PHILOSOPHY

Discipline of the Disciplines

PHILOSOPHY

Discipline of the Disciplines

D F M Strauss

PAIDEIA PRESS
2009

Grand Rapids, MI 49507

A publication of the
REFORMATIONAL PUBLISHING PROJECT
www.reformationalpublishingproject.com

ISBN 978-0-88815-207-7

Foreword

This work aims at investigating the way in which academic disciplines are influenced by philosophy, while at the same time acknowledging the dependence of philosophy on developments within the special sciences. The emphasis will be on the comprehensive and encompassing scope of philosophy. In order to advance this view of philosophy as a comprehensive discipline ("totality science"), i.e. as a discipline that precedes the differentiation and the specialization found in the natural and social sciences, it will be necessary to investigate the implications of basic systematic philosophical distinctions for various academic disciplines. Naturally pursuing this aim requires an analysis of the distinctive features of the scholarly enterprise in the light of developments within the domain of the philosophy of science during the 20th century.

Investigating the various aspects of reality in their uniqueness and mutual coherence will constitute the core focus of this work. It will turn out that such an analysis of the aspects of reality intrinsically coheres with the dimension of concretely existing natural and social things or entities and processes. In Chapter 1 a number of seemingly unrelated problems will be discussed in order to highlight what will be called the *philosophical frame of mind*. However, these problems will be presented in such a way that the interplay of the history of philosophy and the disciplines, systematic distinctions and the connection with particular disciplines are elucidated.

Generally the most fruitful discussion of any philosophical problem will therefore need to take three interconnected perspectives into account. First of all one should understand a problem against the background of its history within philosophy and the particular discipline within which it occurs. Secondly an attempt should be made to advance a distinct systematic philosophical perspective on the problem. Finally, one should investigate the implications of such a systematic perspective for various academic disciplines.

The presence of diverging trends of thought even within the ('exact') natural sciences contains a challenge to the traditional rationalistic overestimation of the 'rational' capacities of human beings. Equally problematic is the alternative relativistic implications of postmodern stances in which the emphasis is shifted to the assumed uniqueness, "poly-interpretable" and changeful nature of reality – stances often making universal claims in order to deny the existence of universality! In the second chapter an alternative argument is advanced aimed at accounting for the role of supra-rational (ultimate) commitments in scholarship – once more followed up in Chapter 9.

Whereas Chapters 5 and 6 enter into a fairly extensive analysis of the implications of inter-modal coherences, the issues discussed in Chapters 7 and 8 further explore the preceding insights in connection with the dimension of many-sided natural and social entities and processes.

Throughout this work references relevant to the philosophical and special scientific sources are given. It is concluded with Chapter 9 in which the main contours of the overall perspective are brought together.

By means of numerous cross-references the reader is constantly reminded of explanations found in other places within this work – often accompanied by a brief summary of the connection. For the sake of ease of reading key distinctions and insights are frequently succinctly repeated – both for the sake of continuity and to serve those readers who first may want to read certain subsections before reading more entensively.

Danie Strauss
(October 2009)

Note 1:

Two unfamiliar conventions are employed:

(i) When single words appear in parentheses single smart quotation marks are used (such as 'single'), while combined words or longer phrases are demarcated by double smart quotation marks ("double quotations marks").

(ii) In the theory of inter-modal connections those analogies that reflect the coherence of an aspect with aspects positioned later within the cosmic order of aspects are designated – not as anticipations – but as *antecipations* (aimed at emphasizing their "forward-pointing" nature).

Note 2:

I want to thank the following people for their valuable comments on earlier drafts of the manuscript: Roy Clouser, Piet Cronje, Rudi Hayward, Jeremy Ives, Kerry Hollingsworth, Martin Jandl, Marty Rice, Paul Robinson, Dick Stafleu, Piet Schoeman, Don Sikkema, Chris Van Haeften and Albert Weideman. In particular I want to thank Alan and Gillian Cameron for their final stylistic editing and proof reading of the entire manuscript.

Note 3:

I also want to express my gratitude to the professors who helped me to digest and further develop the legacy of reformational philosophy, both here in South Africa (proff. P. de B. Kok and my father, H.J. Strauss) and abroad (proff. H.J. Hommes, M.C. Smit, A. Troost, J. Van der Hoeven and H. Van Riessen); the fellow students with whom I interacted during my studies at the University of the Free State (Elwil Beukes, Frederick Fourie and Albert Weideman) and at the *Free University of Amsterdam* (between 1969 and 1974) and later on, namely Bram Bos, Elaine Botha, Phil Brouwer, Henk Geertsema, Sander Griffioen, Bas Kee, John Kok, Piet Schoeman, Egbert Schuurman, Tony Tol, Harry Van Dyke, the late Peet Van Niekerk, Ponti Venter, Al Wolters and Uko Zijlstra; and my colleagues from the Philosophy Department at the University of the Free State for more than three decades, Kobus Smit and Johan Visagie as well as Dirk van den Berg (Dept. of Art History). Not to forget the ever-responsive sound-board, my wife, Tharina.

Contents

Chapter One

Preliminary examples

Chapter Two

The uniqueness of science

Chapter Three

The uniqueness of modal aspects

Chapter Four

Being human

Chapter Five

Inter-modal coherence

Chapter Six

The inter-disciplinary significance of modal analysis

Chapter Seven
Things

Chapter Eight

Human society

Chapter Nine

Philosophy is more than merely the "Discipline of the Disciplines"

Chapter One

Preliminary examples

"Philosophy at large can dispense with Universities, but Universities, that try to dispense with Philosophy will be found in the long run to tamper with the mainspring of their own constitution" (Land, quoted in Spruyt, 1889:127).

1.1 The philosophical frame of mind

Since there are just as many definitions of philosophy as there are philosophers, it may be more instructive to commence our considerations by highlighting a few issues that played a role in the history of philosophy and the academic disciplines. Three examples will be discussed:

(i) The problem regarding our awareness of the boundaries of our experience and the urge to contemplate what lies beyond these limits.

(ii) The way in which the problem of human knowledge, truth and insight touches upon the boundary questions of science.

(iii) Unexpected influences from the humanities upon the natural sciences (with a brief reference to Darwin's theory of evolution).

In all three instances of the issues provisionally discussed, our account of these examples will reveal the *philosophical frame of mind* when, eventually, it will be subjected to a more extensive analysis later on in this work. While highlighting the second example we will mention the remarkable influence exerted by the work of a philosopher from the 18th century on the three main schools of thought within twentieth century mathematics.

Before we subsequently discuss one of the holy cows of our Western intellectual tradition, namely human rationality, a brief overview of diverging standpoints within a number of academic disciplines will be given in order to show that as a matter of fact "universal reason" in no way secured uninamity within the special sciences, not even within the supposedly most exact one of them, mathematics.

It should also be mentioned in advance, however, that particularly during the past few decades, the encompassing scope of a philosophical orientation was challenged by what has become known as *postmodernity* and its rejection of so-called metanarratives. Yet, an explanation of the intellectual background of postmodernism will have to wait until we discuss the complex nature of nominalism in Chapter 6 (see pages 370-379 below). Let us therefore now first pay attention to the three problems mentioned above.

(i) *The boundaries of our universe and of our knowledge*

Human beings have always been puzzled by the *boundaries* of our experience and by the *frontiers* of human knowledge. We are even tempted to think about what lies *beyond* the horizon of what we can experience, in that we are urged to contemplate the *unbounded* and what is supposedly *infinite*.

It is true that, during the early days of Western civilization in ancient Greece, the world was viewed initially as limited and finite. Nevertheless, Anaximander, one of the prominent pre-Socratic thinkers who lived during the 5th century BC, accepted as the principle of origin (*Archē*) something that he called the unbounded-infinite, the *apeiron*.

Yet, to the Greek intellectual world, *space* was always concrete and bound to the presence of bodies as material entities. After the Pythagoreans discovered what was called "incommensurable" proportions[1] – actually the discovery of numbers that cannot be expressed as the ratio between any two integers, i.e. of 'irrational' numbers that are not fractions – the position or place of a material body was seen as its *essence*. 'Place' is a property or predicate only applicable to a body – and in the absence of a body, there is no subject to which the feature 'place' could be attributed. Therefore Greek scholarship did not contemplate an empty space – if the body, which is supposed to be identical to its place, is absent, then it stands to reason that there simply is no place (= no body).

This assumption inspired Zeno, one of the Greek thinkers of the school of Parmenides, to question the reality of multiplicity and *movement*. He argued, for example, that at any given moment of an arrow's flight, it can only be at one place. But, so he continued his argument, if at every moment of its 'flight' the arrow is at *rest*, then it follows that it actually does not move at all! In the third *Fragment* that we inherited from Zeno, he starts his argument by first granting that something moves and then proceeds to the denial of motion. He says, "Something moving neither moves in the space it occupies, nor in the space it does not occupy" (Diels-Kranz, B Fr.3).

Thus the scene was set for an ever-expanding search after the interconnections between number, space and infinity, up to the modern era, where Einstein formulated his answer to the problem by stating, in his theory of relativity, that, since gravitation causes space to be curved, we have to affirm that although space is unbounded, the universe is finite. Whatever continues to move in a "straight line," i.e. a gravitationally 'curved' line, eventually will return to the initial point of departure!

(ii) *Knowledge, truth and insight*

On a more general level the philosophical frame of mind is revealed in our human concern for insight, knowledge and truth. Traditionally these issues were considered to be intimately connected to the thinking human being as a member of society with an awareness of what is good and bad, normally associated with the ethical or with morality. For example, Socrates, Plato's teacher, argued that insight and wisdom ought to be guided by *virtue*, for otherwise it

1 Uprooting their claim that "everything is number."

may end up in service of evil purposes. During the Enlightenment of the 18th century, Immanuel Kant (1724-1804) added another dimension when he exclaimed that he was always intrigued by the starry sky above, governed by the universal law of *causality* (cause and effect), and the *moral law* within me.[1] Implicit in his wonder and awe is an awareness of the difference between *natural laws* and a *domain of normativity*, of *ought to be*, but unfortunately he turned this difference into a dualism, into a separation of what he called the domains of *Sein* and *Sollen*, of *is* and *ought to be*. (The view of reality advanced in this work aims at avoiding such a dualistic understanding of the relationship between is and ought to be.)

The enduring concern for issues at the *borders* of human experience and knowledge not only characterizes the history of philosophy and its connection with the special sciences, but also permeates the various disciplines themselves. The Dutch philosopher, Henk van Riessen (1970), even proposed that we should circumscribe philosophy as that discipline constantly involved in wrestling with *boundary questions* ('grensvragen'). If this view touches something essential and typical of the philosophical frame of mind, it entails that the special sciences – both the natural sciences and the humanities – cannot escape from an ever-present philosophical thrust just as little as philosophy can ignore the enriching influence of on-going special scientific investigations.

Without providing any detail, we only have to mention the *fact* that the development of scholarship demonstrates the direction-giving influence of philosophical stances throughout the entire history of every single discipline, mathematics not excluded.

Just consider the remarkable influence exerted by one of Kant's main works on modern mathematics. In 1781, the first edition of his influential work, "Critique of Pure Reason" (CPR) appeared. The main systematic subdivisions of this work provided the springboard for three diverging trends in 20th century mathematics, namely *intuitionism,* exploring the "transcendental aesthetic" of the CPR, *logicism*, oriented to the "transcendental analytic" and *axiomatic formalism*, affirming the "transcendental dialectic." This effect caused fundamental differences within the discipline of mathematics, such that what is true within intuitionistic mathematics may be false within formalism, while what is mathematically accepted by formalism, such as Cantor's theory of transfinite numbers, is rejected as a phantasm by intuitionism (see Heyting, 1949:4) and as non-existent.[2]

This example immediately alludes to the title of this book: *philosophy as the discipline of disciplines*, as the *science of the sciences*. Our next example will explore the influence of philosophical conceptions operative in the interaction between various special sciences, with special reference to Darwin's

1 "Der bestirnte Himmel über mir, und das moralische Gesetz in mir" – Kant, 1790, A:289.

2 Compare Weyl (1946:6) and the constructivism of Markov that rejects the so-called actual infinite (see Kushner, 2006:561 ff.)

theory of evolution.

(iii) *The influence of the humanities upon the natural sciences*

The various disciplines are constantly confronting philosophy with new problems and insights. Yet this does not mean that the special sciences are operating *without* philosophical presuppositions. For example, even within the natural sciences it may be one of the neglected perspectives that our conceptions about nature and the human body are strongly influenced by our social-political views and practices. A German medical scholar asks the question why there are so many different medical practices in diverse cultures, given the fact that the human body in its organic functioning is the same in all these cultures. He argues for a direct link, for example, between the social, economic and political situation and the conceptions of the human body.[1]

Within this context some background assumptions of Darwin's theory are quite instructive in connection with the influence of the humanities upon the natural sciences. Let us go back to the 17th century, to the political philosophy expressed in a work from the year 1651. The title of this work, written by Thomas Hobbes (1588-1679), is *Leviathan* (1651). It portrays the hypothetically assumed state of nature as one within which there is a constant battle of everyone against everyone (*bellum omnium contra omnes*). Combined with a publication from T.R. Malthus, *An Essay on the Principle of Population* (1798), this perspective inspired Darwin to formulate his idea of a *struggle for existence.*[2] The 19th century British philosopher-sociologist, Herbert Spencer, turned this idea into the well-known phrase: "survival of the fittest." Yet, the contemporary social-political thinker, P.A. Kropotkin (1842-1921), has already pointed out that Darwin presents to us a skewed image of nature, ignoring the fact that next to phenomena of struggle there are also numberless examples of *peaceful* and *harmonious co-existence* (see Kropotkin, 1903; Kropotkin, 1995:117 ff. and Gould, 1996:144).

Although the greater part of the 18th century did not have a positive appreciation of *historical change*, the transition from the 18th to the 19th century is characterized by the emergence of *historicism*, an intellectual movement surfacing during the first half of the 19th century, that twisted a sound historical consciousness by claiming that everything is taken up in a process of historical change. This opened the way for Darwin to employ the idea of *change* in his theory of variation through natural selection (1859). A contemporary of Darwin, Wallace, had already brought it to his attention that the idea of 'selection' presupposes a *choosing intelligence*, and can therefore not be appreciated as a purely natural factor. Darwin is explicit in this regard: "I have called this principle, by which each slight variation, if useful, is preserved, by the

1 He does this on the basis of distinguishing between what is perceivable and invisible in the human body (see Unschuld, 2003:74 ff.).

2 However, Sober remarks that "the degree to which Malthus *changed* the direction of Darwin's thought remains controversial" (Sober, 1987:15).

term of Natural Selection, in order to mark its relation to man's power of selection" (Darwin, 1968:115).

As we now proceed with a discussion of the longstanding trust in the capacities of human reason, the primary question is whether or not it is possible to find a sufficient foundation for intellectual endeavors in human rationality itself?

Is it not true that rational insight testifies to the objectivity and neutrality of theoretical thought? If this is indeed the case, the last thing to be expected is to find diverging and oftentimes radically opposing trends of thought in all the academic disciplines that are supposed to proceed in a neutral and objective fashion. Briefly mentioning diverging orientations found in a large number of basic academics disciplines will support two perspectives at once: (i) human 'reason' or 'rationality' is not the ultimate judge, and (ii) more comprehensive philosophical perspectives caused a proliferation of standpoints in all the special sciences – just consider the brief overview given below.

- *Mathematics*: Axiomatic *formalism* (Hilbert), *logicism* (Russell, Frege) and *intuitionism* (Brouwer, Heyting, Troelstra, Dummett). When someone like Salmon merely refers to "intuitionistic philosophers of mathematics" he implicitly denies the truly *mathematical character* of this trend in 20th century mathematics (see Salmon, 2001:23 – he refers to Körner's work *The Philosophy of Mathematics*, 1968). That this proliferation of viewpoints does not merely occur within the *philosophy* of mathematics but within the discipline itself is confirmed by Stegmüller and Beth: "The special character of intuitionistic mathematics is expressed in a series of theorems that contradict the classical results. For instance, while in classical mathematics only a small part of the real functions are uniformly continuous, in intuitionistic mathematics the principle holds that any function that is definable at all is uniformly continuous" (Stegmüller, 1970:331; see also Brouwer, 1964:79). Beth also highlights this point: "It is clear that intuitionistic mathematics is not merely that part of classical mathematics which would remain if one removed certain methods not acceptable to the intuitionists. On the contrary, intuitionistic mathematics replaces those methods by other ones that lead to results which find no counterpart in classical mathematics" (Beth, 1965:89). We shall return to some of these quotations below (see pages 213 ff.).
- *Physics*: Classical determinism (Einstein, Schrödinger, Bohm and the school of De Broglie) and the mechanistic main tendency of classical physics (last representative Heinrich Hertz)[1] versus the Kopenhagen interpretation of quantum mechanics (Bohr and Heisenberg); the contemporary ideal to develop "a theory of everything" (Hawking and super string theory:

1 Mario Bunge says: "It is now generally understood that mechanics is only a part of physics, whence it is impossible to reduce everything to mechanics, even to quantum mechanics." Although he holds that the physicalism of the Vienna Circle and the Encyclopedia of Unified Science is dead, "the sharp decline of physicalism has not been the end of reductionism" (see his "The Power and Limits of Reduction" in Agazzi, 1991:33).

Greene).

- *Biology*: The mechanistic orientation (Eisenstein), the physicalistic approach (neo-Darwinism), neo-vitalism (Driesch, Sinnott, Rainer-Schubert Soldern, Haas, Heitler), holism (Adolf Meyer-Abich), emergence evolutionism (Lloyd-Morgan, Woltereck, Bavinck, Polanyi) and pan-psychism (Teilhard de Chardin, Bernard Rensch); recent complexity theory (Behe's notion of "irreducibly complex systems" – see Behe, 2003) and the idea of "intelligent design" (see Dekker *et al.*, 2006).
- *Psychology*: The initial atomistic association psychology (Herbart), the stimulus-response approach, *Gestalt*-psychology [the Leipzig school (Krüger and Volkelt) and the Berlin school (Koffka and Köhler)], depth psychology (Freud, Adler, Jung), the logo-therapy of Frankl, phenomenological psychology, contemporary system theoretical approaches (under the influence of von Bertalanffy), and postmodern trends under the spell of the "linguistic turn."
- *The science of history*: Compare the conflict between *linear* and *cyclical* conceptions of history, the Enlightenment ideal of linear accumulative growth, the recurrence of the Greek conviction that history is *eternally recurrent*, in the thought of Vico, Herder, Hegel, Goethe, Daniliwski, Nietzsche, Spengler and to a certain degree also Toynbee.
- *Linguistics*: Two lines of thought dominated the 19th century – Rousseau, Herder, Romanticism, von Humboldt and the rationalistic trend running from Bopp, Schleicher, and 'Jung-Grammatici' to Paul (with his historicistic conception that views language as language-in-development). Cassirer, by contrast, developed his neo-Kantian theory of language (in which *language* is a thought-form imprinted upon reality), Bühler pursued the stimulus of behaviorism in his theory of signs, at the beginning of the 20th century Wundt dominated the scene, De Saussure contributed to the development of a *structuralist* understanding (followed by Geckeler, Coseriu and others), Reichling explored elements of *Gestalt*-psychology in his emphasis on the *word* as the core unit of language, Chomsky revived the doctrine of the *a priori* (that is what precedes experience) within the context of his transformative generative grammar.
- *Sociology*: The initial organicistic orientation (Comte, Spencer) was continually opposed by mechanistic and physicalistic approaches (cf. L.F. Ward – late 19th century – and in the second half of the 20th century W.R. Catton), the dialectical heritage of Hegel permeated Georg Simmel's formalistic sociology with its individualistic neo-Kantian focus (Park and Burgess explored this direction in the USA), Max Weber developed the sociological and economic implications of the neo-Kantian Baden school of thought, Talcott Parsons made the systems model (based upon von Bertalanffy's generalization of the second main law of thermodynamics) fruitful for sociological thinking, opposed by conflict sociology (Dahrendorf, C. Wright Mills and Rex and by the Frankfurt school of neo-Marxism), a systems theoretical approach was recently revived by J.C. Alexander, A. Giddens developed his structuration theory – and during the past two de-

cades Habermas elaborated his theory of communicative actions.

- *Economics*: The classical school of Adam Smith, the neoclassical approach (from Cournot and Dupuit to Menger, Jevons, Walras and Pareto), the marginalism of Marshall, Keynes's "General Theory," as well as alternative approaches to competition (Chamberlin and Robinson).
- *The science of law*: The historicistic orientation of the Historical School of von Savigny – followed by the Romanist (von Jhering) and Germanistic school (von Gierke), neo-Hegelianism (Binder), neo-Kantianism (Stammler, Radbruch, Kelsen), the revival of natural law theories after the second world war (Rommen, Stadtmüller, Küchenhof, Schmitt, Fuchs, Auer Coing, Maihofer and others), legal positivism (which always seems to remain alive amongst legal scholars), and *Critical Legal Studies* (H.L.A. Hart and Dworkin).
- *Theology*: Dialectical theology (Barth, Gogarten, Brunner) in its dependence upon Kierkegaard and Jaspers, Bultmann (dependent on Heidegger), theology of hope (Moltmann – dependent upon the neo-Marxism of Ernst Bloch), the historicistic design of Pannenberg (dependent upon Dilthey and Troeltsch), the 'atheistic' theology of Altizer and Cox (influenced by neo-positivism), existentialist-hermeneutical trends (Fuchs, Ebeling, Steiger), theology of liberation (influenced by neo-Marxism).

What is particularly striking regarding these philosophically founded schools of thought within the disciplines, is that many of them are entangled in what should be labeled *reductionism* in a pejorative sense. Popper is unambiguous in his rejection of reductionism: "As a philosophy, reductionism is a failure" (Popper, 1974:269). Goodfield remarks: "Reductionist methodology may have been extremely successful, but the history of science abounds with examples where forms of explanation, successful in one field, have turned out to be disastrous when imported into another" (Goodfield, 1974:86).

A *positive* appreciation of reductionism is, for example, found in the thought of Dawkins and Dennett (see Dennet, 1995:80 ff.). Surely there are also positive and largely unrelated connotations attached to the term *reduction* in different special sciences. For example, mathematicians may speak of the construction of numbers from sets and then designate it as 'reduction'. Separating chemical compounds into their simpler constituents is also known as 'reduction', and so on. We shall see that the idea of irreducibility precludes every reductionistic approach. The only meaningful option is to investigate the interconnections of what is irreducible. The challenge of all scholarship is to understand the "coherence of irreducibles" – or more strongly formulated: to understand *in terms of* the "coherence of irreducibles"! Whoever wants to avoid the reduction of what is irreducible opts for what we shall designate as a non-reductionist ontology (see below pages 200, 369).

In order to capture more problematic situations the term *reductionism* emerged by the middle of the 20th century. In 1953 Quine used it in his discussion of "The Verification Theory and Reductionism" (see Quine, 1953:37 ff.) and in the early seventies the work "Beyond Reductionism" appeared (see

Smythies & Koestler, 1972). Smith considers the scientist-philosopher Michael Polanyi to be "perhaps the severest and most comprehensive critic of reductionism" because he "was a major scientist of this century and was drawn into philosophical debate primarily because of the threat to scientific freedom, political democracy, and to humane values that he saw in reductionism." To this he adds the remark:

> His works *The Contempt of Freedom, The Logic of Liberty, Science Faith and Society, Personal Knowledge,* and *The Tacit Dimension* have as a common theme the criticism of reductionism in all its scientific, cultural and moral forms.[1]

Particularly in Chapter 6 we shall investigate the limits of logical discernment – identification and distinguishing – in order to make a plea for the avoidance of real antinomies arising from the attempt to reduce what is truly *irreducible.* This aim is similar to the strategy defended by the physicist Henry Margenau in following some ideas of Mario Bunge. He advances "the strategy consisting of reducing whatever can be reduced without however ignoring emergence or persisting in reducing the irreducible" (cf. Margenau, 1982:187, 196-197). Our assessment of the systematic distinction between antinomy and contradiction will eventually furnish us with a criterion that will serve every attempt to avoid the multiple 'ismic' orientations found within the various disciplines (see below pages 172, 264 ff.).

The picture portrayed in the preceding pages justifies the question mark of the next subsection.

1.2 Human rationality – a divine spark?

Although we have remarked that practically all the academic disciplines are stamped by diverging and often conflicting trends of thought, the discipline of philosophy is even more notorious in this respect.

In a sense this is rather remarkable, because one of the hallmarks of the rich legacy of philosophical reflection is its claim of *rationality*. How is it possible to proceed in a *rational* way and still not be able to sidestep radical *differences of opinion*? Particularly when the underlying assumption is that *rationality* ought to be the final judge in intellectual endeavours, this outcome is *perplexing* and *embarrassing*.

It should indeed be conceded that such a *trust in reason* characterizes almost the entire history of philosophy (and the disciplines). The origins of Western philosophy in ancient Greece had already witnessed the enthronement of *reason*, and ever since that era philosophical claims appear to be made in terms of claims to *rationality*. During the hey-day of medieval thought,

1 See Smith, G.L., 1994. *On Reductionism*. Sewanee, Tennessee – available on the web at: http://smith2.sewanee.edu/texts/Ecology/OnReductionism.html (accessed on 22-01-2005). Putnam holds that scientism and relativism are *reductionist* theories (Putnam, 1982:126). In respect of 'phenomenalism' he remarks: "the idea that the statements of science are translatable one by one into statements about what experiences we will have if we perform certain actions has now been given up as an unacceptable kind of reductionism" (Putnam, 1982:187).

Thomas Aquinas (1225-1774) even attempted to develop purely "rational proofs" for the existence of God.

In the spirit of the age of Enlightenment, the above-mentioned prominent and extremely influential 18th century German philosopher, Immanuel Kant, in the *Foreword* to the first edition of his *Critique of Pure Reason* (1781), explains that his age is that of rational critique. Not even law in its majesty or religion in its sanctity are allowed to withdraw themselves from the critical scrutiny of reason, for reason can only show respect to that which has withstood its critical assessment.[1]

Even today, this legacy is manifest in the widely accepted maxim that students and scholars should be *critical.* Rarely anything is ever said about the status and nature of the criteria involved in the exercise of criticism. Are they (i) *derived from* the rational agent or (ii) do they *hold for* rational pursuits?

Let us briefly look at these two options:

Suppose the criteria for rational conduct are derived from the rational agent itself. This position entails that rational activities *generate* their own norms and that rational behaviour in the full sense of the word is "self-normed," that is, it is *autonomous*. This well-known word derives from two Greek words – *autos* = self and *nomos* = law. This position entails that in its rationality, the human being is supposedly a *law-unto-itself.*

Of course the mere idea of *autonomy* may raise many further questions.

For example: am I merely 'norming' my own (strictly individual) 'rational' activities? If the answer is affirmative, the next concern relates to rational interaction between *different* individuals. If all these individuals produce their *own* norms for rationality, will they ever be able to agree or to reach consensus? Does the affirmation of rational insights not rather require or presuppose universal normative standards that are not reducible to the subjectivity of merely one single rational agent? In other words, are we not all bound by supra-individual and non-arbitrary standards for rational behaviour in the first place?

Yet, as soon as this is conceded, the initial idea of autonomy is seriously threatened, for now we have implicitly accepted given norms for rationality, to which human beings are subjected in their rational endeavours.

We have to ask a more radical question: is the idea of autonomy not self-contradictory? In order to answer this, we have to consider a related age-old legacy. From the earliest times, a *law* is understood as pertaining to the conditions for the existence of something. For example, the conditions for *being an atom* hold universally for all atoms. Yet each individual atom is distinct from the conditions for (law for) being an atom. That is to say: the conditions for

1 "Our age is, in every sense of the word, the age of criticism and everything must submit to it. Religion, on the strength of its sanctity, and law on the strength of its majesty, try to withdraw themselves from it; but by doing so they arouse just suspicions, and cannot claim that sincere respect which reason pays to those only who have been able to stand its free and open examination" (Kant, 1781:A-12 – translation F.M. Müller – see Müller, 1961:21).

being an atom are not themselves an atom, just as little as the conditions for being *red* are themselves red.

Obviously there is an important difference between a law or a norm or principle and whatever is subjected to it – even in the case of rationality. This does not only imply that rationality is normed, i.e. that it is subject to universal normative standards transcending the individual, since this subjectedness prompts us to reflect on the origin and nature of these *norms for* rationality. The moment we start to investigate this matter, we are inescabably confronted with direction-giving ultimate commitments, transcending the realm of rationality itself – since they are embedded in one or another (pre-theoretical) world and life view.

The prevailing implicit trust in reason did not realize that such a trust or faith in reason is not itself rational! Twentieth century philosophers from different philosophical traditions started to acknowledge this fact. The well-known philosopher of science, Sir Karl Popper, radically attacks an uncritical or comprehensive *rationalism* based upon "the principle that any assumption which cannot be supported either by argument or by experience is to be discarded" (Popper, 1966-II:230). He argues that this kind of rationalism is demonstrably inconsistent, i.e. in terms of its own criteria: since "all arguments must proceed from assumptions, it is plainly impossible to demand that all assumptions should be based on argument" (Popper, 1966-II:230). Popper is aware of the fact that behind the idea of an 'assumptionless' approach, a huge assumption hides itself – something eventually also criticized by the prominent hermeneutical[1] philosopher, Hans-Georg Gadamer, in his mocking of the prejudice of *Enlightenment* against prejudices (cf. Gadamer, 1989:276).

Popper's own position unequivocally demonstrates his insight into the *self-insufficiency* of 'rationality'. He knows that the rationalistic trust in reason is not rational itself, and explicitly speaks of "an irrational faith in reason" – which means that, according to him "rationalism is necessarily far from comprehensive or self-contained" (Popper, 1966-II:231).

Stegmüller, an equally formidable philosopher of science from the second half of the 20th century, holds a similar conviction when he says that there is no single domain in which a self-guarantee of human thinking exists – one already has to *believe* in something in order to *justify* something else (Stegmüller, 1969:314).[2] This position is reminiscent of a remark by Max Planck in his rectoral oration of 1913: "One should not believe that it is possible, even in the most exact of all the natural sciences, to make progress totally without a worldview, that is to say, completely without unprovable hypotheses. Also for

1 Hermeneutics is the discipline investigating the nature of *interpretation*.

2 Lennox mentions the immunologist, George Klein, who holds that his "atheism is not based on science, but is an *a priori* faith commitment." In response to the accusation that he is an agnostic, Klein says: "I am not an agnostic. I am an atheist. My attitude is not based on science, but rather on faith... The absence of a Creator, the non-existence of God is my childhood faith, my adult belief, unshakable and holy" (Klein, 1990:203; see Lennox, 2007:34).

physics the statement is true that one cannot attain salvation without faith, at least faith in a certain reality outside ourselves" (Planck, 1913:78).[1] These viewpoints are supported from a different angle when De Vleeschauwer says: "A science without any 'presuppositions' is therefore purely from a rational standpoint impossible. The last reality towards which epistemology drives us, is an act of faith in thinking ...".[2] Clement of Alexandria (150-250) held the view that all theology is founded upon knowledge proceeding from an immediate conviction (Mühlenberg, 1966:73).

The trust in the human capacity to think and argue underscores the fact that, from this perspective, 'rationality' does not have the first and the last word in the life of human beings. Human life indeed knows a multiplicity of relationships built upon *trust* and *confidence*.

A more encompassing account of what is involved in the many-sidedness of human knowing will have to investigate the intricacies of the human personality-structure – a task belonging to the discipline called *philosophical anthropology* – and the embeddedness of "knowledge-communities" within human society. Both these contexts are intimately related to the workings of rationality evinced in the inevitable human urge to conceive, conceptualize, argue and understand. It has become customary merely to refer to 'rationality' without trying to come to a more precise explication of what is really meant with the term. As a consequence, some authors even prefer to avoid this word altogether. The problematic way in which the terms 'rational' and 'rationality' are used inspired McGrath to replace it altogether:

> Given the growing criticism of the uncritical use of the word-group which includes 'rational' and 'rationality,' I propose to avoid using them in this work, and instead deploy more meaningful and appropriate terms such as 'justification,' 'warrant', and 'entitlement' (McGrath, 1999:13-14).

It is significant that Paul Bernays, the co-worker of the foremost mathematician of the 20th century, David Hilbert, in his contribution to the *Festschrift* of Karl Popper, remarks that any account of rationality has to pay attention to concept formation. He states that the "proper characteristic of rationality" is "to be found in the conceptual element" (Bernays, 1974:601).

1.3 The conceptual roots of rationality

An awareness of and references to concepts are common-place elements of the human experience of the world. The way in which we are oriented within diverse cultural milieus crucially depends upon the human ability to classify and categorize. Cultural anthropologists are extremely sensitive to uniquely

1 "Aber man wähne nicht, daß es möglich sei, selbst in der exaktesten aller Naturwissenschaften, ganz ohne Weltanschauung, das will sagen, ganz ohne unbeweisbare Hypothesen, vorwärtszukommen. Auch für die Physik gilt der Satz, daß man nicht selig wird ohne Glauben, zum mindesten den Glauben an eine gewisse Realität außer uns."

2 "'n Volstrekte «voraussetzungslose» wetenskap is derhalwe reeds vanuit 'n suiwer rasionele standpunt beskou, onmoontlik. Die laaste realiteit waartoe die kennisleer ons dryf, is 'n akte van geloof in die denke" (De Vleeschauwer, 1952:244).

differing systems of world-orientation in various cultures. Yet, however different these cultures may be, not only do they all still bear the label 'culture', but each one of these cultures inevitably also has to classify and categorize.

But what is entailed in acts of classification and categorization? Think about our first experiences as human beings at home, coming to self-consciousness within a family and living within our own room. The normal daily routine of going to bed in the evening and getting up again the next morning presupposes our cognitive ability to *identify* the bed and to *distinguish* it from other furniture in the bedroom. Without knowing what a bed is and without realizing that a chair is not a bed, one may find one's pants in the bed and oneself hanging over the chair. Identifying and distinguishing *this* bed from *this* chair presupposes an understanding of the general (universal) categories of *chairs* and *beds*. In other words, observing a bed *as* a bed rests on the *concept* of a bed (implying, amongst others, the property: "something to sleep in"). Likewise, noticing a chair *as* a chair depends on the prior concept of a chair (implying, amongst others, the property: "something to sit on"). In these minimal indications, enough is found to highlight the fact that classifying this chair and this bed into the categories of chairs and beds, requires both similarities and differences. Both chairs and beds are cultural artifacts, in our comparison captured by the anonymous reference to 'something' – the moment of *similarity* between them. But although we are referring in both instances to 'something', the two 'things' are different, for the one is a chair and the other a bed.

At once it is also clear that it is meaningless to say that two things are "absolutely different" – implicitly suggesting that they do not share *anything* – for the claim itself rests upon the assumption that both are 'things', showing the *similarity* between them.

Without elaborating a theory of concepts and concept formation fully in this context, one crucial element must be mentioned. Concepts always unite, in the sense of bringing together (or synthesizing), a multiplicity of properties or features that are constitutive building blocks of whatever is conceived. If one defines a 'circle' as a spatial figure where the delimiting curve is equi-distant from the center, then various terms are involved in this concept – such as: "spatial figure," "delimiting curve," "equi-distant," and "mid-point." To a greater or lesser degree it may be possible to find synonymous terms for those employed, for example when 'center' is replaced with "mid-point." But it is soon evident that every attempt to produce further 'definitions' of the terms entailed in this concept inevitably will run into the production of mere synonyms. In itself, this is a sign of the indefinability of the most basic terms operative in any specific concept.[1]

What is therefore at stake in this regard, is the acknowledgement that concept formation and definition ultimately rests upon the acceptance and employment of *primitive terms*. In order to avoid a *regressus in infinitum*, this

1 We shall return to this in connection with an analysis of the unique nature of the various aspects of reality.

state of affairs ought to be respected. Cassirer has a clear understanding of this when he writes:

> In order not to accept a *regressus in infinitum* a critical analysis of knowledge has to stop at specific original functions which are not in need of genuine derivation and which are also not capable of it (Cassirer, 1957:73).

The upshot of these considerations is that the *key terms* involved in rational conceptual understanding are themselves *not* open to (rational) conceptual definition![1] Mühlenberg points out that already for Aristotle thinking presupposed knowledge that was not mediated by any proof (Mühlenberg, 1966:73).

Rationality in this sense does rest upon a non-rational (or more than rational) basis. Yet, this foundation should not be confused with something irrational. We may designate this basis of irreducible primitives as the *restrictive boundary* of rationality. The acknowledgement of irreducible primitives as such reflects a positive awareness of one of the most fundamental perennial issues in philosophy: the quest to account for the coherence of what evinces itself as irreducible, i.e. what we designated as the *coherence of irreducibles*.[2] For example, one may argue that constancy in the sense of uniform flow is irreducible within the discipline of mechanics (kinematics). The same applies to the notion of change within the discipline of physics. But these aspects intimately cohere because one can only establish change on the basis of something persistent or constant. An experience from everyday life can elucidate this point further. Suppose an ageing person claims that she has changed a lot during the past two decades. This statement only has meaning on the basis of an implicit awareness of persistence, for there is a constant reference to the same person. The relationship between *constancy* and *change* is therefore an example of the "coherence of irreducibles."

Provisional summary

Before we proceed, we may briefly summarize some of the core insights regarding rationality discussed thus far.

> Although much more ought to be said about it, in concept formation, we have unveiled a core element of rationality. Concept formation, in turn, depends on the uniquely human ability to identify and distinguish, and it is always dependent upon the discernment of similarities and differences. What is normally called analysis is nothing but this ability to identify and to distinguish. In other words, without similarities and differences all systematic thought will be impossible, because every instance of classification is nothing but performing acts of identification and distinguishing.

We proceed now by further exploring the meaning of analysis. To be sure, identification itself is precisely what concept formation is all about – in the

1 Ewing mentions that G.E. Moore believed that the good has no definition: "I think we shall see that some terms must be indefinable if anything is to be defined at all" (Ewing, 1962:87).

2 The fact that we also have to acknowledge realities transcending the capacity of concept-formation will be treated in a later context. Such *concept-transcending knowledge* (idea-knowledge) will then be designated as the *expansive boundary* of rationality (see pages 176 ff.).

sense of bringing together, i.e. synthesizing or uniting a multiplicity of logically discerned properties into the unity of a concept.[1] At the same time we have to realize that identifying and distinguishing are equivalent to "lifting out" and 'disregarding'. This entails that the two 'legs' of analysis, namely identification and distinguishing, are equivalent to the two 'legs' of abstraction, namely lifting out and disregarding. Moreover, analysis and abstraction are therefore equivalent – as captured in Figure 1.

Figure 1

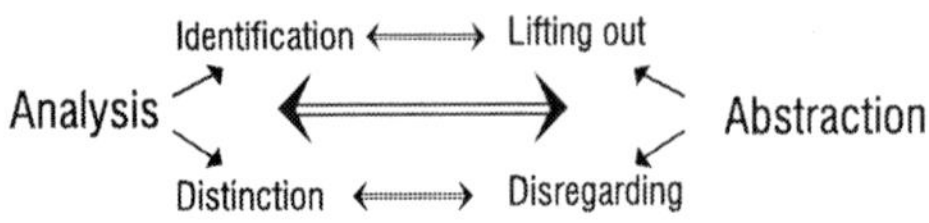

This basic insight was articulated with reference to the presence of similarities and differences. We have seen that cultural artifacts (such as beds, chairs books and tables) display similarities and differences. They are made from natural material in service of particular human aims and needs. However, we have also seen that on the basis of these similarities, we can discern the obvious differences between them: the typical destination of a table is to be used as a place for eating (a social ritual), whereas the typical destination of a book is to be read (a lingual activity). Books are carried around, taken on holiday, sold in book shops, and made available to the public in libraries – all properties that are not applicable to tables. This implies that we have to add a further element to our initial portrayal of the reciprocity of *analysis* and *abstraction* in Figure 1.

Figure 2

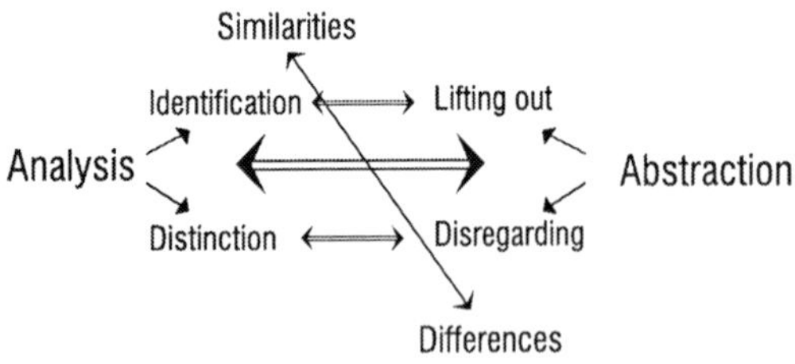

Various philosophical trends, particularly during the past 100 years, emphasized the importance of acknowledging the embeddedness of ordinary human activities within the (inter-subjective) human life-world (German: *Lebens-*

1 That the scope of each one of these properties is *universal* will be argued when the basic thrust of the modern era since the Renaissance has been investigated. We shall see that a proper understanding of the correlation between universality and what is individual is decisive for insight into the difference between *rationalism* and *irrationalism* (see pages 176 and 369 below).

welt).[1] In everyday life, our logical thinking is embraced by an awareness of a more-than-logical diversity, and it is merged within this diversity in multiple contexts. This pre-scientific awareness is not something that ought to be eliminated or denied by scholarly thinking, since it forms the unavoidable basis and starting point of scientific reflection.

Although we have already made the point that analysis and abstraction are formally equivalent (synonymous), it will turn out to be important to realize that our logical-analytical function as such is foundational to our lingual abilities. Throughout their life, humans encounter situations where they have already identified whatever is at stake without always being able to designate it appropriately in a lingual way, or to find the most suitable *word* to designate what has been apprehended.

A striking instance is the following example. A little girl who first notices a *pigeon* and learns its name can abstract 'concretely', for instance when she shortly thereafter refers to a shrike as a **pigeon**. The child actually indicates the concept 'bird' with the *name* (verbal sign) 'pigeon'. This is only possible because, from the concrete sensorially perceived image of a pigeon the girl has lifted out certain bird-characteristics, e.g. a beak, wings, feathers, while simultaneously relinquishing the specific characteristics that distinguish a *pigeon* from a *shrike.*

Figure 3

This kind of abstraction is part of our everyday life, since we argued earlier that we are continually classifying (identifying) all sorts of entities by placing them within certain categories. Otherwise, how would one be able to identify a particular horse as *a* horse (= belonging to the category of horses), or a particular car as *a* car? Without general concepts such as cars and horses (in

1 Particularly the so-called phenomenological school of Edmund Husserl (1856-1938) emphasizes this life-world. Heinrich Schutz, who worked out Husserl's ideas within the domain of sociology and social philosophy, starts with the human being in its naive-natural attitude. The latter is born within a "social world" encompassing 'you', 'we', and 'myself' (see Schutz, 1974:138). In his analysis of "communicative action" Habermas also extensively discusses the concept 'Lebenswelt' (1987-I:102 ff., II:182 ff.).

which the detail of particular cars and horses are relinquished), this would be impossible.

This example shows that within the intellectual development of human beings, logical concept formation *precedes* matching lingual abilities. Viewed from the perspective of modal distinctness and coherence, language use is built upon the basis of logical skills. Within the language *Afrikaans*, a quite interesting example of this foundational relationship is found. The double negation in the Afrikaans language generates a *logic* peculiar to the language itself. It is found that relatively young children (3-5), who display a clear sense of logical consistency and logical soundness, answer questions phrased in terms of the double negation with 'yes', where older children and adults, who matured lingually to such an extent that they are "at home" with the (apparently 'illogical') double negation of Afrikaans, would say 'no'.[1]

The foundational role of the logical function is highlighted not only by the fact that concept formation is a logical(-analytical) activity, but also by the insight that one must carefully distinguish between *concepts* and *words*, keeping in mind that concepts are employed in statements. Yet, the logician Quine does not sufficiently distinguish between the linguistic and the logical side of utterances when he writes:

> The peculiarity of *statements* which sets them apart from other linguistic forms is that they admit of truth and falsity, and may hence be significantly affirmed and denied. To deny a statement is to affirm another statement, known as the *negation* or *contradictory* of the first (Quine, 1958:1).

Concrete utterances are many-sided. Although they are certainly characterized by their linguistic side – from which perspective one speaks of them as *sentences* – their logical-analytical side makes possible what is called statements. In order to discern concepts and statements, one therefore has to adopt a logical-analytical angle of approach. An insight into the basic role of the human logical function in concept formation also explains why one cannot *translate* a concept. The word 'cat' could be translated in many different languages, but the concept *cat* transcends any particular language – it displays an inherent universality.

Moreover, our analytical ability to identify and distinguish enables us to notice the linguistically significant accents, sounds and signs employed in language use (both written and spoken language). Conceding that every concrete act of thinking at once functions both in the logical-analytical and lingual facets of our experience does not invalidate our argument regarding the *foundational position* and role of the logical aspect in its relation to the lingual aspect of reality, but presupposes it.

One further remark needs to be made in this context:

1.4 Reality embraces more than (natural and social) entities

Earlier we alluded to a given, more-than-logical diversity within reality. Al-

1 In Afrikaans one may ask: "Is jy nie honger nie?" ["Aren't you hungry?"] A young child will answer *yes* whereas more mature language users would say *no*.

though this awareness certainly includes different kinds of (natural and social) entities ('things'), much more is intended. What we have in mind is an elaboration and transformation of an age-old distinction in philosophy, namely that between *things* (*entities*) and their *properties*, normally exemplified through a variety of *relations* between them.

Henk Hart opens his work on our understanding of the world with the following illuminating explanation – an explanation focusing on the entities we can experience, the properties (attributes) we can discern and the relations between these entities:

> Our universe, the empirical world of time and space, is populated by little girls, white-tailed deer, yellow slippers, planets and many other things. We can attribute what may be called qualities, or functions, or properties to all of these entities in our world and we can say that they relate to each other. Little girls are cute and have mothers. White-tailed deer are fast and eat leaves. Yellow lady slippers have brown spots on their petals and need light. Planets move around the sun. We can record countless situations that always have these three elements: things with attributes in relation. Little girls feeling warm as they are cuddled by their mothers. White-tailed deer standing motionless as they listen to a sound. Yellow lady slippers hanging low as they bend under the weight of unexpectedly late snow (Hart, 1984:1).

In the subsequent development of his thinking, Hart unfortunately became more sympathetic to the postmodern inclination that emphasizes *change* at the cost of *constancy* (see Hart, 2000). We shall argue below in more detail that *change* can only be detected on the basis of *constancy* (see pages 163 ff.).

Ontologists commonly hold that existent properties are *entities*. Usually they distinguish between 'objects' or things and their properties. Yet, as a general indication of multi-aspectual 'things' we will use the term 'entity' (as in referring to the *entitary dimension* of reality). It is meant to refer to the 'whatness' of reality, to concrete things and events. When we sometimes speak about 'abstract' or 'theoretical' *entities* (in order not to deviate too far from the mentioned general practice), the context will prevent confusing it with the 'concrete' entitary dimension. Later we shall argue that the common use of the term 'object' is inadequate from an ontological perspective (see page 175 below). However, it should be noted that in his German writings, a prominent philosopher like Stegmüller does speak of (concrete) entities ('Entitäten') instead of 'objects' (see for example Stegmüller, 1987:106). The same applies to the contemporary mathematician and physicist Penrose, who refers to physical 'objects' as *entities* (see Penrose, 2004:66).

Our experience of reality always concerns this trio of *entities*, *properties* and the *relations* between these things. The same applies to all events we can experience. Events are always embraced and delimited by these dimensions of reality. Since philosophy is precisely that discipline which reflects on the *limits* of our experience, i.e. exploring the *horizons* of our possibilities, we have to elucidate the reality (in technical philosophical parlance: the *ontic status*) of the various *aspects of reality* – such as the *quantitative* aspect,[1] the *spatial* aspect, the aspects of *movement* and *energy-operation*, the *jural aspect*, and so on.

> *Comment*: The term *ontic* is derived from the Greek word ὄν where it is used to designate the existence of what *is*. It acquired a particular speculative meaning in the thought of Parmenides and his school for according to this school what *is* participates in *static being*. Dooyeweerd prefers to articulate the difference between God and creation by reserving the word 'being' for God alone. The dependent, self-insufficient and relational character of everything within creation entails that every creature is embedded in a *cohering diversity* (embracing what we have called the "coherence of irreducibles"). Dooyeweerd employs the term 'meaning' for this coherent diversity within creation (cf. 1997-I:99). However, while honouring the *intention* to respect the distinction between Creator and creature, one may just as well employ the term *ontic* to designate the creaturely existence of different creatures.

Distinguishing different categories is already found in the thought of Aristotle. Aristotle commences his work *Categoriae* by postulating the existence of a *primary substance* that is purely individual.[2] This primary substance is supposed to lie at the basis of all the *accidental* categories – such as *quantity*, *quality*, *place*, *relation*, and so on. However, this entire scheme is embedded in the primordial and ultimate dualism between *form* and *matter* in his thought. The effect of this dualism is that he distinguishes between *accidentia* related to matter (such as *quantity*) and others related to form (such as *quality*). His statement that whatever is many in number has matter (*Metaphysics*, 1074a33-34) caused serious trouble for the medieval view of the 'immaterial' souls of human beings, for once 'separated' from their "material bodies" the "rational human souls" could only maintain their *number* if they are still connected to matter, for Thomas Aquinas accepted the statement that "whatever is many in number has matter."

During the medieval period, the being of whatever *is* was supposed to be determined by *unity*, *goodness*, *beauty*, and *truth*. However, these determina-

1 The Dutch word 'aantalligheid' refers to a *given* plurality, prior to human conceptualization or lingual designation. Numerals (i.e. "number-names") emerge when multiple things or properties are (successively) named. What is therefore required is an appropriate word that can capture the *ontic status* of a given plurality ('aantalligheid'). In the absence of such a candidate in the English language the term 'quantitative' is chosen to perform this ontic function (the *denotation* ought to be ontic in the sense of something truly *prior* to human naming or conceiving). The denotation of 'quantitative' has (un-counted and un-named) plurality in mind. Since the term 'numerical' is derived from *number* it suffers from the same shortcoming as *quantitative*. Nonetheless, the existing lingual limitations will sometimes cause us to refer to the ontic reality of the first modal aspect also as the *numerical* (arithmetical) aspect.

2 Although references in the text will be to the standard (and generally accepted) numbered system of Aristotle's works, the English translations will be derived from Aristotle (2001) (unless differently specified).

tions were supposed to encompass God and creatures *analogically* alike. In the 13th century A.D., Thomas Aquinas developed his theory of God as the highest being (*ipsum esse*), in this regard with all other creatures participating to a lesser degree in this divine Being.[1]

Early modern philosophy explored new avenues. Descartes (1596-1650), for instance, claims that number and all universals are mere *modes of thought* (*Principles of Philosophy*, Part I, LVII). This conviction holds that only concrete entities are real and that their properties are *human constructions*. The fundamental question is: are there not, *prior* to any human intervention or construction, a *given* multiplicity of *aspects* or *functions* of reality?

Hao Wang remarks that the famous mathematician, Kurt Gödel,[2] is very "fond of an observation that he attributes to Bernays":

> That the flower has five petals is as much part of objective reality as that its color is red (quoted by Wang, 1982:202).

This position suggests that the quantitative side (aspect) of things (entities) is not a *product of thought* – at most human reflection can *explore* this given (functional) trait of reality by analyzing what is entailed in the meaning of multiplicity. Yet, in doing this, theoretical and non-theoretical thought merely explores the *given* meaning of this quantitative aspect in various ways, normally first of all by forming (normally called: 'creating') *numerals* (that is, number symbols, such as '1', '2', '3', and so on). The simplest act of counting has already explored the original meaning of the quantitative aspect of reality – and it happened in a twofold way, because (i) every successive number symbol ('1', '2', '3,'etc.) is correlated with (ii) whatever is *counted*.[3]

The point we want to underscore is actually straightforward: both entities and the various aspects of reality truly 'exist', i.e. entities and aspects are both *real*, they belong to what *in fact* is given (in an ontic sense). What is designated as factual should not be identified with human knowledge of "facts" – in an epistemic (or epistemological) sense. (The Greek word for *knowing* is *episteme*.) What we shall call the *factual side* of reality is correlated with its *law side*. What is therefore needed is a word (or the meaning-nuance of a word) capturing both, the subject-pole of the law-subject correlation, and subject-subject and the subject-object relations. When the word 'factual' is used for what is given in an ontic sense, prior to human cognition, then this mean-

1 It is remarkable that the contemporary German (social) philosopher, Jürgen Habermas, uncritically and without any historical account of his preference, still discusses human society in terms of *truth*, *beauty*, and *goodness*.

2 At the young age of 25 Gödel astounded the mathematical world in 1931 by showing that no system of axioms is capable – merely by employing its own axioms – to demonstrating its own consistency (see Gödel, 1931). Yourgrau remarks: "Not only was truth not fully representable in a formal theory, consistency, too, could not be formally represented" (Yourgrau, 2005:68). The devastating effect of Gödel's proof is strikingly captured in the assessment of Hermann Weyl, quoted below on page 315.

3 Frege correctly remarks "that counting itself rests on a one-one correlation, namely between the number-words from 1 to n and the objects of the set" (quoted by Dummett, 1995:144).

ing-nuance from the semantic domain of the term *factual* meets the required yardstick – therefore in this work it will be used in this sense.

Factual reality embraces both entities and aspects. Although he appraises it perhaps not as clearly, this is exactly what a prominent modern mathematician has in mind. Paul Bernays, the co-worker of the famous mathematician David Hilbert, is convinced that mathematical axiom systems are not created out of thin air, because ultimately they relate to the *nature* of the subject matter of mathematics. He therefore explicitly questions the dominant conception that ascribes reality only to entities (he does that by saying that this conception only acknowledges *one kind* of *factuality* – namely that of the 'concrete'; Bernays, 1976:122).[1] With Cantor's theory of *transfinite numbers* in mind, Titze says that "the domain of the transfinite" cannot be "applied to reality" and therefore "merely represents [an] ideal but not meaningless construction of the imagination" (Titze, 1984:149). This statement denies what we call the *ontic status* of the *functional modes* (or aspects) of reality. In other words, it *rejects* the second kind of 'factuality' Bernays has in mind, namely the **reality** of *ontic modes* or *functions*.

Another person who is on the verge of assigning an ontic status (an existence prior to human cognition) to those facets of reality with which mathematics is concerned, is Kattsoff. He also makes a plea for the acknowledgement of both physical and mathematical factuality, although he holds that "mathematical objects" are "quite different from physical objects": "They are clearly not the sort of things that can be observed by means of the senses" (Kattsoff, 1973:30). He argues that it is through intellectual involvement that "mathematical objects" come into sight: "In analogy to physical objects which are called sensory objects because they are observed by the senses, mathematical objects may also be called intellectual objects (or rational objects?) because they are observed by the intellect" (Kattsoff, 1973:33). Later he calls his approach "quasi-empirical" (Kattsoff, 1973:40).

Perhaps the plea for acknowledging the aspects of reality as truly existing (and not merely products of human thought) found its most impressive advocate in the thought of Gödel who introduces the idea of 'semiperceptions' when it concerns "mathematical objects." Next to a physical causal context within which something can be 'given', Gödel refers to data of a second kind, which are open to 'semiperceptions'. Data of this second kind "cannot be associated with actions of certain things upon our sense organs" (quoted by Wang, 1988:304). In terms of our own approach these 'semiperceptions' relate to the functional aspects of reality (see page 45 and page 206). Gödel says:

> It by no means follows, however, [that they] are something purely subjective as Kant says. Rather they, too, may represent "*an aspect of objective reality*" (my emphasis – DS), but, as opposed to the sensations, their presence in us may be due to another kind of relationship between ourselves and reality (quoted by Wang, 1988:304).

1 Note that Bernays employs the term "factual" in the sense in which we have introduced it – referring to what is given in reality prior to human cognition.

Wang is "inclined to agree with Gödel," but he does "not know how to elaborate his assertions" (Wang, 1988:304). The awkward language used by Wang in this regard[1] is actually a sign of lacking theoretical distinctions. In terms of the distinction between aspects and entities, one can simply say that the various aspects (including the numerical facet) belong to reality (i.e. that they have an ontic status), but they do not concern the concrete 'whatness' of the dimension of entities, because they manifest the 'howness' of the functional modes of reality. The aspects as functional modes of being are just as 'real' as the concretely existing entities in reality.

Yet it is clear that Gödel and Wang do contemplate the 'reality' of 'ontic' (designated by them as 'objective') "aspects of reality," which are not like "concrete entities" occupying "a location in spacetime." It is not accidental that, earlier in the 20th century, the eminent German neo-Kantian philosopher, Ernst Cassirer, effectively captured the legacy of Western philosophical thinking in the title of a significant book, *Function Concept and Substance Concept*, focused on the distinction between *function* and *entity* ('substance'). The first edition of this book appeared in 1910, and it covers both the Greek-Medieval and the modern (post-Renaissance) eras. The former was pre-occupied with concrete entities – just recall our remark (on page 2) about the idea of "empty space" and the identification of an entity with its 'place' – while the latter increasingly started to appreciate *functional* relations within reality.

Cassirer, as we have seen, did have an understanding of "original functions that are not in need of genuine derivation." In addition he also realized that there are relations between different functions (aspects) of reality, for in a later work he speaks of "original functions" and their interconnections. In particular he refers to the similarity and difference between a *logical* identity and diversity and a *numerical* unity and difference.[2] Unfortunately, in spite of these seminal ideas in his reflections, he did not develop a theory of functions or modal aspects and their mutual coherence.

1.5 Functions, aspects or modes of reality – the contribution of Dooyeweerd

It was in connection with the distinction between (natural and social) entities and the various functions of reality that the highly original 20th century Dutch

1 Wang says that he "used to be troubled by the association of objective existence with having a fixed 'residence' in spacetime," but he now feels "that 'an aspect of objective reality' can exist (and be 'perceived by semiperceptions') without its occupying a location in spacetime in the way physical objects do" (Wang, 1988:304).

2 "In der Tat ist nicht einzusehen, warum man lediglich logische Identität und Verschiedenheit, die als notwendige Momente in den Mengenbegriff eingehen, als solche Urfunktionen gelten lassen und nicht auch die numerische Einheit und den numersichen Unterschied von Anfang an in diesen Kreis aufnehmen will. Eine wirklich befriedigende Herleitung des einen aus dem anderen ist auch der mengentheoretischen Auffassung nicht gelungen, und der Verdacht eines versteckten erkenntnistheoretischen Zirkels blieb gegenüber allen Versuchen, die in dieser Richtung gemacht werden, immer bestehen" (Cassirer, 1957:73-74).

philosopher, Herman Dooyeweerd, introduced his general theory of *modal aspects* (or *law spheres*) during the third decade of the 20th century. Dooyeweerd grew up in a house familiar with the ideas of the influential Dutch statesman and theologian, Abraham Kuyper, who founded the Free University of Amsterdam in 1880, with an oration discussing "sphere-sovereignty."[1] In this presentation, Kuyper made a plea for acknowledging the inner sphere of academic pursuits free from interference by the state and the church. He applied the principle of sphere-sovereignty to human society – though not in a consistent way – and he also emphasized the diversity of laws found within creation. Kuyper had a significant understanding of the fact that *all* human beings have a faith function and that the human self-hood, the *heart*, in terms of the figurative speech of the Near East during Old Testament times, plays a central and directing role in the lives of human beings. This insight was worked out in his view of the Christian world and life view as an encompassing orientation differing in principle not only from the traditional Roman Catholic world and life orientation but also from that of modern Humanism.

Unfortunately the theological account of the basic orientation of the Free University soon took refuge in what was called "reformed principles" – largely understood in accordance with the idea that intrinsically scholarly problems and issues could be settled by a *direct* appeal to the Bible and certain texts within Scriptures. Most of the time the unique historical circumstances of quoted Bible texts evinced such a large historical distance between *then* and *now*, that this method lost its dependability. Such a biblicistic or fundamentalistic use of the Bible in scholarship distorts both the Bible and the nature of scholarship. A new approach to Christian scholarship is found in the thought of Herman Dooyeweerd.

Dooyeweerd finished his PhD in law in 1917 with a dissertation on the place of the cabinet within Dutch constitutional law.[2] By the early twenties he and his brother-in-law, Hendrik Vollenhoven, arrived at their first rudimentary ideas regarding a radically *new* understanding of created reality. Dooyeweerd started to work at the Kuyper Foundation in the early twenties on the explicit condition that he be allowed to spend half of his time on the elaboration of this new philosophical view of reality. He immediately established the scholarly journal known as "Antirevolutionaire Staatkunde" and started to publish extensively in it.

His series of articles on *The Struggle for a Christian Politics* (1924-1926 – see Dooyeweerd, 2008) demonstrates to what an extent he involved himself with the history of philosophy in general. But it also contains the emerging ar-

1 Before Kuyper, Groen van Prinsterer has already coined the expression "sphere sovereignty." The first historical point of connection, however, is already found in the work of the German jurist Johannes Althusius (in the year 1603): *Politica Methodice Digesta* (a discussion of sphere sovereignty and sphere universality is found in Chapter 5 – see pp. 161 ff. – in Chapter 8 we shall return to the societal significance of the principle of sphere sovereignty).

2 *De Ministerraad in het Nederlandsche staatsrecht.*

ticulation of an entirely novel and innovative theoretical view of reality. The development of this new approach benefited from his struggle with the dominant schools of thought within the domain of the *science of law*, particularly with those of (neo-Kantian) Baden and Marburg.[1] Rudolf Stammler, Gustav Radbruch, Emil Laski and Hans Kelsen were prominent neo-Kantian *legal philosophers* of the time and their challenge for Dooyeweerd's development is mainly found in their unsuccessful attempts to derive the basic concepts of the discipline of law from supposedly purely *logical thought forms*. Hans Kelsen even conjectured that, disconnected from other disciplines, something like a "pure theory of law" is possible.

In response to these approaches, Dooyeweerd first elaborated his new philosophical insights within the field of law in order to test their fruitfulness, but soon he started to explore their general philosophical implications.[2]

In his investigation of the basic concepts of the science of law, Dooyeweerd discovered that some of the most basic concepts primarily relate not to *things* and *events* (i.e. not to the concrete *what* of reality), but to the *modes of being*, i.e. to the *ways* in which things and events exist and *function* (i.e. to the *how* of reality). For example, answering the question: "what is this?" calls for a response in which some concretely existing 'thing' such as a chair, is identified. However, once something is pointed out, subsequent questions about this entity relate to its aspectual *properties*, embedded in questions about the *how*, about the *modes of existence* of such an entity: "how many legs does it have?"; "how expensive is it?"; "how comfortable is it?" and so on. Clearly these last three questions address *aspects*, *functions*, or *modalities* of chairs, respectively the quantitative (how many), the economic (how expensive) and the mode of sensitive feeling (how comfortable). Diemer speaks of an "aspect discipline" (*Aspektediziplin*) and shows an awareness of the multi-aspectual nature of 'objects' of everyday life, such as a coin (*Münze*), which can be something physical-chemical, historical, aesthetic, a means of payment and eventually even a cultic object (see Diemer, 1970:219).

When a dead person is found, this fact has to be reported to the police. Why? Because the integrity of the human body constitutes a *public legal interest* protected by the legal order of the state as a public legal institution. The discovery of a dead body is therefore an *effect* relevant to *jural* considerations. For example, the first question is: what or who *caused* this *jurally relevant legal effect*? Since the discipline of physics also speaks of (physical) causes and effects (causality), one can step back and ask the more fundamental (philo-

1 Both these trends of thought worked within the legacy of Immanuel Kant. The former, with figures such as Wilhelm Windelband, Heinrich Rickert and Max Weber, introduced the idea of 'values' into scientific discourse for the first time, while the second neo-Kantian school of thought counted amongst its representatives thinkers such as Herman Cohen, Paul Natorp and Ernst Cassirer, Emil Lask and Hans Kelsen.

2 In a later context, attention will be given to the critical questions raised by Van Woudenberg in connection with the nature of 'aspects' (see Van Woudenberg, 2003 and pages 393 ff. below).

sophical) question: is there a *difference* between *jural causation* and *physical causation*?[1]

Consider for example the case of a naturalistic understanding of human actions, where a jural act is defined as a "willed muscle movement." In connection with train signals Dooyeweerd, in an article on jural causality, writes:

> The person controlling the signals who disregards the duty to switch the signal from safe to unsafe, causes a dangerous condition on the railway lines through this neglect (Dooyeweerd, 1997a:61).

This person did not *move* the *muscle* needed to make the switch and therefore, in terms of the definition of a human action as a "willed muscle movement," did not *act*. But because of the *obligation* to switch the signal from safe to unsafe, that person jurally *caused* the derailment of the train and the damage flowing from it. In other words, both a *commission* and an *omission* from a juridical perspective are seen as *jural acts*! Consequently, the person not 'doing' anything in a physical sense is still held responsible (liable) for the accident. This concerns the legal issue of *accountability*.

Considerations like these opened Dooyeweerd's eyes for a two-fold perspective.

(i) First of all, one has to distinguish between *different aspects* or *functions* (modes of being) of events and (natural and social) entities. For example, as we have just now seen, the *jural* aspect differs from the *physical* aspect, and also from the *quantitative* aspect, the *emotional* aspect and the *economic* aspect, to name a few other facets of reality as well.

(ii) Secondly, these modes are *interconnected* in a peculiar way, because without such connections, it will be impossible to articulate *compound phrases* expressing the *coherence* between different modes, such as evidenced in the expression *jural causality* briefly discussed above.

As noted, the *uniqueness* of each aspect is captured by referring to its *sphere-sovereignty*. This term was first employed by the 19th century Dutch statesman, Groen van Prinsterer, in his political thought. Kuyper then explored the significance of this principle for human society, and eventually Dooyeweerd applied it to an understanding of reality in three of the four dimensions he distinguished. Furthermore, the *interconnectedness* between the various aspects, through which every aspect displays *moments of coherence* with the others, is designated as the principle of *sphere-universality*. These distinctions are worked out in a *general theory of modal law spheres*. At this stage, one important implication of this theory ought to be mentioned. The *meaning* of each aspect only comes to expression in its *coherence* with all the other aspects. Therefore, as soon as an attempt is made to lift any aspect out of this *inter-modal coherence*, the elevated (absolutized or deified) aspect will *lose its meaning*. When Albert Schweitzer brought his *vitalism* into practice in hospitals in Africa, the motto *live and let live* did not allow the nursing personnel to kill flies, and consequently many people died as an effect of the lack of

1 A more extensive analysis of the meaning of jural causation is found in Chapter 6 (see pages 281 ff. below).

hygienic circumstances. Marxist communism in Russia promised an economic heaven on earth, in that everyone will possess everything – yet the irony in Russia was that no subject possessed *anything*. The irony of all *isms* in the disciplines (and in practical life) is that one always achieves the opposite of what was aimed for.

Before we can explore a systematic understanding of reality in more detail, we have to return to the nature of the scientific enterprise and the nature of scientific concept formation.

1.6 Concept formation in scholarly reflection

Practically within all disciplines, we find concepts of things and events as well as function concepts. Our analytical ability soon prompts us, within hierarchical contexts, to come to a classification starting from similar entities on one level and then proceeding to higher-level concepts uniting only what is common between the different entities of the previous level. During the Middle Ages, the Latin formulation of this (entitary-directed) method of classification became well-known in the form of the distinction between a *genus proximum* and *differentia specifica*. This method of concept formation is well at home within domains where a *typological* classification is required, as is the case within biology as a discipline. Within the domain of biological classification a *Genus* encompasses various *species*. Yet different *Genera* belong to one or another *Family*, a number of *Families* to one or another *Order*, and so on.[1]

Although this kind of hierarchical progress does lead to higher levels of abstraction, it does not move away from the *entitary dimension of reality* as such. Being a mammal is not less real than being *this* or *that kind* of mammal or even this *individual mammal* (except, once again, if one joins nominalism in rejecting all universal properties *outside* the human mind).

Every *typological* classification is directed at (one or another) kind of entities, but it is of no help in any attempt to identify modal aspects. It is nonetheless important to have a proper understanding of the difference between aspects and entities, particularly since the difference implies that there are two distinct kinds of law – (i) "laws for aspects" (universal modal laws) and (ii) "laws for entities" (type laws). We may call (i) *modal laws* and (ii) *type laws*.

Re (i): When we direct our theoretical attention toward the modal aspects or functions of reality – such as the spatial aspect, the physical facet or the social function – we are no longer involved in the classification of entities according to the *kinds* or *types* to which they belong, and are therefore also not interested in the "kind laws" or "type laws" for entities. The mere distinction between economic and un-economic, for example, is not specified in any *typical* way. Both a state and a business can *waste* their money (and thus act un-economically) and both are called to function under the guidance of economic considerations of *frugality*. But it is only possible to phrase these perspectives when

1 The modern neo-Darwinian theory of evolution, given its *nominalistic* orientation (according to which "organisms are not types and do not have types" – Simpson, 1969:8-9), of course operates with a different understanding of these systematic biological categories.

the economic aspect is understood in its *modal universality*, i.e. when the typical nature of the business and the state is disregarded. Modal laws hold *universally* without any specification – universities, businesses, states, families and sport clubs all have to observe the general meaning of economic norms. In his discussion of "theories about everything" Breuer approximates the idea of modal universality when he states that a *theory* is universally valid if it holds for the "entire material 'world'," i.e. when "no part of the material world is excluded from its domain of validity" (Breuer, 1997:2). However, he does not realize that the intended universality of a theory presupposes modal universality (see below pages 421 ff.). The physicist Von Weizsäcker is therefore correct in his reference to the universal validity of physical laws (see page 423 below).

Re (ii): The law holding for a specific *kind* or *type* of entities still has its own universality, but this universality is *specified*. The law for *being a state* is universal in the sense that it holds for *all* states. But because not everything is a state, this type law is specified – it only applies to states. Likewise businesses and states belong to different *kinds* of societal entities, and this typical difference is seen in the typical differences between the function of a state and the function of a business within the economic aspect – business economy differs from state economy (a business cannot 'tax' its clients, but the state can *tax* its citizens). In general, one can say that *modal laws* encompass all possible entities, whereas typical laws (type laws) only hold for a *limited class of entities*.

> *Comment*: Natural and social entities function in a 'typical' way within every modal aspect. The word 'typical' actually refers to the *typonomic* specification of entitary functions (*typos* = type and *nomos* = law). Therefore *typical functions* can also be designated as *typonomic functions*.

The (universal) conditions for being *this* or *that* type of thing must be distinguished from the (universal) way in which particular entities evince their conformity with these conditions (laws). In being an atom or being human, this or that atom / human being shows that it meets the conditions for what it is (see our remark earlier on page 5 of this chapter). Sometimes the word 'structure' is used both for the "law for" an entity and and for the "actual composition of" an entity. The structure (composition) of the latter reveals what is correlated with (and therefore distinct from) the order for entities. A *structure for* has the meaning of a *law for*, while a *structure of* represents the universal way in which individual entities reveal their conformity with the given law for their existence (also known as their *law-conformity*). Unfortunately Dooyeweerd did not properly distinguish between *law* and *lawfulness* (law-conformity) – he simply used these expressions interchangeably.

> *Comment*: Frege's restriction of abstraction to *entitary abstraction* and the difference between *ordinal* and *cardinal numbers*. (This comment and the subsequent one may be skipped without losing the main thrust

of the general argument).

Where the mentioned German mathematician and logician from the late 19th century, Gottlob Frege, phrases his objections against the idea of *abstraction*, his arguments are restricted to this kind of *entitary-directed abstraction*.

His intention is to show that we cannot arrive at the concept of number with the aid of *abstraction*. He states that the properties through which entities distinguish themselves from each other are indifferent in respect of their number (Frege, 1884:40 ff.). He explicitly asks the question from what one should abstract in order to arrive at the number 'one' when one starts with the moon as an entity. By abstraction, he proceeds with his argument, one only arrives at (more general) concepts such as: "attendant of the earth," "attendant of a planet," "celestial body without its own light," "celestial body," 'body', 'object' (*Gegenstand*) – and nowhere in this series the number '1' will occur (Frege, 1884:57; §44). Frege also uses the example of a white cat and a black cat in order to highlight the shortcomings of 'abstraction': "The concept 'cat', that has been obtained through abstraction does indeed contain no particulars, but precisely for that reason it is only one concept" (Frege, 1884:45-46; §34; translation by Dummett, 1995:84).

In his remark related to Cantor's definition of a subset (see Cantor, 1962:282), Zermelo also refers to the attempt to introduce the notion of "cardinal number" with the aid of a process of 'abstraction', which would imply that a cardinal number is to be seen as a "set composed of pure ones." The cardinality or power of a set disregards any order-relation between its elements. When such an order-relation is kept in mind, ordinal numbers are at stake. Counting the 'first', the 'second' and so on therefore employs ordinal numbers. Cantor holds that the concept of a cardinal number emerges when, with the aid of our active thought-capacity, we abstract from the character of the different elements of a given set M and also disregards the order in which they are given (Cantor, 1962:282). But according to Zermelo it follows from the fact that these 'ones' are still mutually distinct, that they simply provide the elements of a newly introduced set equivalent to the first one, which means that the required abstraction did not help us (cf. his remark in Cantor, 1962:351). An alternative understanding is made possible by what we shall designate (in Chapter 2) as *modal abstraction* (see pp.48 ff.).

Comment: *Ordinal numbers and cardinal numbers*
Cardinal numbers relate to the general question: *how many*? Ordinal numbers, by contrast, do incorporate such an order relation (where it is known that a particular number is smaller than, equal to or larger than

another one). In order to establish that I have 14 marbles in a bag (i.e. in order to find the *cardinality* of the marbles in the bag) no *specific* counting order is required – one can start by counting anyone of them *first*, any of the remaining marbles *second*, and so on. However, if each one of the marbles is successively marked, one can count them afterwards in their ordinal arrangement, the one marked '1' firstly, the one marked '2' secondly, and so on. In this case one cannot just start with any one of the marbles. In modern mathematics this distinction relates to an interesting debate: which one of these two notions is more fundamental, that of *cardinal number* or that of *ordinal number*? The axiomatic-formalist and logicist trends (Hilbert, Gödel, Russell and others) tend to affirm the primacy of *cardinal numbers*, whereas the intuitionist school (Brouwer, Weyl, Heyting and others) give preference to the foundational role of *ordinal numbers*. That ordinal numbers are indeed foundational is clearly seen from the fact that every instance of the cardinality of a set presupposes an order of succession (a more extensive discussion of ordinality and cardinality is found on page 276 ff.).

However, is it possible to distinguish different (modal/functional) properties of one and the same entity? In terms of Frege's example of the moon, we may be more specific: does the moon have any *numerical properties*? Frege indeed realizes that 'number' is an answer to the question "how many?" and explicitly discusses this question in connection with the moon (Frege, 1884:57; §44). He even asks whether or not the moon is 'one' or "more than one"? But in the absence of a *theory of ontic functions* (*modalities*), he cannot relate the *numerical properties of entities* to the (universal) *ontic meaning* of the quantitative aspect of reality – and consequently categorically *denies* that 'number' is the "property of something" (Frege, 1884:63; §51).

Questions about "how many?" point in a different direction – the direction of what may indeed provisionally be called "aspect abstraction." This kind of abstraction, surely, is fundamentally different from *entitary directed abstraction*.

Yet there is an even more fundamental issue at stake, because the question "how many?" requires a *human response*. Are there (universal) ontic features presupposed in our answer to this question, which are quantitative in nature? Or, alternatively, do we have to revert to the position that number and all universals are *creations* of human thinking (as Descartes asserted)?

Keeping in mind that "relational concepts" are for Cassirer the same as *function concepts*, his following statement is significant for the distinction between *aspects* (functions) and *entities* (things, so-called 'objects'): here Cassirer highlights what we have just called *aspect abstraction* (or as we shall argue below: *modal abstraction*).

The function of 'number' is, in its meaning, independent of the factual diver-

> sity of the objects which are enumerated. ... Here abstraction ... means logical concentration on the relational connection as such (Cassirer, 1953:39).

To our mind the theory of modal aspects indeed opens up a new avenue in this regard. For this reason, the phrase we have used, namely "the numerical aspect of reality," implicitly refers to an *alternative view of reality*, which is foreign to Cassirer's thought and to Frege's understanding of the nature of number, because, according to the latter, "specifying a number contains a statement about a concept" (Frege, 1884: §74 – page 81).

Although conceiving the numerical qualities of entities requires a *human response*, this response is made possible by what is *given* in reality (as ontic aspects).

Likwise, the human experience of societal relationships is always connected to an awareness of the reality of social entities and functions or aspects of reality. Within the context of a capitalist understanding of economic relationships and free enterprise, it is often said that *businesses* ought to be *profitable*. The term 'business' refers to the dimension of (social) entities, and the term 'profitable' to that of aspects (modal functions). When a state is accused of being *unjust*, the same distinction is evident: a societal collectivity (namely the *state*) is assessed in terms of the normative perspective of one modal aspect, namely the *jural* function.

Moreover, the concrete existence of societal collectivities, such as the business and the state, is never *exhausted* by any aspect of reality in which they function. Each one of these social entities displays a many-sidedness (or multi-functionality) that enables numerous functionally distinct modes of speech (see Chapter 8 below). A business may be profitable, but in addition it is known by its *trade name* (its function in the sign mode), it has a *unique history* (its function in the cultural-historical aspect), it operates as a context for *social interaction* between its employer and employees (social aspect), and so on. Similarly a state has its own *national symbols* (sign mode), its *unique history*, its delimited cultural *domain* (*territory* – spatial aspect), its typical function in social intercourse (a visit of the police is something different than a visit by a friend), and so on.

1.7 Summary

In this introductory chapter we have illuminated the *philosophical frame of mind* with reference to some borderline questions and by provisionally highlighting the scope and inherent limitations of rationality. In the course of our argumentation, a number of important basic distinctions surfaced. The most significant distinctions are those related to the analytical (or abstracting) abilities of human beings, focused on the nature of identification (lifting out) and distinguishing (disregarding). The last part of our considerations introduced the basic distinction between the various aspects of reality (the *how*) and the diverse concrete entities surrounding us (the *what*). This distinction entails epistemological implications as well as all-encompassing ontological consequences. Epistemology investigates the nature, scope and limitations of human knowing, while ontology accounts for the structure of concrete reality.

Against this background, we can now move ahead in the next chapter and investigate the uniqueness of scholarly (scientific) activities. This analysis, actually exploring basic epistemological distinctions normally falls within the domain of the philosophy of science. It will also serve a more articulated understanding of the 'nature' of philosophy and its foundational position in respect of all the special sciences. In the later chapters we shall return in more detail to the ontological problems, embracing both the dimensions of aspects and entities.

Chapter Two

The uniqueness of science

2.1 The problem of 'demarcation' in the philosophy of science

In North America and Britain, it is customary to restrict 'science' to the domain of *physics*. This practice is mainly the outcome of a particular *philosophical* tradition, known as *positivism*, but its roots go back to the Renaissance and the rise of the modern era. According to positivism, genuine *science* is based upon so-called *empirical observation*, and *experimentation*. What is meant by *empirical observation* is that true science has to start from what could be experienced through the senses, i.e. it proceeds on the basis of *sensory perception* and *sense data*. From sense data, science is supposed to construe its concepts and derive its laws. The *scientific method* consists in formulating *hypotheses* and in *testing* them (experimentally) in order to arrive at *theories* (successfully 'verified' hypotheses).[1]

This strict delimitation of 'science' provides an instance of what is known as the problem of *demarcation* – how does one differentiate between what is *sound science* and what falls outside that intellectual domain? In the context of his constructive realism, Wallner phrases this issue as follows: "It was always attempted anew to set limits to science, in order not to give up what is typically human" (Wallner, 1992:63).[2]

Although this formulation of the problem may sound extremely *innocent*, its historical roots testify to the contrary. In Chapter 1 we have alluded to Kant's fascination with the *law of causality* governing the outer world, and the *moral law* governing the inner life of the human being. This fascination is actually rooted in his struggle with the basic motive of modern philosophy.

What is this basic motive? In order to answer this question, we have to go further back. Kant lived in the 18th century, known as the Enlightenment era because during this period, the exceptional appreciation of the rational capacities of the human being was triumphant. Yet the first half of the 17th century had already clearly manifested the basic motive of what the Renaissance initiated during the 14th and 15th centuries. It concerns the ideal of an all-encompassing natural science (physics) on the one hand and the ideal of an autonomous human person on the other.

Since Descartes, the ideal of such an encompassing natural scientific *con-*

1 Because normal mathematical investigation does not leave any room for *experimentation* in the customary sense of the term, the positivistic criterion actually *excludes* it from the realm of science proper. (The requirement of 'verification' was eventually relativized to 'confirmation'.) A number of issues featuring in this Chapter are treated in the excellent article of Rice on what science is (see Rice, 2000:239-270).

2 "Man wollte der Wissenschaft immer wieder Grenzen setzen, um das typisch Menschliche nicht aufgeben zu müßen."

trol of all of reality started to dominate the scene.[1] In order to proclaim its *autonomy* (being a *law unto itself*) and its *freedom*, the human being had to master reality with the aid of the newly developing *natural sciences.*

In the mould of this new spiritual climate, the 'world' no longer encloses the human being. Rather, the world is recovered as an 'object' at the disposal of the *autonomously free rational human being* with its all-determining natural scientific abilities. Husserl (1856-1938) characterizes this in terms of what he calls the *rationalistic science ideal* since Galileo (Husserl, 1954:64 ff., 119), and Dooyeweerd in a similar fashion speaks of the modern humanistic *science ideal* in its opposition to the humanistic *personality ideal*.[2] The basic motive of modern Humanism is that of *nature* (science ideal) and *freedom* (personality ideal) (cf. Dooyeweerd, 1997-II:215 ff.).[3]

2.1.1 *The infallibility of mathematical thought – Descartes and the classical science ideal*[4]

In his *Discourse on Method* (Part IV), Descartes affirms the motive of control of the new personality ideal, where he holds that the use of our knowledge should "render ourselves the lords and possessors of nature" (Descartes, 1965:15). He praises the success of algebra (Descartes, 1965:18), the clarity and certainty of geometry (Descartes, 1965:33), and believes that with his principles he can satisfactorily explain everything he has observed (Descartes, 1965:51). In his *Discourse on Method* (Part I), Descartes mentions that he "was especially delighted with mathematics, on account of the certitude and evidence of their reasonings" (Descartes, 1965:7). However, in order to justify his trust in reason he was entangled in a *circulus vitiosis* (vicious circle) in his attempt to prove the existence of God. This attempt in fact merely demonstrates his ultimate *trust* in *human reason* as well as the underpinnings of the modern science ideal.

1 In his discussion of the thought of Descartes, Von Weizsäcker reveals a penetrating understanding of this orientation: "This state of affairs is characteristic of modernity. It is not the world in which I find myself that guarantees my existence. This guarantee is not lost, for when I recover the world then it is as the object of my self-assured thinking, that is to say, as an object which I can manipulate" ["Dies ist ein charakteristisch neuzeitlicher Sachverhalt, Nicht die Welt, in der ich mich vorfinde, garantiert mein Dasein. Diese Garantie geht nicht verloren, und wenn ich die Welt wiederfinde, dann als Gegenstand meines selbstgewissen Denkens und darum als Objekt, das ich hantieren kann" (2002:130-131)].

2 Georg Simmel (1858-1918), who viewed himself primarily as a philosopher but in fact, alongside Weber (1864-1920) and Durkheim (1858-1917), is seen as one of the founders of the discipline of sociology, already spoke of (a person's own) *personality ideal* (see Lotter, 2000:188).

3 Note that the expressions "naturalistic science ideal," "rationalistic science ideal" and "science ideal" are employed as synonyms – all expressing what is intended by the humanistic science ideal (the nature motive of modern Humanism). Herold also refers to the way in which Hobbes carried through the *science ideal* in its consequences (Herold, 1974:503).

4 The title of this subheading ought to be appreciated against the background of a book written by a prominent mathematician, Morris Kline: *Mathematics, The Loss of Certainty* (1980).

Starting from the *cogito* (I think),[1] he proceeds under the guidance of the maxim that rational thinking ought to be *clear* and *distinct*. In his *Meditations* III, he takes as a "general rule, that all that is very clearly and distinctly apprehended (conceived) is true." However, the fundamental question is: "What guarantees the truth of clear and distinct thought" (Descartes, 1965:95-96)? Descartes answers that, since God "is wholly superior to all defect" he cannot be a 'deceiver', and therefore guarantees that whatever I clearly and distinctly observe must be true (cf. Descartes, 1965:110).

But for God to be able to guarantee the truth of clear and distinct insight, he must exist. The following question is then: how do we know that God really exists? In order to answer this question, Descartes once again appeals to the maxim of *clear* and *distinct* thinking. The idea of God as an "eternal, infinite [immutable], all-knowing, all-powerful" being, contains "more objective reality than those ideas by which finite substances are represented" (Descartes, 1965:100).

As long as one thinks *clearly* and *distinctly* (and does not allow the will to distract one from this path), one cannot be deceived, and whatever is apprehended is always *true* – because God will not deceive us. Of all the ideas in the human mind, the idea of God is therefore the *clearest* and *most distinct*. Hence God must exist.

The *vicious circle* is 'clear (!)': The existence of God is dependent upon the truth of clear and distinct thinking, while the truth of clear and distinct thinking is dependent upon the (existence of the) non-deceiving God!

This vicious circle actually unveils the fact that Descartes merely used his *idea of God* to impregnate his new *mathematical method of analysis* with the feature of *infallibility*. Underneath the *methodical doubt* leading to the conclusion: "I think, therefore I exist," one finds his deeply rooted modern *trust in* ("faith in"/"*certainty* about") the *rationality* of 'reason'. Unfortunately his argument is *self-defeating*. While *doubting* whatever otherwise seems to be true, he 'discovered' that he cannot doubt that he actually *is* caught up in doubt – which is a *form of thinking* – and from that basic fact he came to the affirmation of his own existence as a thinking being. But let us look closely at his argument:

> Accordingly, seeing that our senses sometimes deceive us, I was willing to suppose that there existed nothing really such as they presented to us; and because some men err in reasoning, and fall into paralogisms, even on the simplest matters of geometry, I, convinced that I was as open to error as any other, rejected as false all the reasonings I had hitherto taken for demonstrations; and finally, when I considered that the very same thoughts (presentations) which we experience when awake may also be experienced when we are asleep, while there is at that time not one of them true, I supposed that all the objects (presentations) that had ever entered into my mind when awake, had in them no more truth than the illusions of my dreams. But immediately upon this *I ob-*

1 Compare his well-known statement: *cogito ergo sum* (I think therfore I am).

> *served* (I am italicizing – DS) that, whilst I thus wished to think that all was false, it was absolutely necessary that I, who thus thought, should be somewhat; and as *I observed* (my italics – DS) that this truth, I think, hence I am, was so certain and of such evidence, that no ground of doubt, however extravagant, could be alleged by the sceptics capable of shaking it, I concluded that I might, without scruple, accept it as the first principle of the philosophy of which I was in search (Descartes, 1965:25-26).

His argument disqualifies every possible perception (observation) and all reasonings formerly taken to be reliable and true. Then he says: "But immediately upon this I observed that, whilst I thus wished to think that all was false ..." – a remark demonstrating that he suddenly elevated **one observation** amongst others *above all doubt* and promoted it to be the first principle of his new philosophy. This indubitable starting point, the infallibility of human reason, served his trust in mathematical reasoning.

This rationalistic trust in in the capacities of mathematical reason inspired the science ideal that dominated early modern philosophy to such an extent that it became a true Frankenstein, threatening its maker. If reality in its totality is reduced to fully determining natural laws (such as the law of causality),[1] human freedom cannot be safeguarded. Without entering into a detailed analysis of the way in which this impasse worked itself out, we shall directly move on to the first serious challenge to the science ideal found in Rousseau's transitional restoration of the freedom motive of modern Humanism. In the thought of Rousseau, the effect of the science ideal is manifest in his contract theory, but in his understanding of the human personality the freedom motive gained the upper hand.

2.1.2 *The road back to autonomous freedom*

Regarding the continued influence of the humanistic natural science ideal in the thought of Rousseau, we have to realize that his theory of the *social contract* still proceeded from the 'atoms' of society, the 'individuals'. This theory attempts to provide a *rational* explanation of an ordered society by reconstructing it from its supposedly *simplest elements*, the *individuals*. But within the on-going dialectical development of humanistic thought, Rousseau forms a *transitional* figure in the sense that he also attempted to liberate himself from the grip of the natural science ideal.

The initial primacy of the science ideal, which 'conquered' all of reality in natural functional categories, such as *spatial extension* (Descartes and Spinoza), the *discrete* and *continuous* [Leibniz with his discrete *monads* and his law of continuity (*lex continui*)], and *perception* (Locke, Berkeley and Hume – the British empiricists), managed to cancel human freedom as well. The deterministic consequences of this humanistic natural science ideal threatened the humanistic ideal of freedom – and Rousseau was the first philosopher who called Humanism back to a radical reflection on its truly deepest motivation: the Renaissance ideal of *free* and *autonomous* humanity.

1 Causality, as we have noted, concerns the relation between *cause* and *effect*.

His reaction to this freedom-threatening (mechanistic) natural science ideal has already surfaced in his treatise on the *origin and inequality between human beings*, where he writes:

> Nature commands every animal, and the brute obeys. The human being experiences the same impulse, but recognizes the freedom to acquiesce or to resist; and particularly in the awareness of this freedom the spirituality of humankind manifests itself. ... but in the capacity to will, or much rather to choose, and the experience of this power, one encounters nothing but purely spiritual acts which are totally inexplicable through mechanical laws (Rousseau, 1975:47).

It is also clear in the essay with which Rousseau won the prize-question posed by the *Academy of Dijon* in 1750. The issue to be dealt with concerned the question whether or not the re-institution of the sciences and the arts contributed to a purification of the morals. The answer that Rousseau gave was *totally negative*. With reference to the classical tradition going back to Egypt and Greece, according to which the god who invented the sciences is to be seen as an enemy of the well-being of humankind, Rousseau explains that the arts and the sciences owe their birth to human defects (Rousseau, 1975:12). The aims of science, as well as its effects, are bad and dangerous for humankind – as product of vanity, science once more merely generates vanity (Rousseau, 1975:13). The progress of science did not contribute anything positive to the well-being of humankind. In fact it corrupted their morals (Rousseau, 1975:22).

The focus of this negative attitude *vis-à-vis* the domination of the humanistic science ideal in the rationalistic culture of the day, resounds in his question:

> Why would we build our happiness on the opinion of others if we can find it within ourselves? (Rousseau, 1975:23).

Immanuel Kant also realized that an *unrestricted* use of the category of *causality* (understood in a *deterministic* natural scientific sense) necessarily leads to an abolition of all freedom.

2.1.3 *Remark about determinism*

The prevailing conception of *physical causality* is convinced that physical entities ought to be seen merely as the extension of physical laws determining their existence exhaustively – explaining why this view is indeed known as *deterministic*. The 20th century physicist, Werner Heisenberg,[1] explains that, according to this deterministic approach, exact knowledge of nature or a particular section of it will suffice to comprehensively and accurately predict the future. He continues:

> If one interprets the word causality in such a strict sense, one also speaks of determinism and means by it that there exist laws of nature determining univo-

1 He is best known for his famous wave formula, constructed in 1927, as a second-order differential equation.

cally from the present the future condition of a system (Heisenberg, 1956:25).[1]

2.1.4 *The basic thrust of Kant's* Critique of Pure Reason

Kant's *Critique of Pure Reason* (CPR) could be seen as an attempt to *restrict* the preceding natural science ideal in order to open up a safe domain for *human freedom*. His aim was to confine the application of reason to sensory phenomena only, in order to leave open a super-sensory domain for the *ethical autonomy* and *freedom* of the human being.

Once this is realized, it is no longer unclear why he introduced the distinction between appearances (sensory impressions) and that which is concealed *behind* the sensory impressions, designated as the "thing-in-itself." This basic distinction between 'Erscheinung' (appearance/*phenomenon*) and "Ding an sich" (thing in itself/*noumenon*), is completely in service of his fundamental aim to safeguard a separate (and *super-sensory*) realm for the human being as an autonomous ethical being, an ethical *aim-in-itself* (*Zelbstzweck*). Kant holds that the category of *cause and effect* (together with all the other categories of human understanding) is only applicable to *appearances* and not to *things in themselves* (such as the *free will* of the *human soul*). In the *Introduction* to the second edition of the CPR (1787), Kant emphatically explains:

> Now let us suppose that the distinction, which our Critique has shown to be necessary, between things as objects of experience and those same things as things in themselves, had not been made. In that case all things in general, as far as they are efficient causes, would be determined by the principle of causality, and consequently by the mechanism of nature. I could not, therefore, without palpable contradiction, say of one and the same being, for instance the human soul, that its will is free and yet is subjected to natural necessity, that is, not free. For I have taken the soul in both propositions in one and the same sense, namely as a thing in general, that is, as a thing in itself; and save by means of a preceding critique, could not have done otherwise. But if our Critique is not in error in teaching that the object is to be taken in a twofold sense, namely as appearance and as thing in itself; if the deduction of the concepts of understanding is valid, and the principle of causality therefore applies only to things taken in the former sense, namely, in so far as they are objects of experience – these same objects, taken in the other sense, not being subject to the principle – then there is no contradiction in supposing that one and the same will is, in the appearance, that is, in its visible acts, necessarily subject to the law of nature, and so far not free, while yet, as belonging to a thing in itself, it is not subject to that law, and is therefore free (Kant, 1787-B:vii- viii).

Kant's ultimate concern to safeguard the (*autonomous*) freedom of the human being indeed necessitated this distinction between *appearance* and *thing in itself*. This is most evident from the entire third part of his CPR (the *Transcendental Dialectic*), where we read:

> The common but fallacious presupposition of the absolute reality of appear-

1 "Wenn man das Wort Kausalität so eng interpretiert, spricht man auch von 'Determinismus' und meint damit, daß es feste Naturgesetze gibt, die den zukünftigen Zustand eines Systems aus dem gegenwärtigen eindeutig festlegen."

> ances here manifests its injurious influence, to the confounding of reason. For if appearances are things in themselves, *freedom cannot be upheld* (my italics – DS; B:564).

The final remark in this subsection reveals the *basic motive* of Kant's whole *Critique of Pure Reason* unambiguously (B:565):

> My purpose has only been to point out that since the thorough-going connection of all appearances, in a context of nature, is an inexorable law, the inevitable consequence of obstinately insisting on the reality of appearances is to destroy all freedom. Those who thus follow the common view have never been able to reconcile *nature* and *freedom* (my italics – DS).

In Kant's thought, the separation of the domains of the science ideal and the freedom ideal (personality ideal) coincides with the separation of 'science' and 'faith':

> Ich musste also das Wissen aufheben zum Glauben Platz zu machen" (Kant, 1787-B:xxx).[1]

This division of *science* and *faith* eventually turned into a kind of *common sense view* within the intellectual development of the West. About a century later, Ernst Haeckel summarizes this dualistic position by claiming that faith *commences* where science *terminates*.[2]

In his Introduction to the first volume of the *Works* of Meinecke, we find a striking summary of alternative expressions for the basic dualism in modern thought. Hofer speaks of the 'discrepancy' between "spirit and nature, values and causalities, ought to be and is, power and culture" (Hofer, 1957:xv).

2.1.5 *The mixed legacy of the 19th century – positivism and its collapse*

Although prominent thinkers after Kant, such as Schelling (1775-1854), Fichte (1762-1814) and Hegel (1770-1831) (early 19th century), further explored the primacy of the modern freedom ideal, Auguste Comte (1798-1857) fell back into the grip of the classical science ideal. The only difference was that he replaced mathematics with *sociology*. Inspired by the motive power of the humanistic natural science ideal, reflection upon human society at this early stage of the 19th century proceeded under the flag of *social physics*. Comte himself also initially used the term *physique sociale*. In a letter to Valat, dated December 25, 1824, Comte for the first time uses the term sociology, although he only made it public in 1838.[3] In his extensive work on *Posi-*

1 "I had to restrict science in order to make room for faith."

2 See the excellent biographical work written by Hemleben (1974). In this context, the penetrating book written by Roy Clouser should be mentioned: *The Myth of Religious Neutrality: An Essay on the Hidden Role of Religious Belief in Theories* (1991, 2005[2]).

3 Cf. *Lettres d'Auguste Comte à Monsieur Valat*, Paris (1870:158), quoted by Horkheimer & Adorno (1973:11-12). Maus remarks that Comte's choice for the term sociology was influenced by his resentment towards the application of *statistical methods* in the science of *social physics*. After the publication of a work on *social physics* by the Belgian statistician Quételet (1835), Comte decided to make his new term publicly known (cf. Maus, 1956:7).

tive Philosophy, Comte accounts as follows for his new term:

> I believe that at the present point I must risk this new term, which is precisely equivalent of the expression I have already introduced, *physique sociale*, in order to be able to designate by a single word this complementary part of natural philosophy which bears on the positive study of the totality of fundamental laws proper to social phenomena.[1]

Comte's positivism was continued both within the humanities (by the British sociologist, Herbert Spencer) and in the natural sciences (particularly in Germany by the physicist, Ernst Mach – who initially influenced Einstein). During the third decade of the 20th century positivism experienced a revival in the well-known *Vienna Circle*, which used as its guideline the *verification principle* – according to which the truth of a statement depends upon its empirical verification (through sense experience).

2.1.5.1 *'Truth' and 'meaning' in logical positivism*

With the rise and development of logical positivism, both the terms 'truth' and 'meaning' played a crucial role, but not to the exclusion of each other. Already in his work, "Allgemeinen Erkenntnislehre" ("A General Theory of Knowledge"), Schlick has been concerned with nothing but the *truth* and *untruth* of propositions and their relation to facts. What was needed, according to him, was another criterion to account for the constitution of scientific knowledge. Stegmüller characterizes Schlick's position as aimed at "obtaining a unique designation of all the facts of the world with a minimal set of concepts" (Stegmüller, 1969:271). In the second impression of Schlick's mentioned work, he explicitly states that there is no other way to the establishment of truth, except by *verification* (Schlick, 1925:151). And eventually, when Schlick published his famous article, "Die Wende der Philosophie" (*The Turning Point in Philosophy*) in the newly established Journal 'Erkenntnis' (1930-1931), he added the notion of *meaning*, but not at the cost of the notion of *truth*. In fact, here he used these two terms in order to distinguish between *philosophy* and *science*: "Every science ... is a system of cognitions, that is, of true experiential statements. ... philosophy is that activity through which the meaning of statements is revealed or determined. By means of philosophy statements are explained, by means of science they are verified" (Schlick, 1959:56).

Even later figures, such as Ayer and Hempel, although certainly concerned with the *meaning* of statements, did not stop speaking of the *truth* of (non-analytical) "empirical propositions": "We may conclude then that the attempt to lay down a criterion for determining the truth of empirical propositions which does not contain any reference to 'facts' or 'reality' or 'experience', has not proved successful" (see Ayer, 1959a:234). Although it is true that a more *explicit emphasis* on 'meaning' is found in the thought of Hempel (see his article

1 *Cours de philosophie positive*, Vol.4, *La partie dogmatique de la philosophie sociale*, identical edition of the first edition, Paris (1908:132 note 1), quoted in Horkheimer & Adorno (1973:12).

on "The Empiricist Criterion of Meaning" – contained in Ayer, 1959, pp.108 ff.), he is not merely focused on meaning, for without any hesitation he speaks of the *truth* of sentences (Hempel, 1959:111, note 6). It appears that logical positivists actually merged their understanding of *sentences* and *statements*, because they do not properly distinguish between the ontic dimensions of the *what* and the *how*. Any *utterance* is constituted by a multi-aspectual human act. Viewed from the modal perspective of its logical-analytical function, an utterance appears as a *proposition* (a logical-analytical configuration) and viewed from its *lingual structuration* it appears as a *sentence* (a configuration evincing the context of the sign mode). Owing to this confusion, Hempel speaks of a criterion of "cognitive meaning" (see Hempel, 1959:116 ff.), where one would have expected something like "semantic meaning" or "lingual meaning"!

2.1.5.2 *Probing the restrictions of sensory perception*

However, positivism did not realize that *sensory perception* relates to *things* and *events*, to the concrete *what* of experience, but that it does not give access to the *terms* employed in *describing* the *how* of what has been observed, for these terms actually stem from the various *modal aspects* of reality. And these modal functions as such are never open to sensory perception. Yet, these aspects provide theoretical thinking with *modal terms* (aspectual terms) that are indispensable for the formation of scientific concepts and theories. As soon as the inevitability of employing modal terms is acknowledged, the Achilles' heel of positivism is laid bare (see the more extensive discussion of the impasse of positivism below, p.44).

In addition, it turned out that in the formation of physical theories, it is never the case that a particular theoretical statement as such is confronted directly with a single experience or sensory experience. The formidable mathematician, Hermann Weyl, subscribes to the conception of Hugo Dingler in viewing physical theories in terms of what is called the principle of *symbolical construction*. Weyl holds the view that the "constructive character of the natural sciences, the situation that their individual propositions do not have a verifiable meaning in intuition (Anschauung), but that truth builds a system which can only as a whole be assessed" (Weyl, 1966: 192). Max Planck[1] states a similar perspective in a concise way:

> Strictly seen it is totally impossible to find any physical question which can be assessed directly through measurements without the aid of a theory (Planck, 1973:341).

Although Popper is sometimes associated with positivism, he considers himself a major figure in causing the *death* of positivism. One of his main con-

1 Planck became famous when he discovered the minimal unit of energy, the quantum h (in 1900). In order to account for the *discrete* nature of the omission or absorption of energy, Planck postulated that radiant energy is *quantized*, proportional to the frequency ν in the formula $E = hn\nu$ – where n is an integer, ν the frequency, and h the quantum of action (*Wirkungsquantum*) with the value 6.624×10^{-34}.

cerns was also the problem of *demarcation*, given in the question: is it possible to elevate the isolated domain of 'science' to the level of being the sole source of reliable knowledge of reality, or is it rather the case that even science as such is dependent upon assumptions that cannot be 'verified' by science itself? As it turned out, this question intimately coheres with the problem of *induction* (i.e. the nature of *generalizations*): is it possible to obtain knowledge with a claim to universality merely by investigating a *limited number* of instances?

Hacohen points out that Popper "sought to overcome the gap the *Wiener Kreis*[1] had opened between science and philosophy" (Hacohen, 2002:195). By attempting this, Popper once again opened the avenue to historical perspectives, because there is no single discipline (special science) which does not, in one or another way, mirror within its own confines the successive trends manifested in the history of philosophy. But we have to explore the historical background of this development a bit further in order to appreciate what Popper achieved.

2.1.6 *Science embedded in a supra-rational commitment*

At this point, we can return to the remarks in Chapter 1 regarding the remarkable *trust* in reason (Popper, Stegmüller, De Vleeschauwer – see page 10 ff.), for this awareness clarifies that it is impossible to separate or demarcate 'science' and 'faith'. It belongs to the very constitution of human rationality that rational pursuits are always already in the grip of more-than-theoretical ultimate commitments, *basic motives*, or *ground motives*. The latter direct theoretical investigations and determine the contents of the theoretical view of reality operative within philosophy and every scholarly discipline.

We have noted that the currently still influential view separating *faith* and *science* originated in the thought of Kant as an outcome of his ultimate ground motive of *nature* and *freedom*. And we have also seen that the disqualification of 'faith' is itself dependent upon an implicit *trust* (i.e. *faith*) in reason which actually stretches back to Greek antiquity.

> *Comment*: At this point we come across the many root-symbols that signify these supposedly created places of rest for the restless heart of the human being. Think of the drawing power of such 'shelters' as happiness, prosperity, wealth, success, freedom, and so forth. P.J. Visagie talks about "pastoral shelters" – i.e. places where humankind in its deepest insecurity and unrest apparently can come to *rest*. It reminds us of the great Dutch historian, Huizinga, who asks in his work, *Geschonden wereld* (*Desecrated world*) whether or not art could bring about renewal in the sunken Western culture. Malherbe recognized this as an overestimation of the aesthetic aspect of reality, "Art cannot be a lasting city for the restless heart of humankind. Art can give passing satisfaction, momentary joyous experiences, but art itself is caught in turmoil, by nature referring us to Him who is the Origin of all things" (Malherbe, 1947:85). In his *Confessions* (written in 400 A.D.) Augustine has already emphasized

1 The *Vienna Circle*.

this foundational biblical perspective: the human heart knows no peace before it comes to rest in God.

At this point we have to refer to the unity and goodness of creation. Literally all world and life views throughout the history of the world *divided* reality by separating a domain considered to be *good*, from the rest of creation, supposed to be inherently *bad*. The only exception is found in the new Testament perspective on the kingdom of God.

The Bible does not localize evil in a specific domain of reality, but in the *apostate direction* of the human heart, while salvation is equally a *directional* matter (first seek the Kingdom of God – in every sphere of life). If we look at philosophy (and the different existing special sciences) from the depth perspective of a world and life view, the most remarkable fact is that we are constantly confronted by what may be called a *surrogate salvific appeal*. In other words, we are confronted with a *way of liberation*, with a call to *move away* from one terrain of creation to "the kingdom of freedom/virtue/self-perfection/goodness/autonomy" etc.

When the good-evil opposition (*antithesis*) is identified with distinct domains within creation, *structure* and *direction* is confused.[1] This confusion leads us to consider a closer look at a fundamental life and worldview issue, the *normativity of human life*.

2.1.7 *The normativity of human life*

Humans are *normative* beings by nature. They are called to respond to the *normativity* of human life either by conforming to or by rejecting the norms guiding human endeavours. Humans are able to discern *truth* from *falseness* and what is *logically sound* from what is *illogical*, just as they are able to know the difference between what is *beautiful* and what is *ugly*. This normative fibre of our shared humanity naturally spans across multiple dimensions of normativity, exemplified in considerations such as:

- Humans are extremely sensitive to the difference between *justice* and *injustice*.
- They are aware of the benefits of *frugality* as opposed to the sorrows of *wastefulness*.
- Their experience of lingual *ambiguities* is filled with examples of correct and wrong *interpretations*.
- They know what the value of *courtesy* is and what the effects of *impoliteness* may be.
- Similarly, humankind has heroic and heartbreaking stories to tell about what is norm-conformative in a *historical* sense and what is historically antinormative or *un-historical* (for example: what is *reactionary* or what is *revolutionary* as opposed to what is *reformational*).

1 Since God's creational law-order strictly speaking gives direction to creaturely subjects, one should speak about its *direction-giving* structure. Similarly the directional antithesis between good and evil is embraced by God's Law-Word – hence one should speak about *structured* direction.

Every inter-human encounter brings to expression this *normative dimension* and takes place under its 'supervision'; is played out within this cosmic theatre of human beings as norm-observing agents. Although individuals oftentimes have *diverging understandings* of what *truth, logicality, justice, love, frugality, interpretation, courtesy* and norm-conformative *historical* actions are, they cannot avoid this "norm-determinedness" of human life. For this reason even in every *antinormative action*, the human being is constantly haunted by the underlying and presupposed normative awareness of what "ought to be" – aptly captured by an age-old legacy that designates it as the uniquely human *conscience*.

But precisely for this reason, it belongs to the very *constitution of human life* and to the intricate *fabric of human society* to have a vital concern and interest in the *normative orientation* of human beings – whether they are fellow citizens or even closer to home, *children* and *students,* who are dependent upon decent educational institutions on their path towards responsible adulthood. Because the multi-dimensional existence of the human being is not *absorbed in* or *exhausted by* any single societal institution (such as the state, one or another ecclesiastical denomination, business enterprise or a particular social club), it is wrong in principle to restrict or narrow down the process of education to be of service to a single or even a number of specific societal institutions only. Therefore, human beings are never (exclusively) educated for 'citizenship', or for "church-membership," or for 'partnership' ('friendship'), or whatever. They are educated to fulfill a *multiplicity of roles* within diverse societal institutions, and throughout one's life, these functionally distinguishable social roles are constantly and concurrently acted out.

Human beings, in their actions and societal institutions, are therefore guided by *norms* and humans constantly give shape to basic *principles.* This at once also explains why humans functioning in diverse societal relations do not *cease* to be *norm-oriented* – for in these instances they have to observe *collective norms*. When a *just state*[1] acts in the pursuit of *public justice*, it has to observe collective norms. Furthermore, when a just state strives to observe *basic rights*, it assumes a task that could be performed in a *better* or *worse* manner. But we shall return to all these issues in more detail once we have entered into an analysis of the different dimensions of a differentiated society. For the moment, we have to return to the distinction between structure and direction.

2.1.8 *The distinctness of structure and direction*

The preceding line of argumentation regards normativity as based upon an acknowledgment of the *distinctness* of *structure* and *direction.* The moment

1 When the rights and freedoms of people are guaranteed within a state such a state is often characterized as a 'Rechtsstaat' in German (Dutch and Afrikaans). Although the term 'Rechtsstaat' is normally translated as a constitutional state under the rule of law one may also opt for the concise alternative: *just state*. Dworkin, for example, also employs this phrase (cf. Dworkin, 2004:376).

these two are confused, the directional antithesis between *good* and *evil* is understood in structural terms, inevitably resulting in a *dualistic worldview*, where 'good' and 'evil' are respectively identified with *specifically opposed areas of life* within creation.

For the later Greek philosophers, *evil* is found in the *material* world; for the existential philosopher of the 20th century, it is found in societal structures that threaten the *individual freedom* of a person; for the neo-Marxist and the social conflict theorist (cf. Hegel, Simmel and Dahrendorf), it is found in the *authority structure* of societal life-forms; for other thinkers, it is found in the supposed inevitability of *natural causality*, and for others, by contrast, in the appearance of *freedom* that an individual is supposed to possess. This style of practising science – in philosophy and in special sciences – still indicates the way to what is supposed to be *good*, to the *meaning* of life and to *freedom*. According to Wolters, these concerns are all directed towards the path to salvation – as the escape from one terrain of creation to another terrain of creation: for example by moving to *rationality*, to *forming*, to the *collective whole* (of the *nation*, the *state* or the *church*), to *freedom* etc.

Each one of these ways to salvation rests on a *misdirected* evaluation of a well-created part of creation. With an inner necessity this leads to a *depreciation* of something else *within creation* (a fundamental characteristic of the ancient heresy of *Gnosticism*), while at the same time coming to the idolization (absolutization, deification) of something else within creation – a point of departure of all those instances where a creature is honoured instead for the Creator. Dualistic worldviews have their *holy cows* and *golden calves*, correlated with their *black sheep* and *scape-goats*.

Sometimes the traditional nature-grace split simply results in an unfortunate two-realm perspective. Francis Collins, for example, grew up as an agnostic and eventually became a committed atheist. However, during his medical studies his worldview and religious orientation was turned around. Alister McGrath refers to his book on the "*Language of God*," in which Collins "argued that the wonder and ordering of nature pointed to a creator God, very much along the lines of the traditional Christian conception. In this book, Collins describes his own conversion from atheism to Christian faith" (McGrath, 2007:20). Unfortunately the account given by Collins himself is still in the grip of a two realm (nature-grace) perspective, for he holds: "In my view, there is no conflict in being a rigorous scientist and a person who believes in a God who takes a personal interest in each one of us. Science's domain is to explore nature. God's domain is in the spiritual world, a realm not possible to explore with the tools and language of science. It must be examined with the heart, the mind, and the soul—and the mind must find a way to embrace both realms" (Collins, 2007:6). Many books within this domain, owing to the absence of a non-reductionist ontology, do not succeed in entering into the field of theorizing of the natural sciences – by and large these works still accept 'science' as a factual given without different schools of thought.

2.1.9 *Once again the problem of 'demarcation'*

In the opening section of this chapter, we have referred to the "problem of demarcation." At this point of our argument we may point out that the problem of demarcaton carries with it a subtle but significant legacy in which the distinctness of structure and direction is confused. Within the dominant philosophical orientation amongst the special sciences, during the first half of the 20th century, the "positive facts" were assumed to be the sole guide to "objective scientific truth." "Sense data," as we have seen, were supposed to be the only source of reliable knowledge, and it supported the postulate of the *neutrality* of human rational endeavors. The latter conviction (!) erroneously labeled any ultimate commitment (conviction) operative within the domain of rationality as a *disturbing factor* that should be *eliminated* from science.

However, as we have seen, without an implicit *trust* or *faith* in reason this postulate itself cannot be maintained. All human beings are endowed with the capacity to think and to argue rationally, but they do this from *diverging direction-giving orientations*. Consequently, in spite of acknowledging the universality of the structural conditions making possible human thinking in the first place, no single human being can escape from one or another deepest conviction.

An analysis of the *structure* of scientific activities therefore does not aim at *securing* a domain of the *good* by protecting it from the evil influence of direction-giving ultimate commitments, for any such analysis can only proceed by implicitly proceeding from a particular life-orientation.

There are not simply 'scientific' people *liberated* from any and all supra-rational convictions, and "non-scientific" people blurred by the 'evil' of adhering to one or another *conviction*. Whatever the life-orientation of thinkers may be, they all equally share in the dimension of *rationality* (or: logicality) and all of them are inevitably in the *grip* of a more-than-rational *ultimate commitment*.[1]

In an earlier context, we have mentioned the fact that the nature of so-called "sense data" is problematic, because the theoretical 'tools' employed in the *description* of what is observed always utilize *terms* that are not susceptible to "empirical observation" themselves.

2.1.10 *The impasse of positivism*

Let us explore this issue in some more detail. In order to highlight the limitations of the senses in the acquisition of knowledge, we may consider the concept of *matter* in terms of some of its main conceptual transformations – anticipating the more detailed analysis that will be given in Chapter 7 (see pages 402 ff. below).

We have referred to the fact that the Pythagoreans adhered to one statement above all else: *everything is number*. After the discovery of irrational numbers (see Von Fritz, 1945) – revealing within the seemingly *form-giving* and *delim-*

1 A penetrating analysis is given in the above-mentioned work of Clouser in terms of the hidden role of religious belief in theories (see Clouser, 2005).

iting function of number something *formless* – Greek mathematics as a whole was transformed into a *spatial mode* (the *geometrization* after the initial *arithmetization*). As a consequence, *material* entities were no longer described purely in *arithmetical terms*. The aspect of *space* now provided the necessary terms required to characterize material entities. This spatial angle of approach remained in force until the rise of modern philosophy, since philosophers like Descartes and Kant still saw the 'essence' of material things in their *extension*. Particularly through the work of Galileo and Newton, the main tendency of classical physics eventually underwent a shift in perspective by attempting to describe all physical phenomena exclusively in terms of (kinematic) *motion*.

Since the introduction of the *atom theory* of Niels Bohr in 1913, and actually already from the discovery of *radio-activity* in 1896 and the discovery of the energy quantum *h*, modern physics realized that matter is indeed characterized by *physical energy-operation*.

From this brief explanation it is clear that different aspects served to characterize matter – starting with the perspective of *number* and then proceeding to the aspect of *space*, the *kinematic* aspect and eventually the *physical* aspect of reality.

The key question is whether these modal aspects could be observed in a sensory way. Can they be *weighed*, *touched*, *measured* or *smelled*? The answer must be negative, for they are not *things* but *aspects* of things (or rather aspects within which concretely existing things function).[1] The first step positivism had to take in order to *digest* "sense data" theoretically has already eliminated the restriction of reliable knowledge to sense data!

The renowned physicist, Max Planck, who eventually became sharply critical of Mach's positivism, distinguished between the real outside world, the world of the senses and the (theoretical) *world of the science of physics*, which he equates with the "physikalisches Weltbild" (*the physical world picture*) (Planck, 1973:208). The abstractions that belong to the 'Weltbild' are not sensorily perceptible – they embrace, according to him, the known law-conformities and concepts such as space, time and causality (see Vogel, 1961:149).

Against this background, it must be clear that the undertaking to delimit scholarly activities and distinguish them from other human practices is not to be equated with the antithesis between *good* and *evil*, but simply has to take on the task of making distinctions *within* the good order of creation.

2.2 What is unique about science?

We may now proceed to determine the uniqueness of science while keeping in mind that 'science' is here understood in the broad sense of scholarship embracing both the natural sciences and the humanities. In order to do that, it is required to identify those properties peculiar to science. Any act of identification involves two kinds of properties: (i) shared properties; and (ii) distinct

1 See our more elaborate discussion of this issue – page 393 ff.

properties.

Suppose we had the task of differentiating between a *plant* and a *stone*.

(i) There are fundamental *similarities* between plants and physical things. A plant, for instance, has a characteristic mass, that is a characteristic shared with material things. Similarly, both plants and material things share a certain spatial *extension*, a certain *durability* and a certain *unity*.

(ii) Only when we pay attention to the fact that plants are *alive* do we come into touch with the difference-side in the comparison of a stone and a plant, that is, with the *distinctive characteristic* of being-a-plant, of *plantness*.

A starting point for the investigation of what science is may appear to be given in the observation that every scientific activity is a *thought-activity*. Yet, this characteristic still does not *distinguish* science from non-scientific activities, because someone who is not engaged in scientific practices certainly can also *think*. We may add a more precise focus to this question by asking: what *sort of thought* is scientific thought? What *kind* of thinking is entailed in scientific thought?

2.2.1 *Thought activities involved in doing science – shared properties*

(a) Is scientific thought *systematic thought*? Certainly, but this is by no means a *distinctive characteristic* of science. The judge who is preparing a verdict in court must similarly work systematically in her argumentation, but this does not mean that a legal verdict changes into a legal scientific treatise.

(b) Is it *verifiable thought*? While the answer to this must also be in the affirmative (taking into account the controversy in contemporary theory of science over the *meaning* of this characteristic), it is still not a distinctive scientific characteristic, since the judge mentioned above must also verify that every bit of evidence she considers is *trustworthy*. In fact, Popper challenged this feature with his criterion of *falsifiability* while the empiricism of Hume has already questioned the strict verifiability of law statements, because a limited number of observations can never justify the universality entailed in such statements. Leibniz, more than thirty years before the first appearance of Hume's *A Treatise of Human Nature* – (1739),[1] wrote as follows in his *New Essays*:

> Now all the examples which confirm a general truth, whatever their number, do not suffice to establish the universal necessity of that same truth, for it does not follow that what has happened will always happen in the same way (quoted by Stich, 1975:45; see Kant, 1787-B:5). It is known that early in his intellectual development, Kant was strongly influenced by Leibniz. Therefore, the claim by J. Bennett, namely that we should "not credit Leibniz with any Kantian insight about the need for intellectual structure" is a bit farfetched (see Bennet, 1974:37). Paul Robinson pointed out that "what is controversial is that any generaliza-

1 His *An Inquiry Concerning Human Understanding* appeared in 1748.

tion cannot even be confirmed to any degree greater than zero: we have no more reason to think that the sun will rise tomorrow than we do that it will not!"[1]

c) Is it *methodical thought*? Since there are also non-scientific methods of doing things, it will always be necessary to distinguish between *scientific* and *non- scientific* methods. Therefore *methodology* as such cannot be the distinctive characteristic of science, for it presupposes the disctintion sought. Positivism even viewed the so-called *scientific method* (of observation, the formulation of hypotheses, and the erection of theories) to be the *sole* entrance to reliable knowledge. But it did not contemplate the option that the nature of what is investigated is decisive and not the method designed to know it, since the latter is dependent upon the former. Method at most has the role of a *servant*.

Neemann, for example, characterizes the prominence of method in positivism analogous to the opening verse of the Gospel according to John: "In the beginning was the Method " ("*Am Anfang war die Methode*" – Neemann, 1986:70).[2]

(d) Particularly tired and disseminated is the notion that science is centered in the relationship between a scientific researcher (the *knowing subject*) and that which is studied, namely the so-called *study object*. In the first place we need to note that the subject-object relation is common to non-scientific human experience as well. Merely think about the way in which human subjects are using *social objects* (like furniture), or *technical objects* (tools), *economic objects* (money), *semiotic objects* (books), *aesthetic objects* (paintings), *ethical objects* (engagement or wedding rings), *legal objects* (property), and the like. All these 'objects' actually indicate object-functions of concrete entities, which can *equally* be studied by various special sciences (each from their own distinctive perspective). Suppose we say that we are interested in the scientific investigation of the business, does that mean that our "study-object" is found in the nature of a business enterprise? Not at all, because different academic disciplines may focus on the business, albeit from *different angles of approach*. If the three disciplines of *industrial psychology*, *industrial sociology* and *business economics* all study the business, merely saying that the *business* is their *study-object* will not suffice to distinguish them from each other. In order to do that, one has to identify the specific angle of approach employed by each.

These considerations make it plain that the cardinal question is not: with what **object** (natural or social entity) or event does *this* or *that* science engage itself? but rather: from what *perspective* (*aspect*, *way of being*, *mode*, *modality*, *function*, *facet*, *mode of explanation*) of reality are certain things, events and societal relationships studied by a particular *academic discipline* (special sci-

1 E-mail remark, November 26, 2005.

2 The relativity of method is emphasized by Putnam: "Today, virtually no one believes that there is a purely formal scientific method" (Putnam, 1982:25).

ence)?

2.2.2 *The distinctive feature of scientific thinking*

At this point, where it is clear that we have not yet been able – through an analysis of possible ways of thinking – to discern the *distinctive feature* of scholarly activities, another option may be to claim that theoretical (scientific) thinking is stamped by the feature of *abstraction*.

In Chapter 1, we discussed Frege's arguments against the appropriateness of (entitary-directed) abstraction, given in his example that higher levels of abstraction starting with the 'moon' does not yield the number one, but only more general categories – such as *attendant of the earth, attendant of a planet, celestial body without its own light, celestial body, body, object* (entity). This capacity is inherent in all the everyday (non-scientific) categorizations and classifications we are used to.

Therefore it must be clear that *abstraction* in the context of entities cannot produce the kind of functional concepts within a scientific universe of discourse. Those of us who had to use an *abacus* in order to be taught how to go about with the most basic arithmetical operations, such as addition and subtraction, will understand this straightaway.

When we learn to calculate with the help of the abacus, we begin by involving *different aspects of reality* at once: we take into account the *color*, the *movement*, the *shape* and the *quantity* of blocks on the abacus, that is, we initially include the many-sidedness of these blocks by leaving their physical, kinematical, spatial and numerical properties *intact*. However, gradually we had to ignore the color, movement and shape, and concentrate on the quantitative side only, i.e. we had to elevate the numerical aspect in order to simultaneously ignore the non-numerical aspects (namely the spatial, the kinematic, the physical and the other aspects not yet mentioned). Keep in mind that the abacus is a *cultural* artifact (its formative aspect), that it has a *name* (its function within the sign mode), that it *belongs to someone* (the jural aspect evinced in the accompanying property right), and so on. We shall see later that the first four aspects (numerical to the physical) serve as *subject* functions, whereas the others are what is called *object* functions. At the same time we will question the popular practice amongst academics to refer to all entities as 'objects'.

With that, we have moved to *theoretical (scientific) thought* – i.e. we have accounted for the *abstraction* of certain *aspects of reality*. The chief point to be observed is that merely referring to *abstraction* as such is not sufficient, because what is required, is the specification: *modal abstraction*.[1] In the absence of a theory of aspects modal abstraction is sometimes designated as *idealizing abstraction* (see Diemer, 1970:213 and Bernays, 1976:37, 196).

Whoever is engaged in *modal abstraction* relinquishes the non-relevant aspects and focuses her theoretical-logical attention on one particular aspect.

1 This expression, namely "modal abstraction," accounts in a different way for what Cassirer calls the "relational connection" (Cassirer, 1953:39; see Chapter 1, p.19).

The distinctive characteristic of theoretical-logical (= scientific) thought, therefore, is *modal abstraction* (see Figure 4 on the next page). Because all concrete entities function in all the various aspects of reality, the abstracted modalities (aspects) provide access to the analysis of the structures of such entities.

Although any person not involved in doing science certainly has an *analytical awareness* of the diverse aspects of reality, this does not mean that in our non-scientific experience of reality we ever achieve *modal abstraction*. One would not, for instance, in one's non-scientific experience, reflect on the nature and structure of the numerical aspect when one notices *six* people walking past one's house, just as little as one would develop an economic price theory when one notices that the price of a car is $30,000. Without an analytical awareness of the diversity of aspects in reality, one would also not have any conception of what is meant when someone comments that a certain car is *beautiful*, yet *expensive*. *Beauty* (aesthetic aspect)[1] and *price* (economic aspect) are facets of one's total experience of cars, although we would maintain that these aspects are generally noticed in a non-abstractive manner *of* and *in* the entities we experience.

In the absence of a theory of modal aspects, Frege did not contemplate the nature of *modal abstraction* in connection with the number concept. When a multiplicity of concretely existing things (or entities) also exhibit a function within the (universal) modal structure of the quantitative mode of reality, the decisive issue is not whether, through (entitary directed) abstraction, higher-level (type) concepts are formed, but rather whether the core meaning of number (discrete quantity) applies to these or any of the entities functioning *within* this aspect. Number, in the first place, concerns the *multiplicity* of entities and not the entities *as* entities. What is shared by a multiplicity of things is the quantitative property of *being distinct*. Even if Frege may have it his way in arguing about abstracting from the typical differences between entities, this line of argumentation does not account for the way in which entities function within the framework of the quantitative aspect. Frege focuses on the familiar conception that the combination of different 'units'

Figure 4

The *modal abstraction* of the social aspect

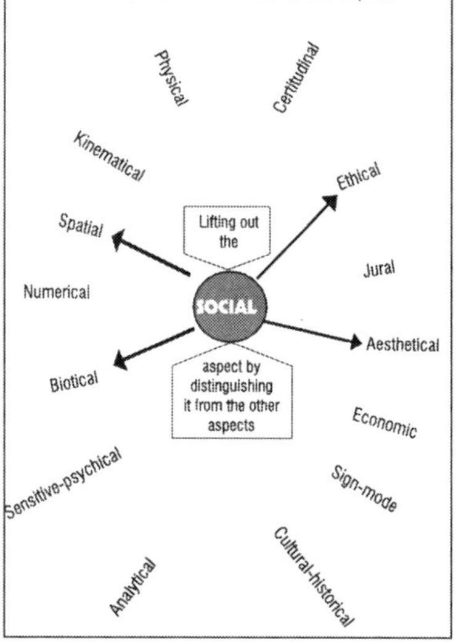

1 Calvin Seerveld and Henk Hart defend alternative views of the unique meaning of the aesthetic aspect. We shall return to this issue in a later context (see page 248 ff.).

successively generates different numbers. And he seems to have sound arguments against this kind of approach. If the term 'unity' is meant to designate "objects to be counted," then, so Frege argues, one cannot define number as (multiple) 'unit(ie)s' (*Einheiten*). In general, he claims that "a plural is only possible of concept-words."[1]

A crucial part of Frege's argument is that, as long as different entities are brought together, the *differences* between them remain intact – and *number* does not constitute these differences. But note that these differences are *typical* differences! He argues that, when one abstracts from the peculiarities of individuals in a "collection of objects," or when one disregards the properties that distinguish separate things, then it is not the case that what is left is the concept of their number (*Anzahl*), for instead one arrives at a *general concept* embracing every one of these things. In other words, the result is, according to Frege, one general *type concept*.

Against this background, Frege considers two equally unacceptable options. When we want to generate number through the *collection of different entities*, then we obtain an accumulation in which the objects with precisely those properties through which they are distinct are found – and that is not number. When, on the other hand, we want to construct number through the union of what is the same (*gleich*), then everything always collapses into *one* – in which case we are unable to arrive at the many (*Mehrheit*).[2] Whenever no differences are left, multiplicity shrinks into *oneness*.

Frege's argumentation is burdened by two mistakes. (i) The first is the focus on different entities (i.e. *typical* differences) instead of accounting for a numerical plurality (*quantitative distinctness*); and (ii) the second is found in his inability to distinguish properly between modal abstraction and entitary directed abstraction – to be seen in his lack of distinguishing between *modal equality* and *entitary identity* ('gleich' in the sense of 'similar' and in the sense of 'identical').

Re (i): It is certainly correct to claim that number (a modal, functional concept) does not constitute the (typical) differences between entities, but this insight does not warrant the conclusion that the multiplicity (plurality) of entities involved in such a process of (entitary-directed) abstraction is nullified by

1 "Ein Plural ist nur von Begriffswörtern möglich. Wenn man also von »Einheiten« spricht, so kann man dies Wort nicht gleichbedeutend mit dem Eigennamen »Eins« gebrauchen, sondern als Begriffswort. Wenn »Einheit« »zu zählender Gegenstand« bedeutet, so kann man nicht Zahl als Einheiten definieren" (Frege, 1884:50). For this reason, he substituted the notion of a set with that of the *domain of a concept* (*Begriffsumfang*). In his review of Frege's 1884 work, Cantor criticizes this view (Cantor, 1962:440).

2 "Wenn wir die Zahl durch Zusammenfassung von verschiedenen Gegenständen entstehen lassen wollen, so erhalten wir eine Anhäufung, in der die Gegenstände mit eben den Eigenschaften enthalten sind, durch die sie sich unterscheiden, und das ist nicht die Zahl. Wenn wir die Zahl andererseits durch Zusammenfassung von Gleichem bilden wollen, so fließt dies immerfort in eins zusammen, und wir kommen nie zu einer Mehrheit" (Frege, 1884:50).

this very process. This consideration naturally leads to our second point.

Re (ii): Frege first summarizes the outcome of his argumentation by pointing out that *number* is not a *property* that can be abstracted from *things* in the way in which colour, weight and density could be abstracted from such things – which leaves us with the question: what is meant when a number is specified?[1]

Frege holds that number is not something *physical*, nor is it something *subjective* like a *representation* – just as little as it is the union of one thing with another. He claims that expressions like *multiplicity*, *set*, and *plurality* are indeterminate and therefore not appropriate to serve as an explanation of *number*. Features like *being delimited*, *being undivided*, and *being unanalyzed* are not useful for what we express with the word 'one', either. For Frege the notion of unities (Einheiten) seems to unite two contradictory properties: identity and difference.[2]

The two words 'identity' and 'difference' implicitly appeal to the dimensions of entities and aspects: different instances of some or other kind of entity are subsumed under the general (universal) entity-concept involved, and Frege correctly argues that this concept is *one*. Yet in our everyday experience, counting always concerns both dimensions, for the first (entitary directed) question is: *what* is counted? whereas the second pertains to the (modal) quantitative question: *how many* are there? Every counted (or: countable) specimen is *similar* ('gleich') to every other one in the *numerical sense* of just being another 'one' to be counted.[3] An entitary perspective on the previous sentence yields the unified "what-concept," while a modal functional arithmetical perspective on it generates the "number concept" of how many entities there are. In other words, the conceptual identity of a multiplicity of entities cannot eliminate this (modal, functional) numerical multiplicity!

Although he is not acquainted with the distinction between the entitary dimension and the dimension of modal aspects, Tait has a very clear understanding of the above-mentioned states of affairs. He claims that Frege tends to confuse the following two questions: "What are the things to which number applies? And, what are numbers?" (Tait, 2005:241). In order to highlight the difference between the *what* and the *how* (i.e. the difference between entities

1 "Die Zahl ist nicht in der Weise wie Farbe, Gewicht, Härte von den Dingen abstrahiert, ist nicht in dem Sinne wie diese Eigenschaft der Dinge. Es blieb noch die Frage, von wem durch eine Zahlangabe etwas ausgesagt werde" (Frege, 1884:58).

2 "Wenn man die zu zählenden Dinge Einheiten nennt, so ist die unbedingte Behauptung, daß die Einheiten gleich seien, falsch. Daß sie in gewisser Hinsicht gleich sind, ist zwar richtig, aber wertlos. … So schien es, daß wir den Einheiten zwei widersprechende Eigenschaften beilegen müßten: die Gleichheit und die Unterscheidbarkeit" (Frege, 1884:58). Of course it is more appropriate to translate the words 'Gleichheit' and 'Unterscheidbarkeit' as 'being the same' and 'being distinguishable/being distinct'. Chihara recently claimed that questions like these have not been addressed by most mathematicians (Chihara, 2004:50 ff.).

3 Numerical similarity means that anyone is as good as another – for the purposes of counting they could be interchanged.

and aspects) we prefer to phrase the second question in terms of 'how' by asking "*how* many are here" instead of "what are numbers"?[1]

A theoretically articulated understanding of the meaning of number is therefore only possible on the basis of modal abstraction (denied by Frege).

From the general meaning of modal abstraction, some important implications follow, for a superficial knowledge of the idea of modal (functional) aspects often caused the misunderstanding that the academic disciplines (special sciences) are restricted to 'aspects' – understood as 'parts' of reality.

2.2.3 *Are the disciplines 'restricted' to certain 'parts' of reality only?*

By calling *modal abstraction* the uniquely distinguishing feature of scientific thought, we have in no way built in a *restriction* regarding the study of concrete things and events of our everyday life, because the different aspects of reality still act as the *gateway* to our experience of those different things and processes within reality. Therefore we can never say that a special science (that is, a theoretical discipline that is delimited by a particular, modally abstracted, aspect of reality) is *restricted* to a 'section' or 'part' of reality.

This misunderstanding is even found amongst those who are acquainted with the general theory of modal aspects. Since every (natural and social) entity and every event or process in principle has a function within *every* different modal aspect, a special science with its theoretical investigations delimited by a particular aspect, still has access to reality in its *totality* – viewed from the angle of that modal perspective. The physicist, the biologist or the economist are not investigating the *modal structure* (the "aspect-structure") of respectively the physical aspect, the biotical aspect or the economic aspect *as such*, since they are actually looking through the 'glasses' of their respective delimiting aspects to whatever is functioning *within* them. The investigation of the modal structure of an aspect as such forms a part of the task of *philosophy*. In the first chapter, we explicated this issue when we alluded to the quest to account for the coherence of what evinces itself as irreducible, that is, the *coherence of irreducibles* (see pages 8, 13, 52, 403).

Consequently, we have to emphasize that being *delimited* by a certain aspect of reality does not entail that any discipline is *restricted* to merely one kind of entity. But what then about something like *atomic physics* or "business economics" (*industrial economics*)? Are these not instances of disciplines restricted to *one* entity only?

The mere fact that the general name of a special science (such as *physics* or *economics*) is here focused on one entity only, does not entail that the discipline as such is restricted in its universal scope. It simply indicates that for practical purposes, one of all the many entities functioning within such an as-

1 Tait also points to the confusion present in Frege's thought regarding the meaning of the word 'gleich': "His reading seems to me to have been misdirected by two related things: his interpretation of '*gleich*' to mean 'identical' and his failure to understand the historical use of the term 'number' to mean what is numbered" (Tait, 2005:242).

pect has been selected, without broadening the scope of investigation in order to include any *non-physical* or *non-economic* angles of approach as well.

Because concretely existing entities surely do function at once in more than one aspect, a responsible understanding of the nature of such entities and processes does require a *multi-disciplinary* approach. But the term 'multidisciplinary' does not eliminate the existence of a multiplicity of (modally delimited) disciplines, since it *presupposes* them.

The full concrete reality of our everyday life experience falls within the field of study of every special science – with the single qualification: *observed from its modally-abstracted angle of approach*. A number of disciplines entered the academic scene in close connection with a prominent aspect of particular entities. In addition to *atomic physics* and *industrial psychology*, we may also think about *political science* (closely connected to the jural aspect of the state as a public legal community), *social anthropology* (closely related to the cultural interaction of peoples), and so on. Even if it may be true that these intellectual traditions actually embrace more than one special scientific viewpoint at once, it still does not cancel the givenness of these primary aspects of reality, or the diverse special sciences delimited by them. The decisive criterion is simply: is the aspect of reality claimed to form the modal delimitation of a specific special science indeed a *unique* and an *irreducible* aspect of reality?

2.2.4 *Modal abstraction entails that every special science has philosophical presuppositions*

If it is true that, in spite of a variety of *shared properties*, scholarly thinking indeed finds its distinctive trait in *modal abstraction*, then it stands to reason that no single special science can account for its modally delimited approach purely in a *special scientific manner*.

This conclusion actually directly flows from Figures 1 and 2 (Chapter 1, page 14). Modal abstraction requires the *lifting out* (*identification*) of an aspect by *disregarding* the other aspects. In order to do this, the existence of *more than one aspect* is required. But being limited to the scope of one aspect only, no single science can accomplish this task in a special scientific way.

If reality contained one modal aspect only *modal analysis* as such would have been impossible, plainly because we can only *identify* an aspect by *simultaneously* distinguishing it from all those aspects which *differ* from the identified aspect. *Theoretical analysis* (*modal abstraction*) must therefore always simultaneously consider *at least two differing* aspects. In itself, this insight questions the nature of *reductionist* modes of thought.

Throughout the history of philosophy and the disciplines, reductionist modes of thought consistently manifested themselves in what became known as *monistic isms*. We have noted that the Pythagoreans believed that *everything is number*, but that through the discovery of *irrational numbers*, Greek philosophy switched to a *spatial* point of orientation.[1] Early modern philosophy advanced under the influence of the classical *mechanistic* worldview that also dominated the main tendency of modern physics up to the end of the 19th century.

Comment: Heinrich Hertz is the German physicist who did experimental work about electromagnetic waves more than a hundred years ago. This work not only established him as the founder of wireless telegraphy and the radio, but also immortalized his name in the unit of frequency (*Hertz*) named after him. Soon after his death in January 1894, his comprehensive theoretical work appeared: "The Principles of Mechanics developed in a New Context."[2] Restricting himself to number, space and movement (the first three modal aspects) only – represented by the concepts mass, space and time – he rejected the concept *force* (a physical concept)[3] as something *inherently antinomic* (cf. Katscher, 1970:329). This shows how consistently he carried through the *mechanistic* approach in modern physics (to which we shall return in Chapter 7).

This mechanistic orientation has played a crucial role in the development of modern biology, since Descartes viewed the human body as a machine. Opposed to this a *vitalistic* inclination, originally found in Greek thought, continued. An early Greek philosopher, Thales, who lived in the 6th century B.C. and became known for his prediction of the eclipse of 585 B.C., said that *everything is alive*. Of course this *pan-vitalism* had a problem with *death*, just as pan-mechanism had a problem with *life* (see below pages 172 ff.).

Anyone acquainted with present-day molecular biology and the incredibly advanced knowledge we currently have about the physico-chemical substructure of living entities, certainly may find Thales' thesis implausible. Yet in spite of its seemingly insignificant proportions, the thin layer of living entities in our world exercises a biotical influence on the entire biosphere of the earth. The biologist, Jones, remarks:

> Through the key processes of respiration and photosynthesis (the world's most important redox reactions!), organisms entirely renew the carbon dioxide in the air every few years, and even the much larger volume of oxygen is renewed in about 2000 years. Most incredibly of all, even the 1.5 billion, billion metric tonnes of water on the Earth are eventually split and reconstituted by the activities of living things. It would seem incredible, because the biosphere is such a very thin film on the Earth's surface (it is spread throughout the oceans of course, but incredibly thinly). Compared with the depth and volume of the atmosphere, hydrosphere or crust, the biosphere is insignificant. For every atom in the biosphere, there are about 700 in the atmosphere, 400,000 in

1 See Chapter 1, page 1.

2 *Die Prinzipien der Mechanik in neuem Zusammenhange dargestellt* (312 pp.).

3 The core meaning of the physical concerns the operation of energy – physical forces cause certain physical effects. One may therefore see the term 'force' as a shorthand for physical causality.

> the hydrosphere and 2,000,000 in the crust! Yet this insignificant scum maintains most of the rest of the world in a steady state adjusted optimally to its own needs. So accurate is the biosphere's system of balances and adjustments that, for example, the oxygen concentration in the air has not measurably varied during the 80 years for which accurate measurements have been available (Jones, 1998:42).

Although everything is not alive, what *is* living indeed effectively *interacts* with almost every non-living thing on the earth! By the end of the 19th century and the beginning of the 20th century, *neo-vitalism* even reigned the day for a few decades, owing to the experimental work done by the German biologist Hans Driesch. He did research on phenomena of *regeneration* and discovered that animals are capable, when divided at an early stage of their development, regenerating the entire living entity. Later on it was shown that, in the case of certain animals, even a part as tiny as $\frac{1}{280}$ can regenerate the entire animal. In general, the mere occurrence of processes of *growth* seem to contradict the second main law of thermodynamics, stating that within a *closed system* the most probable condition would be an increase in *chaos*, i.e. *disorder*.[1]

Since the thirties of the 20th century, after Ludwig von Bertalanffy (the founder of modern *general system theory*) expanded the second main law to *open systems* (constantly in interaction with their environment), it was clear that living entities did not violate the law of entropy: every living entity building up an increasing internal order while growing still 'extracts' food from its environment – and by doing this causes more *disorder* outside itself than the order generated inside it.

The mentioned examples of reductionistic isms attempted to reduce all of reality to one (modal) principle of explanation, such as *number* (the Pythagoreans), *space* (the subsequent development of Greek thought), *movement* (the mechanistic main tendency of classical physics up to Hertz), *atoms* and (macro) *molecules* (the physicalistic orientation of the mechanistic legacy within modern biology), and *life* (classical vitalism and modern neo-vitalism). We can continue to list further examples of this monistic inclination within other disciplines, but the point we wanted to underscore has already been made. Each *monistic ism* is uncritical in a two-fold sense:

(i) The monistic attempt to eliminate the entire cosmic diversity of (modal) aspects by reducing them to the perspective of one (absolutized; deified) aspect, in effect destroys the meaning of the absolutized aspect. By contrast, we shall argue that the *meaning* of an aspect only comes to expression in and through its *coherence* with other modal aspects.

(ii) In its reductionist attitude, every monistic approach is circular, for the choice of an aspect implicitly presupposes a *given diversity* within creation before it can attempt to *reduce* this diversity to *one* principle of ex-

1 This law could be designated as the law of *non-decreasing entropy* – within a closed physical system *entropy* can only remain constant or increase. *Entropy* indicates the tendency towards the *most probable* situation within a closed system.

> planation only. That *reductionistic* modes of thought indeed proceed on the basis of an awareness of *different* aspects, is amply demonstrated in the "nothing but" argumentation they pursue.

Compare the position of Descartes:

> At all events it is certain that I seem to see light, hear a noise, and feel heat; this cannot be false, and this is what in me is properly called perceiving (sentire), which is nothing else than thinking (*Meditation* II.)

Hume opts for the opposite conviction:

> To hate, to love, to think, to feel, to see; all this is nothing but to perceive (*A Treatise of Human Nature*, Book I, Part II, Sec. vi).

That different isms in fact opted for different aspects in their attempted reductionist modes of thinking also testifies to the fact that there are indeed more than one aspect available for absolutizations. Nonetheless, it should be kept in mind that monistic isms normally arise in reaction to other one-sided isms and one should also always appreciate what was indeed discovered by such a stance, even though its distorted perspective needs not to be supported.

The upshot of these considerations supports the view that, by its very nature modal abstraction implies that the various academic disciplines – implicitly or explicitly – operate on the basis of one or another *philosophical orientation*.

In other words, theoretical thinking is only possible on the basis of an idea of the *cohering diversity of aspects* (and entities) within reality. What is at stake, is the necessity and inevitability of a *theoretical totality view* of the cosmos. Since a special science, by definition, is limited to the angle of approach of one modal aspect only, it stands to reason that such a *totality view* of reality cannot be special scientific in nature.

Agazzi realizes that the "conceptual framework adopted by the single sciences never has the dimensions of the *whole*" (Agazzi, 2001:9). He mentions for example the "discussions concerning the nature of space and time, determinism, indeterminism and causality, the continuous or discontinuous characteristics of matter and energy, the infinity or finiteness of the universe that have been produced by the developments of contemporary physics" and adds that "even when they have been led by professional and outstanding physicists, they were of a genuine philosophical nature." He continues that this "can be easily seen if we consider that they also occupied the mind of several outstanding professional philosophers of our time" (Agazzi, 2001:9). Agazzi is aware of the limitations of the special sciences and he understands the totality character of philosophical endeavours:

> Moreover, they count among the most typical and classical questions of the philosophy of nature of all times, and the fact that they are still the object of lively discussions also after the acceptance, say, of relativity and quantum physics clearly indicates that they have not been *solved* by such theories, but simply further problematized as a consequence of them. ... Therefore, what emerges is the ineliminability of the *point of view of the*

whole … Well, the point of view of the whole is precisely that which characterizes the philosophical attitude at large" (Agazzi, 2001:10).

The reaction of postmodernity against such totality perspectives evinces a lack of understanding of the *conditions* of human thinking. It became fashionable during the latter part of the 20th century to downplay the emphasis on (conceptual) rationality that was indeed overemphasized in the age of Enlightenment (the 18th century) – on the basis of acknowledging the *historical situatedness* of every thinker – sometimes referred to historicity – and the dependence upon a *linguistic community*, necessarily implying the task of *interpreting* and *reinterpreting* whatever has acquired the form of a (philosophical or special scientific) *statement*.

In addition to the condition of *historicity* this is sometimes designated as the condition of *linguisticality*. But these two (*historicity* and *linguisticality*) are not the only conditions for human thinking! Within the diversity of aspects, the quantitative and the spatial aspects, for example, play an indispensable role as well, alongside all the other modal aspects. The former provides us with an awareness of *multiplicity* and the latter, as we shall argue below, with an awareness of *wholeness* or *totality* (mere synonyms for *continuous extension*) (see pages 61, 225, 236, 307, 353, 353, 391, 405, 407).

Since *all* the modal aspects simultaneously *co-condition* human thinking, no meaningful thinking is possible without (implicitly or explicitly) employing (amongst others) spatial notions such as *wholeness* and *totality*. This insight alone is already sufficient to counter the accusation of postmodern thought, that the employment of the notion of totality and wholeness represent a fallback into "totalizing metanarratives."

Although our argument in support of the idea that human thinking is conditioned in a *multi-aspectual* way has already answered the postmodern challenge, we shall return to this issue in a later context, when an assessment of the hidden philosophical assumptions of postmodernity are discussed in more detail (see pages 370-379).

2.2.5 *Philosophy and the special sciences*

The most important *historical* argument for the claim that the various disciplines operate on the basis of philosophical presuppositions has already surfaced in Chapter 1, where we pointed out that the history of every special science demonstrates this dependence highlighted in the diverse (often conflicting) trends of thought within it (see Chapter 1, pages 3 and 5 ff.).

From a different approach, Max Planck underscores the interconnection between philosophy and the academic disciplines in 1926 (particularly physics). He says that the era of opposing philosophy and physics is behind us, for philosphers realize that they cannot prescribe to experimental research which methods to pursue, while those involved in physical research realize that sense perception alone is not sufficient, for they have to proceed from a sure dose of metaphysics.

According to him, the new physics particularly once again impresses upon us the old truth that there are realities independent of our sense-impressions, and that there are problems and conflicts in which these realities assume for us a higher value than the richest propositions, or our entire sensory world.[1]

In addition to the argument based upon the nature of modal abstraction, there is a second way to make the same point. Suppose we ask a mathematician, what is mathematics, and a theologian, what is theology, and suppose they provide us with definitions such as: (i) "mathematics is the discipline encompassing algebra and topology,"[2] and (ii) "theology is the discipline ...".[3]

Re (i): It is immediately clear that the 'definition' of mathematics is not *mathematical* in nature, for the definition as such is not an *axiom*, *theorem* or *deduction* either in algebra or topology. It shows that *defining* mathematics is not the same as *doing* mathematics. In order to define mathematics one therefore has already taken a step back. Instead of being involved in the practice of mathematics, one is talking *about* mathematics.

Re (ii): The same applies to theology. The moment one starts to talk *about* theology one has already left the special scientific field of investigation of this discipline. In theological faculties where sub-disciplines such as the bibliological group, the ecclesiological group, the dogmatological group, the missionary group, church history, and so on are found, the striking fact is that these faculties also accommodate a discipline assigned with the task of informing theological students about the nature and the sub-disciplines of theology. This discipline is sometimes called the "encyclopedia of theology." Yet, the name "encyclopedia of theology" is never found in the list of *theological sub-disciplines*!

This shows that the discipline telling us what theology is all about, is not itself a *part* of theology – and this fact is implicitly acknowledged by the discipline itself. What then is the nature of this latter basic discipline?

Our claim is that this basic question, recurring within every special science, is *philosophical* in nature. There are in fact two kinds of disciplines: (i) those

1 "Es hat Zeiten gegeben, in denen sich Philosophie und Naturwissen-schaft fremd und unfreundlich gegenüberstanden. Diese Zeiten sind längst vorüber. Die Philosophen haben eingesehen, daß es nicht angängig ist, den Naturforschern Vorschriften zu machen, nach welchen Methoden und zu welchen Zielen hin sie arbeiten sollen, und die Naturforscher sind sich klar darüber geworden, daß der Ausgangspunkt ihrer Forschungen nicht in den Sinneswahrnehmungen allein gelegen ist und daß auch die Naturwissenschaft ohne eine gewisse Dosis Metaphysik nicht auskommen kann. Gerade die neuere Physik prägt uns die alte Wahrheit wiederum mit aller Schärfe ein: es gibt Realitäten, die unabhängig sind von unseren Sinnesempfindungen, und es gibt Probleme und Konflikte, in denen diese Realitäten für uns einen höheren Wert besitzen als die reichsten Schätze unserer gesamten Sinneswelt" (Planck, 1926:205).

2 In other words, the discipline of number and space.

3 The dots simply represent whatever a particular theologian may want to insert, for our argument can accommodate all possible variations. One "filling in" could be: "consisting of sub-disciplines such as the bibliological group, the ecclesiological group, the dogmatological group, the missionary group, church history and the internal law and practices of the church."

which can only account for what they are doing by moving *beyond* their own universe of discourse, and (ii) those which are actually destined to treat (boundary-exceeding) questions like these. By definition, we can designate those belonging to (i) as *special sciences* and (ii) as *philosophy*, as *the discipline of the disciplines*.

If we combine arguments (i) and (ii) (flowing from the nature of *modal abstraction* and the necessity of moving beyond the confines of a discipline in order to define it), the following statement seems justified.

> Special scientists have two options (but just one choice!):
> (i) either they give an account of the philosophical presuppositions with which they work – in which case they operate with a philosophical view of reality, or
> (ii) implicitly (and uncritically) proceed from one or another philosophical view of reality – in which case they are the victims of a philosophical view.

The first natural reaction of a special scientist is to state that the philosopher cannot explain what a specific discipline is about, for in order to do this, one needs to be informed about what is going on in that discipline. However, nothing argued above contradicts this objection. Indeed it does form part of the relationship between philosophy and the disciplines, that the philosopher ought to be informed by the specialist about what is going on within her particular special science. The point above is not a *personal* one – it does not concern the question: *who* gives the answer (the *person* who is a special scientist or the *person* who is a philosopher)? What is at stake is the *systematic question*: what is the *nature* of the answer given to the question? Is it *special scientific* or *philosophical* in nature? And we have seen that it cannot be special scientific in nature. Ergo ...

Even within the "most exact" of all the sciences, namely mathematics and physics, the intimate link between philosophy and the special sciences is discernable. Within modern mathematics there are different kinds of mathematics (see our remarks in a later context, page 213), while modern physics itself did not succeed in liberating itself from philosophical assumptions. Max Planck, for example, admitted in his presentation on the genesis and subsequent development of quantum theory at the reception of the Nobel Prize, that his initial positivistic philosophical orientation prevented him, in his research of black radiation, from investigating the real problem, namely the relationship between *entropy* and *probability* (see Vogel, 1961:151).

In our provisional discussion of *rationality* in Chapter 1, the perennial philosophical question about the *coherence of irreducibles* surfaced (see page 8). Against the background of our analysis of the implications of *modal abstraction*, the foundational position of this issue is better understood. But many more equally perennial issues highlight other crucial philosophical questions.

At this stage of our analysis we have not yet developed the necessary distinctions and arguments required to deal with them in an appropriate manner, but a brief indication of what is at stake will be helpful in order to arrive at a better understanding of the foundational role of philosophy within the entire scholarly enterprise.

Systematic philosophical considerations always had to account for the relationship between *individuality* and *universality*, for the relationship between *unity* and *diversity*, for the connection between *constancy* and *dynamics*, and for what is *conceptually knowable* in distinction from the kind of knowledge *exceeding the boundaries* of conceptual knowledge. The history of philosophy and the disciplines bear witness to the existence of many stances in which a one-sided approach to these problems is advanced.

2.2.6 *The problem of unity and diversity within various disciplines*

Let us briefly look at the problem of *unity* and *diversity* and then succinctly investigate the traces it left within a number of disciplines. The existence of *monistic isms* alluded to above, such as *physicalism*, *vitalism*, *moralism*, and *historicism*, are all (negative) answers to the problem of *unity and diversity*, also referred to as that of the *coherence of irreducibles* (see page 8). Russell refers to Hegel in respect of the difference between a so-called continuous magnitude (wholeness) and a discrete magnitude – as "*different*" instances of the "class-concept" and then proceeds by saying that he 'strongly' holds "that this opposition of identity and diversity in a collection constitutes a fundamental problem of Logic – perhaps even *the* fundamental problem of philosophy" (Russell, 1956:346).

The two most dominant philosophical orientations operative in the history of the disciplines in this regard are those of *atomism* and *holism*. An understanding of the world avoiding any effort of reducing what is irreducible, in other words a *non-reductionist ontology*, will be inclined to affirm the uniqueness and irreducibility of diverse aspects of reality.[1] By contrast, an *atomistic* thinker will employ the meaning of the *one* and the *many*, of a *discrete multiplicity* in the quantitative sense of the term (or at least *analogical usages* of this quantitative meaning within the context of other modes of explanation), in order to comprehend all of reality. Applied to human society, every social collectivity is simply reduced to its simplest 'elements', the *individuals*. All variants of *holism* (universalism), on the other hand, proceed from the employment of the concept of a *whole* (*totality*) with its *parts*. Thus the *whole-parts relation* (or analogies of this relation) serves as the guiding star, dictating that reality ought to be understood in terms of wholes and their parts (sometimes referred to as *systems* and *subsystems*). Even social relations among human beings have to be captured by this schema. Particularly in connection with reflections on the nature of human society, the opposition between *atomism* and *holism* also became known in terms of the opposition between *individualism* and *universalism*.

There is literally no single special science, from mathematics to theology, that is not burdened by the effect of these two orientations, pervading the entire history of the disciplines. In wrestling with the problem of unity and diver-

1 Already in 1931, Dooyeweerd advocated the scholarly importance of not subsuming any aspect of temporal reality under any other one (Dooyeweerd, 1931:93).

sity a philosophical position is assumed in respect of what may be called the quest for discovering a *basic denominator*. Inevitably the choice of such a basic denominator – as explanatory device employed in the comparison and explanation of whatever there is – always entails an account of the mutual coherence and diversity within reality, since it invariably concerns a perspective on the "coherence of irreducibles." A brief visit to a number of natural sciences and some of the humanities will illustrate our claim.

2.2.6.1 *Mathematics*

Mathematics received its first intellectual stimulus from the mentioned *atomistic* Pythagorean thesis that everything *is* number, but soon had to revert to a *spatial perspective*, in terms of which the whole-parts relation dominated the subsequent development of mathematics up to the 19th century. Eudoxos has already approximated the discovery of the modern calculus (later independently attributed to Leibniz and Newton), but clinging to a *spatial perspective* prevented him from conceiving the *numerical* concept of a *limit*, which turned out to be a necessary tool for the development of the calculus and the modern theory of real numbers ('irrational' numbers – see Boyer, 1959:8 ff.). Laugwitz mentions that Cantorean set theory advances an "atomistic conception" of "the continuum" (not even intuitionism succeeded in decisively deviating from this atomism) (Laugwitz, 1997:266).

Although modern (axiomatic) set theory (Cantor, Zermelo, Fraenkel, Hilbert, Ackermann, Von Neumann) pretends to be a purely atomistic, arithmetic in the structure of set theory actually implicitly (in the undefined term 'set') *borrows* the whole-parts relation from space.[1] This explains why Hao Wang informs us that Kurt Gödel speaks of *sets* as being 'quasi-spatial' – and then adds the remark that he is not sure whether Gödel would have said the "same thing of numbers" (Wang, 1988:202). Paul Bernays (the author of a successful axiomatic system for set theory, who published until the mid-seventies of the previous century), started his mathematical career as the young co-worker of David Hilbert. He emphatically affirms that it is the *totality-character* of spatial continuity that will resist a *perfect arithmetization* of mathematics (see Bernays, 1976:74). Without contemplating the existence of a modal aspect of space as well, Bertrand Russell concedes: "The relation of whole and part is, it would seem, an indefinable and ultimate relation" (Russell, 1956:138).[2]

1 Having resolved the continuum into a set of (isolated) points set theory then super-imposes upon it, with the auxilliary set theoretical construction of environments and open sets a 'topology' within which it is once again possible to speak of 'continuity' ('Stetigkeit') (see Laugwitz, 1997:266).

2 By acknowledging that "greater and less are undefinable" Russell in addition implicitly accepts the primitive meaning of *numerical succession* (see Russell, 1956:194; see also page 167). Later on he remarks that "progressions are the very essence of discreteness" (Russell, 1956:299). In an earlier context he also criticizes Bolzano for not distinguishing the "many from the whole which they form" (Russell, 1956:70).

2.2.6.2 *Physics*

The classical mechanistic worldview, in an atomistic way, reduced the universe to the notion of *particles in motion*. Van Melsen says that, in "most forms of atomism, it is a matter of principle that any combination of atoms into a greater unity can only be an aggregate of these atoms." By contrast, he refers to *holistic* tendencies within the discipline of *physics*: "In modern theories atomic and molecular structures are characterized as associations of many interacting entities that *lose* their own identity. The resulting aggregate originates from the converging contributions of all its components. Yet, it forms a new entity, which in its turn controls the behaviour of its components" (Van Melsen, 1975:349).

2.2.6.3 *Biology*

Similarly, the mechanistic theories in modern *biology* proceed from atomistic assumptions – up to neo-Darwinism. According to Smith, such an atomistic view – regarding genes and what they 'code' – is indeed a "problematic component of the neo-Darwinian outlook" (Smith, 1992:439). Process structuralists, such as Lambert and Hughes are critical of the fact that neo-Darwinians "invariably treat organisms as loose collections of discrete parts" (Smith , 1992:439).[1] Eventually atomism was opposed by the holistic orientation of vitalism and neo-vitalism. The holistic biology of Smuts (1926) and Meyer-Abich (1964) explicitly operates with the whole-parts scheme in the way they have framed their basic concepts. The same applies to the organismic biology of Von Bertalanffy, for in this approach the concept of wholeness also acquired a central role, as opposed to all forms of atomistic understanding. Von Bertalanffy considers the organismic worldview to be a step beyond the mathematical *more geometrico* ideal and also beyond the mechanistic worldview:

> First came the developments of mathematics, and correspondingly philosophies after the pattern of mathematics – *more geometrico* according to Spinoza, Descartes and other contemporaries. This was followed by the rise of physics; classical physics found its world-view in mechanistic philosophy, the play of material units, the world as chaos ... Lately, biology and the sciences of man come to the fore. And here organization appears as the basic concept – an organismic world-view taking account of those aspects of reality neglected previously" (Von Bertalanffy, 1968:66).[2]

The traditional *idealistic morphology* in biology is also intimately attached to a holistic orientation. What is considered an 'ideal' plant or an 'ideal' leaf is understood in a Platonic sense as a-temporal static forms of being (see the extensive botany text book by Troll, 1973, Chapter 1).

Even in respect of the assessment of what constitutes a species, the difference between an additive (atomistic) approach and a whole-parts (holistic)

1 Lambert and Hughes remark that 'trait' thinking, or "the atomizing of wholes into their hypothesized component parts, is an intimate and inseparable part of the neo-Darwinian approach" (see their 1987 work, quoted by Smith, 1992:439).

2 Beckner comments elsewhere that "[E]ven though in fact many biologists agree with the organismic position, they will say they disagree" (Becker, 1971:60-61).

view, still causes divergent views. Grene points out that Ghiselin and Hull propose "that species taxa be considered, not as classes with members, but as individuals (wholes) with parts (see Grene, 1986a:440 and also Sober, 1986). Sterelny points out that the neo-Darwinian synthesis was "individualist, microevolutionary, gradualist, and selectionist" (2009:314).

2.2.6.4 *Psychology*

Within the discipline of *psychology*, the dilemma of atomistic and holistic theories is also discernable. The legacy of an atomistic association psychology prevailed in the 19th century during the rise of psychology as a distinct academic discipline. However, holistic theories soon entered the scene, particularly in the *Gestalt*-school (the Berlin school, Krüger; and the Leipzig school, Köhler and Koffka). More recently, the influence of general *systems theory* – which operates in a holistic way with the whole-parts relation (in the shape of the idea of systems and subsystems) – also had its effect on this discipline.

2.2.6.5 *Logic*

Modern *logic* also did not escape the 'fate' of atomism and holism. Whether or not one is willing to accept the existence of an infinite *totality* is decisive for the *scope-of-validity* of the logical *principle of the excluded middle* (also known as the *tertium non datur*).[1]

2.2.6.6 *Linguistics*

Within the discipline of *linguistics* an example from the sub-discipline of semantics illustrates the dilemma between atomism and holism. Antal considers a *word* to be the primary "sign-unit" in language. He actually dismisses the idea of multiple meaning nuances of a word by transferring them to what is *denotated* (Antal, 1963:53, 54, 58). This atomistic approach was left behind in the development of *semantic field theory*, that has already been initiated by Trier during the first half of the 20th century. This trend asserts that the multiplicity of meaning-nuances of a word are bound together in order to form an authentic *whole* (*Ganzheit*). A word is a genuine totality, embracing its parts fully, while in turn it can only signify because opposing words within its environment act in a meaning-delimiting way (see Trier, 1973:1, 5 ff., 15, and also Geckeler, 1971).

2.2.6.7 *Sociology*

Anther example is found in the field of sociology. Initially this relatively modern discipline pursued a so-called *organicistic* paradigm, for its founder, Comte, viewed society *as an organism* in a *holistic* sense. Although the British sociologist, Herbert Spencer, continued this *organicistic* line of thought,

1 Intuitionistic mathematics accepts the applicability of this principle only in a finitistic sense, i.e. only when the successive infinite (the *potential infinite*) is given, rejecting its validity in the case of the *actual infinite* (preferably designated as the "at once infinite"). This issue, to which we shall return below, is treated in more detail in an article on "The ontological status of the principle of the excluded middle" – see Strauss 1991:73-90 and see page 303 below.

his own orientation reverted to an *atomistic* (individualistic) approach:

> So far from alleging, as M. Comte does, that society is to be re-organized by philosophy; it alleges that society is to be re-organized only by the accumulated effects of habit on character. Its aim is not the increase of authoritative control over citizens, but the decrease of it. A more pronounced individualism, instead of a more pronounced nationalism, is its ideal (Spencer, 1968:22).

The remarkable situation here is that, although both thinkers advocated *organicism*, Spencer did it in an *atomistic* manner and Comte in a *holistic* way! Alexander casts this opposition in the following terms: *rational-individualistic* versus *rational-collectivist* (Alexander, 1987:12).

For the logical positivist, Ayer, "the English state, for example … [is] a logical construction out of individual people" (Ayer, 1967:63). Karl Popper designates his own approach as "methodological individualism":

> It rightly insists that the 'behavior' and 'actions' of collectives, such as states or social groups, must be reduced to the behavior and to the action of human individuals (Popper, 1966-II:91).

Max Weber also explicitly denounces the idea that societal collectivities could be genuine wholes or totalities. In terms of his atomistic conviction, he states:

> Concepts such as 'state', 'club' ... signifies specific kinds of communal human actions ..., that could be reduced to 'understandable' (*verständliches*) actions, and that means that it can, without an exception, be reduced to the actions of the individual human beings (*Einzelmenschen*) concerned (Weber, 1973:439).

Similar to the way in which a misunderstanding of the relationship between unity and diversity gave rise to *isms* such as *atomism* (*individualism*) and *holism* (*universalism*), the relationship between universality and individuality gave rise to *isms* such as *rationalism* and *irrationalism*. The nature of both these pairs of *isms* is further complicated when the distinction between *constancy* and *dynamics* as well as the difference between *conceptual knowledge* and *concept-transcending knowledge* (*idea-knowledge*) is brought into the picture as well – but an explanation of the latter distinction and these isms will have to be postponed to a later stage of our analyses (see pp.176 ff.).

2.2.6.8 *Economic Theory*

Gunnar Myrdal discusses the speculations of German political philosophy found in the writings of Spann. He refers to the aprioristic category of the tax obligation flowing from citizenship and then mentions two opposite poles in the field of political economy: the utilitarian and liberalistic *atomism* on the one hand and and the more heterogeneous organic or juridical orientation on the other. The notion of the *common good* (*Allgemeinwohl*) is the 'metaphysical' element serving both approaches. "In the first case it concerns solely a purely arithmetical sum, while in the other instance we encounter an immediate totality, a whole that is not simply the sum of its parts" (Myrdal, 1932:87).

He holds that the atomism of the utilitarian political economists is less mistaken than the social theory of mutually contradicting constructions of the metaphysical German conceptions of the state. This latter position is also characterized by Myrdal as "romantic, organic, state-absolutizing and not less oriented to natural law than the English utilitarianism" (Myrdal, 1932:86).

2.3 Intellectual creativity and the acquisition of new ideas

The preceding brief case study of the foundational influence of atomism and holism as opposing theoretical ideas regarding the cohering diversity within reality should be appreciated in terms of the broader general perspective mentioned above (see page 59), namely that every special scientist (coming from the domain of the natural sciences or from the humanities) has two options: either (i) such a special scientist accounts for her basic theoretical assumptions – in which case she *has* a philosophical view of reality, or (ii) a special scientist implicitly and uncritically proceeds from one or another philosophical view of reality – in which case that scientist is the *victim* of a particular philosophical view of reality.

One of the remarkable facts concerning the history of intellectual creativity is that most of the decisive and influential theoretical insights advanced by the famous names registered in the intellectual history of philosophy and the disciplines were acquired at a relatively early stage of the intellectual development of those scholars – and Newton is not the only example. A few more will suffice to elucidate this.

In his *Autobiography*, A.A. Fraenkel mentions the fact that some of the most famous mathematicians died before reaching the age of 30 – such as Évariste Galois (1811-1832), N.H. Abel (1802-1829) and F.G.M. Eisenstein (1823-1852). The creativity of gifted scholars in the fields of mathematics, physics, astronomy and symbolic logic in most cases is only manifested when those scholars are relatively young – between the ages of 15 and 35 and sometimes up to 40 years of age (Fraenkel, 1967:147). The same applies to C.F. Gauss, known as the 'Prince' of mathematics, who achieved his classical results between the age of 19 and 37 (Fraenkel, 1967:148). Einstein published his special theory of relativity at the age of 25, and according to Fraenkel, he did not achieve anything truly original after the age of 35.

Johann (later: John) von Neumann (1903-1957), who eventually contributed substantially to our current concept of the computer,[1] published an *Introduction to Transfinite Numbers* at a very young age (eventually published in 1925). However, his remarkable youthful talent as a mathematician is clearly reflected in the account given by Fraenkel of an extensive study of 84 pages submitted by von Neumann to the *Mathematische Zeitschrift*. As nestor of *Set Theory* in Germany at that stage, Fraenkel has already received this manuscript with the request to assess it. He writes that with great effort (*großer Anstrengung*) he succeeded to work through the study. It in fact deviated from everything written up to that point on the axiomatization of set theory, amongst others, in that it did not, as in contemporary views, start from sets,

1 The acronym MANIAC was invented by Von Neuman: *Mathematical Analyzer Numerical Integrator and Computer*.

but from functions. Far from having understood everything, Fraenkel did realize that the author must be exceptionally gifted and invited him to visit him in Marburg, Germany. Fraenkel reports about this visit as follows: "His appearance made a enormous impression, not only upon me but also upon my wife; the slim, almost 20 year old young man exactly looked like one imagined a young genius to be like" (Fraenkel, 1967:169).

From another field, a slightly different assessment is made of the hold of ideas acquired at a fairly young age, in a remark by John Maynard Keynes: "Practical men, who believe themselves to be quite exempt from any intellectual influences, are usually the slaves of some defunct economist. Madmen in authority, who hear voices in the air, are distilling their frenzy from some academic scribbler of a few years back. I am sure that the power of vested interests is vastly exaggerated compared with the gradual encroachment of ideas. Not, indeed, immediately, but after a certain interval; for in the field of economic and political philosophy there are not many who are influenced by new theories after they are twenty-five or thirty years of age, so that the ideas which civil servants and politicians and even agitators apply are not likely to be the newest" (Keynes, 1939:383-384).

Dooyeweerd's discovery of the idea of unique modal aspects and their mutual coherence, at a relatively young age (probably before he was thirty years old), is indeed comparable to the above-mentioned examples of intellectual creativity at a fairly young age.

2.4 Concluding remarks

From the first two chapters, we can infer that human rationality is embedded in a *more-than-rational* commitment, ultimately giving direction to all of life, including scholarly activities. This all-permeating life-orientation manifests itself in a theoretical totality perspective on reality, embracing all the academic disciplines. As the distinguishing feature of scholarly investigations, *modal abstraction* entails that one or another philosophical perspective invariably underlies and directs the work done in the special sciences.

In order to articulate a better understanding of the way in which the "discipline of the disciplines" interacts with the orientation of the special sciences Chapter 3 will provide a first exploration of the general theory of modal aspects, in anticipation of the more detailed analyses of Chapters 5 and 6.

Chapter Three

The uniqueness of modal aspects

3.1 The relation between aspects and entities

From Chapters 1 and 2, it has already been clear that our experience of the world is embraced by two distinct but intimately cohering *dimensions* of reality, namely *modal aspects* (functions, facets) and concretely existing and functioning (natural and social) *entities*. The latter dimension includes concrete *events* as well. Since these dimensions at once *delimit* what we can experience, they serve as the frontiers of what we can *see*, as the *horizon* of our experiential perspective. Therefore an apt metaphor conveying this meaning is given in the saying that the dimensions of modal aspects and of concrete entities and events constitute the *horizon of our experience*.

Implicitly, an important problem has been addressed in our preceding investigations, namely that of the relationship between the 'what' and the 'how' of reality. It entails the question of whether or not the aspects are merely *properties* of *entities*. There seems to be two options:

(i) Modal aspects are merely *aspects of entities*, that is, they exist only as *properties* attached to *entities*.

(ii) Alternatively, one may view the modal aspects of reality as *universal, constant, functional frameworks* at the *basis* of all entitary functioning and thus co-conditioning the occurrence of all kinds of events.

3.1.1 *Are modal aspects merely properties of entities?*

Let us look at an example of a view that amounts to an assertion that aspects are properties of entities. Such a view is similar to the position taken by Greek thinking in connection with the nature of *place* (the idea of an "empty space"). In Greek philosophy, aspects are viewed as *inhering properties*, dependent upon the nature of the entities to which they are 'attached'. But both Plato and Aristotle acknowledged *universality*: the former in the supposedly supra-sensory ideal forms and the latter in the universal substantial forms of entities. Aristotle actually discerned the *orderliness* of things,[1] for in his *Metaphysics* he remarks that when fire terminates the existence of a house it is not *houseness* that burnt down (cf. *Metaph.*, 1035 b 32; *De Anima*, 412 b 16).[2]

Medieval philosophy by and large adhered to this Aristotelian view, but since the Renaissance modern philosophy abolished the *reality* of universality *outside* the human mind. This introduced a new orientation, which holds that whatever exists outside the human mind is *purely individual*. This implies that *aspectual properties too*, in their attachment to *entities*, are purely *individual*.

1 That is, their *universal side*. Every concretely existing entity displays an individual side (*this* entity) and a universal side (it is *an* entity).

2 In addition to this so-called *primary* substance, Aristotle here introduces a *secondary substance* as the *universal substantial form* of entities (see also Aristotle, *Categoriae* 1 ff.).

Suppose the chair I am sitting on weighs 12 kg and the desk in front of me 30 kg. Then, as properties pertaining to distinct individual things, they are *individually* different. But is it then still meaningful to say that both have the same property, called 'mass'? Is it not the case that the *individuality* of the chair and the *individuality* of the desk impress a *uniqueness* upon each of their properties, including that of their *mass*? In other words, does it not follow that the *distinctness* of these two entities constitutes a *gap* between their properties that can only be *bridged* in thought, that is, by *human conceptual thinking*, subsuming these properties under one general (conceptual) category, namely the category of *mass*? This argument proceeds from the assumption that an individual entity is truly *individual*, that is, that it does not evince any *universality* whatsoever. The only problem with this view is that the property of *being individual* applies to *all* entities, and therefore holds *universally*. Is there an alternative view capable of avoiding this *impasse*?

3.1.2 *Modal aspects: universal, functional conditions for the existence of concrete entities*

In ancient Greece and during the Medieval period, it was believed that the laws governing entities on the earth are *different* from those governing celestial bodies. Incorporated in this view was the conviction that motion can only be explained through direct 'contact' between bodies – 'pushing' each other so to speak. But then Newton introduced his law of gravity, positing the idea that entities anywhere in the universe attract each other according to a force directly proportional to their respective masses and indirectly proportional to the square of the distance between them.[1] The force of gravity in the formula exercises its effect despite the fact that the attracting bodies may be separated by a vast empty space. Suddenly it appeared that Newton's formulation brings to expression a physical law that holds *universally* for all physical entities, locally and in outer space. It was simply impossible to explain the effect of this law in terms of the physical mechanism of bodies in "contact with" or 'pushing' each other. Kline remarks that

> the abandonment of physical mechanism in favor of mathematical description shocked even the greatest scientists. Huygens regarded the idea of gravitation as 'absurd' because its action through empty space precluded any mechanism. He expressed surprise that Newton should have taken the trouble to make such a number of laborious calculations with no foundation but the mathematical principle of gravitation. Many others, including Leibniz, objected to the purely mathematical account of gravitation. . . . The attempts to explain "action at a distance" persisted until 1900 (Kline, 1980:55).

Yet the incredible scope of this discovery, which actually unveiled an insight into what we have called the *modal universality* of the physical aspect of reality (see page 25), is best demonstrated by the story about the discovery of

1 The law of gravity states that $F = G\frac{mM}{r^2}$, where F is the force of gravity, G a constant, m the mass of the one physical body, M the mass of the other, and r the distance between the two bodies.

Neptune.[1]

With the aid of his newly designed powerful telescope, W. Herschel discovered the planet Uranus in 1781. The problem was that the path of this planet did not obey the predictions made for it. Alexander Bouvard conjectured that this deviation was the effect of another – yet unknown – planet. Various attempts were made to observe or calculate the possible size and path of this unknown planet. In 1845 a young student from Cambridge, J.C. Adams, on the basis of Newton's law of gravity, calculated a highly accurate estimate of the mass, position and path of this unknown planet and sent his calculations to sir George Airy at the "Royal Astronomical Observatory" in Greenwich. The latter failed to appreciate the significance of these calculations – which gave another student, the Frenchman J.J. Leverrier, the chance to come up with approximately the same calculations independently. He communicated his findings to the German astronomer, Johann Galle. Galle received them on September 23, 1846, and discovered Neptune that same evening – with a deviation of only 55 minutes off the path, as predicted by Leverrier. Kline is certainly justified in asking how one can doubt the *predictive power* of a mathematical-astronomical theory that manages to make predictions accurately up to one ten thousandth of a percentage point. This additional historical perspective may also help to come to a better appreciation of Alexander Pope's assessment:

> "Nature and Nature's laws lay hid in night:
> God said: Let Newton be! and all was light."[2]

This remarkable story is indeed an example of the *universality* of the modal aspects of reality. This universality entails that it underlies and makes possible the *functioning* of every entity *within* this aspect. Philosophically speaking, the technical term designed to capture the nature of those conditions *making possible* the existence and functioning of something, is *transcendental*. This term acquired a particular meaning in the philosophy of Kant, for in his *Critique of Pure Reason* the word *transcendental* is employed to account for that which provides the basis of all experience in the sense that it makes possible what we experience.[3]

Unfortunately Kant did not accept any *ontic universality*, for according to him the formal source responsible for the *ordering* of the chaotic sensory impressions presented to us in experience, is found *within* the human subject itself – in what he calls the (*a priori*) forms of *intuition* (*space* as outer and *time*

1 Cf. Kline (1980:62-63).

2 Apparently the story of the apple is a myth, but it did inspire Lord Byron to write of Newton: "[T]his is the sole mortal who can grapple, Since Adam, with a fall, or with an apple" (see Dunham, 1994:130).

3 In his *theory of knowledge* (*epistemology*), Immanuel Kant aspired to show that, in mathematics, physics and metaphysics there are present what is called synthetic judgments *a priori*. The opposite of *a priori*, that is, of that which precedes and lies at the foundation of experience, is *a posteriori*, that is, that which derives from experience. His basic question is: *How are synthetical propositions a priori possible*? (Kant, 1787 B:19).

as inner form), and in what he designates as the *twelve categories* of understanding (arranged in four groups, namely *quantity*, *quality*, *relation* and *modality*) (see Kant, 1787-B:104 ff.). We shall return to Kant's view of so-called synthetic a priori propositions below (see pages 421 ff.)

Without the recognition of *functional modes of existence*, given in an *ontic* sense, scientific thinking will constantly be burdened by an inability to account for the *applicability* of functional scientific insights. The latter will appear to be 'miraculous', for how is it possible that through rational insight we can formulate 'laws' describing the functioning of entities we can experience? Von Weizsäcker frames Kant's epistemological problem in terms of the question: What is nature, that it must obey laws which a human being could formulate with his/her understanding? (see Von Weizsäcker, 1971:128).

In his work on "Warrant and Proper Function," Plantinga calls upon various authors to illustrate this 'miracle' or 'mystery':

> This hasn't been lost on those who have thought about the matter. According to Erwin Schrödinger, the fact that we human beings can discover the laws of nature is "a miracle that may well be beyond human understanding" (*What is Life*? [Cambridge: University of Cambridge Press, 1945], p.31). According to Eugene Wigner, "The enormous usefulness of mathematics in the natural sciences is something bordering on the mysterious, and there is no rational explanation for it" ("The Unreasonable Effectiveness of Mathematics in the Natural Sciences," in: *On Pure and Applied Mathematics*, [13, p.2]) and "It is difficult to avoid the impression that a miracle confronts us here, quite comparable in its striking nature to the miracle that the human mind can string a thousand arguments together without getting itself into contradictions, or to the two miracles of the existence of laws of nature and of the human mind's capacity to derive them" (p.7). And Albert Einstein thought the intelligibility of the world a "miracle or an eternal mystery" (Lettres à Maurice Solouine [Paris: Gauthier-Villars, 1956], p.115) (see Plantinga, 1993:232, note 2).[1]

Once the *ontic universality* of modal aspects – such as the quantitative, the spatial and the kinematic – is properly understood, all these views will have to concede that these aspects are not mere "modes of thought," but indeed in an *ontic* sense co-conditioning the existence of concrete entities and processes functioning within them. Therefore it is not mysterious at all that a theoretical insight into the nature of arithmetical laws, spatial laws, laws of motion, and physical laws (such as the law of gravitation) relates to "the real world."[2]

1 Grünfeld also remarks: "The formalist tenet, in particular, that the laws of thought are strictly formal, that is, altogether unrelated to experience, made their application to actual phenomena a mystery." He continues by saying that it is more plausible to assume that the invented system of notations will "reflect some aspects of human experience." The following statement exhibits the need for a theory of *ontic modes*: "Any hypothesis of why nature conforms to our mathematical calculations takes some form of preestablished harmony for granted" (Grünfeld, 1983:43).

2 We shall presently continue the argument of Chapter 1, namely that what is 'real' embraces both the dimension of concretely existing *entities* and the ontic dimension of *modalities* (see pp. 16 ff.).

Although the subsequent epistemological developments after Kant continued his stance in various ways, we have noted in Chapter 1 that prominent 20th century mathematicians contemplated a *different kind of factuality* (distinct from the 'concrete' – Bernays), and opted for accepting *aspects of objective reality* (Bernays, Wang and Gödel – see page 20).

These mathematicians in fact reacted to a long-standing *one-sidedness* in the history of Western thought – a legacy in which the term 'existence' is restricted to the reality of *concrete* entities only, such as material things, plants animals and human beings. These things supposedly constitute the domain of 'experience'. Whenever modal functions (sometimes designated as "abstract entities" or 'properties') are contemplated, they are either transposed to a *suprasensory* "intelligible realm" (as Platonism did in its various forms, traditionally also known as *realism*), or they are embedded in the *creative powers* of the individual (and sometimes collective) *human mind* (*intuitionism* in mathematics and other variants of *nominalism* in philosophy and the various scholarly disciplines).

The German mathematician, Gottlob Frege, actually deserves the credit for opening the way to an insight into the nature of *modal universality* by emphasizing the 'applicability' of mathematical insights, in spite of his denial that number is a property of things outside the human mind (see Frege, 1884:27 ff. §17 ff.).[1] When he employs the term 'logical' (such as in the phrase: "logical objects"), he always has in mind a kind of *generality* (universality) that is not dependent on the specific nature of any kind of entity.[2] Yet, if this universality does not pertain to an *ontically given* aspect of reality, the *applicability* of arithmetical insights – such as is evidenced in the calculations of J.C. Adams Cambridge and J.J. Leverrier of Paris, regarding the path of the newly discovered planet Neptune – indeed will remain a *mystery*.

3.1.3 *Modal aspects are not "modes of thought"*

But what is the *ontic foundation* of the question: *How many*? First, we have to observe that, although all the modal aspects of reality are *given* in an ontic sense, their *cognition* depends upon a *constructive human activity*. Thus, in the case of mathematics, the search for something 'ontic' in its subject matter

1 He does hold the view that *a number* is required in order to answer the question: *How many?* (Frege, 1884:57; § 44). Unfortunately he sold 'number' out to *logical objectivity* by claiming that it attaches to a *concept*. With reference to Frege (1884: § 38), Dummett explains that numbers are for him [as *abstract, non-actual logical objects* (entities)] determinate objects of scientific enquiry. Frege writes: "We speak of 'the number one', and indicate by means of the definite article a single, determinate object of scientific enquiry. There are not distinct number ones, but only a single one. In 1 we have a proper name, and, as such, it is as incapable of a plural as 'Frederick the Great' or 'the chemical element gold' ... Only concept-words can form a plural" (Frege, 1884:49; translation by Dummett, 1995:87).

2 Dummett says that this generality "does not relate to any special domain of knowledge, for, just as objects of any kind can be numbered, so objects of any kind can belong to a class" (Dummett, 1995:224). Frege and Dummett do not realize that their argument stumbles upon the modal universality of the quantitative aspect of reality.

does not exclude the *constructive role* of the mathematician.[1]

Therefore, the acceptance of *ontic quantitative properties* may serve as a starting point for mathematical reflection – but then the comprehension of the *meaning* of such ontic quantitative properties requires *human intervention*; it needs the reflective and constructive disclosure and opening-up of the meaning of the quantitative aspect through the formation of *mathematical concepts* and *theories*.

What then is the *relationship* between these *ontic quantitative properties* and the *particulars* of our (numerically specified) answer to the question: *How many?* To phrase it in a different way: In order to relate to a given unity or multiplicity, humans have to grasp this quantitative meaning with the aid of numerals (*numerical symbols*), for only then the *one* and the *many* will be intellectually *articulated*. Naturally this entails that we can *identify* and *designate* ontic quantitative properties by assigning *numbers* to them through acts of *identification* and *designation*, that is, by actively – *as humans* – functioning within the *analytical* and the *sign modes* of reality.

Unfortunately we have seen that the early development of modern philosophy has already witnessed the emergence of the idea that 'universals' are mere "modes of thought."[2] During the 17th and 18th centuries, this idea was radicalized. Particularly, the procedure followed by Galileo in formulating his famous *law of inertia*[3] exerted a strong influence on the way in which Kant elaborated this idea in his view of *thought categories*. In his famous 1638 treatise on "two new sciences," Galileo used a historically significant *thought experiment* (see Galileo, 1973). He contemplated the movement of a body in *motion*, on a *path* extended into the *infinite*, and then argued that this body will *continue* its motion on this path indefinitely if nothing impetes its motion (*friction*, for example).[4]

Against this background, Kant accredited human understanding with an incredible capacity. If it is indeed possible for Galileo to formulate a *thought experiment* merely on the basis of the *spontaneous subjectivity* of human theoretical thinking, and that the subject subsequently succeeds in deducing a *natural law* from this mere *thought experiment* – namely the kinematical law of *inertia* – then this *must* imply that there are elements in our knowledge *before* all experience (i.e. *a priori*). What is even more 'astounding,' according to Kant, is that these *a priori* elements[5] in the human mind actually make possi-

1 This view side-steps the one-sidedness present both in a *platonistic* and a purely *constructivist* approach.

2 For example in the thought of Descartes – see Chapter 1, page 19.

3 Maier points out that the formulation of the law of inertia was anticipated by almost two centuries in the transitional period from the Medieval era to the early modern period (see Maier, 1949).

4 Holz argues that the formulation Galileo gives to this principle of *inertia* strongly influenced Kant (cf. Holz, 1975:345-358).

5 I.e., *previously present* elements.

ble our knowledge of reality in the first place.

The 'solution' suggested by Kant is therefore that these laws of nature are not *derived from* nature, since as categories of thought, they are *prescribed to nature* in an *a priori* way. This indeed represents what has become known as the crucial *Copernican turn* in modern epistemology (theory of knowledge) – in ascribing the primacy no longer to the 'object', but to the thinking human *subject*.[1]

In a somewhat different context, Kant wrote about the difficulty involved in this *turn*, namely how "*subjective conditions of thought* can have *objective validity*, that is, can furnish conditions of the possibility of all knowledge of objects" (Kant, 1787-B:122). Kant advanced the radical *humanistic* conclusion: the laws of nature are *a priori* contained in the *subjective understanding of the human being*:

> the categories are conditions of the possibility of experience, and are therefore valid *a priori* for all objects of experience (Kant, 1787-B:161);
>
> Categories are concepts which prescribe laws *a priori* to appearances, and therefore to nature, the sum of all appearances (Kant, 1787-B:163);
>
> Understanding creates its laws (*a priori*) not out of nature, but prescribes them to nature (Kant, 1783, II:320; § 36).

Human understanding is thus promoted to become the (*a priori*) *formal law-giver of nature* in a *universally valid* way. But soon this position had to suffer from the *relativity* introduced by the emphasis on unique historical events during the early 19th century. The supposedly universally valid construction of reality by the human subject eventually fell prey to the *relativistic* consequences of this *historicism*. Combined with the so-called *linguistic turn* (by the end of the 19th century and the beginning of the 20th century) the outcome of this process settled for a *personally* or *socially constructed* world – each person or each society *constructs* its own unique life-world. If the emphasis falls upon language, the additional qualification is that every person or society gives its own *meaning* to the world.[2]

As a consequence, we can speak about a general shift from *concept* to *meaning*, from *thought* to *language*. Introduced by Wilhelm Dilthey before the end of the 19th century, this transition is still popular a hundred years later. In a book on *Knowledge and Postmodernism in Historical Perspective*, the combined 'Introduction' says that the most recent spiritual climate is marked by a 'shift' away from "documentation to interpretation, away from reconstructing a chain of events to exploring their significance." They continue:

1 Frege accused Kant of having become stuck in the subjective psychological side instead of pursuing the avenue of logical objectivity (Frege, 1884:37; § 26: "Ist die Zahl etwas Subjectives?" – 33 ff.). We shall return to this issue below.

2 Consider the titles of the following books: *The Social Construction of Reality; a Treatise in the Sociology of Knowledge* (Luckmann & Berger, 1969); and *Der sinnhafte Aufbau der sozialen Welt* ("The meaningful construction of the social world" Schutz, 1974).

"Using a conceptual shorthand, we could say that *meaning* has replaced *cause* as the central focus of attention" (my italics – DS; Appleby *et al*, 1996:1).

This shift merely introduced additional functional conditions to an understanding of the human person and human society, namely those labelled above as *historicity* and *linguisticality* (see Chapter 2, page 57). Contrary to the postmodern unease with the idea of *universality*, these two conditions apply universally. Any (postmodern) statement containing the term 'all' in its negation of universality is self-referentially incoherent, because what it wants to deny is presupposed by its mode of formulation. For example, when Van Huyssteen speaks of the shortcomings of 'foundationalism', he remarks:

> As will soon become clear, nonfoundationalism, in a strong reaction against modernist and universalist notions of rationality, highlights the fact that *every* (my italics – DS) historical context, and *every* (my italics – DS) cultural or social group, has its own distinct rationality (Van Huyssteen, 1999:63).

Universality is here questioned in the name of universality, because the logical function of the word 'every' is equivalent to what modern symbolic logic calls a *universal quantifier*. This impasse is evident in another context, where he writes:

> Postmodernity challenges us to deal with the fact that we have been robbed of any general, universal, or abstract ways to talk about the relationships between religion and science today (Van Huyssteen, 1998:2-3).

We have to note that Van Huyssteen can only formulate this statement on the *basis* of employing three *concepts* (each with its own *universal scope*): namely *generality*, *universality* and *abstraction* ("general, universal, or abstract").[1]

Exploring the idea of modal aspects in an *ontic* sense furthermore naturally requires *criteria* when it comes to the identification of modal aspects that are supposed to be unique and irreducible. But let us first articulate the structural features of a modal aspect (see Dooyeweerd, 1997-II:75).

3.1.4 *The structure of a modal aspect*

We noted that Cassirer has already realized that, in order to avoid a *regressus in infinitum* in scientific concept formation, the acceptance of "specific original functions" is required. Such functions cannot genuinely be derived from *other* functions – and there is no need to attempt such a derivation (Cassirer, 1957:73 – see page 12). Throughout the history of philosophy thinkers wrestled with this problem, but the first *systematic*, *consistent* and *reality-oriented* treatment is found in the *general theory of modal law-spheres*, as developed by Herman Dooyeweerd (see in particular Dooyeweerd, 1997-II).

The irreducibility of every modal aspect is guaranteed by its core meaning, designated as its *meaning-nucleus* or *meaning-kernel*. It is not only *irreduc-*

1 Kreitzer formulates a similar critique in respect of Van Huyssteen's attempt to side-step both foundationalism and relativism: "Unfortunately, Van Huyssteen does not see the logical contradiction. He 'knows' (a universal truth claim) that all knowledge is interpreted merely within a parochial group" (Kreitzer, 2007:7).

ible and *indefinable*, but also *irreplaceable*. As such, it founds the *sphere-sovereignty* of every aspect – a feature already alluded to in Chapter 1 in connection with a *jural action* (a *commission* or an *omission* in relation to jural causality) (see page 22).

The flip side of the coin is that the various aspects are related to each other through "moments of coherence," that is, through *analogies* of other aspects reflected within the structure of a particular aspect. All such modal analogies are *stamped* or *qualified* by the nuclear meaning of an aspect, whether they are pointing backward to aspects that are *foundational* to the aspect concerned, or forward to succeeding aspects.[1]

Comment: Initially Dooyeweerd designated the backward-pointing moments of coherence between modal aspects as *analogies* or *retrocipations*. The coherence between an aspect and aspects positioned later within the cosmic order of aspects was called *anticipations*. Stafleu used to accept this earlier position of Dooyeweerd, but in his 2002 work decided to substitute the idea of an analogy (or: *retrocipation*) by using the word 'projection' (see Stafleu, 2002:20 ff., 53). Since this work nowhere addresses the nature of an *analogy* as such (an analogy is present when two aspects or entities differ in that respect in which they are similar – see pages 144 and 151 below), no argument is provided for this new preferential designation of the idea of (retrocipatory) inter-modal coherences. However, the absence of such an account turns out to suffer from a complication: while analogies referring backwards to cosmic earlier modal aspects are no longer designated by Stafleu as retrocipations, he continues to designate the inter-modal coherences between modal aspects and *cosmic later aspects* as 'anticipations' (see e.g. Stafleu, 2002:53)! Apparently without being aware of it, Stafleu thus merely continues a view found during an earlier phase of Dooyeweerd's intellectual development, when he still identified analogies with *retrocipatory analogies*. It was only later, in the English translation of his *magnum opus* (*A New Critique*), that he realized that an *analogy* embraces both these moments of coherence. As a result Dooyeweerd at this stage made the systematic distinction between those analogies pointing backwards ('retrocipations') and those analogies pointing forward ('anticipations') within the cosmic order of time (see Dooyeweerd, 1997-II:75). In other words, his general point of view is therefore that the nature of an analogy embraces both 'retrocipations' and 'anticipations' – yielding a distinction between "retrocipatory analogies" and "anticipatory analogies." It is clear that Stafleu did not realize that by employing the term 'anticipation' he implicitly employed the idea of an analogy as it is present in the mature conception of Dooyeweerd.

In discussing the nature of *jural causality*, we dealt with a (retrocipatory) modal analogy, for within the jural aspect, the configuration of *jural causality*

1 The idea of modal aspects entails an order of succession amongst them – argued for in more detail in Chapter 5 (see page 162 ff.)

points backward to the physical aspect where the relation between cause and effect appears in its original modal context (see page 24). In our everyday experience, such analogies are just as frequent as they are unnoticed. Young people may refer to their "social life" without realizing that this phrase expresses a connection between the biotic and the social aspects of reality; neighbours may talk about the *social injustices* of international policies without contemplating the interconnection between the social and the jural aspects evinced in this expression; and so on.

Our language is simply flooded by such composite expressions – just think about "emotional strength" (sensitive and physical aspects), "economic trust" (economic and certitudinal aspects), "aesthetic expression" (aesthetic and lingual modes), "emotional vitality" (the sensitive and the biotic facets), "social distance" (the social and the spatial), and so on.

The most important implication of this interconnectedness of the various aspects is that the *meaning* of an aspect only comes to *expression* in this intermodal coherence with other aspects. Complementing the sphere-sovereignty of modal functions, this feature is known as the *sphere-universality* of every aspect.

Moreover, within each modal aspect, there is a strict *correlation* between its *law side* and its *factual side*. Whereas the (modal) laws for the functioning of natural things are *valid* (applied), the (modal) laws for human functioning (in the post-sensitive modes) are not yet applied or made valid, for they are given only *in principle*. Without the intervention of human functioning, they do not obtain any validity. It belongs to the formative cultural task of humankind to give a positive shape to these principles in varying historical circumstances. Within each typically human aspect of reality, we therefore have to distinguish between its 'principle'-side or *norm side*, and its *factual side*.

Both Dooyeweerd and Vollenhoven initially distinguished between *law side* and *subject* side. This caused an ambiguity, because on the *subject*-side, they discerned *subject*-subject and *subject*-object relations – implying that the term 'subject' acquired an ambiguous meaning. From a systematic perspective, the most general approach is to differentiate between law side or norm side and *factual* side – where the latter then comprises factual subject-subject and factual subject-object relations.[1] This view represents Dooyeweerd's mature conception, as found in his *Encyclopedia of the Science of Law*. In the Systematic Part, he for example writes: "With respect to jural ground and jural effect the subject or factual side of this interconnection expresses itself in the relation of the factual jural cause to the factual jural effect" (Dooyeweerd,

1 After my promotion in December 7, 1973, Vollenhoven one evening invited me to his home. Amongst the issues discussed was the nature of 'law'. He explained to me that he gives preference to the systematic distinction between a 'norming' and a "non-norming" law.

1987c:15). See also pages 31, 35,[1] 50, 56 and 62 in this connection. Occasionally the phrase "factual side" already appears in *A New Critique of Theoretical Thought* (see 1997-I:96).

On the factual side of every aspect, there are subject-subject and subject-object relations, with the exception of the first modal aspect,[2] where we only find subject-subject relations – exemplified in operations of *addition*, *multiplication*, and so on. When we add or multiply two numerical subjects – such as 2 and 2 – the outcome ('4') is still just another natural number (numerical subject). The same applies to the subtraction of one number from another one, for once again the result is another numerical subject.[3]

Finally, every modal aspect is fitted into a *cosmic time order*,[4] while bringing to expression within its own modal structure a functionally 'coloured' manifestation of what Dooyeweerd calls *cosmic time*. On the law side, it expresses itself as *time order*, and on the factual side as *time duration*.

These structural moments present in every modal aspect are captured in Figure 5.

Figure 5

The structure of a modal aspect

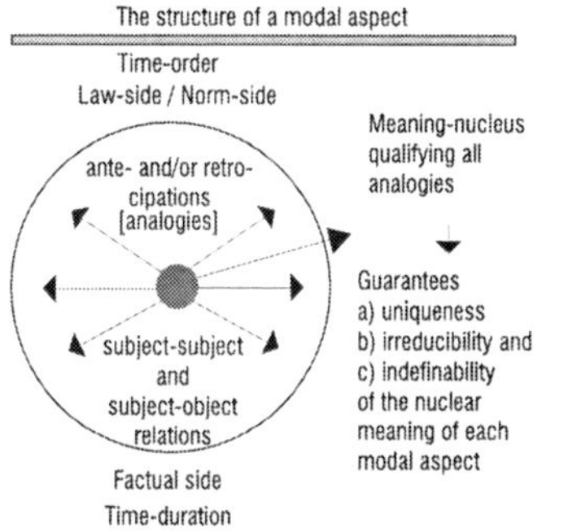

3.1.5 *Criteria for the identification of modal aspects*

Before we elaborate the points below in extensive arguments in Chapters 5 and 6 we may give an indication of criteria to be applied in the search for unique and irreducible modal aspects:

(1) Throughout its entire history, Western philosophy always had to account for a given diversity within reality. This awareness is an *indirect* indication of the existence of *distinct* aspects (this claim is implicit in the footnote on page 70 regarding the reality of *aspects* and *entities*).

(2) Also, in our non-scientific (so-called 'naive') experience, we find this diversity – as reflected in the shared human analytical awareness of this diversity.

(3) The great variety of isms in philosophy and the special sciences reflects

1 "Naar de *wetszijde* doet ze zich voor als de in de *rechtsnormen* vervatte relatie van *rechtsgrond en rechtsgevolg*, naar de *subjects-* of *feitelijke zijde* in de relatie van *feitelijke juridische oorzaak tot feitelijk juridisch* gevolg, ..."

2 That the numerical aspect is foundational to the spatial aspect is argued in Chapter 5, pages 218 ff.).

3 In the case of addition and multiplication only natural numbers are implied. When subtraction is introduced, new numerical subjects are envisaged, namely zero and the positive and negative integers. See the explanation below (pages 214 ff.).

4 See pages 206-211 for an account of the dimension of cosmic time.

the modal diversity within reality. At least in the case of *monistic isms* a different aspect of reality is elevated to provide the all-encompassing theoretical perspective towards an understanding of the universe, also *indirectly* implies distinct aspects of reality.

(4) Reflection on the various realms in nature (material things, plants and animals), as well as on the various human societal collectivities (such as the state, church, sports club, school, cultural society, theater group, marriage, business firm or language association) directs us towards the various modalities (aspects) that provide access to the modal function uniquely characterizing those social entities. An analysis of these characteristic (or: qualifying) functions may be helpful in the search for unique aspects.

(5) The occurrence of *antinomies* in theoretical thought is an indication that certain aspects of reality are confused. Introducing the appropriate modal distinctions should then be able to resolve the antinomies concerned.

(6) The development of independent special sciences, delimited in their area of study by a particular aspect of reality, indicates the variety of aspects of reality.

(7) Another aid in the identification of a particular aspect is given in the appeal to our *immediate intuition* (experiential insight), when reference is made to the meaning of any distinct aspect.

(8) All the special sciences use typical *entity concepts* (such as: atom, molecule, plant, animal, tool, book, money, painting, murder weapon, engagement ring, church building), as well as *functional concepts* unmistakably appealing to the modal aspects of reality (such as *mass*, *volume*, *life*, *feeling*, *control*, *meaning*, *exchange*, *beauty*, *lawfulness*, *love*, *trust*, *faith*).[1]

(9) An indirect method of analysis, such as the indication of an analogical structural element in the modal structure of an aspect, can lead to the identification of the original, non-analogical nature of a particular aspect. The *fact* that something like jural *agreement* and *disagreement* – legitimacy and illegitimacy – exist, refers to the logical aspect in which agreement and disagreement first appear.

(10) In the case of the normative aspects of reality, a negative indication, or even the negation of a negative indication, can sometimes help to express our insight into the nature of a core of meaning.[2] The core of meaning of the economic aspect can be captured as an "avoidance of excess." Economic normativity requires non-excessive actions. The negation of this negative formulation highlights that it refers to a manner of *having*

1 This distinction between typical concepts of entities and modal concepts of function is equivalent to that between modal laws and type laws (see page 25 ff.).

2 Note that one cannot conceptually comprehend it, because every meaning-nucleus is conceptually indefinable and primitive – and concepts are formed on the basis of terms that ultimately are primitive. However, what is *primitive* can always be *designated* or *named*. A detailed discussion of primitive terms is found in Chapter 5 (see below pp. 170 ff.).

enough (and how many large businesses, with their incredible striving for excessive profits, do not know when they have earned enough). Without sensitivity to the modal demand of *having enough*, a person may simply ignore her responsibility for *economic stewardship*.

3.2 Aspects and entities: modal laws and type laws[1]

Once the *universality* of every modal aspect is acknowledged, another avenue for the distinction between these two dimensions of reality is provided. The immediate implication of the modal universality of the aspects of reality is that no single (natural or social) entity or process can 'escape' from *functioning* within every aspect of reality. It will turn out that this statement is only sound if a distinction is made between *subject* functions and *object* functions (see Chapter 6, pp. 345-392). For example, although a material entity cannot buy and sell, it may have an object function within the economic aspect as a *coin*, or simply as a *precious stone* found in nature.

In this sense, the modal laws of every aspect apply to *every possible* entity and process. Modal laws therefore apply *universally*; their *scope* is not limited to any *class* or any specific *kind* of entities. Quantitative laws do not merely hold for apples or human beings, they apply to any multiplicity (plurality) of whatever kind.

Conversely, the laws for any specific kind of entities are solely applicable to those types of entities. The law for being an atom does not apply to mammals or states alike. Similarly, the structural principle for the state as public legal community does not apply to other kinds of societal institutions, such as ecclesiastical communities or business firms.

In general, it can be said that modal laws hold for *all possible classes* of entities, whereas type laws hold for *a limited class* of entities only. This explains why even Kant was compelled to make a distinction between his (supposedly universally valid *a priori*) *thought categories* on the one hand, and so-called *empirical laws of nature* on the other:

> We rather have to distinguish empirical laws of nature, which always presuppose particular perceptions, from the pure or general natural laws, which, without having a foundation in particular perceptions, only contain the conditions of their necessary connection in an experience. In respect of the latter nature and possible experience are entirely the same; and since within these the law-conformity of the necessary connection of appearances in an experience (without which we are totally incapable of knowing any object of the world of sense), actually is based upon the original laws of the understanding, so it initially does sound strange, but it is nonetheless certain, when I state with respect to the latter: understanding creates its laws (a priori) not out of nature, but prescribes them to nature (Kant, 1783 par.36:320).

This distinction runs parallel with the one we have drawn between *modal* laws and *type laws* (see page 25). Although misdirected by the rationalistic as-

1 See our initial discussion of the distinction between modal laws and type laws on page 25 of Chapter 1.

sumptions of his epistemology,[1] we noted that Kant, in his search for the *synthetic a priori* (see page 69 above and pages 421 ff. where the issue is discussed in more detail), actually struggled with the nature of *modal universality*. Positivism and neopositivism ought to be acknowledged for their emphasis on *experimental testing* and *confirmation* (which is not the same as *verification).* Only through studying the *orderliness* or *law-conformity* of entities, is it possible to arrive at an understanding of the *type laws* holding for that limited class of entities conforming to their peculiar type laws. In the case of physics, it requires empirical research through *experimentation.*[2]

In order to appreciate Kant's position better in this regard, we have to look at the historical background of the distinction between *modal laws* holding for whatever there is and *type laws* applicable to a limited class of entities only.

Whoever modally abstracts a particular aspect, gains access to the (unspecified) *universality* of modal-functional relationships. Since modal aspects are not concrete entities or events, they cannot be treated *as if* they are entitary in nature, because this would simply amount to a *reification* of modal functions.[3] A widespread and well-known example of such a reification is the reference to the origin of 'life'. Of course the *intention* is to refer to *living things*, yet no single living entity is exhausted by its biotic (life) function, since, amongst others, living entities also display a *physical* aspect, and both physicists and biologists know that the physical-chemical constituents of living entities are not alive.

If one really wants to gain an understanding of the *type law* of any particular *kind of entities*, one has to investigate those entities *empirically*. One cannot derive the *typical nature* of different kinds of physical entities merely from modal analysis or abstraction – what is required, is empirical testing *through experimentation* – the legitimate facet of the positivist claim.

The fact that modal laws – such as those of quantum physics – hold for all possible 'objects', is clearly seen by Von Weizsäcker: "Quantum theory, formulated sufficiently abstract, is a universal theory for all *Gegenstandklassen* (classes of objects)" (Von Weizsäcker, 1993:128). When he explains, on the next page, that one cannot deduce the kinds of entities of experience from the universal scope of quantum theory, he implicitly alludes both to universal modal laws and type laws (the latter with their *specified* universality).

Weyl also implicitly makes an appeal to the distinction between modal universality and typicality:

1 In terms of which human understanding is viewed as the formal (*a priori*) law-giver of nature.

2 Of course this does not free physics from an overarching and underlying paradigm (theoretical perspective), in which *modal properties* are also accounted for. Sometimes this dimension of theory formation is implicitly acknowledged when reference is made to *theoretical terms* that cannot be *tested* directly against actual experiences ("sense data").

3 Whenever a *function* or *aspect* of reality is treated as if it is an *entity*, such an aspect is *reified* (or: 'hypostatized'). Conversely, if an entity is treated as if it is an *aspect* or *function*, it is *functionalized*.

But what is connected with the *a priori* construction is *experience* and an *analysis of experience through the experiment* (Weyl, 1966:192).

Discussing the nature of an *a priori* synthetic element in the "empirical sciences," Stegmüller raises the following possibility – alluding to the same issue (Stegmüller, 1969:316):

Surely, this cannot imply that the totality of law-statements present in a natural science could be of an *a priori* nature. Much rather, such an apriorism should limit itself to the construction of a limited number of *a priori* valid law relationships, while, furthermore, all more specific laws of nature should be dependent on empirical testing.

Note the similarity between Stegmüller's statement and the following explanation given by Stafleu (related to the distinction between *modal laws* and *typical laws*):

Whereas typical laws can usually be found by induction and generalization of empirical facts or lower level law statements, modal laws are found by abstraction. Euclidean geometry, Galileo's discovery of the laws of motion ..., and thermodynamic laws are all examples of laws found by abstraction. This state of affairs is reflected in the use of the term "rational mechanics," in distinction from experimental physics (Stafleu, 1980:11).[1]

Another example of what modal universality (a modal law) is all about, a remark from the 19th century British philosopher, John Stuart Mill, is illustrative in respect of the scope of 'numerosity'.

All numbers, must be numbers of something: there are no such things as numbers in the abstract. But though numbers must be numbers of something, they may be numbers of anything. Propositions, therefore, concerning numbers, have the remarkable peculiarity that they are propositions concerning all things whatever; all objects, all existences of every kind, known to our experience" (quoted by Cassirer, 1953:33-34 – with reference to Mill, *A System of Logic*, Book II, Chapter 6, 2).

Particularly widespread is the confusion of modal universality and typicality when it comes to art works and an appreciation of the 'aesthetic'. Entities characterized by their aesthetic aspect should certainly be known to us as *art works* – the universal scope of the aesthetic also embraces those entities that are not characterized by the aesthetic mode, but still have a function within this aspect, owing to its modal universality. Therefore aesthetics should not be retricted to those things dominated by their aestheticity, but also leave room for the aesthetic function of those entities that are not typically aesthetic.

Zuidervaart lucidly explains this state of affairs:

I can not avoid viewing the structure of aestheticity from both typical and aspectual angles. In the first place, much of that which is called aesthetic is typically artistic. There exist countless things and manners of functioning that can not be explained except as members of an artistic type of existence that is dominated by its aestheticity. For example, the concrete existence of an art work is

1 In the light of the distinction between modal abstraction and "entitary-directed abstraction" (see pages 15 and 48 ff.) the phrase "found by abstraction" should have been "found by *modal* abstraction."

experienced as a typically aesthetic object. On the other hand, much aestheticity is not typically aesthetic. For example, a human being and a 'beautiful' landscape are present to one another through certain aesthetic relations and functions. Neither the human nor the landscape is typically aesthetic. Their aesthetic relationship is certainly not artistic. Rather it is one instance of the fact that an aesthetic aspect permeates all creaturely existence (Zuidervaart, 1977:6-7).

3.3 The various modal aspects

Naturally everyone introduced to the theory of modal aspects would like to know how many there are and which are the ones that qualify as unique modalilties. While investigating this issue, a number of related problems and questions will surface as well. Yet, we shall attempt to restrict our discussion in this chapter mainly to an account of the *nuclear meaning* of the various aspects, reserving a more detailed analysis of their interconnections for later. However, the first aspect will require more extensive treatment, due to the complexities surrounding the relationship between what is ontically given – as a plurality – and the way in which its human articulation gives birth to arithmetic through the employment of number symbols.

3.3.1 *The quantitative aspect*

This aspect constitutes the "lower limit" of all the modal aspects, in the sense that disregarding the various aspects of any entity or process, invariably terminates in *quantity* as the last (or rather: *first*) modal function of reality. The example of the *abacus* showed that one has to disregard the physical aspect, the aspect of movement and the spatial aspect of the blocks, in order to be able to focus on the numerical question: *how many*?[1]

Specifying any number does not presuppose the meaning of the aspects of space, movement or energy-operation. By contrast, if any spatial figure is specified number is always presupposed. Referring to a *triangle* brings with it the number "3"; speaking of points, lines, surfaces and volumes is accompanied by an awareness of a number of dimensions (we shall argue that a point is a subject in zero dimensions, a line a subject in one dimension, a surface in two and a volume in three (see pp. 146 ff.). The meaning of number is therefore *constitutive* for the meaning of space since the meaning of space is built upon the primitive meaning of number (as the latter comes to expression in everyday acts of *counting*). We shall see that magnitude not only reflects *an order of extension* (length is a one dimensional magnitude, surface a two dimensional one, and so on), but that specifying any magnitude inevitably requires a measure of extension (such as *inches*) which is always given by a *number*. Simple acts of counting do not involve any spatial notions, such a dimension or extension. The numbers 1, 2, 3, 4, and so on are solely constituted as a (re-

1 Of course an *analysis* of the meaning of the numerical aspect cannot escape from employing terms derived from other aspects, for example when the *domain* (spatial) of number is mentioned or when certain numbers are depicted as *constants* (kinematic) and others as *variables* (physical). Regarding the meaning of constitutive and the distinction between constitutive and regulative, see below pages 182 ff., 228 ff., 312 ff., 316 ff., 361391 ff.

versible) *succession*.

Likewise *motion* entails a *path* and it occurs at a certain *speed* (expressed by a number). Therefore the meaning of motion is built upon the foundation of space and number and not the other way around – it does not make sense to speak of the speed of a number, or of its form or its shape. Consider the natural number 'one'. In spatial terms *oneness* is specified as 'wholeness' and in kinematic terms as *uni*-formity. Although physical events in a causal relation succeed each other, the relation between cause and effect is more than a mere succession. Kant already realized that the succession of day and night (and night and day) does not entail that the day is the *cause* of the night or that the night is the *cause* of the day (see page 208).

An awareness of the *one* and the *many* indeed lies at the basis of our human intellectual capacities. It has already been implicit in the Pythagorean legacy, although in a distorted way, because they did not merely recognize this foundational role of number, but actually attempted to elevate it to the level of an all-embracing perspective, manifested in the claim: *everything is number*.

We have frequently referred to this aspect as the *quantitative* aspect. The reason is that as soon as we speak of *numbers*, we have already involved human reflection in our terminology, because the human response to a given *multiplicity* is to introduce *numerals* (number symbols) in service of actual processes of counting. In cases where cultures did not develop a very articulate sense of counting, they nevertheless had a rudimentary understanding of *succession* and they often evinced the ability to establish a *one-one correlation* between groups of entities.[1]

The co-conditioning role of the quantitative aspect explains why all entities functioning within this aspect of reality are indeed (numerically) *distinct*. Yet, as Frege has shown convincingly in his criticism of "entitary-directed" abstraction, more general *entity concepts* merely result in *one* general concept. In so far as entities are *different* it will not help us in any way to attain a quantitative understanding of their *number*, for any combination of entities maintaining their differences does not constitute a number.[2] Does this mean that number would be obtained when we *abstract* from these differences? This will not help either, for the procedure simply results in a higher level concept of the kind of entities involved. Suppose we start with *different cats*. Once all the differences between cats are disregarded, there is no *multiplicity* left, but only the *general concept* 'cat'.[3] Conceiving different things merely as 'units'

1 The story of a farmer who never attended a school illustrates this ability, for when counting his sheep he moved small stones *one by one* from one bag to another one. Cantor explored this idea of a one-one correspondence in his mathematical set theory.

2 Frege remarks that the properties through which entities distinguish themselves from each other are irrelevant and strange to their number (Frege, 1884:45; § 34).

3 Angelelli explains this by pointing out that, through "abstracting from the particular differences and natures of the given objects no plurality can be attained, but only one thing (the concept 'cat', for example)" (Angelelli, 1984:467). See our earlier remark about Frege's reference to a white cat and a black cat (page 27).

or 'ones' avoids the issue, because when they are made *equal* or *similar* through the (abstractive) elimination of all their *differences*, only the general concept of "being the same thing" (that is, of "being one") is left – without any *plurality* at all (like the general concept *cat*).

The typical nature of "entitary abstraction" involved in any similar process amounts to *logical addition* only, without producing any (new) number(s). Where Kant argues for the *synthetic* nature of mathematical judgments in his *Critique of Pure Reason* (CPR), he clearly realizes that such a *purely logical addition* (a merely *logical synthesis*) cannot give rise to a *new* number (cf. CPR, 1787:15, where he considers the proposition that 7+5=12). In his own way, Frege simply elaborated the same point in his argument about the *moon* (see page 27), showing that *entitary-directed abstraction* can only proceed to *more abstract entitary levels*, but that it can never yield any *number* as such. Likewise, the logical addition of 'ones' or 'twos' cannot but end with the repeated identification of another specimen of the same kind: having identified a 'one' and another 'one' or a 'two' and another 'two' and so on, still results only in the 'abstract' notion of 'oneness' or 'twoness'. Frege remarks: "If we want to construct number, on the other hand, through a combination of what is similar, then everything continues to flow into one, preventing us forever to arrive at a plurality."[1] Frege quotes Jevons who says: "It has often been said that units are units in respect of being perfectly similar to one another; but though they may be perfectly similar in some respects, they must be different in at least one point, otherwise they would be incapable of plurality" (cf. Dummett, 1995:86). Or to put it in the words of Frege himself:

> When the things to be counted are called units, then the unconditional assertion that these units are similar is false. That they are similar in certain respects is correct but worthless. The things to be counted necessarily have to be different if their number is to be greater than 1 (Frege, 1884:58; §45).

It is rather tempting to follow Frege in the *solution* he proposes for this difficulty. Suppose we want to ascertain the number of many *different* or even various *similar* things. Is it not true that we first have to conceptualize *what* it is that we are counting? Think about a zoologist observing a whale and an elephant and in the process she realizes that they are *both* mammals. In other words, the zoologist arrives at the (quantitative) insight that there are *two* mammals. This statement seems to be dependent upon the general concept *mammal*.[2] The same seems to apply to *similar* entities, such as in the observation *there are four cats*.[3] Apparently it is only on the basis of the concept *cat*

1 "Wenn wir die Zahl andrerseits durch Zusammenfassung von Gleichem bilden wollen, so fliesst dies immerfort in eins zusammen, und wir kommen nie zu einer Mehrheit." (Frege, 1844:50; § 39 – see page 50 above).

2 If it remained uncertain that both are indeed mammals, one can retreat to a more general level, such as simply stating that there are two *animals*, for then the concept *animal* still guides the attachment of the number '2' to it.

3 Remember that whatever is *different* at a certain level, may be seen as *similar* at a higher level of (entitary-directed) abstractrion.

that we can come into touch with this *plurality* of cats. Consequently, it seems as if the assignment of a number is always attached to a *concept*. Therefore Frege says: "Assigning a number contains an expression about a concept."[1]

Although everyday life employs number in an *attributive* sense, Frege argues that a careful scientific practice should always avoid this usage. For example, instead of saying that Jupiter has four moons, one should rather say: "The number of Jupiter's moons is four" (Frege, 1884:69; § 57).[2] Frege believes that, in this way, we have liberated ourselves from the mistaken conviction that number is a *property* of anything. For it is not a property of things similar to the way in which we can abstract properties such as colour, mass or impenetrability from entities.[3]

We noted earlier that Frege indeed understood something about the *universality* of modal aspects in his employment of the term 'logical' – for, in the words of Dummett, *objects of any kind can be numbered* (see Chapter 3, page 71). This view of Frege actually pertains to a common feature of every *aspect* of reality, not only to the *logical aspect*. It is part of the nature of an aspect or *modus* that its scope transcends the multiplicity of individual things merely *functioning* within it. This constitutes what ought to be designated as the *modal universality* to which we alluded above – amongst others, also holding for the *quantitative aspect of reality*. This modal universality of the numerical aspect clearly surfaces in Frege's account, as explained by Dummett:

> A correct definition of the *natural numbers* must, on his view, show how such a number can be used to say how many matches there are in a box or books on a shelf. Yet number theory has nothing to do with matches or with books: its business in this regard is only to display what, in general, is involved in stating the cardinality of objects, of whatever sort, that fall under some concept, and how natural numbers can be used for this purpose (Dummett, 1955:272).

At this point Frege (and Dummett) was on the verge of the alternative theory which we have in mind when we speak of *modal universality* – were it not the case that Frege identified the *ontic meaning* of the quantitative aspect of reality with the nature of a *concept*. We have seen that the theory of modal aspects *first of all* eludicates the *uniqueness* of functions as distinct from *entities* – and that, within conceptual thought, this difference surfaces in the distinction between the *concept of number* and *type concepts*. The quantitative aspect of reality *co-conditions* the universe in which we live. If we disregard non-quantitative modal functions of entities, such as the physical (their *mass* or *impene-*

1 Frege's original wording reads: "daß die Zahlangabe eine Aussage von einem Begriffe enthalte" (see Frege, 1884:59; §46). On the basis of extensive argumentation, Dummett renders this phrase as follows: "The content of an ascription of number consists in predicating something of a concept" (Dummett, 1995:87, 88).

2 The copula 'is' does not denote attribution or predication, but functions in service of equality, in the sense of stating the *identity* of what is found on both sides. Thus the 'is' has the meaning of asserting that both sides are 'similar' or "the same as" (Frege, 1884:69; § 57).

3 "Die Zahl ist nicht in der Weise wie Farbe, Gewicht, Härte von Dingen Abstrahirt, ist nicht in dem Sinne wie diese Eigenschaft der Dinge" (Frege, 1884:58; § 45).

trability), the spatial (their *shape*), and so on, they turn out to be totally similar, or *gleich*, as Frege argued. But if it comes to the numerical question: *How many*? Frege correctly points out that non-quantitative properties are worthless.

Yet, what he did not consider is the co-conditioning role of the *ontic quantitative mode*. That different entities, through abstraction made similar according to all their non-quantitative properties, are indeed still *distinct* (and therefore a *plurality*), demonstrates the *conditioning role* of the *quantitative aspect*, for the universality of this modality entails that no single entity can 'avoid' having a (subject) function within it.

As soon as one acknowledges that "objective reality" displays an *ontically given* numerical (arithmetical) *aspect* (compare the ideas of Cassirer, Gödel, Bernays and Wang mentioned in Chapter 1, page 20), the quoted phrase: "Yet number theory has nothing to do with matches or with books" has to be corrected. Material things – and whatever else belonging to the dimension of entities within reality – invariably *function* within the numerical (and other) modes or aspects of reality. In so far as 'matches' or "books on a shelf" therefore *function within* the quantitative mode, this *functioning* has everything to do with the meaning of this aspect and with its *conditioning role* in the existence of *matches* (plural!) and *books* (plural)! The ultimate Fregean question concerning the necessity of at least one last 'difference' accounting for the plurality of what has to be counted, finds its answer in the acknowledgment of the co-conditioning effect of the quantitative mode on the existence of all entities, underlying their *distinctness* as entities. Nothing more or less is required than the fundamental insight into the nature of this unique (and irreducible) aspect of our experience. Denying this foundational ontic mode in favor of the logical objectivity of concept formation only apparently solved the problem, for, as we shall explain in connection with the meaning of the logical aspect, the entire *logicistic* course taken by Frege resulted in an unresolvable *impasse*.

The distinctness of entities showing the co-conditioning role of the quantitative aspect is ultimately dependent upon the meaning-nucleus of this aspect, which is to be designated as *discrete quantity*. However, acknowledging this role of the quantitative aspect does not imply that the *individuality* of entities is derived from this mode; that this mode serves as the *principium individuationis* as it was called in Aristotelian Scholasticism. In our discussion of the nature of entities, we shall return to this point (see pages 301 ff. below).

This nuclear meaning also determines the most basic awareness of infinity we have, that of one, another one, yet another one, and so on *indefinitely*, *without an end*, *endlessly*, *infinitely*. However, a proper account of this fact does require further distinctions not yet explained.

Although it is strictly speaking incorrect to talk about the *numerical* or *arithmetical* aspect of reality – owing to the fact that these designations incorporate a contribution from the human subject, for the sake of convenience we shall nevertheless frequently employ these terms in our subsequent discussions; also in cases where the primitive meaning of multiplicity (a given plurality) within the quantitative mode is actually intended (see page 18).

3.3.2 *The spatial aspect*

The spatial aspect succeeds the numerical aspect, but presupposes the meaning of the *one* and the *many*. How else will it be possible to understand the meaning of one or more dimensions (such as length, width and height) or different magnitudes (such as 9 centimeters, 5 square meters or 3 liters)?

Our awareness of space is always related to an understanding of *extension*. Extension in any dimension – such as that of a *straight line* (one-dimensional extension), that of a *surface* (two-dimensional), etc. – is at once attached to an awareness of the *connectedness* of whatever is extended in a spatial sense. What is connected *hangs together*, that is, *coheres*, and this entails that extension cohering in such a way embraces every connected *part*. When *every part* is given, it is understood as a *whole*, a *continuous* whole.

This shows that the relation between a *whole* and its *parts* originally belongs to the modal aspect of space. Yet it seems quite difficult to *define* the meaning of *continuous extension*. Interchanging 'continuity' with terms like 'uninterrupted', 'connected', 'coherent', and so on, simply repeats what is meant with the term *continuity*, instead of really *defining* it![1] In connection with a purported implicit definition, Shapiro mentions the feature of being *coherent*, but adds that "coherence is not a rigorously defined mathematical concept, and there is no noncircular way to characterize it" (Shapiro, 1997:13).

Without entering into a more extensive investigation of the interconnections between number and space, at least *one* outstanding feature of the spatial aspect ought to be mentioned, namely the fact that continuity allows for an endless (sub)division. Aristotle, in following up certain insights by Anaxagoras, holds it to be self-evident that "everything continuous is divisible into divisible parts which are infinitely divisible" (*Physica*, 231b15 ff.). Already the way in which Parmenides has characterized *being*, illuminates important features of continuity and the whole-parts relation. The B Fragments 2 and 3 of Parmenides, contained in Diels-Kranz (1959-1960), hold that being "... was not and will never be because it is connected in the present as an indivisible whole, unified, coherent" (B Fragment 8, 3-6).

Modern intuitionistic mathematics made an appeal to these insights of Greek thinking in developing their alternative to the atomism entailed in the thought of Cantor, and the formalism of Hilbert. The intuitionist Hermann Weyl, for example, points out that the fact that it "... has parts, is a basic property of the continuum," and adds: "... it belongs to the very essence of the con-

1 When we discuss in more depth the interconnections between number and space it will turn out that much more can be said about the *meaning* of continuous extension, because its meaning indeed only comes to expression in its *coherence* with other aspects. Within this context the *formal similarity* between the criteria stipulated by Aristotle for continuity and those specified in the modern mathematical analysis of Cantor and Dedekind will be highlighted.

tinuum that every one of its parts admits a limitless divisibility" (Weyl, 1921:77).

The best way to capture the nuclear meaning of the spatial aspect is therefore given in the phrase *continuous extension*.

3.3.3 *The kinematic aspect*

Already in his *Isagogè Philosophiae* from 1930, Vollenhoven has distinguished between the *mechanical* and the *physical* aspects[1] of reality. However, in the 1936 edition, this distinction no longer appears. Dooyeweerd, on the contrary, initially maintained the order *numerical, spatial, physical* – thus identifying the kinematic aspect with the physical aspect. However, already in 1926, with reference to the idea of a *logical creation* of *nature* (found in the thought of Cohen, Natorp and Cassirer), Dooyeweerd explicitly mentions the four primary aspects he eventually distinguished: "In this way the concept of the infinitesimally small, of number, of space, of mechanical movement, and, finally, the specifically natural scientific concept of energy as logical constant in functional causality have already been achieved" (Dooyeweerd, 1926:27). Equally remarkable is his appreciation that the modern natural sciences discovered the "simplest law-spheres of nature: those of number, space, movement, matter (energy)" (Dooyeweerd, 1928:24). Once in 1930 Dooyeweerd also mentioned the numerical, spatial, movement and energy functions of a tree guided by the biotic function (Dooyeweerd, 1930:249). During 1950 he once again realized that a distinction between the kinematic aspect (of uniform movement) and the physical aspect is necessary in order to account for the fact that kinematics (*phoronomy*) can define a *uniform motion* without any reference to a *causing force* (compare Galileo's law of inertia) (see Dooyeweerd, 1997-II: 99).

The Aristotelian view that whatever moves requires a *causing force* in order to continue its movement was reversed in Galileo's mentioned thought experiment[2] and derivation of the law of inertia. The issue is whether or not one can *deduce* the meaning of uniform motion from any other facet of reality, such as the physical aspect. Since physics always deal with dynamic forces operative in the interplay of energy transformation, and since a uniform motion can indeed be envisaged without making an appeal to a *cause*, it is clear that a unique and irreducible aspect is at stake here.

In a functional sense movement is something *original*. It is clear from Galileo's above-mentioned thought experiment that whatever moves will continue its motion endlessly.[3] The awareness of a *uniform motion* should not be confused with *flexibility*, since the latter requires *change of motion*. Whereas a uniform motion does not need a cause, a change of motion *does* need a cause,

1 In Vollenhoven 1968:3 the physical is designated as the *energetical*.

2 See Chapter 3, page 72.

3 Note that the 'endless' refers back to the original meaning of the numerical (the quantitative) aspect, where we first encountered an endless succession (the infinite in the literal sense of "without an end").

for both in the case of *acceleration* or *deceleration* an energy-input (that is, a *physical cause*) is required.[1] Our immediate experiential awareness of a uniform motion allows for another appropriate synonym capturing its unique meaning: *constancy*. If something in motion *continues* its movement endlessly, its movement will be *constant*. Therefore movement – as the mode of *constancy* or *uniform flow* – is indeed given in an original sense, just as it is the case with *number*, *space*, the *economic* or the *ethical*. Of course the term flow is a metaphor that can be specified with the aid of numerical, spatial or kinematic terms. In a kinematic sense an enduring motion remains unaltered, i.e. it is constant (a kinematic specification). Exploring our awareness of space may generate the qualification *continuous flow*. The term *uniform* suggests the *smoothness*[2] of a constant movement and it may even reflect something numerical in referring to *one* motion. Yet the idea of constancy itself is not found in the quantitative and spatial aspects.

Although terms like continuation, uniform flow and constancy may all be viewed as synonyms referring to the kinematic aspect as a unique modal function of reality, we may give preference to the term *constancy* when it comes to the task of designating the meaning-nucleus of this aspect.

3.3.4 *The physical aspect*

Both the concept of *force* and that of *acceleration / deceleration* do not belong to the domain of the kinematic aspect, as we have argued. Their location within the physical aspect flows from the fact that *energy* in all its forms always *causes* physical changes, that is, it generates certain *effects*. Therefore *energy-operation*, which may be seen as a proper designation of the meaning-nucleus of the physical aspect, entails the *cause-effect* relation, that is, it lies at the basis of *causality* in its original physical sense. Words like *change* and *dynamics* are once again mere *synonyms* of this nuclear meaning of the physical mode.

According to Sikkema the kernel of the physical aspect is 'interaction' (Sikkema, 2005:20). "While some use force and/or energy to characterize the physical aspect, these concepts lose their meaning and relevance at the quantum level, while interaction does not" (Sikkema, 2005:30). However, the expression *interaction* is composite, and the first element, 'inter', represents a spatial analogy within the physical aspect and therefore cannot serve as a part of the characterization of the meaning-nucleus of this aspect. What is left is the second part, 'action', which is the equivalent of 'operation' in the expression *energy-operation*. If the term 'energy' fails to apply to the quantum level, the expression energy-operation within such a context may be interpreted just to refer to 'operation' or 'action'. In an e-mail (May 24, 2008) Stafleu states his preference for *activity* (equivalent to the Greek word *energeia*). We shall continue to use the phrase *energy-operation* – understood as the equivalent or

1 Stafleu points out that Aristotle's view that motion *is* change, obstructed the development of the modern view of motion that distinguishes between *inertial motion* and accelerated (changing) motion (Stafleu, 1989:66).

2 Another metaphor reflecting space in its coherence with the meaning of the kinematic aspect.

incorporation of the idea of "*energeia-activity*."

The modal universality of the physical aspect entails that its core meaning pertains to *all* forms of matter. Interestingly, a leading physicist of the 20th century such as Heisenberg, found it difficult to distinguish between modal functions and concretely existing physical things. His conception was influenced by Ostwald's 'Energetik', which led him to affirm that the "basic stuff" ('Grundstoff') of everything 'real' ought to be designated as *energy*.[1] Unfortunately Heisenberg here confuses *matter* with the universal meaning of the *physical aspect*.

In his critical reaction to this view, Vogel points out that Heisenberg simply exchanged matter with one of its universal properties, namely *energy*. Elementary particles, as forms in which matter appear, display the property of energy, that is to say, the ability to cause effects. But, so Vogel continues, it is mistaken to reify this feature, as done by Heisenberg, in order to provide a foundation for his 'Energetik'.[2] "Energy is not a form of appearance of matter, but a universal property of *all* appearance forms of matter, whether planets, photons or whatever else."[3]

Vogel's criticism at once highlights both the modal-functional nature of energy (that ought not to be reified) and its modal universality (whatever there is, has a function within the physical aspect).

3.3.5 *The biotical aspect*

Note that we start our discussion of the unique meaning of this aspect by using the term 'biotical' instead of 'biological'. This confusion of *biotical* and *biological* follows the long-standing rationalistic legacy of the belief that the world *itself* has a 'rational' structure. In explaining the views of Davies, one finds various statements made by Van Huyssteen explicitly stating that the world is *rational*:

> What is astounding, however, is to what a great extent our world is truly rational, that is, in conformity with human reason (Van Huyssteen, 1998:68);
> It is indeed fascinating to see, precisely through the fact that the rational nature of our universe is reflected in its basic mathematical structure, that Davies ultimately comes to the point where he has to acknowledge the limits of this reasonableness (Van Huyssteen, 1998:71).

1 "So wie die alten Griechen es sich erhofft hatten, so haben wir erkannt, daß es wirklich nur einen einzigen Grundstoff gibt, aus dem alles Wirkliche besteht. Wenn wir diesem Grundstoff einen Namen geben müssen, so könnten wir ihn heute nur 'Energie' nennen" (Heisenberg, 1973:97).

2 We explained that when an aspect is treated as an entity it is *reified* and when an entity is treated as an aspect it is *functionalized* (see page 80 above).

3 "Hier liegt eine ganz simple Verwechselung der Materie mit einer ihrer allgemeinen Eigenschaften, der Energie, vor. Elementarteilchen sind Erscheinungsformen der Materie, die ebenfalls die Eigenschaft der Energie haben, d. h. die Fähigkeit, Wirkungen auszuüben. Es ist falsch, die Eigenschaft zu verselbständigen, wie es Heisenberg hier tut, um seine Energetik zu begründen. Die Energie ist nicht Erscheinungsforin der Materie, sondern ein universelle Eigenschaft *aller* Erscheinungsformen der Materie, sie seien Planeten, Photonen oder anderes" (Vogel, 1961:37, note 14).

However, Van Huyssteen does not question the deeply rooted *rationalistic assumption* in this view of the 'rational' (even 'mathematical') structure of the world – a conception that is on a par with the Kantian view of human understanding as the formal law-giver of nature, discussed above (see page 73). The universality of a modal (or typical) law makes it possible for human beings to conceive (form a *concept* of) such laws, but it does not mean that these *ontic* laws themselves are *rational* in nature.

Biotic phenomena are given – only the theoretical investigation of these phenomena ought to be called *biological*. Biology is a scholarly discipline investigating (from the perspective of the modally abstracted biotical aspect) those entities and processes occurring (functioning) within this aspect.

From the outset, we have to question another widely accepted practice in this regard, namely the habit of speaking about 'life' as if it is a 'something', an 'entity'. The best known expression in which this confusion appears is in biological references to the origin of 'life'. We have to question this *reificatory* mode of speech.

If 'life' is treated as 'something', the first question is whether or not *life* is really *alive*, that it, is it thoroughly *living* – as a *whole* and in *all its parts*? It does not need much argumentation to demonstrate the untenability of this position, for the obvious question is whether the atoms, molecules and macro-molecules within such an entity are *alive*? Of course the answer is: *no*, for even the well-known neo-Darwinistic thinker, George Gaylord Simpson, had to concede that, since biology studies *living* entities, the term *molecular biology* is self-contradictory, for molecules are not alive (Simpson, 1969:6).[1] Therefore the reification of the biotic aspect in an unqualified reference to 'life', results in the self-defeating situation that 'life' is not fully *alive*, since it turns out to be a 'mixture' of 'life' and 'death' (we shall return to this issue in Chapter 7).

The flip side of this problematic coin resides in the habit of biologists differentiating 'living' from "non-living" *matter*. The physicist does know what 'matter' is, because our universe is populated by material entities, such as those just mentioned (atoms, molecules and macro-molecules). But none of them are alive. It is not even proper to refer to such physical entities as *dead matter*, for then the implicit assumption is that those entities, during some earlier stage of their existence, were indeed alive. The German physicist, Von Weizsäcker is therefore fully justified in introducing the new term 'unbelebt', designating that which is not, and has never been, alive (1993:32).[2]

Living entities have a modal function within the biotic aspect, but in princi-

1 "Since biology is the study of life (read: "living things" – DFMS) and molecules, as such, are not alive, the term 'molecular biology' is selfcontradictory."

2 He writes: "Stones are 'unbelebt'. But one should not say that they are dead. Only something that actually once lived could be dead" ("Die Steine sind unbelebt. Man sollte aber nicht sagen, sie seien tot. Tot sein kann eigentlich nur etwas das gelebt hat"). The term 'unbelebt' intends to say that physical entities (matter) were never alive and therefore should not be designated as *dead* – for then it is assumed that once they were alive.

ple they also function within all the other aspects of reality. Surely the biotic aspect presupposes the aspects of number, space, movement and energy-operation, but it cannot be reduced to these foundational aspects. The modal concepts of function employed by physics do not include terms derived from the original meaning of the biotic aspect. Just consider biological terms such as *life*, *growth*, *differentiation*, *integration*, *adaptation*, *goal-directedness* (*finality*), and so on.

Since the term *organism* is derived from 'organs', and the latter are indeed *alive*, it is not appropriate to refer to living entities (embracing what is alive – their organs – and what is 'unbelebt' – physico-chemical constituents – alike) as *organisms*. The meaning-nucleus of the biotic aspect could best be designated as *life* instead of *organic life*, since the latter rather reflects an interconnection with the numerical aspect in the presence of a *multiplicity* of organs.

3.3.6 *The sensitive-psychical aspect*

It is unfortunate that the term 'psychic' acquired an esoteric connotation in our contemporary culture. It would therefore be wise to replace it with the term 'sensitive' or with a reference to the *feeling aspect*, or at least to use the composite phrase *sensitive-psychic*. *Feeling* is also an ambiguous designation, because it may refer to one of the senses (alongside seeing, hearing, etc.) but it can also be used to refer to *inner emotional states*. Yet all these phenomena are found within the structure of this aspect. Particularly because the term 'psychological' is frequently applied to refer to sensitive or emotional phenomena, it would have been convenient to use the term 'psychic' without any esoteric associations.

To be sensitive entails the presence of *consciousness*. In the biotic mode, consciousness is absent. The animal and human *will*, in the sense of *desire*, is part and parcel of this aspect, just like *drives*, *instincts* and *needs* appear within the modal structure of this aspect. If the term is intended in a general functional (modal) sense, *sensitivity* or *feeling* may capture its nuclear meaning. In his excellent work on psychology Ouweneel suggests that this aspect should be two distinct modalities, namely the *perceptive* and the *sensitive* facets (Ouweneel, 1984:42-44). His distinction depends upon the presence or absence of a (complete) limbic system which, to my mind, merely argues for a subdivision within the sensitive mode (it seems to be a matter of degree and not of kind).[1]

3.3.7 *The logical aspect*

Particularly in connection with our analysis of the distinctive feature of theoretical thinking, namely *modal abstraction*, we already have entered into a preliminary investigation of the meaning of analysis and abstraction – see in particular Figures 1 (page 14) and 2 (page 14) in Chapter 1.

Analytical activities are intimately connected to our sensitive-psychic abili-

1 In passing it should be noted that Ouweneel's work the human being at once provides an encompassing orientation in the systematic philosophy of Dooyeweerd (see Ouweneel, 1986).

ties and they give rise to different kinds of *analytical representations*. The term used to capture this state of affairs is 'apperception'. Non-scientific modes of conceiving result in *conceptual representations*. Frege accuses Kant of confusing *subjective representations*, which may be widely differing for different people, with *objective representations* (which are the same for all people). His own alternative is to reserve the term 'representation' (*Vorstellung*) for what is *subjective* and to note that it is to be distinguished from what is objective in a *logical* sense (Frege, 1884:37; § 27). Yet we still have to investigate more fully the nature of scientific concepts, because the relationship between different aspects – captured in *concepts of function* and analogical relations between them – differs from what is normally designated by *metaphors*. Once we have embarked on a more detailed analysis of the principle of sphere-universality (giving expression to the interconnectedness of the various cosmic aspects) the relationship between *function concepts*, *analogical concepts* and *metaphors* will be considered.

At this point it may be sufficient to state that the meaning-nucleus of the logical aspect is *analysis*. But we cannot leave our discussion of the meaning of this aspect without returning to the fact that it is a *normative aspect*, as we have briefly done in Chapter 2 in our first discussion of the normativity of life (see pages 41 ff.). It means that human thinking can either *obey* logical principles or *violate* them, thus making possible the distinction between *logially sound* and *illogical* thinking.

Figures 1 and 2 highlight the congruency between *analysis* and *abstraction*, since *identification* is the same as *lifting out* while *distinguishing* is *equivalent* to *disregarding*. These two corresponding pairs contain the key to an understanding of the nature of *logical concept formation*. Confusing entities or modal states of affairs inevitably leads to *illogical concept formation*. Bertrand Russell provides a classical example in the (illogical) concept of a "square circle." Cassirer uses a similar example: a "rundes Viereck" (a "round square") (Cassirer, 1910:16).[1] However, figurative speech in the use of language seems to 'transcend' the demands of logicality, for nobody has a problem with a *square circle*, such as a "boxing ring"!

3.3.8 *The cultural-historical aspect*

In the face of modern *historicism* – which, by and large, emerged during the age of Romanticism and acquired its full-grown form during the 19th and 20th centuries – Dooyeweerd distinguishes between the concrete *process of becoming* – encompassing *all* aspects of reality – and the *historical aspect* as merely *one* of its *modal functions*. In German, Dutch and Afrikaans, one can designate this difference by employing *different* words for this many-sided (genetic) process and its historical aspect: 'Geschichte', 'Geschiedenis', 'Geskiedenis' on the one hand, and 'historisch' and 'historie' on the other.

1 This example is actually derived from Immanuel Kant (1724-1804) – see his *Prolegomena zu einer jeden künftigen Metaphysik die als Wissenschaft wird auftreten können* (1783:341; § 52b).

During the 1920s Dooyeweerd – particularly in the development of his *legal philosophy* – wrestled with the problems of *natural law* and *historicism*. The former orientation accepted universal principles as (supposedly) *valid* for all times and places, rooted in human reason, whereas the latter emphasized *historical changefulness* and *contingency* at the cost of *constancy* and *universality*.

The human *competency* to give a positive *shape* or *form* to underlying universal, constant principles in varying historical circumstances indeed acquired a prominent role in the attempt to overcome the one-sidedness present in natural law conceptions and in historicism. Being mere products of human *cultural formation*, no single particular positive shape (designated as a *positivization*)[1] may be elevated to the level of the underlying *conditioning principle*. This insight overcomes the shortcoming of natural law theories – which (rationalistically) postulated principles that are already applied or *valid* (in the sense of being positivized) for all times and places – and it also avoids historicistic approaches, i.e. approaches denying the existence of *universal* and *constant* principles by only acknowledging historically varying shapes and form(ation)s.

Strangely enough Dooyeweerd did not fully escape from the *terminology* of the natural law tradition, because he continued to speak of *universal validity* as a characteristic feature of underlying principles – without realizing that, insofar as principles are *universal* and *constant* (i.e. *pre-positive*), they are not yet *valid* (i.e. not yet *positivized*), and insofar as principles *are* given a positive shape and form (i.e. are *positivized*), they have lost their *unspecified* (pre-positive) *universality*. Habermas implicitly continues the same legacy, for his legal and political philosophy is presented in terms of the distinction between *validity* ('Geltung') and *positivity* ('Faktizität') (see the title of Habermas' 1998 work).

McIntire questions the existence of the historical aspect, but unfortunately, amidst many other questionable interpretations, misunderstood the nature of a *special science* in Dooyeweerd's thought. He believes that, according to Dooyeweerd, historical study "examines the historical aspect of anything" (1985:89). This is a fairly common misunderstanding of Dooyeweerd's position, for many scholars took him to say that each of the various disciplines studies a particular aspect of reality. But in fact Dooyeweerd holds that an analysis of the *modal structure* of any aspect belongs to the *philosophical foundations of a discipline*. Special sciences, as we have argued (see page 47), merely look at reality through the *point of entry*, *angle of approach* or 'glasses' of particular aspects. Moreover, Dooyeweerd writes that an investigation by a special science "does not focus its theoretical attention upon the modal structure of such an aspect itself; rather, it focuses on the coherence of the actual phenomena which function within that structure" (Dooyeweerd, 1996:11).

1 The term *positivization* is also used by other philosophical orientations (see for example Luhmann, 1985:147 ff. and Luhmann, 1986:26).

Since concrete things, events and processes function in all aspects of reality, their multi-aspectual nature cannot be used to argue against the existence of what we call the cultural-historical aspect. Formative cultural activities are intrinsically related to the historical mode or aspect.[1] *Cultural historical formation* or *formative control* is an effect of the free formative imagination of human beings. Through it, things (artifacts) are brought into existence in dependence upon the *formative power* of human beings. Yet this formative power is not exclusively directed towards subject-object relations, for in the case of inter-human relationships, there are numerous instances of people exercizing power over other people. Such instances concern *subject-subject* relations in combination with the nature of an *office* and the *competency* to positivize principles valid for other human subjects – such as those promulgated by the parliament of a constitutional state under the rule of law – therefore also entailing the spatial analogy of above and below (super- and subordination).

The meaning-nucleus of the cultural-historical aspect is therefore to be designated as *formative control* or simply as *power* – as long as it is remembered that power differentiates into power over objects and power over subjects. Moreover, this last consideration eliminates the possibility to call this aspect the *techno-formative* mode, for then its meaning is restricted to subject-object relations only.

3.3.9 *The sign aspect*

Normally this aspect is known as the *lingual* mode. Yet there are important considerations suggesting an alternative formulation. The human ability to *express* meaning and *interpret* it is a response to the normative demand to *assign* meaning, to *signify*. Within inter-human contexts such an expressive assignment of meaning or signification always calls forth the *interpretative* response of another human subject. Suppose we restrict ourselves to the normative sense of our (human) calling to signify, leaving aside whether or not it is done with the aid of verbal language.[2] Then it may be justified to refer to this aspect as the *semiotic aspect*. But other options are also open to us, for we can just as well focus our attention on the subjective acts of signification, in which case the *formation of language* (and linguistic structures) acquires a prominent place, apparently once again justifying the designation *lingual mode*. Finally, if we focus on the objects intended by acts of signification in their relation to the meaning of the words, we may designate this aspect as the *semantic mode* (consider the discipline of semantics studying the meaning of words).

Combining these three options – *semiotic*, *lingual* and *semantic* – while at the same time avoiding the relative one-sidedness contained in each in isola-

1 Interestingly, Diemer, as a philosopher of science, without hesitation designates the science of history as an "aspect discipline" (Diemer, 1970:224).

2 In addition to language, all other non-lingual entities, processes and actions also function within the sign mode. Just think about the significance of a *polite greeting* or of what is known as *body language*.

tion, the entire modal structure of this aspect may simply be designated as the *sign mode*. But because the limitations of normal language use in some cases may give preference to alternative usages, we sometimes will have to use the term 'lingual', even when the *total structure* of the sign mode is intended.

Given his view that the bond between the signifier and the signified is arbitrary, and that therefore the "linguistic sign is arbitrary," De Saussure remarks that the word 'symbol' is not appropriate as a designation of the "linguistic sign," because

> [O]ne characteristic of the symbol is that it is never wholly arbitrary; it is not empty, for there is the rudiment of a natural bond between the signifier and the signified. The symbol of justice, a pair of scales, could not be replaced by just any other symbol, such as a chariot (De Saussure, 1966:68).

In a letter to J. Gallois (by the end of 1672), Leibniz pursued the tradition of Aristotle and Boethius when he used the term 'symbol' as synonymous with 'nota'. As an arbitrary sign, it serves as a *genus concept* for linguistic expressions and written signs, including mathematical signs. The epistemology of the 18th century combined this view of an arbitrary sign with the theory of symbolical knowledge (*cognitio symbolica*). Meier-Oeser mentions the 'organon' of Lambert (1764) in which an ambivalence can be observed regarding the purely arbitrary nature of the sign on the one hand and its co-determination by relations or analogies (sensory image and symbolic knowledge). He also refers to Kant, who conceived the symbol as "a sign of signs" – although Kant opened up another avenue by associating a *symbol* with *allegory* and *metaphor* (see the Meier-Oeser, 1998:718-720).[1]

Typical semantic phenomena, such as *synonymy*, *antonymy*, *metaphoricity*, *metonymy*, the use of an *oxymoron*, and so on, all originate within the sign mode of reality. Dooyeweerd depicted the meaning-nucleus of this aspect by employing the expression *symbolical signification*. Given the vast legacy involved in distinguishing between *signs*, *symbols*, *similes*, and so on, one might have expected that Dooyeweerd would have motivated his choice in confrontation with it. It may be the case that his intention was to employ the traditional connotation of the term 'symbol' in the linguistic sense of an *arbitrary sign*. Of course there are also other thinkers in the 20th century who used the term symbol in the sense of freely chosen signs (just compare the three criteria specified by Von Bertalanffy in the next chapter – see page 120). However, this connotation practically duplicates what is intended by *signification*. Therefore one may refer to the meaning-nucleus of this aspect by describing it with the verb 'signifying' or 'signification.'

3.3.10 *The social aspect*

Because all the modal aspects are bound together within a unique dimension of reality, delimiting our experiential horizon in a specific way, it is impossi-

1 Roelofse points out that *symbolism* differs from ordinary connotative meanings as well as from myths "in that it allows only for [a] specific interpretation. ... It is, one may say, totally culturally determined" (Roelofse, 1982:89).

ble to assume that human beings function first in some aspects and later in others. For example, it does not make sense to start a reflection about the nature of the 'social' by first contemplating a supposedly lonesome individual, for it belongs to the very constitution of every individual human being that it functions *at once* (*simultaneously*) within every modal aspect, including the social aspect. There are modern theories proceeding from postulating abstract (pre-societal) individuals. Such a view actually denies that human existence is equally determined by every modal aspect, including the *social function*. A person is not an "abstract individual" that has a "social function" only in the second place. Every modal function – in a primary and fundamental sense – *inherently* co-conditions the existence of human beings. Particularly compare the abstract construction of a "state of nature" in modern theories of *natural law* and in *social contract theory* (Thomasius, Pufendorf, Wolff, Hobbes, Locke, Rousseau and Kant – recently revived by Rawls).

Social interaction within human society gives expression to phenomena such as *courtesy*, *tact*, and *politeness*. Within this aspect, we find a point of orientation for a first classification of different kinds of social intercourse between human beings.

Yet, particularly within the English language, it seems to be quite difficult to find a suitable word or phrase capturing the meaning of its nucleus. The Dutch expression "omgang en verkeer" does not have a direct translational equivalent in English. The term *interaction* is not appropriate, because it merely (analogically) reflects the meaning of the spatial ('inter') and the physical aspects ('action') – implying that without the qualification "social interaction," the term *intercation* can refer to any kind of non-social interaction (such as the 'interaction' of atoms in a chemical reaction). A term such as 'socializing' carries with it the suggestion that there are a-social beings that ought to be brought to the arena of the social, thus wrongfully denying the *transcendental* (ontic) status of the social aspect, i.e. the fact that being human *inherently* entails a function within the social aspect.

In the English translation of the works of Georg Simmel, the German sociologist, the term 'sociation' is employed, suggesting that the term 'association' may be quite close to the core meaning of the social. Unfortunately it has also acquired a specific connotation, referring to certain kinds of social collectivities called 'associations'. If it is not possible to attach a more general and basic meaning to the term 'association', we may still opt for the mentioned term 'sociation' as the equivalent of the Dutch (and Afrikaans) "omgang en verkeer."

3.3.11 *The economic aspect*

In the discussion of the preceding aspects, references to the etymology of words were pushed to the background, mainly because in most cases they derive from concrete metaphors that originally were directed towards analogies between concrete entities, while we are currently in search of terms with a modal (aspectual) focus. This situation is particularly clear in the case of the 'economic', for the etymology of this term goes back to the Greek word for *household*. Originally the latter is actually an undifferentiated societal totality embracing the lives of the people living within its confines in a totalitarian sense. Modern economic theory, under the influence of the modern humanistic natural science ideal, starts from the guiding idea of a (physical) equilibrium and the acceptance of free and equal individuals (atomistically understood).

This explains why the classical school in economic theory saw the prime manifestation of the operation of the basic economic 'law' of *supply and demand* in the market. This approach in fact used the functioning of the *market* to explain *everything* within economic life. Both Coase (1937:386-405) and Williamson (1975 and 1985) start from the assumption that markets are basic and historically original: *in the beginning there were markets*. The *firm* is a secondary phenomenon, emerging through the attempt to economize and internalize the cost of market transactions between individuals. Fourie correctly asks the question whether it is legitimate "to attempt to explain the typical and distinguishing inner nature of the firm exclusively in terms of another, typically different relation, i.e. the market (transactions, contracts)" (Fourie, 1993:41).

Markets not only include exchange relations, but also rivalry relations. However, markets do not 'produce' anything – this task belongs to the *firm*. Since markets are constituted by an interaction between buyers and sellers on an equal footing, that is, by being positioned next to and over against each other, they do not display any relation of authority and subordination. The typical totality character of the older dwelling places of production – such as ancient villages and medieval guilds – indeed historically *preceded* the development of *inter*-relationships within which the market could emerge (cf. Chandler, 1979). The assumption of "markets in the beginning" is therefore just as **fictitious** as the hypothetical *state of nature* assumed in the *social contract theories* of humanistic natural law.

Economic considerations always relate to a relative scarcity of means and the need to act in *non-excessive* (*frugal*) ways, which then constitute its meaning.

3.3.12 *The aesthetic aspect*

Similar to the important distinction between the concrete, many-sided historical process and its normative *cultural-historical* aspect, and similar to the distinction between actual language use and its *sign function*, we have to avoid the identification of art with its aesthetic aspect. A work of art functions in other aspects as well, albeit the case that the aesthetic aspect is dominant.

Finding appropriate terms to capture the unique meaning of this aspect, caused serious differences of opinion, also within the tradition of reformational thinking. In our discussion of the normativity of human life, we have alluded to the many contraries present in our everyday experience of life (see Chapter 2, pp.41 ff.), such as those between *logical* and *illogical*, *kind* and *hostile*, *legal* and *illegal*, *thrifty* and *wasteful*, and of course *beautiful* and *ugly*.

The latter contrary, known to us from early childhood, cannot be located in any other aspect than the aesthetic. Closely connected to this understanding of *beauty*, is that of *harmony*, explaining why one of the initial formulations of the meaning-nucleus of the aesthetic is *beautiful harmony*.

Calvin Seerveld suggested that this designation should be discarded because it is burdened with distortions from the Greek culture and by the 19th century ideal of beauty.

In Greek culture, we indeed find a pagan view of beauty embedded in a particular understanding of harmony. It found expression in the Pythagorean conviction that everything is number, because the mutual relationship between the integers 1, 2, 3, and 4 apparently provides the basis for all harmony, such that the cosmos itself ought to be appreciated *as* a work of art. This *formative harmony* is opposed to the *formless corruptibility* and *transience* of the visible world of change. Ultimately, the Greek view of beauty and harmony is rooted in the radical opposition between *form* and the *formless* operative within Greek culture – an approach that cannot do justice to the unity and goodness of creation.

However, without becoming a victim of this Greek ideal of beauty (or even of that of the 19th century), it cannot be denied that the contrary *beautiful – ugly* belongs to the horizon of our experience of modal properties and that therefore it has to find a 'home' in a specific modal aspect. What should be guarded against, is selling out the nuclear meaning of the aesthetic to any specific art forms or positivizations of aesthetic style. When we discuss the complexities of the interconnections between the various aspects of reality below, we shall return to the development in Calvin Seerveld's thought in this regard – from his initial suggestion of the "law of coherence" and 'ambiguity' up to his eventual preference for *allusivity* and *imaginativity* (see pages 250 ff.).

3.3.13 *The jural aspect*

Within the universal scope of the jural aspect, *retribution* should not be restricted to *revenge* or to the typical configuration of *penal law*. The meaning of the jural aspect is based on the insight that, whatever belongs to a person ought not (illegally) to be taken away – and if it happens, the restorative meaning of this aspect entails that it ought to be given back (*re*-tribution). The classical Roman jurists (Ulpianus and others) captured this meaning with their time-honored phrase: *give* every person his or her *due* (*ius suum quique tribuere*).

The crux of this formulation is given in the positive meaning of 'tribuere', 'tribute' – which provides the basis for our understanding of *retribution*. What is therefore required is to find a suitable term (or word) that can designate this core meaning – the same applies to the way in which the meaning nuclei of all the other aspects are captured as well. "Giving a person his or her due" is simply synonymous with the 'tribution' part of *retribution*. An infringement or violation of that which rightfully belongs to a person (either in a civil legal sense or in the context of penal law), gives rise to the required restoration signified by the term 'retribution'.

The English term *retribution*, interestingly, brings better to expression the meaning of the jural aspect than the equivalent terms in Dutch and German ('vergelding'; 'Vergeltung'), thus reversing the difficult situation encountered in finding an appropriate rendering of "omgang en verkeer" within the English language.

3.3.14 *The ethical aspect*

It is general practice to identify all forms of normativity, requirements of *ought to be*, with the *ethical* or *morality*. Within the context of academic reflection, it often seems unproblematic to reason about business ethics, medical ethics, professional ethics and so on – and in addition a distinction is drawn between theological and philosophical ethics. Traditional Roman Catholic moral philosophy incorporates a *theory of virtues* and since the last two decades of the 19th century, Western civilization became accustomed to speak about *values* – a theme introduced by the Baden school of neo-Kantianism (Weber, Windelband, Rickert and others).

The 'basket' conception of normativity, putting all its manifestations within the category of the 'ethical' or 'morality', simply denies other *normative* aspects their proper right of existence, but then only at the cost of running into serious problems. The example of an illogical concept mentioned in an earlier context (see page 93), presupposes the existence of *logical* normativity, to be distinguished from ethical normativity. The examples given in our initial portrayal of the normativity of life (see pages 41 ff.) also clarified that human life is embedded in various normative contexts, amongst them that of morality (ethics).

Once released from the impossible burden of being the sole bearer of human normativity, ethics may start its investigation into the core meaning of the moral aspect. Similar to the relevance of contraries within the aesthetical aspect, we also in this case have experience of an undeniable opposition between *love* and *hate*, suggesting that the meaning-nucleus of the ethical aspect is none other than *love*. The general ethical contrary between what is considered to be *moral* or *immoral* testifies to the universality of the moral aspect of reality. Although there may be differences of opinion, within the same or within different cultures, concerning what is moral and what is immoral, the universality of the contrary *moral-immoral* is not threatened by these differences, for the latter presuppose this universality of the moral aspect. In the absence of an understanding of the modal universality of the moral aspect one will be inclined to be in doubt about the true universal scope of this aspect. For example, see the ambiguity in the argument of Ferns in this regard (Ferns, 2007:157 ff.). The reality of differences of opinion about the nature of morality apparently conceals the fact that the universality of moral phenomena makes possible such differences of opinion.

As an addition to appreciating love as the core meaning of the moral aspect one can refer to *love in temporal relationships*, but then one has already embarked upon an analysis of the intermodal connections between the moral and other aspects of reality, in this case the spatial and social aspects.

Within a differentiated experience of *love*, certain other nuances evince themselves, for example when we associate with the dedication accompanying genuine love qualities such as *sincerity*, *honesty*, and *integrity*. This explains why the interconnections between the ethical aspect and other aspects of reality sometimes employ these terms as well, in order to capture the connection with the moral aspect. When the philosopher of science, McMullin, speaks about *intellectual honesty*, he actually deals with a so-called epistemic value based upon the ethical antecipation[1] within the structure of the logical-analytical aspect.

Stafleu also highlights the ambiguity present in the two terms normally employed in the designation of this aspect which, according to him, has "loving care" as its core meaning (see Stafleu, 2007:23). Since Stafleu fully acknowledges a unique aspect having as its core meaning "loving care," his argument boils down to a terminological preference.[2] Whereas Stafleu does not want to designate this aspect as the ethical or moral aspect of reality, owing to its traditional more-than-this-one-aspect connotation, someone else may decide to stipulate that instead of loading all forms of normativity into the undifferentiated 'basket' of morality (ethics), it is preferable to restrict these two terms to the aspect under consideration.[3]

3.3.15 *The certitudinal aspect*

Within this aspect, the terms *certainty*, *trust*, *confidence*, *reliability*, and *faith* are facets of the same diamond. As an after-effect of the traditional Thomistic and modern humanistic dualistic separation of 'faith' and 'science', it is customary to restrict the former term to "religious belief," such as found in Roman Catholic or Protestant church denominations. This practice, however, was unsuccessful in eliminating the multifarious forms *faith* can assume (in the sense of certainty, trust, or confidence). A tremendously significant and powerful instance is found in what we have met as a *faith in reason* (see Chapter 1, p.10). In this context we have also mentioned that Derrida affirmed that *faith is absolutely universal* (Derrida, 1997:22). He specifically uses *credit* – economic faith (trust) – as an example. In our analysis of the interconnectedness of the various aspects of reality, we shall return to this example of economic trust, for then it will become clear that, whereas the avoidance of certitudinal excesses is a reference from the faith aspect to the economic as-

1 In order to avoid any misunderstanding attached to 'anti' in the sense of 'against', we shall not speak of 'anticipation' but rather of 'antecipation'.

2 Dooyeweerd considers the term *love* – without any further qualification – to constitute the core meaning of this aspect (see Dooyeweerd, 1997-II:152, 156.

3 The phrase "loving care" rather looks like a cultural-historical analogy within the moral aspect. *Love* is sufficient as an indication of the core meaning of the ethical aspect.

pect, credit (economic trust) does the reverse, because it is a reference from the economic mode to the aspect of faith.

When faith (trust) is directed towards other human beings and even focused on the *reliability* of God's laws for creation, also operative in the existence of artifacts such as chairs (evinced in the confidence we have in the ability of a chair to bear the person who intends to use it as a seat), then the term to designate this modal aspect can also be derived from terms designating *faith* ('fides'). From the Latin word *fides*, the derivation *fiduciary* is possible. Therefore, alongside the designation *certitudinal mode*, the name *fiduciary aspect* may also be used.

Some theological exegeticists are of the opinion that we do find a 'definition' of faith Hebrews 11. What is actually done here is only to repeat what has been said – and understandably it is done by using synonymous terms. When we say that faith concerns something about which we are sure (convinced) and in addition to that it is claimed that it is what we are certain about, then there exists no ground for holding that a *definition* instead of a repetition (tautology) is given. This is indeed the case in the New Testament:

> Now faith is being sure of what we hope for and certain of what we do not see (Hebr.1:1).

To say that you are *sure* is the same as to say that you are *certain* (or *confident* or *convinced*). Similar to the way in which terms like *connectedness* and *coherence* are mere synonyms for (spatial) continuity, terms like *certainty*, *trust*, *confidence*, and *reliability* are (synonymous) designations of the core meaning of the faith aspect.

3.4 Diversity and the quest for an 'origin'

Every world and life view brings to expression some or other account of the *origin* of the universe. Present-day physicists find the "Big Bang" theory reasonably attractive and even assume that the actual "original moment" can be calculated or measured. The "Big Bang" is sometimes portrayed as proceeding from a primordial initial hot and condensed condition that precedes time and space. In some respects the mode of speech attached to the *Big Bang* hypothesis closely imitates the theological tradition of a *negative theology* – where one cannot positively say what God is like, but only state what God is *not* (see pages 192 ff. below). Of course in all instances of a negative theology one always finds one or another last remnant of a *positive* characterization. In the case of the *Big Bang* it is therefore not surprising that *size* and *heat* play a crucial role – just consider terms such as *density* and *temperature*. Hubble and Lemaître contributed to the idea of the *expanding* universe.

The equations formulated by Alexander Friedmann are based upon Einstein's *General Theory of Relativity* and they presuppose the *conditioning role* of the first four aspects of reality – number, space, the kinematic and the physical.[1]

---

1 We shall see that the history of time measurement reveals a general awareness of four modes of time: *earlier and later*, *simultaneity*, *time-flow* and *irreversibility* are well-known modalities of time (see below page 209).

However, Gentry questions the expansion postulate, arguing that "the universe is relativistically formatted in accordance with the Schwartzschild static spacetime solution of the field equations, not the Friedmann-Lemaître spacetime" (see Gentry, 2001:1). Whatever the outcome of this controversy may be, it cannot qualify as an account of *creation*. Dating this primordial event to almost 14 billion years ago appeals to time measurement and time measurement always involves the *time duration of a process*. Any time duration is always delimited by and subject to a specific (correlated) *time order*. Therefore it will always be a circular undertaking to attempt to determine ('date') the origin ('creation') of the presupposed time order. For that reason it is in principle impossible to date creation.

In the discussion of the uniqueness of the various aspects merely focusing on a first approximation of their meaning-nuclei, the all-important perspective in this regard is that we have to accept this radical diversity as it presents itself to us without attempting to *reduce* the rich diversity to one or a few modal perspectives employed as devices for explaining whatever there is.

The insight that everything is unique – including the various modal aspects of reality – only tells the first half of the story (that part governed by the principle of "sphere-sovereignty"). The second part emphasizes the unbreakable coherence between all the aspects, implying that any and every attempt to pursue a reductionistic strategy in scholarship will inevitably result in a distortive understanding of the given diversity of reality. In other words, every attempt to elevate any one (or a combination of more than one) aspect(s) to the level of the 'origin' of all the other [in that they can be derived from or reduced to this (these) privileged one(s)], simply amounts to a search of finding the origin of the rich diversity of reality *within that diversity itself*. In the subsequent chapters, we shall see that this aspiration is self-defeating, because the unique meaning of each aspect *presupposes* the unique (that is, irreducible) meanings of all the other aspects. In fact, we shall argue that the meaning of every aspect only comes to expression in its *coherence with every other aspect*. This investigation will require that further systematic distinctions are introduced, constantly staying in touch with both the *history of philosophy* and developments in the various *special sciences*.

But before we proceed to a more detailed analysis of the complexities involved in an in-depth investigation of the *interconnectedness* of the various aspects of reality, the eluding and ever intriguing question about the *human being* will occupy our attention first.

Chapter Four

Being human

4.1 The outward search turned inward

The human interest in the diversity of creation thus far mainly surfaced in our discussions in connection with the dimension of modal aspects. But amidst these considerations, the central question regarding the unique nature of humanity always seems to return unanswered. This fundamental puzzle calls forth such an immediate response from scientific reflection, that early Greek thinkers already held the opinion that there is no meaning to the attainment of knowledge about all else if humankind does not know itself. As Heraclitus declares: "I investigate myself" (Diels-Kranz, B. Fragment 101).[1] His reflection is situated within the context of an aspiration to discover a *cosmic order* which is valid for everything (cf. B Fr. 30). What he says about the nature of a human being, furthermore, is formulated with an account of the relation between God and humankind – and the negative limit of the relation between *a human being* and a *beast*. Note his simile:

> The most beautiful ape is despicable in comparison to the human race. The most wise human being, however, stands to God as an ape ... (B Fr. 82, 83).

Humankind, for Heraclitus, is situated between *beast* and *God* – a problem echoed even into the 20th century in the title of a book by the eminent German zoologist, Bernard Rensch: *Homo sapiens, from Animal to Demigod*.

Socrates deepens and internalizes the Greek question concerning the nature of being human. He wants to know who he is himself: is he related to the many-headed animal TIPON (the mythological symbol of the *flowing stream of life* without any *set limit* or *form*), or does he share in a more *measured*, simple *divine* nature (the prominence of the *form motive* in Greek thought)? The maxim guiding Socrates' thought is found in the urge: "Know Yourself."[2] The term *know* acquires a new significance, since it no longer refers to the *acceptance* of a pre-existent truth, but to *investigation* and *searching* (cf. Landmann, 1962:67).

In pursuing this search Plato realized that *distinctive characteristics* would have to be found. In one of his later periods, he is of the opinion that a person might be described as a "bipedal living being without feathers." This leaves little room for the distinctive nature of *being human*. Landmann mentions the anecdote that Diogenes plucked a cock as an example of Plato's human being, upon which Plato had to add another qualification: "with flat toenails."

1 The Nobel prize winner, Walter Gilbert (lecturer in biochemistry at Harvard University), claims that the instruction "know thyself" actually refers to the (biological) knowledge of the human 'genome'! (Cf. Elseviers Weekblad, September 5, 1987: 87ff.)

2 *γνῶθί σεαυτόν*.

In the *Phaedo* – the first dialogue in which Plato's famed theory of ideas comes to fruition – the traditional Western dualistic view of being human finds expression. In this view, a person is seen as the union of two entities: a *rational soul* and a *material body*. Plato introduced the existence of static ideal forms[1] in an attempt to explain how we *know* things.

From Heraclitus he learned that all things accessible to *sensory perception* are in an *ever-fluctuating* state. It is therefore impossible to *know* these things. However, this conclusion rests on the presupposition that everything is subject to *change*. Yet, if this was true, the possibility of knowledge would be precluded, because if the claim is made that something is known at *this* moment, it ceases to be valid in the next moment, for then something previously known already has *changed* into something else, i.e. into something *not yet known*, into something *unknown*.

For the sake of upholding the possibility of knowledge, Plato postulates the *essential being* of things, their static *eidos* – which is not subject to change (*Cratylus*, 439c-440a).

In his dialogue *Phaedo*, Plato explains that what which is invisible and constant can only be thought about rationally, while that which is visible and changeable can only be observed through the senses. When the soul investigates without the mediation of the body, it is directed at the world of the pure and eternal, immortal and unchanging, constant and equally-natured things (79d). The *soul* exhibits the greatest similarity to the divine, immortal, conceivable, simple, indissoluble, constant and "self-identical," while the *body* bears the greatest similarity to the human, mortal, multifarious, non-conceivable, dissoluble and never-constant (80b:1-6).

In his greatest dialogue, *Politeia* (*The Republic*) – representing the culmination of the first phase of his theory of ideas – Plato defends, in preparation of his view of the ideal state with its three classes, a tripartite understanding of the soul (cf. 436 ff.). Via the Middle Ages, these three parts of the soul[2] continued to exert an influence on the traditional understanding of the 'faculties' of the soul (even in 20th century Reformed theology): *thought*, *will* and *feeling*.[3]

Not even the classification found within the discipline of biology escapes its influence. It has been assumed of old that a person has something that is *missing* in animals: rational insight (wisdom/sapiens) – hence the typification *Homo sapiens*. Since Darwin, admittedly, this biological classification has been placed within a climate of thought that links humankind in a continuous line of descent to its supposedly animal forebears – a line which has to extend back (via lower animals, plants, pre-organic systems, macro-molecules, at-

1 According to Plato these ideas are foundational to the transient sensory world of perceivable things, for they exist as *invisible*, *unchanging*, *essential* (ontic) forms.

2 Namely the *logistikon, thumoeides* and *epithumétikon*, i.e. thought, fervor and desire.

3 Compare also Hitler's estates in Nazi Germany and the *id*, *ego* and *superego* in the depth psychology of Sigmund Freud.

oms and elementary particles) to some supposed primal configuration – at which point an end must be made so as to prevent the continuation of the material-physical 'origin' into nothingness or the acceptance of some sort of creation idea.

Owing to the all-pervasive influence of neo-Darwinism, hardly any scientific discipline today manages to escape from its claims. A shortcoming in this claim, not usually recognized for what it is, is found in the pretence that the scholarly investigations involved in searching the origin of humankind, are purely biological in nature.[1] Yet the presence of diverse trends of thought within the discipline of biology can only be explained in terms of alternative philosophical views operative within them. A brief consideration of the claim that human beings indeed really descended from animal forebears may illustrate this point and also provide further insights into the distinctiveness of being human.

4.2 Philosophical assumptions operative in theories of evolution

One may distinguish 'evolution' (gradual development within certain limits) from 'evolutionism' (gradual development across all barriers). Evolutionism assumes a continuous development from non-living ('unbelebte') material things like atoms and molecules, to plants, animals and eventually human beings (supposedly the culmination of the evolutionary process).

Nonetheless, viewing evolution as gradual, continuous development is by no means a *new* concept. The Greek philosopher Anaximander had claimed, six centuries before Christ, that living creatures came into existence in an ascending line one after another. This theory was only elaborated in a modern biological-scientific manner by Lamarck (1744-1829) during the 19th century. It was in reaction to Lamarck's work that Charles Darwin published his famous book on "The Origin of Species" in 1859.

Diametrically opposed points of view emerged in biology since the end of the 19th century. The still prominent *mechanistic* approach assumes that all living entities can be completely understood in terms of physical, non-living material particles – particularly the interactions of atoms, molecules and macro-molecules constituting them. Alternatively, *vitalism* ('vita' means *life*) teaches that all living things exist by virtue of the presence of some or other *immaterial* "life force" or "vital force" (*entelechie*). The mechanistic approach exalts the physical aspect of created reality as the explanatory principle of origin – and sees everything as *transformations* of material particles that continuously and completely by chance caused all forms of life, whereas (neo-)vitalism starts with the biotical aspect of reality as principle of explanation. Since living things exhibit a remarkable *purposiveness* and *finality*, neo-vitalism emphasizes this (*teleological*) *purposiveness*, and rejects the blind fate of neo-Darwinism as an effect. Richard Dawkins emphasizes this idea of randomness and chance in his work on the blind watchmaker (see Dawkins, 1986).

1 By applying the yardstick of modal abstraction (see Chapter 2), biology is to be seen as the science that studies living things from the perspective of the *biotic aspect* of reality.

The multiplicity of opinions in modern biology makes it nonsensical to speak of *evolutionary theory* as a single, uniform and coherent theoretical stance. It is undeniable that a number of different evolutionary theories exist, while even non-evolutionary views are found in contemporary biology.

Of course speaking of 'evolution' is truly misleading, because we have noted in Chapter 1 that the past 100 years of biological thinking embraces diverse special scientific paradigms (both 'evolutionary' and 'non-evolutionary'). Amongst these (mutually exclusive) orientations, we have mentioned 20th century biological standpoints such as (classical) *mechanism* (Eisenstein), *physicalism* (neo-Darwinism – Huxley, Simpson, Dawkins, Bennett), neo-vitalism (Driesch, Reiner-Schubert Soldern, Sinnott), organismic biology (Von Bertalanffy), holism (Smuts, Adolf-Meyer Abich), pan-psychism (Theilhard de Chardin, Bernard Rensch), emergence evolutionism (accepting continuity in descent and discontinuity in existence – C Lloyd-Morgan, Alexander, Richard Woltereck, Bernhard Bavinck, Michael Polanyi) (see also Strauss, 2005 for details).

Here we limit ourselves to an examination of some of the pitfalls and problem areas in the conceptual world of the best-known evolutionary theory – physicalist (mechanistic) neo-Darwinism.

4.2.1 *The mystery of the genesis of the first living entities*

Initially and independently from each other Haldane (already in 1928)[1] and the Russian Oparin (cf. Oparin, 1953, chapters 4-7: pp.64-195) developed a hypothesis regarding the origination of the first living entity. The assumptions of the Oparin-Haldane approach eventually turned out to be questionable, namely that the initial atmosphere of the earth was mainly composed of hydrogen, methane, ammonia and water vapor. In particular Oparin holds that carbon "made its first appearance on the Earth's surface not in the oxidized form of carbon dioxide but, on the contrary, in the reduced state, in the form of hydrocarbons" (Oparin, 1953:101-102).

Silver points out that there is at present "no evidence that the atmosphere was reducing (methane and hydrogen)" and remarks that "the prevalent opinion at the moment is that the Earth's atmosphere, at the time that life emerged, was mainly carbon dioxide and nitrogen" (Silver, 1998:344). The role of methane is also unacceptable in the Oparin story since it is one of the components of natural gas which is produced by the "effect of millions of years of pressure and heat acting on prehistoric plant material" (Silver, 1998:344). Although the Haldane-Oparin conjecture was kept alive for a considerable time, supported by the experiments done by Stanley Miller (from Chicago) in 1953, it does not bring us closer to an understanding of the mystery of the genesis of the living cell. With regard to Miller's experimentation Silver remarks:

1 Names associated with the "New Synthesis" introduced by Julian Huxley in 1942 (*Evolution: The Modern Synthesis*) are: R. A. Fisher, Theodosius Dobzhansky, J.B.S. Haldane, Sewall Wright, E.B. Ford, Ernst Mayr, Bernhard Rensch, Sergei Chetverikov, George Gaylord Simpson, and G. Ledyard Stebbins

> The Haldane-Oparin hypothesis is out of fashion. Of the forty or so simple molecules that would be needed to form a primitive cell, the experiment produces two. It is worth bearing in mind that glycine contains only ten atoms and alanine, thirteen. The simplest nucleotide contains thirty atoms. The probability that a given large molecule will be produced by chance from small molecules, by sparks, falls drastically as the molecular size increases. It has to be realized that even if heat, radiation, and lightning, on the young Earth, had produced all the amino acids and nucleotides needed for present forms of life, the gap between an aqueous solution of these molecules and a living cell is stupendous. It's a question of organization: in the absence of a guiding intelligence, presentday scientists are not doing very well. For the moment, let's show the Miller experiment to the side door and see who is next in line in the waiting room (Silver, 1998:345).[1]

The past 50 years witnessed a tremendous increase in our knowledge of the micro-dimensions of living entities. On the one hand these developments opened up a domain that cannot be reconstructed from fossils[2] and on the other it revealed such an astonishingly complex picture that questions now arise – not because we know too little – but because we know so much! Darwin honestly stated:

> If it could be demonstrated that any complex organ existed, which could not possibly have been formed by numerous, successive, slight modifications, my theory would absolutely break down (Darwin, 1859:219).

The general assessment of Behe is quite remarkable in this context: "The story of the slow paralysis of research on life's origin is quite interesting, but space precludes its retelling here. Suffice it to say that at present the field of origin-of-life studies has dissolved into a cacophony of conflicting models, each unconvincing, seriously incomplete, and incompatible with competing models. In private, even most evolutionary biologists will admit that they have no explanation for the beginning of life" (Behe, 2003a:292).

1 Behe remarks: "Of course, if conditions on the ancient earth actually resembled Miller's unsuccessful attempts, then in reality no amino acids would have been produced. Moreover, joining many amino acids together to form a protein with a useful biological activity is a much more difficult chemical problem than forming amino acids in the first place. The major problem in hooking amino acids together is that, chemically, it involves the removal of a molecule of water for each amino acid joined to the growing protein chain. Conversely, the presence of water strongly inhibits amino acids from forming proteins. Because water is so abundant on the earth, and because amino acids dissolve readily in water, origin-of-life researchers have been forced to propose unusual scenarios to get around the water problem" (Behe, 2003:169-170).

2 It appears to be hopeless to provide an evolutionary explanation for the origination of the eye: "Anatomy is, quite simply, irrelevant. So is the fossil record. It does not matter whether the fossil record is consistent with evolutionary theory, any more than it mattered in physics that Newton's theory was consistent with everyday experience. The fossil record has nothing to tell us about, say, whether or how the interactions of 11-cis-retinal with rhodopsin, transducin, and phosphodiesterase could have developed, step by step" (Behe, 2003:292).

From the perspective of the genetic information content of the genome Yockey mentions another serious problem: "The reason that there are principles of biology that cannot be derived from the laws of physics and chemistry lies simply in the fact that the genetic information content of the genome for constructing even the simplest organisms is much larger than the information content of these laws" (Yockey, 2005:2).

Particularly in respect of what is known as the *molecular basis* of vital phenomena the current scholarly predicament lacks a sound scientific basis. Some of the most prominent biochemistry text books do not even have an index entry *evolution*. Behe lists about thirty editions of standard biochemistry text books with none or almost none *Index Entries* referring to *evolution* (the total number of index entries varies from 100 up to 10,000). Add to this the fact that there exists a Journal with the name *Journal of Molecular Evolution* (JME – established in 1971). The underlying *rationale* of this Journal is to explain how living entities came into existence on the molecular level.

A closer look at the articles that appeared in JME reveals that this "self-contradictory" name did not produce anything substantial. Behe remarks that "*none* of the papers published in *JME* over the entire course of its life as a journal has ever proposed a detailed model by which a complex biochemical system might have been produced in a gradual, step-by-step Darwinian fashion" (Behe, 2003:176). He continues on the same page by formulating a number of relevant questions that should have been addressed in this Journal, only to conclude that *none* of these problems in fact had been addressed, "let alone solved". According to him this strongly indicates that "Darwinism is an inadequate framework for understanding the origin of complex biochemical systems" (Behe, 2003:176). His general concluding is devastating for Darwinian theory: "There has never been a meeting, or a book, or a paper on details of the evolution of complex biochemical systems" (Behe, 2003:179). He remarks:

> As a result, evolutionary biology is stuck in the same frame of mind that dominated origin-of-life studies in the early fifties, before most experiments had been done: imagination running wild. Biochemistry has, in fact, revealed a molecular world that stoutly resists explanation by the same theory so long applied at the level of the whole organism. Neither of Darwin's starting points – the origin of life, and the origin of vision – has been accounted for by his theory. Darwin never imagined the exquisitely profound complexity that exists even at the most basic levels of life (Behe, 2003:173).

Without any doubt the ontic distance between the physical and the biotical (the non-living and the living) is fully mirrored and matched by the absence of any feasible conjecture about the origination of biochemical systems. Those who have respect for scientific modesty may do well to reflect upon a remark made by Haldane in a discussion with Silver: "I had a long conversation with J.B.S. Haldane, which started off with politics and ended with science. When I questioned him about evolution, one of his remarks sparked my interest, and sent me to the library that evening: 'Evolution's not the problem. Life is' Then he said, 'Oparin and I once had an idea about that, but we'll never know the real answer' " (Silver, 1998:353).

4.2.2 *Neo-Darwinism as a theory of change: are there any constants in the bio-world?*

'Mutation' is the conceptual term for the supposed phenomenon of sudden drastic, and subsequently inheritable, changes in the biotic structure of living things.[1] It has to serve as an explanation for the origin of more developed types. Unfortunately, all known mutations have a negative effect. It is supposed that "natural selection" is the 'power' capable of rescuing this shortcoming. Neo-Darwinists therefore suppose that these disadvantaged mutants (i.e. the transformed individuals), through natural selection, will turn out to have a better chance of surviving and thus will manifest further evolutionary developments.

With "natural selection," Darwin had in mind the constant struggle for survival in which only the *fittest* survive. As a result, mainstream (neo-)Darwinist evolutionary theory holds that these two phenomena, mutation and natural selection, always act in tandem. This makes it possible for the *disadvantaged* living entity to emerge as the advantaged. In this manner, all transformations and links between different living things can be explained: from the lowest form of plant and animal life up to humankind. Note that the *protista* are grouped apart as a result of their basic biotic organization – including algae, bacteria, fungi, slime, and protozoa. Since *Protista* generally possess both plant and animal features, it seems that we are not quite able to *classify* them properly. Each is *either* a plant *or* an animal, but for the sake of convenience (that is, owing to our *insufficient criteria*) we group them together.[2]

Apparently the role of mutation and natural selection support the main thesis of neo-Darwinism, namely the assumption that there is but one reality within 'living' nature: *change*. But in fact the contrary is true, for if there is no constancy attached to the (combined) *conditioning role* of mutation and natural selection, no single evolutionary change is conceivable. In other words, only when *mutation* and *natural selection* are operating as a constant (condi-

1 The two strings of the nucleic acid are ordered in a double spiral structure and can duplicate themselves. Nucleotides are present in the nucleic acid – DNA: Desoxyribonuclein Acid – and are formed by the link between a sugar with a nitrogen-inclusive base on the one side and a phosphoric acid complement on the other. Every *nucleotide* attracts its complement from the nucleotides freely present in the environment, leading to the formation of two new DNA-spirals that are faultless copies of the original. It could happen, as a result of chemical influences, cosmic or Röntgen radiation, that one or more of the nucleotides fall away or are added, changing the genetic information of the DNA-molecule. This 'fault' can then again be faultlessly copied. Mutations can bring about changes in individual genes, chromosomes, or even a number of chromosomes (e.g. in the case of *polyploide*).

2 We return to contemporary biological conceptions of *realms* within nature in Chapter 7 below – where we consider the distinction between prokaryote and eukaryote cells as well as the proposal of *five* 'kingdoms'.

tioning) *law for* the existence of living entities, is it possible to account for the changefulness of the latter – thus implying that it is not true that everything *changes* within the bio-world, for the "law-effect" of mutation and natural selection is supposed to remain *constant*! Yet, when the combination of mutation and natural selection is applied in order to account for the genesis of first living entities a vicious circle lurks. Von Bertalanffy remarks:

> In contrast to this it should be pointed out that selection, competition and 'survival of the fittest' already presuppose the existence of self-maintaining systems; they therefore cannot be the result of selection. At present we know no physical law which would prescribe that, in a 'soup' of organic compounds, open systems, self-maintaining in a state of highest improbability, are formed. And even if such systems are accepted as being 'given', there is no law in physics stating that their evolution, on the whole, would proceed in the direction of increasing organization, i.e. improbability. Selection of genotypes with maximum offspring helps little in this respect. It is hard to understand why, owing to differential reproduction, evolution should have gone beyond rabbits, herring or even bacteria, which are unrivalled in their reproduction rate (Von Bertalanffy, 1973:160-161)

Depew underscores this objection by pointing out that natural selection "depends for its operation on the very sort of variation and heredity that exists only in organisms and so can hardly be used to explain how organisms came into existence in the first place" (Depew, 2003:448).

The zoologist Thorpe highlights another instance of *fixity* (constancy) that is problematic for the neo-Darwinian emphasis on *change* – keeping in mind that the historicistic emphasis on *change* – at the cost of constancy – had a clear impact on the mode of speech employed by Darwin in his 1859 work on the origin of species.[1]

> [i]t seems to me that there is an outstanding problem raised by our discussion – namely the problem of fixity in evolution. What is it that holds so many groups of animals to an astonishingly constant form over millions of years? This seems to me to be the problem now – the problem of constancy; rather than of change. And here one must remember that the genetic systems which govern homologous structures are continually changing. Thus the control system is continually changing but the system controlled is constant, and constant over millions of years. This problem seems to me to stick out like a sore thumb in modern evolutionary theory.[2]

1 The term 'constancy' appears twice and 'persistent' (or: 'persistently') three times, whereas the term 'change' occurs 268 times, 'variation' 281 times, and 'variations' 162 times.

2 A discussion comment after the contribution of Ludwig von Bertalanffy (*Change* or *Law*) in the collection edited by A. Koestler and J.R. Smythies (1972:77). In passing it should be noted that that Gavin de Beer posed serious questions in his study on *Homology: An Unsolved Problem* – Mills *et al.* remark: "homologous structures do not necessarily derive from similar positions in the embryo or parts of the egg, nor do they share the same organizer-induction processes, nor are they even necessarily controlled by corresponding genes" (Mills *et al.*, 2003:215).

In Chapter 7 we shall return to the problem of constancy and change in paleontology.

4.2.3 *The scope and limitations of genetics and "molecular biology"*

It is scientifically clear, however, that no single molecule, however complex its structure, is *alive*. We have noted that Simpson had to admit that the expression "molecular biology" is self-contradictory. His earlier quoted statement reads (see page 91 above):

> Since biology is the study of life (it would have been better to say "living things" – DS) and molecules, as such, are not alive, the term 'molecular biology' is selfcontradictory (Simpson, 1969:6).

Neo-Darwinism extends the effect of mutations across all existing boundaries. In this manner, it attempts to establish a theory with a scientific status, aimed at eliminating every possible *uncertainty*. The openness to possible misunderstandings and an awareness of the provisionality of human scholarly insights, so prominently emphasized both within the contemporary philosophy of science and postmodern circles, seem to have been traded in for the 'security' of a "scientific face." Perhaps the extreme fundamentalistic views advanced by the "Bible Science Association" and other biblicistic creationists contributed to the equally fundamentalist attitude amongst the atheistic evolutionists of our day.

Extensive and widely-known studies of the fruit fly[1] have contributed considerably to our knowledge of micro-evolution.[2] Practically, this has brought about the current situation, in which the artificial selection of plants and improved animal breeds have become an everyday occurrence. The previously-mentioned geneticist, Theodosius Dobzhansky, nonetheless observes that all the mutations of the fruit fly still belong to the species *Drosophila* – as their ancestors did.

The eminent Swiss biologist, Adolf Portmann, questions, with good reason, the neo-Darwinist attempt to take the long and uncontrollable step from micro-evolution to macro-evolution.[3] He claims that the knowledge we currently possess, based on experiments, is far too little to explain such awesome phenomena as found in paleontology in its study of fossils. Therefore he finds no justification for attempting to derive the larger animals from simpler earlier forms (Portmann, 1969a:30).

Explanations of evolution by means of adaptation commonly refer to true and controllable instances of *adaptation*. An example is the white moth in

1 Known as *Drosophila melanogaster*, it has a life cycle of ten days, which means that great quantities can be bred with great success.

2 Micro-evolution is development within the nature of a single species – "small-scale evolution" – in distinction from macro-evolution, which conjectures development across all typical barriers, and across the barriers of more common systematic units such as genera, families, orders, classes and phyla.

3 Cf. his discussion of the matter in his work on the problems of life (Portmann, 1967:113-121).

England which "turned black" in very polluted areas during the industrial revolution. In terms of adaptation and natural selection this means that since birds could catch the *white* moths more easily against the dark background created by pollution, it increased the chances of survival of the black moths. If this is true, it still does not, however, provide proof of macro-evolution. After all, black moths still belong to the *same species* as the white moths. In addition, the entire story about the moths turned out to be fundamentally flawed (see Sargent *et al.*, 1998 and Majerus, 1998). It proceeded from the assumption that peppered moths normally rest on tree trunks, in 1980 it emerged that they rest predominantly in the vicinity of narrow branches in the canopy. Wells points out that most "textbook pictures of peppered moths ... show specimens that have been manually placed on tree trunks" and from this he concludes that "the classical example of natural selection is actually an example of unnatural selection!" (Wells, 2003a:190).[1]

Every biological theory calling upon the support of mutation must still consider the remark by Dobzhansky:

> Mutation alone, uncontrolled by natural selection, could only result in degeneration, decay and extinction (Dobzhansky, 1967:41).

If mutation alone results in "degeneration, decay and extinction," i.e. in *devolution*, then the *magic wand* turning devolution into evolution is *natural selection*!

In the meantime, new directions in biochemistry have begun to investigate the dimension of other possible relationships, with particular attention to the molecular building blocks of living entities. These investigations concern the nature of proteins – including hemoglobin, albumin, etc. It also concerns the nature of enzymes, built up out of 20 different amino acids. Enzymes perform a catalytic function in metabolic processes (building up and breaking down processes) of cells (sometimes as many as 100,000 are found in a single cell). Finally, this new direction in biochemistry investigates blood group antigens (that cause the formation of antibodies) and makes a comparative study of various kinds of DNA (think about the impressive genome project, completed in 2003).

To arrive at intricate "family trees" on the basis of this information is impossible, since this sort of analysis does not provide information on the time factor – essential for any theory of descent. Soon after the apparent successes in applying biochemical analyses to the problem of human origins, Henke and Rothe has already mentioned in 1980 that all efforts to draw "family trees" on the basis of biochemical research have been unsatisfactory due to the numerous unproven presuppositions regarding evolutionary tempo at the molecular level. They also add the remarkable statement regarding family trees based upon biochemical information:

1 Wells also mentions the discredited "biogenetic law" of Haeckel and in particular Heackel's "distorted drawings" that ignored "groups that did not fit nearly nearly into Haeckel's scheme" (see Wells, 203: 180 ff.).

> It indicates furthermore quite prominent deviations from those "family trees" constructed in terms of morphological measures (Henke & Rothe, 1980:17).[1]

More recent studies comparing human DNA with those of the anthropoids, found that in some cases the overlap exceeds 99% – for example in the case of the chimpanzee. The situation seems to be the following: if there are truly wide-ranging and important differences between the anthropoids and human beings and if it is the case that their respective DNA structures are almost identical, then the unconsidered option suggests that the DNA similarity is unable to account for the tremendous differences that are still present.

Particularly in the light of the argument pursued in the subsequent paragraphs of this chapter, focusing on these remarkable *differences*, the implication is indeed that merely comparing DNA molecules does not account for the real *differences* between human beings and animals. The neo-Darwinian paradigm makes biologists prone to diminish the distance between animals and humankind and thus misapprehends the unfathomable complexity involved in these differences. In an earlier work, by the biologist Overhage, the over-optimistic conclusions drawn from morphological and anatomical similarities are questioned in a way to remind us of the *logical* point, namely that similarities always *presuppose* differences (see Figure 2, p.14). Faced with the uniqueness of human characteristics, one of the (above-quoted) champions of the neo-Darwinian theory of evolution during the 20th century, remarkably enough, does recognize the *qualitative differences* distinguishing human beings from animals:

> Man has certain basic diagnostic features which set him off most sharply from any other animal and which have involved other developments not only increasing this sharp distinction but also making it an absolute difference in kind and not only a relative difference of degree (Simpson, 1971:270).

This insight clarifies why some biologists hold that the issues involved actually transcend the scope of biology as a discipline:

> To reduce the whole question about human origins simply to the biotical-bodily (morphological-anatomical) facet, witnesses an astonishingly one-sided approach and imply a radical simplification of the total depth of the problem (Overhage, 1959a:5).

Later on, Overhage returns to this problem by taking into account the normative functions of being human (which he designates as human 'spirituality'):

> Even if, more or less exactly, the fossils could be shown that represent the stadia through which the human body passed during a long process of evolution, the problem of the origin and the phylogenetic becoming of humanity would still not be solved. Whatever the value of this knowledge may be, it provides only a partial perspective on the genesis of the human being, since it cannot encompass the total life-form of a person. Because, in the fossil findings, the boundaries between the bodily morphological features of a human being and that of an animal fade away, and since the spirituality of a person does not ex-

1 See also the remarks of Mills *et al.* on sequence similarities and ancestral descent (Mills, *et al.* 2003:216-217).

> press itself univocally in the mere form, we no longer have any certainty about the question whether somatic characteristics – such as the erect gait, the free hand and the cranial capacity (all of which are exactly determinable in fossil findings) – are standing in a strict correlation to the spiritually stamped behaviour of humankind, such that it allows us to make claims about the nature and way in which the primates experienced and viewed the world ... The riddle of human genesis is not solvable simply by transforming and recombining animal proportions (Overhage, 1973:374-375).

4.2.4 *Will the fossils ever be able to 'tell'?*

The responsibility for fostering the credibility of the neo-Darwinist evolutionary hypothesis rests largely on *paleontology* (the study of unearthed fossils).

At its deepest, evolutionary theory attempts to answer the question of the origin of living entities during a virtually inconceivable past. Its pretence is to explain in a satisfactory way events in the process of biotic development over a period of some three billion years (three thousand million years). It is obvious that such a pretence cannot be bolstered by means of direct 'verification', 'observation' or 'experimentation'. The acceptability of the "family trees" sketched by paleontologists is additionally dependent on the fossils that are found. Since the publication of Darwin's controversial writings, much evolutionistic hope was placed on the conclusiveness of such finds. Much trusting expectation was spent on the discovery of "missing links." The hope was cherished that paleontology would clear up the mystery of the major moments in the historical descent of plants, animals and human beings.

> *Comment:* The following gives a summative history of the paleontological appearance of a few relevant plants, animals, and human beings: unicellular algae are the most ancient (3,100 million years: *Archaeosphairoides babertonensis*). A few invertebrate animal phyla are known from the pre-cambrium (such as *Trilobita, Porifera,* and *Coelenterata*). In the *Paleosoicum*, different kinds of fish appeared: Agnatha (jawless fish), *Placodermi, Chondrichtyes, Actinopterygii, Crossopterygii*, as well as amphibians and reptiles; in the *Mesozoicum* mammals as well as the first primeval bird *Archaeopteryx* (discovered in 1861). The supposed ancestors of human beings are: the southern apes (*australopithecines* 5-1 million years), *Homo habilis* (3-2 million years), Java- and Peking-apeman (currently *Homo erectus* – 1 million years), Neanderthal people (about 100,000 years) and *Homo sapiens recens* (40,000 years).[1]

In the 1960's most evolutionists still believed that modern human beings descended from the southern apes, with the Java and Peking forms as links. The

1 In spite of all uncertainties and in the absence of decisive fossils, modern evolutionary scientists believe "that our early hominid relatives arose in Africa, but disagree on when the direct ancestors of living humans left Africa to populate the globe" (quoted from http://www.geneticorigins.org/geneticorigins/mito/intro.html – accessed August 1, 2003).

latter had been dated back some 500,000 years. This is now dated to 1 million years ago. Subsequent discoveries, however, upset these hypotheses.

Since the beginning of the 1960's, L.S.B. Leakey has made known several fossil finds that belong, according to some specialists, to a separate species within the genus *Homo* – *Homo habilis*. This form, however, was supposed to be two million years old, while being contemporary with humankind's supposed ancestors, the southern apes. In 1972, Richard Leakey found skull fragments (given the registration number 1470) which, though almost three times older than the Peking and Java forms (grouped together by Leakey as the *Homo erectus*), still has a brain volume almost as large, and without the prominent brow of the erectus-forms.[1] But although skull 1470 is currently considered to be a *Homo habilis* type (cf. Henke & Rothe, 1980:95), it is not quite clear what the comparative relationship between it and modern human beings amounts to.

According to a web article discussing the status of *Homo habilis*, it turned out that "although 1470 is usually placed in the genus *Homo*, it is definitely not a modern human" (see the web reference to *Homo habilis* 1997 in the literature). This article refers to a remark by Leakey (1973), that the upper jaw and facial region are "unlike those of any known form of hominid." Brace (1979) is quoted saying "that ER 1470 retained a fully *Australopithecus*-sized face and dentition." He also mentions a remark by Cronin (1981) stating that KNM-ER 1470, "like other early *Homo* specimens, shows many morphological characteristics in common with gracile australopithecines that are not shared with later specimens of the genus *Homo*." It goes on to mention the more recent assessment of Walker and Shipman (1996): "Ignoring cranial capacity, the overall shape of the specimen and that huge face grafted onto the braincase were undeniably australopithecine." Although the author of this web article concedes that "[S]orting out the exact relationships of these fossils is very difficult," he is convinced that the various *Homo habilis* finds discussed are all similar to "a mixture of *Homo* and *Australopithecus* features." He claims that "there is no 'significant gap' separating 1470 from the others."

Strangely enough, another perspective on this issue came from an unexpected angle. To appreciate this information, we have to remember that the artificially created category of *proto-hominids* is supposed to contain the "tree dwelling" forebears of humankind, as Zeitlin writes: "The proto-homonoids were predominantly tree dwellers" (Zeitlin, 1984:17). From this assumption it is 'natural' to say: "The single most important condition that accounts for the beginning of this process is the fact that they were forced to leave the trees and to make their way permanently on the ground" (Zeitlin, 1984:18).

1 Leakey, R.E.: *Skull 1470, Discovery in Kenya of the earliest suggestion of the genus Homo – nearly three million years old*, National Geographic, Vol.143, no.6, June 1973, p.820. Cf. also pp.822,823,828. Later Kamoya Kimieu, a colleague of Richard Leakey, discovered a well-preserved *Homo habilis* skeleton on the west side of Lake Turkana in Kenya – it is about 1,6 million years old and is probably that of a young boy of about 12 years old (cf. *Newsweek*, October 29, 1984:39).

Some years ago a Dutch paleontologist, Fred Spoor, who is particularly interested in the supposition that human forebears descended from trees to an erect posture on earth, did research in this domain and came up with the modest confession that we do not know what is really going on.

Combined with the expertise of an ear, nose and throat specialist, and utilizing the CT technique of Wind and Zonneveld (CT = Computer Chromotography), De Burgh started to investigate the *balance organ* – located about three centimeters inside the human ear. It consists of the semi-circle like channels equipped with membranes, capable of containing fluid. Any head movement is registered by the nerve cells, enabling the balance organ to send the required signals to the muscles controlling the erect posture of the head. In the case of human beings, the two *vertical* channels are *large* – given the erect human bodily posture – whereas the *horizontal* channel is *small*. Since it is possible to investigate these channels in fossil findings, the method raised considerable interest, because it may help us find information otherwise inaccessible to paleontologists.

Spoor and his friend also visited South Africa, where the CT tests were performed on a specimen of *Homo habilis* found at Sterkfontein. The result was straightforward: this type of *labyrinth* is characterized by an exceptionally large *horizontal* channel, clearly indicating that this *Homo habilis* type never walked upright.

> What is merely suggested by the labyrinth is that *Homo habilis* was not more and also not less bipedal than the australopithecines. Its structure looks like that of gibbons or apes, but in any case is not human (De Burgh, 1995:21).

In the last couple of decades, the history of the emergence of the (human-like) hominids experienced so many alterations as a consequence of new discoveries, that it can be assumed that the situation will only become more complex. L.S.B. Leakey (with Napier and Tobias) abandoned, for example, brain volume as a characteristic of the genus Homo. It has become increasingly clear that the features regarding the human build and form (i.e. anatomical and morphological features) are inadequate to define a human person.

It is interesting to note that, some time ago, one of the world's authoritative pro-evolutionist journals, 'Evolution', published an article (in 1974) written by the paleontologist D.B. Kitts. He wrote that the spatial distribution and succession in time of living entities with which paleontologists work, are founded in the ordering principles of geology, and not in any *biological theory*. Paleontology therefore provides information that is inaccessible by means of biological principles alone. For this reason, paleontologists cannot substantiate evolution:

> We can leave the fossil record free of a theory of evolution. An evolutionist, however, cannot leave the fossil record free of the evolutionary hypothesis" (Kitts, 1974:466).

Thomson points out that "factual patterns of *change over time*, particularly as seen in the fossil record, can be studied in the absence of theories of how these patterns came to be" (in the words of Mills, *et.al.*, 2003:213). A leading paleontologist is therefore saying explicitly that evolution is a provisional (theoretical-hypothetical) presupposition. Kitts also remarks that many biological thinkers become convinced evolutionists on the grounds of a theory that is already inherently evolutionistic. With regard to "missing links," Kitts says:

> Evolution requires intermediate forms between species and paleontology does not provide them (Kitts, 1974:467).

With regard to Darwin's hope of a continuous line of descent without gaps, he declares:

> Most of the gaps were still there a century later and some paleontologists were no longer willing to explain them away geologically.[1]

In proposition 30 of his theses on being human, Dooyeweerd refers to a lack of evidence regarding the bodily descent of humankind from animal ancestors. Stafleu quotes Dooyeweerd, saying that neither 'paleontology' ... has [not] provided evidence for the bodily descendence of mankind from animal ancestors." Stafleu holds that this statement does not make any sense: "Scientific *evidence* differs from logical *proof*. Science does not require logical *proof* for a hypothesis. It requires scientific *evidential material* that does not contradict the hypothesis but sustains it" (Stafleu, 2002a: 10, note 25). Nonetheless Dooyeweerd is not the one who speaks of "logical proof" – in Stafleu's quotation he only uses the term 'evidence'. Stafleu introduced the notion of "logical proof" and then rejects it in favour of "evidential material" – as if he thereby opposes what Dooyeweerd has said (but in fact did not say). Stafleu therefore reacts to his own construction and not to the view advanced by Dooyeweerd. Stafleu and Dooyeweerd in fact apply the *same yardstick* – 'evidence' and "evidential material," although they differ in their respective interpretations of the 'evidence' concerned.

In terms of an overall perspective, Dooyeweerd holds that the

> structural types of plants and animals as such are indeed not individual subjects that originate in the temporal process of becoming for much rather they are ordering types belonging to the law side and not the factual side of our empirical world. They can only realize themselves in transient individual living beings, but as ordering types they necessarily bear a constant and foundational character in the time order. This is the case because they *make possible* our experience of the plant and animal world irrespective of the way in which we theoretically envisage the process of origination of living beings (Dooyeweerd, 1959:132).

His account of the central biblical perspective regarding humankind reads as follows:

> The Scriptures reveal God's act of creation. In their statement of this basic truth, which transcends all theoretical thought, they do not primarily appeal to certain temporal cognitive functions of man, but to ourselves in the religious root of our existence. They do not use theoretical scientific concepts, but by

1 The so-called 'punctuated equilibria' introduced by Stephen Gould is nothing but an attempt to come to grips with the overall image of *discontinuity* increasingly presented by the paleontological 'record' (see Gould, 1997:68 ff.).

> means of their central basic motive they appeal to the heart of man in the language of naïve experience.
>
> And then they impress two things in our minds: man does not make his appearance in time until the whole foundation for the normative functions of temporal reality has been laid in the creation; and at the same time: in man the whole 'earthly' temporal cosmos finds its religious root, its creaturely fulness of meaning. Adam's fall into sin is the fall into sin of the whole 'earthly' world, which is not independent of the religious basic relation between God and the human race (in any of its temporal functions).
>
> For that very reason the metaphysical conception of a natural reality in itself, independent of man, is un-biblical. The religious basic motives which gave rise to it, are incompatible with the Biblical one (Dooyeweerd, 1997-II:52-53).

By briefly investigating facets of the uniqueness of the human being we may gain a better understanding of the differences between animals and human beings and at the same time underscore the scope of a philosophical view that transcends the confines of the various disciplines (special sciences).[1]

4.2.5 *The uniqueness of the human being*

Young children learn that people and animals differ. An animal is an animal and a human is a human. This knowledge, which the child can check with its senses, is challenged for the first time in about grade four, when the child is taught that people are actually *mammals*. Consequently, conflict and doubt grows in the mind of the child. On the one side is the growing reality-conforming experience of life, and on the other the scientific knowledge with which children come into contact.

There are of course many *similarities* between human beings and animals, particularly between humans and *mammals*, the latter being a class of *Vertebrates*, which is a subphylum of the *chordata*. Whenever similarities are indicated, however, it implies (or stronger: *presupposes*) differences, as we have noted. If there were no differences, entities would simply belong to the same kind of entity, that is, they would be *identical*. But as long as there are *differences*, we are not dealing with exactly the same (kind of) thing. To emphasize similarities exclusively – and subsequently draw 'identity' conclusions – is scientifically indefensible.

From antiquity, it has been accepted that a human person possesses something *lacking* in animals: *rational insight* or *wisdom*. Thence the name ascribed to a person in biological classification: *Homo* (the genus) *sapiens* (the species name, where *sapiens* means 'wisdom'). Darwinism, however, links this wise person with animal ancestors, and prior to them, with the lower animals, sub-organic systems (such as viruses), macro-molecules, atoms, and elementary particles, all the way back to the supposed primal mass. By choice this was the end of the process, since an extension through the primal matter into nothingness would come too close to the (biblical) idea of *creation*.

1 For a more extensive and better documented treatment of this problem, see Strauss (2005, Chapters 4 and 5).

What, however, is unique about humankind?

Some thinkers are of the opinion that *language* is the particular characteristic that distinguishes humankind from animals. By means of language humanity owns and utilizes a consciousness of the *past* and the *future*, a consciousness including the knowledge of the individual person's limited lifespan. It is interesting, understandable and noteworthy that the evolutionist Dobzhansky considers the *awareness of death* as typifying the distinctive characteristic of human beings. Some thinkers are even of the opinion that the ability to *commit suicide* typifies the unique nature of being human.

4.2.5.1 *The eccentricity of the human being*

Cassirer (cf. 1944) introduces the well-known distinction between *signals* and *symbols*. The former belongs to the physical world of being and the latter is a part of the human world of *meaning*, the world of human culture. Von Bertalanffy says that *symbolism* "if you will, is the divine spark distinguishing the most perfectly adapted animal from the poorest specimen of the human race" (1968:20). In order to identify symbols, he uses three criteria: (i) Symbols are *representative*, that is, the symbol stands in one way or the other for the thing symbolized; (ii) Secondly, symbols are *transmitted by tradition*, that is, by the learning processes of the individual in contrast to innate instincts; (iii) Finally, symbols are *freely created* (Von Bertalanffy, 1968:15, cf. 1968a:134).

Helmut Plessner wants to transcend the self-contradictory notion of an "*entelechie*," presented to him by his tutor Hans Driesch. As an alternative, he introduces the notion of *positionality*. Physical entities are delimited by the surrounding environment. In the case of organic entities, this delimitation belongs to the entity itself (for example, the *membrane*), and thus evinces *positionality* (Plessner, 1975:291). This concept provides the possibility of viewing human beings as belonging to the last level of living beings. Animals are considered to be closed and *centric*, distinguished from human beings who are *eccentric* living beings (and who are relatively 'Weltoffen' – "world-open") (Plessner, 1975:292). The first anthropological 'Grundgesetz' (fundamental law) mentioned at the end of his book: "*Die Stufen des Organischen und des Menschen*" (Plessner, 1928, reprint 1965) states the "*vermittelte Unmittelbarkeit*" (mediated immediacy) valid for all eccentric positions (cf. Plessner, 1975:297).

Language positions itself between the grasp of the hand and the view of the eye – the eye as the "organ of making-something-immediately-present." Thus, in various respects, the hand and the eye become *dispensable* (cf. Plessner, 1964:38 and Hofer & Altner, 1972:203). Animal communication, according to Plessner, does not know a "mediation through objects" (Plessner, 1975a:380, cf.379). Surely, this phenomenon is particularly remarkable, since, in the domain of human sensitivity, the sense of seeing and the sense of touching dominate that of smelling (cf. Haeffner, 1982:16). Language provides human begins with a *mediated immediacy* in the world. Animal communication does not refer to the past or the future. It refers to the vital *here* and *now*. For this reason, animal signs strictly have *one* content for every single sign.

All human utterances, by contrast, can signify a number of different things, depending on the context, intention, or even, in the case of written language, the punctuation. Compare this with the famous dance of the bees that always indicates by means of the (i) tempo, (ii) direction and (iii) angle of the figure eight executed, the (i) distance, (ii) location, and (iii) direction of the source (see Eibl-Eibesveldt, 2004:258 ff.). Human language, on the other hand, presupposes a *freedom of choice* and the concomitant multiplicity of meaning, requiring *interpretation*, which in turn requires further interpretation from the addressee (cf. Nida, 1979:203; De Klerk, 1978:6). It presupposes the responsible free activity of the human being, which requires responsible choices.[1]

4.2.6 *Animal 'speech'*

The order of primates, under which humankind is classified evolutionistically, is noticeably poor in nuanced sounds – with the obvious exception of the human being. Mammal's sounds simply do not compare with, for example, *birdsong*.

Ever since Descartes, it was believed that the uniqueness of the human brain is responsible for human language. The result was that anatomists insisted that anthropoids have the 'machinery' available to articulate speech. The order of primates – including humans, according to the prevalent classification – is nevertheless, of course with the exception of *human beings*, unable to *vocalize*. The ability to reproduce human speech sounds, as found in birds, is totally absent in mammals. The vocal potential of the gorilla and orangutan is exceptionally poor. The chimpanzee is somewhat better, and the gibbon can produce sounds covering almost an octave. All these anthropoids, however, completely lack the playful sounds produced by the human suckling. The unprecedented possibilities of human sound production transcends that of the anthropoids by far. In addition this sound production displays an exceptionally rich modifiability (Overhage, 1972:242).

Post-mortem studies of the upper respiratory tract in mammals, as well as cineradiographic studies, have shown that the position of the larynx is crucial in determining the way in which an individual breathes, swallows and vocalizes (Laitman, 1985:281). This implies that there are certain anatomical peculiarities that go hand in hand with the contribution of brain functioning in the production of human speech; in particular the gradual descent of the larynx after the post-natal period (cf. Portmann, 1973:423).

The 'humanlike' apes (*anthropoids*, i.e. the orangutan, gorilla, chimpanzee, and gibbon), are, as a result of anatomical shortcomings, born incapable

1 This is why there is a difference in principle between the learning of certain signs by chimpanzees and gorillas and all human language usage – these animals are simply not free to react *responsibly* to norms.

of speech. It is interesting to note that at birth the human *larynx* is positioned in exactly the same way as that of all other mammals. One reason for this is that the human infant needs a way for milk intake that is *separate* from the windpipe. The baby can breathe calmly while drinking. Exactly because of this the human infant is incapable of speech, like all mammals. Only by means of the gradual removal of this division, caused by the downward movement of the larynx – freeing the larger pharynx cavity – is the human person eventually enabled to speak. Only human beings possess an intermediate area between the nasal cavity and the larynx where air and food channels cross.[1] As Laitman observes:

> This high position permits the epiglottis to pass up behind the soft palate to lock the larynx into the nasopharynx, providing a direct air channel from the nose through the nasopharynx, larynx and trachea to the lungs ... In essence, two separate pathways are created: a respiratory tract from the nose to the lungs, and a digestive tract from the oral cavity to the esophagus. While this basic mammalian pattern – found with variations from dolphins to apes – enables an individual to breathe and swallow simultaneously, it severely limits the array of sounds an animal can produce ... While some animals can approximate some human speech sounds, they are anatomically incapable of producing the range of sounds necessary for complete, articulate speech (Laitman, 1985:282).

In order to provide the newborn human suckling with a milk tract separate from the respiratory tract, the position of the human larynx at birth is the same as that of mammals. In the period between the first and second year, this highly positioned larynx starts its descent in the neck. This downward movement creates the pharynx cavity, necessary for the articulation of the richer voice disposition in human beings. Laitman declares that the precise time this shift occurs, as well as the physiologic mechanisms that underlie it, are still poorly understood (Laitman, 1985:282). As soon as the larynx reaches its destined low position, it can no longer lock into the nasopharynx. Consequently, the respiratory and digestive pathways cross above the larynx. This creates the possibility of suffocating, which surely is, evaluated in itself, something negative. However, it is precisely this *expanded pharynx* that provides human beings with the unique potential to produce a rich variety of speech sounds. The palate between the mouth and nose cavities serves as basis for the resonance of the sounds produced. Goerttler even mentions the fact that, in the third month after conception, a distinctively human structural element develops (the vocal chord 'blastem' – Goerttler, 1972:250).

It is interesting in this connection that Laitman informs us that the basicranial similarities between the australopithecines and extant apes suggest that their upper respiratory tract was also similar in appearance. Consequently, as with living non-human primates, the pharynx portion for sound modification in these early hominids would have been greatly restricted:

1 When the mobile epiglottis does not handle the 'traffic' effectively, we suffocate. Cf. Goerttler 1972: 249 and Portmann, 1973: 397-424.

> As a result, these early hominids probably had a very restricted vocal repertoire as compared with modern adult humans. For example, the high larynx would have made it impossible for them to produce some of the universal vowel sounds found in human speech patterns (Laitman, 1985:284).

His conjecture is that the first instances of full basicranial flexion similar to modern humans does not appear until the arrival of *Homo sapiens* (estimated at 300 000 to 400 000 years ago): "It may have been at this time that hominids with upper respiratory tracts similar to ours first appeared" (Laitman, 1985: 286).

If we define a speech organ as that bodily part which exists solely in service of the production of speech sounds, then a surprising fact is that there are no human speech organs. Let us enumerate possible candidates: the *lungs*, *larynx*, *mouth cavity*, *palate*, *teeth*, *lips* and *nose cavity*. Without exception, all these organs perform primary functions that would continue to function in their normal way even if human beings never uttered a single word (Overhage, 1972:243)! Human language simply takes hold of all these different organs in the production of speech sounds.

This highly developed and subtle cooperation, especially of three organs so heterogeneous in character as the mouth, the larynx and the brain, integrated in the production of human speech sounds, makes it rather difficult, if not hopeless, to provide us with a causal evolutionistic explanation of this astonishing phenomenon. The question arises, What number of miraculous changes should have occurred to produce the articulation conditions necessary for truly human language formation?[1]

> Such an unfathomable process of change affecting so many differently structured organs and organ complexes, closely correlated with each other, should have proceeded harmoniously as a total change, if it was to come to the unprecedented perfection of human speech (Overhage, 1972:250).

4.2.7 *The absence of logical concept formation and argumentation in animals*

The German zoologist, Bernard Rensch, who believes that animals can form *a-verbal concepts* (concepts without words), admits that only a human being can form a concept of *causal relationships*. Only a human being can make deductions, accompanied by parts of speech such as "in consequence of," 'because', "in case," etc. The human ability to come to logical conclusions is lacking in animals.

If it is possible to show that animals can *locate* similar entities and afterwards act accordingly, are we then justified in concluding that they have *averbal concepts*? What do we mean with the term: **concept**? This question becomes more urgent when Rensch stresses that his use of the term 'averbal' is meant to emphasize that "averbal concepts" in animals do not proceed from *logical operations* (1973:118). For this reason, he claims that *the* trait determining the gap between humans and anthropoids is *logical thinking* (Rensch,

1 Cf. the similar comments by Overhage in his work of 1972:250, as well as his work of 1977: 109-112.

1968:147). Rensch says that, although animals approach causal 'concepts' (concerning relations in different situations), human beings transcend it in their unique ability to grasp truly *logical* relations, expressed in concepts like "*as a result of,*" '*because*', "*in case of,*" and so on. What Rensch says in this regard is on a par with the standard way in which logicians define their subject matter. In his well-known *Introduction to Logic*, Copi, for example, believes that the arguments involved in making inferences occupy the main task of logic as a special science (Copi, 1994:7; also compare page 2). If we distinguish between *premiss-indicators* and *conclusion-indicators* with Copi, we may expand the phrases mentioned by Rensch as follows:

> *Premiss-indicators*: since (as indicated by); because (the reason that); for (for the reason that); as (may be inferred from); follows from (may be derived from); as shown by (may be deduced from); inasmuch as (in view of the fact that). *Conclusion-indicators*: therefore (for these reasons): hence (it follows that); thus (we may infer); so (I conclude that); in consequence (which means that); accordingly (which entails that); proves that (which implies that); as a result (which allows us to infer that); for this reason (which points to the conclusion that) (Copi, 1994:8-9).[1]

Leakey's announcement that the "ability to see similarities between objects of the same type – classes such as trees, fruit, predators, birds, etc. – is a crucial step in creating conceptual order in what otherwise might be an overwhelming perceptual chaos" (Leakey & Lewin, 1978:204) ascribes, in a typical Kantian fashion, a truly formal creativity to concept formation. The capacity of the anthropoids to recognize perceived objects (a limited ability in animals) and to associate them with each other (even in different contexts), does not provide conclusive evidence that they are able to function subjectively in the logical-analytical aspect of reality.

Examples on the basis of empirical research always show that animals are concretely bound to specific perceived *forms*. On the basis of their *sensitive intelligence*, animals are capable of *seeing* similarities and differences. In the case of the signs taught to Sarah, Washoe, Moja and Lana it is clear that the way in which they use them is always found in sensory *sound-like* and *image-like* modes of locating the relevant similarities. Certainly human beings share this *perceptive* dimension with animals – but humans are not confined to or qualified by this *sensitive way* of dealing with similarities and differences.

Leakey and Lewin also refer to the use of *signs* by chimpanzees and gorillas. With the aid of different *sign-labels*, these animals are said to be able to generalize to cognitively economical concepts (essential for language), "For instance, Lucy calls a watermelon a *drink fruit*; Washoe refers to ducks as *water birds*, and she invented the name *rock berry* for a brazil nut when she first encountered one" (Leakey & Lewin, 1978:174). Even if we leave aside the inability of animals to act in a *norm-conformative* (or: anti-normative) way, an-

1 Given his definition of logic as the study of the methods and principles used to distinguish correct from incorrect reasoning (Copi, 1994:2), it is understandable that we do not encounter a discussion of the nature of a *concept* or a *word* anywhere in this book!

other way of acknowledging their lingual limitations would be to point out the fact that they are not concerned with or interested in any *normative dimension* of reality (such as the cultural-historical, economic, the aesthetic, the juridical or the ethical).[1]

The perception of a multiplicity of objects, the sensitive delimitation of particular perceptual objects or events (capable of exerting a controlling influence on behaviour in later situations – such as avoiding fire), due to the continuity provided by the associative perceptive ability of animals – all of this is enclosed within the domain of *sensitively qualified beings*. In support of such limitations, Portmann characterizes animals as *Umweltgebunden* (constrained by environment) and *Instinktgesichert* (protected by instinct) (Portmann, 1990:79).

Precisely because our human capacity to judge is foundational to every act of identification and distinguishing, it differs in principle from a merely *sensitive delimitation* and *association*. Koehler explicitly claims the contrary: "In the absence of verbal language, we call such an operation with representations, concepts and judgments, founded on intuitions without bearing any names, unnamed thinking" (Koehler, 1973:119). Overhage, on the other hand, stresses that the practical intelligence of animals never exceeds the sensory-perceptive domain (Overhage, 1977:117). A correct logical concept must entail the multiplicity of identified characteristics, united in the concept in such a way that logically justified judgments could be inferred from it. Anything logically known is, by means of subjective logical conceptualization, *logically objectified*.[2]

The capacity of anthropoids to distinguish between entities in sense perception, and even to associate these with one another (compare the signs taught to gorillas), still do not provide conclusive evidence that these animals can function actively – i.e. reason (il)logically – in the logical aspect of reality.[3] For example, animals are not able to distinguish between *logical* and *illogical* concepts because they lack an accountably free will.

Portmann typifies the peculiar human freedom of choice as follows:

> The narrow limitations of animal interests is opposed to our freedom of choice and direction. Animals can escape the bonds of their urges only to a limited extent, while I myself can, in every moment, in accordance with my entire attentativeness, turn my entire inwardly participative dedication to some or other matter, however insignificant it may appear to be (Portmann, 1974:102).

4.2.8 *Sensitive and rational intelligence*

Does this imply that we cannot ascribe any form of *intelligent* behaviour to animals? Buytendijk refers to ethological research in order to substantiate his

1 We are provisionally proceeding from the perspective that considers all post-sensitive aspects of reality – that is the logical-analytical, the cultural, the lingual, the social, the economic, the aesthetic, the legal, the moral and the certitudinal – as *normatively structured*, i.e., meant to apply to the accountable freedom of correlative human responsibility.

2 This objectification is nothing but the opening of the logical object-function of entities.

3 Cf. e.g. the arguments of R.E. Leakey and R. Lewin (1978:202 ff.).

conclusion that the animal world merely shows gradual differences in this respect. "Every species has its own practical intelligence, limited by disposition and experience" (Buytendijk, 1970:98). This conclusion presupposes his basic distinction between animal and human intelligence. When, in a given situation, human beings and animals pursue a similar goal, they will experience similar emotional drives. However, what is absent in the case of the animal is action on the basis of *judgment*:

> Therefore, one defines animal intelligence as the concrete experiential and senso-motoric structuring of practical behaviour, whereas human intelligence displays itself as a rational-logical, categorically judging conceptualization of the task-setting nature of the concrete situation and the discovery of a solution which does not follow from the immediate sensory effect of the situation.[1]

Overhage, rejecting the anthropomorphic mode of speech in the writings of Rensch, Koehler and Lorenz, emphasizes that animal form perception does not result in genuine concept formation, since it remains encapsulated within the sensory-perceptive sphere (1965:307; cf. Overhage, 1972:251-276). According to the evolutionist Bernard Rensch, animals lack the capacity of insight. At most, he speaks of "proto-levels" of insight (cp. Overhage, 1977:116). Perhaps this is the reason why Sarah, Washoe, Moja and Lana never asked a *question* (Overhage, 1977:107)!

What should also not be forgotten, is that many attempts to 'prove' the lingual abilities of animals are fundamentally *circular* at least in one respect: they argue on the basis of what could be achieved when humans teach animals a certain kind of sign language. With reference to chimpanzees, Grant points out the following crucial issue: "Furthermore, chimpanzees do not pick up a simple symbolic language by themselves. They learn such a language only after being tutored by humans in intensive tutoring sessions" (Grant, 1985:425)!

Experiments with chimpanzees attempted to teach them to copy a given drawing of a square or a triangle, but without any success (Rensch, 1968:148). The question is: if these animals are not even able to *draw* these figures, how are we going to be convinced that they truly have the *concept* of a square or the *concept* of a triangle? We must remember that a concept is something different from a *sensory picture* that can be associated with something else (cf. Overhage, 1972:252) – as is the case in the so-called name-giving mentioned by Leakey. Consequently, we have to deny the possibility of non-logical (a-logical) concepts. Concept formation is always a *normatively qualified* activity, which is constantly subjected to universally applicable logical norms or principles. The sound-like and image-like way in which animals cope with similarities and differences should be designated with a different term than *concept* if we want to avoid an unacceptable anthropomorphism.

1 "Man definiert daher der tierische Intelligenz als die konkret-erfahrungsmässige und sensomotorische Gestaltung des praktischen Handelns, während die menschliche Intelligenz ein rational-logisches, kategorisch urteilendes Begreifen der konkreten Situation als Aufgabe ist und die Entdeckung einer Lösung, die nicht durch die umittelbare sinnliche Wirkung der situation erfolgt" (Buytendijk, 1970:97; cf. Overhage, 1977:118).

The decisive point, however, in showing that animals possess concepts would be to show that they can think *illogically* by forming, for example, the self-contradictory concept of a triangular square or a square triangle! This has never been shown by any of the experiments referred to by the mentioned authors. In other words, animals are simply not able to function *subjectively* in the *analytical* aspect of reality. Consequently, it should not be surprising that they are unable to *think* – be it logically or illogically!

What is truly human is apparently evinced in a person's erect stature, free hand, opposing thumb, and spiritually characterized facial expression. Konrad Lorenz says that a human being is a *specialist in non-specialization*.

Gehlen is inclined to see the typically human functions as a *compensation* for the lack of instinctive certainty and environmental fixation.[1] The opposite is however the case. The physical, biotic and sensitive-psychic dimensions of human existence stand in service of and are directed towards the normative character of a person. A person can think logically, speak, interact socially, and form culture. The freedom of decision and the need to reflect rationally (expressed in the great variety evident in the formation of culture) characteristic of human existence, requires a non-specialized and relatively instinct-poor foundation. Portmann speaks of our "second nature," the transformed formation of a world of culture in this regard. From the perspective of the normative-cultural character of our human activities, we should perhaps rather speak of our "first nature."

4.2.9 *The formative imagination in human tool-making*

Initially the use of tools was seen as a distinctive human feature, but since it has been shown that animals also *use* tools, this criterion had to be changed. With reference to Oakley's definition of a human being, Overhage emphasizes our distinctive ability to *make*, rather than merely *use*, tools (Overhage, 1974:359). Despite the continuing placement of humankind in the animal kingdom, Simpson defines a human being summarily as "the only living animal that uses tools to make tools" (Simpson, 1969:91).

This description, however, typifies the nature of technique, since, differently from other widely divergent cultural products such as money, cars and test tubes (respectively economically, socially and academically qualified), tools are the only artificially made cultural products (their technical formative foundation) made to make something else (their technical formative qualification).

The importance of technical cultural products (tools) as a distinctive criterion has increased as it became clear that *anatomical* and *morphological* criteria fall far short in the evaluation of fossils. There is an increasing dependence on evidence of typically human cultural activity, which has increasingly brought archaeology into the picture. The archaeologist Narr has indicated in 1959 that "largely descent researchers with a natural scientific bent have sought the border between the human being and an animal anew where the

1 This typification derives, as we saw, from the thought of the Swiss biologist, Adolf Portmann.

particular spirituality of a person appears in singular indications of cultural activity" (Narr, 1959:393).

The obvious and distinctive human cultural activities are particularly closely bound to the free formative imagination of a person, which is the foundation of all technical inventions. As Von Königsberg states, with reason, a person is a cultural being; "without culture no *Dasein* (concrete human existence – DS) worthy of being human can be contemplated" (Von Königswald, 1968:150). Mentioning the fact that human tools are conceptualized particularly with a view to future use, he states explicitly that true invention already took place in the earliest phase of the Paleolithic Age, i.e. the earlier stone age (Von Königswald,1968: 167). The presence of a person's inventive formative imagination provides the foundation for practically useful archaeological criteria, in terms of which typically human tools can be distinguished:

(a) The form of the produced tool might not be suggested or determined by the original raw material (e.g. in distinction from a stick from which irritating leaves and twigs need merely be removed);

(b) The function of the tools might not be suggested (a rock in its natural shape is a strengthening of the fist; a stick an elongation of the arm or fingers), that is, tools may not be merely extended bodily organs;

(c) The manner of production might not be suggested, with appeal to the technical moment that implies that tools must be formed by means of (formed or unformed) tools (cf. Narr, 1974:105 and Narr, 1976:99-101).

These criteria are exemplified in the images of manufacturing *chopper tools* (Narr, 1974: 101). Whereas Immanuel Kant defined the human *Einbildungskraft* (imagination) as the ability to represent an object that is not in our intuition,[1] the free formative imagination of human beings can also accomplish the opposite task, namely it can imagine what *is* given to the senses in a *different* way.

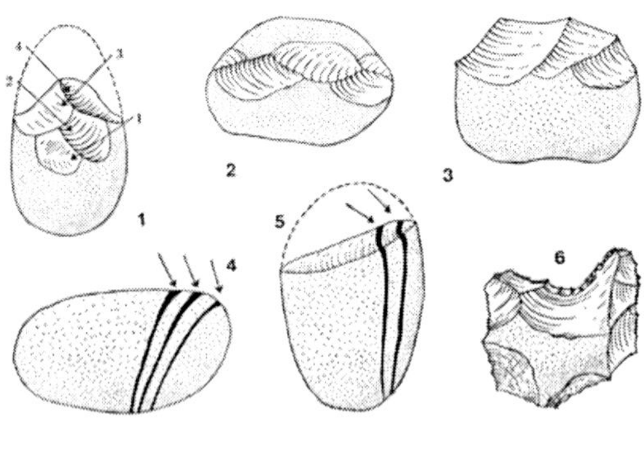

1 "Einbildungskraft ist das Vermögen, einen Gegenstand auch ohne dessen Gegenwart in der Anschauung vorzustellen" (Kant, 1787-B:151). See also Kant, 1800-A:67 where he defines the *Einbildungskraft* (facultas imaginandi) as the power (vermögen) to bring objects to consciousness without their presence to the senses.

The fact that the earliest human tools had multiple purposes and only gained a relative task-specific speciality in due course indicates that the means-end relation is inherent in all tools. The typically human use of tools presupposes a person's analytical ability that enables her to distinguish means and ends.

A philosophical analysis of the unique nature of a human being must advance to the question of the particular manner in which one experiences reality.

4.2.10 *Flexibility and specialization – the difference between human beings and animals*

Portmann considers, as noted above, that the animal nature is *instinctively secured* (*instinktgesichert*) and *bound to an environment* (*Umweltgebunden*).[1] Animals experience reality exclusively from their natural inclination, directed at that which is physically, biotically and psychic-sensitively important to them. In a later context, we shall return to the fact that animals experience reality in terms of that which is negotiable and not negotiable, edible and inedible, in terms of same sex and opposite sex, comforting and alarming (see page 358 below).

Since those facets of reality in which a person functions in a typically human way are not instinctively assured or bound, but directed at a person's normatively qualified, responsible freedom of choice,[2] humankind has a flexibility that makes possible an incredible specialization in differentiated civilizations. Even Simpson emphasizes this: "Such specialization, which is non-genetic, requires individual flexibility and could not occur in a mainly instinctive animal" (Simpson, 1969: 90).[3]

Such normative specialization, however, requires and presupposes an *unspecialized bio-psychic foundation* – a further characteristic unique to being human.

4.2.11 *Is Dollo's law of irreversible specialization universal?*

In contrast to the instinctively assured and environmentally bound specialized way in which animals are adapted to their natural environment, a person enters this world with *unspecialized physical equipment*: that person possesses no natural adaptation to a particular environment, and is distinctively unspecialized, physically and bio-psychically, in comparison to animals. Human teeth are not adapted to either eating plants or animals. The lack of gaps be-

1 "Constrained by environment and protected by instinct: simply and briefly, that is how we can describe the behaviour of animals" (Portmann, 1990:79).

2 We have already referred to this in the brief discussion of opposites such as logical-illogical, historical-unhistorical, social-unsocial, and so forth – opposites that presuppose universal measures, 'ought' demands, and principles (see Chapter 2, p.41).

3 Contrast this with the closed nature of animal existence. As Hart comments, "A worker ant is just that – and all its functions are geared to being a worker ant. A human being, on the other hand, has multiple roles to play and is not exhausted in any of them" (1984:146).

tween the eye teeth and premolars (which is specialized into e.g. fangs in anthropoids) is also an archaic (primitive, in the sense of unspecialized) characteristic of human teeth in comparison with animal teeth. The human hand and foot is equally archaic in comparison with those of the anthropoids (cf. Gehlen, 1971: 86 ff). G. Altner notes that even anthropoid teeth are relatively unspecialized, but cannot deny the general trend of the mentioned data (as emphasized by e.g. Gehlen) (Altner & Hofer, 1972:199-202).

Comment: Since Dollo formulated the law of irreversible specialization, existing anthropoids lost their claim to the ancestry of humanity, since it is impossible to deduce the unspecialized characteristics from the progressively specialized nature of the anthropoids. This leaves two equally limited possibilities:
(i) construct a hypothetical primal form that could serve as a basis and point of departure for the specialization of the anthropoids (but then these would be descended from human beings), or
(ii) negate the law of Dollo with reference to e.g. neoteny (rejuvenation phenomena among animals, L. Bolk) and the theory of self-domestication (K. Lorenz).[1]

Gehlen typifies a human being – in comparison to the natural inclination of animals – as a *defective* creature (Gehlen, 1971:20, 30, 83, 354). He neatly turns around the position that animals have no mind: a human being lacks something, since it is *unspecialized*! Gehlen returns to the position of Herder, who in 1770 (in a work on the origin of language – *Ursprung der Sprache*) said the following regarding humankind: "This instinctless, miserable creature, emerging so lonely from the hands of nature, was from the very first moment a free-acting, inventive creature who had to help himself and could not but do so" (quoted by Altner, 1972:157).

Even though human beings are not entirely without instincts, their natural inclinations do fall considerably short by comparison with the abilities of animals. Humankind is earthbound, unable to soar through the air like a bird. A human being is much slower than many wild animals and lacks a naturally protective hairy hide. Human senses are considerably limited in comparison with the acuity of animal senses. Human beings possess no naturally dangerous weapons, especially in comparison with the muscular strength, claws or jaws of carnivores. There are animals that can register supersonic waves, see ultraviolet rays as light, fish can sense electrical fields, and birds use the magnetic poles of the earth as navigating devices – all senses lacking in a human being.[2]

A human being only appears as an unspecialized and defective creature when the natural inclinations of animals are used as the only basis of comparison. As Hans Freyer objects: the human being is first fictitiously portrayed *as*

1 Cf. resp. Bolk (1926) and Lorenz (1973), as well as the more extensive treatment in Strauss (2001).

2 Portmann discusses this in one of his works: *Der Mensch ein Mängelwesen?* In: Portmann (1970: 200 ff.).

an animal, after which it appears that, as such (that is, *as* an animal), a human being is highly *incomplete* and actually *impossible*!

What picture do we receive when we look at the human being and at animals in terms of common factors – as revealed in the biotic functioning of both?

4.2.12 *The ontogenetic uniqueness of humans*

Portmann's pioneering work in this field has not only indicated that human beings cannot be pigeonholed in either of the two developmental types that he identified in the animal kingdom,[1] but also that, in comparison with the typical animal growth rhythms – which are gradual and continuous – the human growth rhythm has two phases of acceleration.

In comparison with the *Nestflüchter*, human beings are born a year too early. Portmann calls this the "social uterus period," which enables the newborn human to gain by means of cultural contact and transference that which the animal instinctively has at birth (Portmann, 1990: cf. Chapters II, III, V, and VI). During the first year of life, the human baby develops at double the rate of the anthropoids, after which a slowing down in growth tempo takes place until the ninth year. After this period, there is another period of rapid growth, culminating in the fifteenth year (during which puberty stage sexual maturity is reached) – after which the process of growth slows down again until maturity is reached at about twenty or twenty-two years.

Similar to this long period of youth (during which a person must master and internalize the expansive cultural tradition within which that person lives), human beings also possess a similarly long period of adulthood within which to transfer this cultural inheritance of generations effectively and educationally. This biotic developmental dynamic shows that each period of development must be seen as completely interwoven with the characteristic human form of life.

From the data brought to the fore regarding the origin and nature of humankind in the preceding discussion, it has gradually become clear that what is involved is an encompassing *philosophical view* of humankind, transcending the limits of any *specific* discipline. The pretense that a mere biological scientific theory is involved, is unfounded. What is fundamentally involved is a philosophical view of reality that continues to reveal a particular underlying life and worldview, in the grip of some or other direction-giving ultimate commitment.

It is not a simple matter to accomplish a reconciliation between the Christian worldview and the idea of unplanned, accidental evolution transcending all boundaries. We are often told that we may as well believe that God merely

1 Namely the *Nesthocker* (nest-huggers) and the *Nestflüchter* (nest-leavers). The latter are animals who have a way of movement, stature and proportions at birth similar to their adult form, with open eyelids and hearing channels and little dependency on the parents. *Nesthocker*, on the other hand, are born in helpless dependence, with closed eyes and ears and dependent on care in a prepared nest.

created by *means of evolution*. That God created humankind as the crown of creation not only entails that all of creation is oriented towards humankind, but also implies that the human being has a mandate on earth, a cultural *calling* and *task*. The neo-Darwinian combination of mutation and natural selection does not leave any room for the idea of a divine plan of creation. As neo-Darwinism teaches, being human is merely the result of a *meaningless* and completely accidental material-energetic process, which did not foresee this development; the human being was not planned.

The biblical Christian knows that there is nothing in creation, not even a single facet, in which the human heart can find rest. God alone may receive that honour, as the true Creator of all things. He created everything according to its own nature (Genesis 1), with humankind as the crown of creation (Genesis 1:28), crowned with glory and honor (Psalm 8:6). No superficial attempt at reconciliation can bridge the gap between a biblically-founded view of scholarship and the many variations of the modern theory of evolution, and no Christian can settle for a *deification* of something created, and so attempt to serve two lords at once.

4.3 The mystery of being human

At this central point, we are confronted anew by the question of who and what a person actually is. Anyone seriously attempting to ascertain what exactly is scientifically known about humankind today is soon overwhelmed by the sheer magnitude of this knowledge – so much is known that no single individual could hope to be up to date with all of it. Yet, if our exposition thus far gave the impression that science can provide the solution to the riddle of human existence, we have to correct this misunderstanding.

Investigations of the micro-dimensions of human existence have spectacularly expanded the scientific horizon during the past sixty years. We only need to think back to the early 1950's, when biologists and biochemists unveiled the mysteries of the DNA molecule. More and more became known about the complex duplication mechanics in the cell during reproduction and about the human 'genome'. Biological engineering is developing at an astounding rate – so much so that the inhuman possibilities with regard to future genetic manipulation are truly disturbing. These developments probably have as their all-encompassing background the rise of depth psychology during the first half of the 20th century – with such great psychologists as Freud, Adler and Jung in the vanguard. Many previously unexplained phenomena were suddenly thrust into the center of scientific interest. The astounding world of the sub- or unconscious was placed on the table and it became possible to scientifically discuss what has become virtually general knowledge today – e.g. pathological schizophrenia (the personality problem of Dr. Jekyll and Mr. Hyde).

Not only the natural sciences have advanced considerably in recent decades. Owing to developments in abstract mathematics during the 19th century and at the beginning of the 20th century (such as the famous *Principia Mathe-*

matica of Russell and Whitehead during the years 1910-1913), we are on the one hand for the first time in a position to plumb the depths of logical reasoning – that was already accessible to Greek thought[1] – in terms of mathematical logic, while on the other hand we have been enabled by means of the micro-electronic developments of our day – developments entirely dependent on insights in the field of mathematical logic – to construct perhaps one of humankind's most astounding tools as yet, the *computer*. By means of the historical and ethnological sciences, humankind has also gained a considerably enriched perspective on the previously unknown origins of its cultural heritage, while we know more than ever before about the striking cultural configurations that distinguish 20th century peoples from truly undifferentiated cultures living in bygone (pre-)historical eras.[2]

We could continue to highlight further examples of the advances of modern science in this vein – without coming any closer to the elusive riddle of who and what a human being actually is.

The influential personalistic philosopher and ethicist, Martin Buber, developed a dualistic view of reality in one of his works. He places all emphasis upon the personal encounter of human beings *in love*. This personal encounter in love is then dialectically positioned by opposing it to all impersonal relations between human beings and the external world. According to Buber reality reveals itself to humankind in two ways, since the 'I' stands in two fundamental relations: the *I-Thou* relationship and the *I-It* relationship. For Buber no *I-in-itself* exists, since the word 'I' always encompasses one of these two relationships. 'He' and 'She' falls within the "I-It" relationship. The world as we experience it, with the 'It', 'He', and 'She', even with internal experiences or secrets reserved for the initiated, is already constituted by 'Its', by objects. The experiences of this world are not reciprocal, and affect only the human being who experiences them. Thus, the world-as-experience belongs to the fundamental term "I-It." In contrast to this the fundamental "I-Thou" relationship forms the basis for the world-of-relationship that knows no inner barriers since only 'Its' are mutually delimited. The "I-Thou" relationship exists in the presence of an encounter, since only in this relationship does the present reveal itself. The objects of the "I-It" relationship, however, are experienced in the *past*. The individual 'Thou' becomes 'It' after the experience of encounter, and the individual 'It' can become 'Thou' by stepping into the experience of an encounter.

Love is the distinguishing mark of the personal "I-Thou" relationship. Buber develops his approach in a world-historical and religious context (he

1 Euclid, the great Greek mathematician, developed an arithmetical proof stating that there are an infinite number of prime numbers. (Prime numbers are all natural numbers which can be divided only by 1 and themselves). In this proof subtle use is made of means of evidence and conclusions which could only be explicitly accounted for by means of the mathematical logic in the 20th century (cf. Gentzen, 1967:14-25).

2 In Africa there are even tribes who have not yet entered the *stone age*, since they are still living in cultures with *soft* objects of daily use.

was Jewish). Every great culture continually draws its spark of life from an original experience of encounter, from an answer to the *Thou*. When these renewing relational occasions are lost, a culture stultifies and becomes subject to that fate which rests on every human being in the full weight of a dead world mass. Liberation from this situation, to becoming children of God, according to Buber, comes only from new experiences of encounter, a fateful answer from human beings to their *Thou*. Only in this way can a culture renew itself. In the dominant idea of fate, which subjects humankind to social, cultural, psychic, historical and other laws, it is forgotten that no-one can meet fate unless she proceeds from a position of freedom. Notice this internal dialectical tension in Buber's thought: *natural law* and *freedom* reciprocally presuppose and threaten each other.

According to Buber, faith in fate surrenders humankind to the overpowering grip of the *It-World*, whereas a person becomes free in the *I-Thou* relationship, free also from the grip of a rationally obvious system (Buber's reaction against rationalism), a freedom fundamentally indicated by liberation from faith in unfreedom. The meaning of life is to be found, according to Buber, in the embrace of *fate* and *freedom*.

The word love is central in this supposed encounter between person and person. Does it provide an insight into the mystery of human existence? Can we truly say that love is the actual core of human existence – or at least that it should be? Both Classical Greek and Eastern philosophy emphasize the ethical (moral) nature of a person – as can be seen in the typification of an individual as a *rational-moral* creature. Let us look briefly at the possibility of seeing love as the essence of being human.

We are immediately confronted by two problems: (i) it is very difficult to *define* love and (ii) love reveals itself in *many ways*.

Dooyeweerd pays attention to this problem in a lecture given in America. With reference to attempts to typify human beings in terms of love, he said:

> The personalistic and existentialist views of man have tried to determine the I-thou relation as a relation of love, an inner meeting of the human persons. But within the earthly horizon of time, even the love relations present a diversity of meaning and typical character. Does one refer to the love between husband and wife, or between parents and their children? Or is it the love-relation between fellow-believers, belonging to inter-related churches that we have in mind? Or is it perhaps the love-relation between compatriots who have in common the love of their country? Or have we rather in mind the general love of the neighbour in the moral relations of our temporal life? (Dooyeweerd, 1999:125).

It is clear that Dooyeweerd is paying attention to distinctly differentiated human relationships of love – such as *family love*, *marital love*, *patriotism*, and so forth. But none of these differentiated relationships can be reconciled with the central, radical and total bond of humanity transcending all the branches of life. If a distinction is drawn between the **one** encompassing and determinative relationship of a person – referred to as RCT (**R**adical, **C**entral and **T**otal)

– and the various *differentiated* relationships within which people exist and within which they are only engaged in a *partial* and *peripheral* way (that is, DPP relations), then it is clear that love DPP relations are here at stake.

To complicate matters even further, we have to realize that the term 'love' is sometimes used in a radical (root-touching) sense, and sometimes in a differentiated meaning. The Bible regularly uses the word *love* in a differentiated way, referring to a particular facet, among others, of human existence. But it also frequently happens that the Bible uses **love** in an RCT sense.

When love is used in a differentiated sense, it should not be confused with love in its *central* (religious) sense – as it is expressed, among others, in the central commandment of love. This commandment, which demands that we love God and our neighbour with all our heart, belongs to the RCT dimension of creation and therefore contains an appeal to all the facets of our existence. When we talk about marriage, family or patriotic love, however, we are only referring to a specific sector of our existence and not to its *totality*. In biblical usage, this difference is obviously present. In distinction from those sections of Scripture that pertinently refer to the central sense of the commandment of love (e.g. Matt. 22:37-40, Deut. 6:5, Lev. 19:18), we find many places in which love is positioned next to other facets of reality. The statements in Gal. 5:22 and I Tim. 6:11 refer, for example, to the "fruit of the Spirit," and then mentions "love, joy, peace, ..." and so forth, and also mentions that towards which we should be striving: "faith, endurance and gentleness".[1]

In terms of these distinctions, it is clear that the term *love* cannot be used without qualification to indicate the core of the human personality. If it is used, however, to reflect the central (religious) meaning of the commandment of love, we have indeed moved a step closer to the mystery of human existence. Scripture refers to the *heart* of human existence – which is, according to the poet of Proverbs, the wellspring of life (Proverbs 4:23). It is fundamentally a matter of self-knowledge, knowledge of the heart of human existence.

Can a person attain self-knowledge autonomously? As a result of the fall, touching the human *heart* – which is why Christ requires a *reborn heart* – humankind has been tempted, in sinful apostasy, to try and find somewhere within creation a *pseudo-place* of rest – and, as Augustine has already realized, it is only possible to find ultimate rest in God. For this reason, Calvin could emphasize that the true knowledge of ourselves depends upon true knowledge of God. Of course the opposite is also true: fallen humanity designs a self-image in the light of its *idolatry*. In modern times, human beings have been greatly impressed by the machine-like control of reality – with the result that a *mechanical* or *mechanistic* view of the human person necessarily followed. As the second half of the 20th century increasingly stood under the

1 Note, by the way, that the same thing happens with regard to the word *faith:* it is sometimes used in the sense of a total heart commitment to God (this is its central religious sense) and sometimes – as in this instance (I Tim. 6:11) – to indicate a virtue which is valued next to and in distinction from others.

spell of the power of the computer, we find that human beings are understood in computer terms: a person as *super-computer*. David Lyons even demonstrates to which an extent all of society is understood in these terms, as strikingly suggested in the title of his book: *The Silicon Society*.

The human self is nothing in itself; that is, it does not exist separately from the three central relationships in which God has placed human beings. First of all, human beings stand in relation to God, then in relation to their neighbours, and lastly in relation to the totality of created temporal reality. Each of these three relationships is engaged in both the DPP and RCT relationships in which human beings take part. My relation to God is for instance not an esoteric inner room experience that can be divorced from my being a citizen, husband, member of an ethnic community or student, since I live out my love for God or an idol exactly in all these positions. Similarly, every relationship with a fellow human being – however differentiated and peripheral it may be – continues to appeal to the whole self, the heart of that neighbour. Finally, every facet of creation is anchored in a central religious sense. It is with good reason that Paul mentions these kinds of activities – like eating and drinking – when it concerns the honor of God: "So whether you eat or drink or whatever you do, do it all for the glory of God." (I Cor. 10: 31).

It can only be a stumbling block for the centuries-old hubris (pride) of Western people – which has, since the rise of Greek philosophy, developed a limitless trust in the capacities of *human reason* – to be told that humankind cannot come to a true knowledge of the self solely by means of its own rational insight, but only by means of a true knowledge of God. True knowledge of God cannot be a human discovery, it can only be received from Christ in the reborn heart. When it comes to this deepest and most central question of life, fallen and sinful humankind cannot *give* anything. In this regard, as Dooyeweerd notes, humankind can only piously *listen* and *receive*.

4.4 The structural principle of the human being

While material things – atoms, molecules, macro-molecules and macro-systems – clearly belong to the realm of *physically-qualified things*,[1] human existence is by no means excluded from this sphere. Surely our physical existence is bound to the presence of those physical entities necessary for our bodily functioning – from the four 'organic' elements (hydrogen, oxygen, carbon and nitrogen) up to the variety of inorganic substances that make an equally necessary contribution to our existence.

This is not the end of the story, since human beings also have distinctive similarities to the realm of living creatures. Like all living creatures, the human body is also constituted by living cells. When we think about the biotic meaning of the many vital organs in the human body – organs such as the heart, lungs, brain – we realize that human beings take part not only in the realm of physical entities, but also in the biotic realm. This structural biotic di-

1 The meaning of a qualifying function will be explained in more detail in Chapter 7. In this Chapter an intuitive understanding is presupposed.

mension of human existence as a bodily structure has its foundation in the physical-chemical building blocks of living things, since the human body could not be healthy without the necessary foodstuffs.

Both these basic entitary layers are in turn foundational for the sensitive-psychic configuration of the human body, which houses a person's complex sensory equipment and equally complicated emotional life – which are both closely interwoven with the sensory and motoric nervous systems. On this level, human beings are obviously similar to animals.

In our discussion of some of the unique and distinctive characteristics of human beings it has become clear that they are in possession of numerous abilities that animals lack – even if we were to conclude on the shared dimensions that, in comparison with animals, human beings lack a *bio-psychic specialization.*

When human beings act under the guidance of *normative vistas*, they transcend animal abilities. We have repeatedly emphasized that normatively correct or incorrect behaviour is only found amongst human beings. No animal can think logically or illogically, shape historically or unhistorically, act socially or anti-socially, be thrifty or spendthrift, just or unjust. The lack of specialization of the three foundational configurations (the physically qualified, biotically qualified and the emotionally qualifed) goes hand-in-hand with their directedness at the normative nature of a person's bodily existence. Dooyeweerd prefers to speak of a person's 'act-structure'. Since he limits acts to inner inclinations that must be converted into external actions, it is probably necessary to find a broader term for this characteristic human configuration. Since the whole "normative ability" of a person not only indicates the distinctively human-ness of being human, but also qualifies the bodily being of the human person in its entirety, it may be better to refer to this qualifying layer – following the preference of my colleague J. H. Smit – as the *normative 'structure'* of being human.

Comment: The term 'structure' is ambiguous. Its most commonly employed connotation captures the configuration of a concretely existing entity (or process). In this sense, this term refers to the *factual* side of reality and it justifies the employment of the expression "structure of" as an embodiment of the universal side of (factual) creaturely existence (the *lawfulness* of reality). However, one may also use the term 'structure' in order to speak of the *law for* (*order for*) whatever is factually subjected to that law – as in the expression "structure for."

When we want to refer to all four of these *structures* (factual configurations), the best term would be *personality*. This term encompasses the particular nature of each partial *structure of* the human bodily configuration, that is, it encompasses the typical human *tempo* (bound to the physical substructure/configuration), the *inclinations* of a person (known as biotic dispositions – bound to the biotic substructure/configuration), the *temperament* (bound to the emotional-psychic substructure/configuration) and the *character* (bound to the qualifying normative structure/configuration of being human).

Since the variety of human expressions and bodily structures are concentrated in the human selfhood, the human heart (which belongs to the RCT dimension of creation), we can describe a human being conclusively as a *religious personality*.[1]

4.5 Why a comprehensive philosophical view is valuable

At the end of this chapter, we may briefly elucidate once more the *value* of a comprehensive philosophical view of a person. Medical science, for instance, is often accused of having lost a view of the whole and multi-dimensional existential reality of the human being – it easily reduces human beings to mere "biotic organisms," which can be manipulated as *objects*. Even from a nursing perspective, this reduced view is sometimes accepted. The power of medical technique particularly grants apparent credibility to this reduction.

The tendency is, however, to lose sight of the fact that a person is indeed *human*, and that therefore in inter-human relationships, a person always *primarily* appears as a *co-subject*, and never as a *manipulable object* in the first instance. Of course there are many historical examples of societies that degraded human beings to mere utilitarian objects. We only need to recollect the institution of slavery, still common practice in the West less than 200 years ago.[2]

To value and respect a human being *as* human in medical and nursing practice, requires, before all else, a recognition of the position of the human being as *subject*. However, the human being as a *religious personality* is not exclusively *qualified* or *characterized* by any aspect of creation. While we can ascertain that a material thing is characterized by the physical aspect of energy-operation, or that the nature of plants is determined by the biotic aspect of life, it would be meaningless to attempt to use any normative aspect as if it could exclusively account for human existence.

Suppose we were to claim that human beings are *social* creatures, that is, that our entire temporal existence is encompassed by the *social aspect*. This would imply that a person could only act in a *social* manner. But how do we then look at those activities guided by *other aspects* of reality – such as *economical* activities, *analytical* activities, *just* or *unjust* actions, and so forth. It is exactly the complete freedom of a person to choose to act under the guidance of any one of the range of normative aspects which particularly distinguishes the normative abilities of human beings.

1 Remember not to confuse this *root-connotation* of the term 'religion' with the modal functional meaning of faith, as normally done in the English language.

2 The well-known neo-Marxist writer from the Frankfurt school, Jürgen Habermas, has a clear awareness of the difference between subject-subject and subject-object relations – as seen in his distinction between "communicative actions" (regarding subject-subject relations) and "technical actions" (regarding subject-object relations). Van Niekerk indicates that this distinction has deteriorated into a fundamental *dualism* in Habermas's thought – cf. his doctoral dissertation (Van Niekerk, 1982:12-42, 82).

One moment we can be engaged in the scientific analysis of a particular problem or phenomenon, and the next we can act technically by forming something, in creative freedom and with cultural creativity, which could not come into existence by itself, then we can buy something (economic activity), appreciate the beauty of a sunset (aesthetic evaluation), or simply relax with friends (a social activity). We even discussed, in Chapter 1, that this differentiated multiplicity of normative expressions of life correlates with the many societal relationships in which a person is engaged.

If we are to understand the multi-faceted subjectivity of human existence in a meaningful way, it is essential to recognize that human existence cannot be encompassed by or limited to any single aspect of reality. None of these aspects can qualify or finally characterize human existence. It is therefore not appropriate to speak of the 'realm' of human beings – 'realms' are limited to natural creatures: the realm of material things, the plant realm, the animal realm. This usage is linked to the specific qualification of each of these realms by a particular aspect of reality.

Structurally, this means that our temporal, earthly existence is characterized by the richly varied normative structure of the human body – a characteristic structure that is not in itself qualified by any particular normative aspect. Otherwise, a person would be able to act *only* socially, *only* analytically, or *only* economically, as argued above.

The illness of patients normally involves a defect in their biotic functioning. Provisionally, we shall disregard the matter of multiple possible causes – illness can be the result of a shortage of necessary chemical elements, defects in particular biotic organs, or even psycho-somatic (tension, worry, excitement, and so forth). Primarily, the duality illness-health has its origin in the biotic aspect of reality – physics does not even deal with these typically biotic terms.

The presence or absence of particular chemical bonds can without doubt have important implications for normal human functioning. Think of the important role of *iodine* in the nature and function of the *thyroid gland*. The thyroid gland (*glandula thyreoidea*) is placed around the lower part of the human larynx and the beginning of the wind pipe. It is responsible for the secretion of the important thyroid gland hormone (thyroxine) which, probably via an influence on the process of oxidation (oxidative phosphorilation) in the mytochondria[1] initiates the exchange of substances throughout the body's cells. This is essential for normal biotic growth as well as emotional and psychic health. Iodine itself[2] is qualified physically-chemically in terms of its own inner structure. While retaining this inner structure it is however enkaptically bound into the biotic functioning of the thyroid gland.

1 It is one of the important 'organelles' in the cytoplasm of every cell that converts the energy contained in food into ATP – adenosinetriphosphate – in order to produce the necessary energy for various cell functions.

2 Concentrated by glandular cells out of the blood in which it circulates as iodide.

> The term *enkapsis* was introduced by Dooyeweerd, following the biologist Heidenhein, to indicate cases where two differently-natured structures are interwoven in such a way that each retains its unique character. The constitutive physical configuration of living things do not lose their physical-chemical qualification when they are functioning within living entities. Thus we can say that such entities are functioning *enkaptically* – that is, retaining their physically qualified nature – within living things. Similarly both the material components and the biotic organs in a human being are enkaptically interwoven in the total bodily existence of a person.

Only the thyroid gland functions subjectively in the biotic aspect of reality (it is alive) while it depends on the *enkaptically* bound iodine for the production (internal secretion) of the thyroid gland hormone. This biotic function – with its influence on the physical-chemical substructure in the human body – is itself foundationally enkaptically interwoven with the psychic-sensitive substructure and qualifying normative structure of the human being – as proven by its importance for the healthy emotional and normative life of a human being. A hyperactive thyroid gland causes excessive energy use which can lead to a faster heartbeat and a general unease, with accompanying heightened nervous sensitivity. It is clear that the interwoven iodine and thyroid gland plays a role within the integrated functioning of the entire human being. The theory of enkaptic structural wholes attempts to understand this enkaptic functioning of a human being as a whole, keeping in view the complex substructural interweaving also present in the structure of our bodies.

While all four of the human bodily structures have, apart from their enkaptic interweaving, a characteristic internal functional sphere of operation, it is impossible to delimit any one of them *morphologically*, i.e. to localize them in a particular *part* of the human body. The foot, hand, leg or the brain of a human being is never simply physical, biotic or psychic. The whole human personality, in all four of its enkaptically interwoven substructures, is expressed in every part of the body. Precisely for this reason, it is impossible for medical and nursing practice to try and work with a reduced "simply biotic human." This reduction can be directly linked to *technicism*, a force increasingly recognized by present-day philosophers as one of the dominant driving forces of contemporary Western cultural development.

4.6 The problem of the mind-brain identity

In spite of the fascinating detail brought to light by contemporary cognitive science, it must be clear that within this field, one can easily end up with conflicting modes of explanation. An inclination towards the supposedly decisive *biotical* structure of the brain (as an *organ*) might have lost sight of other equally important modes of explanation, such as explored in approaches primarily interested in *sensitive* phenomena, in empirical detail about *logical reasoning*, about *techno-formative skills* (for instance in tool making), in the uniquely human *linguistic abilities*, and so on.

If the distinction between *modal function* and (transmodal) *entity* is neglected, the fallacy of reification becomes actual. The concrete many-sidedness of human subjectivity and human agency transcends any mode of explanation because no single aspect exhausts human functioning. Therefore, if the brain is explained in terms of a mode *different* from the biotical, then the supposed mind-brain identity claim would be highly problematic (and recognized as reductionistic). Glas also proceeds from the idea of modal functions as modes of explanation, when he remarks that he employs the term 'level' with reference to "the qualitative distinctiveness between modes of reality" and to the distinctiveness of modes and entities: "When the modal analysis of an aspect of reality is fundamental in the process of abstraction, then the entitary reformulation of this modal (or functional) point of view may become a first step into the direction of reification and undue substantializing" (Glas, 2002:154). In response to the approaches of Churchland and other neuroscientists, he highlights the shortcomings of both a dualist and a monist account:

> Psychophysical dualism and dualist interactionism may be seen as a result of the unjustified belief in the independent existence of mind and of body as (quasi-) substances. The modal aspect – mental functioning as a mode of functioning that qualitatively differs from biological functioning – is substantialized and changed into a mental substance or a mental entity called "The Mind." Psychophysical identity theories, on the other hand, run the risk of being transformed into one or another form of monism. This occurs when the nature of the 'common ground' of different forms of functioning is defined in terms of one particular scientific discipline, such as biology (Glas, 2002:164).

4.7 The danger of technicism[1]

During a guest lecture at the University of the Free State (18 October 1988) by Egbert Schuurman – the well-known Dutch engineer-philosopher – he referred pertinently to this. The danger of such technicism is that it reduces illness and health to mere scientific abstractions – losing sight of the totality of human existence. Technique can only be of service if it escapes the limitations of this reduced abstraction:

> When medical techniques are used in service of medical care, the physician's responsibility is enlarged while his or her attention is, next to the prevention and cure of illness, directed towards suffering, co-suffering, care and the meaning of all this.

Manipulation of the human embryo in particular easily loses sight of the fact that this embryo is the minimal enkaptic structural whole of a person as a human being. Such manipulation consequently has consequences for all four structures of human bodily existence – effects which, in the light of the limited medical knowledge available in this regard, cannot be foreseen on several vital points. Such experimentation does not only affect particular biotic organs with regard to their internal biotic functioning, but rather a person in its totality.

Apart from the limitations contained in the recognition of the enkaptic interweaving of the human body, medical and nursing practice also has to take

1 Varous books of Schuurman analyze the problems surrounding modern technology and the all-pervasive role of technicism – see Schuurman, 1993, 1995, 2005 and 2007.

into account the variety of societal relationships in which every human being takes part. Whoever enters these professions, must not only have an integrated encompassing philosophical view of a person, but also a balanced encompassing philosophical view of individuals within society – the focus of Chapter 8 below.

Chapter Five

Inter-modal coherence

Starting Points for a Special Theory of Modal Aspects

5.1 The nature of modal aspects

Thus far we have argued for the ontic existence of universal functional modes of reality. In this basic sense, modal aspects are neither functions nor properties of things. Belonging to a distinct dimension of reality, modal aspects cohere with the dimension of entities without being 'dependent' upon those entities for their irreducible 'dimensional' existence. The dimension of modal aspects co-determine the existence of entities (and events) in the sense that the latter invariably *function* within them. But whenever an entity or event "functions within" the universal constant cadre (framework) of any particular modal aspect, the unqualified universal meaning of the latter acquires a specification that is 'coloured' by the entity-structure of the entity function within a mode. One may call these entitary functions *typical* modal properties of entities. In a later context the example of the difference between the fluid, gaseous and solid state of matter is discussed (see page 517 below).

An account of the *epistemic* status of the various dimensions of reality inevitably enters into a complex of epistemological considerations. Therefore we have to explore a few basic distinctions relevant for a meaningful discussion of the nature of a modal aspect.

5.1.1 *Metaphoricity*

Our initial reflections in Chapter 1 were supported by Figures 1 and 2, where we argued that identification and distinction are only possible on the basis of *similarities* and *differences*. We may now move on and look at the peculiar nature of differences that come to expression in what is similar between them and vice versa, at similarities evinced in what is different. Because this formulation may sound *contradictory*, it will be helpful to use some examples.

When we differentiate between life in a *biotic* sense and *social* life, we are confronted with a *moment of similarity* in the term *life*. However, life in a biotic sense differs fundamentally from life in a social sense, for the latter does not need 'watering' and does not produce processes of photosynthesis, simply because it is subject to (social) norms and not to natural (biotic) laws. Similarly, the logical unity and multiplicity of a concept differs from the quantitative sense of the one and the many, for Frege taught us that disregarding differences between entities does not cancel these differences, and therefore at most arrives at a *single* general concept (like the concept 'cat') (see Chapter 2, page 27).

Take another example. Observing the president of a country accompanied by a bodyguard highlights another similarity-in-the-difference, for although they share spatial proximity – the spatial closeness (small distance) between them – in terms of their respective social positions (president and bodyguard), they are *far apart* – that is, there is a *large social distance* between them. Once again, we can discern an element of *similarity* – captured by speaking of *distance* – and an element of difference in the fact that the social has a meaning that is distinct from the spatial. The important point is that the difference is *shown* in the moment of similarity. As soon as one looks at the meaning of the element of similarity, namely *distance*, the distinctive nature of the social and the spatial becomes apparent – *social distance* (being far apart) differs from *spatial distance* (being nearby)!

This kind of a "difference in terms of similarity" may be called an *analogy*. But our examples are all derived from the interconnections between different modal aspects. What about similarities and differences between various *entities*? Language is filled with words and expressions capturing entitary analogies – just think about "foot of the mountain," "the elbow of my finger," the "nose of the car," and so on. Our conjecture is that these entitary analogies are *lingually designated* by the use of *metaphors*. It is the structure of an analogy, where the differences are shown in what is similar, that underlies the apparent *mystery* sometimes discerned in the nature of a metaphor. Max Black writes:

> So perhaps the 'mystery' is simply that, taken as literal, a metaphorical statement appears to be perversely asserting something to be what it is plainly known not to be (Black, 1979:21).

Clearly, understanding a metaphor in a *literal* sense robs it of its suggestive power, lingual significance and truth value. Our suggestion to relate metaphoricity to the nature of an analogy does not entail that either an 'objectivist' or an 'antiliteralist' position is advocated. Lakoff and Johnson are justified in their reaction to the view that our "language consists of words expressing ideas that literally fit the world" (Lakoff & Johnson, 1999:119). Yet they do concede that this "folk theory" ("commonsense theory") is "fundamentally right" for "basic-level concepts," but that in a general sense, this "commonsense theory" would deny that metaphor "serve the central function of language" and that it banishes metaphor from the realm of truth by referring it to uses of "language in which truth is not thought to be at issue: poetry, rhetorical flourish, fictional discourse, and so on" (Lakoff & Johnson, 1999:120).

They portray the traditional theory of metaphor as one where "ideas have to be literal if they are to fit the world," and that such ideas therefore "cannot be metaphorical." As a consequence, "metaphor must be a matter of words, not thoughts. That is why the very idea of *conceptual* metaphor is at odds with this interpretation of the commonsense theory" (Lakoff & Johnson, 1999:120).

The position taken by Lakoff and Johnson proceeds from a peculiar view of the relationship between *word* and *concept*. The *terms* in which the problem is stated concerns 'words' and 'thoughts' – but the truly *modal-functional problem* regarding the relationship between the *logical-analytical aspect* and the *sign mode* is not articulated. The "very idea of *conceptual* metaphor" actually hides it. In order to explicate this, we may consider an example used by Lakoff and Johnson, where the relation between quantity and extension is at stake.

What should be kept in mind is that, from a (ontic) modal perspective, the structural meaning of the spatial aspect *inherently* displays interconnections with the quantitative aspect of reality at its foundation. Of course the nature of these two aspects is manifest in everyday experience, where we implicitly encounter *typically specified* modal properties, such as the *level* (*height*) of water in a glass. The phrase "typically specified" captures the fact that the modal functions of reality are not experienced *in the abstract* but *attached to*, and intimately *connected with*, concrete entities and events functioning within these aspects in a *typical* way. It is only through *modal abstraction* – as the distinctive feature of scholarly thinking – that we can explicitly enter into an analysis of the interconnections between various modal aspects.

But once this dimension of theoretical thinking is opened up,[1] it becomes important to distinguish between modal analogies – evinced in the interconnections between different modal aspects – and entitary analogies designated by metaphors. Does this mean that a *metaphor* has no intrinsic connection with concept formation? (see page 155 below for an alternative account of the ontic conditions of metaphors).

Lakoff and Johnson trace the traditional (and, according to them, currently standard) view of a *concept* back to Aristotle's circumscription of a *definition*. With reference to Aristotle's *Topics* (102a), they mention his view of a 'definition': "A definition is a phrase signifying a thing's essence." They summarize this stance by saying that "it is a collection of necessary and sufficient conditions for an object to be a particular kind of thing." This definition, which is "still commonplace in logic and philosophy," specifies membership of a conceptual category and at the same time expresses "what philosophers today would call a concept." That is, a "definition expresses an idea, which (via Ideas Are Essences), specifies an essence that characterizes a kind of thing existing objectively in the world. Thus, from Aristotle's central metaphor Ideas Are Essences, plus the Folk Theory of Essences, we get the mainstream contemporary philosophical notion of a concept" (Lakoff & Johnson, 1999:379).

According to this approach *metaphors* are constituted by mappings across "conceptual domains." These mappings may derive from experiences, such as when the rising level of water being poured into a glass enables the cross-domain mapping between *quantity* and *verticality*:

> This correspondence between quantity and verticality arises from a correlation in our normal everyday experiences, like pouring more water into the glass and seeing the level go up. Early in development, Johnson hypothesizes, such correlations are 'conflations' in which quantity and verticality are not seen as separate, and associations between them are formed. After the conflation pe-

1 Actually the phrase "opened up" refers to a general feature of subject-object relations. For example, appreciating a sunset as beautiful opens up its aesthetic aspect by objectifying it in this mode, of course correlated to the human subject valuing the beauty.

> riod, according to Grady, the associations between More and Up and [48] between Less and Down constitute a cross-domain mapping between the sensorimotor concept of verticality (the source domain) and the subjective judgment of quantity. Conventional linguistic metaphors like "Prices fell" are secondary manifestations of the primary cross-domain mapping (Lakoff & Johnson, 1999:47-48).

In the absence of a theory of modal functions and concrete (natural and social) entities and processes (in principle functioning in all aspects), and accompanied by the absence of an acknowledgement of the *ontic status* of functional *modes of reality*, the metaphor theory of Lakoff and Johnson suffers from blurred distinctions. In the just-mentioned example of cross-domain mapping between *verticality* and *quantity* the implicit assumption is that these mappings are between *conceptual* domains.

However, as soon as one enters into an analysis of what is *ontically given* (in its "pre-conceptualized" reality), then, as we shall argue in more detail later (see pages 218 ff. below), we immediately realize that *spatial* configurations – such as *vertical lines* or *verticality* – in a constitutive ontic sense, *presuppose* the quantitative meaning of the 'more' and 'less' (that is of the arithmetical aspect). The nature of *dimensional extension* is not *purely* spatial, because there are always *one* or *more* dimensions within which it is meaningful to speak of length (*one*-dimensional extension), surface (*two*-dimensional extension), volume (*three*-dimensional extension), and so on. Consequently, the notion of *verticality* is embedded in that of dimensionality – and the latter collapses into nothingness outside its coherence with the quantitative meaning of *one*, *two* and *three*.

The structural meaning of the spatial aspect is *inherently* connected to the (foundational) *quantitative* meaning of number. Consequently, one should not confuse "conceptual domains" with "ontic functions," because the latter ultimately co-condition both our (integral) experience and our concept formation. This confusion may also tempt us to deny ontic interconnections between modal aspects on the basis of the supposed disconnectedness of "conceptual domains." In a different work, Lakoff is even drawn into a position where his emphasis on "conceptual metaphor" convinced him that *continuity* and *discreteness* are **opposites** (Lakoff & Núñez, 2000:324) instead of belonging to mutually cohering but distinct ontic functions of reality.

Furthermore, as soon as one wants to elucidate the relationship between a *concept* and a *word*, it appears that one first has to account for the implied *modal distinctions*.

The extreme opposite positions taken in this regard often argue that whoever engages in the activity of *thinking*, automatically activates *language*, and whoever uses *language*, is involved in *conceptualization*. These arguments simply state that any concrete human activity necessarily *at once* functions both in the *analytical* and *sign* modes (aspects) of reality, acknowledging that the *ontic modes* of reality are not only *making possible* the concrete functions entities and events may have within them, but also entails that such modes form the (non-subjective, ontic) *transcendental reference point* for human reflection about the uniqueness and coherence between them.[1]

Only *after* this distinction between concrete activities and their diverse *functions* has been drawn, does it make sense to ask *second-level questions* about the *uniqueness* and the *coherence* prevailing between the analytical mode and the sign mode. Only at this point will it be possible to address the problem whether or not the *sign mode* presupposes the meaning of the *analytical mode* or vice versa.[2] Fodor aptly remarks:

> The goal we have been pursuing is the traditional one of reducing meaning to some more basic and better understood entity. But analyticity is too intimately related to meaning to provide such a reduction. In fact, as far as anyone knows, there is no meaning-independent way of characterizing either analyticity or meaning (Fodor, 1977:43).

In Chapter 1, it was argued that *analysis* is the ability to *identify* and to *distinguish* and that *identification* itself constitutes the crux of all *concept formation*, for within a concept a *multiplicity* of (identified and distinguished) properties are brought into a *unity* (they are *synthesized, united*).[3] Lingual phenomena, such as (spoken or written) words, do function in the sign (and other) modes. But we have seen that in young children, the analytical ability to identify and distinguish plays a foundational role in the development of lingual skills (just recall the child who actually formed the concept 'bird', but *designated* it with the *name* (verbal sign) *pigeon* (see page 15).

The view that forming a *concept* is always *language-bound* therefore needs clarification.

(a) First we have to observe that *language* involves a concrete activity embracing *multiple* aspects, explaining why it can never be identified with only one aspect (such as the *sign* mode).

(b) From the first consideration it follows that any *lingual performance* at once functions *both* in the logical-analytical aspect and the sign mode – once again explaining that any such performance will always *simultaneously* exhibit both functions.

Arguing for the *foundational role* of the logical aspect therefore is not in any way intended to deny the actual (inter-modal) connectedness prevailing between the logical and sign aspects of reality. Yet the relationship between analyticity and metaphoricity may serve to highlight the foundational role of the

1 It is needless to say that this kind of reflection will always remain *provisional* and *open to future correction* and *reinterpretation* (owing to the very conditions of *historicity* and *linguisticality* embracing the on-going dynamics of human endeavours).

2 Compare in this regard the analysis found in Strauss (2002a:9-15).

3 Since identification and distinguishing mutually presuppose each other, it should be clear that the counter pole of the *synthetic* act of identification is given in *distinguishing* – the other 'leg' of analysis. Therefore it is a mistake to oppose *analysis* and *synthesis*, for the latter is simply the *other* 'leg' of analysis. Strictly speaking one should say that it is through logical objectification (identification and distinguishing) that the properties of what has been conceived are brought into the logical *unity* of a concept.

logical mode.

5.1.2 *Concept and word*

If we question the nominalistic assumption (see page 25 above) that there is no universality outside the human mind, then it does not need to be problematic to relate the formation of a *concept* to *universal ontic features*. The *unity* of a logical concept is constituted by acts of identification and distinguishing, and is made possible – in a modal functional sense – by the foundational connection between the logical and the quantitative aspect.[1]

Cassirer argues that the determination of a concept as a *unity in multiplicity* belongs to the classical legacy of logic and philosophy as such (Cassirer, 1928a:339). Whatever is logically grasped, cannot fully prescribe in which way the multiplicity of features should be united in the unity of a concept (Cassirer, 1928:134), because it is also a result of the real creative element in our thinking, *the power to discern* or to *observe*.[2] Logical concept formation is always aimed at discerning the *multiplicity* of *universal traits* of that which is conceptualized. As such, it is subject to universal (modal/functional) logical norms, such as the principles of *identity* and *non-contradiction*. Although the construction of each concept is dependent on logical subjectivity – since, as we argued in Chapter 4, only human beings are able to respond with normative freedom to these normative conditions for logicality – no concept is *exclusively* the product of our subjective logical functioning.

Viewed modally, the use of language presupposes the logical function. The lingual intention of symbolical signification is always dependent on the correct lingual *identification* and *distinction* of the relevant words.[3] This identification and distinction is important even in respect of the letters used in words, since the simple switch of letters may result in a complete change in the *meaning* of a (written or spoken) word. Any act qualified by the sign mode must therefore make an appeal to the foundational logical function, and even to the kinematic aspect insofar as an element of *lingual constancy* is required by the possibility entailed in every language of producing endless *lingual* (*semantic*)

1 Distinguishing affirms that, whatever is distinguished, is *not identical* to that from which it is distinguished. This is the nature of analysis assessed in terms of the interplay between its two 'legs', *identification* and *distinguishing*. But the logical meaning of 'sameness' and 'difference' presupposes the original numerical meaning of being distinct, of "more and less" – which explains why Russell states that, in his mathematical philosophy, he will "never use the word *unequal* to mean merely *not equal*, but always to mean *greater or less*" (Russell, 1956:160). *Logical negation* is not the same as *numerical inequality* (bringing to expression the meaning of the *numerical time order of succession*).

2 It is difficult to find a suitable English equivalent for the German word *Aufmerksamkeit* used by Cassirer, "... der *Aufmerksamkeit* als dem eigentlichen schöpferischen Vermögen des Begriffsbildung" (1910:31). "... the power of observation (of being attentive) as the truly creative ability of concept formation."

3 The expression, "lingual *identification* and *distinction*" must be seen as a modal (retrocipatory) analogy within the structure of the sign mode.

variations on the basis of lingual constancy. Think here in particular about the deferral of meaning entailed in Derrida's frequently employed word *différance*. Cilliers remarks that for Derrida the "two concepts: 'trace' and 'différance' ... are actually neither concepts nor words" in the sense that they "cannot be given a full meaning" (Cilliers, 1998:43). He adds that Derrida's "intention seems to be to prevent these two terms from acquiring fixed meanings" for it is "by their very instability that they allow us to say something more general about language" (Cilliers, 1998:44). Yet we have to note that the *instability of varying meanings* still crucially depends on an element of constancy, given in the fact that these two 'terms' – namely : 'trace' and 'différance' – are persistently spelled in exactly the same way in any specific language. Habermas explicitly emphasizes the foundational role (the *constancy*) of alphabet letters: "Already on the level of the sign substrate of meanings, it must be possible repeatedly to recognize the sign-type in the diversity of corresponding sign events as the *same* (my emphasis – DS) sign" (Habermas, 1998: 26-27).[1]

Cilliers further explains Derrida's view of language: "If the system of language is as open as Derrida suggests, if the relationships are always playfully changing in an unpredictable way, how is it possible to say anything about these relationships? In some sense we cannot say anything permanent and specific about them that would apply to language in general" (Cilliers, 1998:43). Clearly Cilliers does not realize that in fact he already said something "permanent and specific" about "these relationships," namely that they "are *always* (my emphasis – DS) playfully changing in an unpredictable way"! The word 'always' reveals the inevitable universality and constancy presupposed in every reference to *change* – confirming the general claim argued for below, namely that *change* can only be established on the basis of an element of *constancy* (see pages 163, 164, 186, 234 as well as Chapter 7 below).[2]

Once a concept is understood in its *logical* structure, it is impossible to deny the possibility of identifying and distinguishing whatever is given within reality. This possibility is related to what should be called the *logical object-side of reality*. Only via this object-side of reality, is it possible to understand the *range* of a concept.[3] A denial of this reality will cause much difficulty in discerning the dissimilarity between the *domain* of a concept and the *set* of ob-

1 "Schon auf der ebene der Zeichensubstrats von Bedeutungen muß der Zeichentypus in der Vielfalt korrespondierender Zeichenereignisse als dasselbe Zeichen wiedererkannt werden können."

2 The just quoted explanation in connection with trace and différance, namely that through "their very instability [that] they allow us to say something more general about language" (Cilliers, 1998:44) came close to an acknowledgement of the foundational position of constancy vis-à-vis change without realizing it. The theory of inter-modal coherences enables an understanding of the foundational meaning of the kinematic aspect (core meaning: constancy) for the physical aspect (core meaning: change).

3 A *spatial analogy* within the modal structure of the *logical* aspect.

jects intended by a concept. Not acknowledging the logical object-function easily generates the incorrect view that a concept *is* merely a group of characteristics of an object, and that the logical extension of a concept is to be equated with the multiplicity (*number*) of objects subsumed under it. Frege clearly recognizes this state of affairs: "The domain (*Umfang*) of a concept does not consist of the objects that the concept encompasses, like a forest constituted by trees, since the domain is only attached to the concept itself" (Frege, 1895:455). Nevertheless the subjective contents of a concept cannot bypass the logical object-side in its reference to *that* which is logically analyzed, i.e. what has been *objectified*. Only in relation to the subjective logical unity of a concept, is it possible to view the *range* of implied objective logical characteristics in its connection to the *contents* of a concept.

Earlier in the 20th century, Cassirer increasingly conjoined the nature of a *concept* with that of *words* and their *meanings* (Cassirer, 1928:130). According to Rickert, the endeavour to think a concept as a *unity* sets an unattainable task for our understanding – a task-setting idea in a Kantian sense – which could only be realized by a *word*, because the unity of a word replaces the unity originally sought in our understanding (Rickert, 1929:50-51).

Closely related to what we said about the letter signs (with reference to Habermas), Reichling, from a semantic angle, also realizes that *lingual activities* do not merely communicate *thoughts*, since knowledge of the *sounds* is also essential (Reichling, 1970:60). In the final analysis, however, the famous author of *Het Woord* (Reichling, 1967) completely reduces his problems with the *meaning* of a word to those which he marked as the "things known" in the different usages of language. For example, many things are known about a *cow* – positively more than those mentioned in a dictionary – and when we identify the sound-form *cow* in a certain context, it is not necessary to think about anything special in our cow-knowledge:

> Therefore, the use of the word *cow* is made possible by the totality of facets of your knowledge of the cow which fits the situation. Such a totality of knowledge-facets in conjunction with the word-form that enables the use of that word, constitutes the meaning of that word in the case concerned (Reichling, 1970:60).

A remarkable indirect-proportional situation is uncovered here. Whereas Cassirer and Rickert try to solve their problems with a concept in the direction of a *meaningful* word, Reichling explores the opposite direction and wants to resolve his embarrassment with the meaning of a word with reference to the *knowledge facets* which are relevant in a specific situation!

Simply by reformulating Reichling's original remark, the foundational role of concept formation, as well as the difference between the logical and the lingual mode can be explained. Instead of saying that the recognition of a specific sound-form (such as *cow*) does not necessarily imply that we simultaneously *think* about every cow characteristic, we prefer to say that we do not necessarily have to *signify* every cow characteristic. For example, it is quite possible that the word *cow* merely means: something *big* and *heavy*.

Lingually, no objection can be raised against the way in which only these (spatial and physical) features are signified (that is, *lingually objectified*). If, however, we now proceed to identify this meaning nuance of the word *cow* with the *concept* 'cow', serious problems are encountered. A correct *concept* of a cow must, analytically seen, imply not only the spatial and physical characteristics of the cow, but also every other essential characteristic, for otherwise the predication of any one of the neglected features amounts to logical contradiction. In a logical sense, therefore, other fundamental traits of a *cow* must also be included in our concept *cow* – such as its health, its appearance, and so on.

A different example may elucidate the logical-analytical nature of concepts. Does the concept of a *chair* appeal only to its opened logical features, or must we assume that all the non-logical characteristics are implied by opening up the logical object function of a chair? If these non-logical characteristics are not implied, then plainly an unbridgeable gap exists between the logical subject-object relation and the non-logical aspects of a chair. We have to conclude, therefore, that in making the logical object-function of a chair *patent* (manifest), the non-logical (modal) characteristics (specified according to the typical entitary uniqueness of the chair) are also logically objectified.

Therefore, the multiplicity combined in the unity of the concept of a chair enables us to make predications like: this chair is beautiful (aesthetic feature); this chair is expensive (economic); this chair is big (spatial); this chair is heavy (physical); and so on. If these (modal) characteristics were not implied in the correct concept of a chair in an analytical sense, all the mentioned statements (explicating them in distinct judgments) would, in a logical sense, be *contradictory*. In other words, if the correct concept of the chair does not *imply* these characteristics in an analytical way to begin with, they cannot be *predicated* afterwards of the chair, except *illogically*: From: P is non-Q; one cannot infer that P is (such-and-such) Q (see Strauss, 1973:139).

Against this background, it therefore seems completely permissible, lingually, to *exclude* various essential features in the meaning of a word in a specific context (without disobeying any lingual norms), while at the same time it remains imperative that every one of the elements of a logical concept should, analytically seen, be *implied* in the applicable *concept*. That which, in a specific *word* (with reference to something signified), is not *semantically* implied, must therefore (without any contradiction) be implied analytically in the corresponding concept.

In general, it may therefore be said that the mentioned semantic phenomena, such as *ambiguity*, *synonymity*, *redundancy* and *metaphoricity*, are foundationally dependent upon our *analytical ability to discern*.

Another manner in which some of the key facets of our current discussion may be captured, is in Figure 6 – taking Figures 1 and 2 of Chapter 1 a step further.

Naturally, there are more interconnections to be considered, because human *analytical discernment* combined with *lingual imaginitivity* enables relationships between modal functions and entities (or events) to be explored.

5.1.3 *The "embodied mind": Conceptual metaphor*

Lakoff and Johnson challenge the Western tradition insofar as it assumed that 'reason' is autonomous in the sense of being independent of "perception, motion, emotion, and other bodily capacities" (Lakoff & Johnson, 1999:17). They are convinced that all "neural beings " 'categorize'. Lakoff and Johnson

Figure 6

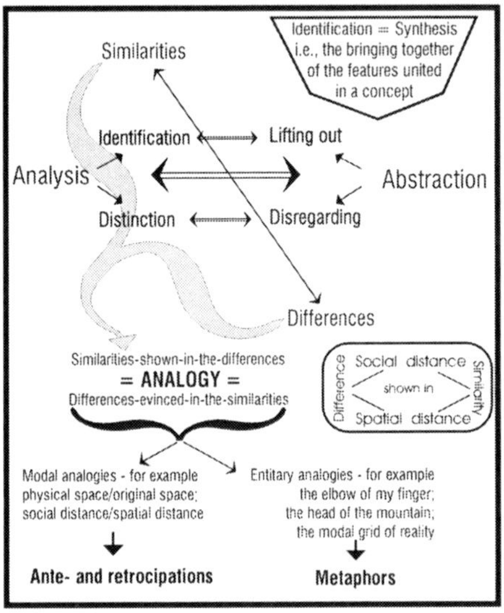

distinguish between 'categories' and 'concepts': the former is the "stuff of experience" and the latter the "neural structures that allow us to mentally characterize our categories and reason about them" – but they emphasize that categories, concepts, and experience are inseparable (Lakoff & Johnson, 1999:19). They refer to the 'categorization' of animals regarding "food, predators, possible mates, members of their own species, and so on." In Chapter 4, we already mentioned the fact that animals experience reality exclusively in terms of their natural inclination, directed at that which is physically, biotically and psychic-sensitively important to them: they experience reality in terms of that which is negotiable and not negotiable, edible and inedible, in terms of same sex and opposite sex, comforting and alarming. This capacity depends upon the sensing apparatus of animals and on their "ability to move themselves and to manipulate objects" (Lakoff & Johnson, 1999:17).

They claim that categorization is a consequence of the way in which we (as "neural beings") are 'embodied', and that most of our categories are formed automatically and unconsciously "as a result of functioning in the world" (Lakoff & Johnson, 1999:18).

Phrased in terms of the distinction between modal functions and the dimension of entitary structures and events, the idea of the "embodied mind" is geared towards the undeniable fact that human beings have *subject functions* in all aspects of reality, including the physical, biotical and sensitive modes. However, within Lakoff and Johnson's approach, the emphasis is on (entitary) concrete functioning of the human bodiliness to such an extent that no consideration is given to the ontic (and *conditioning*) role of the modal aspects of reality.

The effect of this shortcoming is that their argumentation in certain respects tends to be circular. As seen in Chapter 4, the complex nature of the human body is constituted by various kinds of entities in their interlacement, from the physically qualified material constituents, biotically qualified organs, sensitively qualified configurations in which the desires, needs, emotions and the nerve system find their seat, to the guiding normative human orientation supported by these foundational layers. It is only *in* and *through* this complex enkaptic whole of the human body that we are 'embedded' ('embodied') in the world, by participating in all of its dimensions, including the dimension of aspects, entities and events.

From our lived experience of being-spatially-extended, of movement and being subject to physical and biotic laws it is natural to explain the possibility of understanding the functional nature of this reality by means of appropriate functional concepts of spatiality, motion, force and vitality. This functional subjectivity caused Merleau-Ponty to claim that "space is rooted in existence" (Merleau-Ponty, 1970:148) and to overstate his case by saying that *I am my body* (1970:150).

But we have to realize that, since reality at once has both a law side and a correlated factual side, it is simplistic and one-sided to take into account only the *factuality* of our bodily subject functions when it comes to the concepts we form, while neglecting the *conditioning* role of the various aspects (as *law* spheres) and of the *type laws* governing the various layers of the human body. In being subjected to physical laws, such as the law of gravity, our bodily existence experiences the effect of physical forces, for example when the movement of our arms and legs manifest the pull of gravitation. But are we justified in claiming that our *concepts* of movement and causality are derived from our ability to move and from the use of our muscles? Lakoff and Johnson write:

> Our abilities to move in the ways we do and to track the motion of other things give motion a major role in our conceptual system. The fact that we have muscles and use them to apply force in certain ways leads to the structure of our system of causal concepts. What is important is not just that we have bodies and that thought is somehow embodied. What is important is that the peculiar nature of our bodies shapes our very possibilities for conceptualization and categorization (Lakoff & Johnson, 1999:19).

At first glance, the answer seems affirmative. However, the unanswered question underneath the surface is the following one: what are the conditions for this "peculiar nature of our bodies"? The answer to this question has to account for both the *type law* of the human body and for the universal modal (functional) laws *specified* by this type law. Modal physical laws hold universally for all factual physical processes (including the functioning of the human body according to its physical subject function) – therefore they cannot be the *product* of our bodily functioning, but rather must be appreciated as (law side) *conditions for* the physical functioning of our bodies. Consequently causal concepts, though co-dependent upon our physical (bodily) subject function, are also dependent upon the universal modal structure of the physical aspect itself. But then we have to distinguish between our concept of physical relations and the ontic reality of such relations (and their ontic conditions).

It is noteworthy that Lakoff and Johnson stumbled upon the difference between what we have called *modal functions* and concretely existing (and functioning) *entities*. In their discussion of Zeno's paradoxes, they accept for example that "[M]otion is not a metaphorical concept" (Lakoff & Johnson, 1999:157). Similarly they speak of "spatial relations" closely approximating our critique of positivism (see page 44), for we argued that the (modal) terms employed in the description of what is observed by the senses are not themselves entities that can be perceived.

They consider spatial relations concepts to be at the heart of our conceptual system, for they "are what make sense of space for us" in that they "characterize what spatial form is and define spatial inference":

> But they do not exist as entities in the external world. We do not see spatial relations the way we see physical objects (Lakoff & Johnson, 1999:30).

What is significant, however, is that they do not realize that the "part-whole structure" itself is also (in its modal functional sense),[1] an original "spatial relations concept" (see Lakoff & Johnson, 1999:28-29). They understand this relation as something intrinsic to *physical* entities: "the properties that make for basic-level categories[2] are responses to the part-whole structure of physical beings and objects. Gestalt perception is about their overall part-whole structure, as is mental imagery. The use of motor schemes to interact with objects depends significantly on their "overall part-whole structure" (Lakoff & Johnson, 1999:28).

But Lakoff and Johnson add another dimension to their approach by introducing their peculiar view of "conceptual metaphor" in the following way. As their starting point, they look at the scope and richness of our subjective mental life, where "subjective judgments" are made about "such abstract things as importance, similarity, difficulty, and morality" and where we meet "subjective experiences of desire, affection, intimacy, and achievement":

> Yet, as rich as these experiences are, much of the way we conceptualize them, reason about them, and visualize them comes from other domains of experience. These other domains are mostly sensorimotor domains ..., as when we

1 As seen in our remarks about the aspect of space in Chapter 3 (see page 88).

2 "Basic-level categories" are those that are in the 'middle' of hierarchical categories such as *chair* and *car* in the "category hierarchies *furniture-chair-rocking chair* and *vehicle-car-sports car*" (Lakoff & Johnson, 1999:27).

> conceptualize understanding an idea (subjective experience) in terms of grasping an object (sensorimotor experience) and failing to understand an idea as having it go right by us or over our heads. The cognitive mechanism for such conceptualizations is conceptual metaphor, which allows us to use the physical logic of grasping to reason about understanding.

The elaboration of this view of "conceptual metaphor" is developed with great care and systematic finesse, and it particularly applies, amongst other features, the idea of *mappings* between *source domains* and *target domains*. The strong element in this approach is the recognition of the innumerable images generated by cross-domain mappings employed by metaphors in all possible contexts of human endeavours. By and large, these cross-domain mappings concern *similarities* and *differences* between different (conceptualized) *entitary* (including event) *domains*, but also between *modal* domains[1] and *entitary* domains (or vice versa) – and in all these cases, metaphors are instances of a distinct type of *analogies* in the sense defined by us.

Since the choice of source and target domains is relatively arbitrary, it is understandable why metaphors could be replaced by others apparently unrelated to the initial ones. It is only in the case of *purely functional modal analogies* that every attempt at such an exchange is unsuccessful, since the invariable effect is that modal (functional) analogies are simply substituted by or with *synonymous* terms. Look for example at different ways of capturing the spatial analogy within the structure of the social aspect (leaving aside the fact that one has to distinguish between analogies on the law side and the factual side): social distance; social proximity ("next-to-each-other"); social super- and sub-ordination; social position; and so on.

All these expressions are in an important way *connotatively synonymous* – namely insofar as they (analogically) reflect some or other structural feature of the spatial aspect. This ability to 'synonymize' *modal analogies* is absent in the case of analogies between entities (or entities and modal properties/modal domains) as designated by metaphors. One may replace the metaphor "the nose of the car" by referring to the "bonnet of the car." Whereas we do have *denotative synonymity* in this case, *connotative synonymity* is absent.

In fact, we have to expand our argument that aspectual analogies (similarities and differences between aspects of reality) ought to be distinguished from *metaphors*, because there are actually four possibilities (see Figure 7).

Metaphors explore analogies (1) between different entities (E–E: "the nose of the car"), (2) between entities and functional aspects (E–A: such as the "web of belief") and (3) between aspects and entities (A–E: a widespread example is found in evolutionary biology, where the biotic facet – with *life* as its core functional meaning – is treated as if it is an entity, for example when biologists speak of the "origin of 'life' instead of the genesis of *living entities* Another example is when we speak of the "social glue" of society. In Figure 6 (see page 151), as an instance of A–A, we gave the example of *social distance*

1 Not recognized as *ontic functions*, since they are normally discussed as 'concepts'.

and *spatial distance*). Metaphors falling within categories 1, 2, and 3 may be replaced by different ones. But we argued that modal functional (inter-aspectual) analogies cannot be replaced – at most they can be *substituted* with *synonyms* (for example when *continuous extension* – the core meaning of the spatial aspect – is 'synonymized' by words and phrases such as *being connected*, *coherent* or even the expression *the whole-parts relation*).

Figure 7

Analogies - there are four possibilities

1) Similarities and differences between different entities E - E	2) Similarities and differences between an entity and a modal aspect E - A
3) Similarities and differences between an aspect and an entity A - E	4) Similarities and differences between different aspects A - A

In the absence of an articulated theory of modal functions, the nature of inter-modal (inter-aspectual/inter-functional) connections is distorted by the theory of conceptual metaphor. Therefore, instead of analyzing the *ontic meaning* of the moral aspect – with love as its unique and indefinable meaning-nucleus – Lakoff and Johnson embark on the investigation of the "concept of love" by asking whether or not it is "independent of the metaphors for love?" (Lakoff & Johnson, 1999:71).

The reason for this is that the inter-modal connections between modal aspects, evinced in the retrocipations and antecipations[1] through which a specific aspect analogically reflects the meaning of other (irreducible) aspects, are structurally fitted into an unbreakable coherence.

Lakoff and Johnson see in their "theory of conceptual cross-domain mapping" an approach that can account for everyday and novel cases, for "the theory of the novel cases is the same as the theory of the conventional cases," which is "thus best called a *theory of metaphor*" (Lakoff & Johnson, 1999:70).

The cross-domain mappings operative in "conceptual metaphor" do require a truly conceptual understanding of the constitutive elements of the (original) domains. Their guideline states that difficulties emerge whenever such mappings are understood in a *literal* instead of *metaphorical* sense.[2]

1 Remember that instead of the usual designation 'anticipation' we prefer the word 'antecipation' (which acknowledges the difference between 'anti = against' and 'ante = referring forward').

2 Just consider their account of the paradoxes of Zeno, where they hold that "Zeno's paradox of the arrow can also be seen as pointing out the mistake of taking a metaphor to be literal" (Lakoff & Johnson, 1999:157).

It is precisely this difference between the literal and metaphorical, which shows that *metaphor* has its original seat within the *sign mode* and not within the *logical-analytical aspect*. It presupposes the conceptual-logical dimension (foundational aspect), but ought to be distinguished from it. Therefore the expression "conceptual metaphor" – although described in an intelligible way – conflates the sign mode with its foundational logical-analytical aspect. Concepts are not words – and therefore they are also not metaphors.

But although they transcend the logical-analytical mode, the lingual nature of metaphors cannot be understood except *on the basis of* concept formation. Without the foundational role of (analytical) concept formation, the entire distinction between a source domain and a target domain becomes meaningless, as well as the distinction between literal and metaphorical. Precisely because a metaphor is not a concept, it can employ *words* metaphorically without violating the analytical scope of some or other concept to which the word(s) under consideration may refer.

5.2 Meaning requires uniqueness and comes to expression in coherence

The idea of modal aspects has been introduced as a response to the perennial philosophical problem of the "coherence of irreducibles" (see Chapter 1, page 8). We have provisionally examined the two sides of this coin, *coherence* and what is *irreducible*. The idea of *sphere-sovereignty* represents the latter and the principle of *sphere-universality* the former.

Almost invariably, the history of philosophy and the various academic disciplines illustrate the multifarious attempts to distort the given diversity in creation by interpreting it in terms of one or just a few privileged *modes of explanation*. What is known as *monistic isms*, i.e. theoretical orientations such as physicalism, psychologism, historicism, moralism and so on, attempting to explain all of reality merely in terms of one principle of explanation, are the outcome of a denial of the idea of *irreducibility*. But strangely enough, this denial lives by the grace of the accompanying principle of *sphere-universality*, because every monistic ism appears to be justified in its claims with reference to structural elements of an aspect that in fact is based upon the *coherence* between the elevated one and all the others denied by its reification. Since this coherence presupposes distinct and irreducible aspects, every absolutization of one particular aspect actually highlights the impossibility of its claim. Normally this impossibility is manifested in what we shall analyze in Chapter 6 as insoluble *antinomies* (a clash of *irreducible laws* – such as those found in Zeno's arguments against multiplicity and movement – see also Chapter 1, page 2).

5.3 The unbreakable coherence between the various ontic modes within reality

Both in Chapter 1 (see page 24) and in Chapter 3 (see pages 75 ff.), we referred to the example of the concept of *jural causality*. This was done in the context of pointing out the *difference* between *jural causation* and *physical causation*. Jural causation reminds us of physical causation (there is something similar) but at the same time there are *differences*. Later we explained this hand-in-hand appearance of differences and similarities by introducing the idea of an *analogy* – claiming that whenever what is different evinces itself in what is similar, a proper instantiation of an *analogy* is given (see page 144).

We are now able to explain the examples given in Chapter 3 more fully (see page 76). We mentioned analogies between different aspects such as "emotional strength" (sensitive and physical aspects), "economic trust" (economic and certitudinal aspects), "aesthetic expression" (aesthetic and lingual modes), "emotional vitality" (the sensitive and the biotic facets) and "social distance" (the social and the spatial).

The first challenge is of course to decide which aspect refers to which. Does the phrase "emotional strength" bring to expression a reference from the physical aspect to the sensitive mode or the other way around? Since the referring aspect is normally mentioned first, its *qualifying role* is easily identified. This means that, in the configuration of "emotional strength," we meet *within* the sensitive psyhic mode an analogy (retrocipation) to the physical mode, where 'strength' resides in its original meaning.

Similarly, because the expression "economic trust" has 'economic' as its qualifier, "economic trust" represents an (antecipatory) analogy of the certitudinal aspect within the economic mode. The other examples mentioned in the previous paragraph are now easily assessed: "aesthetic expression" is a lingual retrocipation within the aesthetic mode; "emotional vitality" is a biotic analogy within the sensitive function, and "social distance" represents a spatial analogy within the structure of the social facet of reality.

Suppose we switch the terms contained in such composite phrases around and speak instead of "lingual feeling" speak of a "feeling for language," or instead of "economic trust" of "certitudinal frugality"? In cases like these, the order is reversed, for instead of the initial retrocipations, we are discerning antecipations: the analogy of feeling within the sign mode now is substituted with a lingual analogy within the sensitive aspect, and the (antecipatory) analogy of trust within the economic aspect is substituted for an economic retrocipation within the faith aspect.

5.3.1 *Dooyeweerd's confusion of retrocipations and antecipations*

In his work, *The Roots of Western Culture*, there is an instance where Dooyeweerd confuses retrocipations with antecipations. After discussing the retrocipations of the emotional mode to those aspects preceding it, he introduces the discussion of the antecipatory coherence with the words: "[B]ut human emotional life is not limited to a coherence with those aspects that *precede* the *psychical* aspect of feeling. It also unfolds itself within its own aspect in the aspects that follow" – and then he proceeds as follows: "so that we can speak of *logical* feeling, *historical* and *cultural* feeling, *lingual* feeling, feeling for *social* convention, feeling for *economic* value, *aesthetic* feeling, *moral* feeling, and the feeling of *faith* certainty. In other words, the structure of the psychical aspect reflects a coherence with *all* the other aspects" (Dooyeweerd, 2003:46).[1]

Although Dooyeweerd clearly intends to explain the *antecipations* within the structure of the sensitive aspect, he in fact only mentions *retrocipations* from the various (cosmic later) normative aspects to the sensitive aspect. For example, *historical feeling (sensitivity)* represents a *retrocipation* from the historical aspect to the sensitive aspect, whereas *emotional control* illustrates an *antecipation* from the sensitive aspect to the historical aspect. Moral feeling and certitudinal feeling are also both retrocipations, referring back to the sentitive mode within the moral and faith aspects. Interestingly, however, further on in the text, Dooyeweerd correctly speaks of *emotional trust* and *emotional certainty* – although in the context where he uses the correct compound phrases, he is no longer focused on explaining antecipatory analogies within the structure of the sensitive psychical aspect.[2]

The nature of an analogy, as explained in Figure 5 (page 151), is centered in the fact that the difference is shown in the moment of similarity. Figure 6 expresses the peculiar nature of the *modal* analogy involved in the basic concept jural force (causality).

It is clear that every possible *modal analogy* always presupposes the *original meaning* of another aspect distinct from the one in which it appears as an analogy. But the mere fact that, through their analogies, the inter-modal connections between aspects come to expression does not as such settle the following questions – there are four options:

(a) Aspects refer to aspects;
(b) The meaning-nucleus of an aspect refers to other aspects;
(c) Aspects refer to the meaning-nuclei of other aspects; or
(d) Meaning-nuclei refer to meaning-nuclei.

The original ('authentic') theory of a modal aspect conjectures that (c) represents the correct view. However, in some instances, an expression may be employed having the 'face' of (d), but the intention of (c). Expressions such as "love life" and "retributive balance" appear to be instantiations of (d), but in fact speak of nothing but "moral life" and "jural balance." This is simply an effect of the fact that the semantic domains of terms like 'love' and 'retributive' include the meaning nuances 'moral' and 'jural.'

1 The same explanation is found in his little work on the transcendental problems of philosophic thought (see Dooyeweerd, 1948:48). In the systematic volume of his *Encyclopedia of the Science of Law* he also equates 'rechtsgevoel' ('legal feeling/sensitivity'), which is a retrocipation to the sensitive aspect within the jural, with its opposite, a jural antecipation within the senstivie-psychic aspect to the jural aspect: "We have seen that human emotional life exhibits modal anticipations to the normative aspects. In our jural sensitivity we discern this anticipatory connection to the jural aspect of society" (Dooyeweerd, 1967:77).

2 In his significant discussion of Dooyeweerd's idea of development, Griffioen mentions these examples, but does not notice the incorrect reversal in Dooyeweerd's account (see Griffioen, 1986:93).

The importance of a correct understanding of this issue is seen from what C.T. McIntire concluded on the basis of confusing (b) and (c).

Figure 8

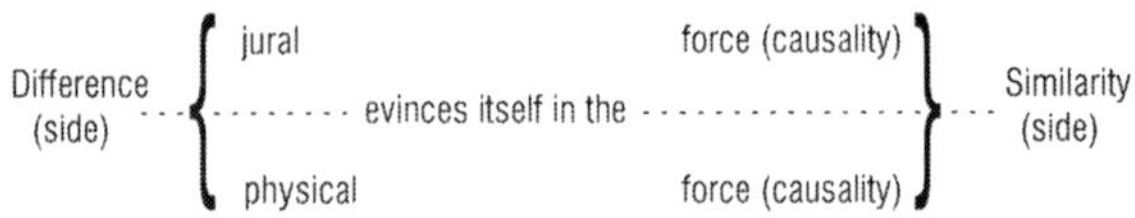

5.3.2 *C.T. McIntire: Turning the theory of inter-modal connections upside down*

In an earlier context, we have discussed McIntire's mistaken view that Dooyeweerd holds the opinion that historical study "examines the historical aspect of anything" (McIntire, 1985:89), while in fact Dooyeweerd holds that an analysis of the *modal structure* of any aspect belongs to the *philosophical foundations of a discipline* (see page 94).

In his discussion of the inter-modal coherence between the historical aspect and the other aspects of reality – in order to substantiate a 'second' argument against Dooyeweerd's theory of a modal historical aspect – McIntire formulates the following one (McIntire, 1985:93).

> I will merely illustrate. As an analogy in the historical mode to the aspect of faith, he [Dooyeweerd] would need something like power-faith, but that would make little sense (McIntire, 1985:93).

Indeed, such a construction does make "little sense." Unfortunately it is not Dooyeweerd's construction, but merely McIntire's misunderstanding of his theory of the modal aspects. In the phrase "power-faith," the term 'power' is supposed to represent the meaning-nucleus of the historical aspect, and 'faith' the certitudinal aspect. But option (c) entails that it is an *aspect* that contains analogical moments (retro- and antecipations) and not the *meaning-nucleus* of an aspect. Furthermore, these analogies point backwards or forwards to the *meaning-nuclei* of other aspects, although they are qualified by the meaning-nucleus of the aspect in which they appear as retro- or antecipations (cf. Dooyeweerd, NC-II, 1997:75). Consequently, the construction presented by McIntire presents a different theory, namely option (b) – not the one advocated by Dooyeweerd, namely option (c).

In this alternatively constructed theory of McIntire, the *meaning-nucleus* refers to another *aspect* instead of an *aspect* referring to the *meaning-nucleus* of a different aspect! If the proper order is restored, and if we adhere to Dooyeweerd's theory, the situation should be phrased as follows: within the cultural-historical aspect, one encounters an antecipation (antecipatory analogy) to the certitudinal (or: fiduciary) aspect in the configuration of *historical certainty* (or: *historical confidence/trust/reliability*) – a perfectly *meaningful* state of affairs – comparable to configurations such as *economic trust* (credit), *aesthetic confidence*, *legal certainty* or *moral trust*.

5.3.3 *Inaccurate account of the core meaning of and social retrocipation within the moral aspect: Stafleu*

According to Stafleu Dooyeweerd holds that "the relation frame of loving care" is designated as "the 'moral or ethical' aspect, with the love for one's neighbour as its meaning nucleus" (Stafleu, 2007:28). In support of this view he mentions certain places in *A New Critique of Theoretical Thought* (NC) (namely NC-I:3 and NC-II:140-163). Unfortunately these references tell a different story. On page 3 of NC-I the moral aspect is mentioned alongside all the other aspects of reality, but nothing is said about "love for one's neighbour" as its supposed core meaning. The reference to the second volume also does not provide any grounds for his claim. The entire section (paragraph 5) from page 140 up to page 163 is concerned with "juridical and social retrocipations in the modal aspect of love" (see Dooyeweerd, 1997-II:140) – suggesting that there is a fundamental difference between the core meaning of this "aspect of love" and the retrocipations found within its structure. In the course of the unfolding analyses of this paragraph Dooyeweerd eventually states: "every serious attempt at an analysis of the modal meaning-structure of the moral relation leads us back to love as its irreducible kernel" (Dooyeweerd, 1997-II:152). Dooyeweerd also explicitly mentions that 'love' is "the meaning-nucleus of the ethical aspect" (Dooyeweerd, 1997-II:156). The phrase used by Stafleu as an indication of Dooyeweerd's alleged designation of the meaning-nucleus of the moral aspect, namely "love for one's neighbour," is used by Dooyeweerd to highlight the *social retrocipation* within the modal structure of the moral aspect. Dooyeweerd speaks of "love of one's neighbour" (pertaining to inter-individual moral relations), alongside the "communal love between parents and children, husband and wife" and concludes that without "this social retrocipation love in its modal ethical sense cannot exist" (Dooyeweerd, 1997-II:158; cf. 1997-III:269, bottom line).

5.4 Sphere-sovereignty and sphere-universality

Implicit in all our argumentations thus far is the acceptance of irreducible (unique) modal functions of reality, guaranteed by the principle of sphere-sovereignty. But this assumption is only meaningful if it is supported by its complement, the principle of sphere-universality. The latter principle supplies an explanation for the *apparent plausibility* of the mentioned monistic isms, that is to those cases where a thinker conjectures that reality in its totality could be understood by only one specific mode of explanation. Modern physics constantly experienced the effect of this reductionistic inclination – from its initial mechanistic orientation up to those who think that something like "string theory" will be able to provide one comprehensive explanation of the entire universe. The principle of sphere-universality precludes the possibility of understanding any aspect *in isolation from* all the others. Clouser calls this the principle of aspectual inseparability: "This means that aspects cannot be isolated from one another since their very intelligibility depends on their connectedness" (Clouser, 2002:257).

Normally, a genuine *ism* is recognizable in some or other "all/every" claim, such as "everything is number" (the Pythagorean school), or "[T]o hate, to love, to think, to feel, to see; all this is nothing but to perceive" (David Hume), or "everything is interpretation" (postmodernism), and so on.

Precisely because one may encounter within an aspect "moments-of-coherence" (analogical connections) with the other aspects, the temptation may be strong to succumb to the distorted view that all reality is "nothing but" this or that facet. These monistic fantasies are produced by human beings involved in theoretical thinking. Yet, as we argued (in connection with modal abstraction as the distinctive feature of scholarly scientific thinking), *if* reality does not contain *more than one* aspect, theoretical thinking (including analysis) as such will be impossible. Therefore every monistic ism, by virtue of its *theoretical* character, is a threat to its own claims! Because the "all/everything" claim is sometimes presented in a "nothing but" format, we may consider the argument that 'love' and 'thinking' are "nothing but" *instinct*. In this case, practically every capacity of the human being could be 'explained' by stating that it is *nothing but instinct*. Sorokin clearly saw through the *emptiness* of this 'all-claim', namely that *everything is instinct*, for he asserts that such a *reductionism* is not only *useless*, but that it is also *tautological* (Sorokin, 1928:607). If everything could be reduced to instinct, asserted in statements such as "love is nothing but instinct," then all scientific judgments will result in the meaningless tautology: *instinct is instinct*!

The inter-modal connections between the various aspects, expressed in the genuine modal analogies binding them together, serve as an important orientation-point for the determination of the *order of succession* between them. Discussing the relation between concept and word has already provided the opportunity to advance arguments for the foundational position of the logical-analytical aspect in respect of the sign mode. Expanding this line of argumentation will also challenge us to account for the order of succession between the other aspects.

5.5 The order of succession between the aspects

The spatial property of 'verticality' enabled us to understand that spatial relations presuppose the meaning of the quantitative aspect, for one or more dimensions as well as spatial magnitudes analogically reflect the original numerical meaning of quantity. The distinction applied in this explanation actually rests upon the fundamental distinction between *law side* and *factual side*. *Dimensionality* constitutes a quantitative analogy on the law side of the spatial aspect, for it highlights an *order of extension*, whereas *magnitude* correspondingly reveals a quantitative analogy at the factual side of the spatial aspect, because magnitude, which can be specified by a number, is attached to factual extension.

It should therefore not be surprising that Greek philosophy first explored the numerical point of entry to reality in its claim that everything *is* number. The eventual discovery of incommensurability (that is, irrational numbers) by

Hippasus of Metapont (about 450 B.C. – cf. Von Fritz, 1945) generated the idea that our intuition of space is more basic than that of number, since it can grasp continuity all at once.

An indication of the foundational position of the numerical aspect is given in the *infinite divisibility* of an extended continuum, such as a straight line – briefly elucidated in our discussion of the spatial aspect in Chapter 3 (see page 88 – a more detailed discussion of related problems is found on page 211). The intuition of one, another one, yet another one, and so on indefinitely, without end, i.e. *endlessly*, indeed constitutes the most basic ('primitive') awareness of *infinity*. In its most primitive numerical sense, infinity makes possible successive *spatial* divisions of factually extended spatial figures.

That both these aspects are *presupposed* within the structure of the kinematic aspect, can be seen from the thought experiment used by Galileo in 1638, where he contemplated the movement of a body on a *path* that is *extended* into *infinity* (see page 72). Clearly the terms 'path' and 'extended' designate the spatial foundation, while the term 'infinity' captures the numerical foundation of the kinematic aspect.

In connection with the relation between the kinematic and physical aspects we promised to provide an argument why *change* can only be detected on the basis of *constancy*.

In our discussion of the kinematic aspect, we mentioned that in his *Isagogè Philosophiae* from 1930, Vollenhoven had already distinguished between the *mechanical* and the *physical* (energetical) aspects, but in the 1936 edition, this distinction no longer appears (cf. Bril, 1973:216).[1] Although Dooyeweerd eventually (round about 1950) once again introduced this distinction, it is noteworthy that, as early as 1930, one finds a couple of instances where Dooyeweerd apparently also contemplated the distinction. In an article where he mentions different aspects, the distinction between the *kinematic* and the *physical* aspects is mentioned in passing. In order to appreciate his remark below, it should be kept in mind that his understanding of the *physical* aspect rests upon its core meaning of *energy*-operation. In his discussion of the *individual unity* of the subject functions of a tree, Dooyeweerd remarks: "Within the internal meaning structure of the tree the numerical, spatial, kinematic and energy-functions are *guided* by the organic function of life (the biotic function) although it takes place while completely maintaining the sphere-sovereignty of each function with its own lawfulness" (Dooyeweerd, 1930:249 – see also page 88 above).[2]

1 In 1968, Vollenhoven says that he has "no objection against the succession of the mechanical as the lower and the energetical as the higher" (Vollenhoven, 1968:3). He holds that, in addition to force (causing acceleration or deceleration), something else is at stake: a change of direction. See our remarks on page 88 ff.

2 "Zoo worden in de interne zinstructuur van den boom de getals-, ruimte-, bewegings- en energie-functies *geleid* door de organische levensfunctie (de biotische functie), doch met volledige handhaving van de souvereiniteit in eigen kring van iedere functie in eigen wetmatigheid."

The distinction between the kinematic and the physical (energetic) is fairly common in natural scientific circles. Max Planck, for example, sharply and correctly distinguishes between a 'mechanical' and an 'energetical' view of nature (Planck, 1973:65).[1] In a different context, Janich also draws a clear distinction between *phoronomic* and *dynamic* statements (Janich, 1975:68-69 – to which we shall return in Chapter 7).

The discovery of the irreducibility of the kinematic function of reality actually occurred in an *indirect* way, because it was concealed in Plato's intellectual struggle with the relationship between *constancy* and *change*. Early Greek philosophy has already challenged the awareness of the *persistence* of entities – the fact that entities *endure over time* – by arguing that, whatever there is, is in constant *flux* (Heraclitus, see Diels-Kranz, B Fr.90). With the intention of distinguishing between a law and what is governed by such a law Avey questions relativism: "There is, however, another aspect of Heraclitan philosophy which should not be ignored, and which relativist theory does not always find it convenient to emphasize. The *law* of change does not itself undergo change in the manner of the changing particulars" (Avey, 1929:521). It was Cratylus, a younger contemporary of Socrates and associate teacher of Plato, who caused the latter to come to a one-sided understanding of the view of Heraclitus. Cratylus interpreted Heraclitus to say that all *perceivable* things are in a process of *change* and that they are therefore *unknowable*. In his dialogue *Cratylus* Plato portays him as claiming that, since we cannot say "is not," we are also unable to say what is *false*. Cratylus recognized *change* in such a radical sense that he even questioned references to what is the 'same' – such as found in the alleged statement of Heraclitus that one cannot step into the *same* river twice: one cannot step into the same river even *once* (see Freeman, 1949:285). Apparently the implication of this radicalized position regarding *change* is that, since there is never something to hold on to, *identity* is impossible as well.

Behind this consequence, we find the school of Parmenides, denying *multiplicity* in its fundamental identification of *thought* and *being*. (Zeno's arguments against multiplicity and movement – Achilles and the Tortoise, the flying arrow, and so on – simply explore the basic position taken by Parmenides in his claim that thought and being are the same.) Thought can only think what is, because it cannot contemplate what does not exist. Veling remarks that Parmenides inspired later thinkers to engage in a rational search for "true reality" amidst all that is *changeful* (Veling, 2000:29).

In order to secure the *possibility of knowledge* Plato then postulated, as we have noted earlier (see page 105) the *enduring essential* (*ontic*) *being* of things (their static *eidos*) – which is supposed not to be subject to *change* (Cratylus, 439 c – 440 a). Without an awareness of *endurance* (*persistence*), the very notion of *change* becomes problematic, for the difficult question is

1 In passing, we can note that during the first half of the 19th century, the philosopher-theologian, Franz von Baader, already distinguished between a *mechanical* and *dynamic* perspective, but unfortunately he constantly identified this *dynamic* element – in line with the romantic spirit of his time – with the *organic* (see for example Von Baader, 1851:52 ff.).

then: 'what' changes? For example, as we briefly noted in Chapter 1 (see page 13 and the discussion on pages 426 ff.), an ageing person can only claim to be ageing because there is a constant reference to the *same* person (who is ageing). One can detect changes only on the basis of constancy – this is the valuable insight of Plato's theory of ideas. While we may distance ourselves from the speculative (metaphysical) construction of transcendent ideal forms (static essences), we still have to account for the brilliant insight that *change* rests on *constancy*. A similar metaphysical element is present in Frege's view, where he correctly points out that there are limits to historical change – just compare his reference to something *eternally* persistent: "If in the continuous flow of everything nothing firm and eternal persists, the knowability of the world would cease to be and everything will collapse into confusion".[1]

In Chapter 2, it was briefly pointed out that, since the development of Galileo's mechanics, classical physics (up to Heinrich Hertz close to the end of the 19th century) by and large attempted to understand all bodies in terms of the denominator of *mechanical movement* (see page 54). The law of inertia assumes the constancy of motion if no *physical forces* interact with this movement. Although there are different modes of explanation, their distinctness and irreducibility entail that there are *set limits* to their *explanatory power*. Not recognizing these limits leads to the (antinomic) position found in the various forms of *reductionism*. The task of science is not to evade unique (and irreducible functional) modes of explanation, but to *recognize* and *explore* them.

Certainly, in ancient Greece and during the period of medieval philosophy, the kinematic aspect ('motion') did not serve as an *original* principle of explanation – rather it was attempted to *explain* motion in other (modal, functional) terms, such as the well-known view that motion is "a change of place." Unfortunately, the term 'place' has its seat within the *spatial* aspect, whereas the term 'change' stems from the meaning of the *physical* aspect.

As soon as the kinematic mode is acknowledged, it is realized that asking about the *cause* of motion is just as meaningless as the classical opposition of *rest* and *motion*, for 'rest' and 'motion' are both "states of motion" (see Stafleu, 1987:58).[2] Alternatively, it will not help to consider 'space' and 'motion' as 'opposites', because two irreducible modal functions (such as the spatial and the kinematic) are never *opposites*. We made the same point in reaction to Lakoff, who mistakenly considers *discreteness* and *continuity* as 'opposites' (see page 146) instead of realizing that two mutually cohering but irreducible aspects are at stake. If motion is 'original' then it does not need to be explained and consequently can indeed serve as an "original mode of explana-

1 "Wenn in dem beständigen Flusse aller Dinge nichts Festes, Ewiges beharrte, würde die Erkennbarkeit der Welt aufhören und alles in Verwirrung stürtzen" (Frege, 1884:VII – *Einleitung*).

2 In a letter to Moritz Schlick (June 7, 1920) Einstein writes: "Rest is a dynamic event in which the velocities are constantly zero, one that for our consideration is, in principle, equivalent to any other event of motion" (Einstein, 2006:186).

tion" – at once eliminating the necessity of searching for the *cause* of motion (for the term 'cause' has its seat in the physical aspect). The only meaningful alternative is to investigate the cause of a *change* in motion (albeit acceleration or deceleration). But then the kinematic mode of explanation is already left behind, for there are physical forces required in order to explain acceleration and deceleration.

The explanatory power of a specific mode of explanation is an expression of what we have designated as its modal universality (see pages 68 ff.). Before the modal universality of the physical aspect was properly understood, the belief was that celestial entities obey laws that are different from those holding for terrestrial entities. The idea of an attractive force – initially contemplated in relation to magnetism – eventually brought Newton to the insight that magnetism is a force not to be explained by *motion*, since motion should rather be applied as a *new principle of explanation*.

The application of this principle of explanation was so successful that we had to note that the initial mechanistic orientation of modern physics continued to dominate practically until the end of the 19th century (we shall discuss this issue more extensively on pages 414 ff. – see Planck, 1973:53).

Similarly the *physical* mode also emerged as a principle of explanation *different from* that of motion. Stafleu explains that the rejection of the Aristotelian distinction between celestial and terrestrial physics, in the case of Newton (following the ideas of Galileo and Descartes), resulted in the assumption that, the same laws apply to both realms (Stafleu, 1987:73). He also points out that just like Kepler, Newton viewed *force* as a "principle of explanation independent of motion" (1987:76), and then summarizes this process in which the physical mode surfaced as *a distinct mode of explanation* as follows:

> In Newtonian mechanics, a force is considered a relation between two bodies, irreducible to other relations like quantity of matter, spatial distance, or relative motion. Though an actual force may partly depend on mass or spatial distance, as is the case with gravitational force, or on relative motion, as is the case with friction, a force is conceptually different from numerical, spatial or kinematic relations (Stafleu, 1987:79).

It is therefore understandable that 20th century physics eventually had to acknowledge that energy-operation is decisive within the material world – aptly captured in Einstein's well-known equation: $E = mc^2$.

Acknowledging the foundational coherence between the kinematic and physical aspects opens up a new perspective of what used to be treated as a mere opposition: *constancy* and *change*. We accredited Plato with the discovery of this momentous insight in spite of the fact that we no longer have to accept his own metaphysical account in terms of his supposedly *eternal* and *immutable* world of transcendent ideal forms. If it is not realized that the terms *constancy* and *change* derive from *different* (irreducibile) modal functions, then the relation between them may turn into a serious problem.

For example, when De Saussure says that time "insures the continuity of language," he considers two "apparently contradictory" influences, for according to him, in "a certain sense, therefore, we can speak of both the immutability and the mutability of the sign" (De Saussure, 1966:74). The editors of this work was uncomfortable with the statement and therefore added the following explanatory footnote: "It would be wrong to reproach F. de Saussure for being illogical or paradoxical in attributing two contradictory qualities to language. By opposing two striking terms, he wanted only to emphasize the fact that language changes in spite of the inability of speakers to change it. One can also say that it is intangible but not unchangeable. [Ed.]"

Remarkably enough just in the following paragraph, De Saussure rephrases his point in such a way that a correct understanding of constancy (continuity) and change is articulated:

> In the last analysis, the two facts are interdependent: the sign is exposed to alteration because it perpetuates itself. What predominates in all change is the persistence of the old substance; disregard for the past is only relative. That is why the principle of change is based on the principle of continuity (De Saussure, 1966:74).

Saying that "the principle of change is based on the principle of continuity" is exactly what we have in mind with the statement that *change* can only be detected on the basis of something lasting, i.e. on the basis of *constancy* (continuity).

Thus far we have formulated arguments to substantiate the order of succession of the first four modal aspects, namely the quantitative, the spatial, the kinematic and the physical. That the biotical aspect succeeds the physical is easily seen as soon as the fact is contemplated that all living entities display their biotic functions only on the *basis* of their physico-chemical constituents. This foundational interconnection provides the basis for the disciplines of biochemistry and biophysics, for they are interested in the *biotically directed functions* of the physical constituents of living entities, and, at least in principle, therefore,[1] should not merely investigate the macromolecular *structures* of the complex conformations found in living entities.

The foundational role of the physical aspect in respect of the biotical function of reality implicitly challenges the most basic claim of vitalistic and neo-vitalistic trends in modern biology, because the claim that there is something 'immaterial' responsible for the uniqueness of vital phenomena is contradicted when this "immaterial entelechie" is designated as a "vital force." In displaying a constitutive interconnection between the biotical and the physical aspects this expression cannot capture an "immaterial reality" simply because the term 'force' analogically reflects the meaning of the foundational physical aspect.

Although the biotical meaning of being *alive*, getting *ill* and eventually *dying* also expresses itself within the physical aspect of reality, the latter cannot in principle capture the unique (and irreducible) meaning of the biotic aspect. For as soon as the *biotical function* of living things is acknowledged, mutually

1 In practice, biochemists actually do what ought to be done by organic chemistry.

enriching modes of explanation are available. For example, there is nothing wrong in asserting that a living thing, considered from the biotical mode of explanation, finds itself in a *stable state* (referred to as *health*), while simultaneously claiming – without any contradiction – that observed from a physico-chemical mode of explanation (with a view to the flowing equilibrium of its physical-chemical constituents – the so-called steady state), it finds itself in an unstable condition. If the physical-chemical basis (substratum) of living things approaches a state of higher statistical probability, the increase of *biotical instability* anticipates the ultimate biotic process of *dying*. Von Bertalanffy has a clear understanding of the shortcomings in all reductionistic attempts aimed at an *elimination* of the *biotic aspect* of reality when it comes to an appreciation of the vital functioning (vital processes) of living entities:

> These processes, it is true, are different in a living, sick or dead dog; but the laws of physics do not tell a difference, they are not interested in whether dogs are alive or dead. This remains the same even if we take into account the latest results of molecular biology. One DNA molecule, protein, enzyme or hormonal process is as good as another; each is determined by physical and chemical laws, none is better, healthier or more normal than the other (Von Bertalanffy, 1973:146).

The absence of a nervous system in plants shows that animals, as *sentient* creatures, are characterized by the sensitive psychic mode of reality. This also supports the insight that the latter aspect succeeds the biotical aspect within the cosmic order of aspects. Within the emotional aspect consciousness comes to expression for the first time and – in the case of human beings – consciousness (also functioning at the deep level of the *unconsciousness*), lies at the *foundation* of all normative functions.

We have already advanced a number of arguments in favor of the view that the logical-analytical aspect is the first *normative* aspect, and that it is foundational to the sign mode. We also argued that concepts are logical configurations and that the logical nature of concepts is also foundational to the cultural historical mode because the distinction between *means* and *ends* inherent in culturally formative activities presupposes the analytical ability to identify and distinguish means and ends.

Since we are accustomed to speak of the *formation* of concepts, the question may be whether or not the historical aspect is not actually foundational to the logical mode? In Figure 5 (see page 151), *identification* is explained as the synthesis of those features united in a concept. The unifying nature of this *activity* of identification of course functions within other aspects of reality as well – notably the historical and sign modes, which explains why we are often tempted to sell out concepts to the formation of the words designating (or implying) them. There are many ways in which words can be shaped and employed in order to *designate* what has been united in a concept, but there is just one (unique) concept underlying all these possible lingual 'assignments' – once more highlighting the foundational nature of the logical aspect. When Derrida applies the mutuality of identity and difference to language, he implicitly underscores this foundational position of analysis vis-à-vis language: "The identity of a language can only affirm itself as identity to itself by opening itself to the hospitality of a difference from itself or of a difference with itself" (Derrida, 1993:10).

A concept comes into existence through a *logical* act of *identification*, and always entails a claim to *universality*. The concept of a triangle, for example, if it is a *proper* concept, applies to any and all triangles – wherever and whenever.

Within every post-logical aspect, the foundational role of the logical aspect is also reflected by the earlier-mentioned *contraries* (such as historical – un-historical, polite – impolite, and so on).[1] In addition, every normative *discernment* always rests on the functional basis of the analytic mode. From the perspective of those aspects themselves, the analytical analogy successively acquires a different meaning. The *nuancefulness* (many-sidedness) displayed by signs (and their lingual usages) is responsible for the peculiar *semantic domains* of words in a specific language. The free formative (human) imagination embodied in language causes similar words in different languages to develop different semantic domains, and this reality is one of the main difficulties involved in translation. The important thing for language is that any given word, with its own semantic domain, is always dependent upon other words 'surrounding' that word, owing to *lingual coherence* making possible language as a *sign system* (spatial analogy).

De Saussure even went so far as to deny the presence of *similarities* in language, for he claims that "in language there are only differences" (De Saussure, 1966:120). What is at stake in this view is his conviction that "[L]anguage is a system of interdependent terms in which the value of each term results solely from the simultaneous presence of the others" – this intrinsic coherence within language as a system opposes the view that "language could be reduced to a simple naming-process" (De Saussure, 1966:114). The latter may appear within the sign mode when it functions in an *undisclosed* way, i.e. not yet opened up towards the modal function succeeding it within the cosmic order, namely the social aspect. In its core meaning, the sign mode is concerned with the as-*sign*-ment of meaning – but only when this assignment is *shared* by a linguistic community (i.e. when the sign mode is deepened under the guidance of the social aspect), do we meet *language* proper.

On the one hand, this shows that the social aspect is founded in the sign mode, and on the other, we can infer that it comes before the economic function since the relative scarcity of economic means always reveals itself within an inter-human context – making possible the concept of economic value (eventually captured by the term 'price').

Avoiding excess in an aesthetic sense, furthermore, shows that the economic aspect is foundational to the aesthetic mode. The jural aspect, in turn,

1 To be more specific: contraries in all post-logical aspects are analogies of the logical principle of non-contradiction.

finds its immediate foundation in the aesthetic mode, because the jural meaning of law in a public legal sense (bound to the typical nature of the state) rests on the basis of a *harmonization* of a multiplicity of legal interests within the territory of the state. Since the quantitative analogy in the jural aspect is given in the nature of a legal order (a unity in the multiplicity of legal norms), and since the legal system is integrated into a ju(ridic)al whole with its jural parts (the spatial analogy), it must be clear that the term 'harmonization' cannot be seen as an analogical reflection of the aspects of number and space within the jural aspect. The only meaningful seat of this analogical term is given within the aesthetical aspect.

Moreover, the jural aspect itself precedes the moral aspect, because those kinds of moral wrongs familiar to human societies, such as theft, murder, etc., always inherently bring with them the (analogical jural) element of *unlawfulness*. The way in which an unethical action differs from an unlawful deed will be explained when we investigate the relationship between *law* and *morality* (see pages 274 ff.).

Finally, the certitudinal aspect, as the limiting aspect of the modal cosmic order, finds its immediate modal foundation in the ethical aspect. The modal structure of the faith aspect presupposes all the foundational cosmic modalities, and at the same time it points beyond them to what is believed to be the ultimate origin and destination of the universe.

5.6 Primitive terms

Since the modal aspects are fitted into a cosmic order such that some are foundational to others, it is clear that this only makes sense if the aspects are irreducible. But because this latter trait crucially depends upon the core meaning of every aspect that transcends conceptual definition, some examples of the indefinability of the meaning nucleus of an aspect will enhance our understanding of aspectual uniqueness, and at the same time it will provide a deeper insight into the fact that conceptual knowledge rests upon a basic cosmic diversity that ultimately cannot be comprehended conceptually.

The acknowledgement of "primitive terms" highlights the one element of the frequently mentioned problem regarding the coherence of irreducibles. Gödel understood this issue in his own way. Yourgrau explains that he "insisted that to know the primitive concepts, one must not only understand their relationships to the other primitives but must grasp them on their own, by a kind of 'intuition' " (Yourgrau, 2005:169). On the next page he adds that "the fundamental concepts are primitive and their meaning is not exhausted by their relationships to other concepts." This is as close one can get to an idea of the (modal) principles of sphere-sovereignty and sphere-universality!

5.7 Multiplicity and meaning as primitive terms

Thinking about the nature and role of concepts soon borders on the *limits* of rationality. Various disciplines in the course of their development had to realize (and concede) that the truly basic terms of their disciplines are conceptually *indefinable*.

Semantics, as a sub-discipline of general linguistics, had to accept 'meaning' as such *a primitive term*. For example, when the distinction by Immanuel Kant between analytic and synthetic propositions (cf. Kant 1787:10 ff.) is pursued, an attempt can be made to *define* a typical semantic phenomenon such as *synonymy* in terms of *analyticity*. Two sentences have the same meaning only if each one of them entails the other one in an analytic sense. Yet Quine highlighted the *circularity* of such an attempt. *Analyticity* is defined in terms of *meaning* (a sentence is supposed to be analytically true if it is true only on the basis of its meaning), whereas *meaning* (in this case: similarity of meaning = synonymy) is defined in terms of *analyticity*.[1]

Similar to the way in which linguistics had to accept (assume) 'meaning' as something *basic* and *primitive*, axiomatic set theory also had to accept *primitive terms*. For example, within the Zermelo-Fraenkel set theory, "*member of*" is introduced as a *primitive term*[2] – and Gödel once remarked that as yet we do not have a satisfactory *non-circular definition* of the term 'set'.

> The operation 'set of x's' (where the variable 'x' ranges over some given kind of objects) cannot be defined satisfactorily (at least not in the present state of knowledge), but can only be paraphrased by other expressions involving again the concept of set, such as: 'multitude of x's', 'combination of any number of x's', 'part of the totality of x's', where a 'multitude' ('combination', 'part') is conceived as something that exists in itself, no matter whether we can define it in a finite number of words (so that random sets are not excluded) (Gödel, 1964:262).

Russell, in his logicistic approach to mathematics, claims that his *class concept* is purely *logical* in nature, without realizing the *circularity* entailed in his argumentation.

> 1 + 1 is the number of a class *w* which is the logical sum of two classes *u* and *v* which have no common terms and have each only one term. The chief point to be observed is, that logical addition of numbers is the fundamental notion, while arithmetical addition of numbers is wholly subsequent (Russell, 1956:119).

Russell speaks about the sum of 'two' classes, where each of them contains 'one' element. This presupposes an insight into the quantitative meaning of the numbers '1' and '2'! Consequently, the number '2', which had to appear as the result of "logical addition," is presupposed by it! In the *Introduction* to this work, Russell says that Hilbert's formalism (by leaving the integers undefined, only having those properties enumerated in the axioms) "have forgotten that numbers are needed, not only for doing sums, but for counting" (Russell, 1956:v-vi). On page 119, it seems as if he himself "has forgotten" what he accused formalism of having 'forgotten'!

1 Recall that Fodor said that "there is no meaning-independent way of characterizing either *analyticity* or *meaning* (Fodor, 1977:43) – see the quotation on page 147 above.

2 In addition to primitive symbols taken from logic, the only set theoretical primitive symbol employed by Zermelo-Fraenkel's set theory is the binary predicate *epsilon*, which denotes the membership relation (cf. Fraenkel *et al.*, 1973:22-23).

David Hilbert points to the circularity entailed in the *logicist* attempt to deduce the quantitative meaning of number from that of the logical-analytical mode. In his *Gesammelte Abhandlungen*, he writes:

> Only when we analyze attentively do we realize that in presenting the laws of logic we already have had to employ certain arithmetical basic concepts, for example the concept of a set and partially also the concept of number, particularly as cardinal number [Anzahl]. Here we end up in a vicious circle and in order to avoid paradoxes it is necessary to come to a partially simultaneous development of the laws of logic and arithmetic (Hilbert, 1970:199; see also Quine, 1970:88).

Singh also states that Russell's attempt makes him a victim of the "vicious circle principle" (Singh, 1985:76) and the same criticism is raised by Cassirer: "Now if we take this process of positing *(Setzung)* and differentiation as a basis, we have done nothing but presuppose number in the sense of the ordinal theory" (Cassirer, 1953:51).

5.7.1 *The distinction between antinomy and contradiction: a provisional account*

We may use another example of the necessity of acknowledging primitive (indefinable) terms, because it demonstrates at once the important difference between a (intra-modal) logical contradiction and an (inter-modal) antinomy. It is found in the briefly-mentioned arguments of Zeno against multiplicity and movement (see page 2). The solution to Zeno's problem of Achilles and the tortoise is not merely given in the claim that Zeno understood the metaphor of the "moving observer" in a *literal* way, as claimed by Lakoff and Johnson (see Lakoff & Johnson, 1999:157-158), since what is ultimately shown by this 'antinomy', is that it is impossible to define *uniform motion* exhaustively (or: exlusively) in *spatial terms*.[1]

When a moving body is, at every moment of its 'movement', in one specific place, it after all is at rest, since "being in one place" simply means "not being in motion." Interestingly, Henk Hart here speaks of a 'paradox' instead of an *antinomy*. Where he explains the relation between *contradiction* and *antinomy* (cf. Hart, 1984:132-133), Hart fails to make this distinction absolutely clear: "A contradiction is illogical or anti-logical, but an antinomy cannot be resolved by logic."

What he needed to say in addition, is that an antinomy always implies a logical contradiction, but that there are many contradictions not presupposing an antinomy. We shall return to this distinction in more detail in Chapter 6 (see pages 262 ff.).

5.8 Primitive meaning: between pan-vitalism and pan-mechanism

Similarly, the traditionally dominant *mechanistic* and *physicalistic* trends in

1 The third B Fragment preserved from Zeno, quoted in Chapter 1 above (see page 2), first grants the reality of movement, but then immediately cancels it again by raising two options denying it: "Something moving neither moves in the space it occupies, nor in the space it does not occupy."

modern biology are not sensitive to the *indefinability* of the *biotic mode* (aspect or function) of reality. Vitalistic, holistic and organismic theories (such as those of Von Bertalanffy) indeed suffer from this shortcoming. (Those biologists who further explored the neo-vitalism of Hans Driesch had to adjust their orientation after Von Bertalanffy generalized the second main law of thermodynamics to cover *open systems* as well – compare the way in which Rainer Schubert-Soldern introduced an "instability factor" in order to account for the health (biotical stability!) of living entities (see Schubert-Soldern, 1959 and 1962).) It should not be surprising that later representatives of neo-vitalism ceased to use the expression "vital force" so dominant in vitalistic thought since Aristotle introduced the notion of an 'entelechie'. The latter was supposed to be *immaterial*, contradicting the term 'force' used in the expression *vital force*. Heitler, for example, simply prefers to refer to a "central instance" (*Zentralinstanz*) (Heitler, 1976:6).

Hans Jonas strikingly typifies the *monistic* forms of *vitalism* and *mechanicism*. Unlike dualists, *monists* do not attempt to reduce reality philosophically to *two* fundamental principles, but rather posit a *single all-inclusive* and *universally explanatory principle*. We may therefore just as well speak of *pan-vitalism* and *pan-mechanicism*.

Whereas biological *physicalism* (with the neo-Darwinian theory of evolution as one of its dominant representatives) aims at *reducing* the biotical aspect of reality to complex material structures and processes, biological *holism* aims at the opposite extreme. Needham gives the following explanation of the position taken by Meyer:

> Thus Meyer, in his interesting discussion of the concept of wholeness, maintains that the fundamental conceptions of physics ought to be deducible from the fundamental conceptions of biology; the latter not being reducible to the former. Thus entropy would be, as it were, a special case of biological disorganization; the uncertainty principle would follow from the psycho-physical relation; and the principle of relativity would be derivable from the relation between organism and environment (Needham, 1968:27, note 34).

Whenever an attempt is made to define what is truly *primitive* (and irreducible), the inevitable outcome is (*antinomic*) **reduction**.[1]

5.9 An example: the meaning of the jural

Within the science of law, the Dutch legal scholar Polak attempted to define 'law' as follows: "retribution is an objective, trans-egoistic, harmonization of interests." The term 'objective' here has the meaning of *not being limited to any particular person*, i.e. it is employed in the sense of what is *universal*. Does it say anything distinctive about the jural? Not at all, because every aspect of reality shares in this feature of *universality*. The term "trans-egoistic" derives its meaning from *moral love*, where it is demanded that a person ought not to be self-centered. Therefore this element of the given definition also does not at all touch on the *jural* meaning of retribution. 'Harmonization' rep-

1 Historicism demonstrates this claim in a very lucid way – the *cul de sac* of historicism is explained below (see pages 271 ff. below).

resents the meaning of the *aesthetic* aspect and therefore fails to capture the core meaning of the jural. Finally, the term 'interests' is multivocal – people may have economic interests, aesthetic interests, legal interests, and so on. This means that *interests* cannot reveal the unique or core meaning of the jural, because it can take on many different qualifications. What Polak has done, is to use non-jural terms in his attempted definition of the jural, which resulted in a formulation totally bypassing the core meaning of the jural. In his (as yet still unpublished *Encyclopedia of the Science of Law*, Vol. III:10-11), Dooyeweerd justifiably concludes: "The result is a general concept fully lacking any delimitation. It could just as well be seen as a moral rule regarding the distribution of alms." The meaning nucleus of an aspect is not only unique and irreducible, but also *indefinable*.

5.10 Implications for rationality

In general, we can therefore conclude that concept and definition ultimately rest upon the acceptance and employment of *primitive terms*. Therefore respecting what is indefinable is the only option if a *regressus in infinitum* is to be avoided in the *theory of knowledge* (*epistemology*).

What we learn from these examples is that the *key terms* involved in a *rational* (*conceptual*) *understanding* themselves are *not* open to (rational) conceptual definition. Rationality in this sense does rest upon a non-rational (or: *more than rational*) basis – but it should not be confused with something *irrational*. We may designate this basis, given in irreducible primitives, as the *restrictive boundary* of rationality. As such, it reflects a *positive* awareness of what we frequntly called one of the most fundamental perennial issues in philosophy, namely the quest to account for the *coherence of irreducibles* (and: '*indefinables*' – see pages 7, 13, 52, 157).

5.11 Rationality: the legacy of an over-estimation of conceptual knowledge

However, unable to oversee its own commitment to reason, modern philosophy pursued the pretension of "universal reason," which led to the modernist Enlightenment conviction that the *world itself* has a "rational structure." In order to elucidate this legacy, closer attention to the nature of concept formation is required.[1]

The legacy of Western philosophical reflection highlights that concept formation is made possible by *two* crucial conditions (see page 148): (i) *universality* and (ii) the capacity to *bring together (identify = synthesize) a logically objectified multiplicity* of traits into the *unity* of a *concept*.

A concept is not simply a *picture* or *image* of something within reality, or of some features of reality. In order to *discern* and *identify*, given possibilities of reality are analytically opened up and deepened through the logical acts of a knowing subject. Opening up the "identifiability" and "distinguishability" of features of reality is only possible through (subjective) acts of *objectification*.

1 Just recall the statement that the world is *rational* and even has a *mathematical* structure (see page 90).

Through logical objectification, concepts emerge as the *unification* (bringing together, synthesizing) of a multiplicity of *universal* traits. The scope of a concept is universal in the sense that it applies to whatever conforms to the *conditions* intended by the concept concerned. A proper concept of a planet, a house, a chair, a human being, and so on makes it possible to place any individual planet, house, chair or human being within the (universal) category of planets, houses, chairs or human beings.

Karl Mannheim accounts for this insight of the universality of concepts in his own way:

> Everything subject to assertion is to be identical for everyone in every assertion of it: and the concept thus [must be] universally valid in two ways: referable to all objects of the same kind (the concept 'table' is thus applicable to all tables that have ever existed or ever will exist), and valid for all subjects who ever will utter it, and who accordingly always understand the same thing by 'table'. That this tendency inheres in every concept formation cannot be doubted; and the creation of such a conceptual plane upon which one concept can be defined by others, with all concepts thereby forming an objective self-contained system, should not be denied (Mannheim, 1982:196).

What is known as Kant's so-called 'Copernican' revolution in epistemology – in no longer ascribing the primacy to the 'object' but to the (formal law-giving) subject[1] – reinforced the notion of things within nature as 'objects'. Someone inclined to defend the neutrality of observation normally would be willing to accept as the most general observation-term the notion of an 'object': all the different things in nature are to be seen as 'objects'. However, this observation-term in itself displays the tremendous *subjectivistic* assumption so deeply impregnating our Western notion of science – as such resulting in an inability to appraise things in nature as genuine *subjects*, i.e. as being (in addition to other kinds of laws – such as quantitative laws, spatial laws and kinematic laws of motion) subjected to *physical* laws for their existence as (subjectively functioning) material things.

Insofar as physical entities are *material*, they are not objects but *subjects* (subject to physical laws), and insofar as they are *objects*, they hold this status because they are considered according to some or other *non-physical* trait – for example, as something perceived (sense-object), as something analyzed (identified and distinguished from something else – logical-analytical object), as something bought or sold (economic object), and so on. Therefore, although such things could be *objectified* by humans, this objectification presupposes their primary existence as (physical) subjects. Speaking of them in all possible contexts as *objects* simply underscores the powerful *subjectivistic*

1 One only has to recall Kant's conviction that human understanding is the formal law-giver of nature – according to him, it does not derive its laws *from* nature, but prescribes them *to* nature: "understanding creates its laws (a priori) not out of nature, but prescribes them to nature" (cf. Kant, 1783:320, § 36 – see below pages346, 382, 382, and 421 where this view is discussed in different contexts).

(human-centred) legacy in Western thinking.[1]

Logical objectification, as a subjective analytical act of identification and distinguishing, does not leave 'reality' untouched – it *opens up* and *deepens* its meaning. Yet concept formation does not *exhaust* the meaning of *knowledge*. The mistaken assumption that knowledge coincides fully with *conceptual* knowledge, which dates back to Aristotle and Greek metaphysics, can best be designated as *rationalistic*. *Rationalism* accentuates universality at the cost of individuality – and thereby simply dismisses the knowledge we have of whatever are unique, contingent and individual. Concepts are *blind* to the unique, contingent and individual – yet it cannot be denied that we do have *knowledge* of what is unique, contingent and individual.

5.12 Concept-transcending knowledge

The kind of knowledge involved in approximating what is *unique*, *contingent* and *individual* transcends the limits of *concepts* and should be acknowledged for what it is: *concept-transcending knowledge*. Because knowing what is *individual* can never escape from terms that have a universal scope, their idea-usages are always merely *approximating*. Nicolai Hartmann once explained the Kantian notion of a 'Grenzbegriff' in a striking way. He says that the notion of an unknowable "thing-in-itself" ("Ding an sich") still requires a *thought-form* through which it is regarded as *unknowable* – and this is what a "Grenzbegriff" intends to capture.[2] Without buying into the role of the so-called "thing-in-itself" in the philosophy of Kant (cf. the critical remarks made in Strauss, 1982:133, 141-143), it is important to leave room for a "form-of-thinking," accounting for knowledge transcending the limits of concept formation.

Comment:

> The translational equivalent usually given for "Grenzbegriff" is *limiting concept*. This is misleading, because it gives the false impression that we deal with knowledge contained within certain limits or confines, whereas the actual intention of the German term is to refer to a kind of knowledge *transcending* the limits of concept formation. For this reason, we prefer to speak of "concept-transcending knowledge."

In Chapter 1 (see page 18), we briefly mentioned that the basic notion with which Aristotle started his work on *Categories* is that of the *primary substance* – which was supposed to be strictly *individual*. However, in order to safeguard the possibility of *conceptual knowledge*, Aristotle had to introduce the so-called *secondary substance*, the *to ti èn einai* (*De Anima*, 412 b 16; cf.

1 Authentic artifacts are designated as *cultural objects*, because their objectified status in human life is intended – but that does not remove their *subject-functions* within the pre-physical and physical aspects of reality.

2 Cf. Hartmann, 1957:311: "Denn bei Kant ist es nicht so, dass etwa das Ding an sich bloss Idee wäre; umgekehrt, da wir das Ding an sich nicht erkennen ..., wohl aber denken können, so muss es eine Denkform, eine Art des Begriffs geben in der es – eben als unerkennbares – gedacht wird. Das ist die 'Idee'."

414 a 9-11 and *Metaph.* 1035 b 32).[1] According to Aristotle, a concept is always involved in conceiving what is general (universal).[2] For Aristotle, true knowledge is therefore always knowledge of the *universal form*. The counter-pole of form, namely *matter* (that lacks any positive determination – cf. *Metaph.* 1029 a 20-26), is therefore outside the reach of conceptualization. As such it is *unknowable*.[3]

The universal substantial form in Aristotle's thought is intended to make the transcendent *eidè* of Plato immanent – inhering within concretely existing substances. It relates to *being-this* or *being-that*. For example, Aristotle holds that a concept (*logos*) is not subject to *coming into being* and *passing away*. It is not 'house-ness' that comes into being, but only *this* house (*Metaph.* 1039 b 22-26).

These two stances, Plato with his transcendent *eidè* and Aristotle with his immanent universal substantial forms (secondary substances), informed the medieval speculation about the *universalia ante rem* (transcendent, seated in the "divine Mind" *a la* Plato) and *in re* (inhering within the created entities as their universal forms – *a la* Aristotle). The additional step in such realistic metaphysics is given in the postulation of *universality* within the human mind, *universalia post rem*. It also underlies the realistic (copy) theory of truth: truth is the correspondence between thought and being (*adequatio intellectus et rei*).

Aristotle actually discovered the (universal) *orderliness* (*lawfulness*) of entities within reality, whereas Plato wrestled with the (universal) *order for* concretely existing entities. The conditions for being an atom are given as the *order for* the existence of any given atom – the latter evinces that it is subject to these conditions (to this *law for* being-an-atom) in its *orderliness*, i.e. in its *being-an-atom* (an element of this insight was explained in Chapter 1, page 20 – we shall return to it again in our discussion of nominalism, pages 370 ff.).

However, during and after the Renaissance, the late Scholastic nominalistic movement (John the Scott, William of Occam) radically questioned Plato's *eidè*, as well as Aristotle's universal substantial forms. The nominalistic orientation actually denies both the order for and the orderliness of entities and accepts universality only within the human mind – as it is clearly seen from the new concept of *truth* it employs: truth no longer relates to a reality outside the human mind, but only concerns the *compatibility of concepts within the human mind*. However, the remarkable fact is that both the traditional Scholastic metaphysics and the modern nominalistic reaction to it continue to adhere to the rationalistic restriction of *knowledge* to what is *universal*! Even Scholasticism was faithful to the conviction that whatever is individual is *inexpressible* (*omne individuum est ineffabile*). It will suffice to quote a well-informed and

1 An extensive analysis of Aristotle's view of a *concept* is found in Prantl (1855:210-362). Prantl translates *ὅρος* with *concept*, *λόγος* with *articulated concept* and *ὁρισμός* with *definition* (Prantl, 1855:211 note 359; cf. 262 note 535). Also compare Prantl (1851:38 ff.).

2 *ὁ δὲ λόγος ἐστὶ τοῦ καθόλου* (*Metaph.* 1035 b 34 – 1036 a 1).

3 *ὕλη ἄγνωστός καθ' αὑτήν* (*Metaph.* 1036 a 8-9).

respected 20th century philosopher from South Africa in order to illustrate the long-standing influence of this rationalistic restriction of knowledge to *conceptual knowledge*. De Vleeschauwer wrote a work on logic and epistemology and explicitly points to the "individual delimitation" (De Vleeschauwer, 1952:213). He illustrates that the domain of the 'individual' is one where our intellectual capacities must fail. Apparently without being aware of it, De Vleeschauwer adheres to the *nominalistic* perspective, which holds that in reality, there are only *individual* things and processes, clearly seen in his explanation that, in spite of all similarities between entities and processes, there will always remain an irreducible kernel of individuality, which causes one thing to be different from another. Science, with its directedness towards the universal has serious difficulties with its inclination to know what is individual – because "knowledge of what is individual is simply impossible" – something about which philosophy, according to De Vleeschauwer, had clarity since its inception (De Vleeschauwer, 1952:213).

Given the awareness emphasized by many contemporary thinkers, namely that we cannot deny the historical and linguistic conditions of being human, one may expect philosophers to be skeptical or at least critical in respect of the acknowledgement of *universality*. As a stance against the long-standing domination of rationalistic notions, this reaction is understandable. Yet, as we have noted in Chapters 2 and 3 (see page 57 and pages 74 ff.), one should not throw out the baby with the bath water, for acknowledging *historicity* and *linguisticality* does not exclude, but rather presuppose *logicality*. Those (postmodern) thinkers who want to deny universality simply continue to reject universality in the name of universality.

What is required is not a denial of the *co-determining* role of *logicality* alongside historicity and linguisticality, but an acknowledgement of the inevitable and inescapable role also played by the *logical-analytical dimension* of being human, accompanied by the critical insight that one should not succumb to the rationalistic restriction of knowledge to (universal) conceptual knowledge.

At this point, we may link our initial analysis of "primitive terms" with the nature of *concept-transcending knowledge*.

The original ontic domains of "primitive-ness" in which irreducible basic terms find their seat, have a further fundamental and complementary side to them. In order to highlight this, we refer to conceptual clusters and the different *ontic domains* to which this complementary side relates (cf. Strauss, 2002:170).

Comment: A rationalistic legacy in everyday language

The use of the term 'ontic' instead of 'ontological' is grounded in the crucial distinction between what is given and what is the result of reflection on what is given. Living entities, such as plants, animals and human beings, are not *biological* phenomena, as we are accustomed to hear. They are simply *alive* and therefore at most the *datum* with which the discipline of biology may concern

> itself. Similarly, a young couple taking a romantic walk on a campus is not a *sociological* phenomenon, but simply a *social* one. The rationalistic legacy, which identifies our ability to know with reality itself – evidenced in the mentioned belief in the "rational (mathematical) structure" of reality – is responsible for this identification of different kinds of phenomena with theoretical thinking about them, thus leading to the practice of employing words with the suffix '-logy' where they do not fit – like *ontological* instead of *ontic*, *biological* instead of *biotic* and *sociological* instead of *social*.

It will suffice to use the following domains: (i) the arithmetical domain, (ii) the spatial domain, (iii) the kinematic domain, and (iv) the physical domain, while keeping in mind that the concrete existence of no single entity, process or societal collectivity is exhausted by anyone of these (or other) *ontic* spheres or aspects of reality. For example, the existence of a chair is not exhausted by its function in the *quantitative* facet of reality – evident in our discernment of its number of legs – since it also exists as a *spatially* extended entity with its relative *movement* (around the axis of the earth, around the sun) and with its relative physical *strength* (suitable for a normal human being to sit on).

Applying our intuitions of number, space, movement and energy-operation in talking of the quantitative properties of a cultural entity (like a chair), of its size and shape (spatial), of its relative speed and of its typical physical characteristics, in every instance inevitably employs notions or terms with a *universal* scope. This means that, whenever any person looks through these different (ontic) points of entry (which are then at once elevated to epistemic *modes of explanation*) at a chair, the terms generated are used in a *conceptual way*. As long as we restrict the use of such terms to the respective ontic domains (modes of explanation), this conceptual focus will always be present – which is actually the case with all our entitary-oriented everyday concepts (just think about the concepts we have of entities such as planets, houses, chairs and human beings).

If we designate such *functional terms* employed in describing the way in which entities function within various aspects of reality as *modal terms* (see Strauss, 2000:26-28, 32-36),[1] then the following distinction should be drawn. When *modal terms* are used to refer to entities that function *within the confines* of particular modes of being, they are employed in a *conceptual manner*. However, whenever a modal term is put in service of referring beyond the limits or boundaries of such an ontic domain, we encounter a *concept-transcending* use of such a term – also designated as an *idea use* of such terms.

For example, while merely exploring our *quantitative intuition*, one can speak of a chair in its totality, including all its properties. In language, this is expressed by referring to its *individuality*, its *uniqueness*, its *being distinct*.

1 Further examples of modal (functional) terms are: quantity, unity, multiplicity (arithmetical); area, coherence, continuity, connectedness (spatial); uniform flow, continuing, constancy (kinematic); life, growth, adaptation, differentiation, integration (biotical); frugality, sparingly, avoiding excesses (economic), and so on. Functional terms such as these relate to the *how* of reality and not to its concrete *what* (see Strauss, 2003:69-74).

The original quantitative meaning of number – captured as a 'primitive' in axiomatic set theory (as we have seen)[1] – is evident in these affirmations, and yet they are intended to refer to much more than the mere arithmetical aspect of the chair. They therefore indeed constitute *idea usages* of modal *numerical* terms. Dengerink stumbled upon this insight without being able to articulate it properly, owing to the absence of our distinction between a conceptual and a concept-transcending use of modal terms. With reference to the quantitative aspect and its analogies within other aspects he adds that it also functions (just like all other aspects) up to the heart of reality ("tot in het hart van de werkelijkheid"), explaining why he alluded to the (central) *unity* of the cosmos ["de (centrale) eenheid van de kosmos"]. He realizes that this central unity is not a mathematical point although it cannot be divorced from the original meaning of number. The next step, not taken by Dengerink, would have been to distinguish between a conceptual use and an idea-use of modal terms. In his final explanation in this context he comes even closer to this view when he explicitly alludes to the referring nature of an idea:

Conceptual knowledge and concept-transcending knowledge

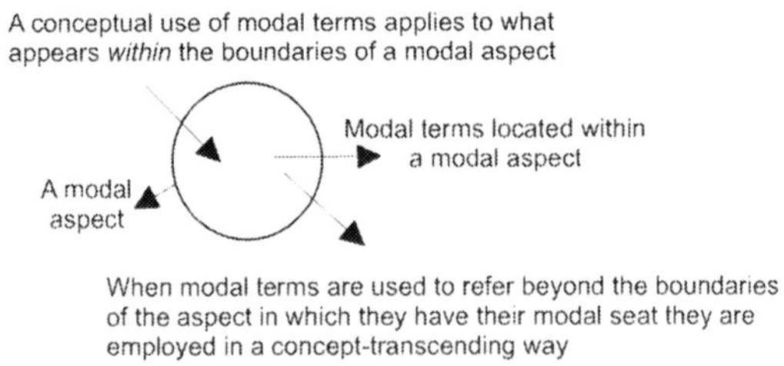

> Also in respect of the numerical we therefore have to avoid a mathematical functionalist reduction, that is to say, of identifying the numerical with what rightfully belongs to the field of investigation of arithmetic. The numerical in turn stretches far deeper than the numerical in its mathematical meaning. As such it is only possible to be understood in a referring idea.[2]

However, it is not the *numerical* that stretches (or functions: 'fungeert') "far deeper" – the issue is that in order to refer to the (central) depth dimension of reality one inevitably has to use *numerical terms* stretched beyond the limits

1 Just recall Gödel's reference to a "multitude of x's," which could be paraphrased but not defined, because every definition would lean upon our awareness of a discrete quantity (distinct multiplicity – see page 171).

2 "Ook ten aanzien van het numerieke moeten we ons derhalve hoeden voor een mathematisch-functionalistische verschraling, d.w.z. voor een identificatie van het numerieke met datgene wat rechtens tot het veld van onderzoek van de getallenleer behoort. Het numerieke reikt op zijn beurt veel dieper dan het numerieke in mathematische zin en is als zodanig slechts in een verwijzende idee te vatten" (Dengerink, 1986:240).

of the meaning of the quantitative aspect. What is approximated in a "referring idea" is not the numerical in its deeper stretching than its mathematical meaning, but the said depth dimension of reality referred to by employing the modal quantitative term 'unity' in a concept-transcending way! (Note that in this explanation two spatial terms are also repeatedly used in a concept-transcending way namely the terms 'central' and 'depth'.)

In his recent encompassing book on the *Philosophy of Theology* Troost, in connection with his idea of the ethos as an ultimate transcendental depth-layer in time, writes: "In both instances we slightly adjusted the meaning of the term 'ethical' to customary language use. Besides, we have to point out that this is often used in a much narrower sense, namely as focused on the moral (or ethical) *aspect* of love" (Troost, 2004:50). He actually employs the term *ethical* in the 'narrow' instance (conceptually) and in the 'broader' sense in a concept transcending way (as an idea).

Similarly, instead of speaking of the sizes and dimensions of a chair, one may use our intuition of the original meaning of spatial extension to speak of all facets of the chair – in which case one may refer to the chair in its *totality*.[1] Once again, it is clear that the term *totality* – in spite of its spatial descent – here refers to much more than merely the spatial aspect of the chair. It therefore constitutes – in terms of the distinction suggested concerning the two-fold usage of modal terms – another example of an *idea use* of modal terms, in this case *spatial* ones.

Modern phoronomy (pure science of movement) understands motion in its original sense as *uniform flow*, without the need of any causes (as Aristotle believed). This kinematic intuition of *constancy*, when used in an idea context, provides us with the idea-knowledge of the *identity* of an entity – its relative constancy amidst all changes – where the latter term finds its seat in the physical aspect of *energy operation*. The operation of energy always *causes* certain *effects* and in *that sense* never leaves anything the *same*, that is, *identical*.

Therefore, the word *change* can also be employed in an *idea context*. But because the *idea-meaning* of *constancy* (consonant with the idea of *identity*) and the idea use of *change* stem from two *irreducible* modes (detecting changes always presupposes constancy), it is not contradictory to use both these ideas.[2]

In a different context we shall expand our view by highlighting the four *most basic* idea statements philosophy can formulate about the universe (see page 455 below). The very nature of (regulative) *idea-knowledge*, referring us

1 In Strauss (2002b) it is explained why the meaning of *continuous extension*, by contrast, entails the awareness of a *gapless connectedness*, which is synonymous with the notion of *coherence* and the original meaning of the *whole-parts relationship*. If all the parts are *connected* (*cohere*), they constitute the *whole* or *totality*. (See also pages 87 and 236.)

2 Although he clearly distinguishes between the kinematic and physical aspects, Stafleu confuses these two when he states that the *dynamic* character of reality meets the kinematic time order (Stafleu, 1989:31). The modal seat of dynamics (change) is found in the physical aspect of reality, not in the kinematic mode.

beyond the limits of (constitutive) concepts, should be seen as the *expansive boundary* of rationality.[1]

5.13 What lies between the restrictive and expansive boundaries of rationality?

We followed up the suggestion by Bernays, namely that the meaning of rationality ought to be related to the nature, scope and limits of *concepts* (see Chapter 1, page 11) and we did so by highlighting the *restrictive* and *expansive* boundaries of rationality. But what is the status of the "in-between," of the *intrinsic domain* of rationality, the domain of *logicality*?

Our conjecture is that this domain of logicality evinces a three-fold *self-insufficiency*: (i) it expresses its unique meaning only in coherence with all the non-logical modal aspects of reality; (ii) it refers beyond itself to grounds transcending the realm of logicality; (iii) it can only function upon the basis of a direction-giving *ultimate human orientation in life*, an *ultimate commitment*.

Let us develop a brief outline of what this entails.

The generally accepted insight that the *validity* of an argument ought to be distinguished from its *truth* left many philosophers with the impression that, as soon as one is involved in the particulars of *logical inference*, one has arrived in a world of *pure thinking*, stripped of any connection with the "outer world." Both predicate logic and propositional (formal) logic seem to operate only on the basis of the logical principles of *identity*, (*non-*)*contradiction* and the *excluded middle*. For example, if one looks at the deductive syllogism in its four modes, while considering that there are four kinds of propositions involved – universal affirmative [A] and universal negative [E]; particular affirmative [I] and particular negative [O] – and then asks two questions: (a) how *many* inference patterns are there? and (b) how many of these constitute *valid* inferences? – then it is clear that, of the 256 possibilities, only a limited number are valid (21). In order to differentiate between *valid* and *invalid* inferences, an implicit or explicit use of the logical principles of identity and non-contradiction is required. Only on this basis is it possible to evaluate a particular inference as being *logically sound* or *illogical*. But this constitutes a *normative contrary* – only human beings with an accountably free will are able to act *in conformity with* logical principles or *violating* such principles. Also, the domain of *propositional logic* presupposes the said logical principles.

The entire distinction between *subject* and *predicate* – in its *logical* sense – is dependent upon the nature of *concepts*. Predicates normally *explicate* traits brought together in the *unity* of a concept. Whatever property is *excluded* from the *logical unity* of the concept cannot be *predicated* of it afterwards, except in an *illogical* way. If it were true that the concept "(material) body" *excludes* the property of *weight* (*mass*) to begin with – as Immanuel Kant asserts

1 The term 'constitutive' is meant in the sense of "building blocks," whereas 'regulative' indicates the deepening and expansive function of concept-transcending knowledge.

in his *Critique of Pure Reason* (1787, B:10 ff.) – then the so-called 'synthetic' proposition: "all bodies are heavy" would be *illogical* since it violates the principle of non-contradiction. It is therefore significant that the prominent German logicians of the 19th century eventually turned this proposition into an 'analytic' statement (Lotze, Windelband, Sigwart).

Comment: Since the discovery of irrational numbers, the switch to space as theoretical point of entry to reality established a powerful and long-standing tradition which is still operative in the thought of Descartes (*res extensa*) and Kant – as it is clearly seen in their shared conviction that a *material body* is exhaustively characterized by *extension* alone. Cf. Descartes *Principles of Philosophy* (Part II, II): "That the nature of body consists not in weight, hardness, colour and the like, but in extension alone"; and Kant: "So, when I separate from the representation of a body what the understanding thinks about it, such as substance, force, divisibility, etc., and similarly what belongs to sensation, such as impenetrability, hardness, color and the like, then in this empirical intuition there is still something left, namely extension and form." (*Kritik der reinen Vernunft*, 1787:B-35 – translation by the author). We shall extensively return to this development in Chapter 7.

As a matter of immanent criticism: Kant did not realize that 'divisibility' is intrinsic to spatial extension (it displays the foundational coherence of the spatial whole-parts relation with the numerical meaning of the successive infinite). Yet his understanding of 'extension' excludes its 'divisibility'!

A *concept* is constituted as a *logical unity in the multiplicity of features* captured in this concept, and every concept is subject to logical principles. The *meaning* of the logical principles of identity and non-contradiction brings to light the intimate connection between the original meaning of a *quantitative* unity and multiplicity, and a *logical* unity and multiplicity (compare the remark of David Hilbert quoted earlier in this chapter, page 172). Clearly, through basic similarities and differences, the *meaning* of the logical mode intimately coheres with the arithmetical aspect. Within the logical mode of reality we find a *logical* unity and multiplicity (particularly exemplified in *concept formation*), which analogically resembles an original *arithmetical* unity and multiplicity. This shows that logic cannot be "contained within itself," since it brings to expression its unique meaning only in *coherence* with non-logical facets of reality (in this case the coherence between the logical-analytical mode and the quantitative mode). Because the status of the logical principle of the excluded middle is dependent upon the notion of infinity that one accepts (either the so-called potential or the actual infinite), a more complicated analysis is required to account for it (see Strauss, 1991).

Since the beginning of the 20th century, the discipline of logic advanced significantly in elucidating the nature of logical inferences. But in spite of all the finesse and intricacies involved in this development, the mere application of the principles of identity and non-contradiction cannot assert the *truth* of any premises or conclusions. The fundamental logical principle referring logic *beyond* its concern with the validity of argumentation, is the one discovered by Leibniz (*principium rationis sufficientis*). Schopenhauer subjected this principle of *sufficient reason* to an extensive investigation in 1813. He called it the principle of sufficient ground of knowledge (*principium rationis sufficientis cognoscendi* – see the quotation on page 285 in the context a discussion of logical principles).

In connection with the nature of energy-operation, we alluded to the original *physical* meaning of the cause-effect relation (causality). Similar to the analogical appearance of numerical unity and multiplicity within the logical-analytical mode, this *physical* meaning of causation is also analogically reflected within the logical mode in the distinction between *logical* grounds and *logical* effects (conclusions). Whenever statements *contradict* each other, the application of the logical principles of identity and non-contradiction can only assert that *both* cannot be true at the same time – but they cannot establish *which one* is true indeed. Yet, the moment an appeal is made to some or other 'reason' or 'ground', logic alone is denied forever from being the *final judge* of rationality! The battle field has now switched, since *divergent views of reality* are suddenly emerging as the *main actors* on the scene of 'rational' interaction. What a particular thinker may consider "sufficient ground" may be viewed by another as "insufficient ground"! This situation implies:

(i) Conflicting views of reality require a *more-than-logical* criterion of truth and
(ii) Such views rest upon an important strategy in order to enable meaningful interaction between thinkers coming from divergent traditions.

Let us make a few brief remarks about these two issues in reverse order, commencing with the strategy required for *meaningful rational interaction*. Perhaps as an effect of the rationalist legacy of the West, our shared intellectual tradition is highly appreciative of *critical* thinking. The Enlightenment period advanced under the assurance of the stance of critical thinking *par excellence*. In the *Preface* to the first edition of his *Critique of Pure Reason* (1781), Immanuel Kant had already appreciated his own time as the *true age of criticism*. This spirit of radically questioning everything resounded quite recently in the stimulating and thought-provoking final address delivered to the *Philosophical Society of Southern Africa* (hosted at Rhodes University in Grahamstown, January 2003), when Graham Priest (from Australia) addressed the final plenary session on the theme "What is Philosophy"? He made a well-argued plea for the view that the ultimate task of philosophy is to be radically critical in questioning whatever there is. But let us first return to Kant by repeating a quote from his *Critique of Pure Reason*, given in Chapter 1 (see page 9):

> Our age is, in every sense of the word, the age of criticism, and everything must submit to it. Religion, on the strength of its sanctity, and law on the strength of its majesty, try to withdraw themselves from it; but by doing so they arouse just suspicions, and cannot claim that sincere respect which reason pays to those only who have been able to stand its free and open examination

(Kant, 1871:A-12 – translation F.M. Müller).

In order to really benefit from the exercise of a critical spirit, one has to observe something more fundamental than critique, namely *showing solidarity*. It is indeed much more difficult to highlight what is *worthwhile* in the thought of a specific thinker, particularly if we accept the challenge to account for it in terms of our own (different) perspective. In other words, if I want to criticize Plato, Aristotle or Kant, I must be able to appreciate positively what they have unveiled *before* it is meaningful to criticize the way in which they account for their discoveries. For example, we appreciated positively the fact that Aristotle discovered that concepts are bound to universal traits, *before* criticizing his rationalistic restriction of knowledge to *conceptual* knowledge. Similarly, we first appreciated Plato's insight that change is founded in constancy, before criticizing his speculative theory of static essences. And we pointed out that the basic epistemological concern of the entire *Critique of Pure Reason* actually highlights the discovery of the nature of *modal universality* (see page 68 above). Although we have already criticized Kant's rationalistic elevation of human understanding to the level of the *formal a priori law-giver of nature*, we shall call upon his epistemological quest in advancing an alternative perspective on the ontic modal universality of the physical function of the world.

On the basis of a genuine sense of *solidarity*, the second best thing to do in service of a meaningful and constructive critical encounter is to exercise *immanent criticism*.[1] This will prevent the conversing partners from ending their interaction with the proverbial: "I say this and you say that, so what?" An inner contact of thought requires the intellectual integrity of immanent criticism. Its aim should always be first to show what is inherently untenable and only *then* to proceed with the formulation of an alternative perspective. At that point, in turn, the conversation partner first has to appreciate the inconsistencies pointed out by *immanent criticism*,[2] then proceed to an appreciation of what is positive in the alternative account *before* renewed criticism is raised.

This brings us to the first point: the requirement of a *more-than-logical* criterion of truth. Observing the logical principle of non-contradiction enables one to discern *contradictions*. But we have seen that real *antinomies* refer to a clash of irreducible law spheres (such as the clash between kinematic laws of motion and spatial laws, in the case of Zeno's argumentation about Achilles and the tortoise). Every attempt to reduce what is *irreducible* inevitably ends in such antinomies – as we also showed above in connection with primitive terms and the limitations of concept formation and definition.

Since a clash of laws concerns irreducible spheres (modes) of laws (or law-spheres), an antinomy always relates to an inter-modal confusion (in the case of Zeno, a confusion of static spatial positions with uniform phoronomic

1 Immediately after his final presentation, I asked Graham Priest about the role of *immanent criticism* in critical confrontations – and his response was that someone else at a different occasion also raised this point – and that he may have to consider it more seriously.

2 Or even, if possible, unmask the purported immanent criticism as not hitting the target!

flow).

The stance taken by a Marxist physicist, Hörz, indirectly demonstrates this issue in a neat way. Acccording to him, classical physics (Newton and his successors) teaches that a moving body finds itself at a specific point in time at a specific place. But if this is the case, he continues, it will be impossible to gain an understanding of true movement. As an alternative Hörz therefore opts for the conception of motion developed by Engels. In the dialectical-materialistic conception of the latter, one can say that a moving body engaged in a *change of place* at the same time *is* and *is not* at a specific place. Hörz explains the inner tension ('dialectics') of this position as follows:

> Insofar as the body changes from one place to another it moves, and it reaches, as a result of its movement, always at a specific time a specific place (Hörz, 1967:58).

This is, according to him, the "dialectical antinomy (Widerspruch)" of *change of place*. A formulation precluding every *logical contradiction* runs as follows: "as the result of movement a body finds itself at a specific place and with regard to the movement itself the body does not find itself at a specific place" (Hörz, 1967:58). The implicit assumption of this assertion of the "non-contradictory" relationship between movement and place is the irreducibility of the law-spheres of *space* and *movement*.

A mere appeal to logical principles does not safeguard anyone from becoming a victim of (antinomically) reducing one of these two spheres to the other. Just consider the account by Von Kibéd of the impossibility of motion in terms of the logical principle of identity:

> The principle of identity, according to which everything is only identical to itself, actually forbids every change, every becoming-different, every stepping-outside of a substance from its being-itself (Von Kibéd, 1979:59).

According to Von Kibéd, the classical metaphysical escape route, namely to distinguish between *essence* and *appearance*, will not help us:

> The difficulties accompanying the concept of the changes of an unchangeable thing are avoided by dividing the entity into an essential and accidental part, thus producing the possibility of associating unchangeability with its essence and changeability with what is accidental (Von Kibéd, 1979:60).

However, this is of no help either, because the accidental features of an entity are also subject to the law of identity: "according to the principle of identity also the accidental must remain identical to itself and cannot abolish its essence, which is given in its accidental nature" (Von Kibéd, 1979:60). His conclusion is therefore to be expected: "The concept of change is therefore logically unthinkable" (Von Kibéd, 1979:60).[1]

What is absent in this argument, are sufficient grounds accounting for the irreducibility of *constancy* and *change* (as unique, mutually cohering, ontic modes enabling their implementation as modes of explantion in scientific

1 What is needed in order to account for change, namely "the concept of causality is, logically seen, non-transparent and shows the limits of logical explanation" (Von Kibéd, 1979:60-61).

discourse). Merely applying the logical principle of identity simply does not solve the problem, and in addition, in the case of Von Kibéd, underscores the *antinomy* entailed in every attempt to reduce *change* and *constancy* to a static 'unchangeability'![1]

The nature of the *more-than-logical* principle needed in this context will be explained in Chapter 6.

5.14 Trust (faith) in rationality

At this point we can return to our initial remarks in Chapter 1 about a *rationalistic faith in reason* (see pages 10 ff.). We mentioned Popper's concession that the rationalistic *trust in reason* is not itself *rational*, for it constitutes *an irrational faith in reason*. To that we may add Gadamer's remarks in which he mocks the prejudice of *Enlightenment* against prejudices,[2] Stegmüller (one already has to believe in something in order to justify something else) and Derrida (on the universality of 'faith' – his slightly different example of credit as economic trust).

The various forms of trust in human life in the final analysis converge in a direction-giving ultimate commitment – what Dooyeweerd calls the central religious ground motives operative in world history.

Our analysis of rationality thus far aimed at bringing to light both the *restrictive* and the *expansive* boundaries of rationality – evinced in the "coherence of irreducibles" and the entailed awareness of *primitive terms* in their *indefinability* on the one hand, and in those kinds of knowledge *transcending* the *limits* of what concepts can achieve on the other. We only briefly paid attention to the place of logical inferences, but had to mention that even the most basic logical principles point out the *self-insufficiency of logical thinking*. The principle of sufficient reason added weight to the dependence of logical argumentation upon the grounds (reasons) that stem from a more-than-logical reality.

What clearly surfaced is the awareness that rationality is *embedded in* and *borders upon* givens that are not open to further 'rational' exploration – givens that both *condition* (in a constitutive sense) and *transcend* (in a regulative sense) the limits of conceptual knowledge. In the last section, it was succinctly argued that the rational human faculty rests upon *trust* – converging in an ultimate life-commitment, which also guides and directs other forms of trust in human life.

Taking into account the many-sided connections with "more-than-rational" *boundaries of* and *conditions for* rationality, we have to conclude that ratio-

1 Prinsloo has a clear understanding of the shortcomings of reductionism. He has a solid understanding of the fact that "we cannot explain movement by reducing it to rest or dynamism in terms of a static state" (Prinsloo, 1989:98). This issue is treated in more detail in the 'Festschrift' dedicated to E.D. Prinsloo – my contribution is entitled: *Dialectics and logicality: Between cultural diversity and ontical universality*.

2 Gadamer speaks of the "discreditation of all prejudices" by Enlightenment and calls for the acknowledgment that "there are legitimate prejudices" (Gadamer, 1998:276-277).

nality as such appears not to be all that 'rational' after all. The meaning of "rational thinking" only comes to expression in the coherence of the logical-analytical function with a more-than-rational (given) *cosmic diversity*. The constructive capacity of human thinking, its ability to acquire knowledge through *normed* subjective acts of identification and distinguishing, constitutes its *creative calling*, which should not be appreciated as the *ultimate source* of order and certainty in the world, but as an *accountable response* to those normative conditions making possible the *important* but *limited* operation of rationality as such.

5.15 Theology and the limits of conceptual knowledge

Not only the etymology of the term *theology*, but also the actual history of the discipline known by this name makes it plain that, in some or other way, the scholarly endeavours of this subject speak about 'God'. Of course there is a rich speculative legacy within philosophy concerned with what philosophers *rationally* construct God to be. Wilhelm Weischedel wrote extensively about this tradition in his work on the "God of the philosophers" (*Die Gott der Philosophen*). Surely the "God of the Theologians" cannot be identified with the "God of the philosophers." Someone like Heidegger even claims that theology is not the "science of God," since it must rather be viewed as *reflection on faith*. For this reason theology, according to Heidegger, "avoids any sort of philosophical system" (Heidegger, 1970:25, cf. p.18).

Pannenberg, by contrast, explicitly returns to Thomas Aquinas in order to restore and maintain the idea that theology does deal with God:

> In proclaiming the one God, Christianity appealed from the start to philosophy and to its criticism of the polytheistic beliefs of other peoples. The reference was first to the Stoic theories and later, above all, to the doctrine of God found in Platonism. Such an appeal to the philosophical doctrine of God must not be interpreted only in an external sense as an accommodation to the spiritual climate of Hellenism. Instead, it reflects the condition for the possibility that non-Jews, without becoming Jews, might come to believe in the God of Isreal as the new God of all humanity. The appeal to the philosophers' teachings concerning the one God was the condition for the emergence of a Gentile church at all. We must therefore conclude that the connection between Christian faith and Hellenistic thought in general – and the connection between the God of the Bible and the God of the philosophers in particular – does not represent a foreign infiltration into the original Christian message, but rather belongs to its very foundations (Pannenberg, 1990:11-12).

Later in this work, he adds the following:

> Christian theology, in contrast to Heidegger's construal of it, is essentially an inquiry [Wissenschaft] into God and his revelation. Everything else that occurs within theology can become a theme for the theologian only "in relation to God," as Thomas Aquinas put it: *sub ratione Dei*. Christian theology would lose not only its specific content but also, and most importantly, the consciousness of truth that is intrinsic to it, if it were to follow Heidegger's advice to stop speaking of God in the realm of thought (Pannenberg, 1990:120).

Suppose we accept for the moment that theology cannot avoid entering into "God-talk" (to be distinguished from the "Divine discourse" discussed by Wolterstorff, 1998). Is it then not unavoidable that theology ought to employ some or other *concept* of God (in the "realm of thought")? Most theologians will affirm this implication. Yet, exactly at this point, inevitable philosophical presuppositions emerge, because reflection on the nature of concept formation necessarily delves into the *epistemological* presuppositions of theology that are *philosophical* in nature, since, as we argued, concept formation always entails an appeal to *universal* properties that cannot be divorced from *universal conditions*. But if God is the origin of all conditions and of being conditioned, can one then still claim that a concept of God is possible?[1]

5.15.1 *The temptation of theo-ontology*

It is amazing to see how easily one can assume a certain philosophical perspective without being aware of the fact that it is indeed *philosophical* in nature. But once this is done, with the best of pious intentions, God is subsequently portrayed in terms of such an implicit philosophical scheme. In doing this, key features of creation are then 'positioned' 'within' God, such that our experience of these properties is subsequently accounted for by deducing them from the elevated 'essence' of God. This is in line with the traditional claim of theology, namely that it merely *listens* to the Bible – a "pure understanding" that should be followed obediently in the scholarly ('scientific') theological reflection about God. We may question this legacy with reference to the attribution of *infinity* to God.

Although the Bible does not explicitly attribute *infinity* to God, the theological tradition deduces God's infinity from his *omnipresence* and *eternity*, with *immutability* and *timelessness* equally important, and related issues (see Leftow, 2005:62 ff.). *Eternity* is understood in terms of two apparently opposing notions: an *endless period of time* or *timelessness*. These two notions, on the one hand, may be related to the so-called Platonic and Aristotelian traditions, but actually should be appreciated in close coherence with the two conceptions of infinity operative in the history of mathematics (and theology). What I have in mind, is the opposition between what is designated as the *potential infinite* and the *actual infinite* (what we labeled as the *successive infinite* and the *at once infinite* – see page 63 above and page 239 below).

Reflecting for a moment upon the assumed property of *infinity* as applied to God, we may be able to uncover a telling example of theo-ontological reasoning.

1 Cf. Strauss (1991a:23-43) where this issue is discussed at length. Since concepts require universal conditions or features, having a concept of God would imply that there is an order for *being-a-God*, i.e. a law-for-being-God, which in fact subjects God to his law for creation and, in an internally antinomic way, turning God into a creature, subject to the conditions for being a God!

5.15.1.1 *God's infinity*

In anticipation of the argument below, we start with a brief summary of what is at stake. When mention is made of God's infinity in a theological context, the (theo-ontological) assumption is simply that infinity originally (*eminently*) belongs to God, and that whenever we employ the notion of infinity in mathematics, it is *derived* from the theological understanding of God's infinity. (Since Cusanus it is also customary to say that infinity in the sense of endlessness belongs to mathematics, but that the actual infinite is reserved for God only.)

Our argument will be that this theological legacy did not start by analyzing the *structural* inter-connections between a fiduciary mode of speech (the "language of faith") and the numerical (quantitative) aspect of reality, and consequently did not realize that infinity is a mathematical notion to begin with which could subsequently be employed theologically only afterwards. If this is not appreciated, a neat theo-ontological *circle* is followed: infinity is first *lifted* from its cosmic 'place' ('seated' in the meaning of number and, in the case of the at once infinite, in its interconnection with space) by assuming that it originally belongs to God – and once this shift is made, infinity can only be (re-!)introduced within the domain of number by taking it over from theology!

In order to illustrate this mode of thought, we look at the stance defended by Chase (1996) as an example. The historical analysis of the examples given by Chase seems to be meaningful for demonstrating the general point that Christian theology did have an impact upon the development of modern mathematics. However, by ignoring the *structural relationships* between (the fields of investigation of) theology and mathematics, certain questions can be raised about the overall perspective in terms of which the historical material is placed.

Perhaps the implicit scholastic definition of theology is responsible for the above mentioned shortcoming, namely that Chase does not explicitly consider the second option open to him: exploring next to the *genetic* perspective also a *structural* analysis of the intended relationship. On a *structural* level, Chase should at least give an account of the fundamental *concepts* (and *ideas*) used by mathematicians and theologians. This *philosophical* issue pertains to the phenomenon that different scientific disciplines frequently use scientific terms which are apparently similar, but still differ in the sense that they are used within the context of diverse scientific universes of discourse – what we will designate as analogical basic concepts of the disciplines (see page 232 ff.).

For example, no one can deny that both mathematicians and theologians use *numerical* terms like the numerals 'one', 'two' and 'three'. The underlying philosophical issue is: are these notions *originally* (that is, in a structural-ontic sense) *numerical* notions that are *analogically* used within a different (faith) context when theologians say that there is but 'one' God, or when they speak of God's "tri-unity"?

A closer analysis of the diversity of modal aspects in reality – and the accompanying *modal terms* – would show that the basic concepts and ideas employed in theological parlance are fundamentally dependent upon the remarkable coherence between the certitudinal aspect of reality, central in the theological concern, and the different other aspects of realty that are, by means of *moments of similarity*, *analogically* reflected within the structure of the fiduciary aspect of reality.

Let us pursue the notion of number further in this context. The awareness of *one*, another *one*, and so on (*indefinitely*), constitutes the most *basic* and most *primitive awareness* of *infinity* one can have. It is only when this *numerical* intuition is deepened by our *spatial* awareness of something (which, as a whole is) given *at once*, that we are able to consider any *infinitely proceeding sequence* (to use a phrase from intuitionistic mathematics) ***as if*** all its elements are given *simultaneously*, that is, as a "completed totality." Our preference is to designate these two forms of the infinite as the *successive infinite* and the *at once infinite*. Both these notions of infinity are originally located within the numerical aspect of reality and can only come to expression on the basis of the integral coherence between the numerical aspect and the spatial aspect (amongst others, which we ignore for the moment). It is therefore surprising to hear Chase asking: "Could infinities such as a completed totality be brought into mathematics without a Christian theological foundation?" (Chase, 1996:209).

He proceeds: "At the very least, some idea of God standing outside of our experience must have been necessary, since apart from God we have no experience of the infinite" (Chase, 1996:209). Chase also mentions the following fact: "Some Scholastics in the middle Ages and Cantor in the nineteenth century believed in an actual mathematical infinity, based on God's infinity" (Chase, 1996:209-210)

Since Chase failed to investigate any structural relationship between mathematics and theology, his historical analysis precludes the option of acknowledging infinity (in both its forms) as 'mathematical' (that is, numerical and numerically deepened) *analogies* operative within the theological universe of discourse. By doing this, the implicit dualism presupposed in his argument could be reversed. Instead of supposing that the notion of "infinities such as a completed totality" originally is a *theological* idea that is completely foreign and external to mathematics, one would then much rather acknowledge that *within* the structural nature and interrelationships between number and space we first of all encounter the notion of infinity – which secondarily could be reflected within the structure of the certitudinal aspect in an *analogical* way. By not tracing the notion of infinity back to its original "modal seat" it can only serve as a notion brought into mathematics "from the outside," that is, as something 'purely' theological that can only bear upon the field of investigation of mathematics in the second place. However, by recognizing the (deepened) numerical seat of the notion of infinity one should rather start from the assumption that theologians could only use notions of infinity as *mathemati-*

cal analogies in their theological argumentation. A shortcoming, in this regard, exists in the fact that Chase does not enter into a discussion of the notion of infinity as it is traditionally employed in Christian theology, for then, at least, he might have taken note of the fact that the Bible nowhere explicitly attributes *infinity* to God. Theologians traditionally extrapolate God's infinity from His *omnipresence* and *eternity*.

If we proceed from a *structural-genetic* perspective as the basis of our historical analysis (something absent in Chase's article), one can argue the point he wants to make as follows:

> Theological reflection and speculation about the 'infinity' of God indeed paved the way for and promoted the eventual mathematical development of a theory of transfinite numbers (Cantor), but in doing that, theology simply digressed into quasi-mathematical considerations which, in the first place, refer to purely *mathematical* notions related to the inter-modal coherence between number and space.

Chase defends a kind of "negative theology": he does not acknowledge the numerical and spatial source of the 'potential' and the "actual/completed" infinite, but argues that these terms are originally *theological* in nature and eventually sends them back to the *domain* from which they were (implicitly) stolen in the first place – in the form of *theological notions* fruitful for the further development of modern mathematics!

The true intent of a theo-ontological approach is best understood when we return to the position taken by Thomas Aquinas and contrast it with a modern theologian like Karl Barth.

5.15.1.2 *Aquinas and Barth*

Thomas Aquinas inherited the opposition of 'essence' and 'appearance' from the metaphysical Greek concept of substance. In his *Summa contra Gentiles* (I,34) and *Summa Theologica* (I,13,1), Thomas Aquinas explains that we can know God through His creatures because, in an eminent way, God bears all the perfections of things *within* Himself. We know God by means of these perfections as they flow from Him into creatures (*procedentibus in creaturas ab ipso* – S.Th. I,13,3).

We have to realize that the emphasis is upon what is supposed to 'pre-exist' within the 'essence' of God. What we consider *good* in creatures, 'pre-exist' in God, albeit in a *superior* and *alternative* way.[1]

Initially, under the influence of neo-Platonism and Augustine we find an inclination towards a negative theological designation of God in the writings of St. Thomas. Sometimes the Plotinian conviction, namely that we could only positively say what God is **not**, almost word-for-word recurs in his writings.[2]

The conviction that being is the *primum notum* (that what is *known first*)

1 "Cum igitur dicitur: Deus est bonus; non est sensus: id, quod bonitatem dicimus in creaturis, praeexistit in Deo: et hoc quidem secundem modum altiorem" – S.Th. I,13,2.

2 S.c.G.III,49; cp. S.Th.I,13,8; I,1,7 and especially I,13,1.

rests on the conception of an *analogy of being* (*analogia entis*) which entails that both God and creation are subsumed under the basic denominator of *being*. According to St. Thomas *being* (esse) and *essence* (essentia) coincide in God (S.Th.I,3,4 and I,13,11).

Although things by themselves are finite and caused, they exist in God in such a way that they are *nothing but* God.[1]

The ultimate epistemological shortcoming in Thomas Aquinas' understanding is that he assumes a *universal condition* for the existence of God, i.e. he deems it possible to form a *concept* of God. With regard to entities, however, concepts always either relate to the *order for* the existence of a particular type of entity (in which case we form concepts of type laws), or they relate to the (universal) orderliness of entities (the *being* an atom of an atom). This approach ultimately levels the difference between God and creature by subjecting God to the (implicitly assumed) *universal order for being a God*. But then there are many instances of 'Gods' conforming to these universal conditions for being a God – contradicting the biblical account that unequivocally claims that there is but *one* God, that God is *unique*. Pannenberg saw this shortcoming in his own way: "The idea of God is destroyed when he is conceived as an application instance (even though it be the highest instance) of some general structure which in its generality is distinct from God and which is asserted as a predicate of God" (Pannenberg, 1990:145).

Karl Barth distinguishes the *oneness* (unity) of God in the sense of *uniqueness* from the oneness of God in the sense of *simplicity* (simplicitas). The uniqueness of any creature, according to him, is only *relative* because it belongs to a *species* that is merely the 'instantiation' of a *genus*. This is just a different way of articulating the nature of concept formation that is directed towards an understanding of universal type laws and the universality (orderliness) of entities 'instantiating' these type laws. The uniqueness of God, on the other hand, Barth claims, is in itself absolutely unique in a way that cannot be grasped in *any* concept (Barth, 1957:447). Clearly, Barth here intends an important distinction needed in order to account for the way in which human beings can speak meaningfully about God.

It is clear that Barth does recognize the cosmic 'residence' of the intuition of *being unique* (*being distinct*), as employed by him. We may raise an immanent-critical question: if it is possible to refer to creatures by using the property of *being unique* does this not entail that we have subsumed both God and creatures under the same 'condition', namely that of being *unique*?

The only way out of this impasse is to further explore the distinction made above, with its deep roots in the philosophical legacy of the West, namely that between *conceptual knowledge* and *concept-transcending knowledge* (*idea-*

1 Cf. Kremer, 1966:399: "Alles Seiende ist so in Gott, dass es in Gott nichts anderes als Gott ist. Die Dinge sind nicht so in Gott wie sie in sich selbst sind. In sich selbst gesehen sind sie nämlich verursacht und endlich, in Gott dagegen unendlich, weil sie in Gott zusammenfallen mit dem göttlichen Wesen... In sich selbst gesehen sind sie Vielheit, in Gott dagegen Einheit."

knowledge). An extensive discussion of the problem of the universal and individual is found in Ouwendorp 1994. In this article he fully accepts and applies the distinction between conceptual knowledge and concept transcending knowledge as well as the alternative understanding regarding *specificity*.[1] One side of this distinction has already been explored by Plato in his dialogue *Parmenides*, which provides the starting point of *negative theology* (where it is only permitted to say what God is *not*), while the other side of the coin in a certain sense reached its apex in the thought of Immanuel Kant. According to him, the acquisition of knowledge occurs in three steps: it starts with the senses, proceeds from thence to understanding, and ends with reason, beyond which there is no higher faculty to be found in us for elaborating the matter of intuition and bringing it under the highest unity of thought (Kant, 1787, B:355, cf. B:730).

Syllogistic inference implies that the conclusion is always subsumed under the condition of a universal rule (the major premise). By applying the rule of reason once more, the condition of the condition must therefore be sought (by means of a prosyllogism) whenever practicable. Thus, according to Kant, the principle peculiar to reason in general, in its logical employment, is: "to find for the conditioned knowledge obtained through the understanding the unconditioned whereby its unity is brought to completion" (Kant, 1787, B:364). The concepts of pure reason are called *transcendental ideas* (Kant, 1787, B:368). These ideas instruct us only in regard to a certain unattainable completeness, and so serve rather to limit understanding than to extend it to new objects (Kant, 1787, B:620). The *unconditioned is* never to be met in experience, but only in the idea – whenever "*the conditioned is given, the entire sum of conditions, and consequently the absolutely unconditioned* (through which alone the conditioned has been possible) is also given" (Kant, 1787, B:436). This means that transcendental ideas are simply *categories extended to the unconditioned* (Kant, 1787, B:436) (this applies only to those categories in which the synthesis constitutes a *series of* conditions subordinated to one another). To Kant, therefore, the transcendental ideas serve only for *ascending*, in the series of conditions, to the unconditioned (that is, to principles; cf. Kant, 1787, B:394).

No *constitutive* usages of these ideas are allowed according to Kant, because then we only arrive at pseudo-rational *dialectical* concepts (the source of which Kant called the *antinomies*) (cf. Kant, 1787, B:672). The three ideas of the *soul* (thinking nature), the *world* and *God* are all to be used in an *as if* way, i.e. *regulatively* (cf. Kant, 1787, B:710-714).

The "thing-in-itself" is not *merely* an *idea*. We noted to the contrary that owing to the fact that we cannot *know* the "thing-in-itself," but nevertheless *think* it, according to Hartmann there must exist a mode of conceptualization in which we can *think* (be it as something *unknowable*) the "thing-in-itself." *This* is Kant's *transcendental idea* (cf. Hartmann, 1957: 311 – see page 176 above). To put it differently: in order to *think* about that which transcends con-

1 See Ouwendorp, 1994:48 ff. and pp.54-57.

cept formation, we still need a "conceptual form" by means of which we can think whatever transcends the *boundaries* of conceptual knowing.

Although the German term 'Grenzbegriff' is normally captured with the English translational equivalent: *limiting concept*, this practice may cause misunderstanding. For that reason we explained that the intention is not to emphasize what lies *within* the boundaries of this limit, but rather to refer to that which *transcends* it. The best way to capture this intention is therefore to view *ideas* are those *forms of thought* through which we approximate that which transcends the limits of a conceptual grasp. Ideas therefore refer to *limit-transcending knowledge*.

In a certain sense the modal (aspectual) dimension of reality conditions *both* the employment of *concepts* and that of (*limit-transcending*) *ideas*. This follows from the fact that the different modalities always serve as *points of entry* to our experience of and reflection on created reality. Modal concepts are always formed in relation to universal features of the different modal aspects – for example the concept *natural number*, *set*, *dimension*, *cause and effect* (causality), and so on. Ultimately, the nuclear meaning of every distinguishable modality is *indefinable*, providing as such the *primitive terms* used for our concept-formation and definitions. In the final analysis, therefore, comprehension is only possible by employing terms that are themselves beyond the reach of concept formation – evincing the *self-insufficiency* of *rational thought*.

But let us return for a moment to the two-fold way in which one can employ modal (functional) terms. Our primitive arithmetical intuition of a *discrete multiplicity* underlies our awareness of *being distinct* of different entities and therefore founds our knowledge of things in their *individuality* or *uniqueness*. Consequently, the concept-transcending idea of *uniqueness* (individuality) ultimately rests upon *numerical terms* in service of an approximating and referring mode of thought – transcending the limits of normal concept formation. In passing we have to note that apparently Barth is unaware of the *speculative metaphysical background* of the notion of the *simplicitas* of God – a legacy dating back to the *simplicity metaphysics* of early Greek philosophy (compare Visagie, 1982:8-9).

Also consider other *modal terms* the Bible uses within the context of referring to God in the said approximating and limiting manner of *idea-knowledge*. God is revealed as *omnipresent* (an idea usage of a *spatial* term); God *acts* (an idea usage of a *physical* term); God is *life* (an idea usage of a *biotical* term); God is *omnipotent* (an idea usage of a *cultural-historical* term); God is *love* (an idea usage of an ethical term), and so on. Surely these idea usages of (modal) terms find their counterpart in the familiar *conceptual* usages of such terms. In the latter case these terms are not employed to refer to something transcending the limits of the aspects in which they have their "modal seat," since they merely capture whatever functions within the boundaries of a particular mode. Saying that there are 20 people present at a meeting employs the numerical property of *being twenty* in a *conceptual* sense. Saying that a body moves *uniformly* employs our *kinematic* intuition conceptually. Saying that a tree is *alive* does the same with respect to the way in which this kind of entity functions within the confines of the biotic mode of reality.

Paul Tillich uses a similar distinction in a different context. He places it within the framework of the distinction between *form* and *dynamics*. Nevertheless, it aims to account for the same difference we have in mind with our distinction between *concept* and *idea*. He argues that *dynamics* transcends a delimited form and, consequently, cannot be grasped in a *concept*. According to him, we nevertheless discover an approximation of this dynamic element almost in all mythologies:

> It underlies most mythologies and is indicated in the chaos, the *tohu-va-bohu*, the night, the emptiness, which precedes creation. It appears in metaphysical speculations as *Urgrund* (Böhme), will (Schopenhauer), will to power (Nietzsche), the unconscious (Hartmann, Freud), *élan vital* (Bergson), strife (Scheler, Jung). **None of these concepts is to be taken conceptually** (my emphasis – DS). Each of them points symbolically to that which cannot be named (Tillich, 1964:198).

At this point, it should be clear that the counter pole of our *concept-idea* distinction is provided by the *negative theological* denial that we can say anything *positive* about God. The latter position, however, simply cannot account for the straightforward *positive* biblical mode of speech about God.

5.15.2 *Turning negative theology upside down*

We may now return to the *necessity* and *inevitability* of a *theoretical* (philosophical) *view of reality* – not only for theology in general – but also for the way in which theology may account for the possibility of speaking (scientifically and non-scientifically) of God.

Given the role of this (mostly implicit) philosophical view of reality, it should not surprise us that theologians could come up with such opposite extremes when they speak of God theologically. For example, when Thomas Aquinas refers to God as the *highest being* (*ipsum esse*), his mode of speech reveals a *philosophical* view of reality differing radically from some prominent theological approaches in the 20th century – approaches that would prefer to refer to God as the *loving Father* who is *close to* us. Apparently complementing nuances easily develop into *mutually exclusive* perspectives – in which case, for example, either the *power* of God or the *love* of God is chosen, without accounting in any way for the *coherence* between these two emphases.

The decisive point to be observed, however, is that there simply is no single scientific discourse dealing with God that is not completely in the grip of and determined by some particular *philosophical* view of reality ('paradigm'). The crucial question therefore is not **whether** such an underlying view of reality is operative in our discourse about God, but much rather whether or not this (mostly *concealed*) theoretical worldview is in the *grip* of the *central meaning* and *radical direction-giving motive-power* of the biblical message. In a slightly different context, Van Huyssteen correctly emphasizes the committed nature of rational thought in the sense that it is rooted in supra-theoretical convictions:

> The high degree of personal involvement in theological theorizing not only reveals the relational character of our being in the world, but epistemologically implies the mediated and interpretative character of all religious commitment, which certainly is no irrational retreat to commitment, but on the contrary reveals the committed nature of all rational thought, and thus the fiduciary rootedness of all rationality (Van Huyssteen, 1997:44).

The point stressed here implies the following "hermeneutical circle": the soundness of a theological call upon specific Bible texts, in the final analysis, is not determined by the pious habit of substantiating a theological argument with quotes from the Scripture, since the outcome of "*Bible-text-support*" could only be reliable when done in obedience to the *integral Kingdom perspective* of the Bible itself. Whenever the central meaning of the Bible is distorted, the unity of God's *good creation* is jeopardized by identifying the *directional antithesis* between *good* and *evil* with distinct terrains/domains within creation – the source of a dualistic legacy easily leading Christians to fight the wrong battles (such as *church* versus *state*, *theology* versus *philosophy*, *faith* versus *reason*, *soul* versus *body*, *calling* versus *occupation*, *direct service to God* versus *indirect service to God*).

Given the contemporary emphasis on the *eschaton*, anyone mentioning the presupposition of creation entailed in the notion of the *eschaton* is questioned. Olthuis correctly remarks:

> The current eschatological orientation in theology which tends to seek even the beginning in the end will need revision. The Bible begins with Genesis and Genesis begins with creation. The Scriptures see the Gospel as the link connecting creation and consummation. And this link between past and future is revealed as the Word which connects the end with the beginning, the consummation with the creation. "I am the Alpha and the Omega, the first and the last, the beginning and the end" (Rev. 22:12). A proper vision of the consummation requires a proper appreciation of the beginning. Without this understanding, the fulfillment lacks substantial content and tends to evaporate into pious words about hope. A non-robust view of creation emasculates the gospel, for it is the creation which is brought to fulfillment in Jesus Christ even as it began in him (quoted by Strauss, 1998:75).

5.15.3 *The philosophical dependence of theology*

The rich *diversity* within creation simultaneously evinces a *coherent unity*. This unity-in-the-diversity is not a *product* of scientific thinking. Much rather, every scientific distinction should seriously take the *given* creational diversity into consideration. By its very nature scientific thinking, as *deepened analysis*, should explain the structural *possibilities* and *limitations* of our human capacity to *know rationally*, without falling prey to the rationalistic limitation of knowledge to *conceptual* knowledge.

A return to this rationalistic position of (Kant and) modernity in a radical form is found in Van Huyssteen's mentioned identification of the *structure* of the universe with human *rationality* and *mathematics*: "What is astounding, however, is to what extent our world is truly rational, that is, in conformity with human reason" (Van Huyssteen, 1998:68 – see page 90 above).

Only the universal *conditions for* creatures and that which behaves in a law-conformative way – as *factually subjected* to (God's) law – could be *conceptualized*. Consequently, the only way to form a "concept of God" would be to subject God to some or other "creational law-order" for "being-a-God," as argued above.

The epistemologically sensitive theologian would, at least at this point, be prepared to support our thesis that humankind is **incapable** of *comprehending* or *grasping* God *conceptually*. It stands to reason that a *pan(en)theist*, believing that everything is (*in*) God, would think differently about God than someone accepting that God *transcends* creation. Traditionally, this boils down to the question whether there is an *essential difference* between *God* and *creation*. Medieval speculation postulates God's *aseitas* – deduced from the Latin expression *a se esse*, which means: *to exist in and of itself*. The *intention* is to honor God's *self-sufficiency*, i.e. that God, in his elevated *aseitas* cannot be grasped in any human concept. The *way* in which this intention is given shape, however, once again illuminates the *philosophical* indebtedness of theological reflection. From God's transcendent *aseitas*, certain *attributes* are deduced, such as God's *eternity*, *omnipresence* and *infinity*. These attributes are denoted as *incommunicable*. They *should* therefore be distinguished from the *communicable* attributes, such as God's *love*, *justice*, *mercy*, *etc.*

This distinction has its foundation in the classical philosophical substance concept that entails a distinction between *essence* and *appearance*.

5.15.3.1 *Inertia and God*

Pannenberg remarks that, when the assumption that movement inherently belongs to the nature of bodies was combined with the principle of inertia, it was no longer necessary to induce the cooperation of God in order to explain natural processes (Pannenberg, 1993:31). Particularly in the light of the *contingency* of natural events, Pannenberg questions the strict interpretation of inertia. According to him, the principle of inertia entailed the emancipation of natural bodies from the creator God (Pannenberg, 1993:20). In spite of his own preference for contingency and historicity as a more encompassing framework in terms of which we have to understand nature, Pannenberg ultimately had to fall back upon the *faithfulness* and *identity* of God: "Yet there emerge regularities and persistent forms of created reality giving expression to the faithfulness and identity of God in affirming the world that God has created" (Pannenberg, 1993:22).

It is not quite clear how he harmonizes this 'identity-'appeal with his conviction that, in the course of time "new patterns of regularity" emerge: "Thus it also becomes understandable that new patterns of regularity emerging in the sequence of time constitute a field of application for a new set of natural laws such that 'the laws governing matter in a higher level of organization can never be entirely deduced from the properties of the lower levels' " (Pannenberg, 1993:21; the last phrase is a quote from A.R. Peacocke).

On the basis of these considerations, Pannenberg argues against the supposed eternity and atemporality of the laws of nature – but he never contemplates the key notion of *constancy* in this regard. Constancy indeed pertains to the core meaning of the phoronomic (or: kinematic) aspect of reality where it is embodied in the notion of a *uniform motion*. In fact, the law of inertia precisely points to this: a body in a state of uniform motion will continue its movement except when some force impinges upon it.

Although Plato has already accounted for the possibility of knowledge with an appeal to *constancy* (elevated to his metaphysical realm of ideas), it was Galileo (*inertia*) and Einstein (the velocity of light in a vacuum) who realized that it belongs to the core meaning of motion.[1] Similar to all other aspects of created reality, the kinematic mode also allows for both conceptual and idea usages of kinematic terms. Consider the idea of *identity* (see page 165 above and pages 425 ff. below). Applied to existing entities the idea of their identity employs the intuition of constancy or persistence – it not merely refers to the kinematic aspect of an entity, but to the full many-sided reality of such an entity.

The basic kinematic intuition of constancy can also be stretched beyond the confines of this aspect when used in an idea context in order to refer to the identity of God – in which case its referring meaning not only transcends the boundaries of the aspect of movement, but in fact it also exceeds creation as such in pointing beyond it to God. Just consider the Old Testament account where God said to Moses: *I am who I am* (Ex.3:14). Clearly, this entails an idea usage of the kinematic meaning of *constancy*, of *persistence*.

The theo-ontological tradition, of course, simply turns this relationship upside down: instead of acknowledging the original "cosmic seat" of the meaning of constancy within the kinematical aspect of reality, it starts with the idea usage of constancy (sometimes twisted into the idea of God's *immutability*) and then in turn attempts to explain *endurance* within created reality by deriving it from God's 'immutability'. This is exactly what Descartes did in a letter to Mersenne (April 15, 1630), where he argues that the foundation for the eternal validity of natural laws is found in the *unchangeability* of God (quoted by Pannenberg, 1993:116).

5.15.4 *Transcendence approached from 'within'*

How can we account for God's transcendence without becoming a victim of *negative theology* and without advocating the other extreme position which simply (theo-ontologically) *duplicates* the creational diversity by projecting it 'into' the 'essence' of God and then derives creaturely properties from God?

1 Heidegger is therefore justified in affirming (in 1924) that the invariance [constancy!] of Einstein's equations in respect of arbitrary transformations represent the positive side of his theory: "Man übersieht leicht über dem Destruktiven dieser Theorie das Positive, daß sie gerade die Invarianz der Gleichungen, die Naturvorgänge beschreiben, gegenüber beliebiegen Transformationen nachweist" (Heidegger, 1992:3)

The constructive service to be rendered by a non-reductionist ontology is to help theologians understand that, in order to speak of God we do not have access to terms that do not proceed from and make an appeal to what is given *within* creation. By using these (creational) terms, however, we precisely want to convey the conviction that God *transcends* all of creation. But how can we continue this claim when we are 'doomed' to do this in a "creational way" by using "creational terms"?

Of course the situation is still further complicated when the creational location of terms employed in our speaking of God is not recognized. Brunner, for example, states: "The omnipresence of God is his elevation above space whereas his eternity and immutability constitute his elevation above time" (Brunner, 1972:272). Brunner does not realize that he has to *use* spatial terms in order to assert that God is 'above' space! Similarly, when we refer to God as *causa sui*, we first have to realize that it is only within the *physical aspect* of reality that we discover the *primitive meaning* of *energy-operation*, *causing* certain *effects*.

It simply seems to be unavoidable that the difference between God and creation can only be explained while inevitably using certain *creational terms* – and we have accounted for this inevitability by introducing the (philosophical) epistemological distinction within knowledge, namely between *conceptual knowledge* and (the limits of concepts transcending) *idea-knowledge* (see page 176 above).

Surely the Bible, as a concrete book, has a function within all aspects of reality and its central appeal touches the religious root-dimension of reality. But when it concerns God *as* God the biblical mode of speech fully employs modal terms and metaphors. God as *Father* and God as *King* are the most familiar metaphors used in a concept transcending way. In addition, the Bible uses terms from every aspect of creation in a similar *concept transcending manner*: God is *unique* (number), God is *omnipresent* (space), God *is* what He *is* (Ex. 3:14 – kinematic: persistence), God *acts* (physical), God is *life* (biotic), God's *will* (sensitive psychic aspect), God's *counsel* (logical-analytical), God's *power* (cultural-historical), God's *name* (sign mode), and so on.[1]

At this point, the theological tradition explored the consequences of the Greek substance concept with its distinction between *being* (essence) and *appearance*. The foundational coherence between the kinematic aspect (with its core meaning of constancy-persistence-endurance) and the physical aspect (with its core meaning of activity-change-dynamics), makes possible our awareness of identity (persistence amidst change). This relation of *constancy* and *dynamics* was distorted by the speculative substance concept, assuming that something exists in itself independent of anything beyond itself. Owing to the initial switch to space in the metaphysics of being of Parmenides and his followers, the coherence between constancy and dynamics acquired a spatial

1 We argued that whenever modal terms are used to designate what functions or appears within the confines of a specific aspect, they are employed in a *conceptual* manner, and when those terms are put in service of designating whatever transcends the boundaries of the aspect under consideration, such modal terms are used in a *concept transcending* way.

projection, embedded in the inside-outside opposition, elaborated by Plato and Aristotle into the *essence – appearance* distinction.[1] The distinction between essence and appearance then informed the theological tradition and inspired it to distinguish between God's self-knowledge (*theologia archetypa* directed towards God as He is "in Himself") and the knowledge through which He revealed Himself to us (accommodated to creation – *theologia ectypa*). All of this flows from the substance concept. In addition this distinction is also closely connected to what is called *communicable* and *incommunicable properties* of God. Bavinck explains that the *theologia archetypa* concerns the knowledge with which God knows himself and that the *theologia ectypa* is the knowledge of God as accommodated and 'anthropomorphized' to be suitable for the finite human consciousness:

> Nonetheless it contains the true conception that the theologia ectypa, which is granted to creatures through the revelation, is not the absolute self-knowledge of God, but that knowledge of God as it is accommodated to and made suitable for the finite consciousness, therefore anthropomorphized.[2]

The assumed (unconditioned) transcendence of God is said to concern the being of God outside time and above all laws, unknowable in this transcendence. This may lead to the idea that God's *accommodation* requires that God subjects himself to the laws for creation. In terms of a biblical Christian orientation, insofar as he was fully human, the human nature of Christ certainly was subjected to the laws for creation, just as God's *revelation* (in Scripture and 'nature' – creation) is subject to God's laws. But instead of 'pulling' God 'down' below His law in order to account for the possibility of our knowledge of Him, one can argue that within creation, God revealed Himself in such a way that we as creatures can realize that, as Creator, Law-Giver, Sustainer and Redeemer, God transcends creation and is not a creature subject to creational laws.

The distinction between conceptual knowledge and concept-transcending (idea-)knowledge first of all enables us to see how the first four modal aspects of reality play a key role in the *theo-ontological* tradition. The meaning of number is implicitly presupposed in the idea of the *unity* of God (speculatively understood in the sense of *simplicity*). The foundational coherence between the kinematic aspect (with its core meaning of constancy-persistence-endurence) and the physical aspect (with its core meaning of activity/change/dynamics), that makes possible our awareness of *identity* (persistence admidst change), serve as the foundation for the distinction between God's self-knowl-

1 It continued to be associated with independence as well – just compare Descartes's definition: "By substance we can conceive nothing else than a thing which exists in such a way as to stand in need of nothing beyond itself in for its existence" (*The Principles of Philosophy*, Part I, LI).

2 "Desniettemin ligt er de ware gedachte in, dat de theologia ectypa, welke door de openbaring aan schepselen geschonken wordt, niet is de absolute zelfkennis Gods, maar die kennis Gods, gelijk ze geaccommodeerd is naar en geschikt gemaakt is voor het eindig bewustzijn, dus geanthropomorphiseerd" (Bavinck, H. 1918. *Gereformeerde Dogmatiek*, I. 6, 4, p.144).

edge (*theologia archetypa* directed towards God as He is "in Himself") and the knowledge through which He revealed Himself (accommodated to creation) to us.

In other words, the distinction between "God in Himself" and "God as revealed to us" is entirely dependent upon an insight into the inter-modal coherence between the kinematic and the physical aspects of reality, embedded into the space-metaphysical opposition of inside and outside (essence-appearance). By and large the idea use of the kinematic meaning of constancy – in the garb of the idea of God's identity (twisted into the idea of "God-in-Himself") – found its strongest ally in the meaning of *space*, for in all instances where the idea of "God-in-Himself" surfaces, God's *transcendence* above creation is always emphasized. What is not realized, of course, is that the terms 'transcendence' and 'above' are *spatial* terms employed in a concept-transcending manner.

At the same time, the theological tradition says a number of specific things about "God-in-Himself" and then it adds the remark that these things are so 'elevated' that in their absoluteness, they cannot be 'communicated' to us as mere creatures – as Bavinck phrases it: "For the knowledge, which God has of itself, is absolute, simple, infinite, and in its absoluteness incommunicable to the finite consciousness."[1] If God's 'essence' is unknowable, then *how* do we *know* that it is *unknowable*?

Since its emergence in Greek culture, informed by the basic dialectic of form and matter, the *substance concept* stood in opposition to the idea of an integral cosmic coherence. Subsequently theological reflections on the relation between God and creation as well as regarding God's revelation itself, bought into the substance concept in order to account for the 'essence' and 'appearance' of God. This process was mediated by neo-Platonism (Plotinus – about AD 204–270 – see also Dionysius the Pseudo-Areopogite) and the thought of the Cappadocians (such as Gregor of Nyssa, about AD 335 – until after 394; Basil of Caesarea and Gregory of Nazianzus – fourth century), further developing the negative theology derived from Plato's dialogue *Parmenides*.

We may focus our discussion on the two options presented by the substance concept (while in both cases holding on to the *negative theological* emphasis on the *unknowability* of God – a legacy also continued in the thought of Dionysius the Pseudo-Areopogite and partially in Augustine and Thomas Aquinas):

(i) Affirm (!) that nothing can be affirmed (!) about God – in His transcendence (*aseitas*, as *causa sui*) – "God-in-Himself" is unknowable. This option then further explored the essence-appearance opposition by applying it to God – some ('essential') properties are *incommunicable* and

1 Bavinck says: "Want de kennis, die God van zichzelven heeft, is absoluut, eenvoudig, oneindig, en in haar absoluutheid onmededeelbaar aan het eindig bewustzijn" (Bavinck, 1918: 144).

other ('appearance') properties are *communicable*. We mentioned that Thomas actually holds that we know God by means of the *perfections* as they flow from Him into creatures (procedentibus in creaturas ab ipso – S.Th. I,13,3) – having commenced from a position where the creational diversity was first duplicated in God (this is typical of the circle entailed in all forms of *theo-ontology*: take something *from creation*, position it within God and then *copy* it back to creation).

(ii) The second option does not differ from the first one regarding the elevated unknowability of "God-in-Himself" – the only difference is that instead of accommodating creation to God (projecting all the creaturely perfections into the esse(nce) of God), the unknowable God accommodated Himself to creation by assuming creational properties (an implication of this stance may add that thus God subjected Himself to the laws of creation).[1] Summarily: in the second option, still as the counter-pole of the *esse(nce)*-part (namely "God-in-Himself"), God accommodated Himself to the creational diversity in order to explain the 'appearance' (revelation) of God to creatures.

My contention is that the entire distinction between what is *unknowable* and *knowable* in God is ill-founded, just as much is the idea that God has to subject Himself to the laws of creation in order to reveal Himself to us in an 'accommodated' form, is misguided, for it inevitably leads to a position where it is asserted that our knowledge of God is not authentic, since it is mediated through an 'accommodated' revelation. By contrast, in terms of the distinction between a conceptual and an idea use of modal terms, the straightforward positive biblical account, saying that God is life, God is love, God is omnipotent, God is onmnipresent, God is just,[2] and so on, could be accounted for in the following manner.

When it is said that God is love, just or wise, then there is nothing "unkown (and therefore different)" *behind* this revelation, for if it were the case, God turns into a *Deus absconditum*, a God that cannot be revealed. God simply **is** the love, wisdom and justice the Bible tells us about, but in His love, wisdom and justice He transcends what we can conceptually know. Therefore the only remaining (biblical) option seems to be that we can take God's Word seriously and accept that God **is** the love, wisdom and justice the Bible asserts, while at the same time acknowledging that through an idea use of ethical (love), logical (wisdom) and jural (justice) terms, our knowledge of God can approximate God's love, wisdom and justice without conceptually encompassing any one of these affirmations. We can *believe* God's Word – God *is* the love He says He is; there is not an unknowable 'essence' behind what is revealed. Rather, in the (trustworthy) love He says He is, God transcends whatever we can conceptualize of God.

1 In this case God's accommodation assumes a functioning within the various modal aspects – which means that all the creational modal terms apply to God in a *conceptual* way.

2 One may add the words of Paul regarding God's eternal power and divine nature as well (Rom.1:19-20).

Therefore, instead of 'pulling' God 'down' below His law in order to account for the possibility of our knowledge of Him, one alternatively can argue that *within* creation we are equipped with the *cognitive ability* to exceed the confines and limitations of conceptual knowledge, namely in the use of *concept transcending knowledge*. Our knowledge of God through His revelation merely explores this "built-in" capacity of (concept transcending) human knowing. As creatures we can therefore realize, in a creaturely way, that, as Creator, Law-Giver, Sustainer and Redeemer, God transcends creation and is not subject to creational laws in any way.

Consequently, it seems as if upholding the distinction between conceptual knowledge and concept-transcending (idea-)knowledge may safe-guard us both from the pitfalls of negative theology and of the accommodation idea subjecting God to His laws for creation.

The difference between the Cappadocian-Reformational (CR) legacy and the Augustine-Anselmus-Aquinas (AAA) tradition seems to be a matter of *emphasis*. Whereas the former (CR) holds on to an elevated *aseitas*, *causa sui* or *essence* of God, elevated above all appearances, the latter (AAA) projects these appearances (the diversity of creational properties – or, as Aquinas calls them, *perfections*) into the 'essence' of God – and then derive the creational diversity once again from God's essence – through the multiple ways in which the essence of God can be copied. We noted that Thomas Aquinas inherited the opposition of 'essence' and 'appearance' from the same metaphysical Greek concept of substance. In his *Summa contra Gentiles* (I,34) and *Summa Theologica* (I,13,1), Thomas explains that we can know God through His creatures because, in an eminent way, God bears all the perfections of things within Himself. We know God by means of these perfections as they flow from Him into creatures (procedentibus in creaturas ab ipso – S.Th. I,13,3).

The two sides of the coin are therefore: either elevates the essence of God into a realm of unknowability (which then needs the idea of accommodation of this unknowable God to creational terms in order to reveal Himself), or projects the creational diversity of 'perfections' into the essence of God before they are copied back into creation.

5.15.5 *Vollenhoven's "negative theology" in his Isagoogè*

Vollenhoven did not realize that modal terms could be employed both in a conceptual and in a concept-transcending manner. According to him, God is *sovereign* and God's law is *valid for* creatures (Vollenhoven, 1967:12, 14; Vollenhoven, 2005: 14, 16).[1] In his discussion of *technè*, Vollenhoven uses the word 'bewerking' (cultivation) in the context of (although in a broader sense than) the historical subject-object relation (Vollenhoven, 1967:135; 2005: 139). However, we argued that the core meaning of the cultural-historical mode is given in *control* or *formative power*. And "having power" is equi-

1 The sections quoted from Vollenhoven (1967) are identical to the corresponding sections in Vollenhoven (2005) – the 2005 page references will be added to the 1967 references.

valent to the notion of 'sovereignty' – from which we can see that affirming the *sovereignty* of God merely employs a modal historical term in a concept-transcending way.

Likewise the idea of the *validity* of the law (its *being in force*) derives from a concept-transcending use of the meaning of the physical aspect. When Vollenhoven accounts for the boundary (*grens*) between God and creation, he immediately provides a qualification: "With 'boundary' one does not think of spatiality: for spatiality itself belongs to what has been created."[1] Once again this idea of (the law as) boundary between God and what has been created is bound to a concept-transcending use of spatial terms. When Vollenhoven says that the law always stands above that for which it holds (is valid – 1967:14), then the same proviso is implied – one cannot be speaking of law in spatial terms when stating that the latter is elevated *above* creation. But in this negation of spatiality an idea use of the spatial term 'above' is present – showing that the negation of spatiality is accomplished by employing the meaning of space!

Van der Walt, in his biblically shaped ontology, entangles himself in the same vicious circle. His claim is that one cannot speak of God in spatial terms – why not? Because "God is elevated above space" (Van der Walt, 1976:128)! Once again the phrase "elevated above" reveals its modal seat in the spatial aspect, implying that it is only through the *use* of spatial terms that he can accomplish his goal, namely to argue that God *transcends* space. As soon as one distinguishes between conceptual knowledge and concept-transcending knowledge, the vicious circle of a negative theology disappears, for then it is no longer necessary to deny what is actually employed in an idea-manner.

5.15.6 *Transcending a metaphysics of Being*: *Jean-Luc Marion*

It is noteworthy that Marion published an article explicitly addressing the issue of a theo-ontology, phrased "onto-theology" (see Marion, 1995). Particularly in a book on *Being without God* (see Marion 1982 and 1995a) he argues against the metaphysics of being in order to arrive at a different understanding of God. However, it appears that he does not realize that the crux of the notion of 'being', as it surfaced in the thought of Thomas Aquinas (with God as the *Highest Being*, the *ipsum esse*), is found in the space metaphysics that originated in the geometrization of Greek mathematics. The idea of *being* (metaphysically) overextends the meaning of space. What is remarkable, however, is that Marion believes that he can escape from this metaphysics of being merely by saying that God's other-directed love *transcends* being (see Bracken, 2001:24 ff.). In order to renounce a metaphysics of being the word *transcendence* is employed, thus showing that he did not succeed in avoiding the spatial point of entry, for in saying that God transcends being the modal meaning of space is still used in a concept-transcending manner.

1 "Bij 'grens' denke men niet aan ruimtelijkheid: de ruimtelijkheid zelf immers behoort tot het geschapene" (Vollenhoven, 1967:13).

5.16 Cosmic time

In our discussion of the Achilles' heel of positivism, we alluded to the inevitability of using *terms* that are not observable in the way in which we acquire so-called "sense data." What is striking in this positivistic tradition is that the idea of time is simply accepted as something obvious, without applying their own criterion of meaning to it. But let us do this by asking the following questions: with which 'sense' do we *perceive* time? Can we *hear* time? Can we *smell* it? Do we *feel* it? or: Do we *taste* it? The absurdity of these questions demonstrates the fact that time is not similar to any observable entity.

That time cannot be identified with any single modal aspect also follows from this consideration. It is perfectly meaningful to speak of *temporal* reality, but it does not make sense to characterize reality exlusively in terms of one aspect only (such as the Pythagorean conviction that everything is *number*, the materialistic belief that everything is *physical*, the historicist claim that reality is *historical*, or the postmodern view that everything is *interpretation*). Dooyeweerd first developed his theory of modal aspects and entitary structures (designated as individuality structures), and only afterwards (probably in 1929) arrived at his first (radically new) understanding of what he calls *cosmic time*.[1] Traditional conceptions of time are constantly identifying time with merely one aspect of time – for example when "true time" is seen as *physical*, *emotional duration* (Bergson), that it is *existential* in nature (where existence is understood in a *historical* sense – Heidegger), and so on. In addition, most views of time are correlated with two powerful and historically long-standing but opposing views of *eternity*. The one equates time with an *endless duration* and the other views time, analogous to *spatial simultaneity* and the *historical present*, as something *timeless*.

The mere fact that we do speak of *temporal reality* rather suggests that time is a unique dimension of reality, cutting across the dimension of aspects and entities in its own way. Every attempt to *define* time invariably results in merely specifying *one aspect of time* – something repeatedly highlighted by Dooyeweerd in his articles on time. "Understandably traditional philosophy constantly attempted to delimit the time problem in a functionalistic manner. Time and again it identified universal cosmic time, which expresses itself at once in all modal aspects of reality because it provides the foundation for them all, with *one* of these modal aspects of time" (Dooyeweerd, 1939:6).

Affirming the temporality of *creation* implicitly assumes the *eternity* of the Creator. But if the biblical view of the eternal destination of the new humanity is accepted, then one may contemplate the idea of a *created eternity*. This was done by Scholastic thinking, in which created eternity (the so-called *aevum*) was distinguished from the *aeternitas increata* of God. Time was supposed to have a beginning and an end, while eternity lacks both. The intermediate posi-

1 In connection with the structure of a modal aspect, we briefly mentioned that cosmic time manifests itself at the law side of the various aspects as *time order* and at the factual side as *time duration* (see p.77).

tion of the *aevum* is then that it does have a beginning, but no end.

Interestingly, before Dooyeweerd developed his philosophy of time he realized that the human being cannot be enclosed within the dimensions of modal aspects and of entity structures. Dooyeweerd *first* realized that the human self-hood is *supra-modal* and *supra-structural*, then he developed his theory of *cosmic time* underlying and embracing the modal aspects and what he calls individuality structures – and on this basis then equated *supra-modal* with *supra-temporal*. From the fact that humanity has an *eternal* destination being human hinges on the boundary-line of *time* and *eternity* – justifying some or other sense of the *time-transcending* nature of the core meaning of being human. Ouweneel and Troost both argued that one should explore the latter direction and defended Dooyeweerd against the accusation of (neo-)Platonic or Scholastic influences. The term "full-temporal" ('tydsvolheid' / 'voltijdelijk') is used as an alternative designation of the "supra-temporal" (see Troost, 2004:79 ff.) It is noteworthy that Dooyeweerd sometimes does speak of the *transcending* concentration-point of human consciousness (Dooyeweerd, 1939:5).

No one will accuse someone who distinguishes between time and eternity (creation and Creator) of being *dualistic*, and the same applies to the distinction between modal and supra-modal. From a biblical perspective, one is certainly justified in contemplating what Dooyeweerd called the creaturely concentration of the temporal on eternity in the religious transcendence of the boundary of time (Dooyeweerd, 1939:5) – although this claim does not of necessity require the idea of an *eternal* selfhood. Dooyeweerd holds: "Yet, the supra-temporal concentration-point in human self-consciousness, which can only actualize itself in the religious concentration of all our functions on eternity, cannot be called *eternal* as such" (Dooyeweerd, 1939:2).

In his response to Van Peursen's critical remarks on *A New Critique of Theoretical Thought*, Dooyeweerd refers to the sense in which we "do transcend time in the center of our existence, even though at the same time we are *enclosed within time*" (Dooyeweerd, 1960:103). Yet he is not wedded to the term "supra-temporal," for in response to the objection raised by Van Peursen to the term "supra-temporal," he explains: "Now I am not going to enter once more into a discussion regarding the question if it is desirable to call the heart, as the religious centre of human existence, supra-temporal. It is sufficiently known that amongst the adherents of the *Philosophy of the Cosmonomic Idea* there is no consensus in this regard. Probably the term supra-temporal, with which I never meant a static condition but merely intended to capture a central direction of consciousness transcending cosmic time, can best be replaced by another one" (Dooyeweerd, 1960:137). The relative clause "transcending cosmic time" refers to the *central direction* of consciousness without claiming that consciousness *itself* transcends time. Although it may mark a change of view, in another respect it does not, because Dooyeweerd never altered his idea of *the transcendent, central religious dimension of reality*, whether or not designated as *supra-temporal*.

A few years later, the same issue surfaced in a discussion of the Annual Meeting of the philosophical association founded by Vollenhoven and Dooyeweerd. In a transcription of the 1964 discussion it is recorded that Steen asked Dooyeweerd about supra-temporality. Dooyeweerd said that sometimes he was inclined to "pull the hair from his head" for *ever having used those words*. Yet he still affirmed that the human being, in the centre of its existence, transcends the temporal cosmic order.[1]

What is of importance for Dooyeweerd, is the centrality of the human selfhood, in the sense that it cannot be identified with any modal aspect or with any "individuality structure" – and *not* the distinction between temporal and supra-temporal *per se*. Those who consider this distinction to be of central importance to Dooyeweerd may want to reconsider their position in the light of his just-mentioned remarks, indeed relativizing those earlier instances where he explicitly speaks of the supra-temporal heart.

5.16.1 *Modes of time*

The first remarkable feature of cosmic time is that it manifests itself within each modal aspect in accordance with the inner nature and unique meaning of that aspect. What is even more remarkable, is that the history of Western philosophy implicitly reveals an insight into different *modes of time* without having been able to relate it to a general theory of *functional modes*. We know that Immanuel Kant believes that time is a *form of intuition*. Yet, in spite of this *psychological one-sidedness*, he distinguishes three 'modes' of time that are strictly correlated with the way in which cosmic time manifests itself within the first three cosmic modalities. His striking remark reads: "The three modes of time are endurance, succession and simultaneity" (Kant, 1787-B:219).[2]

Leibniz juxtaposes time – as "an order of successions," with space – as "an order of coexistences" (Leibniz, 1965:199). Kant also realized that one has to distinguish between *succession* and *causality* – for although day and night *succeed* each other, it is meaningless to say that the day is the *cause* of the night or vice versa. In the 20th century, after modern physics was successful in transcending its mechanistic restriction, it was realized that physical time is intrinsically connected with causation, for the *effect* can never precede the *cause*. The numerical order of succession is *reversible* – manifested in the *plus* and *minus* directions of the system of integers, closed under the operations of addition, multiplication and subtraction. Saying that these operations are closed means that applying them to the set of integers always yield integers from the same set. When any two integers are added, multiplied or sub-

1 The transcription reads: "... waar ik soms de haren uit mijn hoofd trek (you understand?), dat ik deze uitdrukking ooit zo gebruikt heb, ik geloof niet dat ik deze uitdrukking ooit zo gebruikt heb. Ik heb wel dit gezegd, dat de mens in het centrum van zijn bestaan de tijdelijke, de kosmische tijdelijke orde te boven gaat. Dat is wel iets anders" (the *Dooyeweerd Archives* available at the "Historische Documentatiecentrum," Free University, Amsterdam – investigated during March, 2006).

2 "Die drei modi der Zeit sind *Beharrlichkeit*, *Folge* und *Zugleichsein*."

tracted, the result is always another integer. The symmetry of any spatial configuration – allowing being turned upside down or front-backwards – shows the reversibility of the spatial time order, and the same applies to the kinematic time order, for the mathematical description of a constant movement (like the swinging of a pendulum) is equally valid in both directions (a mere switch of the sign provides a description in the opposite direction). Finally, the physical time order is *irreversible.*

Einstein explains the difference between physical *irreversibility* and kinematic (mechanical) *reversibility*:

> On the basis of the kinetic theory of gases Boltzman had discovered that, aside from a constant factor, entropy is equivalent to the logarithm of the 'probability' of the state under consideration. Through this insight he recognized the nature of courses of events which, in the sense of thermodynamics, are 'irreversible'. Seen from the molecular-mechanical point of view, however, all courses of events are reversible (Einstein, 1959:43).

From another angle, the distinct manifestation of cosmic time within the first four modes is evident, particularly in the history of time measurement. It belongs to our general awareness of time: *earlier* and *later*, *simultaneity*, *time-flow* and *irreversibility* are well-known modalities of *time.*[1] In his work on the foundations of physics, Stafleu remarks (1980:16):

> This is most clearly shown by an analysis of the historical development of time measurement. Initially, time measurement was simply done by counting (days, months, years, etc.) Later on, time was measured by the relative *position* of the sun or the stars in the sky, with or without the help of instruments like the sundial. In still more advanced cultures, time was measured by utilizing the regular motion of more or less complicated clockworks. Finally, in recent developments time is measured via *irreversible* processes, for example, in atomic clocks.

What is striking in this whole development is that different *time orders* are used, the one after the other: the numerical time order of succession, the spatial order of simultaneity, the kinematic time order of constancy and the irreversible physical time order, expressed in the relationship of cause and effect.

Both Vollenhoven (see Tol, 1995:99 ff.) and Van Riessen (1970:113, 186) adhere to the view that time implies change. However, the (original) physical meaning of *change* refers to the meaning of those aspects that are foundational to the physical aspect. We have seen that change can only be detected on the basis of constancy (see page 163), and therefore endurance (persistence or constancy) is not only inherent in our awareness of time, for it forms an indispensible condition for change. Constancy and change are on an equal footing, similar to succession and simultaneity. As soon as the meaning of (physical) change is analyzed, its dependence upon these three foundational modes of time is evident, because change presupposes (the modal meaning of)

1 In his work on Einstein and Gödel we find various places where Yourgrau implicitly touches upon the difference between the kinematic and physical aspects (see Yourgrau, 2005:36, 114, 115, 123, 129, 133, 142.).

constancy, *simultaneity* and *succession*.[1]

Within the biotic aspect, the homogeneity of physical time is absent because the time phases correlated with the biotical time order – such as the duration of birth, growth, maturation, ageing and dying – are accelerated in the sense that the older a living entity gets, the quicker the process of ageing occurs. The French biologist, Lecomte du Noüy, confirmed this accelerated process of biotical ageing experimentally. Even the so-called "moment of death" eludes the scope of the physical understanding of time. Whatever criteria are applied by the biologist, only once they *have been* applied and the living entity (plant, animal, or human being) is declared 'dead', the on-looking physicist may look at a physical clock and note the (thus externally correlated) "moment of death."

The sensitive mode adds its own unique modal meaning to the experience of time, for whereas it may *feel* as if a boring event takes hours, something intriguing or capturing one's attention may *feel* as if time passed very quickly. Pursuing an argument in a logical sense is only successful when *conclusions* are reached on the basis of *premises*. Even if the physical sequence of words mentions the conclusion *before* the premises, the *logical time order* (*prius et posterius*) will always be such that, in a *logical sense*, the premises *precede* the conclusion.

Similarly, within each of the post-logical aspects, the dimension of cosmic time "takes on" the original meaning of that specific aspect. Although the awareness of *past*, *present* and *future* rests upon a "more-than-modal-historical" reality (as correctly pointed out by McIntire – 1985:95 ff.), the demarcation of truly historical periods, eras and epochs is dependent on the functional time order within the cultural-historical aspect. Only when truly modal (and typical) *historical* criteria are applied, is it possible to understand the cultural meaning of historical eras. Such an assessment is always related to what are considered the historically *significant* events and tendencies that surfaced, and eventually became direction-giving and dominant within a particular era. If physical time was the only 'real' time, it would have been impossible to speak of peoples who are still living in the age of 'soft' cultures (predating the stone age) today, or about dwelling places of which one could say that time "stood still."

The sign mode in turn reveals the meaning of cosmic time in its own way, for the temporal effects of punctuation marks (or pausing in speech acts), are all relevant to the intended meaning of language users. Likewise an awareness of social priorities is a reminder of social time – even in the case where one will allow an important person to go ahead in spite of one's own haste. Everyone will immediately understand that *interest* is intrinsic to *economic time* (not to mention the well-known expression: "time is money"). Within the aes-

1 Interestingly Heidegger holds the Platonic view that numbers are not *earlier* or *later* because according to him they are not in time at all: "Die Zahlen sind nicht früher oder später, weil sie überhaupt nicht in der Zeit sind" (Heidegger, 1992:18).

thetic aspect, the dimension of cosmic time takes on a nuanced diversity of forms and shapes – depending upon the typical nature of different kinds and genres of art – such as the *performing arts* (bound to a limited duration and filling this time-span with a unique aesthetical expression), *literature*, and for example *painting*. But even in spite of the apparent timelessness of paintings and works of sculpture, they not only objectively *endure* over time, but in an internal sense, also bring to expression their own aesthetic presence.

But perhaps hanging on to the "merely physical" nature of time receives its heaviest blow from the nature of jural time, for within this sphere, one sometimes encounters a different 'calendar', recognizing no public holidays and Sundays in its contractual or legislative "count-down," and one also has to acknowledge laws with a retroactive effect. Through a declaration of age (*venia aetatis*) or as an effect of getting married, the jural time involved in "coming of age" may differ from the generally specified age of majority in the legal order of Western states.

Courtship and eventually getting engaged and married obey the normative time order of the moral aspect – although the duration of these successive events may vary considerably. Finally, within the certitudinal aspect, practically all religions distinguish an order of spiritual growth, correlated with a factual enrichment and maturation in faith. Through the eye of faith, the temporal is appreciated with a view to eternity.

Implicit in the preceding brief discussion of cosmic time, is the difference between the law side and the factual side of cosmic time. Time order at the law side of the various aspects is always correlated with factual time duration. For example, although all living entities are subjected to the *same* biotical time order of birth, growth, maturation, ageing and dying, the individual life-span of living entities may vary from a factual duration of one day up to thousand or more years.

Of course this perspective on the correlation between law side (time order) and factual side (time duration) equally applies to the modal structure of every single modal aspect in general.

5.17 The correlation of law side and factual side within the natural aspects of reality

5.17.1 *Do 2+2 really equal 4?*

Particularly within the natural sciences, the positivistic legacy declared 'facts' to be sacrosanct. Is it not true that being a Christian, a Communist or a Muslim does not make a difference to the 'fact' that 2+2=4?

It is still commonplace to appreciate mathematics as a discipline in which rational argumentation is convincing and compelling. Fern writes:

> Mathematical calculations are paradigmatic instances of universally accessible, rationally compelling argument. Anyone who fails to see "two plus two equals four" denies the Pythagorean Theorem, or dismisses as nonsense the esoterics of infinitesimal calculus forfeits the crown of rationality (Fern,

2002:96-97).

This explicit and unqualified trust in "mathematical reason" apparently did not take notice of Morris Kline's assessment, almost three decades ago:

> The developments in the foundations of mathematics since 1900 are bewildering, and the present state of mathematics is anomalous and deplorable. The light of truth no longer illuminates the road to follow. In place of the unique, universally admired and universally accepted body of mathematics whose proofs, though sometimes requiring emendation, were regarded as the acme of sound reasoning, we now have conflicting approaches to mathematics. Beyond the logicist, intuitionist, and formalist bases, the approach through set theory alone gives many options. Some divergent and even conflicting positions are possible even within the other schools. Thus the constructivist movement within the intuitionist philosophy has many splinter groups. Within formalism there are choices to be made about what principles of metamathematics may be employed. Non-standard analysis, though not a doctrine of any one school, permits an alternative approach to analysis which may also lead to conflicting views. At the very least what was considered to be illogical and to be banished is now accepted by some schools as logically sound (Kline, 1980:275-276).

We may question Fern's claim by stating something different, namely by asserting that $\overline{2}+\overline{2}$ is actually equal to $\sqrt{8}$! One reaction may be to refer to 2 fingers and another 2 fingers, which indeed add up to 4 fingers. But unfortunately the issue is more complicated than it may seem at first, for when the *context* is changed to "spatial addition" – like in vector analysis, where a vector possesses both *distance* and *direction*, indicated by overscoring the vector symbols – the mentioned *geometrical sum* of the two vectors $\overline{2}+\overline{2}=\sqrt{8}$, is encountered.[1] In the first half of the 19th century, Grassmann has already introduced the idea of a vector. He designated such a line segment ('Strecke') with a specific direction and length on a specific straight line as a 'linienteil', which became more generally known in German literature as a *vector* ('Vektor').[2] Hedrick and Noble mistranslated "Richtung und Länge" as "length and sense," perhaps because later on in the same original German paragraph Klein himself used the German words "Länge und Sinn") – also explaining why the word-order of "Richtung und Länge" was reversed to "length and sense" by them (see Klein, 1939:22).

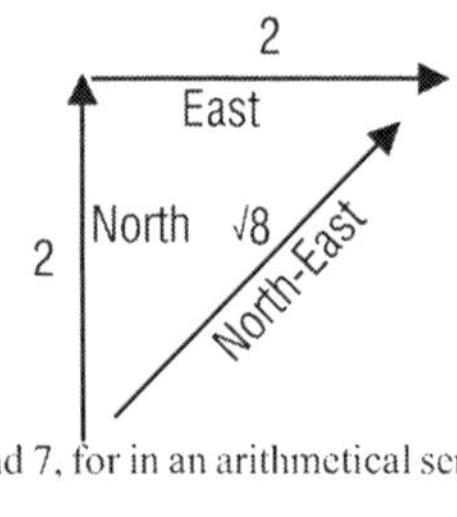

When a person walks 2 miles north and afterwards 2 miles east, then that person will be $\sqrt{8}$ miles away from the initial point of departure. Therefore, we now clearly have two different

1 The same point could be made by using the numbers 3, 4, 5, and 7, for in an arithmetical sense 3+4=7 and in the case of a geometrical sum $\overline{3}+\overline{4}=\overline{5}$.

2 "Graßmann nannte eine solche Strecke bestimmter Richtung und Länge auf einer bestimmten Geraden einen *Linienteil*; jetzt ist in der Deutschen Literatur der Name Vektor üblicher" (Klein, 1925:24).

kinds of facts: a *numerical fact* (2+2=4) and a *geometrical fact* ($\bar{2}+\bar{2}=\sqrt{8}$). Moreover, 'facts' are not simply 'facts', for they are always *delimited* and *determined* by some or other *order*. In our example, the underlying *order diversity* therefore makes possible the said distinction between *numerical* and *spatial* facts.

The identification and distinguishing of unique modal functions in reality, such as the quantitative and the spatial, is not independent of an encompassing (philosophical) view of reality, while the latter, in turn, is ultimately in the grip of a central and direction-giving commitment springing from the core dimension of being human. That the *human subject* is involved in every thought activity is not the issue – the decisive question is whether or not the human subject acts while observing the normative meaning of logical analysis and the underlying cosmological principle of the *excluded antinomy* (see the explanation on page 183). Nonetheless the classical (rationalistic) science ideal is still very much alive when Fern differentiates between morality (dependent upon "subjective faith") and 'mathematics' and 'science' "that allows us to prove – establish beyond a reasonable doubt in universally accessible, rationally compelling terms":

> The frightening thought, so far as moral convictions go, is that we cannot get around the element of subjectivity, that in the end it all comes down to what we see (or fail to see) in the wolf's eye. In part, I want to allow that this is so; there is no method, scientific or otherwise, that allows us to prove – establish beyond a reasonable doubt in universally accessible, rationally compelling terms – that a particular moral outlook is correct. In this sense, morality rests on faith in a way not true of mathematics, logic or, even, at its core, modern science – the affirmation of which serves as a measure of one's basic reasonableness (Fern, 2002:95).

Fern may be surprized to learn that modern mathematics indeed shelters different *schools of thought* (as we have noted in Chapter 1, page 3). For example, Cantor's transfinite arithmetic (with its hierarchy of transfinite numbers) is appreciated by Hilbert as a *paradise* created for us by Cantor,[1] while at the same time it is discarded by the intuitionistic mathematician, Heyting, as a *phantasm* (Heyting, 1949:4). In addition to the above-quoted statement by Kline about the "loss of certainty" in mathematics, we may highlight what Stegmüller remarked in respect of intuitionistic mathematics.

> The special character of intuitionistic mathematics is expressed in a series of theorems that contradict the classical results. For instance, while in classical mathematics only a small part of the real functions are uniformly continuous, in intuitionistic mathematics the principle holds that any function that is definable at all is uniformly continuous (Stegmüller, 1969:331).

The Dutch logician, Beth, underscores this remark in his statement quoted in Chapter 1:

> It is clear that intuitionistic mathematics is not merely that part of classical

1 "Aus dem Paradies, das Cantor uns geschaffen [hat], soll uns niemand vertreiben können" (Hilbert, 1925:170).

> mathematics which would remain if one removed certain methods not acceptable to the intuitionists. On the contrary, intuitionistic mathematics replaces those methods by other ones that lead to results which find no counterpart in classical mathematics (Beth, 1965:89).

It should be reiterated that, what is presupposed in the distinction between *numerical* and *spatial* facts, is the (implicitly) assumed *modal irreducibility* of number and space. Yet this assumption is by no means what is found in the history of mathematics, for we noted that the initial Pythagorean arithmeticism has already beeen replaced by the geometrization of mathematics in the Greek era, until the 19th century eventually claimed – through the work of Cauchy, Weierstrass, Dedekind and Cantor – to have arithmetized mathematics completely once again. Clearly at this point the views of mathematicians depart. That different schools of thought emerged in modern mathematics on the basis of alternative views with respect to the nature of and the coherence between various modal aspects should not surprise us, for an apparently innocent statement of fact, 2+2=4, now suddenly 'deteriorated' into a show-case of the philosophical challenge to account for what we have called the "coherence of irreducibles" (see Chapter 1, page 8).

But there is something else present in this distinction between two kinds of facts, namely the reference to the operation of *addition*. Modern mathematical set theory normally first of all approaches this domain in terms of the algebraic structure of *fields* – where the (binary) operations called addition (+) and multiplication (×) meet the field axioms (specified as *laws*). A *field* is defined as a set *F*, such that for every pair of elements *a*, *b* the sum *a*+*b* and the product *ab*, are still elements of *F*, subject to the associative and commutative laws for addition and multiplication, combined with the presence of a *zero element* and a *unit* (or *identity*) *element*.[1] This definition of a field is then expanded to that of an ordered field[2] and is finally combined with the idea of completeness. Modern *axiomatic set theory* developed in a way in which the corrrelation between law side and factual side is clearly highlighted. The basic operations at the *law side* of the numerical aspect only have meaning when they are correlated with numerical subjects at the *factual side*. If any collection of arithmetical subjects is designated as a *set*, then speaking of a *system* is only possible when both the law side and the factual side are involved.

For example, the set of natural numbers finds its determination and delimitation in the operations of addition and multiplication that are *closed* over the natural numbers – in this sense adding or multiplying any two natural numbers will always yield another natural number.

System of natural numbers N_s	Operations at the law side:	(+, ×)
	Numerical subjects:	N_t = (1, 2, 3, ...)

1 See Bartle (1964:28) and Berberian (1994:1 ff.).

2 "The existence of least upper bounds differentiate the real numbers from all other ordered fields" (Berberian, 1994:11-12).

Once further operations are considered at the law side of the arithmetical aspect, an extended domain of numerical subjects is required at the factual side, because of the strict correlation between law side and factual side. This means that, when the operation of *subtraction* is added to those of addition and multiplication, the correlating set of integers (Z_t) is found, and taken together we have the system of integers. Likewise, extending the operations at the law side of the arithmetical aspect by introducing *division* not only constitutes the correlating set of *fractions*, but also the system of rational numbers.

System of	Operations at the law side:	$(+, \times, -, \div)$
rational numbers Q_s	Numerical subjects:	$Q_t = (\frac{a}{b}, a,b \in Z_t / b \neq 0)$

This explanation, in terms of the strict correlation between operations on the law side and numerical subjects at the factual side, is *formally* similar to the way in which Klein introduces negative numbers and fractions (by means of the reverse operations of addition and multiplication – see Klein, 1932:23 ff. & 29 ff.).

Against this background, we may now point out in conclusion that the *systematic* mathematical statement 2+2=4 does not designate a *brute fact* (a fact *in itself, an sich*), since the factual *relation* it specifies for the numerical subjects involved exhibits the *measure* of the numerical *law of addition*.

Therefore, the expression 2+2=4 is nothing but a *law-conformative numerical fact*, just as $\overline{2}+\overline{2}=\sqrt{8}$ represents a *law-conformative spatial fact*. Law-conformity is always a universal feature at the *factual side* of reality.

5.17.2 *Is a line the "shortest distance between two points?"*

In an earlier phase of his development, Russell 'corrected' the definition: "A straight line, then, is not the shortest distance, but is simply the distance between two points" (Russell, 1897:18). This conception is still found in the thought of contemporary mathematicians. Mac Lane, for example, explicitly says: "The straight line is the shortest distance between two points" (Mac Lane, 1986:17). The three key terms in this statement concern spatial configurations, namely the terms 'line', 'point' and 'shortest'. Yet the crucial element maintained in Russell's 'improved' definition echoes something of our awareness of numerical relations, namely *distance*. If this is indeed the case it may turn out that an analysis of this statement will be entangled in the consideration of arithmetical and spatial issues at the same time, which means that it cannot be analyzed purely in spatial (or geometrical) terms.

The first observation in this connection is to establish that the notion of a 'line' as the 'distance' between two 'points' concerns *spatial* realities. A line is a spatial configuration, not an arithmetical one. Yet the crucial question is: how can one designate the 'distance' between two points? The answer is: by specifying a *number* (for example by saying it is 3 inches long). The problem with this answer is that something *spatial*, namely a 'line', is now apparently equated with something *numerical*, namely 'distance'! (In passing we note that the term 'distance' in yet a different way evinces an intrinsic connection with the meaning of number, because a line is supposed to be the distance between **two** points. Multiplicity is numerical; a multiplicity of *points* is spatial.) Does this mean that the domains of space and number coincide? If this is the case, then the question of priority arises: is space numerical (then a 'line' is identical to 'distance', that is, to number), or is number spatial (then number, that is, 'distance' is identical to space, that is, a 'line')?

The situation is further complicated by the fact that the number specified (such as '3') does not stand on its own, that is, it appears within a *non-numerical context* – one in which the general issue of *magnitude* prevails, with *length* as a *one*-dimensional magnitude. To add insult to injury, we now suddenly have to account for another spatial notion: *dimensionality*! But still new problematic questions are generated, for in our example of "3 inches" – related to the extension of a line – the reference to *length* brought with it the (spatial) perspective of *one* dimension (*length* specifies magnitude in the sense of *one dimensional extension*). On the one hand, this suggests *extension*, which presumably essentially belongs to our awareness of space, while at the same time, just as in the case of the term 'distance', it reveals a connection with number, for one can speak about 1-dimensional extension (magnitude; that is, of length), 2-dimensional extension (magnitude; that is, of area), 3-dimensional extension (magnitude; that is, of volume), and so on.[1] Even if priority is given to the spatial context by admitting that the distinction between different dimensions is indeed something spatial, no one can deny that in some or other way number here plays a foundational role, for without number, the given specification regarding 1, 2, or 3 dimensions is unthinkable.

Clearly, the term 'distance' is embedded within the domain of space and it also evinces a strict correlation between an *order of extension* (the law side of this domain – that is, dimensionality) and *factually extended spatial subjects* – spatial figures (such as 1-dimensional ones, i.e, lines), 2-dimensional ones, that is, areas) and 3-dimensional ones (that is, volumes).

The complexities generated by a consideration of extension in the sense of an order of extension and of factually extended spatial subjects (spatial figures), add weight to the suggestion that, although something like a line has a spatial nature, its extended character cannot reveal its true spatial meaning without showing a dependence upon the (foundational) meaning of number. The reason for this acknowledgement is found in the intrinsic role of numeri-

1 In our discussion of the views of Lakoff and Johnson, we have already highlighted the fact that the notion of *dimension* reveals the inter-modal coherence between space and number (see p.146).

cal terms that are 'coloured' by space, such as *distance* and *dimension*. Within a numerical context, such as what is mathematically known as "real analysis," one can easily dispense with the concept of distance. But text books on real analysis sometimes still acknowledge that the geometric meaning of the term 'distance' may be useful, for "instead of saying that $|a - b|$ is 'small', we have the option of saying that a is 'near' b; instead of saying that '$|a - b|$ becomes arbitrarily small' we can say that 'a approaches b', etc." (Berberian, 1994:31).

Two years after Russell gave his mentioned modified definition of a line as the distance between two points, the German mathematician, David Hilbert, published his axiomatic foundation of geometry: *Grundlagen der Geometrie* (1899). In this work, Hilbert abstracts from the contents of his axioms, based upon three *undefined* terms: 'point', "lies on," and 'line'. Suddenly the term 'distance' disappears. The next year, when Hilbert attended the second international mathematical conference in Paris, he presented his famous 23 mathematical problems that co-directed the development of mathematics during the 20th century in a significant way – and in Problem 4, he provides a formulation that opens up a new perspective on this issue, for instead of speaking of the *distance* between two points, he talks of a straight line as the (shortest) *connection* of two points.[1]

This choice of words completely avoids the traditional view, even found in the mentioned work of the mathematician Mac Lane, who still believes that a "straight line is the shortest distance between two points" (Mac Lane, 1986:17). The German term 'Verbindung' ('connection') does not *define* a line, since it *presupposes* the meaning of continuous extension.[2] Every part of a continuous line coheres with every adjacent part in the sense of being *connected* to it. Although it is tautologous to say that the parts of a continuous line are fitted into a *gapless coherence*, it says nothing more than to affirm that the parts are *connected*. In this sense, the connection of two distinct spatial points also highlights the presence of (continuous) spatial extension between the points that are connected to each other. In other words, two points cannot be connected by a third point, but only by means of a line, i.e. by means of *spatial extension*.

Therefore it is clear that, combined with the primitive terms employed in Hilbert's axiomatic foundation of geometry ('line', 'lies on' and 'point') the term 'connection' no longer equates a line with its distance. Once 'liberated' from this problematic bondage, alternative options emerge in order to account for the meaning of the term 'distance'. If *distance* is the 1-dimensional *measure* of factual (spatial) extension, then one can do two things at once:

(i) acknowledge the spatial context of this measure (1-dimensional magnitude) and

1 "[Das] Problem von der Geraden als kürzester Verbindung zweier Punkte" (see Hilbert, 1970:302).

2 Where Penrose explains Euclid's first postulate, he employs Hilbert's terminology: "Euclid's first postulate effectively asserts that there is a (unique) straight line segment connecting any two points" (Penrose, 2004:28-29).

(ii) allow for the reference to number that is evident both in the '1' of 1-dimensional extension and in the (numerically specified) *length* evident in 'distance', as a specified (factual) spatial magnitude.

The core meaning of space, related to the awareness of extension and dimensionality, now acquires a new appreciation, further supported by the undefined nature of the term 'line' in Hilbert's 1899 work. The message is clear: if the core meaning of space (extension) is indefinable and primitive, then it is impossible to attempt to *define* a line by using a non-original term within space, such as the term 'distance'. Distance as the measure of extension of a (straight) line depends upon and presupposes the existence of the line in its primitive 1-dimensional extension and can therefore never serve as its definition. Therefore the 'definition' of a (straight) line as "the distance between two points" (Russell) presupposes what it wants to define and consequently begs the question.

5.17.2.1 *What is presupposed in space?*

We have seen that every specification of a spatial configuration is unavoidably connected with terms reflecting some or other element of coherence with number (*magnitudes* and the *number of dimensions*). This outcome opens up the way to an alternative: investigate the consequences of the assumption that, although space and number are unique and distinct, they still unbreakably cohere. The new question is then: *what is the interrelation between the spatial and the numerical*?

If the measure of the factual (one dimensional) extension of a straight line could be specified by its *distance*, then the distance of a line not only presupposes its spatial extension, since it also presupposes the intrinsic interconnection between the meaning of space and the meaning of number. Various mathematicians appreciated this state of affairs. Paul Bernays (the mentioned co-worker of Divid Hilbert), for example, says that the idea of continuity is a geometrical idea, which analysis expresses in an arithmetical language (Bernays, 1976:74).[1]

The basic acknowledgement of the distinct domains of number and space does not solve the subsequent problems, for the following two issues are still in need of clarification:

(i) which one of these two domains is more fundamental, in the sense of *foundational*, to the other? and

(ii) how should one account for the interconnections (interrelations) between these two domains?

5.17.2.2 *Which region is more basic – number or space?*

Let us start with Bernays's approach where he considers the way in which one can distinguish between our *arithmetical* and *geometrical* intuition. He rejects the widespread view that this distinction concerns *time* and *space*, for according to him, the proper distinction needed is that between the *discrete* and the

1 "Die Idee des Kontinuums ist eine geometrische Idee, welche durch die Analysis in arithmetischer Sprache ausgedrückt wird" – see page 61 aove).

continuous.[1] But then the question recurs: what is the relationship between the 'discrete' and 'continuous'?

Fraenkel *et al.* even speak of a 'gap' in this regard, and add that it has remained an "eternal spot of resistance and at the same time of overwhelming scientific importance in mathematics, philosophy, and even physics" (Fraenkel *et al.*, 1973:213). These authors furthermore point out that it is not obvious which one of these two regions – "so heterogeneous in their structures and in the appropriate methods of exploring" – should be taken as a starting point. Whereas the "discrete admits an easier access to logical analysis" (explaining, according to them, why "the tendency of arithmetization, already underlying Zenon's paradoxes may be perceived in [the] axiomatics of set theory"), the converse direction is also conceivable, "for intuition seems to comprehend the continuum at once," and "mainly for this reason Greek mathematics and philosophy were inclined to consider continuity to be the simpler concept" (Fraenkel *et al.*, 1973:213).

Of course the modern tendency towards an arithmetized approach, particularly since the beginning of the 19th century, chose the alternative option by contemplating the primary role of number. Although Frege equated mathematics with geometry by the end of his life (consistent with the just-mentioned position of Greek mathematics), his initial inclination certainly was to opt for the foundational position of number. In 1884 he has already asked if it is not the case that the basis of arithmetic is deeper than all our experiential knowledge and even deeper than that of geometry?[2]

From our discussion of the difference between an arithmetical and a spatial sum, and in particular from our remarks about the term 'distance', it may be possible to derive an alternative view on the order relation between the regions of discreteness[3] and continuity. Suppose we explore our initial suggestion that *discreteness* constitutes the core meaning of the domain of number, and that *continuous extension* highlights the core meaning of space (see pp.87 ff.). Then these core meanings guarantee the distinctness (uniqueness) of each domain. The domain of number, with its sphere of arithmetical laws and numerical subjects, is then seen as being stamped or qualified by this core meaning of *discreteness*. Likewise the domain of space, with its sphere of spatial laws and spatial subjects, is then viewed as being qualified by the core meaning of *continuous extension*.

1 "Es empfiehlt sich, die Unterscheidung von 'arithmetischer' und 'geometrischer' Anschauung nicht nach den Momenten des Räumlichen und Zeitlichen, sondern im Hinblick auf den Unterschied des Diskreten und Kontinuierlichen vorzunehmen" (Bernays, 1976:81).

2 "Liegt nicht der Grund der Arithmetik tiefer als der alles Erfahrungswissens, tiefer selbst als der der Geometrie? " (Frege, 1884:44).

3 "Below we shall 'liberate' the idea of 'discreteness' from the arithmeticistic habit of distinguishing between discrete, dense and continuous sets – in order to allow *discreteness* to play its role as meaning-kernel of the numerical aspect that qualifies all kinds of number, even when such kinds of number may imitate in an antecipatory way spatial features (such as wholeness, divisibility and continuity)."

But we have seen that a basic spatial subject, such as a (straight) line, cannot be understood without some or other reference to the meaning of number, for observing the *measure* of the line's extension requires the notion of 'distance' that involves number, and since a line is a spatial figure extended in one dimension, it clearly only has a determinate meaning in subjection to the first order of spatial extension. We have argued that, in both domains (number and space), there is a strict correlation between the law side and the factual side. In the case of space it is therefore possible to discern a reference to number both on the law side and the factual side.

Speaking of one or more dimensions presupposes the meaning of number on the law side, and specifying the one-dimensional extension (magnitude) of something like a line presupposes the meaning of the number employed in the designation of the *length* of the line. The domain of number therefore appears to be more basic, because an analysis of the meaning of space invariably calls upon foundational arithmetical considerations.

5.17.2.3 *The interconnections between number and space*

A metaphorical way to capture this state of affairs is to use an image from human memory by saying that, within the meaning of space (both at the law side and the factual side), we discover configurations *reminding* us of the core meaning of number. We found a key element in all metaphorical descriptions in the connection between *similarities* and *differences*. This enabled us to argue that whenever differences are shown in what is similar, one may speak of *analogies*. We even broadened the scope of an analogy in order to include more than what is normally accounted for in a theory of metaphor. Our first designation has already achieved this goal, for whenever *differences* between entities and properties bring to expression what is *similar* between those entities or properties, we find instances of an *analogy*.

Whenever *entities* are involved in the figurative speech we employ, we advanced the view that such designations are considered *metaphorical*. Similarities and differences between modal functions represent a domain of analogies distinct from metaphors. Yet we have seen that when purely inter-modal connections (analogies) are metaphorically explored, an element of the entitary dimension of reality will always be present, such as found in the current metaphor of a person being 'reminded' of an original domain. Implicit in the nature of an analogy is the distinction between something *original* and something else, which 'reminds' one of what is originally given, but is now encountered in a *non-original* context, that is, within an *analogical* setting.

This is exactly what we have noticed in the terms 'distance' and 'dimension' – for in both cases we are *reminded* of the quantitative meaning of number. In terms of the idea of an analogy one can say that there is an analogy of number on the law side of the spatial aspect (one, two, three or more dimensions) and that there is an analogy of number at the factual side of the spatial aspect (magnitude – as the correlate of different orders of extension: in *one* dimension, magnitude appears as length, in *two* dimensions it appears as area; in *three* as volume, etc.).

This view articulates the general idea of the modal structure of an aspect explained earlier (see pp.77 ff.). Note that any description of modal aspects inevitably employs *metaphors* (involving entitary analogies). For example, one may say that aspects are 'points of entry' to reality, that they provide an 'angle of approach' to reality, and so on. Conversely, the modal aspects provide access to the dimension of entities – they may serve as *modes of explanation* of concrete reality (an expression we used frequently – see pp.60, 140, 157, 179). An account of the basic position of number can now be articulated in terms of the idea of analogies, for since basic *numerical analogies* are presupposed within the domain of space, the original meaning of number is indeed *foundational* for the meaning of space.

Like every other academic discipline, mathematics will always be confronted with the philosophical problems of uniqueness and coherence (unity and diversity). One facet of this foundational philosophical problem is given in the unavoidability of employing analogical terms, that is, in the use of terms reflecting the interconnection between different aspects. This follows from the fact that different modal aspects are interrelated in such a way that each of them, within its own structure, reflects the modal meaning of others. Physical extension, for example, shows some likeness with spatial extension.

However, in this moment of similarity, the modal difference is simultaneously expressed – spatial extension is continuous in the sense that it allows for an infinite divisibility (see p.163), whereas physical space is *not* continuous (since it is determined by the quantum-structure of energy) and is therefore not infinitely divisible (already in 1925, Hilbert has mentioned this difference – see Hilbert, 1925:164). Bernays also distinguishes between physical space and mathematical space: "Only through the contemporary development of geometry and physics did it become necessary to distinguish between space as something physical and space as an ideal multiplicity determined by spatial laws."[1]

Sensitive space, for example the sensitivity for distinct sensations on the human skin, may be experienced as continuous in spite of the fact that the stimuli are physically discontinuous (distinct) (see Gosztonyi, 1976, I:13).

In an e-mail correspondence, the physicist Stafleu (July 27, 2005) remarked that, "as far as I know in concrete physically qualified 'things' (like molecules), energy cannot be infinitely divided, for *within* such a thing energy is always quantized." One can of course revert to a modal functional (mathematical) *description* of processes involving energy (with reference to a *continuous variable*), but then recourse is taken to a *functional mathematical notion* – where it is indeed meaningful to maintain that such a continuous variable en-

1 "Erst durch die zeitherige Entwicklung der Geometrie und der Physik tritt die Notwendigkeit hervor, zwischen dem Raum als etwas Physikalischem und dem Raum als eine ideellen, durch geometrische Gesetze bestimmten Mannigfaltigkeit zu unterscheiden" (Bernays, 1976:37).

tails infinite divisibility. Maddy implicitly alludes to this distinction: "But it is also true that the appearance of, say, a continuous manifold in our best description of space-time does not seem to be regarded as establishing the continuity of space-time; the microstructure of spacetime remains an open question" (Maddy, 2005:455 – she refers to Burgess).

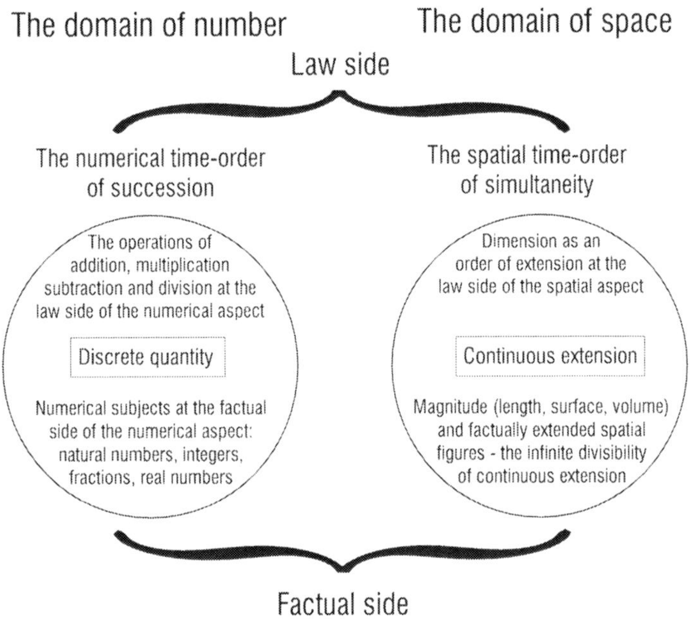

5.17.2.4 *The primitive meaning of space underlying Hilbert's primitive terms*

Within the arithmetical aspect, the factual relation between numbers is constituted as subject-subject relations – as present in the addition of numbers, the multiplication of numbers or establishing the numerical difference between numbers (subtraction). However, at the factual side of the spatial aspect, there are not only subject-subject relations (such as intersecting lines), for there are also subject-object relations, mainly expressed in the idea of a *boundary*.

Already in his abstraction theory, Aristotle has employed the notion of a boundary (or limit) – which is intuitively associated with *spatial* notions (Aristotle used the term *eschaton*). By the 13th century AD, Thomas Aquinas accounts for a one dimensional line by means of a descending series of abstractions. In contradistinction to natural bodies, all mathematical figures are infinitely divisible. The Aristotelian legacy is clearly seen as the *principium* of a line in his definition of a point (cf. *Summa Theologica*, I,II,2), which indicates the fact that a determinate line-stretch has points at its extremities ("cuius extremitates sunt duo puncta" – *Summa Theologica*, I,85,8). This legacy returns in a somewhat more general form in the 18th century (the era of *Enlightenment*). Kant remarks:

> Area is the boundary of material space, although it is itself a space, a line is a space which is the boundary of an area, a point is the boundary of a line, although still a position in space (Kant, 1783, A:170).

In 1912, Poincaré discussed similar problems. Concerning the way in which geometers introduce the notion of three dimensions, he says: "Usually they begin by defining surfaces as the boundaries of solids or pieces of space, lines as the boundaries of surfaces, points as the boundaries of lines" (cf. Hurewicz & Wallman, 1959:3). Although only related to three dimensions, Poincaré here provides us with an intuitive approach to dimension, implicitly stressing the unbreakable correlation between the law side and the factual side in the spatial aspect:

> ... if to divide a continuum it suffices to consider as cuts a certain number of elements all distinguishable from one another, we say that this continuum is of one dimension; if, on the contrary, to divide a continuum it is necessary to consider as cuts a system of elements themselves forming one or several continua, we shall say that this continuum is of several dimensions (Hurewicz & Wallman, 1959:3).

Before 1911, the problem of dimension was confronted with two astonishing discoveries. Cantor showed that the points of a line can be correlated one-to-one with the points of a plane, and Peano mapped an interval continuously on the whole of a square. The crucial question was whether, for example, the points of a plane could be mapped onto the points of an interval in both a continuous and one-to-one way. Such mapping is called homeomorphic. The impossibility of establishing a homeomorphic mapping between an "m-dimensional set and an (m+1)-dimensional set ($h > 0$)" was solved by Lüroth for the case where $m = 3$ (Brouwer, 1911:161, cf. also the footnote on page 161). Brouwer provided the first general proof of the invariance of the number of a dimension (see Brouwer, 1911:161-165). Exploring the suggestions by Poincaré, Brouwer introduced a precise (*topologically invariant*) definition of *dimension* in 1913, which was independently recreated and improved by Menger and Urysohn in 1922 (cf. Hurewicz & Wallman, 1959:4). Menger's formulation (still adopted by Hurewicz and Wallman) simply reads:

a) the empty set has dimension -1,
b) the dimension of a space is the least integer n, for which every point has arbitrarily small neighbourhoods, whose boundaries have dimension less than n (Hurewicz & Wallman, 1959:4, cf. p.24).

Whereas a spatial subject is always factually extended in some dimension (such as a one-dimensional line, a two-dimensional area, and so on), a spatial object merely serves as a *boundary* (in a delimiting way). The boundaries of a determined line-stretch are the two points delimiting it (with the line as an one-dimensional spatial subject). But these boundary points themselves are not extended in one dimension. Within one dimension, points are therefore not spatial subjects, but merely spatial objects, dependent upon the factual extension of the line. Yet a line may serve in a similar delimiting way within two dimensions – for the lines delimiting an area are not extended themselves in a two-dimensional sense.

Comment: Lonergan does not understand the spatial subject-object relation, because he clings to (sensory) space perception. He argues that "points and lines cannot be imagined." No matter how small an imagined dot may be, it still "has magnitude." "To reach a point all magnitude must vanish, and with all magnitude there vanishes the dot as well" (Lonergan, 2000:32). Without the one-dimensional extension of a line (as spatial *subject* – mistakenly designated by Lonergan as 'magnitude'), no point can exist, because a modal object is always dependent upon a modal subject. As a spatial subject, a determined line-stretch is delimited by its starting point and end point. In a similar fashion a surface can assume the role of a spatial object, namely when it delimits three-dimensional spatial figures (such as a cube).

In general, it can therefore be stated that whatever is a spatial subject in *n* dimensions, is a spatial object in *n+1* dimensions. A point is a spatial object in one dimension (an objective numerical analogy on the factual side of the spatial aspect), and therefore a spatial subject in *no* dimension (that is, in *zero* dimensions). In terms of the fundamental difference between a spatial subject and a spatial object, it is impossible to deduce spatial extension from spatial objects (points).

We can now explain the three primitive terms in Hilbert's axiomatization of geometry in terms of the spatial subject-object relation. The term 'line' reflects the primary existence of a (one-dimensional) spatial subject, the term 'point' accounts for the primary existence of a (one-dimensional) spatial object and the phrase 'lies on' highlights the *relation* between a spatial subject and spatial object – in other words, these three terms reflect the *spatial subject-object relation*.

Primitive features at the factual side of the spatial aspect	Subject ↕	Object ↕	Relation ↕
The primitive terms in Hilbert's axiomatization of geometry (1899)	Line	Point	Lies on

Given the foundational position of the numerical aspect in respect of the spatial aspect, one should also consider the distinction between retrocipatory and antecipatory analogies in these two aspects. Distance, as we argued, at the factual side of the spatial aspect points back to the numerical mode. It may therefore be designated as a *retrocipatory* analogy within space. We shall see that the idea of the *at once infinite* (see page 239), furthermore, represents an *antecipatory* analogy on the law side of the numerical aspect pointing towards the order of simultaneity on the law side of the spatial aspect. Likewise, the infinite divisibility of any (factually extended) spatial subject refers back to the law side of the numerical aspect, where the order of arithmetical succession reveals the primitive meaning of endlessness. In this context, one should mention that the continuous extension of any spatial subject embodies the original meaning of the spatial *whole-parts relation* (with its implied infinite divisibility). The *interval* within the system of rational numbers analogically reflects this infinite divisibility of a spatial subject and the latter, as we have just pointed out, represents a retrocipation from space to the primitive meaning of the successive infinite on the law side of the numerical aspect.

Note: *Mereology*

The Polish logicist and philosopher, Stanislaw Lesniewski, under the influence of Twardowski and continuing what Husserl started in the third part of his "Logische Untersuchungen" (*Logical Investigations*), where he analyzes the relation between a *whole and its parts*, was the first to subject this relation to a formal (set theoretical) analysis. Owing to misunderstandings created by calling it set theoretical the term *mereology* was introduced. Its intention is to embrace both the *element-set relation* and the *set-subset relation*. The Aristotelian-Scholastic legacy is reflected in the distinction between *distributive totalities* (related to universality) and *collective* (integrating) *totalities* (cf. Lorenz, 1980:1145-1146). Unfortunately, the development of mereology did not enter into an analysis of the whole-parts relation in terms of the uniqueness and mutual coherence between the aspects of number and space, for then it would have realized that set theory itself is a spatially deepened arithmetical theory that had to rely on the spatial whole-parts relation – concealed within axiomatic set theory in the undefined term 'set' itself!

White remarks that Aristotle and (modern mathematical) topology part ways in respect of "the *collective identification* of sets of points with (proto)topological regions: topology accepts such an identification as fundamental while Aristotle will have none of it" (White, 1988:7). The jump from (atomistic) points to continuity then emerges: "However, topology maintains that when sufficiently large classes of points are considered *collectively*, these can, in some cases, be identified with spatial regions in the intuitive proto-topological sense. The result is that, from the topological perspective, continuity becomes an 'emergent' property" (White, 1988:8). White does not realize that the 'collective' already entails the *wholeness* of spatial continuity and that therefore the attempt to allow continuity to 'emerge' from a collective totality begs the question.

More recent developments highlight the inevitability of primitive numerical and spatial terms in mereology, for example when Obojska intends to introduce a single primitive notion: (primary) relation (see Obojska, 2007:644, 646). The term 'primary' derives from the core meaning of number ('first') while the term 'relation' is derived from the core meaning of space. To be *related* is to be *connected* and to be connected entails *coherence* which is synonymous with *continuity* which in turn is synonymous with the original spatial whole-parts relation. She says that Cantor's theory is *distributive* (starting with points as elements considered as a whole) while Lesniewski's approach is *collective* because "a mereological 'set' is a whole (a collective aggregate or class) composed of 'parts' and the fundamental relation is that of being a 'part' of the whole, an element of a class" (Obojska, 2007:642). However, Fraenkel notes that to Frege we owe the distinction between the membership relation ($\in$)

and the set-subset relation (⊆) (Freankel et al., 1973:26-27).[1]

5.17.2.5 *Law and factuality within the physical aspect*

In Chapter 2, Heisenberg's description of determinism was quoted (see page 35) – claiming that every *effect* is strictly determined by a *cause*. This view represents a *deification* of the law side of the physical aspect – which explains why physical entities (subjects) are merely seen as an *extension* of physical laws. Determinism therefore *reduces* the factual side to the law side. Stafleu correctly points out, by contrast, that the causal relation on the law side of the physical aspect simply states: *nothing happens without a cause – but what the effect of a specific cause may be need not be fixed in advance*. This formulation avoids the one-sidedness of both determinism and indeterminism: it grants *determinism* that the concept of a cause is meaningful and should not be discarded, as claimed by indeterminism; and it grants *indeterminism* that the effect need not be fixed in advance (just think of the half-value of radio-active elements), thus highlighting the untenability of determinism in this regard.

Heisenberg's principle of uncertainty had the effect that the paths of the great 20th century physicists parted – concerning the question whether or not the concept of causality ought to be maintained in the further development of physics. Planck and Einstein wanted to uphold the claims of determinism, whereas Heisenberg and Bohr (the Copenhagen interpretation of quantum physics) opted for the other extreme: indeterminism. If it is the case that determinism absolutizes the law side of the physical aspect, then we have to say that indeterminism absolutizes the factual side of the physical aspect. The alternative approach here is to view the law side and factual side as *irreducible correlates* – an alternative implicitly supported by the necessity to employ statistical laws in physical theories (cf. Stafleu, 1968:304).

5.17.2.6 *The law-subject distinction in biology*

George Gaylord Simpson, the well-known neo-Darwinistic thinker, distinguishes between physics and biology as follows. According to him, the physical sciences are largely *typological* and *idealistic* for they "usually deal with objects and events as invariant types, not as individuals with differing characteristics" (Simpson, 1969:8). This characterization reflects the basic orientation of the classical humanistic science ideal in its *rationalistic* orientation – resolving the factual side of physical reality into the law side, thus turning physical entities into "invariant types." But as soon as truly biotical phenomena are considered, this *rationalistic* inclination makes room for the *irrationalistic side* of modern nominalism – denying, as quoted earlier, any and all *type concepts*: "Organisms are not types and do not have types" (cf. Simpson, 1969:8-9 and Chapter 1, page 25).

This explains why a physical approach is inappropriate when it comes to "phenomena special to the biological levels" (Simpson, 1969:8). What is stri-

1 Obojska mentions that for Cantor a set is "a many considered as a one," whereas Lesniewski's approach defines the 'part' in terms of the 'whole' (Obojska, 2007:644).

king here is that Simpson distinguishes between two *types* of phenomena, namely *physical* phenomena and *biotical* phenomena. In other words, in order to demarcate the domain of biotical phenomena – where a *typological* (and so-called *idealistic*) method is inapplicable, a prior typological classification is required – given in the distinction between the two types of phenomena: physical and biotical. The logical flaw in this approach is evident: biology as a discipline can only proceed in a non-typological manner when it is founded in a typological classification!

The law side of the biotical aspect is completely sacrificed in this irrationalistic nominalism – biology does not investigate types, for according to Simpson "no two are likely ever to be exactly alike" (Simpson, 1969:9). The unique-individual side of biotical factuality is accentuated at the cost of both the universal (factual) orderliness of living entities and by rejecting any order determination, every form of a conditioning biotical law.

Although Darwin refers to a "law of nature" at six places in his "On the Origin Of Species " (1859), he does not specify such a law as a *biotical* law (see Darwin, 1859:143 (2x), 147, 268, 427, 445).[1]

Physics had to formulate its *typical laws* in statistical terms precisely because the unique individuality of entities cannot be seen as a mere extension of the law side, and are therefore also not merely "objects and events" with "invariant types." Simpson has a rationalistic (and deterministic) understanding of physical phenomena, and an irrationalistic (and nominalistic) view of living entities – on both scores, the true law side – factual side correlation is misunderstood.

5.18 Disclosure as an opening-up of modal antecipations

The architectonic configuration of a modal function entails – with the exception of the quantitative mode – that there are always foundational structural elements presupposed by its own structure. These foundational structural moments present themselves as retrocipatory analogies within all the post-arithmetical aspects. These retrocipations are *given*. By contrast, those structural elements through which the inter-modal coherence between a particular aspect and aspects appearing later in the cosmic order come to expression, are not given, since they are dependent upon the 'guidance' of later (aspectual) functioning.

Animal sensory life, for example, is completely closed and contained within its instinctually secured nature. Human feeling life, by contrast, acquires a broadened scope, because it provides the functional basis for every post-sensitive mode, reflected in the sensitive analogies of these aspects, made possible by human consciousness. Since the sensitive aspect constitutes a founda-

1 The WEB version of this work acknowledges Darwin's dependence upon Spencer regarding the use of the expression, "survival of the fittest" as a substitute for his own "struggle for existence": "But the expression often used by Mr. Herbert Spencer of the survival of the fittest is more accurate, and is sometimes equally convenient." The Penguin edition omitted this acknowledgement (see page 115).

tional condition for all the post-sensitive modes, one may say that, through its analogical position within the post-sensitive aspects, it acquires a *constitutive broadening of meaning*, exemplified in configurations such as *logical* feeling, *cultural-historical* feeling, *lingual* sensitivity, *social* sensitivity, and so on. In all these instances, human feeling and sensitivity are expressed in normative contexts, and through these interconnections, human perception unfolds itself – in the sense of a constitutive broadening of scope – with a *norm awareness* that is absent amongst animals.

When the structure of an aspect is opened up in the antecipatory direction of cosmic time, its meaning is not merely deepened in a constitutive sense, but also in a *regulative* sense.

The retrocipatory structural elements within an aspect are like the bricks in a wall – they are constitutive for the very existence of the wall. Functioning within any aspect therefore 'actualizes' every constitutive building block. But those structural moments referring to later cosmic aspects are not automatically realized, for they are always dependent upon a *guiding* and *disclosing* activity.

5.18.1 *The ethical antecipation sphere within the jural aspect*

Since the jural aspect is foundational to the moral (ethical) aspect, it can factually be the case that the structure of this aspect may still only appear in its restrictive, that is, "not yet disclosed," meaning. A system of penal law, in which the meaning of the jural is not yet disclosed, will still display all the constitutive meaning moments within the jural aspect. What is particularly striking in an undifferentiated society with an undisclosed jural awareness, is the dominance of a form of accountability on the basis of the *effects* of an action only (in German known as 'Erfolgshaftung'). A person is held liable for the effects (consequences) of a deed without taking into account the intentions of the actor – the well-known *lex talionis* applied the proportionality of an eye for an eye and a tooth for a tooth. On the one hand, this measure established a certain *jural balance*, because one is not entitled to take a *head* for an eye. In undifferentiated societies, this configuration is intertwined with a *collective accountability* – which is also found in the Old Testament (see Deut. 5:9). Instances of 'Erfolgshaftung' in the Old Testament are found in Gen. 9:6 and Lev. 24:16-21. The legal stipulations of the Old Testament oftentimes instantiate the *lex talionis* – an undisclosed principle of penal law that has already been found in the law books of Hammurabi (almost verbally repeated in the Old Testament).

Later in the Old Testament, we read about cities providing shelter to those who wanted to escape from vengence – once safely within such a city, no revenge was allowed. Yet this did not totally preclude the seeking of vengeance, because the moment such a person left the city, the protection ceased (see Num. 35:9-29 and Num. 35:11, 15 and 27).

It is only when the jural awareness of a society and the jural order of the state is *regulatively* deepened under the guidance of the aspect of moral love, that it is possible to account for the moral disposition of the perpetrator, for the subjective intentions of the person who committed the deed. Only then do the (disclosed) principles of *jural morality* come into play, such as the *fault* principle – in its two forms: *dolus* (intent) and *culpa* (culpability). In Dutch and German, the term 'Schuld' is normally translated as either *fault* or *guilt*. Alan Cameron points out that in English-speaking Common Law jurisdictions, 'fault' is usually reserved for civil wrongs (torts) and 'guilt' for criminal wrongs, but that Dooyeweerd "uses 'schuld' to refer to both types of wrong (i.e. to both civil and criminal delicts)." Therefore it can be "translated as 'fault' in a broader sense, not specific to any particular category of legal wrong."[1]

Aristotle already had an exceptional understanding of another deepened (morally disclosed) jural principle, namely *equity*. He does distinguish between *justice* and *equity*, but ascribes to the latter a higher value (cf. *Nicomachean Ethics*, Book V, Chapter 10). Although equity is just, it is not the justice of the law. Enacting a law necessarily entails a general statement that cannot possibly foresee all particular unique circumstances that may pertain, and therefore it cannot exclude the possibility of error. Applying equity as an effect of the occurrence of an exception to the rule, essentially amounts to a rectification of the law. This modified statement should be what the lawgiver would have done if the special circumstances had been known. Everything cannot be regulated by law – and when the applicable law, in fact, would effectuate an injustice, the original law statement ought to be rectified *ex equitate*, that is, on behalf of equity.

All the deepened principles of jural morality – such as the fault principle, the principle of equity, that of good faith (*bona fides*), and so on – are actually principles of *justice*.

5.18.2 *The disclosure of the sign mode and the aesthetic aspect*

The possibility of assigning certain meanings to particular verbal sounds or letter combinations enable lingual acts of *naming*. But it is only when such signs are *shared* by multiple sign users in *mutual communication*, that language as such emerges. Language is therefore inherently dependent upon the *social disclosure* of the sign mode – as Wittgenstein argued, there is no 'private' language.

Societal differentiation is reflected in the lingual differentiation of various lingual domains, such as *scientific language*, *legal language*, *economic language*, the *language of faith*, and so on. The growing complexity of languages eventually benefit from the deepened meaning of lingual economy – evidenced in "sign frugality," such as found in the employment of *morphemes*.

Since the use of morphemes enables the creation of new words on the basis of existing ones, the lingual economy achieved by doing this relieves human

1 Editorial note added by Alan Cameron to Dooyeweerd's analysis of *jural causality* contained in the *Collected Works* of Dooyeweerd, B Series, Volume 2 (see Dooyeweerd, 1997a:42).

memory of impossible burdens (also see the remarks of Van Heerden, 1965:37). A thorough systematic analysis of this antecipatory moment is found in Weideman 2007.

In the development of *rhyme*, *alliteration* and *assonance*, a disclosure of the meaning of the sign mode under the guidance of the aesthetic aspect takes place. The jural antecipation within the sign mode binds the lingual use of words and phrases to their proper contexts, requiring the choice of *appropriate* options. Choices like these provide the jural grounds for *lingual honesty* in the formation of phrases and lingual expressions, emphasizing that words ought to be *faithful*[1] to the meaning nuances constituting their semantic domains[2] – even setting limits even to the metaphorical use of words.

Within the aesthetic aspect, there are three antecipatory moments. If a work of art does not do *justice* to the nuancefulness of created reality in an aesthetic sense, it displays an *un-aesthetic* trait. Likewise, an artwork lacking *aesthetic integrity* and *aesthetic honesty* (moral antecipations within the aesthetic mode) can never be *convincing* in an aesthetic sense (the certitudinal antecipation).

5.19 Sphere-universality and conceptual links between disciplines

The brief overview of the retrocipations and antecipations within the jural aspect displayed in Figure 7 (below) illustrates the scope of the principle of sphere-universality.

Figure 9

Aspects	Retrocipations and Anticipations
Faith aspect	Jural/legal certainty trust)
Ethical aspect	Jural/legal morality (fault, good faith, etc.)
Jural aspect	[Meaning-nucleus: retribution]
Aesthetic aspect	Jural/legal harmony
Economic aspect	Jural/legal economy (avoiding excess)
Social aspect	Jural/legal interaction
Lingual aspect	Jural/legal signification and interpretation
Cultural-historical aspect	Jural power / Legal competence
Logical-analytical aspect	Jural lawfulness and unlawfulness (consistency)

1 Note the *certitudinal antecipation* within the sign mode.

2 A semantic domain represents a *spatial retrocipation* within the sign mode.

Aspects	Retrocipations and Anticipations
Sensitive aspect	Jural/legal sensitivity (intention, will)
Biotical aspect	Jural/legal life
Physical aspect	Jural/legal dynamics (causality)
Kinematic aspect	Jural/legal constancymovement (transfer, conveyance)
Spatial aspect	Jural/legal sphere, jurisdiction, ambit
Arithmetical aspect	Jural/legal order (unity & multiplicity)

Of course the mere presentation of the overview contained in Figure 9 ought to be understood for what it is – just a brief sketch. What is displayed in this figure presupposes the application of a sound method of analysis, known as the transcendental-empirical method. The aim of this method is to investigate the (i) *structural conditions* making possible our (ii) *experience* of reality. (i) Accounts for the 'transcendental' element[1] and (ii) for the 'empirical' element.

The complexities involved in the application of the transcendental-empirical method are best demonstrated with reference to the illusion in many disciplines, namely that a particular special science possesses a special jargon, set apart from other academic disciplines.

5.20 The elementary basic concepts of the academic disciplines

Special scientists tend to think that their own discipline employs concepts that are peculiar to that specific discipline. This explains why some scholars on the one hand want to get away from certain 'misleading' *figures* or *metaphors*, but on the other hand want to demarcate a *unique* and if possible even *exclusive* universe of discourse. The German sociologist, Fichter, for example, first explicitly treats *typical* concepts, focused upon the investigation of specific entities, societal collectivities or social processes – such as *behaviour*, *role*, *institution*, *culture*, and *society* – and then immediately proceeds with an investigation of *basic concepts*. In this context he refers to the "imaginative analogies" used to explain "social life." He particularly has the "organic analogies" of the 19th century in mind. In a similar vein, but not as totally exclusive, Giddens remarks:

> There are few today who, as Durkheim, Spencer and many other in nineteenth-century social thought were prone to do, use direct organic analogies in describing social systems (Giddens, 1986:163).

Throughout this work, we nevertheless repeatedly find the expression "social

1 We have to keep in mind that 'transcendental' in this context does not have a *constructive epistemic* sense, but an *ontic a priori* meaning, because it is supposed to capture what is *given* prior to any human construction.

life"[1] – without any apology! Fichter also does not critically reflect on the meaning of the phrase "social life" – something clearly seen from his straight-forward *rejection* of biologistic, mechanistic, psychologistic and other approaches to sociology (Fichter, 1968:6). He claims that it is certain that the reality of the social dimension could not be reduced to 'biological', 'physical' or 'psychological' concepts. Remarkably enough, he correctly realizes that an analogy refers to a *partial similarity* and a *partial difference* (Fichter, 1968:5). At the same time he holds the opinion that "the social sciences managed to develop their own terminology so well that these analogies are totally dispensible" (Fichter, 1968:6).

Yet, if this is true, he cannot answer the question why he still employs the expression "social *life*"?! Is it not the case that the term 'life' in the first place refers to *living entities*? If so, is there a *difference* between *social life* and *biotic life*? If the answer is **yes**, then the question arises whether or not only the science of *biology* is allowed to use the term 'life'? What about the other disciplines? Is it not possible to explore *legitimate* and *meaningful* usages of *biotical analogies* in disciplines *different* from biology? May other disciplines use the term 'life' – albeit in an *analogical sense*?

5.20.1 *The unavoidable interconnectedness of scientific terminology*

When we investigate the terms presented by Fichter as belonging to its own distinctive terminology, it turns out that he is not at all aware of the fact that the same terms are employed in a *different sense* by other disciplines as well. In other words, he does not realize that he still uses *elementary* (*analogical*) *basic concepts* which, in different contexts, are also employed by other academic disciplines!

After he emphasized the *dispensability* of *analogies*, Fichter continues by analyzing the problem of *constants* (Fichter, 1968:7). The term *constancy*, however, as we have seen, originally belongs to the domain of the *kinematic aspect* of *uniform motion* (see Chapter 3, page 88). The implication is that it could only be used by other disciplines in a *non-original* way, that is, in an *analogical* way. Fichter believes that the basic concepts analyzed represent "constant and everywhere appearing elements" (Fichter, 1968:7). Once again Fichter does not realize that the term 'everywhere' stems from the meaning of the *spatial aspect* of reality. Just consider the equivalent spatial term: *universal*. Similarly, the term 'elements' reflects the unique meaning of the *quantitative aspect*, since it is related to the meaning of *multiplicity*: the *one* and the *many*. This implies that Fichter necessarily had to *use* numerical and spatial terms in order to explain his employment of the (kinematic) term *constancy*.

What surfaces here is the *complex nature* of an analysis of *elementary basic concepts* of a special science, for it inevitably calls for this kind of procedure: any specific analogical concept can only be analyzed by (implicitly or explicitly) using *other* (analyzed or not yet analyzed) *analogical* structural moments

1 My italics – DFMS.

within the modal structure of the aspect concerned.

It should not surprise us that Fichter, on the basis of his introduction of "social constants" (Fichter, 1968:8), proceeds by speaking of *social dynamics* and *social change* (Fichter, 1968:8). A few pages further, he also speaks of *social causes* (Fichter, 1968:12). The relation between *cause* and *effect* in the first place manifests itself within the structure of the *physical aspect* of reality. Analogous to this *physical* relation, sociology employs the (analogical basic) concept of *social causation* (compare the book with the same title written by McIver in 1942). In other words, although Fichter believes that he can dispense with the former "imaginative analogies" used by sociologists through the development of an "own terminology," he continues (albeit unconsciously and unintententionally) to *use* certain *analogical concepts* – including those from the biotic aspect (such as the mentioned phrase: "social life").

As a result, Fichter's philosophical prejudice, convincing him that sociology does not need any (modal) analogies, did not prevent his continued *unconscious* use of such analogies.

Postmodern thinkers exemplify another negative instance regarding the employment of analogical terms from mathematical and physical theories. Sometimes they claim that 'modernist' thinking is *linear*, while postmodern thought is supposedly *non-linear*. Quantum mechanics would be an instance of *non-linear thinking*, while *chaos theory* is supposed to be in opposition to the deterministic Newtonian legacy in physics.

However, the well-known *second-order differential equation* of Schrödinger in quantum mechanics is absolutely *linear*, while *chaos theory* is nothing but an extension of Newton's *classical mechanics* (see Sokol & Bricmont, 1998:164 ff.). Yet Sokol and Bricmont themselves struggle to come to terms with the connections between different scholarly universes of discourse. Unfortunately they only consider analogies between the different *disciplines* or *theories* (such as between the *theory of relativity* and *social theories*), without contemplating the *ontic interconnections* discernable in truly modal (functional) analogies (retrocipations and antecipations).

What is therefore actually at stake in the employment of (modal) analogical concepts is not the relationships between *scientific disciplines* as such, but the recognition of connections between different "ontic domains" within reality itself ("ontic coherences"). It may be the case that the theory of relativity, for example, does help us understand the nature of the kinematic aspect of uniform movement better by highlighting its core meaning of *constancy* (the velocity of light in a vacuum). When, for example, sociology cannot avoid references to the notion of *social constancy*, we do not discern an analogy between the *disciplines* of sociology and physics (kinematics or phoronomy in particular), but rather with an analogy between two *modal functions* (the kinematic and the physical) of ontic reality, which cohere *prior* to any scholarly reflection.

Of course our argument thus far implicitly rests upon a particular theoretical account of the nature and coherence of the various aspects of reality. Yet, this acknowledgement does not escape the inevitability of *multi-vocal* terms employed by disciplines. In order to illustrate this inevitability, we may also look at the central role of terms like *constancy* and *change* (variability) within (mathematical) logic. This discipline simply introduces the terms *constants* and *variables* intuitively, without even addressing the question about *analogical concepts*. The numerical meaning of *multiplicity* surfaced in the discussions of the role and place of logic in mathematics. Logicism, as we noted, attempted to find a basis for the concept of number in the logical class concept (see page 171).

The inevitability of employing analogical basic concepts is embedded in the other above-mentioned key ontic feature of the various aspects of reality, namely the fact that the unique meaning of every aspect only comes to expression in its coherence with all other aspects. This coherence comes to expression through the analogical structural moments that appear as retrocipations and antecipations within each aspect to the other aspects.[1]

Scholarly thinking throughout the history of philosophy and within all the disciplines constantly stumbled upon this reality, but never attempted a systematic and comprehensive analysis. Although the building blocks for such an investigation are present in various attempts to develop a "categorial framework" (to use a phrase from Stephan Körner) for the understanding of reality, the actual articulation of specific categories by prominent philosophers in the history of philosophy largely pursued a speculative path without really investigating these aspects in a transcendental-empirical way, i.e. without discerning those unique modal aspects that make possible our experience of the diverse kinds of things functioning within these aspects.

This holds from the Aristotelian theory of categories (he distinguished 10), the medieval transcendental determinations of being (*unity*, *truth*, *beauty* and *goodness*), the table of categories of understanding in Kant's *Critique of Pure Reason* (the categories of quantity, quality, modality and relation – each consisting of three concepts) up to Hartmann's account (see Stegmüller, 1969a:237) where he distinguishes 24 principles of being, arranged in pairs: principle – concretum; structure – mode; form – matter; inner – outer; determination – dependence; quality – quantity; unity – multiplicity; unanimity – conflict; antithesis – dimension; discreteness – continuity; substrate – relation; element – system.

The inter-modal coherence between the various aspects unveil a depth layer of ontic inter-connections confronting every possible scholarly discipline. Precisely in the (mostly *implicit*) choice of basic concepts, the philosophical orientation of a discipline comes to expression. A few brief examples may extend our demonstration of the inevitability of employing elementary (analogical) basic concepts and ideas.

1 For a systematic analysis of the elementary basic concepts of the disciplines of law and linguistics see respectively Hommes (1972) and Weideman (2009). An account of analogical basic concepts is also significant for applied linguistics (see Weideman, 2006:241).

5.20.2 *Mathematics and the nature of infinity*

On the law side of the quantitative aspect, cosmic time is expressed as an *order of succession*. This numerical time order of succession makes possible our most basic awareness (intuition) of *infinity* – understood in the literal sense of one, another one, yet another one, and so on, indefinitely and endlessly.[1] This awareness is determined and delimited by what may also be called the *order side* of the numerical aspect. It determines every *infinitely proceeding row of numbers*, as well as the different operations on the law side of the numerical aspect [operations such as addition, multiplication, subtraction, division, as well as the principle of induction – which can be disclosed by (non-theoretical as well as theoretical) mathematical thinking].

5.20.2.1 *Space presupposes the successive infinite*

The primitive meaning of infinity as endlessness is presupposed in the meaning of the spatial aspect. The factual side of the spatial aspect contains *extended figures* such as lines, surfaces and volumes. Every spatial subject entails a "turning inward" of the original quantitative meaning of infinity, for whereas the infinite in its basic arithmetical meaning proceeds beyond all limits, the very nature of spatial extension allows for multiple *divisions*. Something continuous coheres in all its parts; every part is connected with other parts and, taken together, these parts constitute the whole (totality) of an extended spatial subject.

Although the terms 'whole' and 'totality' are closely related to the term 'continuity', it seems difficult to *define* the meaning of *continuous extension* as realized by Dantzig when he writes: "From time immemorial the term *continuous* has been applied to space, ..., something that is of the same nature in its smallest parts as it is in its entirety, something *singly connected*, in short *something continuous*! Don't you know .... any attempt to formulate it in a precise definition invariably ends in an impatient: 'Well, you know what I mean!' " (Dantzig, 1947:167). Synonyms like 'uninterrupted', 'connected', 'coherent', and so on simply repeat what is meant by continuity, instead of *defining* it! Furthermore, a spatial term such as *domain*, for example, could be replaced by others such as *range, scope* or *sphere*. Although, for technical purposes, mathematicians may decide to attach slightly different connotations to these terms, the fact that they all share a *generic spatial meaning* cannot be denied.

5.20.2.2 *Infinite divisibility as an analogical basic concept*

In order to understand what is at stake in the case of truly *analogical basic concepts*, the classical legacy of the *infinite divisibility* of continuity is very instructive. In an article entitled "On the Infinite" – commemorating the mathematician Karl Weierstrass (published in 1925) – David Hilbert commences by looking at this issue from the perspective of the *infinitely small* and the *infi-*

1 Russell broadens the concept of endlessness by specifying it to mean that an endless series has "neither a beginning nor an end" (Russell, 1956:297).

nitely large (Hilbert, 1925:163 ff.). The discovery of *quanta of energy* on the one hand and Einstein's *theory of relativity* on the other, eliminates both possibilities.[1] The distinction between *mathematical space* and *physical space* is therefore necessary, as we have analyzed it in an earlier context (see page 221). We have noted that whereas the former – in a purely *abstract* and *functional* perspective – is both *continuous* and *infinitely divisible, physical space* is neither *continuous* nor *infinitely divisible.* Moreover, since it is bound to the *quantum structure* of energy, physical space cannot be subdivided *ad infinitum*. Energy quanta indeed represent the *limit* of the divisibility of energy.[2]

We argued that an analogy is present whenever differences are shown in what is similar. In this case: both mathematical space and physical space are extended (their *similarity*), but in being discontinuous and not infinitely divisible (their differences), the latter differs from the former.

5.20.2.3 *The modal seat of the whole-parts relation*

We have seen that the most prominent recognition of the spatial 'home' of the terms *wholeness* and *totality* is found in the thought of the mathematician Paul Bernays (see page 219 above), for he holds that the arithmetical and geometrical intuition should not be distinguished according to the *spatial* and the *temporal*, but rather by accounting for the difference between the *discrete* and the *continuous* (Bernays, 1976:81). Being fully aware of the claims by modern mathematics that it is possible to arithmetize this discipline fully, it is all the more significant to know that Bernays questions the attainability of this ideal of a *complete arithmetization* of mathematics – and he does this on the basis of acknowledging the original spatial meaning of continuity and wholeness (totality). (It should be kept in mind – as mentioned on page 11 above – that Bernays worked with Hilbert who from 1912, considered to be the leading mathematician of the world.)

Bernays unambiguously writes:

> We have to concede that the classical foundation of the theory of real numbers by Cantor and Dedekind does not constitute a *complete* arithmetization of mathematics. It is anyway very doubtful whether a complete arithmetization of the idea of the continuum could be fully justified. The idea of the continuum is after all originally a geometric idea (Bernays, 1976:187-188).[3]

Particularly in explaining the difference between what is traditionally known as the *potential* and the *actual* infinite, the difference between *succession* and

1 A more extensive explanation is found in his oration on *Naturerkennen und Logik* (in 1930; it appeared in Hilbert, 1970:380-381). The nature of Planck's *Wirkungsquantum* was briefly explained in Chapter 2 (see page 39 above).

2 In addition, owing to gravitation ("curved space"), the universe is considered *finite* though *unbounded*.

3 "Zuzugeben ist, daß die klassische Begründung der Theorie der reellen Zahlen durch Cantor und Dedekind keine *restlose* Arithmetisierung bildet. Jedoch, es ist sehr zweifelhaft, ob eine restlose Arithmetisierung der Idee des Kontinuums voll gerecht werden kann. Die Idee des Kontinuums ist, jedenfalls ursprünglich, eine geometrische Idee."

at once, as well as the irreducibility of the notion of a *totality* surfaces. The extension of any spatial figure is bound to a specific *dimension* and the spatial time order of simultaneity, for if all the parts (i.e. a *multiplicity* – evincing the undeniable quantitative foundation of space) of such a figure are not present *at once*, the spatial figure concerned is also absent. Only when all three sides of a triangle are present at once do we have a triangle – the succession of its three sides does not constitute a triangle.

5.20.2.4 *The inter-modal meaning of an 'infinite totality'*

Hilbert introduces the difference between the potential and the actual (or genuinely) infinite by using the example of the "totality of the numbers 1, 2, 3, 4, ...," which is viewed as a unity that is given at once ('completed').

> If one succinctly wants to characterize the new conception of the infinite, established by Cantor, one could well say: in analysis we only encounter the infinitely small and the infinitely large as limit concept, as something becoming, originating, generated, i.e. as one says, with the potential infinite. Yet this is not the truly infinite. We have the latter, for example, when we observe the totality of the numbers 1, 2, 3, 4, ... themselves as a completed unity or the points of a line as a totality of entities that are completed at hand. This kind of infinity is designated as the actual infinite (Hilbert, 1925:167 – translation by the author).

According to Lorenzen (whose preference in his constructive mathematics is *not* to accept the 'actual infinite') the understanding of real numbers with the aid of the actual infinite cannot camouflage its ties with space (*geometry*):

> The overwhelming appearance of the actual infinite in modern mathematics is therefore only understandable if one includes geometry in one's treatment. ... The actual infinite contained in the modern concept of real numbers still reveals its descent (*Herkunft*) from geometry (Lorenzen, 1968:97).

Lorenzen highlights the same assumption when he explains how real numbers are accounted for in terms of the actual infinite:

> One imagines much rather the real numbers as all at once actually present – even every real number is thus represented as an infinite decimal fraction, as if the infinitely many figures (Ziffern) existed all at once (*alle auf einmal existierten*) (Lorenzen, 1972:163).

These modes of speech highlight the inevitability of employing terms with a *spatial provenance* even when the pretention is to proceed purely in *numerical* (or *arithmetical*) terms. Lorenzen is also justified in pointing out that arithmetic by itself does not provide any motivation for the introduction of the actual infinite (Lorenzen, 1972:159).

In terms of the inter-modal coherence between number and space, the infinite divisibility of a spatial subject – such as a straight line – analogically (as a retrocipatory analogy) reflects the successive infinite made possible by the numerical time order at the law side of the quantitative aspect. It should be pointed out that the classical designation of endlessness as the "potential infinite" lacks intuitive clarity. In order to regain this loss, it seemed appropriate rather to refer to this (basic) form of infinity as the *successive infinite*. The original numerical meaning of the successive infinite is analogically reflected at the factual side of the spatial aspect in the (successive) infinite divisibility of any spatial continuum (see page 237 above).

When, under the guidance of our theoretical (that is, *modally abstracting*) insight into the meaning of the spatial order of simultaneity, the original modal meaning of the numerical time order is disclosed, we encounter the *regulatively deepened antecipatory* idea of *actual infinity* or *completed infinity*. Any sequence of numbers may then, directed in an antecipatory way to the spatial order of simultaneity, be considered *as if* its infinite number of elements is present as a whole (totality) *all at once*. In other words, following from the unbreakable coherence between the law side and the factual side of an aspect, the modal antecipation from the numerical time order to the spatial time order must have its correlate at the factual side. At the factual side of the numerical aspect, we have distinguished between the row of natural numbers and integers (expressing the primitive meaning of numerical discreteness).[1]

When we employ the antecipation at the law side of the numerical aspect to the law side of the spatial aspect, we unveil the inter-modal foundation for the notion of *actual infinity*. This antecipatory moment also determines the correlative multiplicity of natural numbers, integers and rational numbers, since these can also be viewed *as if* they are present as completed (though infinite) *wholes* or *totalities*.

At this point, we only have to mention Georg Cantor's description of the actual infinite as a *constant* quantity, *firm and determined in all its parts* (Cantor, 1962:401). Throughout the history of Western philosophy and mathematics, all supporters of the idea of *actual infinity* implicitly or explicitly used some form of the *spatial time order of simultaneity* (determining the spatial property of *wholeness* or being a *totality*).

Comment: We mentioned that, although Russell holds that the "relation of whole and part" is "an indefinable and ultimate relation" (see page 61), he nonetheless did not consider its intrinsic link with the *spatial* aspect or the nature of *spatial continuity*. Russell finds it strange that his account of a "continuous series" depends on 'progressions' that "are the very essence of discreteness": "it seems paradoxical that we should require them in defining continuity" (Russell, 1956:299). It only "seems paradoxical" insofar as he does not realize that his account of continuity (using as starting point Cantor's idea of a *perfect* and *coherent* set – see Cantor, 1966:194 and Klein, 1928:105) is a spatially deepened *numerical* approach to the meaning of continuity that presupposes the just mentioned *semi-disclosed* meaning of number (see the text corresponding with the previous footnote).

What should have been used as an *antecipatory* (*regulative*) *hypothesis* (the idea of *actual infinity*), was often (since Augustine) reserved for God or an eternal being, accredited with the ability to oversee any infinite multiplicity

1 That the introduction of the *dense* set of rational numbers manifests an antecipation to a retrocipation while representing the *semi-disclosed* meaning of number, requires a more detailed analysis exceeding the scope of our present discussion.

all at once.

This antecipatory regulative hypothesis of actual infinity does not eliminate the original modal meaning of number, but only *deepens* it under the guidance of *theoretical thought.*

Similar to the new expression that we earlier introduced as an alternative designation of the "potential infinite," namely the "successive infinite," we proposed to introduce another phrase that has an equally immediate intuitive appeal as an alternative expression for what has been known as the actual infinite. Instead of speaking of the *actual infinite*, we introduced the idea of the *at once infinite*. Both these terms were actually already used in the early 14th century disputes about the infinity of God.[1]

These new expressions relate directly to our basic *numerical* and *spatial* intuitions, viz. our awareness of *succession* and *simultaneity* – and their mutual irreducibility is based upon the irreducibility of the aspects of number and space.[2]

Given my suggestion that the rational numbers represents an antecipation to a retrocipation (page 238), we may see in the rational numbers a particular emphasis on the 'part'-moment of the spatial *whole-parts relation*. A spatially disclosed analysis of the real numbers, of course, since it necessarily presupposes the *regulative hypothesis* of the *at once infinite* (actual infinity), particularly emphasizes the 'totality'-moment of the spatial *whole-parts relation*. Being dependent upon an irreducible (antecipatory) structural element within the deepened structure of the numerical aspect, real numbers cannot be obtained from the rational numbers alone – as was clearly emphasized by Russell, observing that "from rational numbers alone, no proof can be obtained that there are irrational numbers at all" and that "their existence must be proved from a new and independent postulate" (Russell, 1956:282).

5.20.2.5 *The 'at once infinite' as antecipatory hypothesis*

Paul Bernays indeed saw something of the nature of this *regulative hypothesis* – and at once he also distances himself from Vaihinger:

> The position at which we have arrived in connection with the theory of the infinite may be seen as a kind of the philosophy of the 'as if'. Nevertheless, it distinguishes itself from the thus named philosophy of Vaihinger fundamentally by emphasizing the consistency and trustworthiness of this formation of ideas, where Vaihinger considered the demand for consistency as a prejudice ... (Bernays, 1976:60).

Paul Lorenzen also sensed something of this approach in his remark that the actual infinite meaning attached to the 'all' shows the employment of a fiction – "the fiction, as if infinitely many numbers are given" (Lorenzen, 1952:593). But what is striking in this case is that the "as if" is ruled out, or at least dis-

1 Compare the expressions *infinitum successivum* and *infinitum simultaneum* (see Maier, 1964:77-79).

2 Dooyeweerd did not accept the idea of the *at once infinite* (the *actual infinite*) due to the fact that he was strongly influenced by the intuitionistic mathematicians Brouwer and Weyl in this regard. (Cf. Dooyeweerd, 1996-I:98-99, footnote 1 and 1996-II:340, footnote 1).

qualified as something fictitious, with an implicit appeal to the *primitive* (undisclosed) meaning of infinity.

Gödel realized that the set concept cannot be effectively defined. He also said that sets are quasi-spatial (see page 61 above). Yet, the so-called antinomies made him cautious for the totality character of sets, at least in the case of infinite "all" claims. He remarks that the 'naively' employed concept of a set "has not led to paradoxes": "This concept of set, according to which a set is anything obtainable from integers (or some other well-defined objects) by iterated application of the operation of 'set of ', and not something obtained by dividing the totality of all existing things into two categories, has never led to any antinomy whatsoever" (quoted by Yourgrau, 2005:137). From this quotation it is clear that Gödel and Yourgrau do not realize the the idea of a set inherently contains an appeal to the whole-parts relation originally given within the spatial aspect.

However, we may characterize the forward-pointing (antecipatory) hypothesis (referring from number to space) as a *disclosed* approach, in which any successively infinite sequence of numbers may be viewed *as if* given *at once*, as an *infinite whole* or *an infinite totality*. The following crucial implications should be highlighted:

(i) The multiplicity in the at once infinite cannot escape from simultaneously *echoing* succession while *transcending* it in being given "at once".

(ii) For example, the mere designation of the initial positions of the set of natural numbers – {1, 2, 3, ...} – is not decisive for the kind of infinity it may capture. In other words, this representation may intend either the successive infinite or the at once infinite. Suppose we actually assume the successive infinite, then this set is literally *endless*. But suppose now that we start this sequence of natural numbers with 11 as the first number: {11, 12, 13, ...}. Of course we can then continue to merely restrict ourselves to the successive infinite. We may now look at the sequence of fractions generated by substituting 11 with 1/1, 12 by 1/2, 13 by 1/3 and so on. Then we have {1/1, 1/2, 1/3, ...} – but still only in the sense of the *successive infinite*.

(iii) At this point we may involve our spatial intuition of simultaneity (at once). We do that by observing the one-one mapping between 'all' the elements of the succession {1/1, 1/2, 1/3, ...} with those points on the line matching these fractional (rational) values ('magnitudes') – as intuitively explained in the Figure below.[1]

(iv) Through the mapping illustrated on the next page we can consider each of the three *successively infinite* sequences [namely {1, 2, 3, ...}; {11, 12, 13, ...} and {1/1, 1/2, 1/3, ...}] as *infinite totalities*, i.e. under the guidance of our spatial intuition, the initial sets are interpreted as "actually infinite" (at once infinite) sets. Without the coherence between

1 See pages 268 ff. below, where the distinction between the succesive infinite and at once infinite is discussed in connection with Zeno's paradoxes.

number and space, the idea of the at once infinite will be *intrinsically contradictory*, for it will simply refer to the *completion* of what is strictly speaking *endless*.

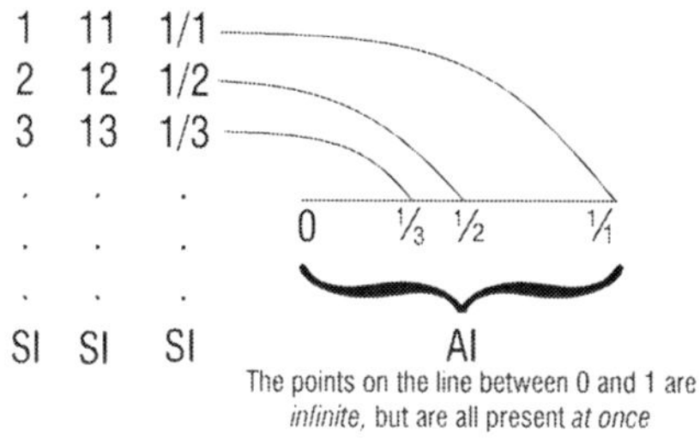

SI = Successive Infinite AI = At Once Infinite

(v) This issue may be 'simplified' with reference to the number 1. Consider the question: is the number 1 equal to 0.999... or not? Suppose we accept the potential infinite only. Then there will always be more "fractional amounts" to be added, since however far one proceeds, there will always be more to come – in which case 0.999... is *not* equal to one. The other option is (in following Weiertrass, Dedekind and Cantor) to accept the "actual infinite" (the *at once infinite*) and straightaway to define the number 1 as the 'totality' of the decimal expansion 0.999...! In a similar way, the approaches of Weierstrass and Cantor had to use the at once infinite in their account of real numbers.

From our analysis of the *meaning* of the *inter-modal coherence* between number and space, it is clear that Cantorean ***set theory*** is in fact a spatially *disclosed number theory*, opened up under the guidance of the regulatively deepened hypothesis of the *at once infinite*. Ultimately the irreducibility of the spatial aspect serves as the foundation for the irreducibility of the at once infinite to the successive infinite and explains why there is no *constructive* transition from the successive to the at once infinite. The human intellect, exploring these interrelations between number and space, is therefore more powerful than any technical device, which is always bound to a limited manifestation of the successive infinite. Consequently, Wachter's assessment – in a slightly different context – is still justified: "our most modern technology even including the 'Super-Protonensynchrotron', compared with our thoughts, is actually a very primitive tool" (Wachter, 1975:19).

A remarkable ambivalence in this regard is found in the **thought** of Abraham Robinson. He explored *infinitesimals* in his non-standard analysis – based upon the meaning of the at once infinite. A number *a* is called *infinitesimal* (or *infinitely small*) if its absolute value is less than *m* for all positive numbers *m* in $\Re$ ($\Re$ being the set of real numbers). According to this definition, 0 is *infinitesimal*. The fact that the infinitesimal is merely the correlate of Cantor's transfinite numbers is apparent in that *r* (*not equal to* 0) is infinitesimal if and only if r^{-1} is infinite (cf. Robinson, 1966:55ff). However, in 1964, he held that "infinite totalities do not exist in any sense of the word (that is, either really or ideally). More precisely, any mention, or purported mention, of infinite totalities is, literally, *meaningless*." Yet he believes that mathematics should proceed as usual, "that is, we should act *as if* infinite totalities really existed" (Robinson, 1979:507). Only when the successive infinite is used as yardstick, does it make sense to disqualify the idea of an infinite totality as *meaningless*.

5.20.3 *The increasing complexity involved in the analysis of elementary basic concepts*

We may use the nature of a *societal collectivity* to illustrate some of the complexities involved in analyzing the meaning of the social aspect. The nature of a distinctly differentiated social form of life could first be characterized by describing it as a *societal unity*. The perspective of the spatial analogy, additionally enables us to say that the *unity* of such a form of life could be seen as a social *whole* or *totality*. This analogical structural moment actually also enables us to speak of the specific social *domain* of a particular societal *sphere*. With the aid of the kinematic analogy we can account for the awareness of the social *continuity* (constancy) of such a life-form – providing the basis for social *changes* taking place within it (the focus of the physical analogy).

The *continuous change* in the ongoing functioning of a social form of life – allowing the individual members of that sphere to come and go without terminating its existence – analogically reflects the thermodynamics of physically open systems. The *continuation*, *maintenance* and *social development* (social growth) of a social collectivity frequently require *competent social organs* capable of structuring societal relationships by means of exercising their *social ordering will* in such a way that the internal functioning of the societal life-form concerned could express itself in constructive manifestations of an *integrated social awareness* (sensitivity/consciousness) – the contribution of the perspective of the sensitive-psychical analogy.

Without looking at other structural moments, it is evident that every analogical moment, reciprocally and with increasing precision, further determines the *meaning* of every other analogy within the structure of the social aspect. This underscores the *inherent dynamics* demanded by an analysis of elementary basic concepts. An analysis of the elementary basic concepts of sociology as a special discipline is therefore a *step-by-step process* that should make it increasingly clear that every analogical element constitutes an indispensable contribution to a progressively deepened understanding of the meaning of the social aspect. This underscores the perspective that the *meaning* of each modal aspect only reveals itself in *coherence* with all the other aspects within reality.

Note that such an analysis should at the same time demonstrate the ideal of showing a sense of *critical solidarity* with those positions in theoretical sociology, which one-sidedly – and often to the exclusion of other equally important modal analogies – want to use a *single* modal analogy (or sometimes a combination of *a few* modal analogies) as definitive explanatory device in their analysis of social reality. The challenge is to show *solidarity* with particular theoretical stances by appreciating the fact that a specific analogical moment has been discovered and explored in an analysis of social phenomena. The *critical distance* that ought to be demonstrated by pointing out to what extent this discovery has been hampered by an inherent one-sidedness (exclusivity) and/or denial of the *integral coherence* of all the analogical structural moments within the social aspect.[1]

5.20.4 *The system concept in economics and sociology*

In the development of modern sociology and economics, the concept of equilibrium – taken in the sense of a *closed physical system* – exerted a tremendous *reductionistic* influence. If a closed system is in a state of equilibrium, it is impossible for that system to produce energy. Furthermore, it does not need any energy input to maintain its equilibrium state. The capacity to perform work is only made actual when the system is *open*. Von Bertalanffy gives the following definition of an open system: "An open system is defined as a system in exchange of matter with its environment, presenting import and export, building-up and breaking-down of its material components" (Von Bertalanffy, 1973:149). Examples of open systems are: a *glacier*, a *fire*, any *living entity*. Whenever a living entity approaches a true state of physical equilibrium, death is in sight.

The "flowing equilibrium" which is thermodynamically present in an open system, is signified in some cases as homeostasis. It is also related to what became known as functionalism. Alexander gives a brief history of the term functionalism. The emergence of this term from a study group conducted by the physiologist L.J. Henderson at Harvard in the 1930s, was influenced by biological functionalism and by Canon's notion of homeostasis (Alexander, 1985:8). Von Bertalanffy also made an appeal to Canon's notion of homeostasis in the development of his system theory (Von Bertalanffy, 1973:10, 14, 21, 78, 169) by Cannon in 1929 (cf. Cannon, 1929:397).

In the autobiographical sketch of his intellectual development (1977), Parsons mentions the fact that he was first confronted with the problem of an equilibrium in the form developed by Henderson and Pareto, and by the way in which it was implemented by Schumpeter in the science of economics. He adds that, at a very early stage, he was influenced by Cannon's physiological conception of homeostasis – a conception showing direct links with the then predominant social-anthropological views of A.R. Radcliffe-Brown and his followers (cf. Parsons, 1977:48).

Parsons frames his analytical distinction of the social system in the following terms:

> The social system-focus is on the conditions involved in the inter-action of actual human individuals who constitute concrete collectivities with determinate membership (Parsons, 1961:34).

1 Strauss (2006) provides an in-depth analysis of the basic (analogical) concepts of sociology and of the way in which they are constitutive for the compound and typical concepts of this discipline.

The primary categories used by Parsons in his functional classification are pattern maintenance (also designated as latency), together with integration, goal-attainment and adaptation (Parsons, 1961:30). He declares that the "function of pattern-maintenance refers to the imperative of maintaining the stability of the patterns of institutionalized culture defining the structure of the system" (Parsons, *Theories of Societies*, 1961:38), and then adds the following:

> Pattern-maintenance in this sense plays a part in the theory of social systems, as of other systems of action comparable to the concept of inertia in mechanics. It serves as the most fundamental reference point to which the analysis of other, more viable factors can be related (Parsons, 1961:39).

When Parsons, Bales and Shills formulate a law imitating Newton's first law of motion (basically Galileo's law of inertia), they characterize it as merely being "another way of stating one aspect of the fundamental postulate that we are dealing with equilibrating systems" (Parsons, 1953:100 – note the influence of Pareto and Schumpeter). Parsons, Bales and Shills do not comprehend the difference between the kinematic and the physical aspects. They also do not adequately distinguish between closed and open systems within the physical aspect. Consequently, they incorrectly identify homeostasis with the analogy of inertia (a kinematic meaning-figure) in their characterization of "equilibrating systems." The same comment is relevant with respect to Parsons' use of the concept of pattern maintenance, which he sees as something comparable with the concept of inertia in mechanics. Maintenance always requires new energy input (into an open system) – which is different from the inertial notion of mere continuation.

5.20.4.1 *Economic theory and the notion of equilibrium*

Modern economic theory (the classical school) was firmly in the grip of the modern humanistic science ideal, which aimed at reducing the normative dimension of creation to deterministic natural laws. Initially a deterministic and physicalistic understanding of causality guided this development, amended by an increasing focus on variants of the physical notion of gravity (an almost inevitable and obvious effect of Newton's successful theory of gravity), eventually crystallizing in the urge to account for economic activities in terms of equilibrium.

Adam Smith believed that there was no relation between the *usability* (practical value or utility) of a commodity and its *exchange value*. At this point, the dominant concept of causality, still prevailing in classical economic thought up to Ricardo, turned out to be untenable. As a substitute, a physicalistic conception of a system in equilibrium was introduced. Wicksell points out that the cost of the production of a commodity and its relative price (exchange value) do not stand in the relation of cause and effect, but determine each other mutually as distinct members of a unitary economic equilibrium system (Wicksell, 1913:73).[1] Myrdal mentions that the theory of marginal utility provides a unified and closed explanation of the context of price establishment given in the equilibrium theory, which is characterized by the replacement of the representation of a one-sided causal connection by that of a mutual functional coherence.[2]

What surfaces here is the decisive difference between the classical theory and the theories of Menger, Jevons and Walras. The latter thinkers managed to transcend the limitation in Adam Smith's approach by drawing on seminal insights by von Mangoldt, Dupuit and Gossen.

The inherent many-sidedness and dynamics of economic life do not warrant the fiction of *static economic equilibria*. Yet economic theorists attempted to introduce the physical concept of equilibrium in order to explain the nature of the free market. Pareto, for example, pursued this avenue explicitly. Before the age of seven, this French-born thinker moved to Italy, where he eventually (in 1896) completed his doctoral studies (in order to become an engineer) at the university of Turin – dealing with the fundamental principles of equilibrium in solids (cf. Bousquet, 1965:391).

Pareto argues that the molecules constituting the social system are individuals (Pareto, 1963, § 2080). Society is therefore to be recognized as a system in equilibrium, with a number of inter-dependent elements which, with the aid of the mathematical concept of function, must be studied in quantitative terms. The resistance of a society to internal and external forces leads to a recovery of the previous situation:

> A society where this occurs can therefore be considered as being in a state of equilibrium, and of stable equilibrium (Pareto, 1966:104).

Although D'Alembert's mechanics allows for the study of the dynamic condition of a system, both economics and sociology must, according to Pareto, "consider a series of *static equilibria* (my italics – DS) rather than the dynamic equilibrium" (Finer, 1966:104). This mode of thought represents a fundamental assumption in neoclassical economics – Walras, for example, argues that successive changes in price would always lead from a state of disequilibrium

1 "Die produktionskosten der Güter und ihre Tauschwerte (relativen Preise) stehen eben nicht 'wie Ricardo annahm, in dem einfachen Verhältnisse von Ursache und Wirkung zu einander..., sondern (bedingen) einander gegenseitig als die verschiedene Glieder eines einzigen wirtschaftlichen Gleichgewichtssystems'. Es ist das große verdienst von Menger, Jevons und vor allem Walras, diesen Nachweis geführt und damit gezeigt zu haben, daß 'das Hinweisen auf die Produktionskosten sogar auf den denkbar einfachsten Voraussetzungen als theoretische Erklärung der tauschwerte der Waren unmöglich ist, so anwendbar es als praktische Regel oft auch sein mag' " (Schneider, 1962:192).

2 "Daruch wurde eine einheitliche, in sich geschlossene Erklärung des großen Preisbildungszusammenhanges ermöglicht, die in der modernen Gleichgewichtstheorie vorliegt und deren Kennzeichnen es vor allem ist, daß die Vorstellung eines einseitigen Usachenzusammenhanges ersetzt worden ist durch die Vorstellung eines wechselseitiges Funktionszusammenhangs" (Mydal, 1932:56).

to that of equilibrium.[1]

The influence of this physicalistic legacy within modern economic theory (dating back to the rise and development of modern physics during the 17th and 18th centuries), however, exhibits an inherent tension with another powerful tradition in the 19th century; the idea of *organic* transformation. The effect of the latter both in economic theories and in the practice of various societies is seen in the contradictory claim that the economic arena, due to the operation of the "invisible hand" (the market 'mechanism'), inherently tends towards a state of equilibrium, while at the same time economic growth ought to improve the quality of life world-wide.

What is remarkable in this regard, is that the notion of a closed system (in equilibrium), as developed in classical physics, stands in contradiction to a deepened (biophysical) understanding of living entities.

Already the discovery of irreversible processes – initially by Carnot in 1824 and eventually through the discovery of radio-activity in 1896 (by Madam Curie and Henri Becquerel) and the quantum of energy *h* (by Max Planck) in 1900 – terminated, as we have noted, the reign of the main mechanistic tendency in modern physics. By 1850, Clausius and Thompson, independently formulated the second main law of thermodynamics. Clausius only introduced the term entropy in 1865. Thomson's formulation in 1852 reads as follows: All the available energy strives at dissipation, it aims therefore at a uniform dispersion (cf. Apolin, 1964:440). A more extensive account of these developments is found in Chapter 7 (see pages 402 ff.).[2]

Von Bertalanffy generalized this law to include cases of a constant interchange of systems with their environments:

> Chemical equilibria in closed systems are based on reversible reactions; they are a consequence of the second principle of thermodynamics and are defined by minimum free energy. In open systems, in contrast, the steady state is not reversible as a whole nor in many individual reactions. Furthermore, the second principle applies, by definition, to closed systems only and does not define the steady state. A closed system must, according to the second principle, eventually attain a time-independent state of equilibrium, defined by maximum entropy and minimum free energy, ... A closed system in equilibrium does not need energy for its preservation, nor can energy be obtained from it. ... the chemical equilibrium is incapable of performing work (Von Bertalanffy, 1973:132).

Hart considers the way in which "a cell exchanges its material components many times over without losing its identity" to be one of a long list of irreducible properties of living organisms (Hart, 1984:399, note 64). However, Von Bertalanffy's generalization not only shows that living entities are thermody-

1 "Walras zeigt, wie ein Prozeß der sukzessiven Žonderung der Preise und der Mengen von einem Zustand des Ungleichgewichts zum Gleichgewicht führt" (Schneider, 1962:264).

2 Clausius only introduced the term entropy in 1865. Thomson's formulation in 1852 reads as follows: All the available energy strives at dissipation, it aims therefore at a uniform dispersion (cf. Apolin, 1964:440). A more extensive account of these developments is found in Chapter 7 (see pages 402 ff.).

namically open, since, as mentioned above, there are also numerous examples of *physical systems* that are thermodynamically open, such as a glacier or fire. This dynamic equilibrium, designated with the term 'Fliessgleichgewicht' by Von Bertalanffy (1973:165), concerns a physical feature of living entities, not a distinctive biotical one! Hart here needed a reference to the well-known work of E. Schrödinger (dealing with the physical aspect of the cell – see Schrödinger, 1955, and Hart, 1984:124, 139).

A deepened understanding of thermodynamically open systems once and for all makes it clear that living processes cannot be explained with the aid of a closed systems model. This attempt caused the neo-vitalism of Hans Driesch to introduce entelechie as an immaterial vital force capable of suspending physical laws, since living entities manage to build up more and more internal order through growth. The apparent 'equilibrium' in a living entity, however, is completely different from any true equilibrium in a physical sense. The latter state is incapable of performing any work. Von Bertalanffy points out that the dynamic pseudo-equilibrium of living entities is kept constant at a certain distance from true equilibrium, enabling them to perform work while requiring a continuous import of energy for maintaining the distance from true equilibrium (Von Bertalanffy, 1973:133). Schrödinger describes this state of affairs by saying that living things feed on negative entropy (Schrödinger, 1955:71 ff.).

The way in which economic life analogically reflects its inter-modal coherence with the physical mode of reality is not by approximating instances of static equilibria, but by analogically evincing the feature of a dynamic equilibrium, i.e. of thermodynamically open systems (Von Bertalanffy). In addition Apolin highlights the coherence between constancy and dynamics present in this *dynamic* pseudo-equilibrium – because living entities *constantly* maintain a certain distance from true (physical) equilibrium.[1]

The notion of static equilibria in the so-called Walras-Pareto-Optimum,[2] cannot be reconciled with a (thermodynamically) unstable process of growth. Within the field of biological contemplation, followers of the neo-vitalistic approach of Hans Driesch had to alter their arguments after Von Bertalanffy's generalization of the second main law. While fully acknowledging the nature of thermodynamic open systems, Schubert-Soldern thus continues the neo-vitalism of Hans Driesch in the following remarkable way. We have seen that in order to explain the (thermodynamic) "state of highest improbability," i.e. instability, present in a self-maintaining system, such as a living entity, he introduced an "instability factor" (Schubert-Soldern, 1962:62, cf. page 68 above).

We are fully justified in saying that the steady flow of building material in a living entity, physically seen, causes it to be in an unstable condition. At the

1 "Auch Von Bertalanffy halt das Lebendige für eine spezifische Einheitsbildung höherer Stufe. Das scheinbare Gleichgewicht aber, in welchem sich der Organismus dauernd befindet, ist ein dynamisches Pseudogleichgewicht, ein Fließgleichgewicht, das in einem bestimmten Abstand vom echten Gleichgewicht konstant erhalten bleibt" (Apolin, 1964:55).

2 Compare the Dorfman, Samuelson, Solow two-fold modification: every competition equilibrium constitutes a Pareto-Optimum and vice versa (Dorfman *et al.*, 1958:409 ff.).

same time, and without contradiction, we may also say that the same entity is, seen from a biotical perspective, in a stable condition! Whenever physical stability is approached, that is true equilibrium, biotical instability is on its way as an inevitable symptom foreshadowing death. This non-contradictory fashion of grasping both the physical instability and the biotical stability of living entities indicates the irreducible nature of the biotical aspect. Evidently, in a purely physical sense, it is contradictory to claim that the same entity can exist both in a stable and an unstable condition – something economists upholding the classical equilibrium concept would have to assert as soon as they want to account for phenomena of *economic growth*!

Dobb aptly remarks that more recent theories of growth effectively transcended the limitations of the assumed prevailing state of equilibrium by introducing a dynamic element in the conceptional scheme of Walras.[1]

Once the issue of economic growth managed to break through the rigid walls of the general equilibrium approach, the myth of an a-normative economic realm is unmasked. Questions about excessive economic growth invariably point to the normative character of economic activities and accounts for the absence of an awareness of having enough within societies dominated by a materialistic life orientation.

Without pursuing this matter further in this context, it must be clear that the true meaning of the economic aspect of creation can only be understood when analyzed in its unbreakable coherence with all the non-economic modes of reality, and when it is seen in its concentric relatedness to the central commandment of love – articulated in the call to *stewardship*. Every humanistic attempt to isolate and 'autonomize' the economic realm of creation is challenged by this articulation of the meaning of God's creational order.

5.20.5 *The meaning of the cultural-historical aspect*

Calling this aspect the technoformative, as Seerveld suggests, harbours an important limitation. The element *techno* emphasizes instrumentality in the sense of subject-object relations. Dooyeweerd wanted to account for these subject-object relations, but in addition he also wanted to account for instances by which human beings have power over other human beings – preferably designated in terms of the idea of *office* and *competence. Competence* is seen as *power over persons*, in other words as an instance of subject-subject relations. Regarding subject-object relations, Vollenhoven also speaks of 'techné' (Vollenhoven, 1968:6) and in general he does not want to assign a modal (functional) meaning to the term *history*.

Instead he prefers to refer to the "function between the logical and the lingual" as the 'form-giving' ('vormgevende') function – but he does not enter

1 "Die neuere Wachstumstheorien jedoch laufen im Gegensatz zu Schumpeter, für den Wachstum eine Störung der vorher bestehenden Gleichgewichtslage darstellte, jedenfals in ihren formalen Aspekten auf eine Dynamisierung des Walrasianischen Begriffs des generellen Gleichgewichts hinaus. ... Dabei ergaben sich die schwerwiegende Zweifel, ob die Gleichgewichte der Wachstumsmodelle Überhaupt stabil seien" (Dobb, 1969:562).

into a discussion of the difference between formative subject-subject and formative subject-object relations (Vollenhoven, 1968:8). Modal subject-subject relations concern *formative control* in inter-human association, organized in societal structures with relations of authority and subordination. If the meaning of *free formative control* is designated by using the synonymous term *power*, then the preceding distinction between subject-object and subject-subject relations could be articulated by differentiating between *power over objects* and *power over subjects*. As mentioned, the legitimacy of power over other human beings requires the notion of an *office* and the *competence* entailed in such an office. The *authority* with which a person occupying a certain office is endowed, legitimizes the competence to concretize principles in the form of rules that other human beings ought to obey.[1]

This competence of an office bearer, enabling the shaping and transformation of principles into rules of conduct that are valid within typical spheres of social intercourse, actually evinces a *subjective moment* functioning on the norm side of reality. Dooyeweerd speaks of the *formative human will* through which "normative principles" are 'positivized', and then adds, "The human formative will is to be conceived of as a *subjective moment* on the law side of these law-spheres themselves" (Dooyeweerd, 1997-II:239). Therefore, restricting the cultural-historical aspect to subject-object relations would be an impoverishment leaving the correlative subject-subject relations unaccounted for.

In addition to the argument above, namely that the normativity of the cultural-historical (or in Seerveld's terms, the technoformative) aspect presupposes the foundational role of the logical-analytical aspect (seen in the contrary historical – unhistorical), we may in this context mention an argument by Vollenhoven in support of the foundational position of the logical aspect. Vollenhoven points out that the technical means-end scheme implies that both the means and the ends have to be *identified* and *distinguished* in advance. He says that this distinguishing rests on an *analysis*, from which it follows that the historical mode is directly founded in the analytical aspect (Vollenhoven, 1948:15).

In a different context, Dooyeweerd advances another argument for positioning the cultural-historical aspect after the logical-analytical aspect. He refers to instances in which the process of meaning disclosure manifests itself within the cultural-historical and post-cultural-historical aspects, without affecting a deepening of non-theoretical thought to the level of the *systematic mastery* of a given cognitive domain. Because *formative control* (mastery) reveals the nuclear meaning of the historical aspect, and since scholarly reflection requires this *deepened meaning* of analysis, it must be clear that the rise of truly *scientific thought* is dependent upon the *disclosure* of the logical-analytical mode, antecipating the meaning of the historical modality as an aspect

1 Within the cultural-historical aspect, this relation of super- and subordination analogically reflects the meaning of the spatial aspect – where *dimensionality* and *position* find their original modal seat.

coming *after* the logical aspect in the order of cosmic time. It is therefore also striking that the *historicistic* mode of thought accepts science as a "cultural factor" – to the exclusion of nonscientific thought (Dooyeweerd, 1938:33; cf. p. 61, footnotes 49 and 50).

5.20.6 *Seerveld's view of the meaning-nucleus of the aesthetic aspect*

The development in Seerveld's thinking on the meaning of the aesthetic aspect is particularly significant when measured against an awareness of inevitable elementary (analogical) basic concepts. This development started with the notion of the "coherent symbolical objectification of meaning" (Seerveld, 1968:45). He particularly reacted against the Platonic view of *beauty* as a "matter of measure and proportion; a thing of beauty is one with appeasing, fitting harmony." Later, he proceeded from *ambiguity* to *allusivity*. In an article on "Modal Aesthetics" the term *allusivity* surfaces (Seerveld, 1979:284 ff.). In a footnote he concedes, positively reacting to the criticism of A.T. Kruijff, that his idea of the "law of coherence" was redundant (Kruijff argued that coherence still made an appeal to the rejected notion of 'harmony'). His account of the "ontic irreducibility" of *allusiveness* mentions related terms, such as "suggestie-rijk" (P.D. van der Walt) and 'nuanceful' (L. Zuidervaart) (Seerveld, 1979:286). The implication of this change is that his well-known formulation of what constitutes art ought to be revised to read: "Art is the symbolical objectification of certain meaning aspects of a thing [better: "meaning-realities" – to accept a corrective comment from N. van Til], subject to the law of allusiveness" (Seerveld, 1979:290; see Seerveld, 1980:132; note 12). The fact that he also started to assign a significant aesthetic function to imagination requires a brief assessment as well – in the light of the conditioning role of the dimensions of aspects and entities in human knowing.

Human knowing indeed seems to be 'geared' towards these two fundamental dimensions of reality, the knowing of *modal aspects* and knowledge of *entities*. The former is known through *functional relations* and the latter through *imaging* that takes on the shape of *imagining* in the uniquely human acquaintance with the world. These two legs of knowing – modally directed and entitary directed – imply each other and open the way to account for our knowledge of *universality* and *individuality* – compare the conceptions of Croce.

> Knowledge has two forms: it is either *intuitive* knowledge or *logical* knowledge; knowledge obtained through the *imagination* or knowledge obtained through the *intellect*; knowledge of the *individual* or knowledge of the *universal*; of *individual things* or of the *relations* between them: it is, in fact, productive either of *images* or of *concepts* (Croce, 1953:1).[1]

Comment: It should be noted in passing that the issue of *invariance* and *constancy*, as opposed to transience and changefulness, discussed by Zuidervaart in connection with his notes towards a social ontology of the arts (cf. Zuidervaart, 1995:41 ff.), needs to take into account

1 Mäckler mentions the following definition of Benedetto Croce: "Kunst ist Intuition, Intuition ist Individualität, und Individualität wiederholt sich nicht" ["Art is intuition, intuition is individuality and individuality does not repeat itself"] (Mäckler, 2000:30).

> Dooyeweerd's dependence on *nominalism*. We argue below that in a nominalistic fashion Dooyeweerd fuses *universality* at the factual side of reality – normally evinced in what is designated as *lawfulness*, *law-conformity* or *orderliness* – with the law side of reality, such that entities are strictly individual (see pages 370 ff., 446 ff.)

However, imaginativity, as the manifestation of a specific directedness of human knowing towards the dimension of individuality structures, extends across this entire dimension and cannot be restricted to *aesthetic* imaginativity *alone*. In addition, the flexibility of human understanding allows for a cross-utilization between the two dimensions of human experience, since the modal aspects serve as points of entry to an understanding of entities whereas the nature of the modal aspects can only be explained with the aid of *metaphors* – the result of imaginatively relating different kinds of entities through predication.

We have seen that the archeologist Narr correctly emphasizes that the human formative imagination must be able to invent something *different* from what is presented to the senses. This view is complementary to Kant, who defines the *Einbildungskraft* (imagination) as the capacity to have a representation of an object without its presence to the senses (Kant, 1787-B:151). This enables human beings to have a *historical* awareness: memory (historical past) and expectations or planning (historical future).

From a historical perspective, one may suspect that both Dooyeweerd and Seerveld, each in his own way, digested too much of the linguistic turn in modern philosophy – as a reaction to the conceptual rationalism of the 18th century and the historicism of the 19th century, Dooyeweerd switched from the idea of *organic coherence* to that of *meaning coherence*, and Seerveld explored his new understanding of *symbolical objectification*, *ambiguity*, and *allusivity*. The title of Croce's work of 1920 is quite significant: *Aesthetic as science of Expression and General Linguistic* (see Croce, 1953 – 'expression' is indeed as "general linguistic" term). Also compare the terms used by Zuidervaart: the aesthetic qualifying function is designated as "interpretable expressions" (purely *semantic-hermeneutical* categories!).[1] Even in 2001, when Seerveld once more argues in favor of 'allusivity', he remarks that it "is more sound for doing justice to the symbolic character of Western as well as non-Western craft and art" (Seerveld, 2001:163).

Croce, in his preface to *Aesthetic* (Naples, December 1901), writes, "If language is the *first spiritual manifestation*,[2] and if the *aesthetic form is language itself*,[3] taken in all its true scientific extension, it is hopeless to try to understand clearly the later and more complicated phases of the life of the spirit, when their first and simplest moment is ill known, mutilated and disfigured"

1 Even his designation of the foundational function reflects his *linguistic preoccupation*: "grounded in expressive design (technical founding function)" (1995:54).

2 The italics are mine, because they highlight Seerveld's positioning of the aesthetic and his dependance upon Croce in this regard.

3 Compare the conception of Hart.

(Croce, 1953:xxvii). One should not be surprised that Rookmaaker's first reaction to Seerveld's Ph.D. thesis (1958) was that in his aesthetics he argued the aesthetical aspect away (cf. Birtwistle, 1996:342).

The linguistic turn may have been more influential than it seemed to be at first. The strange thing about Seerveld's proposal is that, after he 'locates' "beautiful harmony" as being analogies of number and space (something surely **not** found in the nuclear meaning of either of these two aspects), these analogies scarcely form an integral part of his aesthetics.[1] It seems as if he never managed to come to terms with the everyday reality of the normative contrary of what is experienced as *beautiful* and *ugly*.[2] I cannot find any other modal aspect than the *aesthetic* to serve as the "home base" of the *beautiful-ugly* contrary – and that not merely as the numerical and spatial analogies within the aesthetic aspect as Seerveld later on attempted (Seerveld, 2001:175). It should be pointed out, furthermore, that the term *harmonization* – as used in a jural context when reference is made to the *harmonization* of a multiplicity of legal interests within a public legal order – says much more than the mere *numerical* and *spatial* analogies within the structure of the jural aspect. The numerical analogy concerns the *unity in the multiplicity* of legal norms – basic for any legal *order* – and the spatial analogy is reflected in jural relations of *super*- and *sub*ordination, *next-to-each other*, as well as in the understanding of a specific *sphere* of competence.

Comment: One should also keep in mind that the *whole-parts* relation is originally spatial in nature (cf. Strauss, 2002a and page 87 above). Without an awareness of this basic *spatial* relation – or analogies of it in aspects appearing later in the modal cosmic order – the notion of 'fit' (designated by Zuidervaart as a *technical norm* – 1995:54) and 'fittingness' (cf. Wolterstorff, 1987:96 ff.) would be meaningless. Aesthetic 'fit' and 'fittingness' therefore first of all reflect *modal spatial analogies* in a cosmic later modality. Although the meaning of any aspect can only be revealed in its *coherence* with other modal aspects, as we repeatedly pointed out, it must be clear that the unique meaning of the aesthetic aspect cannot be captured merely by referring to *one* or *more* of its retrocipatory analogies.

Duncan Roper provides a good overview of different reformational perspectives on the aesthetic in an article on this theme (cf. Roper, 1992:17). He may be correct in pointing out that the scope of *imaginativity* transcends the aesthetic as such. If imaginativity concerns our epistemic involvement in know-

1 In his later representations, it does turn up in connection with the aspects of number and space.

2 See Seerveld 2001:164 where he attempts to liberate the meaning of 'ugly' from what is antinormative, thus effectively eliminating the normative meaning of the contrary beautiful-ugly. Systematically carried through, this kind of argumentation implies that the other antinormative poles of contraries will also have to give way (such as *illogical*, *illegal*, *impolite* or *immoral*), effectively challenging the idea of normativity as such. When a work of art is *ugly*, in spite of this antinormative trait it continues to be a work of art, similar to impoliteness that continues to be social or an illogical concept that continues to be a concept.

ing the entitary dimension of reality it certainly can take on *any* normative modal qualification. However, where Roper wants to support Seerveld in seeing the "kernel" of the aesthetic aspect in "suggestiveness, nuancefulness or allusiveness" (Roper, 1992:8), we have to pay attention to the following considerations.

First, these three terms are *not* synonymous. *Suggestiveness* and *allusivity* relate to the *ambiguity* of *lingual phenomena*, i.e. they stem from the *sign mode* of reality where meaning presupposes *choice* and requires *interpretation* (see Chapter 4, page 121).

Secondly, the term *nuancefulness* without any doubt analogically reflects nothing but precisely the meaning of the *numerical* and the *spatial* aspects of reality – those aspects in which Seerveld wants to 'locate' and to which he wants to restrict the meaning of *beauty/harmony* and from which he wants to escape in his characterization of the core meaning of the aesthetic! *Nuancefulness* is synonymous with *many-sidedness* – and it does not require much reflection to realize that the term *many* originally appears in the numerical mode while the element of '*sidedness*' refers to *spatial configurations* or *sides*. Let us summarize the situation.

Initially, in order to avoid *numerical* and *spatial* analogies, Seerveld introduced his idea of the "law of *meaning-coherence*" – until it became clear that the term *coherence* still evinces a *spatial* descent. But the same fate befalls the subsequent introduction of *nuancefulness* and *allusivity*. Seerveld (and Roper) in their designation of what they consider to be the supposed unique and irreducible meaning-nucleus of the aesthetic aspect simply do not escape from the trap of numerical and spatial analogies. In her Ph.D. dissertation on the problem of meaning and identity in the art of George Grosz, Magda Van Niekerk explicitly discusses various stances within the reformational tradition regarding the position of the aesthetic aspect. At different places in this dissertation she formulates a variety of arguments challenging Seerveld's postulation of a different modal order (see Van Niekerk, 1993).

All-in-all it seems that a detailed analysis of *analogical concepts* substantially contributes to an understanding of the core meaning of the aesthetic aspect. Acknowledging the numerical and spatial descent of the term *nuancefulness* (many-sidedness) rules out the possibility of using it as a designation of the meaning-nucleus of the aesthetic aspect.

Chapter Six

The inter-disciplinary significance of modal analysis

6.1 The logical function of theoretical thinking

The idea of an inter-modal coherence and its exploration through modal analysis are intimately related to the meaning of *logicality*. One may even suspect that, since theoretical thought (modally abstractive thinking) is qualified (stamped/characterized) by the logical-analytical mode, *logic* ought to be of a *decisive* significance for all the disciplines and should therefore be foundational to the entire scholarly enterprise. Modern mathematics, for example, holds that logic is foundational to it – even though some trends believe that arithmetic is not reducible to logic. If logic is indeed foundational to mathematics, can it then still be defended that the quantitative meaning of number – in a modal functional sense – is foundational to the functional meaning of the logical-analytical aspect?

6.1.1 *The double-sided edge of analysis*

We may start our considerations by referring to the fact that people from all walks of life enjoy the fun of wrestling with 'logical' problems – such as the so-called liar paradox attributed by Diels and Kranz to Epimenides (5th Century B.C.), where it is asserted that one of the Cretans, their own prophet, said *all Cretans are liars*. In the account of Titus 1:12-13, it is reported that the Apostle Paul holds that this testimony is *true*. How can such a statement, made by a liar, be true without being false at the same time?[1]

6.1.1.1 *The limits of conceptual affirmation in Plato's thought*

The mere statement of the above-mentioned contradiction shows that ancient Greek thought has already wrestled with the basic logical ability of humans to *identify* and *distinguish*.[2] The school of Parmenides postulated the primordial nature of being and even identified it with thought.[3] The argumentation found in Plato's dialogue *Parmenides* highlights, in a *negative way*, the *mutuality* (relatedness) of *identification* and *distinguishing*, which we have explained in Chapter 1. In the *Sophist*, it is acknowledged that trying to know what *being* and *non-being* are in themselves presents thought with an *aporia* (*ἀπορία*), that is, an unresolved problem. Logic eventually used the term 'aporia' in connection with the theoretical truth of a statement, where there are grounds *for*

1 Normally a liar is a person who *sometimes* tells a lie, but not necessarily *always*.

2 We have noted that Derrida also applies the mutuality of identity and difference to language (see Derrida, 1993:10 and page 169 above).

3 Diels-Kranz (I, 231); Parmenides, B. (Fr. 3): "For thinking and being are the same": (*τὸ γὰρ αὐτὸ νοεῖν ἐστίν τε καὶ εἶναι*).

and *against* it. In Latin, *aporia* was turned into *dubitation* and *question* (see Waldenfels, 1971:448). In the Sophist (250e-251a) we read:

> So much then for a full statement of our problem. But since reality and unreality are wrapped alike in obscurity [aporia], there is for that reason some hope that any light, however much or little, cast on the one may likewise reveal the other. [251] And again, if we can get sight of neither, we will at least thus be able to investigate as best we can the logical relationship of both at once in some manner or other.[1]

Whenever *being* is thought, *non-being* is also thought.[2] In other words, in order to *identify* something, whatever is *distinct* has to be thought of as well. But within Plato's thought behind the correlation between *being* and *non-being*, the hidden *an-hypotheta* constitute the radical dualism of the *One* (*hen*) and the *Many* (*apeiron*) – which are not accessible to a logical (dialectical) method of investigation. Particularly in the dialogue *Parmenides*, the *logical inaccessibility* of these opposing principles of origin is investigated (see Strauss, 1973:16-19).

As soon as one tries to conceive the One in an absolute ("origin-al") sense, it escapes the grip of logical concepts. The first and fourth paths of the dialectical argument developed in this dialogue proceeds from the original Eleatic thesis that *being* in its absolute sense should be seen as a unity lacking plurality, motion, change and becoming. But then it cannot be said that it is a *whole*, for a whole contains *all* its parts, which would imply that the One is *many* (Parmenides, 137c4-d3). Likewise, the One is not *limitless* (137d7-8), and it is not *formless* (neither round nor straight: 137d8-e1). Furthermore, the one is not anywhere (neither in itself, nor within something else), it cannot be in motion or be at rest, it cannot be identical or different from itself, and so on (138a-142a). Whatever is *affirmed* of the One will inevitably entail that it is fraught with plurality.

The fourth path pursues a similar line of argumentation with regard to the Many (the *apeiron* – see 159e-160b). The intermediate two paths demonstrate that all properties could be *affirmed* and *denied* in respect of both the *One* and the *Many* (142b-157b and 157b-159b)! The final conclusion to all four paths of this dialectical argument reads: "Therefore, if the One is, it is everything and nothing, in relation to itself and to the many" (160b1-3).

From this *negative* argument, the *positive* result is that the *unity* of a concept always encompasses a *multiplicity*. A concept is a unity in the multiplicity of logically objectified features brought together in it. Of course this concerns a *logical* unity and multiplicity. The emphasis on the term *logical* is intended to remind us of the difference between a logical-analytical unity and

1 This remark is made against the background of the supposed *opposition* of *rest* and *motion* (see *Sophist*, 250c and d). Being is supposed to transcend rest and motion.

2 Spivak explains Derrida's view of deconstruction in similar terms: "Deconstruction, as it emerged in Derrida's early writings, examined how texts of philosophy, when they established definitions as starting points, did not attend to the fact that all such gestures involved setting each defined item off from all that it was not" (Spivak, 1999:426).

multiplicity, and a quantitative unity and multiplicity – as distinguished by Cassirer (see page 16 above).

However, we have noted earlier that (according to Hilbert) there is a circularity entailed in the *logicist* attempt to deduce the quantitative meaning of number from that of the logical-analytical mode (see page 172 above). Alternatively the inter-modal coherence between the logical and the quantitative modes will enable us to account for the two most basic logical principles.

6.2 Principles for logical reasoning

Throughout the history of logic, the principles of identity and (non-)contradiction played a dominant role. Roughly speaking, the former entails that, within the analyzable, A is always A (A is identical to itself), while the latter implies that A is never non-A (A is different from what it is not). Plato had a clear understanding of the meaning of the logical principles of *identity* and *contradiction*. The following phrase highlights both principles: "No objection of that sort, then, will disconcert us or make us believe that the same thing can ever act or be acted upon in two opposite ways, or be two opposite things, at the same time, in respect of the same part of itself, and in relation to the same object" (*Politeia*, Book IV, Ch.XIII, 436 – translation by Cornford 1966:130). Aristotle, in addition, had already understood the principle of the *excluded middle* (see *Metaph*. 1057a).[1]

During the middle ages, alongside the continuation of Aristotle's predicate logic, a notable dialectical tradition proceeding from Heraclitus and the above-mentioned dialectical logic of Plato, remained in force. Through the *via negativa* of neo-Platonism (Pseudo-Dionysius, Plotinus) it was eventually articulated by Nicholas of Cusa in his notion of the *coincidence of opposites* (*coincidentia oppositorum*). Nicholas of Cusa explored the so-called *actual infinite*, in terms of which he claimed that God, as the actual infinite, is at once the largest and the smallest (*De Docta Ignorantia*, I,5), i.e. the *coincidentia oppositorum* (see *De Docta Ignorantia*, I,22).

Perhaps the most significant elaboration of this dialectical tradition is found in the thought of Hegel, Marx and those sociologists of the 20th century who are known as *conflict theorists* (Simmel, Rex and Dahrendorf) on the one hand and the philosophy of "As If" of Vaihinger on the other. The significance of the latter is linked to its exploration of the various academic disciplines (including special sciences such as mathematics, physics, linguistics, economics, and the science of law – to name some of them). Vaihinger claims that the use of *inherently antinomic* constructions (designated as *fictions*) may serve human (scientific) thought in surprisingly efficient ways. For example, he characterizes mathematical constructs such as *negative numbers*, *fractions*, *irrational* and *imaginary numbers*, as "fictional constructs" (not hypotheses) with a "great value for the advancement of science and the generalization of its results in spite of the crass contradictions which they contain" (Vaihinger,

1 This principle concerns the case where the totality of what can be analyzed is divided such that any statement either belongs to the one or the other part – there is no third option (Latin: *tertium non datur*). We shall return to the *ontical status* of these logical principles.

1949:57). In general, Vaihinger aims to provide an explanation of "the riddle that by means of such illogical, indeed senseless concepts, correct results are obtained" (Vaihinger, 1949:240). His answer is given in what he terms "the general law of fictions," that is, in the "correction of the errors that have been committed" or in a procedure called "the *method of antithetic error*" (Vaihinger, 1949:109).

If a fiction is illogical, then the existence of logical principles is implicitly *assumed*, for without standards for the acquisition of logical concepts and for logically sound reasoning, one cannot speak of *illogical* fictions. But before we further explore the normative sense of contraries, terminological clarification is needed.

6.2.1 *Terminological considerations*

Both in scholarly and everyday contexts we hear of *contradictions*, *antinomies*, *paradoxes*, *riddles* and even *puzzles*. Particularly since Immanuel Kant explained apparently stringent proofs in his *Transcendental Dialectics* (second Book, second chapter) for a set of four *theses* and *antitheses*, under the category of *antinomies*, the latter term became common knowledge in subsequent philosophical literature and reflection.[1] In 1849, a posthumously published book by the German philosopher-mathematician, Bernard Bolzano, appeared under the title: *Paradoxien des Unendlichen* (*Paradoxes of the Infinite*). More recently, E. Teensma also published a book with the title: *The Paradoxes* (1969), while E.P. Northrop wrote a book on *Riddles in Mathematics* (1944 – also published as a Penguin book in 1964). Sometimes the word 'puzzles' is used – Martin Gardner employed it in a (Pelican) book on *Mathematical Puzzles and Diversions* (1968). Closer to the present, one may add the work of Michael Clark: *Paradoxes from A to Z* (2002).

By and large the legacy of classical and modern logic, as well as philosophy in general, did not develop a *systematic analysis* of the *differences* between these diverse designations, since most of them are normally used as synonyms – at least the terms *contradictions*, *antinomies* and *paradoxes* are used interchangeably. For example, when Faenkel *et al.* discuss the known contradictions and paradoxes, they are called *antinomies*. They distinguish between *logical* antinomies (those of Russell, Cantor and Burali-Forty) and *semantic*

1 See Kant (1787:B-454-483). These are: (1) "The world has a beginning in time, and is limited also with regard to space" versus "The world has no beginning and no limits in space, but is infinite, in respect both to time and space"; (2) "Every compound substance in the world consists of simple parts, and nothing exists anywhere but the simple, or what is composed of it" versus "No compound thing in the world consists of simple parts, and there exists nowhere in the world anything simple"; (3) "Causality, according to the laws of nature, is not the only causality from which all the phenomena of the world can be deduced. In order to account for these phenomena it is necessary also to admit another causality, that of freedom" versus "There is no freedom, but everything in the world takes place entirely according to the laws of nature" and (4) "There exists an absolutely necessary Being belonging to the world, either as a part or as a cause of it" versus "There nowhere exists an absolutely necessary Being, either within or without the world, as the cause of it." We shall return to the first three arguments below.

antinomies (those of Richard, Grelling and The Liar – 1973:5-12). The 'paradox' of Epimenides mentioned at the beginning of this chapter (see page 254) is equivalent to the statement "This statement is false."[1]

In order to provide an argument for distinguishing between antinomies and contradictions, the contrary between logical and illogical may serve as a starting point.

6.2.2 *The normative sense of the contrary: logical-illogical*

The scope of the logical principles of identity and (non-)contradiction applies to the human ability to *conceive* and *argue*. Copi explains that the "principle of contradiction asserts that *no statement can be both true and false*" (Copi, 1994:372).

We have seen that the classical example of an illogical *concept* stems from Immanuel Kant (the concept of a "square circle" – Kant, 1783:341; § 52b – see page 93).[2] Establishing that this concept is *illogical* entails that a *normative standard* is applied and that the said concept does not *conform* to the requirement of the *ought to be* inherent in this normative standard. It is *contradictory* not to distinguish between a square and a circle or, to put it differently, confusing two *spatial figures* violates the demands for *identifying* and *distinguishing* properly – a square is a square (logically correct *identification*) and a square is not a non-square (such as a circle – logically correct *distinguishing*).

Thinking in a logically *contradictory* way, i.e. thinking *illogically*, remains bound to the structure of logicality and does not turn into something *a-logical*, such as the economic, the moral or the jural. These (non-logical) facets of our experience are said to be *a-logical* – but they are not *illogical*. In fact, we have seen that they also have room for contraries *similar to* the contrary between logical and illogical. Just think of contraries like *economic* and *uneconomic*, *moral* and *immoral* and *legal* and *illegal*. Although the history of humankind makes different assessments of what may count as economically, morally or legally proper behaviour, one can hardly deny the *normativity* inherent in these dimensions, as it is manifested in the mentioned contraries. The logical contrary actually lies at the foundation of all these other instances of normative contraries – the latter analogically reflect within their own domains the meaning of logical analysis (identification and distinguishing). In fact we designated contraries as analogies of the logical principle of non-contradiction (see page 98).

In this context Stafleu writes: "According to Danie Strauss, the logical aspect should precede all normative aspects, because normativity presupposes

1 In his discussion of Tarski's semantic conception of truth Max Black explains this by considering "The statement printed within a rectangle on this page is false" (this statement is symbolized by 'c'). He then showed that an apparent contradiction can be deduced from an empirical truth and a statement true by definition and then resolves the issue by distinguishing between *primary* statements and *secondary* statements (see Black, 1949:92).

2 This concept could be seen as shorthand for the *statement* that a given spatial figure is both square and round.

the possibility of distinguishing good and evil" (Strauss 2000). Yet nowhere in this article do I say what he states. I do not even use the words "good and evil" at all. My reference is to the directional antithesis between *good* and *bad* – raised in the context of my argument that "the law-side of the logical-analytical aspect itself is [not] directionally broken apart" (Strauss, 2000:12). My argument is not at all that the normativity of the post-logical aspects "presupposes the possibility of distinguishing good and evil."

The context is a discussion in which these issues surfaced: "In our mentioned discussion of Friday January 19, 1996 I raised the point that normativity – as evidenced in contraries like polite/impolite, frugal/wasteful, and so on – always presuppose the logical aspect and the logical principle of non-contradiction." The suggestion was made "that accounting for contraries with reference to this principle may confuse the fundamental distinction between structure and direction" – explaining why I subsequently proceeded with a discussion of the directional antithesis between good and bad.

However, it is important to note two things in this regard: (i) the mentioned contraries concern *post-logical* aspects; (ii) Three pages earlier, in a footnote, I included the logical aspect and then stated: "Normative contraries, such as logical/illogical, polite/impolite, love/hate, clear/obscure, and so on, are found in all aspects of reality which require an accountable free will. The argument we are going to advance in this subsection does want to acknowledge this uniquely human normative accountability."

It is therefore out of context for Stafleu to state (just after he wrote what I have quoted above ("According to Danie Strauss …"): "As a bizarre consequence, one should assume that either the logical aspect is not normative, or that this aspect precedes itself." Although it is clear that I only mentioned post-logical contraries, it is indeed possible to defend what Stafleu deems to be *bizarre* – taking into consideration my explicit reference to the logical principle of non-contradiction (i.e. a reference to the law side of the logical aspect).

The contrary between logical-illogical (illustrated in my argument regarding the example of a square circle as an *illogical* concept), can only appear at the *factual side* of the logical-analytical aspect, which presupposes its norm side. It is impossible to discern a statement as illogical without (at least implicitly) applying the logical principles of identity and non-contradiction belonging to the law side of this aspect. For that reason it is perfectly meaningful to hold that the (factual) contrary logical-illogical presupposes the normative structure (i.e. the law side) of the logical aspect. This does not entail that the "logical aspect precedes itself" for it simply asserts that the factual contrary between what is logical and illogical is correlated with the law side of the logical aspect.

6.2.3 *The difference between confusing spatial figures and making space an all-encompassing denominator*

What is at stake in an *illogical concept* such as a *square circle* is the confusion of spatial figures. It is a different issue when an attempt is made to explain whatever there is, purely (and exclusively) in *spatial terms*.

Just consider what happened in Greek philosophy after the discovery of *irrational numbers* – an event that led to the *geometrization* of Greek mathematics (after its initial Pythagorean arithmetization – see page 45). This alteration within mathematics inspired the development of a speculative metaphysics, in which material entities were exclusively characterized in terms of their *spatiality*. The result was that the Greeks did not contemplate an *empty space*. According to their mature understanding, space does not exist, only *place*. Place is a property exclusively attributed to a concrete body. In the absence of a body, there is no subject for the predicate place. From this, it naturally follows that an "empty place" is the place of nothing – in other words, it is no place at all! But what then do we have to say about the movement of a material body? Will it be possible to assert that motion is a "change of place"? Surely, given the identification of a *body* with its *place*, motion would then be an impossibility – at least when a body is supposed to be the subject of motion – for a *change of place* will amount to a change of 'essence'! In terms of such a metaphysics of space, the introduction (or 'definition') of motion yields to the 'contradiction' that a body can move if and only if it cannot move – which approximates Zeno's arguments against multiplicity and movement, alluded to above.

The attempt to explain whatever there is exclusively in spatial terms is nothing but pursuing the aim of *reducing* everything to space (similar to the Pythagorean claim "everything is number"). The first 'victim' of a spatially oriented reductionism is found in the elimination of the meaning of *motion* – manifest in the unique contribution of Parmenides and his school. But before we enter into a more detailed analysis of this *space metaphysics*, we briefly have to relate our current discussion to the above-mentioned problem of the "coherence of irreducibles" (see page 8).

6.2.4 *'Primitives' and the problem of 'reduction'*

The claim that a *reduction* is unwarranted implicitly presupposes the conviction that '*irreducibles*' and *primitives* exist. Typical (reductionsistic) **all**-claims, such as that *everything is number* (the Pythagoreans), or *everything is interpretation* (postmodernism), challenge the idea of *uniqueness* and *irreducibility*. All-claims like these are mainly *monistic* in nature – in the sense that they want to find *one* single, all-encompassing perspective or principle of explanation capable of accounting for the entire diversity manifest in our experience of the universe. With reference to Einstein's thirty-year search for a unified field theory, Brian Greene, a specialist in the theory of super strings, believes that physicists will find (have found) a framework fitting their insights into a "seamless whole," a "single theory that, in principle, is capable of describing all phenomena" (Greene, 2003:viii). He introduces Super String theory as the "Unified Theory of Everything" (Greene, 2003:15; see also pp.364-370, 385-386). (Although the most important inspiration of *theories* *about everything* is found in the idea of reductionsism, Breuer warns that a theory of everything does not necessarily need to entail a reductionistic perspective because his own idea of universal validity approximates the nature of modal universality – see page 26).

An argument in favor of the acknowledgement of irreducibility – as one side of the coin (with the mutual coherence of what is unique as the other side) – ought to show that an unwarranted *reductionism* is entangled in unsolved problems (normally referred to as contradictions, paradoxes or antinomies). In following some ideas from Mario Bunge, we noted that a similar strategy is defended by Henry Margenau, in respect of what he calls "moderate reductionism." He takes this to be "the strategy consisting of reducing whatever can be reduced, without however ignoring emergence or persisting in reducing the irreducible" (cf. Margenau, 1982:187, 196-197 – see Chapter 1, page 8).

We noted that Ernst Cassirer, the philosopher from the neo-Kantian Marburg school (perhaps best known for his *Philosophy of Symbolic Forms*), is also explicit in this regard when he claims that a critical analysis of knowledge, in order to avoid a *regressus in infinitum*, has to accept certain *basic functions* that are not capable of being 'deduced' and not in need of a deduction (see Chapter 1, page 12). This insight enabled us to connect the inevitable employment of such basic (and irreducible) concepts to the indefinable meaning-nuclei of the different (irreducible) modal aspects. Precisely because these concepts are *basic*, they cannot be defined straight away. Various disciplines acknowledge this state of affairs by explicitly introducing "primitive terms."

For example, we have seen that in Zermelo-Fraenkel's set theory first order predicate calculus is assumed, and on this basis it introduces as an *undefined term* the specific set-theoretical primitive binary predicate $\in$, which is called the *membership relation* (Fraenkel *et al.*, 1973:23). This approach follows a general pattern: an axiomatic theory (axiomatic theories of logic excluded) "is constructed by adding to a certain basic discipline – usually some system of logic (with or without a set theory) but sometimes also a system of arithmetic – new terms and axioms, the specific undefined terms and axioms under consideration" (Fraenkel *et al.*, 1973:18).[1]

We also noted that in general linguistics, the term 'meaning' is primitive, just like the term constancy in kinematics ('invariance' – normally associated with a *uniform movement*) operates as a primitive. Likewise, in the discipline of law, the term 'retribution' is primitive – and so on.

The upshot of this is that the acquisition of *concepts* and the formulation of *definitions* ultimately rest upon *primitive terms* – they are not defined and they cannot *be* defined. These primitive terms, referring to unique and irreducible modes of existence, ought to be distinguished from terms used to refer to (types of) entities. When Loux discusses the difference between "defined and

1 For the sake of an economy of primitive terms, even the term *identity* need not be taken as primitive, since in the approach to axiomatic set theory explained by Lemmon, it could be defined by the use of the axiom of extensionality (see Lemmon, 1968:124).

undefined predicates" (Loux, 2002:42 ff.), he does not make this distinction (the term 'unmarried' relates to the dimension of entities). Against this background we now return to the *space metaphysics* of the Parmenides school.

6.2.5 *Zeno's paradoxes: a different understanding of antinomies*

In the school of Parmenides, Zeno argued against *multiplicity* and *movement* by assuming an absolutely static *being*. The well-known reasoning regarding the *flying arrow*, *Achilles and the tortoise*, as well as what is known as the *dichotomy paradox*, is reported by Aristotle in his *Physics* (239 b 5 ff.). Aristotle writes:

> The first asserts the non-existence of motion on the ground that that which is in locomotion must arrive at the half-way stage before it arrives at the goal. This we have discussed above. The second is the so-called 'Achilles', and it amounts to this, that in a race the quickest runner can never overtake the slowest, since the pursuer must first reach the point whence the pursued started, so that the slower must always hold a lead. This argument is the same in principle as that which depends on bisection, though it differs from it in that the spaces with which we successively have to deal are not divided into halves (Physica, 239b12-20).

Aristotle's approach proceeds from the assumption that "it is impossible for anything continuous to be composed of indivisible parts" (*Phys*. 232 a 23 ff.) and that "everything continuous is divisible into an infinite number of parts" (*Phys*. 238 a 22).

The basis of the first paradox is in *divisibility* and that of the third in the successive addition of two distinct series of *diminishing magnitudes* both converging to the same limit – but given the different points of departure the first one is *nested* within the second. It seems as if the Aristotelian account of paradoxes one and three collapsed movement into an issue of *spatial divisibility* and the addition of *diminishing magnitudes* from the very start (therefore both cases are related to the *density* of spatial continuity),[1] whereas the account of the flying *arrow paradox* seems to *allow* for movement to begin with, but then 'freezes' it into distinct 'moments' of time – as if something moving from 'moment' to 'moment' has a *definitive place* in space.[2]

It may be worthwhile once again to mention Zeno's fourth *Fragment* known to us, for it explicitly starts by *granting* the reality of movement and then it proceeds with an argument launched from the perspective of the static nature of space in order to cancel the possibility of movement: "That which moves neither moves in the space it occupies, nor in the space it does not oc-

1 The *density* of spatial continuity is constituted by its *infinite divisibility*. This property is analogically reflected in the structure of the system of rational numbers (fractions), since the numerical difference between any two fractions could be 'bisected' indefinitely.

2 If "being in one place" means "being at rest," and if this is "every moment" the case with the "flying arrow," then the arrow is actually only "at rest" – i.e. it is not moving at all. Of course, modern kinematics holds that 'rest' is a (relative) state of motion. Without reference to some or other system, one cannot speak of the motion of a specific kinematic subject (see Stafleu, 1980:81, 83-84).

cupy" (Diels-Kranz B Fr.4).[1] This certainly explains why Grünbaum distinguishes between the "paradoxes of *extension*" and the "paradoxes of *motion*" (Grünbaum, 1967:3) – but he explicitly distances himself from the authenticity of the historical sources by restricting himself to the legacy of Zeno in "the present-day philosophy of science" (Grünbaum, 1967:4).[2]

Unfortunately, in his encompassing treatment of Zeno's paradoxes, Grünbaum does not anywhere pay attention to the problem of *uniqueness* and *irreducibility* in his analysis (with the accompanying problem of *primitive terms* and *indefinability*). Yet this does not mean that in his mode of argumentation, there is not an *implicit* acceptance of the uniqueness of the core meaning of motion. In his work on space and time, for example, he discusses Einstein's "principle of the constancy of the speed of light" (Grünbaum, 1974:376) and points out that it concerns an upper limit that is only realized in a vacuum (Grünbaum, 1974:377). Einstein's theory of relativity proceeds from the hypothesis that one singular light signal has a constant velocity (in respect of all possible moving systems), without necessarily claiming that such a signal actually exists.[3]

We have seen that, although modern physics was dominated by a mechanistic inclination until the end of the 19th century, it eventually realized that a purely *kinematic* explanation of physical phenomena is untenable. In his work "The Principles of Mechanics developed in a New Context" Hertz demonstrates the dilemma of *reductionism*, for in his attempt to restrict the discipline of physics merely to the modes of *number*, *space* and *movement* [represented by the concepts *time*, *space* and *mass* (see page 54 and the more extensive discussion in Chapter 7 below)], he had to reject the (physical) concept *force*, claiming that the concept of *force* is something *inherently antinomic* (cf. Katscher, 1970:329). Bertrand Russell takes the same view in his *Principles of Mathematics* – the only difference is that he speaks of 'force' as a "mathematical fiction": "The first thing to be remembered is – what present day physicists will scarcely deny – that *force* is a mathematical fiction, not a physical entity" (Russell, 1956:482; cf. 494 ff.). Yet, as soon as the *dynamic* physical sense of *force* is acknowledged, as 20th century physics does, what Hertz

1 Without exploring it further, it should be noted that B Fr.3 probably contains the first explicit statement of the spatial whole-parts relation – which makes possible the notion of the *infinite divisibility* of a spatial continuum. In passing it should be noted that the PhD of Verelst contains an analysis of Zeno's paradoxes that will have to be considered by any future investigation of them (see Verelst, 2006).

2 This position closely imitates a similar disclaimer found in Russell's treatment: "Not being a Greek scholar, I pretend to no first-hand authority as to what Zeno really did say or mean. The form of his four arguments which I shall employ is derived from the interesting article of M. Noël, 'Le mouvemont et les arguments de Zénon d'Elée', *Revue de Métaphisique et de Morale*, Vol.I, pp.107-125. These arguments are in any case well worthy of consideration, and as they are, to me, merely a text for discussion, their historical correctness is of little importance" (Russell, 1956:348, note).

3 Stafleu remarks: "The empirically established fact that the velocity of light satisfies the hypothesis is comparatively irrelevant" (Stafleu, 1980:89).

deemed to be an antinomy turns out to require an acknowledgement of two unique and irreducible *functional modes* of reality, namely the *kinematic* and the *physical*.

6.2.6 *The inter-modal meaning of an 'antinomy'*

The kinematic function of uniform motion also differs from the functional modes of number and space. The above-mentioned B Fragment 4 of Zeno actually demonstrates that the irreducibly unique meaning of uniform flow (motion) cannot be captured purely in spatial terms, except in an *antinomic* way. In order to explain what this means, we have to follow Dooyeweerd in altering the customary meaning of the term 'antinomy', and further explore our initial succinct distinction between contradiction and antinomy (see page 172). The most obvious way to accomplish this is to allude to the literal sense of the word *antinomy*, which is intended to designate a *clash of laws*.[1] But contrary to the position taken by Immanuel Kant – who designates an antinomy as tension amongst laws of pure reason[2] – one should honor the *ontical* reference of antinomies, for the latter emerges as soon as theoretical thought confuses what is ontically unique and irreducible. The attempt to explain movement in terms of space therefore results in a (*theoretical*) *conflict* between *kinematic laws of motion* and *spatial laws*. Stafleu suggests that one way to interpret Zeno's arguments 'against' motion is that he in fact demonstrates that motion cannot be explained by numerical and spatial relations (see Stafleu, 1987:61).

The outcome of our argument is that such a conflict or clash between distinct functional modes (laws) indeed demonstrates the nature of a *theoretical antinomy*. After all, in the actual world, these two modes of explanation are unique *and* mutually cohering.[3] Yet the attempt to *reduce* one unique mode to another invariably results in genuine (theoretical) antinomies.

> In this sense antinomies therefore concern an *inter-modal* confusion, that is, a lack of distinguishing properly between *different* modes, functions or aspects of reality.

Furthermore, an antinomy always entails a *logical contradiction*, whereas a contradiction does not *necessarily* presuppose an antinomy. The logical contradiction in a "square circle" exemplifies an instance where two *spatial* figures are not properly identified and distinguished. In other words, such a contradiction has an *intra-modal* character since it is manifested within the modal-functional boundaries of only one aspect. The error is caused by confusing modal-functional figures within the boundaries of one specific aspect (namely within the spatial mode).

We noted that, within the dialectical tradition of Marxism, this difference between an (inter-modal) antinomy and (intra-modal) contradiction surfaced

1 *Anti* = against, and *nomos* = law.

2 Kant (1787:434): "... Widerstreits der Gesetze der reinen Vernunft."

3 The 'path' of a movement highlights the undeniable interconnection between motion and space. The serial order of events reveals a connection with the numerical meaning of succession.

in a striking way – resulting in what was called the "dialectical Widerspruch" of movement (see page 186 above): one can say that a moving body (that is, a body involved in a change of place) at the same time is and is not at a specific place. Hörz's designation of this situation as the "dialectical antinomy (Widerspruch)" of *change of place* is also found in Hegel's thought:

> When we speak of movement as such, we say: a body is at a specific place and then moves to a different place. While it is moving, it is no longer at the first place, but also not yet at the second. When it is at any one of the two it is at rest. When it is said that it is between both, this is not said for between both there is also a place and therefore the same problem occurs. But movement means: to be at this place and at the same time not be there; this is the continuity of space and time and this is it that makes possible motion.[1]

In the case of Zeno's argument, however, two different modal aspects are at stake. The theoretical attempt to reduce the meaning of movement to that of space is *antinomic* – and this antinomy shows itself in the implied logical contradiction: Zeno granted movement to begin with, but then concluded that movement is an impossibility. Something can therefore move if and only if it cannot move – the logically contradictory outcome of his antinomical attempt to reduce movement to static spatial extension.

Similarly, Descartes defines movement as a change of place, thus merely reiterating Zeno's *antinomy*, since according to him, the essence of a material body consists in its extension (its place) (Descartes, *Principles*, Part II, XXIV). However, as we have noted, if something **is** its place, and if movement is 'defined' as a **change** of place, then something can only move if it undergoes a change of *essence*, implying that it cannot move after all.[2]

Since an antinomy results from an attempted reductionism, it is always inter-modal in nature. Yet it always *simultaneously* expresses itself (intra-modally) within the analytical mode as a logical *contradiction*.

Of course this perspective does not eliminate a meaningful analysis of the numerical and spatial aspects of a moving body (as intended in Grünbaum's 1967 work). Material entities and processes display various functional properties without being exhausted in their concrete many-sided existence by any single one of these modes of being and explanation.

1 "Wenn wir von der Bewegung überhaupt sprechen, so sagen wir: Der Körper ist an einem Orte, und dann geht er an einen anderen Ort. Indem er sich bewegt, ist er nicht mehr am ersten, aber auch noch nicht am zweiten; ist er an einem von beiden, so ruht er. Sagt man, er sey zwischen beiden, so ist dieß nicht gesagt; denn zwischen beiden ist er auch an einem Orte, es ist also dieselbe Schwierigkeit hier vorhanden. Bewegen heist aber: an diesem Orte seyn, und zugleich nicht; dies ist die Kontinuität des Raums und der Zeit, und diese ist es, welche die Bewegung erst möglich macht" (Hegel, 1833:337 ff.).

2 Phrased in terms of its illogical implications, this antinomy boils down to what we had with Zeno: a body can move if and only if it cannot move. Note that we initially referred to Zeno's argument by merely designating it as a 'contradiction' (see page 260 above).

6.2.7 *Apparent contradictions/antinomies resulting from a lack of understanding of inter-modal coherences*

An example of what we intend in this context has already been mentioned with reference to the work of Heinrich Hertz. We have noted that his *mechanistic reductionism* only allowed for the employment of the basic concepts of mass, space and time in his understanding of the foundations of physics. Since he denied the original and primitive (irreducible) meaning of the physical aspect of the universe, he believed that the notion of 'force' is internally antinomic.

Another example concerns the mentioned paradoxes of Zeno:

> It is not surprising that the *application* of the principle of the infinite divisibility of time is found to be associated with logical fictions formed, strictly speaking, in violation of the law of contradiction. For the principle itself involves just such a logical fiction [that is, self-contradiction], as is evident when Zeno's Dichotomy paradox – which he appears to have formulated for a moving body – is applied to time itself ... before any interval can elapse a completed infinity of overlapping sub-intervals must have elapsed. One can, therefore, either conclude that the idea of the infinite divisibility of time must be rejected, or else if one wishes to make use of the device, one must recognize that it is, strictly speaking, a logical fiction [that is, self-contradiction] (Withrow, 1961:152; see Grünbaum, 1967:65-66).

The accusation of introducing a (self-contradictory) "logical fiction" rests upon the idea of "a completed infinity." Two issues are involved in this account: (i) the possibility of the *infinite divisibility* of something continuous (Aristotle holds that "everything continuous is divisible into an infinite number of parts" – *Phys*. 238 a 22); and (ii) the idea of "a completed infinity."[1] In its own way, the ambiguity in the distinction between *endlessness* (the literal meaning of the infinite) and the idea of completed infinitude (the at once infinite as an *infinite whole*) plays a role in the observations by Kant about his first antinomy (*the world has a beginning in time, and is also limited with regard to space*).[2] Kant also wrestles with the difference between what, since Greek philosophy and mathematics, is known as the *potential* and the *actual infinite* (Kant, 1787:B:458 ff.):

> I might have apparently proved my thesis too by putting forward, as is the habit of dogmatists, a *wrong* (my italics – DS) definition of the infinity of a given quantity. I might have said that the quantity is *infinite*, if no greater quantity (that is, greater than the number of given units contained in it) is possible. As no number is the greatest, because one or more units can always be added to it, I might have argued that an infinite given quantity, and therefore also an infinite world (infinite as regards both the past series of time and exten-

1 We have designated the deepened meaning of infinity with the phrase: the at once infinite (see page 239).

2 Cassirer refers to a letter from Kant to Garve, in which the former states that the theory of antinomies actually constitutes the centre and starting point of his entire *Critique of Pure Reason* (see Cassirer, 1912:85).

> sion in space) is impossible, and therefore the world is limited in space and time. I might have done this, but, in that case, my definition would not have agreed with the *true concept of an infinite whole* ("*einem unendlichen Ganzen*") (the italics are mine – DFMS). We do not represent how large it is, and the concept of it is not therefore the concept of a *maximum*, but we conceive by it its relation only to any possible unit, in regard to which it is greater than any number. Accordingly, as this unit is either greater or smaller, the infinite would be greater or smaller, while infinity, consisting in the relation only to this given unit would always remain the same, although the absolute quantity of the whole would not be known by it. This, however, does not concern us at present.

Like Withrow, Kant (in this context) initially rejects the idea of "a completed infinity," of an *infinite totality* or *an infinite whole*. According to him, "[T]he true transcendental concept of infinity is, that the successive synthesis of units in measuring a quantum, can never be completed" – to which he adds the footnote: "This quantum contains therefore a multitude (of given units) which is greater than any number; this is the mathematical concept of the infinite."[1]

Rejecting the idea of completed infinitude is also used by Hermann Weyl in connection with his response to Zeno's antinomy of Achilles and the tortoise. "The impossibility of comprehending the continuum in terms of static being cannot be formulated in a more pregnant way than through the known paradox of Zeno between Achilles and the tortoise" (Weyl, 1966:61). He mentions that the current solution of the paradox refers to the successive partial sums of the row $1/2 + 1/2^2 + 1/2^3 + \ldots, 1 - 1/2^n$ ($n = 1, 2, 3, \ldots$) that does not grow beyond all limits (since they converge towards the number 1), and then adds the remark that when the infinitely many partial distances are viewed as a completed totality, the essence of infinity is contradicted in the claim that Achilles completely passed through the 'Unvollendbaren' in the end (that which cannot be completed) (Weyl, 1966:61).[2]

This issue may be 'simplified' by further considering the following problem (see page 241 above). Ask the question whether or not the number 1 is equal to 0.999...? Suppose we only accept the potential infinite. Then there will always be more "fractional amounts" to be added, since however far one proceeds, there will always be more – in which case 0.999... can never be

1 This note reminds me of the definition the professor in mathematics (A.P. Malan) formulated in response to my question in the first-year class (1965): "What is the infinite"? He replied that it is a *number greater than any number, however large it may be*!

2 "Die Unmöglichkeit, das Kontinuum als ein starres Sein zu fassen, kann nicht prägnanter formuliert werden als durch das bekannte Paradoxon des *Zenon* von dem Wettlauf zwischen Achilleus und der Schildkröte. Der Hinweis darauf, daß die sukzessiven Partialsummen der Reihe $1/2 + 1/2^2 + 1/2^3 + \ldots, 1 — 1/2^n$ ($n = 1, 2, 3, \ldots$) nicht über alle Grenzen wachsen, sondern gegen 1 konvergieren, durch den man heute das Paradoxon zu erledigen meint, ist gewiß eine wichtige, zur Sache gehörige und aufklärende Bemerkung. Wenn aber die Strecke von der Länge 1 wirklich aus unendlich vielen Teilstrecken von der Länge 1/2 1/4 1/8, als 'abgehackten' Ganzen besteht, so widerstreitet es dem Wesen des Unendlichen, des 'Unvollendbaren', daß Achilleus sie alle schließlich durchlaufen hat."

equal to one. The other option is (in following Weierstrass, Dedekind and Cantor) to accept the "actual infinite" too and straight away to define the numer 1 as the 'totality' of the decimal expansion 0.999...!

Kaufmann and Fischer consider this account *internally antinomic*. The former says:

> It is generally accepted that an infinite decimal fraction indicates nothing else but a sequence of natural numbers, where ... the term 'sequence' does not bear upon an infinite totality, but merely pertains to the domain of a determined relationship (law-conformity) (Kaufmann, 1968:122-123).

Likewise, with reference to the square root of the number 2, Fischer argues:

> Every 'representation' of the square root of 2, whatever its nature may be, is only to be conceived of as an endless and absolute sequence of rational approximating values incapable of completion. Only when the self-contradictory fiction of completed infinitude is added, are we allowed to view the infinite decimal fraction as a representation of the square root of 2. Without the antinomic concept of 'completed infinity' we cannot form the concept of an irrational number (Fischer, 1933:108).

But what is entailed in the idea of "a completed infinity"? Although his constructivist approach also questions the idea of an infinite totality, we saw (see page 237 above), that Paul Lorenzen does explain its nature with reference to classical analysis (Weierstrass, Dedekind, Cantor) when he says that "... each real number is represented as if the infinitely many decimal figures existed all at once" (Lorenzen, 1972:163).

This formulation actually captures a long-standing tradition that is closely related to the view of eternity as the *timeless present* (Plotinus, *Enneads*, III/7) and that dates back even further to the mentioned position of Parmenides (B Fr.8:3-6), where we find a characterization of *being* in terms of the *present* and of (a unified, coherent) spatial *wholeness* (see page 87 above).

A way out of this apparent impasse is therefore given in the acknowledgement of the uniqueness and irreducibility of *space*, as well as in an acknowledgement of the inter-modal connections between number and space. Whereas our basic awareness (intuition) of number is determined by the quantitative *order of succession*, that of space is determined by the order of *simultaneity* (at the basis of the kinematic order of *uniform flow*).

In Zeno's paradoxes, we have noted that Fraenkel discerned a tendency towards arithmeticism (see page 219 above): "... the tendency of arithmetization, already underlying Zenon's paradoxes, has been impressing its mark upon modern mathematics and may be perceived in axiomatics of set theory ... as well as in metamathematics" (Fraenkel *et al.*, 1973:213). Let us recall here also once again his subsequent statement: "However, the converse direction is also conceivable, for intuition seems to comprehend the continuum *at once* (my italics – DS); mainly for this reason Greek mathematics and philosophy were inclined to consider continuity to be the simpler concept and to contemplate combinatorial concepts and facts from an analytic view" (Fraenkel *et al.*, 1973:213).

If the most basic meaning of infinity is expressed in an *endless succession*, or, as we prefer to call it, in the *successive infinite*, then this quantitative (arithmetical) perspective retains its *relative* validity within a numerical framework – a context within which there is indeed no room for the idea of 'actual' or 'completed' infinity. But if the *totality character* of continuity is something original within the *spatial aspect*, then this perspective opens up the possibility of an alternative *systematic account* of the meaning of the actual infinite. The idea of an *infinite totality* points to a deepening and disclosure of the original primitive meaning of number. It entails that any successively infinite sequence of numbers could be viewed – anticipating the spatial feature of wholeness (totality) – *as if* all the elements in the sequence are *at once* present as a *whole*, as an infinite *totality* (see the Figure on page 240 above).

We argued that the best way to capture this spatially deepened meaning of infinity is to designate it as the *at once infinite*. Phrased differently: through the idea of the at once infinite, the meaning of number is deepened by pointing towards the original *spatial* meaning of wholeness (totality).

Since there is no *constructive* transition from the *successive infinite* to the *at once infinite* (cf. Wolff, 1971), it stands to reason that those who only accept the closed (not yet opened-up) meaning of number, will *by definition* consider the *at once infinite "internally antinomic."* But what seems to be an antinomy turns out to be a meaningful configuration as soon as it is positioned within the context of a proper understanding of the *inter-modal coherences* between number and space!

6.2.8 *Are Kant's 'antinomies' real antinomies?*

Kant's entire argumentation rests on the conviction that both the thesis and antithesis are conceivable without containing a contradiction as such. It is only when they are juxtaposed, that the yardstick of the logical principle of non-contradiction shows them to be (logically) incompatible in the sense that both cannot be true at the same time.

In his discussion of the thesis and antithesis of the first antinomy, Kant argues against the at once infinite by applying the restricted (not-yet-opened-up) meaning of the successive infinite.[1] In this case, the at once infinite ("completed infinitude," an "infinite totality") is considered to be a *self-contradictory* idea, whereas in fact it is not the case if one acknowledges the interconnections between number and space. The 'solution' of the first antinomy takes recourse to 'endlessness' (the successive infinite) as well as to the distinction between the world of phenomena (*mundus phaenoumenon*) and the noumenal world (*mundus intelligibilis*). In a later context, we shall return to the crucial role played by the distinction between *appearance* and *thing-in-itself* in the

1 Remember that the first one reads: (1) "The world has a beginning in time, and limited also with regard to space" versus "The world has no beginning and no limits in space, but is infinite, in respect both to time and space."

solution of the antinomies.

Let us look at the second 'antinomy': "Every compound substance in the world consists of simple parts, and nothing exists anywhere but the simple, or what is composed of it" versus: "No compound thing in the world consists of simple parts, and there exists nowhere in the world anything simple."

In his remarks about the first thesis of the second antinomy, Kant states that space is not a *compositum* since in determining its parts, space is a *totum*:

> We ought not to call space a compositum, but a totum, because in it its parts are possible only in the whole, and not the whole by its parts.

However, this proper insight into the nature of *spatial wholeness* (the totality-character of continuity) is burdened by the legacy of the Greek metaphysics of space, for Kant still (in line with Descartes) holds the view that matter is *qualified* (characterized) by space, that it is essentially *extended* (cf. Kant, 1787:B35). This explains why he continues to speak about the *infinite divisibility* of matter (in his remark about the antithesis of the second antinomy) – without realizing that denying the physical nature (qualification) of material entities is truly *antinomic*. Whereas mathematical space – in a purely abstract and functional perspective – is both *continuous* and *infinitely divisible*, physical space is *neither* continuous nor infinitely divisible. Since it is bound to the quantum structure of energy physical space cannot be subdivided *ad infinitum*. Energy quanta indeed represent the limit of the divisibility of energy – as already emphasized by Hilbert (see Hilbert, 1925:165 and page 300 below).

In other words, although internally *antinomic* views do occur within the argumentation of Kant's first two antinomies (not as such recognized by him), the theses and antitheses of them are merely standing in a relation of *contradiction* to each other.

The *distinction* between discreteness and continuity or between constancy and dynamics derives from irreducible functional modes of reality. Therefore they are not 'opposites'[1] – like *few* or *many* (an arithmetical opposition), like *small* and *big* (a spatial opposition), like *slow* and *fast* (a kinematic opposition) or like *weak* and *strong* (a physical opposition).

6.2.9 *The irony of reductionism*

It is remarkable to note that the internally antinomic nature of every attempted reductionism harbours the irony of achieving the opposite of what was was intended. Classical Marxism, for example, pursued the goal of everyone possessing *everything*, but the tragic irony was that, in Communist Russia, no one really possessed *anything*. Likewise the vitalist motto "live and let live" caused problems in African hospitals where Albert Schweitzer worked, for not killing flies threatened the health (and lives) of patients.

But let us investigate in some more detail the irony of historicism.

1 As Lakoff *et al* assert (see page 146 above).

6.2.10 *The "cul de sac" of historicism*

The name of Friedrich Meinecke is closely connected to an understanding of the rise of *historicism*, since he dedicated a comprehensive work to this theme. An interesting historical fact is mentioned by the editor, Carl Hinrichs, in the *Introduction* to this Volume (Volume III of the *Collected Works* of Meinecke), indeed highlighting an *ironic* development in the life of Meinecke. His father gave him a conservative pietist education and sincerely wished that Friedrich would continue his intellectual life within these parameters.

Hinrichs remarks that excessive pressure from the church led to the opposite result as the one that was aimed for – Meinecke broke all ties with the church and his home. Unfortunately, as a result, he was absorbed by the dialectical humanistic motive of nature and freedom. This explains his reaction against the spirit of the time (the second half of the nineteenth century) – which was dominated by the view that the "universe was caught up in the firm connection of a mechanical causality" (see Hinrichs, 1965:x). In his work on the "reason/necessity of state" ('Staatsräson') this orientation is explicit:

> Nature and spirit, law-conformative causality and creative spontaneity are these poles, which as such are sharply and apparently irreconcilably opposed. But historical life, situated between them, is always simultaneously determined by both, even though not with an equal force (Meinecke, 1957:10).[1]

During the age of Enlightenment (the 18th century), faith in the possibilities of *conceptional rationality* dominated the scene. It was believed that human reason will be able to unravel all the secrets of nature and that humankind will be set on a path of infinite progress. Through an increasing sensitivity to the reality of *individuality* and what is *unique*, romanticism witnessed the rise of *historicism* at the turn of the 18th to the 19th century – a philosophical trend attempting to turn away from the universalities of human reason by taking seriously what was considered inherently *changeful* and *irrepeatable*: historical reality in its *uniqueness* and *individuality*.

Initially it seemed as if this transitional process would succeed in maintaining a sound balance between universality and individuality. According to Meinecke, modern historicism even pursued the task of understanding both the individual structures of historical humanity and the general timeless laws-for-life in their universal interconnections.[2]

When Meinecke characterizes the highest worldview achievement of

1 "Natur und Geist, gesetzliche kausalität und schöpferische Spontaneität sind diese Pole, die sich als solche scharf und anscheinend unvereinbar gegenüberstehen, aber das geschichtliche Leben, das zwischen ihnen liegt, wird immer gleichzeitig von beiden her, wenn auch durchaus nicht von beiden immer in gleiche stärke bestimmt."

2 "Die individuelle Gebilde der geschichtlichen Menschheit, gleichzeitig aber auch ihren zeitlosen Kern, das Generelle in ihren Lebensgesetzen, das Universale in ihren Zusammenhängen zu erfassen, ist Wesen und Aufgabe des modernen Historismus" (Meinecke, 1957: 22).

Goethe in terms of a *balance* between Heraclitean and Eleatic thought (eternal becoming and eternal being),[1] it still seemed as if Goethe was trying to do justice to both universality and what is individual. His need for eternal laws manifests the influence of neo-Platonism and it revealed an element of continuity with the Enlightenment. Yet, in the final analysis, his appreciation of the universal experienced a twist in the direction of the individual. Meinecke writes: "Goethe's concept of law was totally different from that of the Enlightenment, completely free from mathematical ingredients"[2] – and then he quotes Gundlof: "Goethe's laws themselves are individuals, delicately, elastic precisely though constantly mobile, mystical inner form forces."[3]

In our discussion of the significant and influential role of nominalism in modern philosophy (see pages 370 to 379 below), we shall explore this line of development in some more detail. But within the context of our present discussion we are first concerned with the inner antinomical nature of historicism.

Under the heading *Change and Permanence: On the Possibility of Understanding History*, Hans Jonas examines this impasse of historicism. He argues that radical historical skepticism is self-defeating:

> For it is easy to show (but need not be shown in detail here) that the alleged nonintelligibility of the historical past can be advocated only in conjunction with the nonintelligibility of the present as well, and that of whole cultures only with that of every individual – since the arguments from otherness and uniqueness urged for the first are equally valid for the second, that is, for all human existence in general. The thesis, therefore, amounts to saying that there is no understanding of the human other of any sort – historical or contemporary, collective or individual. That is to say, it leads straight into solipsism and its absurdity. Its spokesmen cannot even utter it meaningfully, except in soliloquy, since they cannot hope for its being understood. Strictly speaking, one need not reply to them (Jonas, 1974:241).

Another way of exposing the impasse of historicism is to challenge its claim that everything in human life is subject to *constant change* – moral standards, jural principles, aesthetic norms, epistemic values – all of these are taken up in the ongoing process of historical change and should ultimately be appreciated for what they are: *intrinsically historical*. The consistent historicist claims that law, morality, art, science, and even the human being, are all historical in nature – without an exception.

We have seen, however, that *change* can only be detected on the basis of an *enduring* or *persistent* element, of something *constant*. The term 'constancy'

1 "So besteht Goethes höchste Weltanschauliche Leistung darin, daß er heraklitisches und eleatisches Denken, ewiges Werden und ewiges Sein veschmolz – 'Dauer im Wechsel' " (Meinecke, 1965:503).

2 "Der Goethesche Begriff von Gesetz war also völlig anders als der der Aufklärung, völlig frei von mathematischen Bestandteilen" (Meinecke, 1965:504).

3 "Goethes Gesetze sind selbst Individuen, dehnbar feine, dem immer beweglichen eben geheimnisvoll innewohnende Formkräfte" (Meinecke, 1965:504).

is preferable to 'permanence'. Constancy (the core meaning of the kinematic aspect) underlies all change (the core meaning of the physical aspect – see page 163 above.) For example, if *law* (in its *jural* sense) is intrinsically historical, it is supposed to have 'happened' somewhere in the past – which is not at all the case, for the discipline of law has a solid sense of the historical changes that took place in the on-going development of law. If the jural *is* history, it could not *have* a history. Therefore the mere fact that we still speak of legal history, art history, economic history, and so on, shows that only that which is not intrinsically historical can *have* a history (see Dooyeweerd, 1997-II:223).

The irony of the radical historicist is therefore that she achieves the opposite of what is aimed for – if everything *is* history, there is nothing left that can *have* a history. Jonas refers to the said element of *constancy* as something *transhistoric* in his assessment that parallels the irony just mentioned:

> And so we have the paradox that the advocates of radical historicism must arrive at the position of complete a-historicism, at the notion of an existence devoid of a past and shrunk to a now. In short, radical historicism leads to the negation of history and historicity. Actually, there is no paradox in this. For history itself no less than historiography is possible only in conjunction with a transhistoric element. To deny the transhistorical is to deny the historical as well (Jonas, 1974:242).[1]

One can simply state that constancy and change are *inseparable*. This is particularly seen from the fact that the historicist tends to use phrases in which both elements are present without realizing it. For example, when it is said that society changes constantly, the intention is to emphasize *change* but unfortunately this emphasis invariably brings with it some or other reference to constancy – think about synonymous terms like 'ever', 'always', and "never-ending/ceaseless" in combinations such as "the ever-changing scene of history," "the always changing historical process," and "ceaseless historical change."

Of course the foundational coherence between historical constancy and historical change is founded in the role of the kinematic meaning of constancy (uniform flow) with respect to physical change (dynamics) (in connection with constancy and change see also pages 186 and 234 above).

Oswald Spengler neglects this intra-cultural reality when he attempts to argue that *number as such* does not exist. He holds that there are different worlds of number because there are multiple cultures. In this regard his historicistic relativism clearly anticipates what eventually was claimed by postmodernism during the secod half of the 20th century.

According to Spengler we therefore find Indian, Arabic, Antique, and Western types of number, each with its own distinctive uniqueness and each bringing to expression a different tone of the world – and as an ordering principle each has a limited symbolical validity. There are therefore more than one

1 Clouser wrote an excellent article on the impasse of historicism with particular reference to the shortcomings in Rorty's approach (see Clouser, 2005a).

instances of mathematics.[1]

With good reason this perspective highlights the fact that different cultures indeed developed different number symbols and different types of number concepts. But at the basis of all these variations a *given diversity* and *plurality* cannot be denied. Things are *distinct* prior to their *being distinguished*. It is only once the question "How many?" is asked, that a *human* response is required – and this response results in the use of (culturally determined) *numerals*. Unfortunately Spengler does not realize that the *differences* he has in mind are rooted in a shared ontic reality – the *quantitative meaning* of the *one* and the *many*. Only at this point is it meaningful to argue that in respect of what is ontically given human responses cannot escape the prevailing cultural conditions of a specific society – while upholding the perspective that cultural relativity does not coincide with *relativism*.

6.2.11 *The antinomy involved in collapsing law and morality*

Although the Aristotelian-Thomistic tradition subsumed 'justice' under the category of "moral virtues," the post-Renaissance period increasingly witnessed an awareness of the distinctness and uniqueness of law and morality. Perhaps Grotius was the first to distinguish explicitly between law and morality, although his example was not followed consistently in legal thought. Thomasius introduces the association of the opposition between *inner* and *outer* with morality and law (compare his posthumously published work: *De jure naturae et gentium libri duo*, 1706). For Rousseau, this distinction eventually became instrumental to the inherent tension in modern philosophy between domination and freedom. On the one hand, Rousseau describes law as coercive and on the other, through his contract theory, he attempts to guarantee moral freedom on the basis of this coercion!

Within modern Roman Catholic moral philosophy the conviction is that rules of "natural law" may be derived from the moral law (the 'decalogue'). Thomas Aquinas (1225-1274) holds that such derivations could be made by using commandments like "You shall not murder," "You shall not commit adultery," and "You shall not steal." What he did not realize is that the concepts *murder*, *adultery* and *stealing* presuppose unlawfulness in a *jural* sense. The prohibition of *murder* requires that one ought not to show such a lack of love and care towards one's neighbour that the desire to intentionally slay such a person arises. But when it is attempted to reduce the *moral meaning* of this commandment to the jural, an *antinomy* appears, since the meaning of morality presupposes the jural sense of unlawfulness. In order to avoid this

1 "Eine Zahl an sich gibt es nicht und kann es nicht geben. Es gibt mehrere Zahlenwelten, weil es mehrere Kulturen gibt. Wir finden einen Indischen, Arabischen, antiken, abendländischen Zahlentypus, jeder von Grund aus etwas Eignes und Einziges, jeder Ausdruck eines anderen Weltgefühls, jeder Symbol von einer auch wissenschaftlich genau begrenzten Gültigkeit, Prinzip einer Ordoung des Gewordnen, in der sich das tiefste Wesen einer einzigen und keiner andern Seele spiegels, derjenige, welche Mittelpunkt gerade dieser und keiner anderen Kultur ist. Es gibt demnach mehr als eine Mathematik" (Spengler, 1923, I:78-79).

antinomy, Victor Cathrein suggested that it is forbidden to *murder unlawfully* (Cathrein, 1909:223 – see also Dooyeweerd, 1997-II:162). However, since the concept 'murder' presupposes the jural element of unlawfulness (murder = unlawful killing), this escape route remains antinomic. The possibility of an unlawful 'murder' entails that its opposite is also possible: "lawful murder." But since *murder = unlawful killing*, the construction of "lawful murder" boils down to the contradiction: "lawful-unlawful killing"!

In general the sphere-sovereignty of the jural and ethical aspects entails both that certain legal actions may be immoral while others may be morally right but illegal (within a particular legal order). From the perspective of rhetorical ploys and fallacies Bowell and Kemp illustrate the conflation of law and morality:

> This is the mistake of assuming that anything legal must be moral, or conversely, that anything illegal must be immoral. For something to be legal within the boundaries of a given political entity (nation, city, district, the United Nations) is to say that there is no law in the statutes of that entity that prohibits it. However, the fact that something is legal does not automatically make it morally acceptable. For example, it is not illegal to cheat on your lover, or to be rude to a shy person at a party just out of cruelty, but these things are immoral. More gravely, in some countries, slavery, though obviously immoral, was legal until the nineteenth century. In other countries, even today, people may legally be denied human and civil rights because of their race, ethnicity, gender or religious belief (Bowell and Kemp, 2005:141).

6.2.12 *The logical function is constitutive for (theoretical) thinking – but not for the ontic meaning of the pre-logical aspects*

In the absence of a foundational *ontological* consideration of the interrelationships between the numerical, spatial, kinematic, physical and analytical aspects of reality, an extremely fundamental *circulus vitiosus* is concealed. Mathematical set theory, for example, is appreciated as a purely *arithmetical* theory (as intended by Cantor and most of his successors). Yet, in order to construct an *axiomatic foundation* for set theory, as we have seen, the aid of an underlying *logic* is required. Does this really mean that logic itself provides a *sufficient foundation* for set theory or for mathematics in general? Whereas Russell claims that logic and mathematics are identical (1956:v), and made an attempt to derive the number concept from the logical class concept, we have seen that this entire procedure begs the question (see page 171). Claims of logicism are untenable, because both the logical *class concept* as well as the *Axiom of Infinity* make an appeal to the *basic meaning of number*. In other words, the very *meaning* of analysis (i.e, the logical mode of reality) presupposes the *meaning* of number, and therefore cannot serve as its foundation.

> The pretended foundation of set theory in logic therefore avoids the crucial issue: in order to provide an axiomatic foundation for an analysis of the meaning of number an underlying logic is required which in itself presupposes this very meaning of number!

This follows from the inevitable presence of *quantifiers* and a *multiplicity* of (*constants* and) *variables* assumed in first-order predicate calculus. The intuition of *multiplicity* is made possible by the unique quantitative meaning of the numerical aspect – first accounted for in the introduction of the *natural numbers* and in the fact that *succession* is also inherent within our understanding of natural numbers (the best-known mathematical 'application' of this order of succession is found in [mathematical/complete] *induction*).[1] Already in 1922, Skolem had a firm grip on these issues:

> Those engaged in doing set theory are normally convinced that the concept of an integer ought to be defined and that complete induction must be proved. Yet it is clear that one cannot define or provide an endless foundation; sooner or later one encounters what is indefinable or non-provable. Then the only option is to ensure that the first starting points are immediately clear, natural and beyond doubt. The concept of an integer and the inferences by induction meet this condition, but it is definitely not met by the set theoretic axioms such as those of Zermelo or similar ones. If one wishes to derive the former concepts from the latter, then the set theoretic concepts ought to be simpler and employing them then ought to be more certain than working with complete induction – but this contradicts the real state of affairs totally (1979:70).[2]

6.2.13 *Once again: cardinality versus ordinality*[3]

Those who are convinced that (axiomatic) set theory provides a sufficient basis for mathematics, give priority to the concept of *cardinal number*. This concept seems to be more abstract and general than that of ordinal number, because in the latter case, the relation between the members of a set is taken into account, whereas it is absent in the case of *cardinal numbers*. According to Smart, the main purpose of Cassirer's "critical study of the history of mathematics is to illustrate and confirm the special thesis that ordinal number is logically prior to cardinal number, and, more generally, that mathematics may be

1 According to Freudenthal, Dedekind was perhaps the first (cf. Dedekind, 1887, par. 59, 80) to call the conclusion from n to $n + 1$ complete induction ("vollständige Induktion"). Neither Bernoulli nor Pascal is the founder of this principle. Its discovery must be credited to Francesco Maurolico (1494-1575) (cf. Freundenthal, 1940:17). In a mathematical context, where "bad induction" is supposed to be excluded (as Freundenthal remarked – 1940:37), no adjective is necessary to qualify the term *induction*.

2 "Die Mengentheoretiker sind gewöhnlich der Ansicht, dass der Begriff der ganzen Zahl definiert werden soll, und die vollständige Induktion bewiesen werden soll. Es ist aber klar, dass man nicht ins Unendliche definieren oder begründen kann; früher oder später kommt man zu dem nicht weiter Definierbaren bzw. Beweisbaren. Es ist dann nur darum zu tun, dass die ersten Anfangsgründe etwas unmittelbar Klares, Natürliches und Unzweifelhaftes sind. Diese Bedingung ist für den Begriff der ganzen Zahl und die Induktionsschlüsse erfüllt, aber entschieden nicht erfüllt für mengentheoretische Axiome der Zermelo'schen Art oder ähnliches; sollte man die Zurückführung der ersteren Begriffe auf die letzteren anerkennen, so müssten die mengentheoretischen Begriffe einfacher sein und das Denken mit ihnen unzweifelhafter als die vollständige Induktion, aber das läuft dem wirklichen Sachverhalt gänzlich zuwider."

3 See Chapter 2, page 28.

defined, in Leibnizian fashion, as the science of order" (Smart, 1958:245).

The notion of a cardinal number faces two problems on the way to its foundational claim. First, it is dependent upon the concept of a *whole*, and secondly, *cardinality* can only be established by implicitly applying *ordinality*. Surely the notion of *order* does not exhaust the meaning of number – as realized by Cassirer: "As soon as we proceed from the mere succession of numbers to a specific multiplicity (Vielheit), we encounter the transition from ordinal numbers to cardinal numbers as it was developed by Dedekind, Helmholtz and Kronecker" (Cassirer, 1910:53).

The intimate link between the notion of a *cardinal number* and the supposition of a *totality* is also seen from the perspective of the concept of a one-to-one correspondence – Cassirer uses the term 'Gleichzahligkeit' (Cassirer, 1910:58) – which clearly presupposes the intuition of *at once* (simultaneity; *Gleichzeitigkeit*).[1] The concept of *equivalence* implied in a one-one correlation underlies the difference between a straightforward *multiplicity* (how many / *Wieviel*) and the concept "just as many"/'equalling' (*Gleich-viel*) (Cassirer, 1910:62).

However, as we argued, the crucial question is whether the intuition of a *whole* (*totality*), *given at once*, stems from our numerical intuition of succession or whether it has an irreducible meaning transcending the confines of relations of number? Is it possible to bring *succession* and *simultaneity* under the same denominator, or are they irreducibly different? Skolem mentions that Zermelo pointed out that the expressions *ordinal numbers* and *cardinal numbers* are employed when *sets* are compared (Skolem, 1929:80). But what is required to compare sets? How is it possible to establish the equality of two cardinal numbers?

The only way to achieve this is to 'map' the members of both sets one by one (i.e. *successively*), because otherwise no one-to-one correspondence could be established. In other words, *comparing* cardinals presupposes *some or other* order of succession! On the basis of a similar argument, Weyl (also employing the term 'Gleichzahligkeit') is therefore fully justified in his claim that *ordinal number is the primary notion*:

> Much discussion took place whether cardinal number is the primary and ordinal number the secondary concept. ... This definition is not merely restricted to finite sets; connected to it Cantor developed his theory of infinite cardinal numbers within the context of his general set theory. Yet the possibility of mapping, which is intended in the criterion of *Gleichzaligkeit*, can only be assessed if the act of ordering takes one after the other, in an ordered temporal succession through which the elements of both sets themselves are ordered. ... Therefore it seems to me beyond doubt that the *ordinal number is primary*. Modern mathematical foundational research, which is once again disrupted by

1 In her penetrating discussion of the paradoxes of Zeno Verelst had a clear understanding of this difference. She states that ordinality relates to succession and cardinality to at once: "We zagen echter eerder al dat Zeno's oneindige deling niet ordinaal maar cardinaal (i.e., simultaan) dient te worden opgevat" (Verelst, 2006:119).

dogmatic set theory, confirms this throughout.[1]

Dummett even considers it to be a shortcoming that Frege did not pay enough attention to Cantor's work. Had he done this he "would have understood what it revealed, that the notion of an ordinal number is more fundamental than that of a cardinal number" (1995:293).[2]

Cassirer is therefore correct when he emphasizes that the "determination of number by the equivalence of classes presupposes that these classes themselves are given as a plurality" (Cassirer, 1923:52). Although he holds that the transition from *ordinal number* to *cardinal number* does not entail any new mathematical content, he does discern a "new logical function" in the "formation of the cardinal number": "As in the theory of ordinal number the individual steps are established and developed in definite sequence, so here the necessity is felt of comprehending the series, not only in its successive elements, but as an ideal *whole*" (Cassirer, 1953:42). But Cassirer does not realize that the supposed "new logical function" actually unmasks a new key element transcending the mere awareness of a succession. We have seen that the idea of wholeness or totality has an original spatial meaning.

6.2.14 *The set theoretical attempt to define an ordered pair*

Our argument in support of the foundational role of the quantitative time order of succession (and its significance for acknowledging the foundational role of ordinality), is also substantiated by the well-known definition of an *ordered pair* in set theory (and topology), in the light of our fore-going considerations.

In Zermelo-Fraenkel (ZF) set theory (in connection with which we have mentioned that the membership relation is its only set-theoretical primitive symbol – see Fraenkel *et al.*, 1973:22-23 and page 171 above), the notion of an ordered pair is evident in their discussion of the *Axiom of Pairing*. The latter states that, for any two elements *a* and *b*, there exists the set *y* which contains just *a* and *b*. The initial understanding of an ordered pair states that

for all a, b, c, d, if $<a, b> = <c, d>$ then $a = c$ and $b = d$.

Wiener and Kuratowski suggested that, within ZF, such a notion can be defined by

1 "Es ist viel darüber gestritten worden, ob nicht umgekehrt die Kardinalzahl das Erste und die Ordinalzahl der sekundäre Begriff sei. ... Es ist diese Definition nicht einmal auf endliche Mengen beschränkt; die an sie sich knüpfende Theorie der unendlichen Kardinalzahlen hat G. Cantor im Rahmen seiner allgemeinen Mengenlehre entwickelt. Aber die Möglichkeit der Paarung, von der im Kriterium der Gleichzahligkeit die Rede ist, läßt sich nur prüfen, wenn die Zuordnungsakte einer nach dem andern, in geordneter zeitlicher Folge, vorgenommen und damit die Elemente beider Mengen selber geordnet werden. ... Daher scheint es mir unbestreitbar, daß die *Ordinalzahl das Primäre* ist. Die moderne mathematische Grundlagenforschung, welche die dogmatische Mengenlehre wieder zerstört hat, bestätigt dies durchaus" (Weyl, 1966:52-53).

2 Dummett continues by pointing out that this is also true of the finite case: "after all, when we count the strokes of a clock, we are assigning an ordinal number rather than a cardinal. If Frege had understood this, he would therefore have characterized the natural numbers as finite ordinals rather than as finite cardinals" (Dummett, 1995:293).

$$<a, b> = \{\{a\}, \{a, b\}\}$$

This definition is 'burdened' by hidden assumptions. First-order predicate calculus requires primitive symbols – such as connectives, quantifiers, variables and equality. Both an awareness of *multiplicity* and an intuition of *succession* are concealed in this underlying discipline. The reference to quantifiers and variables exhibits the intuition of the *one* and the *many*. If there are more than one member in ZF (where the general form is "*x* is a member of *y*"), then both ordinality and cardinality are implicitly assumed (and successively explicated in the subsequent axioms of ZF – just consider the combination of the axioms of pairing, of union and of power-set).[1]

The definition of an ordered pair proposed by Wiener and Kuratowski simply begs the question, because, although the sets $\{a\}$ and $\{a, b\}$ as such appear to exhibit *cardinality*, the number of members (respectively 1 and 2) hides an inevitable element of ordinality: there has to be a *first* set of the pair (with one member) and a *second* set of the pair (with two members). It is therefore not at all surprising that, in their initial formulation of the notion of an ordered pair – namely that for all *a*, *b*, *c*, *d*, if $<a, b> = <c, d>$ then $a = c$ and $b = d$ – Fraenkel *et al.* had to insert the (circular!) qualifier "taken in that order": "The ordered pair $<a, b>$ is an element which corresponds to *a* and *b* (taken in that order) such that ..." (Fraenkel *et al.*, 1973:33).

Since the core meaning of number inevitably comes to expression in the primitive notions of *multiplicity* and (the arithmetical time order of) *succession*, every attempt to define *order* (or something more specified like an *ordered pair*) will turn out to be *circular*.[2] Interestingly, the human infant appears to develop a sense of (numerical) order only at the age of 15 months, whereas at an earlier stage children are already capable of observing a given multiplicity at once (subitizing) (cf. Lakoff, & Núñez, 2000:18 ff.).

But let us now consider the antinomies that arise from an insufficient understanding of the coherence between natural and normative modes in reality.

6.2.15 *The antinomous attempt to expel causality from the domain of normativity*

By and large the upshot of the dualistic separation of *nature* and *freedom* (in modern humanistic philosophy[3]) is that, by the end of the 19th century the rise of a new appreciation of the humanities (*Geisteswissenschaften*), as distinct

1 The axiom of power-set makes explicit the dependence of ZF on the original spatial whole-parts relation, for it postulates for any set *a* the existence of a set whose members are just all the subsets of *a* (see Fraenkel *et al.*, 1973:35).

2 Of course this does not entail a "vicious circle" – as the mathematician Johan Heidema pointed out in an e-mail. In his discussion of the construction of ordinal numbers, Chihara also enters into an analysis of impredicativity (Poincaré, Russel, Weyl and Feferman) (Chihara, 1973:171 ff. and Chihara, 2004:19 note 13). Although this issue is intimately connected to the whole-parts relation, we will leave it aside in this context.

3 Immanuel Kant was aware of the motive power of nature and freedom in a self-critical way (see pages 32 ff. above).

from the natural sciences resulted in a *rejection* of the use of the concept of *causality* within the domain of the humanities – except in those instances where the natural science ideal still dominated the scene to such an extent that it (deterministically) reigned *at the cost* of normative accountability. These issues are also discussed in the penetrating article of Cameron (see Cameron, 2000:191-238).

In a different context, Husserl also clearly rejects the naturalistic prejudice in his lectures on psychology (from 1925):

> The naturalistic prejudice must be given up. Solely from this prejudice emerges the determinism that mechanizes the natural world into an on-going machine that in principle is incomprehensible and meaningless. This prejudice at once also blindfolds the naturalist by preventing the recognition of the peculiar achievement of the totality of historical and generalizing sciences focused on personal spirituality and culture.[1]

In the discipline of law, the general assessment of the 19th century regarding the meaning of *causality* was purely in terms of a *physical necessity*, stripped of every element of normativity. The so-called *philosophical* concept of causality in this context is derived from J.S. Mill. It received the label of the (*conditio*) *sine qua non* theory – from which Traeger claims that every juridical theory of causality has to proceed.[2]

Mill specified the *conditio sine qua non* as follows: "philosophically speaking, [it] is the sum total of the conditions positive and negative taken together; the whole of the contingencies of every description, which being realised, the consequent invariably follows" (Mill, 1886:217). Although this formulation is intended to be *philosophical*, Mill emphatically declares: "the causes with which I concern myself are not efficient, but physical causes" (Mill, 1886: 213). His core definition reads as follows: "We may define, therefore, the cause of a phenomenon to be the antecedent, or the concurrence of antecedents, on which it is invariably and unconditionally consequent" (Mill, 1886:223). This definition brings to expression the position of the classical science ideal in its *deterministic* sense – just recall the explanation by Heisenberg (page 35 above).

Hart and Honoré (1985:444) mention the Marxist writer Lekschas, who thinks that there is "a sound core in the notion of a *sine qua non* but that the credit for discovering it belongs to Hobbes" (they refer to Hobbes's *Principles of Philosophy*, IX page 101). They mention *Die Kausalität bei der verbre-*

1 "Das naturalistische Vorurteil muß fallen. Nur aus diesem Vorurteil entspringt jener die natürliche Welt mechanisierende Determinismus, der die Welt zu einer prinzipiell unverständlichen und sinnlos dahinlaufenden Maschine macht. Ein Vorurteil, das den Naturalisten zugleich blind macht für die eigentümliche Leistung der gesamten historischen und der generalisierenden Wissenschaften von der personalen Geistigkeit und Kultur" (Husserl, 1962:143).

2 "Als ein nicht unterchätzender Gewinn für das Verständnis der Kausalitätsfrage im Recht muß die sich immer mehr Bahn brechende Erkenntnis betrachtet werden, daß *jede juristische Kausalitätstheorie von der condition sine qua non auszugehen hat*" (Traeger, 1904:38).

cherischen Handlung (1952:24-25) by J. Lekschas. However already in 1904, Traeger has traced it (via Herbart and Lotze) to Hobbes (see Traeger, 1904:17).

The problem with Mill's definition is that the cause-effect relation is taken up in a closed chain, merely allowing for thinking through an endless succession of prior causes without providing a yardstick to arrive at a point where an *accountable cause* is found – a cause to which a jur(idic)ally relevant effect could be *attributed* (and thus closing the endless succession – before it 'regresses' into infinity). The well-known theories developed on the basis of this (deterministic) concept of causality are those of the *conditio sine qua non* (Von Buri)[1] and the *adequate cause* (Von Kries, Traeger) (see Hart & Honoré, 1985:442 ff., 465 ff.).

These theories did not contemplate the possibility of an inherently *normative* jural concept of causality, i.e. the (normative) concept of *jural causality*. This possibility would have been *antinomic*, since according to the naturalistic, physicalistic prejudice entailed in the deterministic understanding of causality, it can only have an a-normative meaning. A physical concept of *causality* can never explain an *omission* which, in a jural sense, can *cause* a juridically significant *effect* without taking any *physical action*. Within the discipline of law and jurisprudence, the employment of a physical concept of causality is internally antinomic.

The above-mentioned dualism between *physical causality* and *normative accountability* also appears in the thought of Hans Kelsen.[2] He belongs to the neo-Kantian Marburg school of thought, founded by Herman Cohen and elaborated by Paul Natorp and Ernst Cassirer. As a natural (physical) law the *law of causality* applies to whatever happens factually. According to him, the decisive question is not whether our actions are caused by our will, but rather whether or not the will is *causally determined* (Kelsen, 1960:100). And Kelsen indeed considers it undeniable that the human *will* is objectively determined by the law of causality.[3]

Whereas what factually happens (in subjection to the law of causality) belongs to the domain of the 'Sein', Kelsen ascribes to the domain of the 'Sollen' (normativity) the feature of 'Geltung'. The equivalents of this term that he mentions are: "in Kraft" ("in force"). Kelsen argues that all statements

1 When all the others are given, Von Buri holds that every single *conditio sine qua non* may be viewed as a cause as well.

2 According to Kelsen, the difference between causality and normativity "is this: imputation (i.e. the relation between a certain behaviour as condition and a sanction as consequence, described by a moral or legal law)." He holds that it "is produced by an act of will whose meaning is a norm, while causality (i.e. the relation between cause and effect described by a natural law) is independent of any such intervention" (Kelsen, 1991:25).

3 "Mitunter leugnet man zwar nicht, daß der Wille des Menschen, wie alles Geschehen, tatsächlich kausal bestimmt ist, ..." (Kelsen, 1960:98). "Da die objective Bestimmtheit des Willens nach dem Gesetze der Kausalität nicht geleugnet werden kann, ..." (Kelsen, 1960:99).

of the science of law are not 'Seinsaussagen', since they have to be 'Sollaussagen'. He holds "that the statement: a particular legal norm is in force means the same as: a particular jural norm is valid.[1] The irony of this approach is immediately evident, for in order to escape from the determinism entailed in physical causality (the causal law), recourse is taken to the idea that the domain of 'Sollen', being totally separate from the domain of 'Sein', is characterized by 'Geltung'. But the term 'Geltung' is synonymous with the terms 'Kraft' (German) and 'force' (English) – and both have no other *source domain* than the physical function of reality! This view is therefore just as *antinomous* as the (neo-)vitalisic claim that living entities are characterized by an *immaterial*, vital *force*(!), for the term 'force' is derived from a 'material' (i.e. *physical*) context.

The aim is to arrive at an understanding of the 'norm' that, in its *validity*, is separated from physical operations, without realizing that the term *validity* is derived from the physical function of reality. On the one hand, Kelsen emphatically argues that, since the validity of the norm is a 'Sollen', which is not a 'Sein', its validity must be distinguished from its operation (*Wirksamkeit*).[2] On the other hand, he promotes the *operation of the legal order* to be the *condition* (*Bedingung*) of *Geltung* (Kelsen, 1960:82).

Likewise, as a serious *legal positivist*, Kelsen distances himself from *natural law* theories. But once again, his conception of the 'Grundnorm' (*Basic Norm*) had to surrender to the enemy by accepting a (pre-positive) starting point serving as the "ultimate reason for the validity of all the legal norms forming the legal order":

> It is a 'basic' norm, because nothing further can be asked about the reason for its validity, since it is not a posited norm but a presupposed norm. It is not a positive norm, posited by a real act of will, but a norm presupposed in juridical thinking, i.e. a fictitious norm[3] – as was indicated previously. It represents the ultimate reason for the validity of all the legal norms forming the legal order. Only a norm can be the reason for the validity of another norm (Kelsen, 1991:255).

Unless the *antinomic dualism* between *Sein* and *Sollen* is rejected, no single *jural fact* could be established. For example, *stealing* something has no jural meaning apart from the application of *jural norms*. A burglary as *material delict* causes *loss* and this specific *jural effect* also cannot be established apart from the application of *jural norms*. Furthermore, the domain of civil private law (*common law*) is presupposed in recognizing this legal fact, for such an infringement of a *property right* enables one to speak of *unlawfully* laying claim to what *rightfully* belongs to the legitimate owner. In this account, the

1 "daß die Aussage: eine bestimmte Rechtsnorm ist 'in Kraft' ('in force') dasselbe bedeuted wie: eine bestimmte Rechtsnorm steht in Geltung, ..." (Kelsen, 1960:82).

2 "Da die Geltung der norm ein Sollen, kein Sein ist, muß die Geltung der Norm auch von ihrer Wirksamkeit, das ist der Seinstatsache unterscheiden werden ..." (Kelsen, 1960:10).

3 His notion of "a fictitious norm" is directly influenced by Vaihinger's thought.

nature of *jural causality* is inherently *normatively* structured – suggesting that we should not try to avoid speaking of *causality* within the domain of law, but rather that we should account for the difference between *physical* and jural *causality* in order to avoid all kinds of *antinomies*.

6.3 Causality and history in the thought of Gadamer

It is understandable that Gadamer has to refer back to the way in which Kant wrestled with this problem. He mentions that the relation between *cause* and *effect* is introduced by Kant as a necessary *a priori* concept of understanding that makes possible our experience, because *nature* is nothing but "matter subject to law" (Gadamer, 1967:192). Although Gadamer allows for the employment of the concept of *causality* within the domain of history, he points out that, within a historical coherence, events occur which cannot be *predicted*. As Von Ranke formulates it, world history is constituted as "scenes of freedom" (Gadamer, 1967:193). Gadamer is here quite ambiguous about the idea of causality. On the one hand, he alludes to a two-fold notion of causality and on the other he often puts the term *cause* within parentheses (Gadamer, 1967:194, 200). His remark that causality *hampers* human freedom and responsibility in a peculiar way (Gadamer, 1967:193-194), demonstrates that the classical tension between *nature* (causality) and *freedom* is still part and parcel of these reflections. Within the context of this legacy he poses the question: "Does causality presuppose freedom or does it rather exclude freedom?" (Gadamer, 1967:196). His preference is not to use the concept of causality (in the sense of a determined relation between causes and effects) within the domain of history. Yet he still continues to speak of 'causes' within historical descriptions and explanations! But he does this with explicit hesitation, because the 'causes' within the domain of human action are always found within *teleological coherence* (Gadamer, 1967:200).

6.3.1 *The historical background of Gadamer's combination of 'cause' and 'teleology' – the dialectical tradition of necessity and freedom*

With reference to the ideal of logical creation, which became apparent in the thought of Hobbes we have already alluded to the new concurrent ideal of an autonomously free personality. Initially, the first aim of this ideal was merely to master nature rationally with the aid of the newly developing natural sciences. An element of this initial motive of *rational control* of the natural science ideal is still present in Gadamer's view of *technical control* within science. Gadamer holds that science is directed towards what the human being can *accomplish* – it therefore constitutes a *knowing control* of nature, which means that it is constituted as *technique* (Gadamer, 1972:xiii). The inherent dialectics between *natural causation* and *freedom* (teleology in Kant's third Critique) finds its one leg in the deterministic classical natural science ideal and the other in the reaction of the personality ideal initiated by Rousseau and completed by Kant.

This inherent dialectics, enclosed in the basic motive of *nature* and *freedom*, already brought Kant to a *negative* interpretation of human freedom in his *Critique of Pure Reason*: freedom is seen as *being free* from *natural necessity* (Kant, 1967-B:651-652). In his *Critique of Judgement*, Kant develops a most influential formulation of the way in which nature and freedom presuppose each other *dialectically* (that is, both oppose and need each other). Although the human understanding applies the category of causality *a priori*, as a strict natural law to nature, Kant approaches organic nature *teleologically*. It means that nature is thus portrayed *as if* its multiplicity of laws is contained in the unifying basis of understanding (Kant, 1968-B:VIII). The concept of a *natural teleology* is proposed by the capacity to judge, in order to function as a mediating concept between the concepts of nature and freedom. However, the purposiveness of nature only functions as a *regulative principle* to the (reflecting) capacity to judge (Kant, 1968-B:LVI). As guiding principle, this natural purposiveness is never to be used in a *constitutive* way, since our reflecting ability then becomes a *determining faculty* of judgement, implying that once again we are introducing a *new causality* (a final cause; *nexus finalis*; cf. 1968-B:269) into natural science (Kant, 1968-B:270).

The teleological principle merely functions as a subjective maxim in judging nature. Therefore, it cannot be applied to the *objective* reality of things in nature. Consequently, the reconciliation between the *causally determining* and the *teleologically reflecting* view of nature is sought in the unity of a supra-sensory principle, which is supposed to be valid for the totality of nature as a system (Kant, 1968-B:304). This 'solution' did not really reconcile the opposing poles of *nature* and *freedom*, since it simply reinforces the basic dualism between *natural necessity* and super-sensory *freedom* – each with its own law giver (cf. Kant, 1968-B:LIII-LIV).

Fr. Schelling made an attempt to 'synthesize' nature and freedom. Yet, according to him, in the absence of the antinomy (*Widerspruch*) between necessity and freedom, not only philosophy, but also every higher will of the spirit will shrink into insignificance (Schelling, 1968:282). As a result of this commitment he believes that a principle of freedom is concealed in nature itself, while history is founded on the hidden principle of necessity. Clearly, the result is not a real 'synthesis' or 'reconciliation', since it only amounts to a *duplication* of the original dialectic: necessity is present within the domain of freedom, and freedom is present within the domain of necessity!

Gadamer believes that the only option is to assume a *dialectical coherence* between the two ways in which one can apply the idea of causality (within nature and within history) (Gadamer, 1967:198). He remarks that the peculiar dialectic between freedom and natural necessity is given in the fact that the freedom through which the human being can manipulate natural events, is itself made possible by the idea of natural necessity (Gadamer, 1967:199).

This 'solution' does not escape from the inherent dialectic within modern humanistic thought. Perhaps Gadamer should have considered where the *negative* understanding of human freedom – namely being free from natural necessity – ends. Karl Jaspers clearly saw the impasse of this whole dialectical legacy. His confession reads:

> Since freedom is only through and against nature, as freedom it must fail. Freedom is only when nature is (Jaspers, 1948:871).

This dualistic legacy also left its imprint upon various academic disciplines. Before we assess the limitations of logic, we consider the attempt to eliminate causality from normative considerations.

In principle the biblical creation motive liberates us from an inherent antinomic understanding of reality – as if the world order contains inherent tensions within itself. Antinomies arise whenever the *given* order-diversity within reality is *dialectically* torn apart – as the outcome of a dialectical ground motive directing such a dualistic distortion. But whenever antinomies arise we are referred beyond the confines of logic to an *ontology*, in which irreducible facets of reality are not properly distinguished. For this reason antinomies actually demonstrate the *limitations of logic*.

6.4 Antinomies and the self-insufficiency of logic

As long as one merely considers the logical principles of identity and non-contradiction (whether or not amended by the principle of the excluded middle), no *material* criterion of truth is available (i.e. regarding the content of an argument), for in terms of these principles, one can at most *(formally)* affirm that two contradictory statements cannot both be true at the same time and within the same context. Kant clearly understood this:

> Therefore the purely logical criterion of truth, namely, the agreement of knowledge with the general and formal laws of the understanding and reason, is no doubt a *condition sine qua non*, or a negative condition of all truth. But logic can go no further, and it has no test for discovering error with regard to the contents, and not the form, of a proposition (Kant, 1787-B:84).

What refers thought irrevocably *beyond* logic is the *principium rationis sufficientis* (also known as *principium rationis determinantis* and *principium reddendae rationis*) – in English formulated as the "principle of sufficient reason."

This principle, originally formulated by Leibniz, was subjected to an extensive investigation by A. Schopenhauer in 1813. He called it the principle of *sufficient ground* of knowledge (*principium rationis sufficientis cognoscendi*):

> As such it asserts that, if a judgement is to express a piece of knowledge, it must have sufficient ground or reason (*Grund*); by virtue of this quality, it then receives the predicate true. Truth is therefore the reference of a judgement to something different therefrom. This something is called the ground or reason of the judgement ... (Schopenhauer, 1974:156).

The general legacy of Leibniz is captured in the phrase: *there is nothing without a sufficient ground* (*nihil est sine ratione sufficiente*). Of course, Plato affirmed that assertions require a foundation (Timaeus 28a), whereas Aristotle distinguished amongst the *αἰτίαι* four causes: *material*, *formal*, *effective* and *final*.

In his Monadology, Leibniz formulates his view as follows:

> ... and the second the *principle of sufficient reason*, by virtue of which we observe that there can be found no fact that is true or existent, or any true proposition, without there being a sufficient reason for its being so and not otherwise, although we cannot know these reasons in most cases (Leibniz, 1976:646 – see Sections 44 and 196).

People are often tempted to think that *logic* is decisive when a "good argument" is mentioned. Since an argument by itself merely links premises and conclusions – either in a *valid* or in *invalid* way – the 'goodness' of an argument does not convey an assessment regarding the reliability of the premisses or the conclusions. The latter requires proper *distinctions* in respect of the ontic nature of the diversity within reality – and the said distinctions ultimately reflect the worldview of a person.

6.5 The foundational role of the *principium exclusae antinomiae*

If the *principium rationis sufficientis* refers thinking beyond the limits of pure logicality, the logical principle of non-contradiction is founded in an underlying *ontical* principle, namely the principle forbidding inter-modal reductions – where the latter invariably result in *antinomies* (see Dooyeweerd, 1997-II: 36 ff.). This principle is *ontical* in nature and should be called the ontical principle of the *excluded antinomy* (*principium exclusae antinomiae*).

The perennial philosophical problem of explaining the coherence of what is unique and irreducible (the "coherence of irreducibles") therefore opens the way to an acknowledgment of the foundational position of the *principium exclusae antinomiae* in respect of the logical principle of non-contradiction – and at once explains why the distinction between *antinomy* and *contradiction* is not *purely logical*. The *principium exclusae antinomiae* not only depicts the *limits* of logic, but also underscores the significance of a non-reductionistic ontology transcending the confines of logic. Ontological reductionism violates the *principium exclusae antinomiae* and it leads to disastrous consequences, entailing all kinds of logical contradictions. Even if we disregard possible underlying antinomies, a negation of the principle of non-contradiction is equally devastating. Hersh remarks: "From *any* contradiction, *all* propositions (and their negations) follow! Everything's both true and false! The theory collapses in ruins" (Hersh, 1997:31).

The alternative to (antinomic) reductionism is given in an analysis of inter-modal connections presupposing their irreducibility. One of the richest implications enclosed in such an analysis is the fact that it is possible to come to a *theoretical* articulation of *modal norms* on the basis of analyzing analogies (retro- and antecipations) on the norm side of the normative aspects of reality.

6.6 Critical thinking versus critical solidarity

In practice *being critical* more often than not simply means that when you read a scientific article or book or when you listen to a scholarly presentation that you then notice *differences of opinion*. Picking up a book and finding something you do not agree with within the first couple of pages is not all that difficult. However, in order to be able really to benefit from the exercise of a critical spirit, one has to observe something more fundamental than critique, namely what we have designated as *showing solidarity* (see page 185).

The acknowledgment of *supra-individual principles* enabling all forms of rational interaction and scholarly communication. Those who have published on the theme of *critical thinking* normally enter into a discussion of modern (informal and formal) logic. The standard textbook on *Logic* of Copi immediately comes to mind (see Copy, 1994), or even a typical work on critical thinking (such as the one written by Bowell and Kemp, 2005). Implicit in works such as these is the acceptance of the basic logical principles of identity, non-contradiction and the excluded middle (*tertium non datur*). Without these principles both classical predicate (syllogistic) logic and modern symbolic logic (propositional logic) collapses. The logical validity of particular modes of inference depends upon the logical principle of non-contradiction.

6.6.1 *A self-defeating argument against the possibility of Christian scholarship*

Consider the following argumentation against the possibility of divergent standpoints within scholarly disciplines (mathematics included) – an argument rooted in the belief that scholarship ought to be 'objective' and 'neutral'.[1] It runs as follows – with an implicit appeal to the later Wittgenstein's idea of "language games":

> Only those participants who accept the "rules of the game" are allowed to join the realm of *science*. When it is asked which 'rules' ought to be followed, the mentioned three logical principles are specified.

However, we noted that intuitionistic logic does not accept the universal applicability of the logical principle of the excluded middle (see page 63).[2] The first two principles are embedded in the unity and diversity within reality because the latter make possible all *identification* and *distinguishing*. The normative demand of the principles of identity and non-contradiction is to *identify* A with A and to *distinguish* A from non-A. The crucial question is therefore: does intuitionism (with its logic) constitute a valid standpoint in mathematics?

Suppose we apply the yardstick of the three mentioned logical principles to this situation, that is, let us assume that only those who accept *all three* logical principles qualify to *play the game* of science. Then the principle of the excluded middle implies that intuitionism either is or it is not a valid mathematical standpoint – there is no *third* possibility. Yet what is presupposed in this

1 In passing we may note that Hersh also questions the supposed infallibility and objectivity of mathematics (see Hersh, 1997, Part One, Chapter 3 – pp.35-47).

2 It is noteworthy that Wittgenstein followed Brouwer [see Wittgenstein, 1968:112 (par.352); cp. p.127 (par.426) – see page 304 below] and that the well-known analytical philosopher Dummett also supported the intuitionistic approach (see Dummett, 1978) – not to mention prominent mathematicians such as Weyl, Heyting, Van Dalen and Troelstra who continued to work within the legacy of the intuitionistic mathematics of Brouwer.

application is an implication of the principle of non-contradiction, namely that *affirming* and *negating* the scholarly status of intuitionism cannot both be true at once. However, on the basis of the three given logical principles one does not find *sufficient grounds* for the truth or falsity of two contradictory statements. The moment *grounds* are needed we are irrevocably referred *beyond* the boundaries of logic. On the basis of the initial argument, holding on to the first three logical principles, the only other option left (next to disqualifying intuitionism as an acceptable mathematical standpoint), is to accept it as a valid standpoint in spite of the fact that it partially truncates the principle of the excluded middle (thus implicitly applying the principle of the excluded middle, for here there is no reference to an *infinite totality*). In other words, if the answer to the question: whether or not intuitionism is a valid standpoint in mathematics? is affirmative, then the principle of non-contradiction is violated, and when it is negative, a new problem arises. Why it is not the case that intuitionism represents the valid mathematical standpoint rather than the Cantorian (or axiomatic formalistic) orientation? Is it unacceptable because the *majority* of mathematicians are not intuitionists?

Unfortunately this option introduces a *new* 'principle', namely the *majority*.[1] However it is simply impossible to provide a *justification* for the majority principle. At most recourse could be taken to a *regressus in infinitum*.

> Did the majority decide that *what* the majority believe is true?
> And:
> Did the majority decide *that* the majority decide *that what* the majority decide is true?! …
> and so on *ad infinitum*.

Clearly, accepting the existence of universal principles for thinking is inevitable, does not entail that there is no room left for disagreement about specific principles of reasoning. Our argumentation not only demonstrates that the claim concerning the objectivity and neutrality of scholarship is self-defeating, but at the same time it also opens up space for a distinctively Christian approach to scholarship guided by the principle of the excluded antinomy and rooted in the ultimate conviction that nothing within creation ought to be deified or absolutized.

6.7 Modal norms (principles)

In an exploration of the incredibly rich scope of inter-modal connections a new perspective on the nature of modal norms emerges. It follows from an analysis of the modal analogies on the *law side* (*norm side*) of the normative aspects of reality.

We commence by considering everyday references to 'principles' – which forms an integral part of our everyday experience. Political parties like to declare their continued commitment to "basic principles"; churches refer to "scriptural principles," young people are raised to guard sensitively against all

1 Amongst the "rhetorical ploys and fallacies" discussed by Bowell and Kemp the "fallacy of majority belief" is also mentioned (Bowell and Kemp, 2005:131 ff.).

that conflicts with the 'principles' according to which they were raised; in arguments it is often concluded that an unbridgeable "difference of principle" exists. But since the beginning of the 20th century, after the neo-Kantian Baden school in philosophy exerted its influence, we have become accustomed to speak of 'values'. Whereas the neo-Kantian philosophers initially defended a view of ideal (supra-temporal) values that are supposed to be *valid*, the co-determining influence of historicism eventually relativized ('temporalized') the status of values – they were turned into our personal (unique and historically determined) choices of what ought to be, what we hold to be desirable. Consequently, we nowadays often hear of an organization formulating its 'mission' by stating that it has chosen *new values* or that it had to *prioritize* its values *differently*.

Surely the underlying issue is whether or not one can speak of *changing* norms/principles/values? Those who opt for an affirmative answer are inclined to accuse the opposite camp of adhering to a *static* view. Are principles *universally valid*? In other words, is it part of the nature of a principle that it is *valid* at all times and in all places? If so, does any space remain for human freedom to adapt to new situations? Universally valid principles have an obvious concrete significance – but then what about the equally familiar thought that principles must be *concretized* (*made valid*)? If alternative applications of a principle is considered acceptable, can changing historical circumstances lead to alternative applications?

6.7.1 *The distinction between principle and application*

We are often informed that something like the *death sentence* is a *principle*. In reality, however, the death sentence refers to the underlying disclosed Western principle of criminal law, which requires that the punishment fit the crime (taking into account *fault*, both in terms of *intent* and *negligence*). We mentioned that, in following a suggestion by Alan Cameron, the editor of Dooyeweerd's *Encyclopedia of Legal Science* (a part of the *Collected Works* of Dooyeweerd), we should note that 'fault' or 'guilt' is the equivalent of what is known in Dutch as 'schuld'. Although, in English-speaking Common Law jurisdictions, 'fault' is normally reserved for civil wrongs (*torts*), while 'guilt' is used for criminal wrongs, we mentioned that both these meanings will be captured with the term 'fault' in a broad sense (not specific to any particular category of legal wrong), encompassing both civil and criminal delicts (see page 229 above).

This principle of punishment relevant to fault is a deepened *legal-ethical* principle, fundamentally different from the strict responsibility for outcomes evident in undisclosed legal systems (e.g. the *talio*-principle in the Old Testament, known as the "eye for an eye" or "tooth for a tooth" principle – see pages 229 and 289 above). In the *talio* principle the ethical aspect of moral love had not yet deepened the meaning of the jural aspect of reality, since the attitude of the actor was neglected, and only the consequences of the act taken into account. In an ethically deepened, or disclosed,[1] legal system, the death penalty can only be considered as an application (positive expression) of the underlying principle of punishment according to fault. Other applications of the same principle could be, e.g., life imprisonment or even a shorter term, depending on the degree of possible mitigating circumstances.

God's creational will approaches humankind as constant points of departure, and humankind's calling is to give concrete effect to these points of departure as cultural shaper, according to the unique historical circumstances of a particular cultural period. Without foundational constant principles, it would be impossible to speak of *adaptation*, *dynamics*, *concretization*, *application* or *positivization*. Only in the light of the Scriptures does the Christian realize that God set his creation-wide law for being human (his Law-Word), and that its central unity and fullness exist in the law that commands us to love God and our neighbour with all our heart.

In the absence of this perspective, modern Humanism is constantly burdened with the impossible task of deriving the (normative) law side of reality from the supposed *autonomy* of the human *subject* – normally ending up in the (mutually exclusive) positions of *rationalism* and *irrationalism*.

6.7.2 *Between natural law and legal positivism*

It seems as if there are two options available in this connection: either one claims universal validity for normative principles *per se*, or one subscribes to the view that there are no universal or constant starting points for human action since all positive decisions by human beings are *variable*. Traditional theories of *natural law* chose the first option, and *legal positivism* opted for the second.[2]

Hommes summarizes the traditional concept of natural law as follows:

> Natural law in its traditional sense is the totality of pre-positive legal norms (not brought into existence through a human declaration of will in the formation of law) that are immutable, universal and per se valid as well as the eventual subjective natural rights and correlating duties, based upon a natural order (whether or not traced back to a divine origin), such that the human being can derive it from the natural order aided by natural reason (Hommes, 1961:55).

By contrast legal positivism received its most powerful ally in modern (Post-Enlightenment) *historicism*. In 1815 Von Savigny wrote, in the first Volume of the newly established *Journal for Historical Legal Science* (*Zeitschrift für geschichtliche Rechtswissenschaft*), that law is a purely *historical* phenomenon and that *next to* or *above* positive law, there is no immutable and eternal legal system of natural law (see Von Savigny, 1948:14 ff.).

In order to transcend the mutual exclusivity of these two positions, an acknowledgement of *ontic normativity* is required, because, as we argued in

1 Notice that this disclosure regards the "opening up" of antecipatory analogies in a particular aspect. Fault here refers to an antecipation from the jural aspect to the ethical aspect.

2 Compare our provisional discussion of the dilemma between *natural law* and *historicism* (page 94).

Chapter 2 in connection with the normativity of life (see page 41 above), the acceptance of *ontically given normative starting points* for human action has to realize that such starting points are not the *result* of human intervention and construction, since these ontic principles lie at the *basis* of all human *shaping* and *construction*. We have repeatedly pointed out that normative contraries[1] suggest that what we know as *analytical*, *social*, *economic* and *jural* functioning is dependent upon universal, constant starting points (principles). Although we may *disagree* about the account of the meaning of *analysis*, of *sociation*, of *economizing* or of *pursuing justice*, the reality of the mentioned contraries affirm the underlying normative *structuredness* of these human capacities. This understanding of principles transcends the *subjectivistic* inclination of modern philosophy insofar as it accepts the existence of normative principles in a truly *ontic-transcendental* sense. This perspective opposes both the rationalistic position of natural law and the irrationalistic stance shared by different trends of legal positivism.

Once the implicit assumption of human *autonomy* is questioned, it becomes clear that our human *experience* of legal relationships and our human sense of justice are not the *product* of individual or collective (rational) *construction*, since whatever we can observe within the domain of *legal relationships* is founded in and made possible by the normative structure of the jural aspect of reality. Proceeding in a transcendental-empirical way (see page 231 above), entails an investigation into what *makes possible* every positive form of our legal experience.

Natural law saw something of the underlying (universal, constant) structure of our legal experience, but it distorted its meaning by assuming that those underlying principles already have been made valid (*enforced*) for all times and all places. Yet *no principle* in this fundamental *ontic* sense is valid *per se*. Every principle requires *human intervention* in order to *be made valid*, i.e. no (pre-positive) ontic principle *holds* by and of itself. Only human beings are able to 'enforce' them (as Derrida correctly emphasizes),[2] and only *human beings* can give a *positive form* or *shape* to them. The activity of giving form to underlying principles is sometimes designated as acts of *positivizing*, and the result of such acts is accordingly known as *positivizations*. Habermas explicitly uses this term, for example when he speaks of "the positivization of law" (Habermas, 1996:71, and 1998:71, 101, 173, 180). Already in 1930 the word 'Positivierung' was used by Smend (see Smend, 1930:98). Hartmann also employs the idea of positivizing ('Positivierung').[3]

1 Like logical – illogical, polite – impolite, frugal – wasteful, legal – illegal, and so on.

2 Derrida says that there "are a certain number of idiomatic expressions" in the English language that "have no strict equivalent in French," such as the phrase "to enforce the law," or the phrase "the enforceability of the law" (Derrida, 2002:232).

3 "Dagegen ist hier wichtig, daß den Werten die Tendenz zur Realisierung immanent ist" (Hartmann, 1926:154 ff.). "Soll aber ein Wert realisiert, ein Ziel erreicht werden können, so muß das Ziel zunächst erkannt und als solches gesetzt werden. D.h., daß der Wert zunächst positiviert werden muß" (Hartmann, 1926:160 ff.; see Horneffer, 1933:105).

It is this intermediate and dependent position of all positivizations that is reified by the idea of the (logical, lingual or social) *construction* of the world.

The pitfall of traditional natural law theories exists in the *double validity* to which they adhere. In addition to those positivizations constituting *valid positive law*, the theory of natural law also accepts *natural law* as an equally valid (pre-positive) order of law. Although Habermas does not explicitly mention the problem of a *duplicated validity* entailed in the view of natural law, he does mention a conceptual duplication present in modern theories of "natural law, in preserving the distinction between natural and positive law." They thus "assumed a burden of the debt from traditional natural law. It holds on to a duplication of the concept of law that is sociologically implausible and has normatively awkward consequences" (Habermas, 1996:105).

Historicism, by contrast, is justified in questioning the metaphysical idea of *immutable* and *eternal principles* of natural law that are (supposed to be) valid *per se*. But its emphasis on the supposedly intrinsically *changefulness* of 'historical' reality collapses the normative meaning of law and justice into an anchorless *relativism*. In legal practice, it results in a merely *formal* account that actually sanctions putting any arbitrary content in the *form of law*.[1] Fukuyama has a good understanding of the *cultural relativism* that blossomed during the first part of the 20th century. He states that *cultural relativism* believes that "cultural rules are arbitrary," that they are "socially constructed artifacts of different societies," and that "there are no universal standards of morality and no way by which we can judge the norms and rules of other cultures" (Fukuyama, 2000:155-156). However, one should broaden his 'basket' understanding of normativity (identifying the latter with *morality*) and also point out that behind 20th century relativism one finds 19th century *historicism*. Grondin mentions that Dilthey envisioned "a new important task for hermeneutics," namely to defend "the certainty of understanding in the face of historical skepticism and subjective arbitrariness" (Grondin, 2003:15; see Dilthey, 1914:217).

Without an insight into the *foundational* relation between constancy and change (dynamics) – frequently pointed out in our discussions (see for example page 163 above) – it sometimes happens that the recognition of what this entails is accounted for in terms of what is considered *permanent*[2] or *unchanging*. In his analysis of the logical status of the principle of non-contradiction, Avey intuitively uses the word *constancy*, but nonetheless falls back to the terms 'permanent' and 'unchanging': He asks: "Rules of practise undergo revision. Why not rules of intellection?" and he then proceeds:

> Might it not be that even though the value concept has reference to some crite-

1 Kelsen, for example, holds: "Daher kann jeder beliebige Inhalt Recht sein. Es gibt kein menschliches Verhalten, das als solches, kraft seines Gehalts, ausgeschlossen wäre, Inhalt einer Rechtsnorm zu sein" (Kelsen, 1960:201). ("Therefore every arbitrary content can be law. There does not exist any human action for which, according to its quality, it is excluded from being the content of a legal norm.")

2 Permanence may be seen as the analogy of kinematic constancy within the physical aspect.

> rion beyond, by which the particular is estimated, and that life is a constant pursuit of something not yet attained, yet the criteria themselves, the standards in the light of which all judgments are made, may be permanent, so that the things valued in the light of those standards come and go, the standards themselves may be unchanging (Avey, 1929:520).[1]

The problem of constancy and change finds further expression in the extremes of natural law and legal positivism when connected with the dilemma of modern *subjectivism*, as it is given in the question: "how can the (collective) human person at once both be the *constructive author* and the *subject* of justice?"

The modern ideal of *autonomous freedom* (exemplified, amongst others, in the thought of Rousseau, Kant and Rawls) actually *reifies* the freedom of human subjects to give *positive form* or *positivize* pre-positive jural (and other) normative principles. Without the recognition of such (universal and constant) principles, dependent on human intervention for making them valid, the extremes of *natural law* and *historicism* cannot be avoided.[2]

6.7.3 *Are principles valid for all time?*

As constant points of departure, all true principles have an appeal for all times and places – they are universal in the sense that no human being anywhere, ever, can escape their normativity. Contemporary "situation-ethics" attempts to make the uniqueness of every situation determinative, elevating the situation itself to a norm. This is complete normlessness. This universality (that is, the point of departure for action in all situations), however, does not mean that any principle is valid in itself. Our argument is that in order to become valid, to be made effective, human intervention and activity are essential – the human being alone is empowered to give concrete expression to principles in particular unique historical situations.

The mere distinction between principle and application is linked by Hart with those attitudes towards life referred to as legalistic, conservative or traditionalistic. According to him, extreme and excessive traditionalism or conservatism is the result of an inability to understand the meaning of this distinction. He explains his claim in terms of the various expressions of respect in social habits of greeting. While the fundamental principle of social respect remains, its concrete expression in greeting changes.

> In certain cultures, men may express respect by taking off their hats to each

1 Also in connection with change and relativism, we saw that Avey stumbled upon the distinctness of constancy and change, phrased in terms of the distinction between a law and that which is subject to that law: "There is, however, another aspect of Heraclitan philosophy which should not be ignored, and which relativist theory does not always find it convenient to emphasize. The *law* of change does not itself undergo change in the manner of the changing particulars" (Avey, 1929:521 – see page 164 above).

2 In respect of Kant and Rawls, O'Neill remarks: "Constructivism for Kant, as for Rawls, begins with the thought that a plurality of diverse beings lacking antecedent coordination or knowledge of an independent order of moral values must *construct ethical principles* by which they are to live" (my italics – DS, see O'Neill, 2003:362).

> other. Let's say that after some time, people no longer actually raised the hat all the way, but just lifted it slightly. Still later, we see people just touching the hat. In the end, all that remains is raising the hand. We can distinguish between a principle (i.e. expressing respect) and actual patterns of behavior (i.e. various actions with the arm relating to headgear). ... In spite of all that varies, something 'in principle' remains invariant through all this historical development (Hart, 1984:59).

Three pages further, he explicitly rejects the extremes of conservatism and chaos:

> Either only lifting one's hat all the way counts as greeting, or anything I choose is greeting. The recognition of 'greeting in principle' makes it possible to avoid both conservatism and chaos (Hart, 1984:62).

Several contemporary theological currents have, as a result of the historicist emasculation of the biblical creational faith, only the future in view (hence their *eschatological* emphasis), without any sensitivity for the creational points of departure from which our obedience should be directed towards the future. As Olthuis observes,

> The current eschatological orientation in theology which tends to seek even the beginning in the end will need revision. The Bible begins with Genesis and Genesis begins with creation. The Scriptures see the Gospel as the link connecting creation and consummation. And this link between past and future is revealed as the Word which connects the end with the beginning, the consummation with the creation. 'I am the Alpha and the Omega, the first and the last, the beginning and the end' (Rev. 22:12). A proper vision of the consummation requires a proper appreciation of the beginning. Without this understanding, the fulfillment lacks substantial content and tends to evaporate into pious words about hope. A non-robust view of creation emasculates the gospel, for it is the creation which is brought to fulfillment in Jesus Christ even as it began in Him (Olthuis, 1989:32-33).

While the appeal of the central commandment of love is without doubt also present in the commandments of the Old Testament, as confirmed by the fact that Jesus, in his reply to the Pharisees, uses the formulation of Deut.6:5 and Lev.19:18, it is equally true that God's covenantal will for Israel was presented in the form of numerous concrete stipulations. These are a diversity of *positivized principles*, which are *as such*, i.e. in their positivized form, not universally applicable. Consider the following example.

What is the meaning of the covenant word: you shall not commit adultery? Suppose we were to put this question to a number of church goers one Sunday morning. Most likely they would reply: I understand it to mean that a husband must be faithful to his wife and vice versa. They may therefore not have any love relations in the marital sense with other men or women, since this would be adultery. In response, we could ask: does your minister understand it in this way, and what about the members of the congregation; how do they understand this commandment when they hear it? To this also, the answer is most likely to be: *yes*. Now, however, comes the critical question: is this what the

Israelites in the Old Testament understood this commandment to mean?

Not at all! In the Old Testament situation, a married man was not only allowed to have more than one wife and more than one concubine, he was even allowed to have sexual relations with an unmarried woman if he was willing to take her as wife or concubine after his involvement with her! Without doubt, the positive content of this covenant word is different from the way in which we give a positive form to the ethical relationship between husband and wife today. How does one account for the difference? The Old Testament positive form cannot be used as yardstick, except if we were to pursue the absurd *casuistic* path of elevating a particular positive form to become a universal norm for all times. But such an attempt would lead to the following problematic situation. If what we understand under this commandment today is identical to the meaning and content of the Old Testament covenant word, then virtually any situation would be justifiable in its terms. How would we counteract claims that the intention of the covenant word justifies one man to have three wives, or one wife three husbands? In this way, *any arbitrary situation* would be justifiable by claiming that contemporary practice is in conformity with the commandment. This would lead to complete normlessness.

What happened when Jesus was approached by the Pharisees with regard to divorce? Christ held that what God has put together, no person may put asunder, to which the Pharisees replied by asking why Moses prescribed the use of a letter of divorce? Jesus replied, "Moses permitted you to divorce your wives because your hearts were hard. But it was not this way from the beginning" (Matt. 19:8). Jesus appeals to the beginning – in the beginning God created the heavens and the earth (Gen.1:1). This is an appeal to creation: in principle no one may divorce, even though a person's sinful heart and its antinormative acts demands it factually (cf. Matt. 15:19).

6.7.4 *Central appeal and contemporary expressions*

Only with an appeal to the creational principle of marriage do we gain a measure which liberates us from the arbitrariness with which virtually any situation could be seen as conforming to the Old Testamental commandment.

The central unity of God's law and the religious fullness of God's claim on whole-hearted loving service is expressed in a differentiated way in the diversity of (modal and typical) creational structures – linked to the historical level of development (differentiation and integration) and disclosure in effect in a particular civilization (cf. the example of the death penalty discussed above or a plea of clemency as an example of mercy). This explains again why we cannot biblicistically consider any particular positive form of the differentiated expression of the meaning of the central commandment of love as valid *per se* for all times.

In the ten covenant words of God, the central commandment of love is given contemporary expression. The commandment: you may not commit murder, acquired an Old Testamental positive expression that must be understood in view of the *relative undisclosed* and *undifferentiated* legal system of the time. Disclosed, deepened jural-moral principles (fault, fairness, and so forth) were not prominent in this system (see pages 229, 229, and 289 above).

The sabbath commandment is perhaps the most obvious in this regard, since it is completely interwoven with the Old Testamental tabernacle and temple orders of worship, with the particular position of the high priest, all of which is part of the whole people of Israel, which is supposed to be holy as God is holy (cf. Lev. 19:2). The holy cultic days did not exist to make the people holy, since Israel was supposed to be a royal priesthood in all her covenantally obedient activities. Thus the people regularly had to recall cultically God's mighty deeds of care and redemption (this included a variety of festivals). Once Christ, priest-king in terms of the order of Melchizedek, sacrificed himself (differently from the high priests who always sacrificed both on their own behalf and on behalf of the people) (Hebr. 7:27), a change in priesthood required a change of law (Hebr. 7:12). This is why we as Christians celebrate Sunday, the first day of the week, since the new covenant is no longer bound to the celebration of the sabbath (the seventh day of the week). In Christ there is a sabbath rest for the chosen people of God (Hebr. 4:9), a restoration of the paradise-order of peace and obedience in all activities of life in God's kingdom come, and coming.

In the New Testament, we find a continuous central appeal to the commandment of love, even as the diverse concrete situations and commandments of which we read provide us with contemporary positive expressions.

From this perspective, the covenant history of the Old and New Israel can be understood within the context of the all-sided dynamic and disclosure of meaning of God's creational order. Conversely, we cannot deduce the differentiated principles for our richly nuanced contemporary life from the covenant words of the Old and New Testamental positive expressions, which were true to their particular times. The common point of reference remains God's universal order of creation, within which God gave his Word revelation and speaks to us in a central religious sense.

Of course, the religious heart appeal of the Bible is normative for all Christian expressions of life, and not only the narrower life of faith. Only in the Bible, the radical (cutting to the root) religious content of the central commandment of love is encountered. One only finds in the Bible a factual (concrete) expression of the content of the Christian faith. It would be clearer, however, to say that the Bible *determines* the content of the Christian faith, rather than that it is the *norm* for faith. We have seen that the Bible itself refers to the principles of divine creation (cf. Christ's reply to the Pharisees). The Bible itself therefore took on a positive shape *within* the order of creation.

In this regard, it cannot be emphasized enough that no insight into the existence of creational principles, nor any actual theoretical analysis or discovery of these principles can ever take place independently of the radical and central perspective of the Bible, since only the Bible contains the revelation that God created all things and subjected all creatures to His (norming and non-normative) Law-Word. Once the radical and total authority of the Bible is recognized, a (fallible) theoretical attempt can be made to uncover creational principles.

The arbitrary and indiscriminate way in which certain positive expressions in the Bible are *biblicistically* elevated to become universally valid 'principles', is well known. Without realizing the inconsistencies of such an approach, an appeal could for instance be made to Deut. 22:5, that a woman may not wear male clothing and a man not women's clothing, while all other expressions in the same context are ignored.[1] At the same time, the question is not asked whether the prohibition could have had something to do with certain heathen cultic practices from which Israel, as a holy nation, had to distance herself.

This sort of abuse of positive expressions follow a particular "exegetical procedure": when it appears in any way as if a particular positive expression in Scriptures shows a similarity to contemporary positive expressions (e.g. monogamous marriage), it is immediately concluded that we are dealing with a "scriptural principle." Even the way in which a modern marriage comes into existence or an end is absent from the New Testament, since it is dependent in our times on the differentiated civil and non-civil private law (to which we return at a later stage – see pages 548 ff. below), which had not as yet crystalized at the time of the New Testament.

All positive expressions that may obviously and considerably differ from our contemporary situation, are simply ignored, without closer justification, even while we are still supposedly bound by positive expressions that already have been invalidated from a New Testament perspective (such as the mentioned difference between *keeping* the sabbath and *celebrating* the Sunday).

In order to proceed with an exploration of the suggestion that every modal analogy on the law side of the normative aspects opens up a way to come to a theoretical articulation of modal principles (norms), we may consider the following summary formulation of the nature of a principle:

> A principle is a universal and constant point of departure that can only be made valid through the actions of a competent organ (person or institution) in possession of an accountable (responsible) free will enabling a normative or antinormative application of the principle concerned relative to the challenge of a proper interpretation of the unique historical circumstances in which it has to take place.[2]

1 You may not sow two types of seed in your vineyard (verse 9) and you may not wear mixed materials – wool and linen – at the same time (verse 11). Where would this leave the modern clothing industry?

2 Notice that this formulation implicitly uses the gateway of various modal aspects – which indicates that the term principle is a *complex* or *compound* fundamental scientific concept – in distinction from the *elementary basic concepts* in the various disciplines that only appeal to a single particular analogy within the structure of an aspect of reality. The nature of *modal analogies*, viewed within the context of a distinction between law/norm side and factual side, opens up the possibility of a philosophical analysis of modal principles. Every analogy on the law side of a normative aspect provides us with a fundamental modal principle.

At this point, we may return to our remark that the modern ideal of *autonomous freedom* reifies the human freedom to give a *positive form* to pre-positive normative principles.

6.8 The humanistic idea of autonomy

A basic distinction ought to be kept in mind regarding the nature of a principle as a universal and constant starting point for human action in the assessment of the humanistic idea of autonomy.

The mere fact that any principle is solely given as a universal and constant starting point for human action underlies our earlier argument that all such principles are *dependent* upon human intervention in order to be *made valid*, *enforced*, *given shape* or *positivized*.

In addition to this, we have to acknowledge that insofar as the human subject – in exercising a certain competency – is dependent upon a given universal and constant starting point, it remains *subject* to that normed point of departure in the traditional sense of the term. Yet, insofar as no principle can "make itself valid" (positivize itself) it requires the activity of a human subject – causing us to speak of a subjective element on the norm side (see Dooyeweerd, 1997-II:239).[1] Clearly the word 'element' contains the insight that we are not autonomous in the sense of being the *source* of normativity. Nonetheless, in a secondary and derivative sense, we have the power (competency) to positivize principles and then have to be obedient to these positivized principles (in political terms: the government is still subject to the "rule of law").

For this reason, we have said that the humanistic idea of *autonomy* absolutizes the moment of positivization and in doing so, uproots the connection with the pre-positive universal and constant starting points found at the basis of all human shaping and form giving.

6.9 Epistemic values and the "laws of thought"

Since analytical thinking always comprises both an act of *identification* and of *distinguishing*, we have argued that these two legs of analysis are subject to the logical principles of identity and non-contradiction. The other so-called basic principle of logic, understood by Aristotle (*Metaphysics*,1057a33), is the principle of the *excluded middle*. According to this principle, any statement is either true or false (see our earlier reference to Copi on page 258). But before we give an account of the way in which the logical aspect expresses its meaning in an inter-modal coherence with the non-logical aspects, we consider the idea of "epistemic values" as it surfaced in the domain of the philosophy of science.

1 Already in 1932 Dooyeweerd obtained the insight that principles require human intervention in order to be positivized: "For on the law side only norm-principles are given, demanding subjective positivization by competent organs. This [act of] positivization therefore in principle belongs to the law side, although it is a subjective activity" (Dooyeweerd, 1932:182).

6.9.1 *Kuhn and McMullin*

Since the rise of the Baden school of neo-Kantian thought at the beginning of the 20th century (Windelband and Rickert), the philosophical legacy of the West has become accustomed to speak of *values* (often instead of *principles* or *norms*). Kuhn distinguishes between the *application of rules* and the act of *evaluating* (see Kuhn, 1977:331 and Kuhn, 1984:379). He mentions five values influencing theory choice: "accuracy, consistency, scope, simplicity, and fruitfulness" (Kuhn, 1984:373). McMullin follows Kuhn's view of epistemic values and also discusses the choice of a theory in terms of "value-judgements" which differ from the mere *application of a rule* (cf. McMullin, 1983:11). He preferably speaks of "epistemic values" and transforms the values mentioned by Kuhn by referring to them as *predictive accuracy*, *internal coherence*, *external consistency*, *unifying power* and *epistemic fertility*. *Epistemic simplicity* is added to this list (McMullin, 1983:15-16).

It is clear that this way of dealing with "epistemic values" is dependent on underlying coherences between the *analytical facet* of theory formation and diverse *non-analytical aspects* of our experience. For example, 'fertility' first reminds us of a *biotic* phenomenon.[1] Plants need "fertile soil" in order to grow properly and bear fruit. Analogously, theories may turn out to be 'fertile' by bearing 'fruit'. In the value of *epistemic fertility*, we therefore meet a *biotical analogy* within the structure of theoretical thought, i.e. within the structure of (deepened) analysis.

It is clear that these "epistemic values" depend upon the sphere-universality of the analytical mode – which presupposes its modal universality as such.[2] Quine indirectly explores this modal universality of the logical-analytical aspect with reference to the disciplines of logic and mathematics. He claims that "the relevance to all science and their partiality toward none" are "two traits of logic and mathematics," which makes it possible "to draw an emphatic boundary separating them from the natural sciences" (Quine, 1970:98). We have seen that Frege already has employed the term 'logical' in order to designate a kind of generality or universality that is not dependent on the specific nature of any kind of entity (see page 71). But from our remark about epistemic fertility it must already be clear that this modal universality does not entail that the *meaning* of the logical-analytical aspect could be *separated* from its coherence with the other (non-logical, ontic) aspects of reality. The mere fact that Hilbert remarked that it is necessary to come to "a partially simultaneous development of the laws of logic and arithmetic" (Hilbert, 1970:199 – see page 172), shows that the *meaning* of analysis has its ontic foundation in the arithmetical aspect – even though an *analysis* of the meaning of number requires an actual human function within the logical-analytical mode. And this mode intimately coheres with the quantitative meaning of the numerical aspect.

1 *Biology* is the study of ontically given phenomena from the perspective of the (modally abstracted) *biotic* aspect of reality.

2 See our earlier remarks about *modal universality* (page 68 above).

Quine correctly affirms: "Any logic has to come to terms somehow with quantification, if it is not going to stop short" (Quine, 1970:88).

6.9.2 *The logical principles of identity and non-contradiction*

The failure to appreciate this *foundational* position of the numerical aspect (in an ontic sense), caused Frege, Peano, Whitehead and Russell to attempt to reduce the meaning of number to the logical mode, which at once meant that they believed that mathematics ought to be reduced to logic. Quine mentions Frege, who "claimed in 1884 to have proved in this way, contrary to Kant, that the truths of arithmetic are analytic. But logic capable of encompassing this reduction was logic inclusive of set theory" (Quine, 1970:66). Weyl goes one step further when he states that mathematics is totally – also according to the logical form in which it operates – dependent upon the essence of the natural numbers.[1]

An additional difficulty arises with regard to the *basic terminology* of a discipline. Particularly in the case of the relation between mathematics and logic, the dominating tendency seems to be to propose a clear-cut *separation*, by assigning certain terms to logic and others to mathematics. For example, Quine discusses the question whether *identity theory* should be reckoned to belong to logic (Quine, 1970:62), or whether it should be looked upon "as extra-logical" (Quine, 1970:61). At this point the theory of modal aspects conjectures a *third option*, by asking that the uniqueness and coherence between aspects (their sphere-sovereignty and their sphere-universality) be acknowledged. This option entails that those terms that bring to expression the *core (primitive) meaning* of an aspect recur in other aspects in *non-original* (i.e. *analogical*) ways. Whenever modal terms are employed within the modal domain where those terms have their "primitive seat," such terms do not require any further qualification. But when they are used in an *analogical way* an account is required of the context *qualifying* their employment. In Chapter 5, we introduced the issue of the elementary (analogical) basic concepts of the disciplines (see page 232 above). The naturalistic sociologist, Catton almost saw the point made in the text, where he remarks that one does not need to use the adjective 'physical' when *physical force* is intended "because physics got there first and has a prior claim on the word 'force' " (Catton, 1966:234). But to make this remark useful, Catton was in need of a theory of modal aspects and their analogies, for only within such a perspective is it possible to explain that the concept of *force* originally refers to the physical aspect and that it therefore can only appear as a *physical analogy* in the social (and other *non-physical*) aspect(s).

In the case of the term 'identity', however, the situation is more complex. In terms of modal aspects, the awareness of *identity* is related to the interconnected nature of the *kinematic* and *physical* aspects of reality, for we argued

1 "Aber dann muß man mit aller Energie festhalten: die Mathematik ist ganz und gar, sogar den logischen Formen nach, in denen sie sich bewegt, abhängig vom Wesen der natürlichen Zahl" (Weyl, 1966:32).

that *change* (a physical term) can only be detected on the basis of *constancy* (a kinematic term). In everyday experience, *identity* is affirmed whenever there is *continuity* (constancy) amidst discontinuity (change). But since the trans-modal nature of any entity *transcends* every single modal aspect in which such an entity functions, the modal meaning of the kinematic aspect can only be used in a concept-transcending way in order to articulate our awareness of the persistence of an entity amidst changes. For this reason, we argued earlier that the term *identity* represents such a concept-transcending use of a *kinematic intuition*. In other words, the term *identity* primarily manifests an *idea use* of the core modal meaning of the kinematic aspect (constancy; uniform motion – see page 181 above). Surely this does not mean that number (in combination with matter) ought to be seen as the *principium individuationis* (principle of individuation), as done by Thomas Aquinas and Aristotle (see *Metaph*. 1074b33-34).

But because the kinematic aspect itself is founded in the spatial and numerical aspects, the *meaning* of these two aspects is implied in an awareness of identity. First of all, the *distinctness* inherent in the meaning of a discrete quantity ultimately makes it possible to *distinguish* between whatever has been *identified*.[1] Phrased from the perspective of the *analytical mode* the nature of analysis, owing to its quantitative foundation, differentiates into *identification* and *distinguishing*. When a mathematician says that x is *different* from y ($x \neq y$),[2] then both the original meaning of number and its analogical recurrence within the logical analytical mode is present. The contribution of the primitive spatial meaning of continuous extension (synonymous with the whole-parts relation – see page 87 above) is found in the specification acquired by analysis because identification and distinguishing rest upon subdivisions from a given *field*, *domain* or *totality*.

The numerical analogy on the norm side of the analytical aspect presents itself in the configuration of a *logical unity* and *multiplicity*. The positive side of this analogy provides the ultimate (modal-analogical) foundation for the logical principle of identity (whatever is distinctly identified is identical to itself). Based upon what is *distinct* the logical principle of contradiction demands that whatever is distinct ought not to be considered as being *identical*.

In other words, the numerical analogy on the norm side of the analytical aspect explores the two sides of *unity* and *multiplicity*, and thus serves as the basis of the two *most basic* logical principles underlying every analytical act of identification and distinguishing. The freedom of choice in the human ability to identify and distinguish can pursue the option to identify and distinguish

1 In connection with what he calls the "correlative notions of consituent and whole," Loux formulates his *Principle of Constituent Identity* (PCI) in such a way that the quantitative foundation in its intimate coherence with identity is revealed (compare the phrase: *numerical identity*): "Necessarily, for any complex objects, *a* and *b*, if for any entity, *c*, *c* is a constituent of *a* if and only if *c* is a constituent of *b*, then *a* is numerically identical with *b*" (Loux, 2002:113).

2 See for example the remark by Fraenkel *et al.* in their discussion of the foundations of set theory (1973:26).

properly (correctly) or *improperly* (incorrectly). The former is achieved when acts of identification and distinguishing *conform* to the logical principles of identity and non-contradiction, while the latter prevails whenever the normative appeal of these principles is *violated*. The unity and diversity within reality thus *make possible* all identification and distinguishing – guided by the normative demand to identify *A* with *A* and to distinguish *A* from *non-A*. Therefore, taking into account their direct ontic foundation, the primary formulation of these two principles may be phrased as follows:

1) *Identity*: Within what is analyzable *A* is always identical to *A*.
2) *Non-contradiction*: Within what is analyzable *A* is never identical to *non-A*.

The act of identification entails an *affirmation*, and the act of distinguishing entails a *denial* – *affirming* that *A* is *A* is at once *denying* that *A* is *non-A*. This brings truth and falsehood into the picture, and therefore makes possible an alternative formulation of these principles in terms of *truth* and *falsity* – as done by Copi in his standard *Introduction to Logic*:

> The principle of identity asserts that *if any statement is true, then it is true*.
> The principle of contradiction asserts that *no statement can be both true and false*.
> The principle of the excluded middle asserts that *any statement is either true or false* (Copi, 1994:372).

In axiomatic set theory, two classes *v* and *w* are said to be *identical* if and only if they have exactly the same members (Lemmon, 1968:10). One can also take the equality symbol ('=') to denote identity.[1] Lemmon's choice is "to take identity as a primitive notion ... and regard it as part of our underlying logical framework" (Lemmon, 1968:11). Where equality is understood to denote identity it is also regarded as belonging to the underlying logic (Fraenkel *et al.*, 1973:25). Weyl speaks of *logical* identity ('$x = y$') as a two-valued (*zweistellige*) relation (Weyl, 1966:19), and later on, in the context of his discussion of automorphisms, he characterizes identity as a *one-to-one mapping*.[2]

But the idea of a *one-to-one mapping* presupposes the spatial time order of at once (simultaneity) as well as the spatial whole-parts relation. These two elements in turn cohere with the notion of a *cardinal number* where the supposition of a *totality* is also crucial when observed from the perspective of the concept of a one-to-one correspondence. We mentioned that Cassirer uses the term 'Gleichzahligkeit' (Cassirer, 1910: 58 – see page 277) – which clearly presupposes the spatial time order of *at once* (*simultaneity*; *Gleichzeitigkeit*). We noted that the concept of *equivalence* implied in a one-one correlation underlies the difference between a straightforward *multiplicity* (*how many* / *Wieviel*) and the concept "just as many" / 'equalling' (*Gleich-viel*) (Cassirer,

1 Another option is to regard *equality* as one of the *primitive* relations of the system or to introduce it *by definition* (Fraenkel *et al.*, 1973:26).

2 "Die Identität ist eine eineindeutige Abbildung" (Weyl, 1966:97).

1910:62). We also argued that the primitive meaning of *number* is *presupposed* in logic.

From these considerations, it is clear that 'identity' in mathematics and logic cannot be understood apart from the meaning of different modal aspects. Speaking about *two* classes, *v* and *w* is presupposed in saying that they are *identical* if and only if they have exactly the same members. Furthermore, establishing that two different *sets* have "exactly the same members" presupposes the spatial order of at once and the spatial whole-parts relation, because only the existence of a one-to-one correspondence secures the meaning of the term 'exactly'. Finally, the term 'same' presupposes an intuition of "continued existence" – from the initial *identification* of *two* classes up to the ultimate assessment, established by using one-to-one mapping, that these two classes are *identical*.

Although Lowe is hesitant to speak of "persistence in time" in connection with "abstract objects," one finds a remarkable reference to identity (as persistence) and a uniform motion: "One may indeed be called upon to explain the *coming-to-be* or the *ceasing-to-be* of a tomato, but surely not its *continuing-to-be*. Isn't the request for an explanation of the tomato's persistence rather like a request for an explanation of an object's continuing to move with a uniform velocity when not acted upon by any force?" (Lowe, 1998:107). Phrased in terms of *persistence* (the idea of identity), two sets are identical as long as they *continue* to have exactly the same elements. Alternatively, one may observe a numercially deepened (set theoretical) notion of identity in terms of one-to-one correspondence, to be distinguished from a concept-transcending employment of the core meaning of the kinematic aspect found in the idea of persistence.

6.9.3 *The principle of the excluded middle*

It is known that Brouwer (partly), rejects the universal applicability of the principle of the excluded middle because his neo-intuitionism identifies mathematical existence completely with what could be *constructed*. For this reason, Brouwer states that the question concerning the validity of the principle of the excluded middle is equivalent to the question concerning the possibility of solving mathematical problems (Brouwer, 1919:9). He says that whether, in the decimal expansion of π, infinitely many combinations of equal successive digits appear must be viewed as uncertain (Brouwer, 1919:9). Heyting specifically uses examples concerning the occurrence of the sequence 0123456789 in the decimal expansion of π. For instance, the decimal expansion of π is shown as:

$$\pi = 3.1415 \ldots$$

and the decimal fraction

$$\rho = 0.3333 \ldots$$

which breaks off as soon as the sequence 0123456789 occurs in the decimal expansion of π. As yet it is completely unknown if this sequence (that is,

0123456789) does occur in the decimal expansion of π and there is no method known to determine its existence. By accepting the principle of the excluded middle, the following must be correct:

$$\rho = 1/3 \vee \rho \neq 1/3$$

According to the logic of intuitionism, the expression "U ∨ (¬U)" implies that we must be able to construct a proof for every mathematical statement U, or construct, by starting with the assumption that U is valid, a contradiction. But then, the same requirements must apply to the above-mentioned case. That is, however, impossible, for in order to prove one of the statements $\rho = 1/3$ or $\rho \neq 1/3$, we must first be able to decide if the sequence 0123456789 does occur in the decimal expansion of π. Since our present mathematical knowledge does not allow this, intuitionism rejects the universal scope of the principle of the excluded middle – whenever the infinite is at stake, it is inapplicable.

With his arguments, Brouwer convinced Ludwig Wittgenstein and Hermann Weyl (the latter only in respect of questioning the at once infinite).[1] Wittgenstein explains his stance in terms of a similar example. It seems self-evident that the sequence 7777 either occurs somewhere in the decimal expansion of π, or it does not:

> We want, that is, to quote the law of the excluded middle and to say: 'Either such an image is in his mind, or it is not; there is no third possibility!' – We encounter this queer argument also in other regions of philosophy. 'In the decimal expansion of π either the group 7777 occurs, or it does not – there is no third possibility'. That is to say 'God sees – but we don't know'. But what does that mean? – We use a picture; the picture of a visible series which one person sees the whole of and another not. The law of the excluded middle says here: It must either look like this, or like that. So it really – and this is a truism – says nothing at all, but gives us a picture (Wittgenstein, 1968:112, par. 352, cp. p.127, par. 426).

Within mathematical logic, the dividing line between accepting the principle of the excluded middle and rejecting it is given in the kind of infinity that is allowed. If the infinite is merely understood as something without an end, that is, *endless* in the sense of the potential infinite (the successive infinite), the principle of the excluded middle is unacceptable, but when in addition the idea of an *infinite totality* is accepted, the principle of the excluded middle can be applied, for in the case of the actual infinite (the *at once infinite*), a strict either/or is possible (concerning the distinction between the successive infinite and the at once infinite see page 237 above).

Gentzen speaks of the "an sich" view. He designates the acceptance of infi-

1 Quine remarks: "But one can practice and even preach a very considerable degree of constructivism without adopting intuitionist logic. Weyl's constructive set theory is nearly as old as Brouwer's intuitionism, and it uses orthodox logic; it goes constructivist only in its axioms of existence of sets. On this approach, constructivist scruples can be reconciled with the convenience and beauty of classical logic" (Quine, 1970:88). But Weyl did support Brouwer in questioning the principle of the excluded middle in the context of infinity (see Weyl, 1921:70-71).

nite totalities as the "an sich" ("in itself") view. A statement regarding "for all" (an "∀-statement") reads: "For everyone of the infinitely many natural numbers the statement concerned is valid." Accordingly, the meaning of "there exists" (an "∃-stament") is: "In the infinite totality of natural numbers there somewhere exists a number for which the statement concerned is valid" (Gentzen, 1965:32).[1] In Cantor's explanation of the nature of a set, the feature of being a *totality* is also prominent: "We understand a 'set' to be any collection into a whole M of definite and distinct objects *m* of our intuition or our thought (which are called the 'elements' of M)."[2] Since the notion of a *set* (or the membership relation) is *indefinable* in the Zermelo-Fraenkel set theory, the formulation of the "Axiom of Infinity" may give the impression that it only intends the *successive infinite* (cf. Fraenkel *et al.* 1973:46). But when the "Axiom of Infinity" is read in coherence with the "Axiom of Power-Set" (cf. Fraenkel *et al.*, 1973:35), it is clear that, taken together, they (also) cover the case of *infinite totalities* (where all the subsets of an infinite set constitutes a new set).[3]

It must be clear that our argument does not introduce continuity in its irreducible spatial sense into the meaning of number. Much rather, we argued for the need of the inter-modal configuration of an antecipatory *modal analogy*: any arithmetical succession may, under the guidance of our spatial intuition of simultaneity, be viewed *as if* all its elements are present *at once* – in which case we employed the regulative idea of the *at once infinite*, accounted for in terms of this antecipatory analogy of space within the meaning of number. In other words, the use of the *at once infinite* (with its irreducibility to the successive infinite), stands or falls with the irreducibility of the *spatial time order of simultaneity*. It is the determining order for any coherent totality, since in order to constitute any spatial whole, *all the parts* have to be present *at once*.

Against the foregoing background, we may now characterize the principle of the excluded middle by formulating its *ontical status*. As such it is first of

1 "Jede transfinite Aussage habe einen bestimmten *Sinn an sich*; insbesondere sei der Sinn einer ∀-Aussage dieser: 'Für jede einzelne der unendlich vielen natürlichen Zahlen gilt die betreffende Aussage', der Sinn einer ∃-Aussage sei: 'In der unendlichen Gesammtheit der natürlichen Zahlen gibt es *irgendwo* eine Zahl, für welche die betreffende Aussage gilt' " (Gentzen, 1965:32).

2 "Unter einer 'Menge' verstehen wir jede Zusammenfassung *M* von bestimmten wohlunterschiedenen Objekten *m* unserer Anschauung oder unseres Denkens (welche die 'Elemente' von M genannt werden) zu einem Ganzen" (Cantor, 1962:282). In his description of the nature of 'Teilmengen' ('subsets') on the same page, the word "at once" ('zugleich') is used *constitutively*. A good discussion of Cantor's concept of a set is found in Singh (1985).

3 We note in passing that it was the postulation of an infinite 'class' that shows the untenability of Russell's logicistic project. In 1919, he had to admit that all of his earlier proofs for the existence of an infinite class are invalid (Russell, 1919:134-135; cf. Morris, 1929: 456). Fraenkel *et al.* remark: "It seems, then, that the only really serious drawback in the Frege-Russell thesis is the doubtful status of InfAx, according to the interpretation intended by them" (1973:186).

all a part of the arithmetical analogy within the modal structure of the logical-analytical mode, intimately connected with the principles of identity and non-contradiction. To be sure, in the finite case, the bifurcation of *A* and *non-A* clearly excludes any third possibility. However, in order to ensure the universal applicability of this logical principle, that is also in the case of the infinite, the (irreducible) meaning of the *at once infinite* must be acknowledged. Since the latter is itself completely dependent on the irreducibility of the spatial order of simultaneity with its implied correlate, the *whole-parts relation*, the numerical analogy within the structure of the logical-analytical aspect acquires, in the nature of the at once infinite, a deepened meaning, for it is only under the *antecipatory guidance* of the regulative hypothesis of the at once infinite, that we are justified in accepting that the principle of the excluded middle holds in the infinite case as well.

Therefore, via the (retrocipatory) analogy of number within the structure of analysis, this principle finds its ultimate foundation in the numerical antecipation to the meaning of space. This justifies the claim that the *ontical status* of the principle of the excluded middle is found in the fact that it is a *retrocipation* to an *antecipation*! In other words, the meaning of the principle of the excluded middle is in a retrocipation from the logical-analytical mode to the arithmetical mode, which in turn antecipates towards the factual spatial whole-parts relation in subjection to and determined by the spatial time order of simultaneity.

6.9.4 *Kinematic and physical analogies within the logical aspect*

In our discussion on the nature of rationality (see page 184 above) we have pointed out that it is only the logical *principle of sufficient reason* (ground) that directs thinking beyond logic to more-than-logical (ontic) states of affairs. This principle captures the way in which the *physical* meaning of causation analogically appears on the norm side of the logical mode. Without sufficient *logical* grounds, the truth of *logical* conclusions (effects) cannot be served by valid modes of argumentation. *Inference* concerns the movement from logical grounds to logical conclusions. This insight is all the more significant when it is considered that Russell (and many others) portray(s) symbolic logic as the discipline "essentially concerned with inference in general" (Russell, 1956:11). The logical *force* of an argument depends on its (formal) *validity*. Russell points out that *implication* – in the form if *p* implies *q* – cannot be *defined*. In general, the relation of implication – if *p* then *q* – reveals the inter-modal coherence between the logical and physical aspects.

The direct inter-modal foundation of this coherence comes to expression in the basic concepts of *constants* and *variables*. When Russell discusses the *mathematical* meaning of constants and variables, he says that "constancy of form must be taken as a primitive idea" (Russell, 1956:89), and earlier he speaks of "indefinable logical constants" (Russell, 1956:11).

The unbreakable coherence between the kinematic and physical aspects is particularly relevant for modern symbolic logic and mathematics, since both these disciplines are dependent upon the idea of *constants* and *variables*. It is remarkable that Weierstrass, in his employment of the *at once infinite*, in order to provide a new foundation for mathematical analysis, apparently succeeded in liberating himself completely from phoronomic and physical connotations. Boyer remarks that previous "writers generally had defined a variable as a quantity or magnitude which is not constant; but since the time of Weierstrass it has been recognized that the ideas of variable and limit are not essentially phoronomic, but involve purely static considerations" (Boyer, 1959: 286). This interpretation even caused Russell to settle for the idea that Zeno's arrow is truly at rest at every moment of its flight!

> After two thousand years of continual refutation, these sophisms were reinstated, and made the foundation of a mathematical renaissance, by a German professor, who probably never dreamed of any connection between himself and Zeno. Weierstrass, by strictly banishing all infinitesimals,[1] has at last shown that we live in an unchanging world, and that the arrow, at every moment of its flight, is truly at rest (Russell, 1956:347).

Although Russell calls the variable "*the* characteristic notion of mathematics" (Russell, 1956:90), he adheres to the *static* interpretation of Weierstrass by implicitly assuming the at once infinite in what he designates as the "indefinable notion of *any*" (Russell, 1956:91).

Without a sensitivity for the *ontic foundation* of the meaning of *constancy* and *change*, it is difficult to realize that an analysis of the meaning of the (logical-analytic and) numerical aspect(s) is a *complex* undertaking inevitably also involving the use of terms derived from *multiple irreducible* (non-arithmetical) *aspects*. In respect of the logical itself Russell nevertheless maintains an ontic appeal: "Logic, I should maintain, must no more admit a unicorn than zoology; for logic is concerned with the *real world* (my emphasis – DS) just as truly as zoology, though with its more abstract and general features" (Russell, 1919: 169).[2]

This explains why, contrary to their arithmeticistic intentions, both *axiomatic formalism* and *intuitionism* constantly employ *spatial notions*, such as the *whole-parts relation* and the *spatial time order of at once*.[3] Dummett, for example, frequently uses the expression "infinite *domain*" as a substitute for an "infinite totality" (cf. Dummett, 1978:22, 24, 57, 58, 59, 63 and so on).

The notion of *operations* in logic and mathematics shares in this inevitable *complex* nature of an analysis of the meaning of number, and of the logical-analytical aspect. An operation *effectuates* something; it causes whatever it *conditions* to exhibit these *effects*. In one of the earliest lucid discussions of *logical calculus*, Schröder explains the *analogy* between elementary arithmetical operations and the three basic logical operations distinguished by him

1 Of course the new introduction of *infinitesimals* in the *non-standard analysis* of Abraham Robinson outdates this remark of Russell (see the explanation given above on page 241).

2 Gödel quotes the second half of this statement with approval – see Wang (1988:313).

3 In the case of axiomatic set theories, either the *membership relation* or the undefined notion of a *set* makes an appeal to the spatial time order of at once.

(Schröder, 1877:2 ff.). Whereas arithmetic, in its concern for number, deals with *addition*, *multiplication* and their inverses (*substraction* and *division*), logic is concerned with *concepts*. Within logic, addition is called *collection*, multiplication is called *determination*, division is called *abstraction* and subtraction is called *exception* (*Ausschliessung* – Schröder, 1877:4). *Opposition* or *negation* is the third *logical* operation – where the combination of determination and collection constitutes the first operation, and abstraction and exception the second.

6.9.5 *Addition within different modal contexts*

Remembering the above-mentioned problem of Quine regarding where a *theory of identity* ought to be 'placed', we now briefly look at different modal meanings of *identity* and *addition*.

Speaking about *numerical identity* is different from speaking of *spatial* identity, *biotical* identity,[1] *logical* identity, *cultural* identity or *social* identity.[2] Identity does not belong to any one of these contexts *exclusively* – there are similarities and differences (*analogies*) between all of them.

Likewise, *addition* can occur in multiple (modal) analogical contexts. We have seen that Kant and Frege had a clear understanding of the difference between a *logical synthesis* (addition) and the meaning of an *arithmetical plurality* (see page 84). The fundamental distinction between the dimension of modal functions and the dimension of concretely existing, many-sided entities opens the way for an account of the quantitative meaning of number. Frege shows that whoever starts with *concrete entities* is doomed not to arrive at a genuine plurality if this goal is pursued by means of abstraction, for in this way only higher-level (more general) concepts are formed – and the result is always a single concept without any plurality. Apart from the numerical time order of succession, even a supposed multiplicity of 'units' collapses into the single general concept of 'oneness' – without any basis for assessing *plurality*. If nothing is left but 'sameness', that is, when apart from *sameness* the abstract notion of a *unit* does not somehow maintain something *distinct*, no discrete multiplicity or plurality is discernable. Apart from the primitive meaning of "less and greater" (Russell), i.e. from the determining and delimiting quantitative time order of succession, the 'many' collapses into the conceptual sameness of 'one' ('oneness'). In § 3 of his *Paradoxien des Unendlichen*, Bolzano therefore had to take recourse in the notion of 'and' (see Bolzano, 1921:2-3).

1 Merely focusing on the material constituents of living entities can never reveal their *biotical identity*. In terms of Von Bertalanffy's theory of a dynamic equilibrium (*Fliessgleichgewicht*), the second main law of thermodynamics is generalized to *open systems*, and this generalization enables us to differentiate between the highly improbable (*unstable*) state of a living entity which, viewed from its physical aspect, serves its *biotical stability* in a foundational sense. Physical stability (equilibrium) spells biotic instability (= death). Therefore the biotic *identity* of a living entity highlights its function within this aspect (see page 173 above).

2 Both in the case of cultural and social identity, there are multiple identities to be distinguished, owing to the existence of different cultures and the existence of multiple societal collectivities (and the social roles human persons can assume within them).

Numerical addition concerns numbers (*numerical subjects*) whereas *spatial addition* concerns *spatial subjects*, such as *vectors* (with *direction* and *distance*). We noted that in the case of arithmetical addition 2+2=4 but that a vector sum yields something different, for in the case of adding the two vectors $\overline{2}$ and $\overline{2}$ (see page 212 above), we obtain the *geometrical sum* $\overline{2}+\overline{2}=\sqrt{8}$.

Initially, modern physics believed that velocities could simply be added *numerically*. According to Galileo the following four equations represent the transition from one coordinate system to another – where the second coordinate system moves with a velocity v along the X-axis, and where a specific point is observed from both systems after a time t.

$$x' = x - vt;\ y' = y;\ z' = z;\ \text{and}\ t' = t$$

This Galilean transformation still presupposes an absolute simultaneity independent of the relative movement of a coordinate system. But Einstein realized that simultaneity is not independent of the relative movement of a system: "Two instants of time t_1 and t_2, observed at two points x_1 and x_2 in a particular frame, are simultaneous if light signals simultaneously emitted from the geometrically measured midpoint between x_1 and x_2 arrive at x_1 at t_1 and at x_2 at t_2" (Eisberg, 1962:17). It can be shown that two events occurring at the same place and within the same frame, are no longer simultaneous if they are observed from within a frame of reference relative in motion in respect of the first frame of reference, i.e. in which the two events are *spatially* separated. The factor involved is

$$\frac{1}{\sqrt{1-v^2/c^2}}.$$

This inspired the Lorentz transformation in which the new coordinates are:

$$x' = \frac{x - vt}{\sqrt{1-v^2/c^2}};\ y' = y;\ z' = z;\ and\ t' = \frac{t - xv/c^2}{\sqrt{1-v^2/c^2}}$$

Suppose two particles move in opposite directions – each with a velocity of $0.9c$ (where c is the 'vacuum-velocity' of light). In terms of the transformations of Galileo, the one particle will regard the other as moving away at a speed of $1.80c$ (the two velocities are simply added). This outcome contradicts the thesis of the special theory of relativity, namely that the movement of no physical entity can exceed the speed of light (see Einstein, 1982:41, note 1). However, applying the equation $V_x' = \dfrac{Vx - v}{1 - \dfrac{v}{c^2}Vx}$ confirms that the speed of the two particles moving away from each other (as assessed from the perspective of any one of them) is smaller than c:

$$V_1' = \frac{0.9c - (-0.9c)}{1 - \dfrac{(-0.9c)}{c^2}(0.9c)} = \frac{1.80c}{1.81} < c$$

These examples show that there is a *modal-functional difference* between - *arithmetical* addition, *geometrical* addition and *kinematic* addition, as well as *logical* addition (the conceptual level).[1]

6.9.6 *The inter-modal meaning of epistemic values*

We are now in a better position to assess the meaning of the epistemic values mentioned by Kuhn and McMullin (in addition to what has been said about *epistemic fertility* as a biotical analogy within the structure of the logical aspect).

What McMullin calls the value of *epistemic consistency* is dependent upon the meaning of the logical principle of non-contradiction, for *epistemic consistency* requires the absence of contradictions. The epistemic value of *predictive accuracy* links the fiduciary meaning of *trustworthiness* (reliability) and the jurally disclosed lingual meaning of *precision* ('appropriateness' as a jural antecipation within the sign mode) with the logical-analytical meaning of knowledge (*episteme*). The values of *internal coherence* and *external consistency* depend upon the numerical and spatial analogies on the norm side of the logical aspect. The epistemic value of *unifying power* points to a deepening and disclosure of the meaning of analysis in the antecipation to the cultural-historical aspect. Modal abstraction opens the way to an understanding of the *modal universality* of every modal aspect, which enables a *theoretical control* and *mastery* over a knowledge-domain – it therefore accounts for the *unifying power* of scientific theories.

The unifying power of a theory enables its *symbolic articulation*. By making an appeal to Dilthey's view regarding natural scientific theory formation as the construction of reality, with the aid of logical mathematical elements of consciousness (and thus asserting the power over nature of this sovereign consciousness as an effect of the autonomy of the human intellect),[2] we noted that Weyl follows the view of Hugo Dingler in respect of the principle of *symbolical construction* – underscored by Planck's conviction that no physical question could be settled directly without the aid of a theory (see above page 39). The meaning and scope of the (deepened) logical symbolism found in every discipline encompass the basic concepts of such a discipline.[3]

The acknowledgement of *intersubjectivity* in scholarly endeavors high-

1 Grünfeld remarks: "One raindrop added to another does not necessarily make two raindrops (Grünfeld, 1983:41). Therefore, in the case of physical addition, 1+ 1 may be equal to 1. *Biotical* addition has further possibilities, for when the first children of a married couple arrive as twins, the size of the family is at once expanded from two to four (1 + 1 = 4). The *sensitivity* of the human skin may experience a needle prick at the *same* spot, even though two *physically distinct* locations are touched (here, in terms of feeling, 2 = 1 – see Gosztonyi, 1976, I:13). In a *logical* sense, the *logical unity* of the concept 'cat' 'reduces' the multiplicity of cats to 'one', i.e. to the universal concept of 'catness', resulting from *logical addition*.

2 Weyl refers to the second volume of the 1923 edition of Dilthey's *Collected Works* (p.260). Cf. Weyl (1966:192).

3 Normally differentiated into *elementary*, *compound* and *typical* (*typonomic*) basic concepts.

lights the interconnection between the logical and the social aspects, and explains why the intellectual history of the world has always been rooted in 'thought-communities' (schools of thought).

It is interesting to note that, in a presentation in Johannesburg (1982) McMullin distinguished between *espistemic values* and *moral values*, with *epistemic integrity* as an example of a *moral* value. As main respondent to his presentation I questioned this account. If *epistemic fertility* exhibits the interconnection between the logical and the biotical aspects, a similar interconnection between the moral and the logical aspects ought to be acknowledged in the epistemic value designated as *epistemic integrity* – which will then still be just another instance of an *epistemic* (and not a moral) value. The published version of McMullin's speech (1983) no longer called epistemic integrity a *moral* value.

6.9.7 *Occam's razor*

In his *Discussion on Philosophy and Literature*, W. Hamilton refers to the *law of parsimony* and calls it *Occam's razor* (Hamilton, 1852:590; see Cloeren, 1984: 1094-1095).

According to the classical Greek view, the acquisition of knowledge is achieved through the cooperation of the mind and copies from external entities. Both Thomas Aquinas and John the Scott accepted "species intellectualis" – but Occam considered the latter to be an *unnecessary duplication* of the outer world. Therefore he posits the requirement that entities should not be multiplied beyond what is strictly necessary (*entia praeter necessitatem non esse multiplicanda*).[1]

John Stuart Mill criticizes Hamilton for viewing this principle as 'ontological' because, according to him, it is "a purely logical precept."[2] But this distinction is untenable, since it denies the *ontic status* of the logical-analytical aspect. However, we owe the formulation of this principle as the principle of *thought economy* to Ernst Mach (the positivistic teacher of Einstein – see König, 1972). According to the German scholars J. Schultz and R. Hönigswald, this principle is not constitutive, but *regulative*. In order to appreciate this remark in a systematic way, we once more have to digress on another important distinction, namely between what is *constitutive* and what is *regulative*.

1 See Windelband (1935:273). In terms of this conviction sensory knowledge loses its character of copying objects. Where Kant discusses the principle of the formal teleology of nature as a transcendental principle of the capacity to judge, he mentions the rule that "nature takes the shortest route" ("Die Natur nimmt den kurzesten Weg") and then calls it: "lex parsinomiae" (Kant, 1790:XXIX; 1793:XXXI).

2 Mill does this in his work: *An examination of Sir William Hamilton's Philosophy*, Boston (1865:238-240 – quoted by Cloeren, 1984:1096). Hugo Dingler (in his work *Grundlegung der Naturphilosophie*) also considers the principle of thought-economy *purely logical*: it is a "free stipulation at the very beginning of the initial theoretical construction" (Dingler, 1967:228).

6.9.8 *Kant and the distinction between constitutive concepts and regulative ideas*

Kant views human understanding as the capacity, by means of rules, to *unite* appearances. The task of *reason*, by contrast, is to unite the rules of understanding under the guidance of *principles*.

> Accordingly, reason never applies itself directly to experience or any object, but to understanding, in order to give the manifold knowledge of the latter an *a priori* unity by means of concepts, a unity which may be called the unity of reason, and which is quite different in kind from any unity that can be accomplished by the understanding (Kant, 1787-B:359).

This implies that, for Kant, the acquisition of knowledge proceeds in three steps: it starts "with the senses, proceeds from thence to understanding, and ends with reason, beyond which there is no higher faculty to be found in us for elaborating the matter of intuition and bringing it under the highest unity of thought" (Kant, 1787-B:355, cf. B:730).

For Kant, syllogistic inference implies that the conclusion is always subsumed under the *condition of a universal rule* (the major premise). By applying the rule of reason once more, the *condition of the condition* must therefore be sought (by means of a *prosyllogism*) whenever practicable. Thus, according to Kant, the principle peculiar to reason in general, in its logical employment, is: "to find for the conditioned knowledge obtained through the understanding of the unconditioned whereby its unity is brought to completion" (Kant, 1787-B:364). The concepts of pure reason are called *transcendental ideas* (Kant, 1787-B:368). These ideas instruct us only with regard to a certain unattainable completeness, and so serve rather to limit the understanding than to extend it to new objects (Kant, 1787-B:620). The *unconditioned* is never to be met in experience, but only in the idea – whenever "*the conditioned is given, the entire sum of conditions, and consequently the absolutely unconditioned* (through which alone the conditioned has been possible) is also given" (Kant, 1787-B:436). This means that transcendental ideas are simply "categories extended to the unconditioned" (Kant, 1787-B:436).[1] To Kant, therefore, transcendental ideas serve only for *ascending*, in the series of conditions, to the unconditioned (that is, to principles; cf. B:394).

No constitutive use of these ideas is allowed, because then we only arrive at *pseudo-rational dialectical concepts* (the source of which Kant has designated as the antinomies) (cf. Kant, 1787-B:672). The three ideas of the *soul* (thinking nature), the *world* and *God* are to be used in an "as if" way, that is, *regulatively* (cf. B.710-714).

Although Kant's distinction between concept and *Grenzbegriff* is fully in the grip of the dialectical ground-motive of nature and freedom, a different perspective is obtained in terms of the theory of *retrocipatory* and *anticipatory analogies*. Whereas the retrocipatory analogies are indeed *consititu-*

1 This applies only to those categories in which the synthesis constitutes a *series of* conditions subordinated to one another.

tive "building blocks" within the modal structure of any aspect (of course the arithmetical aspect excluded, for it does not have any retrocipations), the antecipatory meaning-moments within an aspect requires an *opening up* or *deepening of meaning* in order to be realized. But because such a deepening of meaning always presupposes the structural foundation of the aspect concerned (inclusive of its retrocipatory meaning-moments), such an antecipating disclosure of the meaning of an aspect is always merely *regulative* in nature.

Without explicitly focusing our attention on it, we have already mentioned the distinction between the constitutive and regulative in connection with the expansive boundaries of rationality (see pages 182, 187, 228, 229, 238, 305 ff.). In connection with the regulative deepening of the structure of number, we have added a further specification to the term *regulative*, for there we have spoken about a *regulative hypothesis* through which successively infinite sequences of numbers could be viewed *as if* they are all given at once.

Returning to our discussion of the principle of thought-economy, we can now attach an expanded positive connotation to the remark by Schultz, namely that this principle is *regulative* in nature, because it indeed belongs to the antecipatory structure of the logical-analytical aspect. As such, it only comes to light when the meaning of analysis is deepened and disclosed under the guidance of the meaning of the economic function. It is noteworthy that Derrida is also sensitive to these states of affairs.

6.9.9 *Credit as economic trust: Derrida*

Modern money economies are guided by the principle of economic trust. In some countries the *Reserve Bank* only requires that 20% of the money invested in a commercial bank should be available for realization at any specific moment. If investors all decided at the same time to withdraw their money, an immediate 'crisis' may result. Of course the bank could put such concerns at rest by explaining the financial arrangements with the Reserve Bank, and everyone would be satisfied. However, when the *credentials* of such an economic institution turn out to be doubtful, the forthcoming *economic crisis* will be more serious. In a *disclosed* economy the guiding nature of economic trust functions as a *regulative principle* for it represents a *fiduciary antecipation* within the modal structure of the economic aspect. However, when this *economic trust* is lost an entire economic system may suffer – as was amply demonstrated by the international economic crisis of 2008.

Although he does not operate with a theory of modal aspects, Jacques Derrida places *credit* against the background of acknowledging the *universality* of 'faith' – he stresses that "faith is absolutely universal" (Derrida, 1997:22) and then, with reference to *credit* [as economic trust] he states:

> There is no society without faith, without trust in the other. Even if I abuse this, if I lie or if I commit perjury, if I am violent because of this faith, even on the economic level, there is no society without this faith, this minimal act of faith. What one calls credit in capitalism, in economics, has to do with faith, and the

economists know that. But this faith is not and should not be reduced or defined by religion as such (Derrida, 1997:23).[1]

6.9.10 *'Reason' and 'faith'*

The long-standing legacy of *separating* reason and faith is incompatible with the modal (functional) nature of the logical-analytical and the fiduciary aspects of reality. The meaning of both of these aspects is dependent upon the inter-modal coherence between them: (i) within the structure of the certitudinal aspect (faith aspect), one of its retrocipations is given in the normative requirement to come to proper certitudinal *identification* and *distinguishing*; (ii) Within the deepened and disclosed structure of the logical analytical aspect, one of the antecipatory meaning-moments is evinced as *cognitive trust* (*logical certainty; epistemic reliability; epistemic confidence*).

Re (i):

Faith implies and demands fidelity in faith and sacrifices of faith, together with knowledge of faith – correct faith distinctions (as emphasized by Calvin), faith sensitivity – not the same as faith directed by *feeling*. It requires a dynamism of faith, perseverance in faith and an integration of faith, it brings about a harmonious and balanced faith, requires the correct interpretation of signs of faith (think about the controversy about the nature of bread and wine during communion), it brings about community in faith, it requires contemporary forms and expressions in response to the new problems and tasks arising out of changing historical circumstances. At the deepest level, faith unifies our life and directs it either in service of something within creation or at the whole-hearted loving service of God and the neighbour.

Re (ii)

Both *assumptions* and *arguments* need to be *convincing* – even regarding *method* within the context of (meta-)mathematics. Grünfeld categorically states: "The ultimate test of whether a method is admissible in metamathematics can only be whether it is convincing, yet equally competent mathematicians find different logical justifications convincing" (Grünfeld, 1983:44).[2]

Furthermore, taking (and arguing for) any intellectual position requires *cognitive trust*, it needs a conviction on the basis of which the intellectual stance is accounted for. Stegmüller and De Vleeschauwer both underscore the guiding role of faith in intellectual endeavors – one needs to believe in something in order to be able to justify something else (see Chapter 1, page 10). *Logical trust* forms the basis of any axiom system, and explains the result of

1 In societies where the meaning of economic life is not yet opened up through antecipatory analogies, *economic trust* in the sense of *credit* is absent. Within such societies, an *exchange economy* is found. Similarly the meaning of the jural aspect is not yet disclosed in the practice of a tooth for a tooth and an eye for an eye (*lex talionis*).

2 Note that the expression "logical justification" relates the physical retrocipation of analysis to the deepened interconnection between the logical-analytical mode and the *jural* mode. Given sufficient grounds, we are logically justified to hold a certain position.

Gödel's incompleteness theorem in 1931, which shows that every axiomatic system requires and presupposes an *insight* transcending the formalism of the system.[1] Hilbert's hope to prove the completeness of mathematics was ruined by Gödel's theorem. After Hilbert died in 1943, his student, Hermann Weyl, who switched to an intuitionistic orientation, wrote: "It must have been hard on Hilbert, the axiomatist, to acknowledge that the insight of consistency is rather to be attained by *intuitive reasoning* which is based on evidence and not on axioms" (Weyl, 1970:269).

In his Encyclical Letter: *Fides Et Ratio* Pope John Paul II treats reason and faith not as modal functions or aspects of being human, but rather as interdependent concrete (entity-like) realities. To him faith does not fear reason but trust it: "Faith therefore has no fear of reason, but seeks it out and has trust in it" (John Paul, 1998). A human *act* indeed does have a multi-aspectual structure – but even in this case the qualification 'human' is decisive. An act of faith and a thought-act are therefore both functioning within all aspects of reality, which means that no act of faith could be envisaged apart from its function within the logical-analytical aspect and that no thought-act is divorced from its function within the aspect of faith. Of course the term *trust* may be used to designate the core meaning of the certitudinal aspect or it may serve to indicate the interconnection between other aspects with the aspect of faith. For that reason we were able to speak of *logical trust* and *economic trust* (credit). Contemplate also composite phrases such as *social trust*, *moral trust* and *historical trust*. In all these cases antecipations of various aspects to the faith aspect are exemplified, showing that we are dealing with an *inherent* structural feature of these aspects. Similarly, no possible human function within the faith aspect can escape from the logical-analytical analogy within the modal structure of this aspect. 'Reason' and 'faith' are therefore not 'foreigners' to each other, because human acts qualified either by the logical or the certitudinal functions display both an internal and an external coherence with the non-qualifying aspects of such acts (see pages 566 ff. where the distinction between internal and external coherence is explained in more detail).

6.9.11 *Constitutive and regulative historical principles*

The normative contrary *historical-unhistorical* uproots the positivistic preoccupation with historical *facts*, for without the implicit application of a historical norm of development, it would be impossible to speak of *reactionary* or *revolutionary* historical events. Reaction and revolution presuppose the normative meaning of *historical constancy* (*continuity*) and *historical change* – revealing on the norm side the coherence between the historical aspect and

1 "Gödel proved that if any formal theory T that is adequate to include the theory of whole numbers is consistent, then T is incomplete. This means that there is a meaningful statement of number theory S, such that neither S nor not-S is provable within the theory. Now either S or not-S is true; there is then a true statement of number theory which is not provable and so not decidable. The price of consistency is incompleteness" (Grünfeld, 1983:45).

the *foundational* role of the kinematic and physical aspects. Reactionary movements cling to the status quo without any flexibility or willingness to face the challenge of *changing* historical circumstances. Revolutionary movements, by contrast, take such challenges so seriously that no room is left for any *historical continuity*.

It is only when a sound application of the (constitutive) norm of historical continuity prevails that constructive *reformation* takes place, avoiding the historically antinormative extremes of reaction and revolution. Historical development is always confronted with a struggle between *progressive* and *conservative* forces, but only through *continuity-abiding reformation* is it possible to bend these opposing forces into the pathway of historical norm-conformity.

Tradition, as the guardian of historical continuity, not only embodies the worthwhile legacy of the past, but also calls for continued reformation. But when an accountable reformation takes place, it only causes *changes* on the basis of historical continuity and not at the cost of it.

The biotic analogy within the structure of the historical aspect is particularly responsible for many controversies. The original biotical meaning of *growth* and *development* provides the source for analogical and even metaphorical usages. In the original, biotic sense, the normal life cycle of any living entity follows the path of the biotic time order of *birth*, *growth*, *maturation*, *ageing* and *dying*. Biotic growth proceeds along the lines of *differentiation* and *integration*. Diverse organs *differentiate* – but if the living entity does not manage to *integrate* this differentiating growth process, it will *disintegrate* and *die*.

Since it is clear that the concept of differentiation and integration is an analogy of *biotic differentiation* and *integration* within all the normative modal aspects,[1] the acknowledgement of this *constitutive role* of the biotic analogy within the normative aspects ought to be specified *both* on the norm side and the factual side of these aspects.

In the context of the historical aspect, the task-setting nature of historical principles entails that the calling to (formative) control – over fellow human beings (the competence vested in some or other societal office) and over cultural objects made by humankind – comes to expression in processes conforming to or violating the fundamental historical principles of historical differentiation and historical integration. These principles are *functional* principles exhibiting the *modal universality* of the cultural-historical mode. Although Griffioen highlights factual mistakes in Dooyeweerd's account of undifferentiated societies, he does not question the constitutive normative meaning of historical differentiation and integration. He points out that Dooyeweerd's view of totem cultures is dependent upon the interpretation of Cassirer. Instead of *identifying* human beings with plants or animals, one rather has to discern a parallelism between a *natural series* and a *cultural series*

1 In 1931 Dooyeweerd still related the term *integration* to the mathematical "infinitesimal calculus"(Dooyeweerd, 1931:65).

within such cultures.[1] He also correctly emphasizes the distinction between a *constant principle* and its *variable form-giving*, although he does not reveal insight into the meaning of *modal principles* as distinct from *structural* (*typical*) principles, for he only discusses *norms* as *structural principles* (Griffioen, 1986: 103 ff.).

Unfortunately, Griffioen nowhere in this illuminating and penetrating article *explicitly* undertakes the task of 'locating' the "modal seat" of the term *differentiation* (and its correlate: *integration*) – which is found solely in the *biotic mode of reality*. In addition, he also does not *explicitly* account for the fact that the inter-modal coherence between the normative aspects and their foundational connection with the biotical (and other natural) mode(s) actually highlights fundamental *modal norms* requiring positivization within the process of cultural development. But not even Dooyeweerd himself realized that *differentiation* and *integration* represent *biotical* analogies within cosmic later aspects (including the historical aspect) – although he did discuss the acknowledgment of vital and dead elements in the tradition in connection with the biotical analogy within the historical aspect (see Dooyeweerd, 2002: 157-158).

Constitutive historical principles are not eliminated when a deepening or disclosure of the meaning of the historical aspects takes place. The first element of deepening the meaning of the historical aspect is found when the awareness of what is historically *significant* materializes in inscriptions, monuments, written historical accounts, and so on. The latter serve as *sources* for the historian. The difference between what is *historically significant* and what is *insignificant* is made possible by the antecipatory coherence between the cultural-historical aspect and the sign mode. Cultures in which this antecipatory moment is not yet disclosed do not, strictly speaking, participate in *world history*, as Hegel realized.

Constitutive meaning-moments within the cultural-historical aspect acquire new meaning under the guidance of regulative moments. For example, an articulated understanding of what is historically significant enables a more nuanced *identification* of a cultural community with its historical past and at once highlights avenues through which what is fruitful in its tradition could be pursued in further historical development. Once the social antecipation is opened up, intercourse with other cultures leads to an equally articulated development of the *national identity* of communities. The uniqueness and individuality of cultures are thus recognized. But since the contours of the normative aspects of reality embrace the multi-faceted nature of all cultures, their uniqueness and individuality can only be manifested within shared dimensions of normativity, for individuality and universality are not *opposites*, but *mutually cohering traits* of every concretely existing creature or societal reality.

Prinsloo (1989) approaches this problem from the angle of the relationship

1 Cf. Griffioen, 1986:91 note 1 – where he refers to more recent ethnological research and to some writings of E. Leach, C. Geertz and Claude Lévi-Strauss.

between *logic* and *culture*. He discusses examples by thinkers such as Peter Winch and Evans Pritchard. Whereas these thinkers want to demonstrate that *consistency* is something different for Westerners and 'primitives', he successfully shows that both actually observe the (underlying, universal, logical) principle of *non-contradiction*.

The applicability of this principle, however, presupposes the nature of *logical concept formation* – and (as we argued – see pages 148 ff.) concept formation, in turn, rests on the nature of *universality*. Yet we have seen that it was exactly this problem – the relationship between universality and individuality – that haunted the new claims by historicism during the 19th century (see pages 271 ff.). Acknowledging cultural diversity and historical uniqueness does not eliminate universality, but *presupposes* it. Only if the phenomenon of culture is something *universally human*, will it be possible to differentiate between the peculiarities of *different* cultures and to gain insight into the process of differentiation and integration occurring within societies. The process of differentiation and integration taking place within a developing society soon discloses the requirement for observing the historical norm of *cultural economy*.

The internal sphere of competence of every newly differentiated societal community and collectivity demands respect, but whenever this is not obtained, history tells the story of the many one-sided abuses of power, leading to situations where one sector of society violates the internal sphere-sovereignty of another. During the middle ages the Roman Catholic Church exceeded the limits of the church as an institution and excessively impinged upon the spheres of competence of the non-ecclesiastical domains of life. Likewise, after the Renaissance, the modern humanistic science ideal breached the integrity of every non-scholarly domain of life – just recall how Kant, in his *Critique of Pure Reason*, even claimed that even *law* and *religion* cannot withdraw themselves from the critical scrutiny of reason (see page 9 above).

These antecipatory moments within the modal structure of the cultural-historical aspect must be distinguished from the *original function* of communal and collective activities of a differentiated society within the post-historical modalities. It is only within the latter context that one can for example speak of "love for culture" or even about the guiding role of faith in the (harmonious or disharmonious) disclosure of culture. Although Dooyeweerd does not properly distinguish between this internal and external coherence, he does provide us with an extensive analysis of the process of disclosure and particularly of the guiding role of faith in this process (see Dooyeweerd, 1997-II: 180-365 and Dooyeweerd, 2003:89 ff.).

Of course it should be kept in mind that modal analogies on the law side of the aspects of nature also reveal fundamental natural laws to us. We briefly mention a constitutive physical law; that of energy conservation.

6.10 'Conservation of energy' – the kinematic analogy on the law side of the physical aspect

Anticipated by Schelling in his philosophy of nature, what became known as the law of the *conservation of energy* was established as a basic physical law in the course of the 19th century. Already in 1847, at the youthful age of 26, Helmholtz presented a formulation of this first main law of physics to the *Physics Society* of Berlin. He began by pointing out that nobody had succeeded in building a successful *perpetual motion* machine. This was a logical consequence of the *indestructibility* of energy. Until the present, physicists have recognized this as the law of energy conservation, which means that energy cannot be created or destroyed. This law fascinated Max Planck, who had a keen interest in the universality and constancy of physical laws. His 1879 dissertation is dedicated to the second main law of thermodynamics, and in 1887 he wrote a work on the law of energy conservation. He considered these to be universal principles of nature.[1]

In opposition to the positivism of Ernst Mach, the lectures of mathematics teacher, Herman Müller (at the "Maximilian-Gymnasium" in München) inspired Planck to accept physical laws – such as the law of energy conservation and the second main law of thermo-dynamics – as firm in their objective existence and operation, and as independent from *human observation*. Later in his life, Planck remarks: "Thus it happened that I absorbed in myself like a message of salvation the first law which possesses independently of human perception an absolute validity, namely the principle of the conservation of energy" (Planck, 1953:1).

In accordance with the core meaning of the kinematic aspect, the law of *inertia* assumes the constancy of motion if no *physical forces* interact with the movement concerned. The mechanistic reduction of the physical to the kinematic (or even its denial of the reality of physical forces) came to an end by the turn of the 20th century, when *irreversible* physical processes had to be taken seriously, such as were found in the processes of decay of radio-active elements (discovered in 1896).

But acknowledging the distinct (and irreducible) meaning of the physical aspect still has to account for the intrinsic (inter-modal) connections between the physical and kinematic aspect at its foundation. For this reason, within the modal structure of the physical aspect there is a structural moment that reminds us of the foundational meaning of the kinematic aspect. *Constancy* in the physical aspect appears as a structural reminder of the meaning of *uniform motion*. In terms of the inter-modal connections between aspects, we may say that, in the configuration of *energy constancy* we find an *analogy* of the kinematic aspect on the law side of the physical aspect.

A transcendental-empirical analysis of the interrelationship between the physical and kinematic aspects thus enables us to arrive at a (more) precise formulation of the meaning of the *first law*. Whereas the accepted physical wording of this law may carry (against its true intention) an element of energy *input* (one connotation of the term *maintaining / conservation*), the designation *energy constancy* makes it plain that this law holds for nothing more and nothing less than the continuation of whatever 'amount' of energy there is. The continued existence of energy does not need any "holding on to" – it sim-

1 He indeed recognized in them what we called *modal universality* (see p.68)

ply *persists* in the sense of remaining *constant.*

We now proceed by considering analogical structural moments on the law side of normative aspects of reality.

6.11 The inter-modal foundation of linguistic communication

We discussed the fact that the core meaning of the sign mode is *primitive* in the sense of being *indefinable* (see page 171). But we have seen that the other side of the "uniqueness coin" is given in the analogical meaning-moments found within each aspect of reality through which its coherence with the other modes comes to expression. The retrocipatory analogical structural elements point backwards to those aspects at the foundation of a given modal aspect and they make possible, as we have seen (page 232), the analogical basic concepts employed by the various disciplines. Particularly in our understanding of *language* and in the use of the term *communication*, this inter-modal coherence surfaces.

A statement by Derrida may be used to start our exploration of this issue. He asks:

> Is it certain that there corresponds to the word *communication* a unique, univocal concept, a concept that can be rigorously grasped and transmitted: a communicable concept? (Derrida, 1982:307).

What is significant, is that Derrida leaves behind the *conceptual context* by switching to the interpretative *lingual context*, reminiscent of the linguistic turn alluded to above (see page 74):

> one first must ask whether the word or signifier 'communication' communicates a determined content, an identifiable meaning, a describable value (Derrida, 1982: 307).

Since Appleby claimed that the focus on 'meaning' replaced 'cause', it may appear that the domain of 'meaning' excludes the domain of "*cause* and *effect.*" But suppose now that *communication* between human beings is described in terms of the communicative *effect* that is *caused* by such an interaction. Is it then permissible to speak of *communicative causality* – as implicitly done, for example, by Austin in his theory of *performative utterances* and *speech acts*?[1] The affirmative answer to this question implies that the *basic concepts* employed by the natural sciences – such as the physical concept of *causality* – might indeed also legitimately be used by disciplines within the *humanities* – as argued above (see pages 279 ff.).

Referring to *causality* and (lingual) *communication*, suggests that both categories reflect facets of reality that are (ontically) unique. This *uniqueness* gains significance when the *interdependence* between communication and causality is examined.

1 His well-known work bears the title: *How to Do Things With Words?* (see Austin, 1962). He distinguishes between locutionary, illocutionary and perlocutionary actions. Locutionary acts are what we have called lingual objectification (the utterance and inscription of words), whereas illocutionary acts are acts taking such words a step further by asking, commanding, asserting, promising and so on in addition to merely uttering.

6.11.1 *Causality versus totality and meaning: Jaspers and Habermas*

The thought of Karl Jaspers, who is particularly known for his extensive discussion of *communication* (see Jaspers, 1948:338-396), reveals a similar tension between the domain of authentic human action (normally associated with or qualified as *freedom*) and what is considered to be causally determined – and something similar is even found in the thought of Habermas.

Jaspers identifies *causality* with its *deterministic* (physicalistic) understanding (see Jaspers, 1948:439) and he opposes *causality* with the I and another I who are, through communication, *ideally* bound by an encompassing *totality* (*Ganzheit*).[1] The historical uniqueness of communication is a non-objective totality without foundation (*grundlos*) – it does not originate in the sense that eventually I acquire something, for in communication, I actually become myself.[2] I am only myself in my *freedom* when the other is and wants to be who s/he is, and this can only be comprehended on the basis of the possibility of *freedom*. For Jaspers, *existential communication* – flowing from freedom – is therefore incomprehensible in an *objective* sense.[3]

In the thought of Habermas a similar opposition is found, though his basic distinction is not between causality and freedom, but between *causality* and *meaning* (cf. Habermas, 1981:145).

6.11.2 *Historical starting points for an understanding of communication*

During the Enlightenment (18th century), the emphasis was fully on *human reason* – on the one hand to liberate it from any external authority and on the other to inspire humans as rational beings not to be shy in employing their rational abilities autonomously. This perspective is emphasized by Immanuel Kant in his discussion of the question "What is Enlightenment?" But the obvious lack of a focus on and appreciation of language as such in Kant's *Critique of Pure Reason* (1781) generate the criticism of Jacobi, Hamann and Herder. The latter even called the human being a *creation of language*.[4]

Although the exploration of this focus had to await the genesis of the "linguistic turn" at the end of the 19th and the beginning of the 20th century, the

1 "... bei der Kommunikation dieses Ichs mit dem anderen Ich als vertretbarer Punkt die weitere Grenze: die übergreifende Idee von Ganzheiten, in denen sie wirken, durch die sie nicht kausal, sondern ideel gebunden sind" (Jaspers, 1948:342).

2 "Das geschichtlich Einmalige der Komnunikation ist ein Ganzes, das nicht entsteht, indem ich selbst schon bin und nun etwas hinzugewinne, sondern worin ich selbst erst eigentlich werde: aber als unobjektives Ganzes ist sie grundlos" (Jaspers, 1948:346).

3 "Der Sinn des Satzes, daß ich erst ich selbst in meiner Freiheit bin, wenn der Andere er selbst ist und sein will, und ich mit ihm, ist nur *aus Freiheit als Möglichkeit zu* ergreifen. Während die Kommunikationen im Bewußtsein überhaupt und in der Tradition erkennbare Daseinsnotwendigkeiten sind, ohne die ein Versinken ins Unbewußte unausweichlich würde, ist die Notwendigkeit existentieller Kommunikation nur eine solche der Freiheit, darum objektiv unbegreiflich" (Jaspers, 1948:345). *Objective* here means: *in causal terms*.

4 "Der Mensch ist ein freidenkendes, thätiges Wesen, dessen Kräfte in Progression fortwürken; darum sei er ein Geschöpf der Sprache!" (Herder, 1978:73).

early 19th century did start to explore the meaning of language. Hegel particularly provided certain starting points for the inter-subjectivity involved in linguistic communication. He conceives language as constituted by self-consciousness, where the latter, in its being there *for the other*, at once also manifests a splitting off of the self from itself, and thus enables a merging of the other with self-consciousness (one hears oneself as one is heard by the other). Hegel therefore qualifies "*language* as the being-there of the spirit" ("die *Sprache* als das Dasein des Geistes").[1] According to Meggle, the way in which Hegel explained the form quality in terms of a theory of predication,[2] indeed provided the design of a universal theory of communication (see Meggle, 1999:703).

But it should be remembered that the background of the linguistic turn is found in the *historicism* of the early 19th century, which eventually led to unbridled *relativism*. In 1922, Troeltsch declared: "We see here everything in the flow of change, in endless and constantly new individualization, in its being determined by the past and in the direction toward an unknown future. State, law, morality, religion and art are dissolved in historical change and they are everywhere only understandable as ingredients of historical developments" (Troeltsch, 1922:573).[3]

After the linguistic turn, Heidegger and Gadamer realized that language itself may be emphasized to escape from the relativism of historicism. Van Niekerk acknowledges this step when he points out that, according to Gadamer, the 'world' should be recognized as a *creation of language* (Van Niekerk, 1993:39). Heidegger also realized that a *new universal* was needed. In *Being and Time*, he focused on "there-being" as a "being-in-the-world" but he still concentrated on historical being ("geschichtliches Dasein"). However, Gadamer points out that Heidegger did not want to introduce once more something essential or divine with his notion of *Sein* (Being). Much rather, his purpose was to introduce something like an event that opens the *space* in which hermeneutics could become (without a final foundation) a *new universal* ("zum neuen Universale wird"). This space is the dimension of language.[4]

The horizon of the past three centuries is filled with the successive epistemic ideals (see below pages 370 ff.) of *universal conceptual knowledge*

1 "Wir sehen hiemit wieder die *Sprache* als das Dasein des Geistes. Sie ist das *für andre* seiende Selbstbewußtsein, welches unmittelbar *al solches vorhanden* und als *dieses* allgemeines ist. Sie ist das sich von sich selbst abtrennende Selbst, das als reines Ich = Ich sich gegenständlich wird, in dieser Gegenständichkeit sich ebenso als *dieses* selbst erhält, wie es unmittelbar mit den andern zusammenfliesst und *ihr* Selbstbewußtsein ist; es vernimmt ebenso sich, als es von den Andern vernommen wird, und das Vernehmen ist eben das *zum Selbst gewordene Dasein*" (Hegel, 1807:458).

2 Compare Hegel (1830:245-269).

3 "Wir sehen hier alles im Flusse des Werdens, in der endlosen und immer neuen Individualisierung, in der Bestimmtheit durch Vergangenes und in der Richtung auf unerkanntes Zukünftiges. Staat, Recht, Moral, Religion, Kunst sind in den Fluss des historischen werdens aufgelöst und uns überall nur als Bestandteil geschichtlicher Entwicklungen verständlich."

4 "Dieser Raum ist die Dimension der Sprache" (Gadamer, 1991:172).

(18th century), of the *uniqueness*, *unrepeatability* and *individuality* of historical change (19th century) and the constitutive role of *human language* (20th century). A *positive* appreciation of this development is given in the acknowledgement of *logicality*, *historicity* and *linguisticality* as indispensable conditions of being human – without isolating and elevating any one of them to become the sole and exclusive condition for being human. The *conditio humana* is indeed just as many-sided as the experiential world in which we live.

An understanding of the many-sidedness of *communicative actions* will benefit from this insight, since it is given in the idea of the *unbreakable coherence* between different (unique) modes of explanation (the coherence of irreducibles – see above pp. 8, 13, 52, 157). Such an understanding should be able to give an account of particular instances of the *interconnectedness* of linguistic communication, revealed within the coherence between lingual communication and every facet of reality distinct from it. In the next subsection, we commence by assuming that the ability to use language is indeed a universal human capacity and that one can capture this universality by employing the term that we have used above, namely *linguisticality*.

6.11.3 *The multi-vocality of the term communication*

The acknowledgement of the fundamental role of (analogical) basic concepts within every discipline entails that no academic discipline can escape from the inevitable employment of *key terms* or *basic concepts*. The problem entailed in this practice surfaces as soon as it is realized that *different* disciplines frequently use the *same* terms or concepts within *different* contexts. This *multivocality* requires a theoretical account of the *meaning* of basic terms in order to explain what is *similar* and what is *different* (the issue of *analogical* concepts).

This approach may help us in transcending the dualism between *nature* (causality) and *freedom* in our understanding of *communication*, for it will have to explore the *interconnections* between the natural sides of reality and the *normative* domain of human communication in a positive sense.

Let us consider the strange position assumed by Habermas in his argument below concerning two kinds of bodily movements instead of simply consistently recognizing that both kinds *at once* display both 'physical' and "semantic/social" features. This will show that the powerful naturalistic legacy of restricting *causality* to *physical* causality still hampers the approach of Habermas, and effectively causes him not to acknowledge what should be called the *modal-functional universality* of the *sign mode* (also embracing what we will describe as *linguistic communication*).

6.11.3.1 *The subtle dualism in Habermas's understanding of communicative action*

In Habermas's discussion of the ideas of Danto, the issue is whether or not bodily movements are *actions*. Danto defines them as "basic actions" and then construes *complex actions* that are performed 'through' such other (basic) actions (such as greeting someone 'through' raising my right arm). By contrast, a basic action cannot be performed by means of an additional act. According to Habermas, this concept is false. The English translation weakens this statement to: "I regard this conceptual strategy as misleading" (Habermas, 1984:97). The original German reads: "Ich halte dieses Konzept für Falsch" (Literally translated: "I hold this concept to be false") (Habermas, 1981:146).

The background of the position taken by Habermas is given in his view of the *three* worlds distinguished by him.

> 1. The objective world (as the totality of all entities about which true statements are possible); 2. The social world (as the totality of all legitimately regulated interpersonal relations); 3. The subjective world (as the totality of the experiences of the speaker to which he has privileged access) (Habermas, 1984:100).

The "objective world" is understood in terms of (objective) *entities*, whereas the social and subjective worlds are respectively described merely in terms of *relations* and *experiences*. Although Habermas does acknowledge institutions within the "social lifeworld" of human beings (see Habermas, 1996:24) he does not have multi-aspectual societal entities in mind, because at most he speaks of "the symbolically structured lifeworld, mediated by interpretations and belief" (Habermas, 1996:36). Culture, society and personality structures are components of the lifeworld that differentiate themselves within the boundaries of a multifunctional language (Habermas, 1996:55). The question not posed by the ontology of Habermas requires an acknowledgement of *ontic modes*, embracing all three worlds distinguished by him. This question is: what is wrong with acknowledging "bodily movements" as *functioning* within aspects of *nature*, such as the physical and the biotical, *as well as* within *normative* functions, such as the logical, the lingual or the social? If such movements are understood merely in terms of their nature context – for example as (physical) *material* events – then they cannot be viewed as being accountable (*normed*) actions of social subjects, but this objection does not as such preclude them from displaying *object-functions* within the modes of the "social world."[1] Besides, even what Habermas would recognize as uniquely human acts, in principle also function within the *natural sides* of reality. This insight explains why there is no need for two different (kinds of) actions (basic actions and secondary actions mediated 'through' the basic actions) – justifying Habermas's claim that the "double action" account is mistaken. But Habermas's own view is dependent upon the basic *dualism* between *instrumental actions* and *communicative actions*, since this dualism sets apart the domain of *subject-object* relations (being 'instrumental') from that of *communicative actions* (restricted to *subject-subject* relations).

Habermas does not articulate a theory of *modal functions* belonging to a

1 Such bodily movements, though assumed to be given in their physical and organic subject-functions, can be objectified in modes such as the logical-analytical, the lingual and the social. Identifying and distinguishing them logically opens up their logical object-function; designating them manifest their lingual object-function; and assigning a social meaning to them instantiates their social object-function. See our discussion of the subject-object relation (pages 344 ff.).

(ontic) dimension of experiential reality *distinct* from (but intimately cohering with) the dimension of (natural and social) *entities*. Certainly, he does realize that his concept of "communicative action presupposes language as the medium for a kind of reaching understanding, in the course of which participants, through relating to a world, reciprocally raise validity claims that can be accepted or contested" (Habermas, 1981:148; 1984:99). But his approach does not realize that all subject-subject relations are actually *founded* in subject-object relations, and therefore no water-tight separation is possible between these two kinds of relationships.

In order to explain this claim, we have to return to the issue at the basis of the problem of uni-vocality and multi-vocality.

6.11.3.2 *Communicative actions in their inter-modal coherence*

Exploring this problem further relates to the idea of the *unity* and *diversity* within reality, and its consequences for an analysis of the interconnections entailed in the nature of *linguistic communication*. Although formulated within a different context, Kaufmann explains the thought of Jaspers by explicitly referring to the idea of *uniqueness* and *coherence* employed in our subsequent analysis of linguistic communication: "Jaspers's view of being shows it as being divided in itself and yet in its various modes mutually related ('a being-torn-apart that is related within itself')" (Kaufmann, 1957:212).[1]

The general idea of the theory of modal aspects, namely that uniquely different modes of being indissolubly cohere, will continue to guide our subsequent reflections.[2]

An analysis and explication of *communication* prompts one to consider interconnections between ontic domains. In respect of the so-called *non-semiotic* meaning of *communication*, Derrida considers the possibility speaking of its "*proper* or *primitive* meaning" (Derrida, 1982:307). Unfortunately he *prejudiced* his view of the relation between a possible "primitive meaning" and a *semantic* meaning to such an extent, that the former is discredited in advance. The self-limitations he imposes in this regard come to light in his more extensive statement:

> ... that consequently the semantic, semiotic, or linguistic meaning corresponds to a derivation, an extension or a reduction, a metaphoric displacement. We will not say, as one might be tempted to do, that semiolinguistic communication is *more metaphor* entitled 'communication', because by analogy with 'physical' or 'real' communication it gives passage, transports, transmits something, gives access to something. We will not say so (Derrida, 1982:307).

1 "Jaspers's Vision vom Sein zeigt dieses als in sich selbst aufgeteilt und doch in seinen verschiedenen Weisen aufeinander bezogen ('eine in sich bezogene Zerrissenheit')."

2 In passing, we mention that Jaspers in a rather speculative way, discerns *seven* modes of being not at all related to the modal aspects of reality. His modes of being are 'Existenz', Transcendence, existence (Dasein), consciousness-as-such (Bewusstsein überhaupt), spirit, world, reason (see Knaus, 1981:152 ff. and his criticism – Knaus, 1981:171-175).

Derrida here confuses the *primitive meaning* of some or other functional domain of reality with truly *analogical instances*, and to this confusion, he adds the inapplicable opposition "more real" and "less real." If it is proper to speak about communication in a *physical context* – constituting a reference to *physical reality* – then it may be just as correct – and *real* – to speak about "semio-linguistic communication." But this issue concerns more than the mere use of *words* – it relates to a fundamental trait of (ontic) *reality* and is connected with what we identified as an inevitable challenge facing *every scholarly discipline* within the encyclopedic diversity of special sciences (encompassing both the natural sciences and the humanities), namely to account for the use of a multiplicity of (*multi-vocal*) *analogical basic concepts*.

6.11.3.3 *Is linguistic communication a transmission and/or sharing of meaning?*

Setting aside for the moment the difference between *concept* and *word* treated in an earlier context (see page 148 ff.), we may look at the following statement by Derrida: "If *communication* had several meanings, and if this plurality could not be reduced, then from the outset it would not be justified to define communication *itself* as the transmission of meaning, assuming that we are capable of understanding one another as concerns each of these words (transmission, meaning, etc.)" (Derrida, 1982:307). Derrida continues by pointing out that the word *communication* opens up what he calls a "semantic field" and that to the "semantic field of the word *communication* belongs the fact that it also designates nonsemantic movements" (Derrida, 1982:308).

Digressing for a moment, we have to note the importance of the term 'field' in this context. Of course this term as such also has *multiple* meanings. It first of all reminds us of *spatiality*, where *continuous extension* makes possible our awareness of a *domain*, an *area*, or a *field*. Yet these terms are frequently found in *different contexts*, such as when the discipline of physics introduces a "field theory," or when biology develops the notion of an *ambient* (*Umwelt*, biotic domain) (see page 352 ff.). The prominent German logicians of the late 19th and early 20th century explored the meaning of the domain ('Umfang') of a concept, and particularly the structural trend in the linguistic sub-discipline of *semantics*, investigated the idea of a *semantic field* (see Geckeler, 1978 and Coseriu, 1978). More recently the phrase "semantic domains" was employed within the context of *cognitive semiotics*. Unfortunately, in the absence of a theory of aspectual interconnections (ante- and retrocipations), Brandt qualifies the expression *semantic domain* as a "spatial metaphor" (Brandt, 2000:11). While Derrida wants to refer to the *semantic field* of the word *communication*, he implicitly employs one amongst a multiplicity of *contexts*, in which the term *field* is put to use! It happens frequently that Derrida enters into a discussion of this phenomenon – but unfortunately he never attempts to develop a *systematic theory* to elucidate the general significance of what is at stake. Our current discussion will explore some crucial elements of such an account.

With reference to "ordinary language and to the equivocalities of natural language," Derrida mentions instances where "what is transmitted or communicated, are not phenomena of meaning or signification" – because one may "*communicate a movement*," or "that a tremor, a shock, a displacement of *force* can be communicated – that is, propagated, transmitted" (Derrida, 1982:308).

Suppose we consider the transfer of magnetism from a magnet inducing a piece of iron and thus turning it into a magnet as well. Then the use of the term *transfer* can be replaced by the word *communication* – the magnet *communicated* its magnetism to the piece of iron. Once this transfer took place, it may be said that the magnet *shared* or *communicated* its magnetism with the iron. But this connotation of the word *sharing* cannot be understood except on the basis of our awareness of *wholeness* and the *division* of such a whole into *distinct* (i.e. a *multiplicity* of) *parts*. As we noted, continuous extension allows for the use of (synonymous) terms or phrases that can be substituted for it, such as *gapless coherence* or *cohering parts* constituting a *whole* (*totality*) (see page 87). We here encounter the *primitive meaning* of *space* – for in its *original* (non-analogical) *meaning*, spatiality concerns a *continuous whole* and the possibility of (endless) *divisions* (see page 163).

In terms of this line of argumentation, it must be clear that the *meaning* of (inter-human *lingual*) communication can only come to expression through its *interconnection* with multiple aspects of reality, owing to the fact that these aspects are *different* from the lingual function. Human communication *loses its meaning* if divorced from the spatial meaning of *wholeness* and *divisibility*, from the physical meaning of *dynamics* and *change*[1] (causes and effects/transfer), from the logical meaning of *consistency* (non-contradiction), and so on. Jaspers claims that communication is actually a truly human phenomenon, because he did not explore an analysis of the (analogical) basic concepts involved in an understanding of the meaning of the lingual aspect of reality. At most, he concedes that one merely refers to other (non-human) instances of communication ("alle andere Kommunikation") by way of comparison ("gleichnisweise ausgesagt werden") (Jaspers, 1957:784).

6.11.4 *Language acquisition – an a priori human faculty: Chomsky*

The lingual capacity of human beings brings to expression the *subject function* they have within the *sign mode* of reality. From Chapter 4 we know that this subject function is embedded both in the functional abilities of human beings and in the (enkaptic) intertwinement of distinct bodily entity structures. As soon as it concerns the acknowledgement of fundamental functional abili-

1 Kaufmann provides the following quotation from Jaspers: "All genuine communication is like the restless Heraclitean flux. Every moment is unique, it cannot be fixed and it cannot be repeated" (Kaufmann, 1957:195). ["Alle echte Kommunikation ist wie das ruhelose heraklitische Fließen. Jeder Augenblick ist einzigartig, er kann nicht festgehalten und nicht wiederholt werden."]

ties, such as thinking or speaking, traditional philosophy takes recourse to the idea of what is supposedly *a priori*. The view that language represents such an *a priori* human faculty derives from a revival of the notion of *idea innate* (innate ideas) during the 20th century in the thought of the linguist Noam Chomsky. Stegmüller employs a neat comparison in order to explain Chomsky's understanding of this issue. He compares acquiring / mastering a language with the complexities involved in studying differential geometry and quantum physics.[1] Whereas it would seem to be far-fetched to believe that a two-year old boy is mastering the said disciplines, no one considers it strange to hear that such a boy is mastering his mother tongue.

The remarkable element in this story is that Chomsky advanced a number of empirical arguments supporting his conviction that learning an ordinary language cannot be accounted for merely in terms of an empirical process. What is at stake is the mastery of a grammatical structure and linguistic rules from an apparently insufficient amount of linguistic data and to this Chomsky adds that even a child can generate more sentences than there are seconds in the life of any average person. Keeping in mind the comparison between mastering a complex scientific theory and learning a language, one should imagine that differences in intelligence would be significant in the former case, but strangely enough the same does not apply to language acquisition, for large differences in intelligence result in negligible differences in linguistic competence. Furthermore, even if the linguistic experience to which the child is exposed is not only limited but also largely degenerate, the child nonetheless masters the principles and rules governing the formation of meaningful sentences and the interpretation of linguistic utterances. Perhaps even more remarkable is the fact that language is learned during a stage in which the child is not capable of achieving anything comparable. The absence of any direct instruction and above all the fact that many children succeed in learning to speak without actively participating in talking activities ought to be mentioned as well. Besides, once the basic linguistic competency is mastered the child can creatively generate meaningful sentences never heard before – ruling out any idea that language merely emerges through acts of imitating what is heard. All-in-all these considerations are used by Chomsky in support of his claim to that an *a priori* element is inherent in the faculty of language-acquisition.

6.12 The semantic domain of words: synonyms and antonyms

The phrase "semantic domain" analogically reflects the meaning of space within the sign mode. This analogical moment lies at the foundation of the linguistic "word-field" theory. Brandt points out that Lakoff and Turner do not provide a satisfactory analysis of the *domain* concept (Brandt, 2000:14 ff.). Brandt views "[D]omains of experience" as "semantic domains" to which, as "kinds of reality," our "beliefs implicitly *refer to*" (Brandt, 2000:15). Although this formulation seems not to acknowledge the possibility of *ontic* co-

1 The explanation in the text is derived from the account given by Stegmüller (1969:530-533).

herences, Brandt explicitly states that "[B]asic semantic domains are neither language-dependent nor culture-dependent" (Brandt, 2000:16), thus implicitly alluding to the constant, ontic nature (and universality) of such domains underlying variable shapes within different languages and cultures. But then he continues by stating this explicitly:

> Semantic domains are constituted by human experience in the richest possible phenomenological sense; languages, cultures, and human semiotics in general are based on experiences and practices in a life-world constituted as a whole, and though it is perfectly possible to divide this whole arbitrarily into comparable segments – a task regularly assumed by natural philosophies and religions – it is also possible to identify genuine parts of it that remain stable under cultural variation; if such parts are identified, they qualify as universally given semantic domains (Brandt, 2000:16).

However, the ontic coherence between the sign mode and the spatial aspect can only be disclosed through actual language use and the concomitant development (growth and decline) of the semantic domains of words.

If the *word-field* (semantic domain) of a word is represented by a *Venn* diagram then the employment of partially overlapping circles can help us to understand the nature of synonyms while antonyms are represented by non-overlapping circles. To be more precise: in the case of a synonym there ought to be at least one shared meaning-nuance and one non-overlapping meaning-nuance within the semantic domains of the words involved. Of course, in order to account for the sign function of synonyms the *semantic unity* of a word is presupposed. This unity is not empty owing to the *multiple* meaning-nuances contained within the semantic domain of such a word. In other words, the spatial analogy within the sign mode is based upon the numerical analogy within it: semantic unity and multiplicity. Moreover, the semantic domain of a word has to endure over time in coherence with other semantic domains for only then is it possible to account for interrelations and the exchangeability of (synonymous) words. Therefore, the kinematic analogy not only intimately coheres with the numerical and spatial analogies for at once it makes possible the actual *lingual changes* taking effect when synonymous words are interchanged. Yet the *differentiation* of meaning-nuances ought to be *integrated* within the semantic field of a word, thus enabling the phenomenon of *linguistic context-sensitivity* (evincing the coherence between the sensory and the sign modes). In Chapter 2 we briefly referred to Antal's atomistic approach (see page 63). This atomism not only distorts the numerical analogy within the sign mode, but actually boils down to a *denial* of the spatial analogy. His view restricts every word to an *identical meaning* in every context: "We said earlier that the meaning of a word is identical in different contexts, because, if it is not identical, we are dealing with different words" (Antal, 1963:59); according to him meaning is "independent of environments" (Antal, 1963: 64).

The semantic domain of words often mediates the task of translation. The translational equivalent of a particular word may not be *direct* – for example when one tries to translate the Afrikaans word 'duister' into English one of the meaning-nuances of the Afrikaans word 'duister' namely *sinister* is the only option. In Afrikaans 'duister' is synonymous with 'donker' for they share one meaning-naunce ("absence of light") and each of these Afrikaans words has at least one distinct (non-overlapping) meaning-nuance.

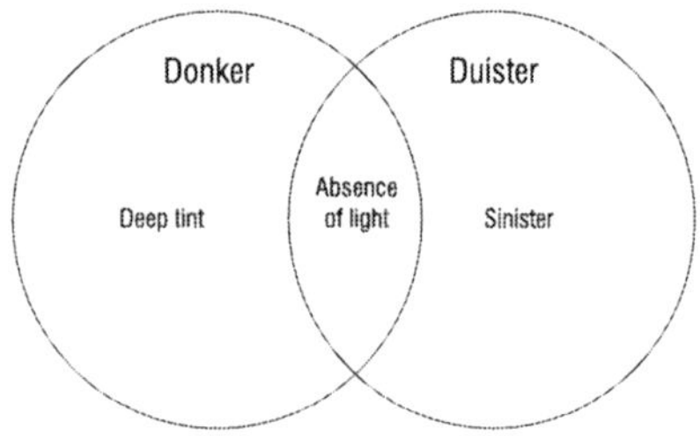

If no single meaning-nuance overlaps an antonym is at hand. More than two words may also be synonymous – just consider the representation of the meaning-nuances of the words *trust*, *confidence* and *certainity*, accompanied by some further 'relatives'.

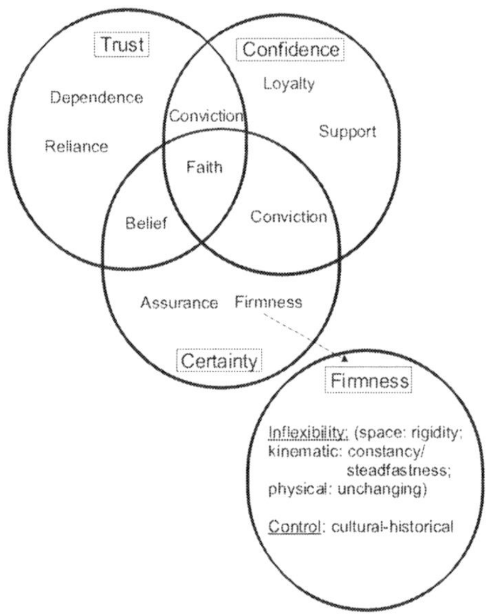

Trust: faith, belief, hope, confide, conviction, confidence, expectation, reliance, dependence [Antonym: distrust]

Trust: (in the sense of *have faith in*): believe, rely on, depend on, confide in, have confidence in, count on, be sure about [Antonym: distrust]

Confidence: (in the sense of *self-assurance*): self-confidence, poise, self-belief, self-reliance, buoyancy, coolness

Confidence: (in the sense of *belief*): faith, trust, support, loyalty [Antonym: doubt]

Confidence: (in the sense of *certainty*): conviction, assertion [Antonym: uncertainty]

Certainty: confidence, conviction, faith, belief, assurance, firmness, sureness, [Antonym: uncertainty]

Note that the semantic field of 'firmness' takes us beyond the limits of the fiduciary mode – *firmness*: inflexibility, steadfastness, control. (The Venn diagram captures some of these meaning-nuances.)

Before we return to the ontic conditions of language the connection of mutability and immutability in the thought of De Saussure requires attention.

6.12.1 *Between the "mutability and immutability" of the lingual sign as medium of communication: De Saussure*

The well-known emphasis on the fact that the "bond between the signifier and the signified is arbitrary" (see De Saussure, 1966:67) may prompt the *mistaken* idea that the use of language as such is *completely* arbitrary. But observe that de Saussure immediately adds important *conditions* and *limitations* applicable to the supposed *arbitrariness of language*. For example, once a signifier is established within a linguistic community, the *individual* cannot change such a signifier *at will* (De Saussure, 1966:69). Soon de Saussure also explicitly had to underscore both the *mutability* and the *immutability* of the sign – although it is *freely chosen* with respect to the idea that it represents, the sign is *fixed* (not free) in respect of the *linguistic community* that uses it (De Saussure, 1966:71). Nonetheless, the ongoing *dynamics* present within each linguistic community (Jaspers calls it the Heraclitean flux), requires a recognition of the multiple processes of *change of meaning* constantly taking place throughout an entire language.

This entails that what appears to be impossible to accomplish by the individual, is indeed possible for the *linguistic community*: "Language is radically powerless to defend itself against the forces which from one moment to the next are shifting the relationship between the signified and the signifier" – but once again this reality is on the shoulders of the *arbitrariness* of the sign: "This is one of the consequences of the arbitrary nature of the sign" (De Saussure, 1966:75). Whereas the *individual* appears to be powerless vis-à-vis the *linguistic community*, the latter now appears powerless vis-à-vis *historical change*!

Because the notion of *change* only has a determinate meaning on the basis of *constancy*, the question may be asked whether or not there are *universal* and *constant ontic conditions* underlying the existence of language, such that they are also *reflected by* and *displayed within* language itself? What Derrida says about the *written sign*, implicitly rests upon the foundational coherence between *constancy* (that "which remains") and *change*[1] (dynamics – "a force of breaking with its context"). Derrida writes:

> [Re *constancy*]: A written sign, in the usual sense of the word, is therefore a mark which remains, which is not exhausted in the present of its inscription, and which can give rise to an iteration both in the absence of and beyond the presence of the empirically determined subject who, in a given context, has emitted or produced it. This is how, traditionally at least, 'written communication' is distinguished from 'spoken communication.'
> [Re *dynamics*:] By the same token, a written sign carries with it a force of breaking with its context, that is, the set of presences which organize the moment of its inscription. This force of breaking is not an accidental predicate, but the very structure of the written (Derrida, 1982:317).

In order to find an answer to the question about the ontic conditions for lan-

1 We have seen that change always presupposes an element of constancy (see page 163).

guage, the distinction between *concrete things* (entities) and *ontic aspects* (within which entities *function*), frequently alluded to above, may help us to clarify an apparent peculiarity within language, as well as obtaining a perspective on the embracing cosmic intertwinement of linguistic communication.

6.12.2 *Implicit ontic conditions for language*

In connection with the nature of *synonyms* and *antonyms*, Geckeler (1971:242) points out that an unresolved problem for the science of linguistics is to explain why it is possible to provide contraries, antonyms or opposites for lexical items such as "old/young," immediately, whereas it is not possible with other items such as 'book'/'?'. De Klerk (1978:114) refers to the fact that most adjectives appear in dichotomous pairs, such as "short-tall, poor-rich, narrow-wide, ill-healthy, and so on." Lyons elaborates this when he says: "The existence of large numbers of antonyms and complementary terms in the vocabulary of natural languages would seem to be related to a general human tendency to 'polarize' experience and judgement – to 'think in opposites' " (Lyons, 1969:469).

This apparently strange feature of language merely reflects the *uniqueness* and *coherence* of the mentioned two *dimensions* of empirical reality, namely that of *modal* (*functional*) *aspects* and that of *concretely existing* (many-sided) *entities*, *events* and *societal relationships*. As we have seen (page 23) one can unveil the former dimension by answering 'how' questions, while the latter surfaces by asking 'what'. Only answers to 'how' questions make it possible to articulate answers in terms of *opposites* or *contraries*, such as the opposites *small-tall* (a spatial contrast), *slow-fast* (a kinematic contrast), *strong-weak* (physical contrast), or such contraries as *logical-illogical*, *moral-immoral*, *legal-illegal*. *Modal functions* always embrace the 'how' of entities and processes, reflecting a *way of being* (*manner of existence*), whereas *concrete (natural and social) entities* specify the basic meaning of these aspects by *functioning within them* in *typical* ways. Entities thus functioning within all aspects of reality always *specify* the general (*universal*) modal functions in a *typical* way. For example, think about distinct modal properties of material things (physical entities), such as *mass* (a physically specified numerical property), *volume* (a physically specified spatial property), *relative movement* (a physically specified kinematic property) or *force* (*strength* – the characteristic physical property of material things – see Chapter 7 below). In other words, these modal aspects provide the *universal* and *constant functional framework* within which the various entities and (human) activities – including acts of communication – *function*.

The emphasis on the term **function** reveals the inherently *dynamic* trait of reality – reflected in all languages by the presence of *verbs*. These functional modes serve as gateways through which one *talks* and *communicates* about entities, events or societal relationships. But since the dimension of (natural and social) entities also inherently (i.e. *ontically*) belongs to our experiential world, its reality within language is reflected by the presence of *substantives* (*nouns*).

The problem raised by Geckeler now has a natural solution in terms of this distinction between *functional modes* and *entities*, for the reason why certain lexical items do not have correlates is that this phenomenon only appears within the "such-and-such" (the 'how') nature of the *aspects* (*functional modes*) *of reality* (like the biotic contrast: "old-young"). But it does not apply to the *dimension of entities* (like 'book'/'?').

This state of affairs is obviously also linked to the fact that some languages display the tendency to be structured by 'substantives' (such as Persian), whereas others (such as old Greek and German) tend to be dominated by a *verb* structure (with multiple stipulations and derivations on the basis of their verbs – see Coseriu, 1978:43). These two dimensions (of *aspects* and *entities*) actually embody the *formal structural conditions* for all lingual activities – not only *formally* mirrored in the *subject-predicate* structure of (logical) propositions, but also in the *noun-verb* structure of *sentences*.

In addition to this *formal conditioning* role, the modal aspects appear as something *non-arbitrary* within all languages, namely in the inevitability of articulating multiple words reflecting the *modal diversity* within reality as such. The other side of the coin lies in the fact that this diversity of ontic aspects may also come to expression within the *lingual field* of particular words – evinced in their respective *connotations* (embedded in the particular *semantic field* of a word) – which also reflect the *aspectual diversity* within reality.

On the basis of the fact that our empirical world itself displays a *quantitative aspect*, all languages are enabled to form *number words* (such as the *one* and the *many*, *singular* and *plural*, *plenty*, *much*, *multiple*, *diversity* and *endless/infinite*). Similarly, the 'givenness' of *space* as an aspect of reality makes possible *spatial terms* (including *diminutives*) in all languages.[1] The same applies to kinematic terms (such as *constancy*, *persistence* and *endurance*), words with a reference to the structural features of the physical aspect (such as *cause* and *effect*, *energy-operation*, *force*, *validity* and *entropy*), words designating (functional) biotical states of affairs (such as *life*, *growth*, *differentiation*, *integration*, *finality* and *adaptation*), words capturing the meaning of emotional relations (such as *feelings*, *emotions*, *perceptions*, *awareness*, *consciousness* and *sensitivity*), logical relations (such as *inference*, *contradiction*, *discern*, *abstract*, *analyze*, *argumentation*, and *proposition*), cultural-historical terms (such as *a free formative imagination*, *power* over persons and things, *control*, *imaginitivity* and *mastery*), phenomena proper to the lingual (or *sign*) mode (such as *expression*, *meaning*, *significance*, *sign*, *allusivity* and *interpretation*), terms related to social relationships (such as *courtesy*, *tact*,

1 Spatial relations, amongst others, encompass terms such as *continuity*, *connectedness*, *coherence*, *division*, *divisibility*, *distance*, *context*, *place*, *simultaneity* (existence 'at once'), and *large* and *small*. Jenkinson explicitly referred to different modal aspects in his account of the interconnectedness of diminutives in ordinary language (see Jenkinson, 1986:55).

kindness, *friendship*, *peer*, *pal*, *partner* and [*as*]*sociation*), economic terms (such as *frugality*, *sparing*, *wasteful*, *stewardship*, *credit*, *profit* and the *avoidance of excess*), words capturing (functional) *aesthetic* realities (*pretty*, *gorgeousness*, *beauty*, *harmony* and *splendor*), terms expressing modal traits of the jural aspect (*retribution*, *fault*, *guilt*, *unlawfulness*, *illegality* and *just*[*ice*]), moral terms (such as *troth*, *love*, *integrity*, *respect* and *sincerity*) as well as certitudinal expressions (such as *reliability*, *trustworthiness*, *certainty*, *fiduciary*, *trust*, *credo* and *faithfulness*).

One of the consequences of *not* realizing that every language is *co-determined* by the underlying and conditioning diversity of modal functions within reality (to the extent that they are *merged into* and *reflected within* its vocabulary), is that it is also not understood that the multiple *interconnections* between these various (lingually designated) ontic domains are also reflected within language itself, particularly in *composite phrases* that bring to expression *analogical connections* between these functional domains.

One may use the metaphor of the *modal grid of reality* in order to capture the foundational role of the dimension of modal aspects with regard to logical thinking and the use of language. In the absence of a diversity of aspects logical analysis – identification and distinguishing – would collapse. As points of entry to reality the modal aspects also enable meaningful classifications, e.g. by distinguishing between different kinds of entities, such as physical things ("matter"), living entities (plants, animals and human beings), sentient creatures (animals and humans), cultural objects – differentiated in multiple categories, such as analytical objects (test tubes), lingual objects (books), social objects (furniture), economic objects (money), aesthetic objects (works of art), jural objects (jails), ethical objects (engagement rings, wedding rings), and so on.

The ultimate issue is therefore whether or not we are willing to acknowledge the foundational role of the dimensions of aspects and entities. The metaphor of the *modal grid* of reality focuses on whether or not there are, prior to any human intervention or construction, a given (ontic) multiplicity of aspects, modes or functions of reality. Furthermore, since these modal aspects co-condition the existence of concrete entities *functioning* within them, language and communication are bound to reflect this functional diversity.

The significance of this insight for an understanding of language and communication is dependent upon another important distinction, namely that between *concept* and *word*. Scholars tend to confuse or even identify concept and word (see for example Rossouw, 2003:17 ff.). A concept has its (logical) *content* and a word has its (lingual) *meaning*. Words may designate whatever there are – aspects, entities and processes and even concepts. But a word is not a concept. A number of years ago I attended an international conference in Vienna (on the comparison of Chinese medicine and Western medicine) where the claim was made that the Chinese do not have the concept *culture*. During discussion time I asked the speaker if the Chinese language does have translations for words such as *power*, *formation*, *control*, and *imagination and a phrase such as the free formative imagination of human beings* – to which the answer was affirmative. Yet all these words are in some respect synonyms for the English word *culture*, indicating that the Chinese do have a *concept* of culture but merely lack a translational equivalent for the English *word* culture! We argued that a concept unites, through logical objectification, a multiplicity of universal features (and it is 'blind' to what is unique and individual – see pages 148 ff. above). Consider the concept *human being* or the concept *triangle*. The universality of the features constituting a human being makes it possible to recognize (identify) a human being wherever one encounters one. Likewise, when the terms *line*, *angle*, (closed) *surface* and *three* are combined in the unity of the concept *triangle* the universality of these traits make it possible to recognize a triangle wherever and whenever it is encountered. A concept in this sense transcends every word and every language. For that reason a concept cannot be *translated*. One understands the concept of a triangle or one does not understand it – irrespective of the lingual sign employed to designate it. The English word *triangle* can be translated into other languages, for example into Afrikaans (*driehoek*), German (*Dreieck*), and so on. There are multiple words for this concept, but the universality of the concept precludes the idea of its 'translation'.

Since the lingual ability to signify presupposes the analytical ability to identify and distinguish – and identifying something amounts to nothing but acquiring a *concept* of it, lingual communication explores this foundational relationship in various ways. First of all it occurs through direct conceptually based interaction. Yet the inherent ambiguity of all language does not warrant a straight-forward claim to *literal* language or the simple distinction between literal and metaphorical language use. By its very nature linguistic expression is *ambiguous* and requires *interpretation*. A well-known case may illustrate this point. As an example of an allegedly *literal* sentence the following one was presented: "The cat is on the mat." The fact that this apparently 'literal' sentence still required *interpretation*, was underscored by the following reaction: "Oh, I know exactly what it means: The poor hippy is once again in the office of the boss!" What is striking about this example is that both sentences contain words making an appeal to familiar *concepts*. There is no doubt about what the concept *cat* or the concept *mat* is all about, just as there is no doubt about what the concept of a (military) *general* or a *lion* entails. Yet, the moment language is at stake the conceptual level is transcended, making possible an expression such as *the Lion of Western Transvaal* (General De la Rey). In a purely logical sense it is contradictory to affirm that a human being is a lion, but within the context of language it is perfectly permissible to generate expressions such as these.

Within ordinary language words reflecting modal qualities (properties) are always embedded in a context where concrete things and events are discussed. To this extent all language and communication, in spite of the richness in variation owing to the operation of the free formative imagination of human beings, is bound to the horizon of modal (aspectual) possibilities. This does not merely imply terms derived from the core (primitive)[1] meaning of particular aspects, but also interconnections between various aspects, evinced in partial similarities and partial differences (in an earlier context designated as *analogies*). It has been pointed out that many composite phrases capture inter-modal analogies – such as *emotional life* (an analogy of the biotic meaning of life within the sensitive mode of feeling), *social distance* (spatial analogy within the social mode), *economic trust* (a fiduciary analogy within the economic aspect – see pages 76, 158, 313 above), *aesthetic integrity* (a moral analogy within the aesthetic facet), *energy constancy* (a kinematic analogy within the physical mode), and so on. What might have seemed, at first sight, to be mere *arbitrary constructions* of human language, in fact turns out to be instances of analogical linkages underlying similar composite phrases found in all languages, once more underscoring the *conditioning* role of the modal grid of reality regarding language and communication.

Let us consider a sentence chosen at random: "Does life in the United States actually show signs of moral and cultural crisis, or does a closer look reveal the continuing resilience of the world's most successful and self-renewing democracy?"[2] In terms of the *modal grid* that conditions what is said, we can identify the following words presupposing different modal aspects: *life* (the biotic mode – here taken as an analogy within human interaction – *social life*); *show* (sensitive aspect); *signs* (lingual mode); *moral* (the ethical aspect); *cultural* (the cultural-historical mode); *continuing resilience* (the equivalent of the intimate connection between *constancy* and *change* – derived from the kinematic and physical aspects); *closer* (reflecting the spatial meaning of *nearby* as opposed to *far away*); *look* (the metaphorical use of an observational term derived from the sensitive mode); *most* (derived from the quantitative meaning of *more* and *less*); *successful* (effective – figuratively derived from the physical cause-effect relation); and *self-renewing* (reflecting the interconnection between persistence and change found in thermodynamically open systems). In addition there are of course entitary-directed words present in the sentence – such as "United States," 'world', and 'democracy' – also reflecting a distinct dimension of reality, namely the dimension of (natural and social) entities, events and processes.

1 The general idea of a modal aspect as developed in this work holds that the core meaning of an aspect brings to expression its *irreducibility*, which is reflected in its indefinability, explaining why it is also designated as *primitive*. Korzybski underscores that one cannot define *ad infinitum*: "We thus see that all linguistic schemes, if analysed far enough, would depend on a set of 'undefined terms'. If we enquire about the 'meaning' of a word, we find that it depends on the 'meaning' of other words used in defining it, and that the eventual new relations posited between them ultimately depend *on the … meanings* of the undefined terms, which, at a given period, cannot be elucidated any further" (Korzybski, 1948:21).

2 Quoted from Skillen (1994:14).

6.12.3 *Habermas and the norms for communicative actions*

Habermas conceives an ideal speech situation, where basic utterances are bound by validity claims, namely *truth* (*Wahrheit*), *rightness* (*Rigtigkeit*), and *appropriateness* (*Angemessenheit*) or *comprehensibility* / *well-formedness* (*Verständlichkeit* / *Wohlgeformtheit*), (see Habermas, 1981: 66). Jacobson and Kolluri take this as a starting point[1] in their aim to reconstruct participatory communication in order to evaluate the interaction between 'experts' and "either an individual or group from a local population" (Jacobson *et al,* 1999:274). They do this by asking four questions:

1. Is the communicator's communication correct, that is, is the information being offered undistorted and reliable? (Truth)
2. Is the communicator's role legitimate, given his or her role and the participation of other interested individuals, social groups, agencies, and nations who are party to the process of development? (Rightness)
3. Is the communication offered sincerely, in good faith without being manipulative, either on the part of the individual, or any organization from which the individual may have been sent? (Sincerity)
4. Is the communicator's communication comprehensible to others, that is, are idiom, cultural factors, and/or message design, adequately accounted for? (Comprehensibility) (Jacobson *et al.* 1999:274).

If we employ – as shorthand for "linguistic communication" – the adjective 'communicative', we may ask whether these four questions are exhaustive of the multifaceted nature of *communicative actions*? What they have captured are the following configurations: *communicative truth*, *communicative rightness*, *communicative sincerity* and *communicative comprehensibility*. A closer look at these four composite phrases reveals interconnections between communicative actions and the *logical-analytical* mode (*truth* and *comprehensibility*), the *jural* (*rightness*) and the *moral* aspect (*sincerity*).[2]

Habermas and Jacobson *et al* are certainly justified in highlighting, for example, the *logical functional substrate* of lingual signification and communication. Viewed from the perspective of modal functions the use of *lingual* communication presupposes the *logical* function. The lingual intention of (symbolical) signification is always dependent on the correct lingual *identification* and *distinction* of the relevant words. Identification and distinction are important even in respect of the *alphabet letters* used in words, since the simple switch of letters may result in a complete change in the *meaning* of a (written or spoken) word. Any act guided by the *sign mode* must therefore make an appeal to the foundational logical function and even to the *kinematic* aspect

1 Though they refer to Habermas 1979 where he still mentioned four claims – the fourth being *sincerity* (see Jacobson *et al*, 1999:271).

2 This slightly 'violates' the preference in the thought of Habermas for *truth*, *beauty* and *goodness* (the moral) – a threefold perspective binding his world-orientation to the Greek-Medieval legacy where these three (accompanied by *unity*) were considered to be transcendental determinations of *being*.

insofar as an element of *lingual constancy* is required by the possibility entailed in every language of producing endless lingual variations on the basis of this lingual constancy. We noted that Habermas explicitly and correctly emphasizes the foundational role of alphabet letters (see page 149 above). In spite of the elusive and ever-referring nature of *différance* in Derrida's writings, the latter consistently had to hold on to exactly the same letters *constituting* this word – once again demonstrating the fact that *différance* itself can only be affirmed on the basis of *constancy* (*identity*)! Habermas also does does not realize that this notion of *lingual* or *communicative constancy* reveals the inter-modal connection between the *sign mode* and the *kinematic mode* of reality, i.e. that the original kinematic meaning of uniform flow (constancy) is *analogically reflected* within the structure of the sign mode.

Of course the core meanings of other modes of reality are also analogically reflected within the lingual mode. Is it not the case that one-sided focal emphases will distort communicative actions? Is it meaningful to support the claim that communicative actions are only meaningful if they are sensitive to *communicative nuancefulness* (*many-sidedness*). It is clear that the term 'many' analogically reflects the original quantitative meaning of number, while the suffix '-sidedness' manifests the original spatial meaning of configurations with 'sides'. Similarly, only on the basis of communicative *constancy* (including *continued* or *uninterrupted* communication) communicative actions are reciprocal in the sense that the *communicative dynamics* involved bring about the desired communicative *effects* – analogically reflecting the primitive interconnected meaning of kinematic *constancy* and physical *change*.

Surely there are just as many interconnections at stake here as there are unique and irreducible aspects to be distinguished within reality. Just consider the following composite phrases capturing the other analogical structural moments not yet incorporated in our analysis – but keep in mind that these interconnections, through which non-lingual terms are analogically reflected within the domain of linguistic communication, could be phrased in many different ways (within the same and within different languages). Yet, since they are *irreplaceable*, they ought not to be confused with *metaphors*.

All these composite designations yield the phrases that reveal an element of coherence (similarity and difference) in each case between *linguistic communication* and a different aspect of reality – and each can be formulated in such a way that its inherent *normative* meaning is displayed (communicative actions can violate these modal-functional communicative principles).

Contemplate the following composite phrases (which include alternative formulations of the four mentioned by Jacobson *et al.*).

Communicative *vitality* and *fruitfulness* (the biotic analogy). Communicative *sensitivity*; communicative *consistency* and *comprehensibility*.

Comment: We have seen that the nature of human knowledge rests on the basis of the acquisition of *concepts* – and that conceptual insights are *normed* in the

> sense of being subject to *logical principles*, such as the principles of *identity* and *non-contradiction*. Logical analysis can only unfold its meaning in acts of *identifying* and *distinguishing*. A succinct way of capturing the interconnection between linguistic communication and the logical-analytical mode would be to speak about *communicative discernment*.

Communicative *mastery*,[1] *care*, *power* and *control*.[2] Communicative *sociation*;[3] communicative *frugality* (and economy – the avoidance of communicative excesses); communicative *harmony*; communicative *retribution* (*justice*); communicative *integrity* (*morality*); and communicative *reliability*, *trustworthiness* (*bona fides*).

The crucial perspective here is that this brief indication of the multi-faceted *meaning* of communicative activities are made possible by acknowledging the multiplicity of *unique* (and *irreducible*), but at once *intimately cohering* (ontic) modes of being (functional aspects also serving as *modes of explanation* in different scholarly disciplines). Within the framework of these modal functions – highlighting universal and constant ontic conditions[4] – creaturely responses are possible in numberless ways (accounting, for example, for the rich variety of different languages, mistakenly thought of as being formed in a completely 'arbitrary' way). If imagination is viewed as the ability to *recollect* what is no longer present to the senses, the idea of Condillac, explained by Derrida, is understandable: "The sign is born at the same time as imagination and memory, at the moment when it is demanded by the absence of the object for present perception" (Derrida, 1982:314).

Because the above-mentioned composite phrases point to fundamental starting points (*principles*) guiding human communicative actions, they call for a positive shape to these principles in the course of history. We argued that they ought to be 'positivized', because different *positive forms* can be given to these principles. In a basic sense, however, they are *unspecified*.

Within the different aspects of reality, there is a difference between natural laws – that are valid per se – and the principles norming human activities, for the latter, as we have pointed out repeatedly (see pages 289 and 297 above), require human *intervention* in order to be *made valid* in specific historical cir-

1 The implicit dualism in Habermas's thought between subject-object relations and subject-subject relations at once explains why he reserves the term *mastery* for *instrumental rationality* and avoids its employment when it comes to *communicative understanding* – see his mentioned distinction between *instrumental mastery* and *communicative understanding* (Habermas, 1984:11).

2 The cultural-historical task of humankind concerns the exercise of formative power and control – not to be confused with an (antinormative) *abuse* of power.

3 The connection with the primitive meaning of the social aspect of reality. Georg Simmel (in his 1908 work *Soziologie*) introduced the idea of 'Vergesellschaftung' here translated as 'sociation'.

4 Of course the *acknowledgment* of universality and constancy does not substantiate any claim to *infallibility* and to (*absolute*) true knowledge – for human knowledge is co-conditioned by the (functional) domains of *historicity* and *linguisticality* – explaining why every human insight in principle is fallible, subject to criticism and open to improvement or reinterpretation.

cumstances. Modal laws and modal principles hold for every possible entity functioning within the aspects concerned. Within the normative functions of reality – such as the *logical-analytical*, the *lingual* or the *social* – natural entities (such as things, plants and animals) do not function as *subjects*, but merely as *objects*. For example, through the employment of a linguistic sign the lingual *object*-function of a plant could be made patent (manifest) in as-*sign*-ing a *name* to it (see page 344 ff. below).

6.12.4 *The modal universality of the sign mode*

When the universality of the sign mode is accepted the mistaken impression may arise that language as *horizon* embraces us in such a way that we do not have an *extra-lingual* access to reality. Is this not intended by Derrida when he states: "*There is nothing outside of the text* [there is no outside-text; *il n'y a pas de hors-text*]" (Derrida, 1976:158)? Eventually, Derrida gave his own interpretation of his initial statement in *Limited Inc* where he writes that "[t]he phrase which for some has become a sort of slogan, in general so badly understood, of deconstruction ('there is nothing outside the text' [*il n'y a pas de hors-texte*]), means nothing else: there is nothing outside context" (Derrida, 1988:136). A re-interpretation of this statement in terms of the theory of modal aspects is that Derrida implicitly stumbled upon the *modal universality* of the sign mode without being able to articulate it in terms of this idea of modal functions. When we substitute 'text' with "sign mode" the view that nothing is "outside the text" simply means that whatever there is in principle has a (a subject- or object-) function also within the sign mode of reality. This understanding does not intend the *reduction* of all of reality to the sign mode (as it is found in the postmodern *all*-claim: "everything is interpretation"), but simply highlights the inevitable function of everything within this unique mode. Within the scholarly reflection of the West, we have noted a similar over-emphasis on the meaning of number in the Pythagoreans assertion: "everything is number." Subsequently scientific thinking was tempted to fall prey to similar one-sided views – such as "everything is matter" (materialism), "everything is alive" (vitalism), "feeling is everything" (*Gefühl ist alles* – Goethe), "everything is historical" (historicism), and so on.

What is known as *conventions* – be they *communicative*, *societal*, *juridical* or *moral* – always relate to the ways in which human beings have given a positive form to underlying *modal* or *typical* principles. Habermas is therefore completely justified in criticizing Wittgenstein (in his use of the semantic concept of *intentionality*) for implicitly *equating* conventions of signifying with societal conventions.[1]

Jaspers expresses the modal universality of the linguistic mode (in its communicative function) in his own way when he, in the words of Kaufmann, underscores the stimulus and guiding force in our "absolute will to communicate

1 "Freilich erhält der semantische Begriff der Intentionalität von Handlungen seine Pointe erst dadurch, daß Wittgenstein Bedeutungskonventionen stillschweigend mit gesellschaftlichen Konventionen gleichsetzt" (Habermas, 1995:308).

with our neighbour."[1]

Within the discipline of hermeneutics, the modal universality of the sign mode is also essential. When Caputo, for example, stresses that "radical hermeneutics cultivates an acute sense of the contingency of all social, historical, linguistic structures, an appreciation of their constituted character, their character as effects" (Caputo, 1987:209), it must be clear that his statement presupposes the *universality* and *constancy* (i.e. *modal universality*) of the *social*, the *historical*, and the *linguistic* facets of reality. This is already seen from the use of the word 'all', that bestows a *universal scope* upon the *social, historical* and *linguistic* dimensions of reality. Where Grondin refers to hermeneutics as the *universal methodology* of the humanities, he actually also highlights the modal universality of the historical and sign modes:

> Dilthey broadened this hermeneutics into a universal methodology of the human sciences, and Heidegger located hermeneutic inquiry on the still more fundamental ground of human facticity. Gadamer ultimately reformulated universal hermeneutics as a theory of the ineluctable historicity and linguisticality of human facticity (Grondin, 1994:3).

However, as soon as the nature of *human society* is explicitly brought into the picture, communicative action assumes (typically) *specified forms* and *shapes*. This does not entail that one has to concede that the "social world" is solely the product of *human construction*. Whatever human beings *construct*, is always in response to *given* modal and typical principles. The theme of "social construction" actually elevates the human ability to give *positive form* to underlying principles to the level of being the *constructing origin* of social reality. Just once again consider the above-mentioned titles: *The social construction of reality*; *a treatise in the sociology of knowledge* (Luckmann & Berger, 1969); and *Der sinnhafte Aufbau der sozialen Welt* ("The meaningful construction of the social world," Schutz, 1974 – see page 73).

A proper understanding of language ought to highlight the *social disclosure* of the sign mode. The fact that a truly *private language* is limited to one person only[2] asks for the acknowledgement of the *communal* character of normal language use and linguistic communication.

6.13 Multiple modal norms

Given our acquaintance with the modern natural sciences it is normal to be aware of "laws of nature" as they are called. These laws are actually *laws for* whatever is correlated with these laws in the sense of being subjected to them. It seems as if the disciplines of mathematics, physics and chemistry discovered (formulated) many distinct natural laws, but that there are not that many 'laws' applicable to humankind. Our first intuition may be to think of the normative meaning of the post-sensitive modes of reality, for within all of them

1 "... durch den Ansporn und die Richtkraft in unserem absoluten Willen zur Kommunikation mit unseren Nächsten" (Kaufman, 1957:206).

2 Wittgenstein remarks: "And sounds which no one else understands but which I 'appear to understand' might be called a 'private language' " (Wittgenstein, 1968:94e, §269).

we encounter *normative* contraries, such as logical – illogical, polite – impolite, legal – illegal, and so on (see page 98 above). Since provisionally we distinguished nine normative aspects, each with its distinct qualifying meaning-nucleus, at least nine norms (or principles) should be acknowledged.

Although it is true that each one of the nine post-sensitive aspects of reality has a normative structure there is much more to be said concerning modal principles. In fact we did analyze many instances where the interconnection between two different modal aspects constituted fundamental modal principles. For example, we related the first three logical principles (*identity*, *non-contradiction* and *the excluded middle* – see pages 300 ff. above) with the interconnections between the logical-analytical aspect and the aspects of number and space. While the meaning of number is analogically reflected within the logical aspect in the configuration of a logical unity and multiplicity, we argued that the logical principles of identity and non-contradiction embody the positive and negative sides of this coin. The principle of the excluded middle turned out to have its foundation in an even more complex interrelationship, for our argument concluded that it rests upon a (numerical) retrocipation to a (spatial) antecipation – on the norm side of the logical-analytical aspect. In the principle of thought-economy the antecipatory coherence between the logical-analytical and the economic aspects of reality is disclosed (see pages 311 to 313 above).

Our discussion of modal norms or principles (earlier in Chapter 6) briefly stated that analyzing the interconnections between aspects sheds a new light on the nature of modal norms. In view of the many instances of modal principles that thus far surfaced in our analyzes we are justified in stating the general perspective, namely that

> *every modal analogy on the law side of a normative aspect constitutes a fundamental modal principle.* A quick calculation of the basic modal norms given in the retrocipatory and antecipatory analogies on the law side of everyone of the nine normative aspects yields 126 possible modal norms (90 constitutive and 36 regulative ones).

Let us elucidate this extremely important insight with two additional examples, the first one highlighting just one modal norm (jural economy) and the other one the full range of retrocipatory and antecipatory analogies within a specific aspect (the aesthetic mode).

6.13.1 *The principle of jural economy*

In his analysis of the meaning of jural causality Dooyeweerd in particular enters into an explanation of the meaning of the economic analogy within the structure of the jural aspect and in doing this implicitly highlights the *modal legal norm* (principle) of *jural economy* (avoiding legal excesses):

> The economic analogy in the legal relation of balance concerns the economical handling of legal means and interests of others within the context of alternative possible choices a person is free to pursue. Every excessive, every unrestrained exploration of one's own legal interest, within legal life, is an inter-

> rupting causal intervention in the legal balance of interests against which the legal order reacts with restorative legal consequences. The driver of a car, who, when another car approaches from a side-street, continues driving on a road that gives the first-mentioned motorist the right of way, does not cause the subsequent accident when the same driver had no reason to expect that the other motorist would not yield. However, if the first motorist still continues to drive on, while having had the opportunity to stop in time after realizing that the other driver had disobeyed the traffic rules, then the loss-causing effect should also be imputed to the former's act since it is in conflict with the principle of jural economy and constitutes as such an excessive pursuit of one's own legal interest. If, according to the theory of the conditio sine qua non, the jural imputation of the effect is only introduced in connection with the question of fault, it would essentially amount to a total elimination of the jural problem of causality. After all, the genuine jural question of who caused the accident no longer has any meaning – and without answering this question the issue of fault could also no longer be posed in a jural sense (Dooyeweerd, 1997a:65).

Of course within the highly differentiated structure of modern systems of criminal law there are many instances in which the general meaning of the principle of jural economy is typically specified. Consider self-defense and excessive self-defense, where self-defense is allowed in turn to the latter or the typical public legal principle prohibiting an abuse of governmental power.

Our second example concerns the modal analogies on the norm side of the aesthetic aspect.

6.13.2 *Modal aesthetic principles*

In Beardsley's standard work on aesthetics basically only *three* aesthetic principles are identified: *unity*, *complexity* and *intensity*. From the perspective of the inter-modal coherence between the various aspects including every analogical moment the shortcomings of this approach are striking. Let us therefore give a brief overview of modal aesthetic principles and then return to the view of Beardsley.

A work of art has to conform to the following (universal) modal aesthetic principles. It ought to display an aesthetic unity amidst an aesthetic multiplicity (numerical analogy). It ought to be structured as an aesthetic whole with aesthetic parts (parts fit within a whole – spatial analogy).[1] These first two modal aesthetic principles underscore that an aesthetic awareness of the nuancefulness or many-sidedness of reality analogically reflects the coherence between the aesthetic aspect and the aspects of number and space. Furthermore, a work of art ought to display aesthetic durability (constancy) (kinematic analogy); it ought to exercise an aesthetic effect (an analogy of the physical cause-effect relation); it ought to display an inherent aesthetic differentiation and integration (biotic analogy); it ought to be aesthetically sensitive to the

1 The neo-Marxist view of literature objects to what Bürger calls an *organic work closed in itself*. Lukács depicts it as realistic and according to Adorno it is ideologically suspect, for instead of rather laying bare the contradictions of contemporary society already the form of an organic work creates the illusion of a beneficial society (see Bürger, 1974:120).

nuancefulness of reality (sensitive analogy); it ought to display an inherent aesthetic consistency (logical-analytical analogy); it ought to take shape within a particular aesthetic style (cultural-historical analogy); in an aesthetic sense it ought to communicate an aesthetic message (sign analogy); in an aesthetic sense it ought to fit into a particular social milieu, it ought to evince an aesthetic 'sociality' (social analogy); it ought to avoid whatever is aesthetically excessive (economic analogy); in an aesthetic sense it ought to do *justice* to the many-sidedness of reality transformed into the aesthetic end-product; it ought to display aesthetic integrity (moral analogy); and finally a work of art ought to be aesthetically convincing, it ought to witness to an element of aesthetic confidence making an appeal to aesthetic trust (fiduciary analogy).

Each one of these modal aesthetic norms could be obeyed or violated – aesthetic activities may be norm-conformative or antinormative, but no work of art can ever avoid the normative appeal of these modal aesthetic principles, although the artistic type law does specify the manner a work of art functions within the various modal aspects in a *typical* way.

Returning to the three aesthetic principles mentioned by Beardsley it is clear that he actually 'loaded' each with structural elements stemming from different analogical moments. Beardsley explains the aesthetic *unity* of an art-work by asking whether or not it is "well-organized," whether it is "formally perfect (or imperfect)" and whether "it has (or lacks) an inner logic of structure and style" (Beardsley, 1958:462). Being well-(dis-)organized or formally (im-)perfect refers to the cultural-historical analogy within the structure of the aesthetic mode and the same applies to the element of *style*, while the "inner logic" obviously refers to the analytical analogy. However, the meaning of the norm of *aesthetic unity* receives its first closer determination from the spatial analogy on the norm side of the aesthetic aspect: *aesthetic unity* ought to be embodied in an aesthetic coherence binding together all *parts* into an aesthetic *whole*. What Beardsley says about *complexity* utilizes merely analogies from the first three modal aspects (format, contrasts and subtlety). Likewise, the norm of *sensitivity* is explained by employing analogical elements derived from the physical, biotic and sensitive modes (compare terms such as *vitality*, *forceful*, *vivid*, and *tender*").

6.14 Disclosure in the sense of opening up object functions

Let us commence by looking at a very deeply rooted misunderstanding almost universally present in the discourse of English-speaking philosophers. What I have in mind is the practice of referring to concrete entities – things in their multi-aspectual existence – as 'objects'. The most familiar expression is the reference to "physical objects." Lowe, for instance, speaks of "changeable, persisting physical objects" (Lowe, 1998:131) and in other contexts repeatedly about "physical objects" with specific properties, such as spatial ones (Lowe, 1998:171), including certain particularized "properties of objects," such as the *shape* of a smile or waves as particular ways "in which the sea's surface is configured" (Lowe, 1998:181).

The implicit assumption of this mode of speech is the primary existence of the human subject, to whom physical entities are related as 'objects'. By putting the human subject in the center, with concretely existing (physical) entities "at hand" as so-called "physical objects," no room is left for the recognition of the true *subject functions* of such physical entities. Of course the *shapes* of physical things indeed pertain to their active function within the universal framework of the (ontic) spatial aspect, that is to the way in which they are *subject* to spatial laws. Likewise, *counting* physical things is only possible due to their quantitative *subject functions*, and the same applies to their movement (as kinematic *subjects*). Since the mechanistic main tendency of modern physics, which attempted to reduce all physical properties to *movement*, was abandoned, 20th century physics also had to acknowledge the *physical* subject function of such entities. Physical things are subject to physical laws, such as the law of energy constancy, as we have argued above.

As soon as the subject functions of material things are acknowledged, it turns out to be a *contradiction in terms* to refer to physical entities as "physical objects": a physical thing (entity) cannot in the same context and the same sense be *both* a physical subject *and* a physical object.

Does this mean it is inadmissible to speak of 'objects' at all? Not at all! This would have been the case only if the existence of physical things were *enclosed* within the first four modal aspects of reality. Our everyday experience of physical things is constituted by numerous *non*-physical (in terms of the order of succession between the different modal aspects of reality: *post*-physical) contexts, in which they appear solely as *objects*. The particular economic value of a diamond comes to expression in its expensive *price*, notwithstanding the fact that a diamond as such cannot act as an economic subject. The assessment of its economic value is exclusively dependent upon the action of an economic subject, of a human being capable of subjectively functioning within the economic aspect. Having a price in economic exchange is a *latent* function of a physical subject like a diamond. When this latent economic function is made *patent*, that is to say, when it is opened up or disclosed, then we may discern its economic *object function* that is always strictly correlated with a human subject function. Another way to say the same thing is to speak of the economic objectification of the diamond by a human subject. The implication is that objectification is always the result of the activity of a subject within the aspect concerned – or at least modal objectivity is always correlated with modal subjectivity (recall the relation between a line as *spatial subject* and a point as a *spatial object* – see pages 222 ff.).

In the light of the foregoing considerations, it is clear that the common practice of speaking of "physical objects" proceeds from a confused ontology.

> *Remark*: Insofar as material things are *physical*, they are *subjects* and not *objects*; and insofar as material things are *objects*, they are considered within some or other post-physical aspect (in which they are *objectified*) – such as the sensitive mode (as something *perceived*), the logical-analytical mode (as something *identified* and *distinguished*), the sign mode (as something *designated* / *named*), the economic mode (as something *bought* or *sold*), and so on.

Initially we introduced the terms/phrases *disclosure* and *deepening of meaning* in order to designate opening up modal antecipations within aspects. But within the context of the subject-object relation a different meaning-nuance is activated – disclosure in the sense of *opening up a latent object function*.

One may argue that it is an effect of the "subject-centeredness" of modern philosophy (sometimes plainly designated as its *subjectivism*) that everything different from the human being is turned into an 'object'. Behind this practice one should observe the return of modern philosophy to the 'subject' – a development that has to be pictured against the background of a long-standing legacy that acknowledged an 'objective' order embracing humankind as well. In their speculation about a world 'logos' and a "chain of being" classical Greek and Medieval philosophy, in this sense, adhered to an order *outside* the human subject. During the late medieval transitional period a turn away from God and a cosmic world order took place. Descartes drew the radical conclusion of this process when he postulated the thinking 'I' (*cogito*) as the new source of certainty. The apparent infallibility of mathematics is used to proclaim the sovereignty of human thought. Kant continued this legacy for in his 'Prolegomena' to every future metaphysics acting as a science, he said that it was David Hume who liberated him from his "dogmatic slumber" and helped him realize that "objective reality" ought to be understood in terms of the subjective categories of thought (cf. Kant, 1783, 'Vorrede':260). This is known as the "Copernican turn" in the epistemology of modern philosophy. Earlier we have seen that Kant actually elevated human understanding, through its *a priori* concepts, to become the *formal law-giver* of nature.[1] A similar return towards the *subject* is found century within the field of biology in the 20th century.

6.14.1 *Subject-object relations in plant and animal life*

The development of the notion of an *ambient* (*Umwelt*) will enable us to understand the coherence between analogical basic concepts and the meaning of the subject-object relation.

Every scholarly discipline certainly has to coin the key-terms necessary for its theoretical endeavors. But particularly as an effect of misguided attempts to elevate certain basic concepts to be more encompassing than they really are, special sciences not only tend to coin their own terms but sometimes also want to shield them from the terms employed by other disciplines. According to Thure von Uexküll, it was the intention of Jakob von Uexküll to pursue this path by designing a new method in the development of his 'Umweltlehre', equipped with its own peculiar conceptual system that cannot be derived from any of the existing sciences. He did this by employing the idea of *purposefulness* ('*Planmäßigkeit*') and *natural plans* ('*Naturplänen*').[2]

1 "Understanding creates its laws (*a priori*) not out of nature, but prescribes them to nature" (Kant, 1783:320, § 36). Also see Kant (1787:163).

2 "Um den unbekannten Inhalt von Naturplänen zu erforschen, bedarf die Umweltlehre einer Methodik, die aus keiner der bestehenden Wissenschaften entliehen werden kann. Sie muß dafür ihr eigenes Begriffssystem entwickeln" (Thure von Uexküll, 1970:XXVII).

6.14.1.1 *Mutually exclusive special scientific terms?*

Is it really possible to develop a conceptual system that does not have any linkages to other domains of knowledge? The negative answer to this question is shown by instances of *partial overlapping* in the terminology of different disciplines. Sometimes an attempt is made to account for this by means of categories with varying degrees of generality, as for example assumed in the approach of the sociologist Catton. He aims to develop a truly *naturalistic* sociology, but does not want to become a victim of *reductionism*. He believes that sociological axioms could be formulated to be parallel to Newton's axioms. He dismisses as irrelevant the charge that they are *metaphorical*. The sociological concepts defined by these axioms and their counterpart physical concepts should simply be viewed as *special cases* of more abstract concepts that are generally applicable to all fields of (naturalistic) inquiry (Catton, 1966:237). In other words, in terms of the classical Aristotelian mode of concept formation (particularly employed in the classification of plants and animals), he appears to advocate the distinction between a *general concept* (*genus proximum*) and more *specific concepts* (*differentia specifica*). He writes:

> If a force is that which produces an acceleration, then a physical force is that which accelerates material bodies in physical space, and a social force is whatever accelerates social processes. It makes sense to use the term 'force' in both contexts because both physical forces and social forces are special cases of the general concept (Catton, 1966:233-234).

Clearly, Catton elevates the concept of *force* to the level of a *genus*-concept, encompassing various *species* of this *general concept* as *special cases* (see our remarks on page 300 above). The crucial question, however, is: if *physical forces* and *social forces* are mere specifications of a general (genus-) concept of *force*, what then is the *original experiential domain* of this "general concept" of 'force'? As a *genus* concept, this general concept of force must transcend the diversity of properties we can experience, for if it does not find an original 'home' or 'seat' within the physical aspect of reality, where else can we locate it?

Alternatively, we have opted for a different approach by acknowledging *analogical* concepts. Although *energy-operation* – with the mentioned associated connotations of *cause* and *effect* (i.e. activity) – surely in a crucial sense pertains to *physical* phenomena, this does not mean that physics has a *monopoly* on the employment of the terms *energy* and *force* – or that it is forbidden for other disciplines also to use them.

If different disciplines employ the term *force* in the *same* sense, it would be difficult to maintain the view that the disciplines are indeed *different*. It seems as if the only other option is to acknowledge that each discipline uses the term 'force' in its *own peculiar way*. But there is more to this: is there a *primary* or *foundational* use of such a term? If so, then other usages simply remind us of its original or primary sense, while at the same time they differ from its original meaning.

When one speaks about the 'force' or the 'validity' of an argument (i.e. employing the terms 'force' and 'validity' in a *logical* sense) the context is evidently different from references to *physical* forces. In his examination of the "place of force in the operation of social systems" Parsons is ambiguous for on the one hand he considers the "primary reference for the concept of force" to be "an aspect of social interaction" (Parsons, 1967a:265)[1] and on the other he sometimes does specify the term 'force' by adding the qualification 'physical' in the phrase "physical force" (Parsons, 1967a:266). In the latter case one should expect that he would have said that the "primary reference for the concept of force" is to the *physical aspect*.

Likewise, in 1942 the sociologist MacIver in 1942 wrote a book that deals with "Social Causation" (see MacIver, 1964). In this work, he discusses all kinds of *social forces*. When economists deal with *economic forces*, they are not using the term 'force' in a physical sense. Similar to physical forces, both social and economic forces *cause* certain *changes* within social and economic life – an unmistakable *similarity* with physical forces and physical causes and effects (*causation*).

Yet social and economic forces are attributed to *persons* or *societal institutions* (*collectivities*) that are considered to be *accountable* in a normative sense, i.e. in the sense that they can also act in *anti-normative* ways, violating certain social or economic principles – which clearly indicates that there are important *differences* between the various *modes* of forces.

It is precisely in the employment of the term 'force' – which manifests the element of *similarity* – that the actual *difference* between the two universes of discourse becomes apparent. The alternative way in which we have captured this state of affairs is to say that, in these instances the differences are shown within the element of similarity, or that the similarities are shown in the differences. Both formulations are nothing but alternative ways to define an *analogy*. Initially, this term was used by mathematics in an arithmetic (10-6 = 6-2) and a geometric (8:4 = 4:2) sense, whereas the combination of both gave birth to a *harmonic* connotation. In the transfer of meaning from a genus domain to that of the species (and vice versa), Aristotle used the notion of analogy to account for the lingual level of metaphor formation (cf. Kluxen, 1971: 214-227).

6.14.1.2 *The interdisciplinary conceptual foundations of the term 'Umwelt'*

The attempt to reflect upon the interdisciplinary conceptual foundations of the term 'Umwelt' soon unveils comparable considerations, because it evinces connections, particularly to terms with *spatial* connotations. Within the context of a scientific study the term *space* can also take on different forms, more or less synonymous with *extension*. Mathematics struggled with this issue throughout its history, initially within Euclidean geometry, eventually in the context of analytic geometry and finally within the more general discipline of *topology*. Mathematical space – or, as we prefer to say, the original *aspectual*

1 He underscores the fact that "[W]e have stated the problem of force in the frame of reference of social interaction" (Parsons, 1967a:267).

meaning of space – provides us with the most basic (foundational) sense of *continuous* (dimensional) *extension*. This claim entails that the other (analogical) usages *presuppose* the original spatial meaning of the term *extension* (and other analogical synonymous equivalents of extension).

Catton approximates this perspective when he remarks that one does not need to use the adjective 'physical' when physical force is intended, "because physics got there first and has a prior claim on the word 'force' " (Catton, 1966:234). Yet, in order to make this remark useful, Catton needs a theory regarding analogical linkages between various aspectual domains. Then it will be possible to explain that the concept of *force* originally refers to the *physical aspect* of reality and can therefore only appear as a *physical analogy* in the *social* (and other non-physical) aspect(s). Any attempt to subsume analogical concepts under a highest genus-concept must inevitably result in the *eradication* of the uniqueness of the different aspects of reality.

An account of the first employment of the term 'Umwelt' shows that it surfaces during the 19th century and subsequently developed in close connection to the notion of *milieu*. Schröder refers to the contribution of Spitzer regarding "Milieu and Ambience," which appeared in his work on "Essays in historical semantics" (1948, 1968²), where he points out that up to the first decades of the 19th century the term 'Umwelt' emerged as a newly formed (undefined and concept-free) word predominantly related to human beings in a topographical sense. The phrases mentioned are: the "surrounding world," 'environment', "outer world" and "surrounding neighbourhood" (Schröder, 2001:99).

Haeckel introduced the term 'ecology' in 1866 (Schröder, 2001:100). During the 19th century, biologists mainly used the term 'milieu'. Related terms apparently intending the same reality are 'environment' and 'habitat'.

Since the rise of modern evolutionary biology, the interaction between living entities and their 'Umwelt' gave birth to the distinction between 'phenotype' and 'genotype'. The latter brings to expression the inherent genetically determined structure of a living entity, whereas the former captures the way in which a living entity appears on the basis of its adaptation to the environment.

The fact that the further development of the term 'Umwelt' was accompanied by a fundamental reassessment of 'object' and 'subject' is best demonstrated by exploring the biological thought of Jakob von Uexküll in more detail.

6.14.1.3 *The theory of von Uexküll*

It was Jakob von Uexküll in particular who explored the concept of an 'Umwelt' in his biological thought. His general work on *Theoretical Biology* explains his views, and they are also articulated in a very accessible way in his work "*Streifzüge durch die Umwelten von Tieren und Menschen: Bedeutungslehre.*"

The traditional scene of academic disciplines dealing with various aspects of living entities encompasses special sciences such as physics, chemistry, physiology, anatomy, the study of animal behaviour (ethology), and even sociology. Yet, as Thure von Uexküll remarks, no one of these disciplines answered to the question of how the individual living being *experiences* its environment, how it orders these entities in its own world, and how this world is connected to the worlds of other living subjects. The greater the distance between human beings and animals, the more problematic the answers provided by psychology become (Thure von Uexküll, 1970:XXIV).

Jakob von Uexküll is critical of the predominantly mechanistic view of the 'organism' as a *machine*. Portmann points out that he respects the mysterious fact that the full-grown organism presents itself as a purposeful structured whole.[1] His theoretical biology, supported by extensive biological investigations, aims at substantiating certain basic features of Kant's philosophy (cf. von Uexküll, 1973:12 ff.).

> *Comment*: In passing we have to note that the recent project of Lakoff and Johnson, in analyzing "conceptual metaphor," stems from the same Kantian legacy. They claim that the world as we *know* it does not contain any so-called "primary qualities," "because the qualities of things as we can experience and comprehend them depend crucially on our neural makeup, our bodily interactions with them, and our purposes and interests" (Lakoff & Johnson, 1999:26). However, this "crucial dependence" actually shows that objectification is always in need of an agent functioning subjectively within the aspect where objectification takes place. What is called our "neural makeup" is a crude substitute for the *subject function* human beings have in the various aspects of reality. Disclosure in the sense of an opening up of (latent) object functions always occurs within the *ontic* context of subject-object relations.

Jakob von Uexküll introduces the notion of a "functional circle" ('*Funktionskreis*') by which he intends to capture the given structural coherence between the animal body and its environment – whether or not the latter is nonliving, belongs to a fellow species member, or even to what may be dangerous. The 'features' ('*Merkmale*') of the environment are co-dependent upon the *sensory organs* (and neural structures) of animals – these structures in advance co-determine the quality and intensity of the relationship between the animal and its environment. In order to penetrate to the 'inner' side of what animals experience ('erleben'), quantitative methods fall short. Suddenly the world of colors, forms, sound tones, and odors, with their joys and pleasures appears as the worthwhile 'object' of scientific biological research (cf. Portmann, 1970:XII).

With great care, Jakob von Uexküll describes the introduction of the 'subject' in biological research. He points out that, although we do not *know* the *experiencing tone* ('*Erlebniston*') of things from the environment according to the subjective quality they have for animal subjectivity, we can deduce their *effects* from the actions of animals. By lifting out this 'toning' ('*Tönung*') of 'objects', biological thought is brought to its limits in the acknowledgement of an *inner mood*. Portmann remarks that the theory of Jakob von Uexküll regarding the peculiar *Umwelt* of every animal species indeed became a chapter

1 "Von Anfang an lenkt Uexküll den Blick des Beschauers auf die übermaschinellen Eigenschaften des Lebensstoffes, auf die geheimnisvolle Tatsache, daß im reifen Organismus ein planmäßig gefügtes Ganzes vor uns ist" (Portmann, 1970:X).

of modern biology (1970:XIII). Whereas the science of ethology and physiology treat living organisms as 'objects', the 'Umweltforschung' (*Umwelt* research) of Jakob von Uexküll focuses on animals as *subjects*.

Jakob von Uexküll constantly argues that we as human beings are never able to see, hear, smell or feel what a foreign subject sees, hears, smells or feels (Thure von Uexküll, 1970:XXV). The relationship with the environment is given in an intricate and intimate connectedness, which causes Jakob von Uexküll to consider this reality in terms of a true *totality* ('*Ganzheit*').

In addition to the 'Merkwelt' of animals, he introduces the notion of their 'Wirkwelt' (action-world). The subject is endowed with both 'Merkorgane' (organs open to 'features') and with 'Wirkorgane' (organs for taking action). The 'Merkorgan' have features inhering in the bearer ('object') of these 'marks', while through the 'Wirkorgane' and the 'Wirkwelt', the "Wirkmal-Träger" (bearer of the effects of the actions of the animal subject) is affected. The 'functional circle' ('*Funktionskreis*') encompasses the coherence, adaptation and interaction between subject and object as a purposeful whole ("*planmäßiges Ganzes*") (Jakob von Uexküll, 1970:11).

His classical example concerns the way in which an oak tree at once functions as the central reference point for different kinds of animals that respectively disclose specific parts of the tree as constitutive for their life-world ('*Umwelt*'). In Chapter 13 of '*Streifzüge*', he discusses this issue under the heading: "The same subject as object in different Umwelten" (1970:94-103). The fox explores the roots of the oak tree in order to build its hiding place, safely underneath the oak tree, which functions as its roof. The tree only acquires a protective tone for the fox, which is similar to that of the owl. For the squirrel the tree has a climb tone, for the bird that builds its nest in the tree it has a supporting tone. Jakob von Uexküll writes:

> Each Umwelt isolates out of the oak tree a particular part whose characteristics are appropriate to be the bearer both of the properties and activities of their functional circle. In the *Umwelt* of the ant the whole of the oak tree diminishes in its cracked bark which, with its valleys and heights, becomes the hunting field of the ant. … In all the various Umwelten of its various inhabitants the same oak plays a widely diverging role, sometimes with particular and then again with none of its parts. The same part can be large or small, the same wood hard and soft, it can serve as a means of shelter or attack (Von Uexküll, 1970: 98, 99, 100).

As we have mentioned, although all these 'Umwelten', in an *objective* sense, coincide with the concrete many-sidedness of the tree, and in this sense overlap, the 'Umwelt'-*experience* of these different kinds of animals do not overlap.

If we invert the perspective and look at the animal subject in order to understand more about the way in which each subject "cuts out" what is required for its own 'Umwelt', the difference between our human ambient and animal 'Umwelten' becomes more striking. The paramecium as a unicellular organism only reacts by fleeing (known as *phobotaxis*). Yet, in the absence of any specific sense organs, this reaction is sufficient to guide the animal to live in optimal conditions. Von Bertalanffy remarks that the "many things in the environment of the paramecium, algae, other infusoria, little crustaceans, mechanical obstacles and the like, are nonexistent for it. Only one stimulus is received which leads to the flight reaction" (Von Bertalanffy, 1973:241). He points out that this example demonstrates how the *organizational* and *functional plan* of a living entity is decisive in selecting what can become a 'stimulus' and a 'characteristic' to which the organism will react in a specific way:

> According to von Uexküll's expression, any organism, so to speak, selects out from the multiplicity of surrounding objects a small number of characteristics to which it reacts and whose ensemble forms its 'ambient' (*Umwelt*). All the rest is nonexistent for that particular organism. Every animal is surrounded, as by a soap bubble, by its specific ambients, replenished by those characteristics which are amenable to it. If, reconstructing an animal's ambience, we enter this soap bubble, the world is profoundly changed: Many characteristics disappear, others arise, and a completely new world is found (Von Bertalanffy, 1973:241).

Jakob von Uexküll portrays the world of a tick, which reduces its environment in an astonishing way. It hangs motionless at the edge of a branch in the bushes and by virtue of its nature, it is capable to explore the possibility of falling onto a passing mammal. No stimulus from the total environment is registered. But then suddenly a mammal approaches and the tick requires blood for its offspring. Of course it may happen that the tick needs to wait for quite a time before a mammal passes. But it is equipped to cope with this obstacle, for ticks have been reported to subsist for 18 years without food (cf. Jakob von Uexküll, 1970:13-14)! Von Uexküll adds that there is a difference between the 'Umwelt' and the 'environment' ('Umgebung') of an animal. Whereas the conditions in the former are optimal, the picture may be more negative ('pessimal') in the latter. Yet, if this state of affairs did not prevail, the optimality of a specific 'Umwelt' may overpower other species (cf. Jakob von Uexküll, 1970:13, note 3). And then something quite remarkable happens: of all the effects that proceed from the mammal body, only three, and in a determined succession, are turned into stimuli. From the majestic world surrounding the tick, three stimuli light up like light signals in the dark, and serve as a guide for the tick to reach its goal with certainty. Its world is reduced to three characteristics and three actions (*Merkmalen* and *Wirkmalen*) (Jakob von Uexküll, 1970:12-13).

Implicit in Jakob von Uexküll's account of the 'Umwelt' of distinct animals, we have to recognize a new assessment of the importance of *subjects* in nature. Although the notion of a subject could be understood in the sense of *activities*, one cannot do without the correlated concept of a *law* to which entities in nature are *subjected*. Physical entities are subject to *physical laws*, living entities to *biotical laws*, while sentient creatures, such as animals, are subject to their *typical structural laws*.

6.14.1.4 *The original spatial foundation of a theory of an ambient ('Umwelt')*

We have argued that space, in its original mathematical sense is characterized by *continuous extension*, and that the arithmeticistic claim of modern set theory, namely that it has succeeded in fully arithmetizing continuity, entails an inherent circularity, since it had to 'borrow' from space the order of "at once" (implicit in the notion of the *at once infinite*) and the *whole-parts relation*. Although the Greek atomists, Leucippus and Democritus, believed that there are indeed last indivisible units, which they called *atoms*, modern philosophy since Descartes switched to the conviction that physical space is both *continuous* and *infinitely divisible* – a view that already has been found in the thought of Anaxagoras and Aristotle. However, by the end of the 19th and the beginning of the 20th century, it was necessary to distinguish between *mathematical* space and *physical* space. Whereas the former – in a purely abstract and functional perspective – is at once continuous and infinitely divisible, physical space is neither continuous nor infinitely divisible (see our earlier remarks on page 221). Being bound to the quantum structure of energy, physical space cannot be subdivided *ad infinitum*. Energy quanta indeed represent the limit of the divisibility of energy. In his theory of relativity, Einstein also has a clear insight into how the original meaning of space (simultaneity) is relativized. Neither the spatial point at which something occurs, nor the temporal moment in which it happens has a physical reality. It is only the *event* itself that has physical reality (Einstein, 1982:33).

6.14.1.5 *The irreducibility of the spatial whole-parts relation*

Although the meaning of space comes to expression in its coherence with other aspects, we have seen that its core characteristic, *continuous extension*, is indefinable. The same applies to our concept of number, where the intuition of a *distinct multiplicity* could be paraphrased, but never 'defined' in a non-circular way. This explains why we noted that modern axiomatic set theory, for example the system of Zermelo-Fraenkel, had to proceed from the acceptance of the *undefined* term "member of" (which is equivalent to the notion of a 'set' – see above page 171 where Kurt Gödel is quoted in connection with the circularity entailed in the attempt to 'define' a set).

Furthermore, we have also established that the term 'totality' ('Ganzheit' / 'Totalität') is merely synonymous with continuity (see page 87). Whatever is continuously extended, such as a straight line, *hangs together* in the sense that all its parts are *connected*. With disconnected parts, it would not be continuous. But if *all* the *parts* are present, then the line is given as a *whole*, in its *totality*. The *whole-parts* relation is therefore derived from the core meaning of *continuous extension*.

In passing we may return to our above-mentioned note (see page 61) that it is precisely this totality feature of continuity that forms an obstacle to the attempt to arithmetize continuity completely. Paul Bernays aptly writes: "The property of being a totality undeniably belongs to the geometric idea of the continuum. And it is this characteristic which resists a complete arithmetization of the continuum" (see page 61 above)."

In a discussion of the foundational issues of mathematics, Hermann Weyl also alludes to the irreducibility of continuity. In following the Dutch intuitionist, L.E.J. Brouwer, it is Weyl's conviction that mathematics should acknowledge the crucial whole-parts relation whenever continuity is treated. He says that the fact that it has parts is a fundamental property of the continuum. The 'atomistic' understanding seriously violates this intuition, as emphasized by Brouwer.[1] It belongs to the essence of every continuum that it permits an infinite division: that every part allows for an unlimited number of further subdivisions (Weyl, 1921:77 – see page 88 above). Mathematical analysis that tries to do without this crucial place of the whole-parts relation when it treats issues of continuity had to employ the notion of an *environment* (*Umgebung*).

> In order to account for the continuous coherence of points, contemporary analysis, which indeed broke apart the continuum into isolated points, had to take recourse to the environment concept.[2]

In spite of the fact that mathematics did not explore a third option – as opposed to both arithmeticism and geometricism, which dominated the history of mathematics – namely to acknowledge number and space in their uniqueness (irreducibility) and mutual coherence, it proves worthwhile to use this insight in the characterization of *atomism* and *holism* (compare our initial characterization of *atomism* and *holism* on page 60).

6.14.1.6 *The one-sidedness of opposing 'element' and 'totality' (Ganzheit)*

Jakob von Uexküll constantly argued against the *atomistic* assumption of the prevailing mechanistic trend in biology. Since Democritus, atomism acquired two senses, one indicating the attempt to explain the material world in terms of last indivisible *elements* ('atoms'), and the other broader conception using the term to designate different forms of *pluralism*. In the case of human society the last units were seen as the *individuals*. Since 1825, Saint-Simon and his followers (amongst them Comte) employed the term *individualism* in order to capture the social philosophy of the 18th century as a whole – the view in which society was broken apart into isolated individuals.

In opposition to both mechanistic monism and vitalistic dualism as biological theories, the term *holism* was introduced by J.C. Smuts in 1926. In this narrow sense, it aimed at a dialectical synthesis that can do justice to the supposedly highest concrete totality (in the case of Adolf Meyer: *Ganzheit* – cf. Meyer, 1964). An expanded connotation is given to holism when it is used in the sense of a *universalism* which, in opposition to (sociological) individual-

1 "*Daß es Teile* hat, ist aber die Grundeigenschaft des Kontinuums; und so macht die Brouwersche Theorie (im Einklang mit der Anschauung, gegen welche der heutige 'Atomismus' so arg verstößt) dieses Verhältnis zur Grundlage für die mathematische Behandlung des Kontinuums" (Weyl, 1921:77).

2 "Um den stetigen Zusammenhang der Punkte wiederzugeben, nahm die bisherige Analysis, da sie ja das Kontinuum in eine Menge isolierter Punkte zerschlagen hatte, ihre Zuflucht zu dem Umgebungsbegriff" (Weyl, 1921:77).

ism, wants to account for the meaningful coherence and mutuality within societal institutions, i.e. for wholeness / totality as an essential trait of societal collectivities (as introduced by the German philosopher, sociologist, and economist, Othmar Spann, in the twenties of the previous century and also applied in his nature philosophy – cf. Spann, 1937). The German biologist, Richard Woltereck, a contemporary of Jakob von Uexküll – in his "Ontologie des Lebendigen" – and Othmar Spann also developed a position that emphasizes *wholeness* and 'Ganzheit' (see Woltereck, 1940 and his reference to Spann on page 476, note 13).

From a theoretical point of view, one can say that atomistic (individualistic) stances in philosophy over-emphasize the explanatory potential of the quantitative mode of reality (or its analogies in other modalities), whereas holistic (universalistic) orientations do the same with the spatial mode and its inherent whole-parts relation or analogies in other aspects of reality.

Comment:

Organicism as a theoretical position does not necessarily need to be *holistic*, because one can just as well explore the *numerical analogy* in the biotic aspect instead of the spatial whole-parts relation. Herbert Spencer, for example, advocates an organicistic view, but clings to an individualistic perspective. Spencer repeatedly speaks about an *organic aggregate*, which is basically the same as a *social aggregate*. But he loads his conception of a biotic organism with an atomistic (individualistic or aggregate) perspective in order to be consistent with his individualistic liberal political convictions. Compare in this regard Spencer's statement concerning a decrease of authoritative control in society: "A more pronounced individualism, instead of a more pronounced nationalism, is its ideal" (Spencer, 1968:22).

Living entities integrate the multiplicity of their vital functions in order to persist. They function in a *unified* way, thus exhibiting a function within the *quantitative aspect* of our experiential horizon. Note the subtle difference in the distinction between the concrete function of a living entity within the numerical mode of reality and the inter-functional connections between the biotical and the arithmetical modes as expressed in the numerical analogies within the biotical aspect. A living entity is a *biotical unit in the multiplicity* of its vital functions. This state of affairs is well reflected in the uniqueness of biotical *growth* phenomena, for if a living entity does not continuously succeed in *integrating* its vital functions, it *disintegrates* and *dies*.

Within the discipline of physics, we encounter the concept of *mass*. It represents a numerical analogy within the physical aspect, since it relates to a "quantity of matter." Similarly, a biologist may speak about *biomass*. Jones, for example, refers to the fact that plants constitute 99% of the world's biomass, while fungi are estimated to have *twice* the total biomass of animals (Jones, 1998:54).

But living entities certainly also function within the *spatial mode* of reality – they are spatial subjects and they are related to other (surrounding) spatial subjects (that is, those in their environment sharing in the mode of being spatially extended). Even the smallest independent "unit of life" – the living cell – exhibits a remarkable constant proportionality in respect of the ratio between the nucleus and the cytoplasm.[1] The size and shape of living entities (a manifestation of their concrete function within the spatial mode) provide a foundation for the inner coherence between the spatial and biotic aspect, captured in the well-known expression, *bio-milieu*, and the equally familiar concept, *biosphere*.

The way in which living entities are adapted to their environment received a peculiar designation in the *holistic* biology of Adolf Meyer. With the aid of extensive empirical information, he formulates a "basic typological law":

> There is no group of existing organisms belonging to any taxonomical category of the Natural System, whose members possess all group characters in their most primitive or in their most progressive phases only. Rather are primitive, intermediate and progressive character phases thus combined with each other in each real member of a group that an organismic holism suited for living in any real existing ecological biotope results from it. Forms which possess all their morphological characters in their primitive or in their progressive phases only are neither living holisms nor suited for existence in ecological biotopes and are, therefore, but purely ideal constructions (Meyer, 1964: 59-60).

Yet one should question the mode of speech advocated by Jakob von Uexküll when he attributes the feature of a true *totality* (*Ganzheit*) to the subject and the object encompassing nature of an 'Umwelt'. The various entities involved in these interrelations are indeed, in their own right, genuine totalities (wholes) that escape any atomistic attempt to reduce them to last constitutive elements ('atoms'). But in their interrelations, these entities do not terminate their wholeness, however intimate their interwovenness with other entities may be. Dooyeweerd employed the term 'enkapsis' (borrowed by Theodor Haering from the German anatomist Heidenhain – as mentioned on page 140 above) in order to explain the interlacement intended in these instances. This term intends to account for the maintenance of the *internal structural properties* of an entity, even when it is intertwined with other entities – like the atoms bound through a chemical bonding into the enkaptical whole of a molecule. A similar interlacement is present when the relationship between physico-chemical constituents are explained as *foundational* to the typical *vital* functions of living entities. A living entity necessarily functions on the basis of material configurations (largely macro-molecular in nature), but the phrase *living organism* should actually be reserved for the biotically characterized dimension of a living entity that is alive through and through – just think about the *organelle* within the organism of a living cell: lysosomes, mitochondria, ribosomes, and so on, are all *alive*. The same could not be said in respect of the living entity as a whole, because although they are also present

1 Hertwig speaks about the "nucleoplasmic index," which is equal to the volume of the nucleus divided by the volume obtained when the volume of the nucleus is subtracted from the volume of the cell (cf. West & Todd, 1962:208).

within the total form in which living entities exist, the atoms, molecules and macromolecules as such (in which the functioning of these *organelles* are founded), are *not* alive.[1]

6.14.1.7 *Objectification in living and sentient creatures*

Compared to the mass of 'unbelebte' matter (as Von Weizsäcker prefers to refer to physical entities – cf. 1993:32)[2] in the universe, living entities indeed appear to be a negligible layer at most. Yet living entities ensure that the earth is kept in a dynamic state, which is not in chemical equilibrium but in a *steady state*. Although biologists tend to speak of *adaptation* – as if the environment fully determines what happens to biotical subjects – the guiding and qualifying *biotic activities* involved should be kept in mind, for otherwise one may lose sight of the fact that the organic function of living entities is not merely dependent upon environmental conditions – the legitimate side of the meaning of 'adaptation' – for this biotic function plays an active and guiding role in respect of the inorganic surroundings of living entities. For example, the splitting up and reconstituting of the water on earth, mentioned above (see page 55), can only occur when living entities absorb water into their life cycle. Through their enkaptic bonding within living entities, water molecules are *structurally objectified* within the biotic aspect of reality, i.e. they function in service of the *biotic needs* of living entities.

Similarly, in the life of animals, the surrounding physical environment becomes partially objectified into the 'Umwelt' through and under the guidance of *subjective animal functioning*. Objectification as such is always the activity of a *subject*.

What should also be kept in mind is that Jakob von Uexküll's 'Umweltlehre' finds a solid basis in the fact that each different species has its own (distinct) 'Umwelt'. However, his attempt to expand its scope in order to incorporate human beings does not withstand the scrutiny of critical assessment. Analogous to the way in which he expanded his 'Umwelt' theory to human experience, we may consider the way in which the oak tree is opened up through human acts of objectification (cf. von Uexküll, 1970:94 ff.). The human experience of the oak tree transcends the natural aspects of reality to which animal experience is restricted. Animals experience reality exclusively by means of their natural inclination, directed at that which is *physically*, *biotically* and *sensitively* important to them. As noted in Chapter 4, they experience reality in terms of that which is accessible and not accessible, edible and inedible, in terms of same sex and opposite sex, and in terms of what is comforting and alarming (or endangering).

The human experience of a tree opens up vistas beyond the grasp of animal capabilities. The natural scientist sees the tree as an object of analytical study, the hiker as something with a particular aesthetic attraction, the criminal as a hiding place from the law, the woodworker as material from which to make

1 Cf. Dooyeweerd (1997-III:634 ff.) and Chapter 7 below.

2 Max Planck also speaks of the "unbelebte Natur" (Planck, 1973:285).

furniture, and so forth. This human experiential perspective with its rich variety, is linked to the cultural calling which enables a person to be variably settled in any environment by means of cultural formation. But in the case of humans, none of these domains of objectifiability is *exclusive*. Human beings share them, and precisely in doing this, they differ fundamentally from animals and their non-overlapping 'Umwelten'. Portmann understands this shortcoming by mentioning that Jakob von Uexküll neglects the fact that all these different 'Weltansichten' (world perspectives) share a communal species world, which enables a mutual understanding, as well as interaction on the basis of opposing views. The communal world in which humans live constitutes a shared domain that is not broken apart even by the most severe differences in potential or cultural traditions. Poetry thrives on differences in world perspectives, but depends on a last horizon of mutual understanding. The term 'Umwelt' should be reserved for the separation of the different worlds of animal species, but it should not be applied to different ways in which human beings view the world.

Gostonyi attempts to account for the difference between animal 'Umwelten' and the domain of human experience in terms of the opposition between *constancy* and *dynamics* (*change*). He holds that the 'Umwelt' of animals is largely stabilized through processes that are directed by instincts that cause them to be by and large *unchanging*. By contrast, as an effect of the 'Instinkt-Ungebundenheit' of human beings (the fact that humans are not determined by their instincts), the 'world' of humans is subject to continuous change, that is, it is constantly broadened and transformed. The animal is borne as a part of its 'Umwelt', while the human being 'conquers' its 'world.'[1]

The term 'Umwelt' emerged and eventually found application mainly within the domain of biology and the discipline of animal behaviour. But given its original *spatial* connotation, it turned out that an account of its meaning had to consider the interconnectedness of terminology transcending the confines of a single discipline. This interdisciplinary insight opens the way for the systematic considerations that we have developed in confrontation with the relevant historical perspectives.

Although we have argued that animal subjectivity is presupposed in every animal process of *objectification*, our subsequent argument had to relativize the "Copernican turn" by reverting to the acceptance of *ontically given* modal aspects of reality, which ultimately *condition* every kind of animal (and human) subjectivity.

The merit of Jakob von Uexküll's 'Umweltlehre' is primarily in its aim to give animal subjectivity its due place. His understanding of the 'Funk-

1 "Der entscheidende Unterschied zwischen der menschlichen 'Welt' und der Umwelt der Tiere ist – vereinfacht formuliert – der, daß letztere durch weitgehend gleichbleibende, weil von Instinkten gesteuerte Verhaltensabläufe stabilisiert und somit im großen und ganzen *unveränderlich* ist, während die 'Welt' des Menschen – dank seiner Entwicklungsfähigkeit und seiner 'Instinkt-Ungebundenheit' – ständiger Veränderung, Wandlung, d. h. Erweiterung und Umstrukturierung unterworfen ist. Das Tier wird in seine Umwelt hineingeboren, der Mensch 'erobert' sich seine 'Welt' " (Gostonyi, 1976:902).

tionskreis', which highlights the interdependence between animal subjectivity and whatever is objectified by animals into their respective 'Umwelten', indeed unveils the constitutive domain of *objectification* underlying the very existence of living entities (in a biotic sense) and animals (as sentient creatures). Although this subjective capacity to *objectify* also inherently belongs to human beings, we had to point out that applying the 'Umwelt' concept in Jakob von Uexküll's sense to this domain as well, neglects the mutuality of human understanding.

In order to come to a more articulated systematic account of the original and analogical meaning of spatial (and numerical) terms, we had to address related issues such as the uniqueness and mutual coherence between number and space, and the analogical concepts found in other disciplines, as being based upon the inter-modal connections between various modal aspects of reality.

These ontic modalities turn out not to be merely *functional conditions* for concretely existing entities, since they also serve as *modes of explanation* within different, but mutually cohering scientific universes of discourse. Through the inter-modal coherence between the spatial and other aspects, we are fully within our epistemic rights to employ (analogical) concepts of space in diverse disciplines. Although we have highlighted some of these analogical phrases – such as *physical* space, *biotical space* (*bio-sphere*, *bio-milieu* and so on), and *sensitive space* (the content given by Jakob von Uexküll to his notion of 'Umwelt') – there are still many others not receiving attention. To mention a few unexplored contexts, we refer to *logical space* (Rudolph Carnap wrote a work in which he reduced space to so-called *formal space* – see Carnap 1922), *lingual space* (also encompassing theories regarding the *semantic domains* of words), *social space* (think about notions of *social stratification* and *social distance*), and *jural space* (intricate questions regarding the *location* of a legal fact).

6.15 Knowledge and the logical subject-object relation

The universal modal structure of the logical aspect finds its core meaning in *analysis* and we have seen that analysis rests on the two 'legs' of *identifying* and *distinguishing*. What is presupposed in analysis, is a *given* cosmic diversity (of aspects and entities) that can be identified and distinguished.

Something that can be *named* is objectified within the sign mode when a name is as-*sign*-ed to it. Something that is *perceivable* is objectified within the sensory mode when it is *observed*. Likewise, when what is *identifiable* and *distinguishable* is indeed identified and distinguished, then logical-analytical objectification takes place. In the subjective act of logical objectification, *concepts* are borne, but we have to remember that the nature of a concept not only involves that a multiplicity of characteristics (modal or typical) are *united* (see page 174), but also presupposes that these features are *universal* (see pages 148, 169).

A concept is acquired through the "bringing together" (the 'synthesis') of a multiplicity of identified hall-marks. This entails that *synthesis* cannot be opposed to *analysis* as normally done. Uniting identified features is one of the key elements of a concept. Consequently, the counterpart of synthesis (*identifying*) is not analysis, but rather *distinguishing*.

We have shown that *abstraction* and *analysis* are synonymous insofar as "lifting out" and 'disregarding' are equivalent to identifying and distinguishing. Furthermore, since *modal abstraction* or *modal analysis* constitutes the distinctive feature of theoretical activities (see Chapter 2, page 48 ff.), theoretical knowledge is obtained whenever any dimension of reality[1] is analyzed through the gateways of identified and distinguished aspects. The delineation of the field of investigation of a (modally delimited) academic discipline is only possible through modal abstraction, i.e. through the *identification* (lifting out) of that particular aspect while distinguishing it from other aspects. The identification of an aspect, is exactly what it says – the identification of (the structural features of) one aspect.

The discipline of biology, for example, owes its existence to the ontic reality (givenness) of the biotic aspect and our human logical ability to focus our theoretical attention on the phenomena evinced within this aspect. Since the features of a modal aspect in general (and a specific aspect in particular) are united in the concept thus acquired, the only meaning of 'synthesis' in this context relates to the "bringing-together" of the characteristics of *one particular aspect*, as they are *logically objectified* within the analytical aspect. It does not entail a synthesis between the logical aspect of our act of thought, and a non-logical aspect.

What is known as the opposition between a *correspondence theory of truth* and a *coherence theory of truth* actually one-sidedly emphasizes either the factual object side or factual subject side of the logical analytical aspect. In its modal universality, this aspect embraces whatever there is and therefore underlies the ability we have to logically objectify whatever is identifiable and distinguishable. Although the conceptual framework within which knowledge is embedded co-determines our knowledge acquisition, it is always at once related to what is logically objectified (compare the difference between the connotation and denotation of a word or sentence occurring within the lingual subject-object relation).

If it was not the case that the entire universe had either a subject- or an object function within every aspect, human beings would have been opposed to things with which nothing is shared. Stafleu therefore correctly holds that "human thought is subject to the same kind of laws as the creation as a whole; this is even a condition for the achievement of knowledge" (Stafleu, 1999:100).

1 This includes the dimensions of time, modal aspects, and multi-aspectual entities as well as the central direction-giving dimension, where human beings surrender to their ultimate commitments. In respect of all dimensions, theoretical thought has its constitutive foundation in *conceptual knowledge* and its regulative foundation in *concept-transcending knowledge* (idea-knowledge).

6.15.1 *The Kantian background of Dooyeweerd's idea of a Gegenstand-relation*

In what is known as his transcendental critique of theoretical thought, Dooyeweerd aims at gaining an insight into the transcendent and transcendental pre-requisites of theoretical thought – in other words, he aims at laying bare those universal conditions that make possible theoretical thought as such. For this purpose, he distinguishes between our non-theoretical experience of reality, embedded in and characterized by the subject-object relation, and the "theoretical attitude of thought" that is stamped by what he calls the *Gegenstand*-relation.

In his epistemology, Immanuel Kant proceeds from the *isolation* of sensibility (Kant, 1787:33) that is strictly *separated* from our pure understanding (Kant, 1787:89). He never succeeded in giving a satisfactory account of their 'synthesis'. Dooyeweerd thought that he could rectify Kant's view by expanding the scope of the *Gegenstand* to all non-logical aspects of reality. Unfortunately he continued something essentially Kantian in his belief that theoretical knowledge originates in the theoretical synthesis between the logical aspect of our real act of thought and some or other non-logical aspect (cf. Dooyeweerd, 1997-I:39).[1]

This idea lies at the basis of the first two problems formulated in his transcendental critique. The first concerns the question of what is *abstracted* (subtracted) in the antithetic attitude and "how is this *abstraction* (my emphasis – DS) possible?" (Dooyeweerd, 1997-I:41).[2] The second question poses the problem: "From what standpoint can we reunite synthetically the logical and the non-logical aspects of experience which were set apart in opposition to each other in the theoretical antithesis?" (Dooyeweerd, 1997-I:45). Clearly, *abstraction* and *synthesis* are opposed to each other by Dooyeweerd. But we have seen that abstraction rests on the same two legs as analysis, namely *lifting out* (identifying) and *disregarding* (distinguishing) (see Figure 6 on page 151), and that *synthesis* (identification), being just one leg of analysis, is not opposed to analysis itself but to the *other* leg of analysis, namely *distinguishing*.

Opposing an aspect to the logical aspect, setting aspects apart in the "theoretical antithesis" or dinstinguish between the logical aspect (of the real act of thought) and an "intentionally abstracted non-logical Gegenstand," entails the *identification* of such an aspect. But identification is equivalent to the acquisition of a concept, of uniting or 'synthesizing' a multiplicity of (logically objectified) universal features. Therefore setting aspects apart or opposing them, i.e. identifying and distinguishing them, is exactly what the acquisition of a concept (as part of analysis) is all about. Nonetheless, the formulation of the *Gegenstand*-relation is based upon the assumption that in spite of the fact

1 For a detailed analysis of this issue, see Strauss (1984).

2 Dooyeweerd alternatively uses the terms 'abstraction' and 'subtraction' – see Dooyeweerd 1958:139.

that the non-logical pole has been 'opposed' to the logical pole of this antithetic relation it has not yet been conceptualized, for Dooyeweerd explicitly states that we "must proceed from the logical *antithesis* to the theoretical *synthesis* between logical and non-logical aspects, if a logical concept of the non-logical 'Gegenstand' is to be possible" (Dooyeweerd, 1997-I:44). From my first reading of the transcendental critique, I was unable to understand how it is possible to *think of* (!) an opposed non-logical aspect (opposed to the logical aspect of our thought-act) without (conceptually) *knowing* what is opposed to the logical aspect! Only after a more serious analysis of the meaning of analysis I eventually not only realized that identifying and distinguishing constitutes the meaning of analysis, but also that it is constitutive for the acquisition of concepts as such – i.e. for bringing together (synthesizing) various universal features of whatever is conceived. Moreover, in the first presentation of his transcendental critique (in 1941) Dooyeweerd actually proceeds in a way that comes very close to his own view of logical objectification (further explored in my alternative explanation below). He writes:

> In the theoretical, scientific attitude of thought, by contrast, logical analysis is directed first of all upon the modal aspects themselves, which are theoretically pried asunder into a theoretical discontinuity by means of this analysis and abstracted from their given, continuous, systatic coherence (Dooyeweerd, 1941:5).

In NC-I:39 it is said that "we oppose" the logical to the non-logical aspects, but initially (in 1941) it is "logical analysis" that is directed at the modal aspects as they are "pried asunder" by "means of this analysis." So here it is not the '*we*' performing the distinction (analysis) between the logical and the non-logical aspects, i.e. that oppose them, but logical analysis itself.

It should be kept in mind that Dooyeweerd understands logical analysis merely in the sense of distinguishing (the traditional meaning of *analyzing*), and therefore not yet in the sense of resting on the two 'legs' of identification and distinguishing – and for that reason he also does not acknowledge the fact that analysis (identification and distinguishing) is synonymous to abstraction (in the sense of lifting out and disregarding) (see pages 14, 151). However, he does add something to his view of "analytical distinction," namely the remark: "distinction in the sense of setting apart what is given together" (see Dooyeweerd, NC-I:39, note 2). Surely, identifying and distinguishing do not eliminate the inter-modal coherence prevailing between the aspects that are identified and distinguished.

Compare in this regard his remark:

> The 'Gegenstand of understanding' arises by means of a conscious *setting apart* of systatic reality, by performing an *analytic epochè to the continuity of the cosmic order of time*, by which the non-logical meaning-structures are fixated in logical discontinuity. This deepening of systatic, naive thinking, along the road of "being placed in opposition," can be explained from the analytic structure of the logical meaning-function itself. Only through logical discrimi-

> nation can the analytic meaning-structure reveal its universality in its own sphere! (Dooyeweerd, 1931:103)

What is obtainable "through logical discrimination" conforms to the meaning given to *analysis* in NC-I (page 39, note 2): "the aspect of analytical distinction; distinction in the sense of setting apart what is given together." Distinctions are also drawn in naïve experience – Dooyeweerd only believes that naïve experience does not enter into the making of *modal* distinctions, for the latter is reserved for the *Gegenstand*-relation.

At most Dooyeweerd would say that in naïve experience we have "an implicit awareness of" the aspects but no explicit concept (Dooyeweerd, 1996:9). Keeping in mind that *aspect* and *function* are synonyms his explanation of naive experience as characterized by subject-object relations is inconsistent. He says: "Naive experience makes a distinction between subject-functions and object-functions" (Dooyeweerd, 1996:8)! If naïve experience does not have a concept of functions (or aspects), how does it distinguish between subject and object *functions*!? Dooyeweerd also says: "Furthermore, these subject-object relations are grasped, in naive experience, as structural relations within empirical reality itself" (Dooyeweerd, 1996:8). How it is possible to distinguish between subject functions and object functions in naïve experience without being able to distinguish these functions *as* functions (aspects)?!

In other words, Dooyeweerd holds that the *analytical function* as such can perform the *setting apart*, but it cannot accomplish the *re-uniting*, the bringing together (the inter-modal synthesis), for then we suddenly require a supra-modal central point of reference (see Dooyeweerd, 1996:15).

Opposing modal aspects in the *Gegenstand*-relation is accomplished by our logical function in its analytic (for Dooyeweerd: distinguishing) activity, without the need for a supra-modal focal point, whereas the theoretical intermodal meaning-synthesis can only bring together what was analytically set apart in a more-than-analytical way, in a way transcending analysis!

At the same time this hidden circularity in Dooyeweerd's argument, according to which something already conceptualized has not yet been conceptualized, explains why the entire account of the *Gegenstand*-relation is burdened by internal inconsistencies.[1] I summarize them against the background of the more extentive account in my PhD, keeping in mind that in general Dooyeweerd considers theoretical thought to be characterized by an opposition between the logical thought-aspect and the non-logical aspects of reality.[2]

1 For the sake of clarity we repeat: the logical-analytical aspect does not merely concern *distinguishing* since it embraces *both* identification and distinguishing. Moreover, *identification* is exactly what *conceptualizing* means. Therefore, whenever a non-logical aspect is opposed to the logical (or any other aspect) it has already been identified (conceived) – and this makes redundant the argument that only afterwards we have to proceed (the second problem of the transcendental critique) to obtaining a *concept* of it!

2 See Dooyeweerd, 1931: 102-103; Dooyeweerd, 1935:365; Dooyeweerd, 1941:5-6; Dooyeweerd, RS-II:82-83; Dooyeweerd, 1962:19; Dooyeweerd, 1948:30-31; Dooyeweerd, 1996: 6-7; Dooyeweerd, 2002:28.

i) Dooyeweerd talks of an antithetical relation between the actual logical thought function and its intentionally abstracted non-logical *Gegenstand*, and about an antithetical relation between a specific non-logical aspect *x* and the remaining non-logical aspects *y* (although in the latter case no inter-modal synthesis is possible).

ii) The restriction of the Gegenstand to the non-logical aspects is contradicted by statements in which Dooyeweerd explicitly speaks of the modal structure of the logical aspect *itself* as being a *Gegenstand* of our "actual logical function" ("of theoretical analysis").[1]

iii) Relating to i) and as a consequence of ii), we must conclude that an antithetical relation between the abstracted modal structure of the logical aspect and the abstracted non-logical aspects is also possible.

iv) (a) Is the theoretical concept of the modal structure of the logical aspect characterized as inter-modal or intra-modal? If intra-modal, then the universal validity of the theory of an inter-modal meaning-synthesis is disqualified. Usually Dooyeweerd suggests that the logical law sphere can only be theoretically analyzed by opposing it to all non-logical aspects of reality, i.e. by means of an inter-modal synthesis with all the non-logical aspects (cf. NC-II:461-462). But then the same applies to all the non-logical aspects. The juridical aspect, for example, can only be analyzed by opposing it to all non-juridical aspects (including the logical one!) which are grasped in an inter-modal synthesis of meaning. Consequently, the logical aspect cannot be analyzed in an inter-modal synthesis, if and only if it can be analyzed in one! (b) The other possibility implicit in the conception of non-logical Gegenstand-aspects, is that *only* non-logical states of affairs are (theoretical-)logically analyzable, i.e. identifiable and distinguishable. For example, the judgment of identity: *legal is legal*, is only possible due to the non-logical nature of this juridical state of affairs. The immanent criticism applicable to this explanation is obvious: Simply consider the identity judgment: *logical is logical*. If all identification and distinction are always directed only to non-logical realities, it stands to reason that this identity judgment also pertains to non-logical states of affairs. Consequently, the identity judgment *logical is logical* is valid if and only if *logical is non-logical*. This implicit possibility actually came up in a discussion I had with Dooyeweerd and Hommes in June, 1973. On this occasion, I discussed my problems with and criticism of the *Gegenstand*-relation with them on the basis of a 14-page succinct statement of my considerations. When I raised the problems connected to the '*Gegenstand*-relation'-account of analyzing the logical mode itself, Hommes argued that our logical function is always directed towards *non-logical* states of affairs. This constitutes the antecedent of the conditional statement: "*If* only non-logical states of affairs can be (theoretically) analyzed (identified and distinguished in an "inter-modal meaning-synthesis") *then* whatever is logical in nature cannot be (theoretically) analyzed." Since Dooyeweerd in fact does

1 Respectively Dooyeweerd, NC-I:40, footnote 1, and NC-II:463.

want to maintain that one can arrive at a theoretical analysis of the logical aspect (see Dooyeweerd, 1997-I;40 note; 1997-II:463), the consequent may alternatively be phrased as follows: "*If* only non-logical states of affairs can be (theoretically) analyzed (identified and distinguished), *then* the logical can be (theoretically) analyzed if and only if it is non-logical in nature." He mentioned the example of the identity judgment legal is legal ('recht is recht'). My immediate reaction was to formulate the mentioned formal contradiction. Neither Dooyeweerd nor Hommes succeeded in providing a tenable defense of the *Gegenstand*-relation against this method of immanent criticism. While Hommes was still trying to get out of the impasse, Dooyeweerd frankly admitted that I may have laid my finger on a problem not solved by his own theory.

v) The sharp distinction that Dooyeweerd upholds between the *Gegenstand*-relation and the logical subject-object relation is contradicted by himself, where he states that the pre- and post-logical aspects of reality can be *logically objectified* (cf. NC, II, pp. 390, 472).

vi) The actual and inter-modal nature of the intuition cannot, as Dooyeweerd assumes, establish the inter-modal meaning-synthesis between the actual logical thought-function and the abstracted non-logical aspects for the simple reason that it cannot actually function in an abstraction.

vii) The question: how is it possible to obtain an implicit knowledge of the modal aspects in our naive pre-theoretical experience (by means of systatical logical objectification) *without* the need for an inter-modal synthesis, is not answered by Dooyeweerd.

viii) When the cosmic meaning-systasis is acknowledged, the first basic problem of a critical theory of knowledge is not given with the question regarding the possibility of an inter-modal meaning-synthesis, but in the question: how is it possible to arrive at an analytical awareness of the modal and structural meaning-diversity in our non-theoretic experience?

What should be kept in mind is that Kant and neo-Kantianism presented Dooyeweerd with the pseudo-problem of an opposition between the logical and the non-logical – Dooyeweerd merely expanded the scope of the latter to more than the *sensory* mode. Paul Natorp actually left open the option taken by Dooyeweerd – just consider his statement from a work frequently quoted by Dooyeweerd:

> It is now not any longer possible to speak about a multiplicity of sensibility, which should synthetically be united by our thought in an act of knowledge performed only afterwards. How will it be possible for the multiplicity of our senses to be combined in the unity of a concept? In this case, the basic elements, unity and multiplicity, would not both have been thought-moments; consequently, also their combination would not have been performable within pure thought as such. Much rather, in order to achieve this synthetic unity, our understanding should first of all unite itself with something external to it, something foreign, 'sensibility'. Alternatively, there must be a third instance, transcending both sensibility and understanding, which can perform this syn-

> thesis. However, in reality this is not the case, since in one and the same thought-act multiplicity originates simultaneous with the consciousness of unity, as both thought-structured modes of determination which are in their thought-character strictly connected with each other (Natorp, 1921:48).[1]

The alternative of "a third instance" "transcending both sensibility and understanding" was indeed explored by Dooyeweerd in his idea of the necessity of a *supra-modal* (central religious) starting point for the supposed inter-modal meaning-synthesis. That Dooyeweerd's acceptance of a supra-modal starting point, residing in the central religious dimension of reality, is not unique, is clearly seen from the fact that Alexander Varga von Kibéd also defended a similar solution. Von Kibéd is strongly influenced by neo-Kantianism (the Baden school) and especially by Nicolai Hartmann. The Kantian opposition of understanding and sensibility, and the neo-Kantian opposition of value and reality (thought and being) influenced his formulation of the unbridgeable gulf between truth and reality. The principle of identity forbids the connection or unification of truth and reality (Von Kibéd, 1979:37). However, according to him, this synthesis is possible in the "metaphysical-religious sphere": "The solution of the unsolvable problems, recovering the lost unity of our being and our thought is only to be expected from this deepest layer" (Von Kibéd, 1979:38).

Dooyeweerd criticized Kant for identifying the human selfhood with the 'cogito', the "I think," by emphasizing its transcendence in respect of all modal functions. Kant's focus on the universal structure of logicality embodies his *rationalism*. Dooyeweerd did something similar in his formulation of the *Gegenstand*-relation, because he identified the full modal structure of the logical-analytical aspect with the human subject-function within it. Once this happened his account of theoretical thought eliminated the logical subject-object relation by pushing this logical object-side of reality into the domain of the *non-logical*. This is an implicit left-over of Kant's rationalism in Dooyeweerd's epistemology, generating the antinomous idea that two irreducible aspects could be 'synthesized' instead of realizing that it is only through a deepening (disclosure) of the analytical aspect that identifiable and distinguishable features could be logically objectified, i.e. united on the basis of their distinctive, objective logical standing apart – ironically enough an insight explicitly advanced by Dooyeweerd himself.

In direct contradiction to his general notion of the *Gegenstand*-relation (as distinct from the logical subject-object relation) Dooyeweerd states that, in theoretical thought, the law-spheres "which precede the logical law-sphere" are "*distinctly* objectified in the latter" (Dooyeweerd, 1997-II:472). In respect of the deepening of the "logical object-side of reality" Dooyeweerd therefore

1 The other prominent contemporary of Natorp within neo-Kantianism is Heinrich Rickert – from the Baden school. He holds the view that within the theoretical sphere a distinction between the logical and a-logical is required: "Without a distinction between the logical and the a-logical within the theoretical sphere we do not even succeed in formulating the problem of number" (Rickert, 1924:9).

indeed involves *all* the non-logical aspects: "At the same time the logical object-side of reality is deepened in the subject-object relation. It changes from an objective logical systasis, merely embedded in temporal reality, into an objective logical 'standing apart', the objective dis-stasis of a functional multiplicity in the analytical aspect" (Dooyeweerd, 1997-II:471).

A comparison of the perspective of the deepened logical subject-object relation tested against the immanent criticism raised against Dooyeweerd's view of the *Gegenstand*-relation, provides the following picture:

i) No ambiguity is present in the notion of an antithetical relation any longer: within the disclosed logical subject-object relation, the modal aspects are distinctly opposed (i.e. logically objectified) to each other. And it is only within this relation of disjunction at the logical object side of reality, that we can meaningfully speak of an antithetical relation in the mentioned meaning of *an objective logical standing apart*.
ii) No restriction to the non-logical aspects is necessary any longer.
iii) The implication of Dooyeweerd's formulations, namely that the abstracted logical aspect may also be antithetically opposed to the non-logical aspects, may be accepted on condition that these abstractions are acknowledged as appearing at the distatic logical object side, where they are *distinctly objectified.*
iv) It is also no longer meaningful to ask for a *synthesis* between our actual logical thought function and some abstracted non-logical *Gegenstand.* Furthermore, in the sense of the notion of logical objectification, it is clear that two distatically objectified abstractions cannot be *synthesized* by our analytical activity. The term *synthesis* can only be used to indicate the way in which the multiplicity of identified structural features of a specific (i.e. *one*) aspect is conceptualized.
v) Claiming a difference between the *Gegenstand*-relation and the logical subject-object relation is no longer necessary. Distinguishing what is distinguishable takes place *within* the logical subject-object relation that refers to a more-than-logical diversity within reality.
vi) In terms of this perspective on the logical subject-object relation, there is no need to claim that intuition must function actively in any abstraction.
vii) Since the modal aspects are implicitly logically objectified in our pre-theoretical thought, their theoretical abstraction and analysis are not in need of any *inter-modal meaning-synthesis*. The implied meaning diversity is only made explicit by means of distatical logical objectification.
viii) This approach confirms that the first basic problem of epistemology is not an account of *theoretical knowledge*, but an explanation of the possibility and limits of our everyday experience and knowledge of things and events. It is only on the basis of our integral everyday experience that we are able to articulate the acknowledged meaning diversity scientifically.

It is unfortunate that Lambert Zuidervaart discussed crucial elements of the transcendental critique without taking into account the above-mentioned immanent criticism of the idea of a *Gegenstand*-relation. The effect is that he uncritically continues to refer to the *Gegenstand*-relation as if this idea is not torn apart by inner antinomies (see Zuidertvaart, 2004:77 ff.; Meynell, 2004 and Strauss, 1973). Dooyeweerd's unsuccesful attempt to defend his position is disussed in Strauss 1984.[1]

6.16 The archeological discourse theory of Visagie

During the middle of the 1980s, Visagie started to develop an account of what he called the "epistemic deep structures of philosophical discourse" (Visagie, 1988:52). He points out that Derrida sharply discerned "the foundational role of a class of conceptual deep structures" (Visagie, 1988:55). The name initially attached to an unveiling of these deep structures is "archeological discourse analysis" (Visagie, 1988:59). He speaks of "Governing Instances" (GIs) crucially significant for "knowledge of the self and the way reality is structured" and then mentions examples of GIs such as "matter, reason, God, freedom, sense-impression, system, historical flow" and so on (Visagie, 1988:59).

Against the background of Derrida's notion of a worded world that is affected by the play of *texts* or *writing* (Ecriture), Visagie highlights the formal archeological predicates at stake in a systematic way, namely *independence, persistence, transcendence, changeability, infinity, complexity, unknowability, causality* and *universality*. A closer analysis reveals certain *pairs* of archeological analysis, such as the binary attributes of *persistence* and *changeability, finite* and *the infinite*, and so on. *Operators* are needed to describe the relation of the *Governing Instance* to a particular *domain* (Visagie, 1988:67).

More recently, he employed the idea of "key-formulas," where "some aspect of the world like economic relations, or power, or physics/biology, or cultural context, etc. is postulated to play an explanatory role regarding other domains of reality." A governing relation is established between a *key factor* – such as *power* – and its domain – such as *knowledge, morality* or *politics*. Similar to the initial idea of a *Governing Instance* and its *operators*, the governing involved in the key factor concerns a "causing, preceding, founding, unifying, constituting" of "its domain" (Visagie, 2005:139). Through "key promotion," something can obtain the status of a "key factor" and then rule a domain by *uniting, grounding, structuring, centering* or *enclosing* the elements of the domain concerned.

But it should be kept in mind that certain *attributes* play a role in such a key, such as pairs that serve "as possible descriptions of the key factor and its do-

1 More recently Glenn Friesen once more attempted to remedy Dooyeweerd's position. However, his first attempt was completely external without any immanent criticism, while his second effort right from the beginning was misguided by losing track of the fact that Dooyeweerd's theory of an inter-modal synthesis precludes theoretical (i.e. inter-modal) knowledge of the logical aspect itself (the relation "logical-logical" is certainly not inter-modal – at most it is *intra-modal*). It is fortunate, however, as mentioned above, that Dooyeweerd was inconsistent and in a different context did open the way to allow for the *distinct* logical objectification of *all* modal aspects.

main":

> unitary/multiple (or simple/complex); finite/infinite; constant (immutable) / changing; knowable/unknowable; universal/individual; necessary/contingent (Visagie, 2005:143). What is striking is that "these particular attribute pairs have featured as such a *constant* in philosophical conceptualization from the Greeks until postmodernism" (Visagie, 2005:144).

In addition, Visagie also distinguishes between certain *ethical postures* (such as *meaninglessness*; *suffering*; and *guilt* – on the dark side – and "*ordinary-everyday living amid work and relaxation*; *contemplative reflection*; even *ecstatic experiences*; *giving up things*; *knowing humility*; *feeling joy*; *finding hope*; *caring for others*; and so on"), *root metaphors* (such as "*servant / worker*; or a *traveler*; or a *warrior*; or a *fully matured being*; or a *child*; or a *player*; or a *lover*; and so on"), *ideological power relations* (*discourse domination*, *types of group domination*), and *macro-themes* (such as "*nature, power, knowledge, personhood, society, humanity*") (Visagie, 2005:146). He relates *rationality* to the "two different horizons of experience: the one enclosing *structure, law, principle, norm, universality, generality, abstractness*, and so on; the other enclosing *subjectivity, factuality, process, event, individuality, contextuality, concreteness,* and so on" (Visagie, 2005:149) and finally accounts for *normative assumptions* related to society and its features.

These distinctions expand the scope of a systematic analysis of the conditions underlying scholarly discourse and at once account for problems arising from negative postures, reductive metaphors, distortive ideologies and misguided expectations from rationality.

The challenge to a non-reductionist ontology is indeed to generate a theoretical account of reality in which what Visagie designates as *attributes* are understood in such a way that we are not tempted to understand them in an exclusive "either-or" sense. For as soon as this avenue is pursued, it is realized that the *one* and the *many* are not in conflict with each other, but co-constitute our experience of reality. Likewise, the *finite* and the *infinite*, *constancy* and *change*, *universality* and *individuality*, as well as *necessity* and *contingency* are overall, mutually cohering traits of empirical reality.

In our preceding analyses, some of the most basic and influential distortive philosophical schemes of thought derive from elevating just one element of the mentioned pairs at the cost of the others. For example, we have argued that, since concepts are always constituted by uniting universal features, *rationalism* results from the absolutization of conceptual knowledge (see page 176 above), whereas *irrationalism* follows from an absolutization of concept-transcending idea-knowledge. In Chapter 2 we have looked at the influence of another well-known *ismic* opposition, namely that between *atomism* (individualism) and *holism* (universalism) – concerned with the quest to find a (explanatory) basic denominator for the diversity within reality – and we have seen that an understanding of isms like these rests upon a more detailed analysis of the meaning of (arithmetical and spatial) aspects and their mutual interrelations.

In the light of the all-permeating influence of nominalism in modern philosophy, it is now necessary to give an account of its basic tenets with a view to the currently well-known distinction between modernism and postmodernism.

6.17 The more complicated challenge to characterize *nominalism*

What became known as the opposition between *realism* and *nominalism* is a conflict of thought that originated within Greek antiquity where it is primarily associated with Plato's theory of ideas and the subsequent Aristotelian-Thomistic metaphysics.

6.17.1 *The historical importance of universality*

At the cradle of Western philosophical speculation, one certainly has to acknowledge Parmenides with his fundamental identification of *thought* and *being* (see page 254 above).

In the meantime, the core issue in this legacy is centered in the notion of *universality* and the nature of *truth*. According to the ripened conception of medieval realistic metaphysics, as defended by Thomas Aquinas during the 13th century, *universalia* have a three-fold existence: (i) as *archè*-typical forms in the divine mind (*ante rem*), (ii) as the immanent (universal) essential forms inhering in entities (*in re*) and (iii) as the subjective universal concepts within the human mind (*post rem*). *Truth* was supposed to be based upon the correlation between *thought* and *being* (*adequatio intellectus et rei*).

6.17.2 *The rise of modern nominalism*

This realistic metaphysics assumed a *chain of being* in accordance with the Platonic scheme of *ontic form* and its *copy* ('Urbild' and 'Abbild'), with God as the highest being ("*ipsum esse*," "Die Form aller Formen") at the top of the hierarchy of being.

Early modern philosophy soon replaced the classical legacy of metaphysical realism with a *nominalistic* approach. The artificial synthesis between Greek antiquity and biblical Christianity – sustained by the societal power of the medieval Roman Catholic Church with its ecclesiastically unified culture – did not survive the disintegrating effects of the *nominalistic* movement emerging during the late 13th and early 14th centuries. John the Scott and William of Occam denied the primacy of the human intellect as opposed to the will and particularly Occam opened up an avenue for an *arbitrary creativity*, by means of which the human intellect can acquire *control* over the surrounding world. Beck correctly points out that modernity caused a transformation in the understanding of human rationality. Human reason no longer *accepts*, but rather *logically controls* nature as an object in service of the human spirit with its self-determination and self-understanding as pure *subject*, directed at the experience of its own power and freedom (Beck, 1999:3).

Instead of looking at the world from the perspective of a pre-ordained hierarchical order of being, with God as the highest being, the *nominalistic attitude* stripped reality (outside the human mind) of any and all forms of *order-determination* and its entire *orderliness*. It thus leaves open a new domain of exploration manifested in the Renaissance urge towards the *rational control* and *mastery* of the world – which soon found a powerful ally in the rise of modern *natural science*. However, we have seen that this development was in the grip of the modern humanistic ground motive of *nature* and *freedom* (the natural science ideal and the personality ideal – see page 32 above).

6.17.3 *Universality and its connection with order and orderliness*

In order to understand this development, it should be noted that Plato and Aristotle respectively unveiled two closely related features of our life-world (*Wirklichkeit*), namely the role of an *order for* or a *law for* entities (Plato) and an understanding of the *orderliness* or *lawfulness of* entities (Aristotle). The former discovery was transposed into the speculative and transcendent realm as *static ideal forms of being* in Plato's thought, while the latter was absorbed in Aristotle's problematic notion of *substance*. We have seen that Aristotle also realized that *universality* plays a key role in human understanding, although he starts with the strictly individual primary substance (*proten ousian*) in his *Categories*.[1] However, in its *individuality* it precludes conceptual knowledge – something Aristotle did not want to sacrifice. As a consequence, he introduced the *secondary substance*, which is supposed to be the *universal substantial form* of an entity. This secondary substance is designated as the *to ti èn einai* (*De Anima*, 412 b 16 and *Metaph.* 1035 b 32). We noted that according to Aristotle, a concept is always focused upon what is *general* or *universal*.[2] In this way, he wants to safeguard the universality of theoretical knowledge. But at the same time, he restricted knowledge to *conceptual knowledge*. For Aristotle, true knowledge is therefore always knowledge of the *universal form*. The counterpole of form, namely *matter* (which lacks any positive determination – cf. *Metaph.* 1029 a 20-26) is therefore outside the reach of conceptualization. As such, it is *unknowable*.

Nominalism took a radical stance in opposition to this realistic metaphysical legacy by denying the existence ('reality') of *universalia ante rem* and *universalia in re*. The place where nominalism allowed universality to prevail was *inside* the human intellect – either as *concepts* or as *words*. The world as such lacks any universal traits; it is populated by *individuals* only.

6.17.4 *Descartes and Hobbes*

Particularly Descartes, in his methodical skepticism, exemplified the new spirit of the Renaissance and the post-Renaissance era by affirming the *autonomy* of the thinking subject as the ultimate starting point for philosophical thought. He carried through the consequences of denying any universality outside the human intellect. The most important implicit consequence of this

1 Cf. Metaphysics (1031 b 18 ff.): *πρώτεν ούσιαν*, and *Cat.* 3 b 10 ff. and see pages 18 and 176 above.

2 The Greek of this claim reads: *ὁ δὲ λόγος ἐστὶ τοῦ καθόλου* (Metaph. 1035 b 34–1036 a 1). Compare also Metaph. (1036 a 8): *καθόλου λόγου*. (See the earlier references mentioned on page 177.)

nominalistic orientation is that it does not acknowledge any *order* transcending the human being as such. A universal law order for creatures and also the orderliness of such creatures (in their subjection to such laws), are transposed into the *human mind*. The seemingly innocent remark that "number and all universals are only modes of thought" (Descartes, *The Principles of Philosophy*, LVIII), demonstrates the radical reorientation caused by modern nominalism.

Without entering into a detailed analysis of the philosophical development from Descartes up to Kant, one key element ought to be highlighted, for it explains the background of the modern idea of *construction*. In the thought of Hobbes we have already found a *new* motive, namely that of *logical creation* (*construction*). In his work on *Corporeal Things* (*De Corpore*), he introduces a thought experiment by imagining that reality in its totality is broken down into a heap of chaos and that human understanding, with the aid of well-defined concepts, *creates* a newly ordered cosmos. Being acquainted with Galileo's mechanics, it is understandable that Hobbes employed the concept of a *moving body* as the basic tool in the logical *reconstruction* of reality. Cassirer explains that this stance no longer accounted for *truth* in terms of the relationship between *thought* and *being*:

> Truth does not inhere in the things, but is attached to the names and their comparison as they are employed in statements (Cassirer, 1971:56).[1]

Before Hobbes, Galileo also formulated a thought experiment and without taking account of any real sense-experience, he arrived at his law of inertia. If a body is in motion and the path is extended into infinity, the body will continue its motion indefinitely, unless some force affects it. Galileo realized that it is not *motion*, but only a *change of motion* that requires a *cause* (see page 88 and 72 above). According to Drake, the astronomy of that period had the task of improving the "methods of describing and calculating the observed positions and *motions* (my emphasis – DS) of heavenly bodies, rather than to explain such motions physically" (Drake, 1999:68). To this, he adds that physics evolved along two distinct paths, a philosophical and a mathematical part, and that astronomical calculations "which fell to the mathematicians, remained strictly *kinematic* (my emphasis – DS)" (Drake, 1999:69).

6.17.5 *Once again the Corpernican Revolution in epistemology*

Eventually Kant took the outcome of Galileo's thought experiment further: Galileo derived a law and prescribed it to moving entities out of the pure understanding of the human being in its *spontaneous subjectivity* (see page 72 above). *This* represents, as we have noted, the crucial modern *epistemological turn*, known as the *Copernican turn*, in ascribing the primacy no longer to the object, but to the *subject*. Of course the problem was now to explain how the human subject inherently furnishes us with *universal* a priori forms, making

1 "Die Wahrheit haftet nicht an den Sachen, sondern an den Namen und an der Vergleichung der Namen, die wir im Satze vollziehen: veritas in dicto, non in re consistit" (cf. *De Corpore*, Part I, Chapter 3, Par.7 & 8).

possible our knowledge of the phenomena. Just compare the context in which Kant wrote about the difficulty involved in this turn, namely to explain how "*subjective conditions of thought* can have *objective validity*, that is, can furnish conditions of the possibility of all knowledge of objects" (Kant, 1787, B:122). Clouser aptly captures the impasse of this subjectivist stance: "Unless there were already laws governing the mind that were not its creations, what would explain the uniformity of the ways the mind imposes laws on experience" (Clouser, 2005:368).

The solution to this problem by Kant must be seen against the background of the ambiguous nature of modern *nominalism*. In spite of the radical differences between realism and nominalism, both orientations restrict knowledge to *universals*. This restriction of knowledge to *universals* is typical of what is known as *rationalism* (see our initial characterization on page 176). By contrast, one can define *irrationalism* as that epistemological approach that accentuates what is *individual* and *unique* (contingent) at the cost of any universal traits outside the human intellect. Of course, this irrationalistic side of nominalism is self-contradictory, at least when one rejects the influential Aristotelian dichotomy between *quantity* (as a category of matter) and *quality* (as a category of form). A chair with *four* legs unmistakably possesses this definite *numerical* (quantitative!) *quality*! Therefore, the *being individual* of each of a plurality of (be it perceivable) things, manifests an inherent structural trait (i.e. something *universal*) of every particular entity.

When Kant sruggles with the problem of how "*subjective conditions of thought* can have *objec*tive validity," his attempted solution illustrates that, in line with the thought experiment of Galileo, he simply drew the radical modern *subjectivistic* conclusion of Humanism: the laws of nature are *a priori* contained in the subjective understanding of the human being.[1]

It was merely a matter of pursuing this approach consistently that the so-called absolute freedom idealism after Kant (Hegel, Schelling, Fichte) believed that logic and dialectic embraces reality in its fullness and totality, and brings it forth. Cassirer remarks that only at this point does it seem as if the circle of philosophy is closed by reaching its aim in the identity of reality and reason – that is the point where Hegel believes his "Science of Logic" stands.[2] In his early development, Ludwig Feuerbach, the Luther of modern Humanism, continued this emphasis on the identity of thought and being. In the spirit

1 We noted repeatedly that Kant elevated understanding with its categories to become the (formal) law-giver of nature (see CPR, B:161, 163 and Kant, 1783, II §36, 320).

2 "Der kritische Idealismus Kants wird zum absoluten Idealismus umgebildet. Dieser absolute Idealismus will erfüllen, was Platon versprochen hat, wozu er aber auf Grund des dualistischen Charakters seines Weltbildes nicht vorzudringen vermochte. Logik und Dialektik sollen jetzt kein bloßes Organon der Wirklichkeitserkenntnis mehr sein, sondern sie sollen diese in ihrer Fülle und Totalität enthalten und aus sich hervorgehen lassen. Damit erst schien der Kreis des philosophischen Denkens geschlossen und sein Ziel, das Ziel der Identität von Wirklichkeit und Vernunft, erreicht zu sein. An diesem Punkt glaubte Hegels 'Wissenschaft der Logik' zu stehen" (Cassirer, 1957:10).

of Hegel Feuerbach holds that thinking is not a capacity of the spirit, but is itself the spirit (see Rawidowicsz, 1964:16). And in his dissertation, Feuerbach asserts that only the universal, the whole, the absolute is completed and perfect (see Rawidowicsz, 1964:27). Philosophy is simply concerned with what is universal and unconditioned necessary ('unbedingt Notwendig') and whatever is individual and particular disappears as unreal in the "One and All" infinite reality (see Rawidowicsz, 1964:34).

These consequences can only be properly understood when it is realized that *nominalism* (also in its *conceptualist* variants, including Kant's own position), in fact *transposes* the universal side of entities (accounted for by Aristotle as the supposed *universal substantial form* of things) into the human mind (*understanding*).

Yet, in our experience of the world, we always encounter the universal and individual sides of entities as being inseparably present at once. *This* atom is *an* atom – the individual side ('this') and the universal side ('an') are simply two sides of the same coin (see page 5 above). The *being an atom* of this atom is nothing but the *universal way* in which this individual atom evinces its subjection to the *law-for-being-an-atom*. Consequently, by stripping an entity of its *orderliness* (its *universal side*), it is simultaneously stripped of being subjected to the applicable *universal order for* its existence, i.e. both its *orderliness* and the *order for* its existence are eliminated.

6.17.6 *The vacuum created by nominalism*

What is left is *factual reality* in its (chaotic) *individuality* and *particularity*, *structureless* and without any *order-determination*. This very feature of nominalism enabled modern philosophy, from Descartes onwards, as driven by the new subjectivistic motive of *logical creation*, to *reconstruct* all of reality (in terms of natural scientific thought). Only the extreme consequences of this natural science ideal, also destroying in principle *human freedom*, were questioned by Kant. But for Kant, within the restricted area of human understanding, limited to sensibility and appearances, the position of understanding had to acquire a firm foundation in the claim that it is the a *priori* law giver of nature.

Clearly, nominalism created an important *vacuum*, by leaving factual reality in its individuality *unstructured*. Kant indeed drew the ultimate subjectivistic conclusion to fill the gap of determination thus created. Human understanding took hold of this vacant position, since Kant promoted it to function as *the formal* (a priori) law giver of *nature* (given as chaotic sensory *material*). Kant's position may be seen as the ultimate *rationalistic* consequence of nominalism that commenced with Descartes and Hobbes.

Yet the permeating effect of nominalism surely takes us beyond Kant. We may even jump to a key figure within the scene of 'postmodernity', Richard Rorty. Richard Bernstein defines the rationalistic tradition (designated by him as 'objectivism') as "the basic conviction that there is or must be some permanent, a-historical matrix or framework to which we can ultimately appeal in determining the nature of rationality, knowledge, truth, reality, goodness, or rightness" (Bernstein, 1983:8). Mary Hesse sees scientific revolutions as "metaphoric rediscriptions" (cf. Rorty, 1989:50). In following her, Rorty remarks: "This account of intellectual history chimes with Nietzsche's definition of 'truth' as 'a mobile army of metaphors' " (Rorty, 1989:17). Rorty (1989:16) views "intellectual history" as "history viewed as the history of metaphor." "Old metaphors are constantly dying off into literalness, and then serving as a platform and foil for new metaphors."

6.17.7 *What caused the shift to language?*

This emphasis on metaphor reveals the effects of what became known as the *linguistic turn* – prompting us to ask what happened after Kant that lead to this position? The key to answering this question is found in the transition from *universality* to *change* and *individuality*. Whereas, roughly speaking, one can say that the 18th century is the period of extreme (conceptual) *rationalism*, the transition to the 19th century can be designated as an acute awareness of the *historical dimension* of reality. By the end of the 18th century this, first of all, was due to the pioneering work by Johann Herder, a contemporary of Immanuel Kant. Korff calls Herder the *German* Rousseau and Cassirer praises Herder as the Copernicus of the (science of) history (Ernst Cassirer, 1957:226.). Pross sees in Herder the key figure who, in rejecting the 'Aufklärung' (*Enlightenment*), prepared the rise of *romantic historicism* (see Cassirer, 1957:226 ff. and the introductory remarks of Proß as the *editor* of Herder, 1978).

Although early romanticism transposes the *universal* to the *unique*, it did not distance itself from the inherent *atomism* (indvidualism) of the 18th century. The step to *holistic irrationalism* was eventually taken by Schelling, Fichte and Hegel – three prominent post-Kantian philosophers in Germany during and after the rise of *romanticism*. We should observe that, although Herder believes that society is subject to thorough *historical change*, he does not want to advocate an *anchorless relativism*. To curb this unwanted consequence, Herder upholds the *ideal of humanity*, which guarantees, as a universally binding rule, the *unity* and the *meaning* of history (Cassirer, 1957:228).

Niebuhr, the tutor of Leopold von Ranke (perhaps best known for his statement that the science of history studies the past as it actually happened to be), demonstrates the transition from the 18th to the 19th century in a remarkable way. From the *romantic movement* – including Goethe and Schiller (Germany), Bilderdijk and Da Costa (The Netherlands), and Shelley and Keats (Britain) – Niebuhr received his appreciation of *mythical thought*. Without relinquishing the imaginative exuberance present in myths and sages, Niebuhr wants to treasure the *historical way of thought* in its own right.

With an obvious hint to Plato's classical allegory of people living in a cave (*The Republic*), Niebuhr compares the historian to a person whose eyes adapted so effectively to the dark, that it is possible to observe things that would be invisible to the newcomer. Where Plato appraises these "shadow-images" *negatively*, Niebuhr assesses them positively – for on occasion he characterizes the work of the historian as "work done under the earth" (cf. Cassirer, 1957:237).

In opposition to Plato, who acknowledges only knowledge directed at the true (static) *being* of things as worthwhile, Niebuhr is convinced that only *historical change* provides genuine knowledge. This kind of knowledge is the most appropriate type of knowledge for humanity, comprising the vital self developing of human beings.

6.17.8 *From historicism to language as new horizon*

Over against the deification of universal (conceptual) knowledge during the 18th century, we are here brought into contact with the importance of *historical change*. However, this *irrationalist* and *historicist* reaction against Enlightenment *rationalism* contains hidden problems that would only become explicit during and at the end of the 19th century. It is noteworthy that this process was anticipated by the first critical reactions to Kant's *Critique of Pure Reason*. It was in particular Jacobi, Hammann and Herder who pointed out that Kant neglected the nature of *language*.[1]

Although Feuerbach in his early development still continues the rationalistic line of nominalism by emphasizing universality at the cost of individuality, he had a clear understanding of the fact that the sensory capacities of the human being exceed the possibilities of the acquisition of concepts, because the former (concepts) are strictly based upon universality, while the latter (the senses) cannot transcend what is individual (see Rawidowicsz, 1964:18).

Karl Mannheim, one of the prominent sociologists of the first half of the 20th century and the founder of the sociological subdiscipline known as *sociology of knowledge*, had a solid understanding of the romantic roots of Dilthey's *irrationalistic historicism*:

> Dilthey is borne by, and may be the most important exponent of, that irrationalistic undercurrent which first became self-aware in Romanticism, and which, in the neo-Romanticism of the present, is on the way, in altered form, to effecting its attack on bourgeous rationalism (Mannheim, 1982:162).

Only what can be experienced in the context of a historical, world-encompassing coherence could serve as the *immediately certain* basis of knowledge acquisition – and only by means of empathy one can attain a genuine understanding (*Verstehen*) of spiritual reality. The natural sciences *know*, the humanities *understand* (Dilthey, 1927:86). Dilthey no longer supports the positivistic science ideal seeking the typically human in some facet of nature. The historical aspect now occupies this vacancy: to be human means to be *historically conditioned* (1927:275, cf. Diwald, 1963:38 note 11). Habermas furthermore mentions the implied *linguistic framework* in Dilthey's hermeneutics:

> We don't understand a symbolic expression without an intuitive prior-understanding (*Vorverständnis*) of its context, because we are not capable of freely

1 See the quotation from Herder given on page 321 above (Herder, 1978:73).

> transforming the presence of an unquestioned background knowledge of our culture into an explicit awareness.[1]

In an article on hermeneutics and the Dilthey school Gadamer explains the transition to language as a *new universal* in the thought of Heidegger: " 'Being' certainly does not mean the being of something, also not the authentic or divine, for it is rather like an event, a pathos, that opens up the space in which hermeneutics – without a final foundation – turns into a new universal. This space is the dimension of language" (Gadamer, 1991:172).[2]

6.17.9 *The hybrid nature of nominalism*

We mentioned that *rationalism* entails the absolutization of knowledge in terms of *universal* features, that is, it deifies *conceptual* knowledge, whereas *irrationalism*, on the other hand, focuses upon whatever is unique, individual, unrepeatable and contingent, thus restricting knowledge to the *approximating* understanding of concepts stretched beyond the limits of their natural application (*concept transcending knowledge*) – that is, to *idea-knowledge*.

The perplexing fact is that **nominalism** comprises **both** these elements: In respect of the *typical structure of entities*, nominalism does not accept any *conditioning order* (*universal structures*) for, or any *orderliness* (*universal structuredness*) of such entities. Every entity is *strictly individual*. In terms of our distinction between *rationalism* and *irrationalism*, nominalism surely represents an *irrationalistic* view of the nature of entities, since every individual entity is completely stripped of its *universal orderliness* (*law-conformity*) and *conditioning order*. This characteristic applies to both *moderate nominalism*, viz. *conceptualism* (Locke, Occam, Leibniz and others), and to *extreme nominalism*, that rejects all general and *abstract ideas* and only accepts *general names* (Berkeley and Brentano).

This *irrationalistic* side of nominalism, however, does not exhaust its multi-faceted nature, because *universals* are acknowledged fully *within* the human mind, at least as *general words* in the case of Berkeley's and Brentano's *extreme nominalism*. This restriction of knowledge to *universals* is typical of *rationalism* in the sense defined by us. Therefore, it is possible to see *nominalism* as being simultaneously *rationalistic* in terms of the universals – concepts and words – in one's mind, **and** *irrationalistic* in terms of the strict individuality of entities outside one's mind. Just compare the way in which Habermas captures the stance taken by Rickert in this regard: "Rickert presupposes – and here he is covertly in accordance with *Lebensphilosophie* – the irrationality of a reality that is integrally present only in nonlinguistic experience" (Haber-

1 "Einen symbolischen Ausdruck verstehen wir nicht ohne das intuitive Vorverständnis seines Kontextes, weil wir das fraglos präsente Hintergrundwissen unserer Kultur nicht freihändig in explizites Wissen verwandeln können" (Habermas, 1983:17).

2 " 'Sein' meint eben nicht ein Seiendes, auch nicht das Eigentliche oder Göttliche, sondern ist eher wie ein Ereignis, ein Pathos, das den Raum öffnet, in welchem Hermeneutik – ohne Letztbegründung – neuen Universale wird. Dieser Raum ist die Dimension der Sprache" (see page 322 above).

mas, 1988:4).

The inability of conceptual knowledge to grasp what is unique and individual caused philosophers to look at the *senses* (cp. the development of positivism and neo-positivism) and at *language* to bridge the gap. It seems as if *language* can indeed mediate between *universality* and *individuality* in a way that transcends the limitations of *concept formation*. Mannheim understood these issues for he clearly grasped something of the *two-fold* nature of nominalism: "Nominalism proceeds from the unjustifiable assumption that only the individual subject exists and that meaningful contextures and formations have being only to the extent that individual subjects think them or are somehow oriented toward them in a conscious manner" (Mannheim, 1982:196-197, cf also p.224 above). As a consequence, we can speak of a general (and currently widely acknowledged) shift from *concept* to *meaning*, from *thought* to *language*.

It should now be clear that 'postmodernity' and its supposed 'new' features are actually 'old' *humanistic* ones. The key *historicistic* claims of postmodernity derive from *post-Kantian Romanticism*, and its *lingual emphasis* was anticipated by *nominalism* from its very inception (cf. Occam and Hobbes), and was also suggested by Jacobi, Hamman and Herder, even before the end of the 18th century! The key figure in the genesis of the linguistic turn, insofar as we may see it as an attempt to overcome the limitations of concept formation with respect to what is unique, contingent and individual, Wilhelm Dilthey, actually lived the greater part of his life in the 19th century. To be sure, what is called *postmodernity* merely constitutes a new *power concentration* of the irrationalistic side of nominalism. This basic orientation even predates *modernity* – the latter taken in the sense of the 18th century Enlightenment.

Yet, acknowledging these historical roots should not mislead us to underestimate the vastly permeating (and uprooting) effects of contemporary postmodernism. Although the features are not new, their current hegemony surely *is*. The claim that in a fragmented and ever-changing world every person is entitled to his or her own 'story' – while negating any and all grand meta-narratives (Leotard) – has the pretension of being just one amongst many other 'stories'. Yet, without realizing it, this new orientation over-emphasizes *historicity* and *linguisticality* at the cost of other dimensions of creation equally co-conditioning human existence. In fact, these postmodern claims operate as an alternative grand meta-narrative, namely the universal claim that holds that *everyone* only has his or her partial story without any 'universal' truth.

From the fact that this statement itself rests upon a *universal claim* – 'enabling' it to apply to 'everyone' – its inherent self-uprooting nature is manifest in its very formulation. Without an inherent *constancy* and *universality*, even the exclusively elevated conditions of historicity and linguisticality lose their meaning.

The enemy of scholarship and culture is not *universality* and *constancy*, but the internally antinomic attempt to assert historical change and lingual ambiguity at the *cost* of constancy and universality. It is only when we take seriously the liberating biblical perspective that creation cannot be explained merely in terms of some or other aspect that we in principle can escape from the one-sidedness of orientations such as rationalism, irrationalism, historicism and 'linguism' – all of them combined and fused in the contemporary fad of *postmodernism* (for a critique of the distinction between so-called linear and non-linear thinking see Sokal and Bricmont – 1999:164-167).

The over-estimation of *rationality* in the legacy of the West cannot be divorced from the all-pervasive *nominalistic* conviction that reality itself supposedly has a 'rational' structure. Since nominalism denies both the God-given order for (law for) the existence of creatures *and* the universality of creaturely responses to those laws (evinced in their lawfulness or orderliness), it is quite 'understandable' why modern secular Humanism 'loaded' the human subject with the additional 'responsibility' of becoming the law giver / (constructive agent) of its own world.

We may now return to and continue our discussion of the general significance of the modal subject-object relation.

6.18 The economic subject-object relation

Within the structure of all the (post-arithmetical) aspects we have discerned various subject-object relations. We have seen that they appear at the *factual side* of an aspect – in subjection to the delimiting and determining law side.

6.18.1 *Economic price theory*

Menger, Jevons and Walras advanced a position, stating:

(a) a commodity 'in itself' does not have any (economic) value – it always 'only' has a "subjective value" for a determined group of commodities;
(b) the (economic) value of a group of commodities is dependent upon the marginal utility of the mentioned group, i.e. upon the increase or decrease of the marginal utility of the commodities;
(c) that the marginal utilities of a commodity decrease when its number increases (Schneider, 1962:193).

It is of critical importance to realize that the extremes of an objectivistic and subjectivistic price theory flows from breaking apart the economic subject-object relation. The statement that something does not have an economic value in itself simply affirms that nothing can objectify itself within the economic aspect of reality. However, the act of objectifying it, is not itself 'objective'. It requires the activity of an economic subject, since all forms of objectification result from subjective[1] actions. This indispensable subjective element in the act of objectification explains that human behaviour is always subject to economic principles and therefore invalidates any notion of *arbitrary* marginal utilities. The relative price of commodities and their marginal utility are always 'positioned' within the universal framework of the economic modality

1 The word 'subjective' here simply means: the activity of a person who has a *subject-function* within the economic aspect of reality correlated with economic principles.

– explaining why concrete economic interactions can always be evaluated as being economically sound or uneconomical (i.e. economically anti-normative). Although dependent upon the economic (objectifying) activity of an economic subject, the economic value of a commodity is not exclusively determined by that subject, since it is fitted in the structural economic subject-object relation.

As soon as one enters the domain of economic actions and transactions, mediated by the interaction of differently structured economic subjects within the economic domain – individuals or societal collectivities, according to their economic subject-function – it turns out that we are dealing with a complex subject$_1$-object$_1$-object$_2$-subject$_2$ relation. Subject$_1$ is the primary subject (that is, the producer/supplier); object$_1$ is some (produced) commodity not yet (fully) objectified in the economic aspect; object$_2$ is the economically objectified commodity, an objectification embodied in its price (i.e. its economic object-function); and subject$_2$ represents the demand for this commodity (by those economic subjects who function as consumers in this context).

6.19 The jural and ethical subject-object relations

The increasing process of globalization gave prominence to the issue of human rights. The threats to the sustainability of 'life' on earth caused by the excesses of modern technology in addition also brought the survival of plant and animal life into the picture – focused on what became known as plant and animal rights. This background explains why animal life in Africa turned into a matter of global concern.

For example, the issue of *culling* elephants in Southern Africa generated international debate and action. From a systematic philosophical perspective, it concerns jural and moral subject-object relations. A brief look at some of the complexities of this problem will provide us with a basis upon which we can come to a more nuanced assessment of the relations between plants, animals and human beings, and the nature of the 'rights' involved in this context.

Hardliners claim that the *Kruger National Park* can only accommodate 8000 elephants if one does not want to threaten the 'rights' of other animals (and the ecosystem). But since the culling program was terminated in 1994, the elephant population in the *Kruger National Park* increased to number of about 12000. Steve Smit, the chair-person of "Justice for Animals," nonetheless holds that it is a *myth* that the elephant population is exploding (*Die Volksblad*, Saturday 30, 2005:21). Michelle Henley supports this view by emphasizing that there is no "scientific proof" that the size of the *Kruger National Park* can only accommodate 8000 elephants. She contends that problems are due to managerial mistakes dating back to the sixties of the previous century. For example, by closing some of the artificially created water points the ecology is recovering and at the same time a change is already observed in the movement patterns of the elephants (*Die Volksblad*, Saturday 30, 2005:21).

A general misconception prevails that elephants are "living bulldozers" that destroy the natural environment in which they live. But in fact they have a tremendous positive influence and effect on their bio-milieu (ecosystem) as a whole.

MacKenzie notes that "as much as 80 percent of what elephants consumed is returned to the soil as barely digested highly fertile manure:

- Elephants provide a vital role in the ecosystem they inhabit.
- They modify their habitat by converting Savannah and woodlands to grasslands.
- Elephants can provide water for other species by digging water holes in dry riverbeds.
- The depressions created by their footprints and their bodies trap rainfall.
- Elephants act as seed dispersers by their fecal matter. It is often carried below ground by dung beetles and termites causing the soil to become more aerated and further distributing the nutrients.
- Their paths act as firebreaks and rain water conduits.
- An elephant's journey through the high grass provides food for birds by disturbing small reptiles, amphibians or insects" (MacKenzie, 2005).

The plans made and the actions envisaged for the meaningful management of plant and animal life in a case like that of the *Kruger National Park* is surely only possible because such a large conservation area was established in the first place.[1] In the meantime, the human factor complicated the scene further: The *Mail&Guardian* reports that up to March 2005, the South African government received a total of 37 land claims in respect of the *Kruger National Park*. It further states that at least a quarter of the land in the *Kruger National Park* is claimed by African communities (including its "headquarters at Skukuza and prime tourist attractions such as Letaba and Pretoriuskop" – *Mail & Guardian*, February 18-24, 2005:6). The heading of this report highlights the serious issue at stake: "Land claims 'could kill Kruger'."

This picture may be broadened by looking at the animal life in Africa in general and at the history of the way in which humankind interfered with the flora and fauna of this continent. A quotation from Douglas Chadwick (in his work on "The Fate of the Elephant") reads: "Conserving elephants, then becomes much more than an issue about how to protect a single species. It is about protecting one of the forces that shapes ecosystems and helps sustain the wealth of wildlife found across much of the continent. It is about saving the creative power of nature."[2] Since the 17th century, firearms enabled hunters to kill animals far beyond any reasonable measure – of course co-determined by the economic value of the ivory and skins. To provide some examples: during the seventies of the 19th century, a hunting party of three shot

1 The Park was established by President Paul Kruger in 1898 and covers 8000 square miles (or: 21000 square kilometers – an area of about 320 x 80 kilometers). It consists of a variation of open field with its sandy flats as well as brush and forests. Here the well-known *acacia*, *marula* and *combretum* species (to name but a few) provide a habitat to an enormous diversity of African animals (such as elephants, lions, leopards, cheetahs, buffalo, rhinos etc).

2 See the web-site: http://elephant.elehost.com/ (accessed on 30-07-2005).

down 10 elephants in a single day (see the account in Selous, 1985:40-41).

A.C. White records the case of Prince Alfred's visit to Bloemfontein in the year 1860. When the Prince arrived at Bloemfontein on the 21st of August, his party entered into "an impromptu hare-hunt" (where the old Town Hall stood). But the truly shameful event concerns "probably the largest big game hunt ever staged in South Africa," which was arranged on the farm of Mr. Bain, the person after which *Bainsvlei* is named:

> Thirty thousand head of game were surrounded by many hundreds of natives. Although the Royal party and friends shot 1,000 head, 5,000 head of game were killed, mostly by the natives. The hunt became in fact a massacre, and the slaughter one for which Bloemfontein must have been ashamed (White, 1949:69).

If we are indeed called not to act in total *arbitrariness* towards animals, a foundation must be established for understanding the nature of the *normative care* that we, as human beings, have to exercise in relation to nature – for only within such a context will it be meaningful to assess the issue of (human and animal) *rights*.

6.19.1 *Moral normativity*

The question concerning the 'rights' of animals (and 'plants') is derived from the emergence of the idea of *human* rights. But even before the problem of human rights can be stated properly, we have to rely on what has been said in Chapter 1 about the *normativity* of (human) life – because it concerns the *interconnection* between multiple human beings and their acknowledgement of underlying *norms* or *principles* guiding human activities in the various domains of societal life. Particularly when the nature of principles (or norms) is considered, the advocated views reveal *supra-theoretical commitments*. Different world and life views diverge radically in this regard.

Modern philosophy since Descartes is impregnated with the *humanistic* idea of *human autonomy*, of the human person as being a *law unto itself*.[1] And we saw that Immanuel Kant carried this starting point through to its ultimate rationalistic consequences when he assumed a position according to which human understanding (with its categories) actually serves as the *a priori* (formal) law giver of nature (see Kant, 1783:320; § 36).[2] Of course we also noted that the crucial problem is given in the *jump* from human *subjectivity* to the supposedly *universal validity* of these (categorical) determinations – something clearly felt by Kant when he addresses the problem in his *Critique of Pure Reason* (CPR), in which he asks the question how "subjective conditions

1 The collective version of this autonomy-idea (and human freedom) is found in the earlier quoted words of Rousseau: "freedom is obedience to a law which we prescribe to ourselves" (Rousseau, 1975:247).

2 It is worth-while to keep reminding the reader about this humanistic view of Kant by once more quoting his significant statement: "der Verstand schöpft seine Gesetze (a priori) nicht aus der Natur, sondern schreibt sie dieser vor" ("understanding creates its laws (a priori) not from nature, but prescribes them to nature" – see pages 73 and 373).

of thought can have objective validity" (Kant, 1787-B:122 – see pages 73 and 373 above).

The on-going development of philosophical reflection never escaped from an equally strong awareness of the necessity of acknowledging a *more-than-merely-individual* normativity. The words of Kuhlen, directed at the distinction between law and morality, touch upon a facet of this insight.

> The delimitation of law and morality does not concern an arbitrary definitional determination, but in respect of certain factual and normative starting points it aims at a reasonable conceptual solution (Kuhlen, 1981:223).

Particularly through the immense growth in the technical mastery of nature during the past two centuries, and the unforeseen consequences for the *conditio humana* flowing from it, humankind as well as scholarly reflection currently find themselves confronted with life-sustaining issues of *normative care* and *control*. Given the long-standing view of the human being as a *rational-moral* creature, it should therefore not be surprising to see how the contemporary scene is flooded with concerns about the differentiated and collective responsibilities of humankind towards nature (and natural resources) and thereby towards itself.

But the fact that these concerns are constantly put into the mould of some or other kind of *ethics* demonstrates at least three issues:

(i) It continues the underlying (above-mentioned) view of the human being as a *rational-ethical being*;
(ii) As a consequence of (i), it collapses all forms of normativity into *morality*, and
(iii) With their *human-centered* point of departure considerations focused upon the issue of plant and animal rights are suffering from a lack of clear understanding of the coherence and difference between subject-subject and subject-object relations.

6.19.2 *The universal scope of the moral aspect*

In this context, the scope of the moral concerns the question: which entities partake in the domain of the ethical?

Within the contemporary scene, wide-ranging differences of opinion regarding so-called *legal rights* for (of) plants and animals are found. One direction pursues the expansion of the scope of morality and ethical subjectivity, whereas the opposing inclination expels animals from the dimension of morality, or at least disqualifies plants and animals as "moral agents."

Stone discusses the development of Roman law, where the father initially had the "power of life and death" over his children. As in ancient Greece,[1] deformed and female children were subjected to the widespread practice of infanticide. Stone quotes Maine, saying that the Roman *paterfamilias* (*Patria Potestas Romana* – with an *ius vitae necisque*) had the power of "uncontrolled corporeal chastisement; he could modify their personal condition at pleasure;

1 In ancient Greece the aged were also treated as *rightless*.

he could give a wife to his son; give his daughter in marriage; divorce his children of either sex; transfer them to another family by adoption; sell them" (VanDeVeer, 1998a:148).[1]

Through a long process of legal development the assignment of rights was then broadened to encompass human beings in various capacities. It is rather amazing to note that throughout the development of law every successive extension of "rights to some new entity has been, …, a bit unthinkable": eventually it was extended to "prisoners, aliens, women (especially of the married variety), the insane, Blacks, foetuses, and Indians" (VanDeVeer, 1998a:148). Such an expansion of the scope of morality eventually somehow aimed at incorporating plants and animals into the domain of "moral agents" – in the attempt to portray them as bearers of legal rights.

On the basis of the shared capacity of animals and humans to 'suffer' the next step is, for example, to argue with Tom Regan "that the same essential psychological properties – desires, memory, intelligence, and so on – link all animals with humans and thereby give us all equal intrinsic value upon which equal rights are founded. These rights are inalienable and cannot be forfeited" (quoted by Pojman, 2000:396). Singer also refers to *moral equality*: "A liberation movement demands an expansion of our moral horizons and an extension of the basic moral principle of equality" (Singer, in Pojman, 2000:400).

However, Pojman himself, by contrast, questions the *moral accountability* of animals, because they "cannot make moral decisions" and because animals are not "members of the moral community." In order to substantiate his claim, he refers to the 'contractualist' Hobbesian tradition, which holds that, for their lack of communicative abilities, animals can never enter into a *contract* (Pojman, 2000:396). Another way is to mirror these duties in the sense that our moral duties towards animals are nothing but *indirect duties* towards humanity – a position that was already taken by Immanuel Kant. The essential element in this view is continued by Martin Buber in his work *Ich und Du* (1923), and we even find it (via the Dutch scholar in theological ethics, Aalders) in the thought of the South African philosopher Stoker and the South African theologian Heyns.

Stoker and Heyns restrict the domain of ethics to *inter-human* (subject-subject) relations. Applied to human action towards animals, this view argues that the meaning of morality is found in 'person-care' ('persoonsbehartiging'). This immediately raises the question whether or not the human love for animals, plants and the land fall *outside* the domain of morality? Heyns explicitly says that a *mere love for nature* does not reveal the ethical or moral dimension, since the latter only occurs when the former is viewed in the light of 'per-

1 Stone refers to "H. Maine, Ancient Law 153 (Pollock ed. 1930)." Maine (who lived from 1822-1888) initially published this work under the title: *Ancient law: its connection with the early history of society, and its relation to modern ideas* (published by J. Murray in London). With a Foreword by Lawrence Rosen, the most recent edition appeared in 1986, published by the University of Arizona Press (Tucson).

soonsliefde' ('person-love') (Heyns, without date:9). In order to explain his intention Heyns gives the example of torturing an animal. Such an act is not immoral because it is an *animal* that is tortured, but because a *human being* performs the act. By torturing an animal, the human being displays a lack of *self-love* and does not fulfill the calling we as human beings have to rule over creation (Heyns, without date:9).

It is striking that Heyns nonetheless refers to love that is *non-moral* in nature, namely love for entities in nature (things, plants and animals). The logical consequence of his restriction of the moral domain to the care between *persons* ('persoonsbehartgiging'), is that since animals are not *persons*, they by definition fall *outside* the *ethical domain*.

When Warren argues that, although animals do "not have *rights*, we are, nevertheless, obligated not to be cruel to them" (Warren, 2000:450), the overall picture increasingly becomes one in which the status of moral agents and their relation to non-moral entities need elucidation.

6.19.3 *Ethical subject-subject and subject-object relations*

One of the merits of the legacy of reformational philosophy is that it developed an account of subject-subject and subject-object relations based upon the theory of *modal aspects* displaying a truly *ontic* universality. This insight entails the difference between the two kinds of laws we distinguished: *modal* and *typical laws* (type laws – see page 25). The former holds universally for all possible entities and processes in an *unspecified* sense, whereas the latter evinces a specified universality, holding for a limited *type* of entities only. The modal universality of the moral aspect entails that nothing in the universe escapes from the scope of this modal function. This holds particularly for entities in nature, although this does not mean that natural entities – things, plants and animals – ought to be appreciated as moral *subjects*. In order to discover a connection between plants and animals and the ethical aspect, there is no need for expanding the scope of ethical subjectivity. It is sufficient merely to acknowledge what Dooyeweerd calls the latent object-functions of natural entities. Of course this theory of subject-object relations does not deny the original *subject-functions* of natural entities (and processes). Modern Humanism has such a strong appreciation of the human *subject*, that it appreciates all non-human entities as 'objects'. Just recall the widespread practice to speak about *physical objects*. The theory of subject-object relations suggests that, insofar as material things are *physical*, they are not objects but *subjects* (see page 175), and insofar as they are *objects* they are not physical, for in the latter case, they ought to be understood in accordance with their *non-physical* properties (*object-functions*). Likewise, plants and animals in the first place are (biotical) and (sensitive) *subjects*, and only in the second place they ought to be considered according to their relevant *object-functions*.

From a systematic perspective, the ethical subject-object relation is fitted within the distinction between the law side (norm side) and factual side of the moral aspect, i.e. it functions at the factual side in subjection to moral principles. This implies that natural entities and natural processes in a *structural sense* display *possible* (latent) moral object-functions, and that the latter can only be *disclosed* or made *patent* through the activities of ethical *subjects* (human beings as *individuals* or the functioning of *societal collectivities*). The normative ethical demands of *caring* for nature embraces both ethical subjects and ethical objects – but the moral aspect should not be viewed in isolation from other aspects of reality, not least if one wants to understand what the true meaning of *rights* is. Since the concept of a *right* – such as in the expression "human rights" – is best known (but the latter should be distinguished from *subjective rights*) as a *juridical* term, a discussion of the nature of "animal rights" is dependent upon an account of the relationship between *law* and *morality*. In view of the fact that the current debate about animal rights is normally phrased in terms of *moral* categories, this issue indeed justifies an investigation of this basic relation.

6.19.4 *The distinction between law and morality*

Understanding *rights* requires insight into the nature of the *jural* aspect of reality, of the *state* as a public legal institution, of the international (global) concern for the environment, and also of the nature and place of *rights* within human society (beyond the state). The emphasis on environmental ethics, in turn, requires a view on the relationship between *law* and *morality*. But these relationships cannot be treated in isolation, because both the meaning of the *jural* and that of the *moral* reveal themselves only in an interconnectedness between these aspects and all other facets of reality.

The first step towards a meaningful differentiation between law and morality is given in a fundamental questioning of the 'basket' view of morality, in terms of which every form of normativity is placed within the domain of morality. In our preceding analyses we have implicitly questioned the immense *simplification* present in this prevailing view. In fact, the contrary between logical and illogical has already manifested the meaning of normativity in the sense that it presupposes the logical principles of *identity* and (*non-*)*contradiction* as they apply to the human ability to *conceive* and *argue*.

Establishing that a concept is *illogical* (we use the example of a square circle) entails that a *normative standard* has been applied and that the said concept does not *conform* to this logical (and not *ethical*!) requirement of *ought* inherent in this normative standard. It is contradictory not to distinguish between a square and a circle – an example used by Kant (see page 93 above). Phrased in a different way, one can say that confusing two *spatial figures* violates the demands for *identifying* and *distinguishing* properly – a square is a square (logically correct *identification*) and a square is not a non-square (such as a circle – logically correct *distinguishing*).

However, thinking in a logically antinormative way, i.e. thinking *illogically*, remains bound to the structure of *logicality* and does not turn into something *a-logical* (non-logical), such as the economic, the moral or the jural. Although one may call these (non-logical) facets of our experience *a-logical*, they are not *illogical*. In fact, we noted that they also have room for contraries *similar* (or analogous) to the contrary between logical and illogical, namely *economic* and *uneconomic*, *moral* and *immoral*, and *legal* and *illegal*. The logical contrary actually lies at the *foundation* of all the other instances of normative contraries – the latter *analogically reflect* the meaning of logical analysis (identification and distinguishing) in an *unspecified* sense.

This perspective uproots the above-mentioned long-standing legacy of identifying the domain of *normativity* with *morality*. All kinds of norms, principles and 'values' cannot be identified with the moral. Moreover, without an acknowledgement of (God-given) *ontic normativity*, the eclipse of a solipsistic subjectivism seems *unavoidable*.

The theological thought of Johan Heyns was influenced by the transition from the medieval era to the early modern period where *nominalism* postulated a *despotic arbitrariness* on behalf of God (*potestas Dei absoluta*). This idea subjected God to His law for creation, for only when a normative standard is present can one meaningfully speak of *arbitrariness*. Heyns likewise believes that the prohibitions contained in the ten commandments are not prohibitions for God. For example, the "inter-trinitarian essence" of God entails that what is forbidden for a human being is allowed in the case of God, because Father, Son and Holy Spirit are "in front" of each other (they therefore have "other Gods" in front of them). He places every one of the Ten Commandments in this perspective, and then concludes that what God is allowed to do is forbidden for human beings. For example, when human beings commit adultery, they sin, but when God demands that husband and wife should love Him more than each other, God indeed turns out to be the big 'Adulterer' (Heyns, 1970:88). His remark that God's adultery is not *sin* camouflages the real problem, for within the *ethical* meaning of adultery *as such* the jural meaning of *unlawfulness* is *presupposed*. Therefore, if adultery is inherently *immoral* (i.e. *sinful*) then the defense that as the great *Adulterer*, God does not sin, boils down to the following contradiction: when God commits adultery (that is, when God *sins*) then He does not *sin*! This argumentation in fact subjects God to His own creational law.

We have highlighted a similar problem in modern Roman Catholic moral philosophy, where it is believed that from Commandments such as "Thou shalt not murder," "Thou shalt not commit adultery," and "Thou shalt not steal," one can derive rules of natural law – an attempt that resulted in the implied antinomy expressed in the contradiction that "lawful murder" is possible. The mere fact that murder is nothing but *unlawful killing* entails that the construction of "lawful murdering" boils down to the following implied contradiction: "lawful unlawful killing" (see page 275 above).

Just recall how Polak attempted to 'define' retribution as the core meaning of the jural (see page 173): "retribution is an objective, trans-egoistic, harmonization of interests." We argued that this 'definition' completely misses the target, because all the elements employed in it are either non-jural or trans-jural. Polak simply set aside *retribution* as the core meaning of the jural aspect of reality – resulting in a general concept lacking any modal (aspectual) delimitation. The meaning nucleus of an aspect is not only unique and irreducible, for it is also *indefinable*. When the term *retribution* is employed to designate this core meaning of the jural, alongside the term love designating the core meaning of the ethical, then it must not be confused with *revenge*. It is based on the insight that what rightfully belongs to a person ought not to be taken away from that person – and if it happens, the core meaning of the jural aspect entails that whatever has been taken away ought to be given back to that person (*re*-tribution). We have noted that the classical Roman jurists (Ulpianus and others) gave expression to this meaning where they emphasized that every person should be given his or her due (*ius suum quique tribuere* – see page 99 above).

In the light of what we have said in connection with the disclosure of the jural aspect (see page 228) we know that these deepened principles are *principles of justice*. Against this background, we can now ask:

6.19.5 *Can animals (and plants) be bearers of subjective rights?*

Perhaps one of the most important contributions of Dooyeweerd's legal philosophy is given in his systematic account of the nature of a *subjective right*. First of all, he realized that both the notion of a subjective right and that of a juridical object (legal object) are not *thing concepts* but *modal functional* concepts (see Dooyeweerd, 1997-II:405 ff.). Most importantly, it should be realized that the juridical subject-object relation constitutes the framework within which both these notions ought to be analyzed (namely that of subjective rights and of a legal object). Once this is acknowledged, the meaningless construction of absolute and inalienable human rights in modern humanistic natural law is unmasked as untenable. A subjective right is always correlated with a legal object and the subjective competence of disposition of a legal object must be distinguished from the competence to do away with a legal object (for instance by *selling* it).

But human subjectivity as such cannot be objectified in a juridical sense. One cannot truly sell one's freedom or the biotical integrity of one's body. Therefore the "rights" in respect of the human body and life are not truly *subjective rights*. Dooyeweerd correctly points out that at most they constitute "subjective legal interests which are so closely connected to human legal subjectivity that they do not qualify to be objectified over against a legal subject. It is not possible to dispose of these interests as rights in a juridical sense" (Dooyeweerd, unpublished *Systematic Volume* III of the *Encyclopedia of the Science of Law*, page 207 – in the original Dutch text page 208).

Animals (and plants) lack *normative accountability*, because they do not function as subjects within the post-sensitive aspects – including the jural and the moral aspects. Therefore they are neither "moral agents" nor "legal subjects." Recalling the above-mentioned example of a "square circle" and the inability of chimpanzees to copy a given drawing of a square or triangle, we noted that the crucial question is: if these animals are not even able to *draw* these figures, how are we going to be convinced that they truly have the *concept* of a square or the *concept* of a triangle? We must remember that a concept is something different from a *sensory picture*, which can be associated with something else (cf. Overhage, 1972:252) – as is the case in so-called 'name-giving' mentioned by Leakey. The decisive point, however, in showing that animals possess over these concepts would be to show that they can think illogically by forming, for example, the self-contradictory concept of a "triangular square" or a "square triangle." This has never been shown by any of the experiments referred to by the mentioned authors. In other words, animals are simply not able to function subjectively in the analytical aspect of reality. Consequently, it should not be surprising that they are unable to think – be it in a logically correct or incorrect way.

Stone approaches the status of natural things from the angle of the current "rightlessness of natural objects" and by asking what is entailed when something is a *holder of legal rights*? For the latter, "something more is needed than that some authoritative body will review the actions and processes of those who threaten it." According to him, a "holder of legal rights" must satisfy "each of three additional criteria." The underlying question is, what makes a thing count jurally? The three criteria are: "... first, that the thing can institute legal actions *at its behest;* second, that in determining the granting of legal relief, the court must take *injury to it* into account; and, third, that relief must run to the *benefit of it....*" (Stone, 1998a:150). The objection that streams and forests cannot *speak* is not sound, for Stone argues: "Corporations cannot speak either; nor can states, estates; infants, incompetents, municipalities or universities. Lawyers speak for them, as they customarily do for the ordinary citizen with legal problems" (Stone, 1998a:150). With reference to incompetent people requiring a *guardian* he holds that "[N]atural objects *can* communicate their wants (needs) to us, and in ways that are not terribly ambiguous," for one can judge with more certainty and meaningfulness whether and when my "lawn wants (needs) water" than whether or not the judge can decide when the "United States wants (needs) to take an appeal from an adverse judgment by a lower court" (Stone, 1998a: 152-153).

Stone's overall aim is to make a plea for the protection of the rights of unborn generations. "Indeed, one way – the homocentric way – to view what I am proposing so far, is to view the guardian of the natural object as the guardian of unborn generations, as well as of the otherwise unrepresented, but distantly injured, contemporary humans" (Stone, 1998a:152).

However, when grass 'says' that it is 'dry' and needs 'water', it requires a human being to *interpret* these *signs* correctly – just like in the last quote the decisive element is the role of the *guardian* of *natural objects*. Outside the structural context of normatively qualified subject-object relations, the realm of plants is unable to obtain the *care* it 'needs'. It is therefore only from the perspective of ontic normativity that we can escape from the impasse enclosed in all humanistic attempts to find a foundation for universal and constant principles with an appeal merely to human *subjectivity* (compare the above-mentioned dilemma of Kant).

This dilemma is intensified through the relativizing effect of postmodernism, evident in its denial of every form of universality and constancy (of course with the exception of the implicit universal validity assumed for its own view). When Regan attempts to account for "the ideal moral judgment," he mentions elements such as *conceptual clarity*, *information*, *rationality*, *impartiality*, *coolness*, and the *validity* of *moral principles* (Regan, 1997:126-130). He then posits criteria for the evaluation of moral principles, such as consistency, adequacy of scope, precision, and conformity with our institutions (Kegan, 1997:131-136). He does not realize that these terms stem from diverse irreducible modal domains – *consistency* resides in the logical aspect (non-contradiction); *adequacy of scope* is equivalent to (specified or unspecified) *universality* which reflects the *spatial* meaning of *everywhere*; *precision* comes from the domain of non-ambiguous linguistic meaning; and *conformity with our institutions* makes an appeal to a particular form of social interaction of individuals within societal collectivities.

As a complex basic concept of any discipline, it is indeed only possible to articulate the meaning of a norm or principle by 'borrowing' terms from every (constitutive) modal aspects. On the basis of our above-mentioned argumentation (see page 297), we suggested the following brief description of a principle. "A principle is a universal and constant point of departure that can only be made valid through the actions of a competent organ (person or institution) having the disposition of an accountable (responsible) free will, enabling a normative or antinormative application of the principle concerned, relative to the challenge of a proper interpretation of the unique historical circumstances in which it has to take place."

The cultural task of caring for nature differentiates in various contexts. The universal scope of all the normative aspects makes it impossible for human beings to withdraw from this normative appeal, although the *typical* way in which particular social entities (coordinational relationships, communal relationships and collective relationships)[1] function within the normative aspects does *specify* the meaning of the implied universal modal norms.

It is neither meaningful to restrict morality to ethical subject-subject relations, nor correct to expand the domain of *normative subjectivity* such that the ('discloseable') object functions of natural entities (concerning plants and animals) and processes are transformed into normative *subject functions*. However, the human calling to care for nature is not merely enclosed in the 'basket' of moral normativity, since in its richly varied diversity it is 'spread' over all the normative modal aspects and the normative calling of all coordinational, communal and collective relationships within human society. Only an analysis of the structural principles of the latter will open up the required scope within which *local societies* and the *global village* can account for their

1 These distinctions are explained in Chapter 8 below.

responsibilities in keeping up the Genesis appeal of *building* and *conserving*. Governments may benefit from a deepened perspective on the requirements of a public justice in which the legal interests of societal institutions and environmental needs are harmonized, for within the context of the disclosure of the jural and moral object functions of natural things, plants and animals ought to be given their *due* and be *cared* for.[1]

Before we enter into an analysis of the significance of the dimension of (natural and social) entities for an understanding of the foundational role of philosophy in respect of the special sciences, we will briefly consider a number of misunderstandings pertaining to the dimension of modal aspects.

6.20 Some misunderstandings regarding the nature of modal aspects

At the Annual Meeting of the Association for Reformational Philosophy,[2] held in the *Hotel Americaine* (Leidse Plein, Amsterdam, January 1970), Dooyeweerd at a certain stage participated in the discussion concerning the invention of the theory of modal aspects. He remarked that although it certainly is the *best known* section of his philosophy it certainly is also the *least understood* part of it.

We will focus on the following misunderstandings (some of them did receive attention in different contexts).

1) Aspects are viewed as 'slices' or 'layers' within reality, in the sense that they are seen as a way in which reality could be 'divided'.
2) Aspects are interpreted as properties of entities.
3) It is confusing to equate modalities, aspects and functions.
4) There are ambiguities regarding the relationship between law and subject and universality and individuality
5) Aspects are mental constructs.
6) Aspects are designated as the field of study of the various disciplines.

6.20.1 *Do the aspects 'subdivide' reality?*

The idea that modal aspects subdivide reality into as many 'pieces' as there are aspects is built upon multiple confusions. (a) The mistaken view that reality (the universe) is one big 'thing' (whole) that can be divided and that the result of such a division is exemplified in the different modal aspects; (b) It rests upon an uncritical employment of the idea of a whole (totality) and its parts; (c) It does not account for the "entitary-natured" (that is 'non-aspectual') 'parts' of reality. (d) It continues a long-standing nominalistic legacy. (e) It does not understand the basic modal sense of the whole-parts relation.

Let us analyze these confusing views in more detail:

(a): Although human thought inherently displays a 'totalizing' tendency, explaining why in everyday parlance we unhesitatingly speak of the *universe*, this tendency does not warrant the conclusion that the universe indeed is noth-

1 Only within such a broad context will it be possible to also do justice to the cause of the elephants in (South) Africa.

2 *Calvinistic Philosophy* as it was still known at the time.

ing but one big entity which in its totality (as a whole) could be divided like cutting a cake. In our everyday experience we are fully aware of multiple (differently-natured) *kinds* of things, processes (events) and societal entities. The idea of a modal aspect is based upon the assumption that *distinct* from the dimension of natural and social entities we can identify the *dimension* of modal aspects. This entails that no single aspect can ever be a 'part' of (entitary) reality. Entities are entities and therefore they can only have *entitary parts*. The general assumption of the theory of modal aspects is that every possible entity (and every possible part of an entity) in principle has a function within each modal aspect of reality. The only condition added to this general claim is that one has to differentiate between subject functions and object functions. Physical entities, as we have argued, are said to have *subject functions* in the first four modal aspects (number, space, movement and the physical), but they have object functions in all the post-physical aspects.

It is worth mentioning in this context that a number of critical appraisals of Dooyeweerd's philosophy suggested a connection between his philosophy and the theory of layers (*Schichten*) developed by Nicolai Hartmann. In addition to a brief remark in Dooyeweerd 1997-II:51 note 3 a more extensive defense against this misunderstanding is found in Dooyeweerd 1960:122-124.

(b) & (c): The theory in terms of which we accounted for the interconnections between different kinds of entities – where each entity maintains its sphere-sovereignty – was designated as the theory of enkaptic interlacements (see pages 139 and 356 above). Enkaptic intertwinements are different from a straight-forward whole-parts relation because in the latter case all parts share – in their being-a-part – structurely determined by the whole.[1] Whenever we look at a whole-parts relationship – whether 'normal' or 'enkaptic' – the parts concerned cannot be *aspects*. In his critical discussion of the expression "aspects of things" Van Woudenberg correctly points out that having a physical aspect does not entail a composition of physical material (Van Woudenberg, 2003:4).

(d): When Descartes declares that "number and all universals are modes of thought" his position bears witness to the influence of modern nominalism. The latter ascribes reality only to concrete entities, implying that the various aspects of reality could only be appreciated as being seated within human thought (reason). Therefore, if the aspects are to partake in 'reality' they have to be appreciated in *entitary terms* – as manifested in the mistaken idea that reality is *divided* by distinguishing its aspects. Surely, it requires some-'thing' with an entitary nature to 'cut' or 'divide' concrete reality.

(e): The whole-parts relation in a modal functional sense, as we have pointed out, for the first time appears within the spatial aspect (see pages 87 and 263 above. A continuously extended spatial subject, such as a one-dimensional line-stretch, must be connected in all its parts, for if some parts are *disconnected*, i.e. if they are not *cohering*, then the gaps will eliminate its *continuity*.

1 It may look as if Dooyeweerd contradicts himself by still speaking of an enkaptic structural *whole* – but this follows from the "cross-fertilization" to be discussed below.

Therefore, in ordinary parlance, one sometimes speaks of a *gapless* connection or coherence. Furthermore, if all the parts are present and fitted in a gapless coherence, they constitute the meaning of the *whole* embracing all its parts.

6.20.2 *Once again: are aspects properties of entities?*

Although it is possible to speak of properties of statements or of syntactical properties, the general awareness of entities with their properties is what is at stake in the formulated question.[1] In his discussion of this issue Van Woudenberg explains Dooyeweerd's view as follows: "So, a stone, or tree, or cat, or person is not a modal aspect. Modal aspects don't, so to speak, exist, in their own right but are aspects *of* individual things. They require a 'bearer', or 'substratum'." He explains that Dooyeweerd does not advance a 'bundle' theory but a 'substance' theory, for "individual things are not just bundles of aspects" because "aspects are aspects *of* such things" (Van Woudenberg, 2003:1). Throughout his discussion he maintains the assumption that Dooyeweerd understands aspects as aspects *of* individual things – even where he [Van Woundenberg] explicitly deals with the possibility of speaking of things as *functioning* within an aspect (Van Woudenberg, 2003:6).

Let us suppose for a moment that it was indeed Dooyeweerd's intention to see (modal) aspects as properties of things and then suppose that we imagine multiple entities having a certain property – for example the property of being *square* ('squareness' – such as a square room, a square table, and so on). The key question then is what "bridges the gap" between these multiple entities, that is, what makes it possible to speak of the *same* spatial property in *different* instances? If the nature of *spatiality* is not more than its individual instantiations, if it does not exceed any (individual) possible case of 'squareness', then it seems impossible to assign the same property uniformly, or, as one may immediately say, *universally*, to all possible square entities. In other words, the first attempt to envisage what the expression "property of an individual thing" means already suggests that something more fundamental is at stake, something with an inherent universal scope transcending merely being a property of an individual thing.

But the fact of the matter is that Dooyeweerd considers the different modal aspects as belonging to a *distinct dimension* of created reality. The modal aspects are seen as the *a priori ontic conditions* making possible the many-sided existence of concrete (natural and social) entities. They form the *universal cadres* within which concrete entities and events function. In Chapter 3 we stated that each modal aspect displays certain universal characteristics – such as having a unique, indefinable and irreplaceable core meaning (meaning-nucleus), (retrocipatory and antecipatory) analogies referring backwards and forwards to the other modes, having a law side and a factual side accompanied

1 Of course as soon as one accounts for configurations like numbers, sets of numbers, spatial points, statements and sentences, it can be argued that we are dealing with 'abstract objects' or 'abstract entities'.

by the correlation of time order and time duration, and on the factual side each one displays both subject-subject and (in the post-numerical aspects) subject-object relations. As such each one has a universal scope, best captured by what we designated as its *modal universality* (see pages 68 and 421).

Modal universality underlies and makes possible the functioning of every entity within the aspect concerned. Once the ontic universality of modal aspects – such as the quantitative, the spatial and the kinematic – is properly understood, they cannot simply be seen as "modes of thought," for in an *ontic* sense they co-condition the existence of concrete entities and processes functioning within them. Therefore it is not at all mysterious that a theoretical insight into the nature of arithmetical laws, spatial laws, laws of motion, and physical laws (such as the law of gravitation) relates to "the real world," because these *modal laws* co-condition the *real world.*

The modal universality of the various aspects enables all entities to function within all these aspects of reality. This universality explains why modal laws hold irrespective of the peculiar nature of different kinds of entities. By contrast, as we argued, the law determining the nature of entities is always limited to a specific *class* or *group* of entities. Our example was that the law for an atom is only applicable to atoms and not to anything else; the structural principle for marriage is only applicable to marriages and not to states or business enterprises. Whereas modal laws encompass all possible entities (and therefore hold *universally* in an unspecified sense), typical laws (type laws) only hold for a *limited class of entities*. *Typical* functions of entities are designated as *typonomic* functions (see page 26).[1]

Van Woudenberg investigates the possibility that an "aspect of a thing" is a point of view belonging to the viewer and not to what is viewed. To such a point of view, for example the physical, there belongs a family of concepts such as "atom, molecule, electron, mass, charge and impulse" (Van Woudenberg, 2003:7). His whole argument here is that *if* aspects are points of view belonging to the subject, they cannot be seen as ontic aspects as well. But the counter question is: why not? The ability to 'look' at reality through the 'gateway' of the physical (or any other aspect) would vanish if reality itself did not display such an ontic mode or aspect. Modal aspects are therefore always at the same time *modes of being* **and** *modes of explanation*. In his discussion of *categories* Loux highlights the latter perspective: "They [categories] are not screens or barriers between us and things; they are, on the contrary, our routes to objects, our ways of gaining access to them" (Loux, 2002:11). In their ontic sense they are modes of being and once (modally) abstracted, they can also (without any contradiction) serve as modes of explanation.

Modal concepts always refer to the universal spheres in which concrete entities and processes *function*, whereas typonomic concepts refer to the dimension of concretely functioning entities, events and societal totalities. The distinction between the dimensions of universal spheres and concretely functioning entities corresponds to that between modal functional concepts and type

1 Belonging to a specific kind therefore entails the feature of *typonomicity*.

concepts – a distinction not considered by van Woudenberg.

In all special sciences the use of both modal and typonomic concepts is encountered. The science of physics employs modal (functional) concepts, such as a uniform (constant) motion, the concept of force (a specific cause with its particular effect), concepts such as volume, pressure, entropy, and so on. In addition physics as a discipline employs numerous typonomic concepts – such as atom, molecule, macro-system, physical processes, and so on. In biology the classification of plants and animals (in phyla, classes, orders, families, genera and species) represents typical biological entitary concepts (type concepts) whereas concepts such as growth, differentiation, integration, adaptation, finality, and so on all manifest biological function concepts. By enumerating physical typonomic concepts Van Woudenberg therefore did not advance any argument against the ontic reality of the physical modal aspect or against the inevitability of physical concepts of function.

In the preceding chapters the terms mode, aspect and function were used as synonyms. Yet for Van Woundberg only the term 'function' appears to be useful. He holds that the idea of an aspect does not make sense. Of course, as we noted, this conclusion stands and falls with the mistaken idea that aspects are *properties* of individual things. Once this erroneous view is abandoned, it is not difficult to explain why the term *modality* and *function* can be understood as synonyms. The original Latin root of the term mode is after all modus – still recognizable in phrases such as *modus operandi* and *modus vivendi*. A modality is simply a *mode of being* – exactly what is meant by Dooyeweerd when he employs the term function.[1]

If we assume that Van Woudenberg might concede that the terms *function* and *modality* are acceptable, we still have to account for the legitimacy of the term 'aspect' as an indication of the different modal functions of reality. At this point a broader perspective is needed.

6.20.3 *The cross-fertilization of the dimensions of functions and entities*

One of the most effective ways to introduce the idea of a modal aspect (function) is to make a distinction between two different questions, that concerning the *concrete what* and that related to the *how*. Asking questions about the concrete *what* of entities and processes does not highlight any modal aspect, simply because the aspects reflect the *way* or *manner* in which such entities and processes *function* – that is, they relate to the *how* of concrete entities and processes. Already in 1910 Cassirer highlighted the importance of this distinction between *entity* ('substance') and *function* (see Cassirer, 1953). When entities and processes are resolved into functions we meet the distorted view of *functionalism*; and when modal functions are treated as if they are entities then those aspects are *reified* ('hypostatized' – from the Greek: *hypostasis*). An

1 In a work published in 1931 (see Dooyeweerd, 1931) Dooyeweerd consistently employs the term *function*. He never refers to functions as aspects, although the term 'aspect' is amply found in publications before (as early as in his Inaugural Address of 1926) and after 1931.

in-depth analysis of the decisive role of functionalism in the development of the modern natural sciences is found in an important work of Rombach (see Rombach, 1965-66).

Ultimately one can only speak of the modal aspects by (implicitly or explicitly) using terms derived from or reflecting the dimension of entities and vice versa, one can only talk of entities (and events) by employing terms derived from the dimension of modal aspects. Van Riessen used to say that the modal aspects are *points of entry* ('toegangspoorten') to concrete reality. This mode of speech highlights an element of the earlier-mentioned difference between modal analogies and entitary analogies, designated by metaphors.

Modal functional terms and metaphors are complementary in the sense that they serve, in a cross-fertilizing way, the (imaginative) options we have of elucidating the nature of the dimensions of aspects and that of entities. This explains the legitimacy (and unavoidability) of using (or creating) some or other image in order to articulate what is meant by the dimension of aspects – ultimately based upon the fact that our experience of reality is embedded in our functioning within it. When Van Riessen frequently speaks of aspects as *points of entry* or *gateways* to our experience of and reflection upon (entitary) reality, he merely explores one example amidst many other possibilities. When modal aspects are seen as *modes of explanation* it is often said that a specific discipline has a particular *angle of approach* – and once again the image of a concrete directional orientation serves the purpose of articulating something about the nature of modal aspects.[1]

The same applies to the term 'sides' that is also often used to talk of modal aspects. From our experience of concrete physical entities we are all acquainted with their (typical) spatial properties. A cube, for example, as a three-dimensional physical entity, has six *sides* – it is in other words *many-faceted* ('nuanceful'). Likewise, the world in which we live displays a rich diversity of *functions* and *things* (the *how* and the *what*). Exploring typical spatial relations makes it possible to speak of multiple modal functions within reality. Since "typical spatial relations" or "typical spatial properties" concern the connection between the universal modal structure of a function of reality and the *entitary* way, that is the *typical* or *typonomical way*, in which concretely existing things function within a particular modality, we have to observe that such a *typical property* reflects the relation between an entity and a modal function and consequently it falls within the scope of what we defined as a *metaphor*. Speaking of aspects as *sides* (or: *meaning-sides*; Dutch: 'zinzijden') of reality represents one amongst many metaphorical options we have

1 Van Woudenberg phrases four sentences in which he uses the term 'property' and then reckons that the untenability of the expression 'modal aspect' is evident when one attempts to replace 'property' in each one of the sentences by the phrase 'modal aspect'. However, recalling our above-mentioned distinction between *modal* and *typical* (typonomical) the proper substitutions would have had to use the following two phrases: 'modal property' and 'typical property' – which in each instance yields a meaningful sentence (see Van Woudenberg, 2003:2).

in order to express our understanding of the nature of modal aspects. The term 'aspect' itself is merely another word used to speak of the many-sidedness of reality.[1]

6.20.4 *Is it confusing to equate modalities, aspects and functions?*

The inevitability of employing such metaphors is not sufficiently appreciated by Geertsema in his critical assessment of Van Woudenberg's article. Geertsema writes:

> It can be conceded that the word 'aspect' does not lead to a proper understanding of what is meant by modal aspect. The word as such suggests something external, especially because of its visual connotations, irrespective whether it points to something real, as side of a diamond, or refers to subjective viewpoints. Modal aspects imply something else. They refer to something much more intrinsic, especially in the case of Dooyeweerd. Modal aspects are 'modes of being' which make possible and determine the specific nature and kinds of individual things. This, of course, implies some understanding of being, but by itself it might explain why an analysis that starts with the common use of the word 'aspect' does not lead to a proper understanding (Geertsema, 2004:61).

The alternative for referring to an 'aspect' presented by Geertsema does something similar – it speaks of modalities by at once taking into account the dimension of functions (*modes of*) and the dimension of entities (*being*). The only difference is that whereas the terms 'side' and 'aspect' are derived from a *specific instance* of "modal typicality" (exploring "typical spatial relations"), the expression "modes of being" explores the *general case*, namely the fact that whatever entity there is, it will have a function within all modes. We should not shy away from the imaginativity of creative metaphors – we only have to avoid a simplistic reliance on some or other 'literal' element present in the semantic field of words serving metaphorical designations. If the word 'aspect' or 'side' carries with it "visual connotations," as Geertsema remarks, then as such it does not *prohibit* a metaphorical designation in which the reality of multiple modes is accentuated.

The distinction between *connotation* and *denotation* (originally from Frege: *Sinn* and *Bedeuting*), 'liberates' a word in order to serve many different "connotational purposes" – including metaphorical designations such as those found in the perfectly meaningful (metaphorical) reference to *sides* or *aspects* of reality. However, the use of the phrase "the word as such" by Geertsema, suggests a connotative atomistic restriction. No single word has a privileged meaning nuance or connotation that legitimizes a view of the supposedly sin-

1 It should be remembered that Dooyeweerd initially experimented with a expressions like "domain-category" ('gebiedskategorie'), "field of view" ('gezichtsveld') (see Verburg, 1989:67) and "modal categories" ("modale kategorieën") (see Verburg, 1989:56). All of them constitute metaphorical explorations – the first one almost in terms of a metaphorical duplication, because both the terms *domain* and *category* derive from the core meaning of the spatial aspect.

gular meaning of a word "as such." The semantic field of a word opens up multiple options both in respect of selecting some or other meaning nuance of a word in a specific context, and with regard to the creation or deletion of meaning nuances (in which case the semantic field of a word is *expanded* or *reduced*). Speaking of modal functions as *aspects* or *sides* of reality therefore does make a meaningful contribution to a proper understanding of the nature of modes of being.

Furthermore, when Geertsema remarks that "[M]odal aspects are 'modes of being' which make possible and determine the specific nature and kinds of individual things" he overextends the meaning of modal aspects. The modal universality of modal aspects at most co-determine the existence of entities, but the modal aspects themselves are not responsible for the "kinds of individual things" – this honor is reserved for the dimension of entities, "individuality structures" and their type laws.

6.20.4.1 *Frege's implicit understanding of the difference between* modal *and* typical

This distinction is similar to the way in which Frege employs the word *quantity*. Dummett writes: "Frege so uses it that a phrase like '2.6 meters' designates a specific quantity of one kind, '5.3 seconds' a quantity of another kind, and so on. He thus takes quantities to be objects, distinct from numbers of any kind. There cannot be two equal quantities, on this use: if two bodies are equal in mass, they have the *same* mass. Quantities fall into many distinct types: masses form one type, lengths another, temperatures a third" (Dummett, 1995:270). Frege therefore implicitly distinguishes between the general (modally universal) meaning of *number* and the *specifications* number can receive when it is attached (within non-numerical contexts) to different *types of quantities* – in which case he does not speak about *number* but about *quantity*. In other words: *number* is a modal term and *quantity* a specified term (sometimes specified within a non-numerical context where numbers appear analogically, and sometimes where an entity functions within the arithmetical aspect – in which case we can speak about its typicality – for instance in respect of the *mass* of a physical body).

6.20.4.2 *The meaning of the term 'function'*

In presenting an alternative to the terminology of aspects (modal aspects or sides) Van Woudenberg shows a preference for the term *function* (Van Woudenberg, 2003:8 ff.). Yet also here he and Geertsema do not realize that this term stems from just another metaphorical depiction – equally dependent upon the "cross-fertilization" currently under discussion. Our human awareness of 'functioning' arises from the interconnection between entities and modes of being, but this interconnection is now not specified with reference to *typical spatial relations* but rather in terms of *typical physical* phenomena. The core meaning (meaning-nucleus) of the physical aspect according to Dooyeweerd and Stafleu is *energy-operation* (activity). When *energy oper-*

ates some or other entity or process *causes* the occurrence of *changes* (alterations) and we capture this connection by speaking of *causality*: the relation between *cause* and *effect*. And according to the general theory of modal aspects no single entity or event can *by-pass* its 'operation' within the physical aspect – a mode of speech synonymous to saying that no single entity or event can *by-pass* its 'function' within the physical aspect. By speaking of modal aspects as *functions* we have simply (metaphorically) explored the typical requirement that every entity or process displays (amongst others) also a physical aspect. Without an awareness of physical *typicality* (*typonomicity*) – involving both the dimension of entities and the dimension of aspects, the designation of aspects as *functions* would not make any sense.

6.20.5 *Aspects caught up in the confusion of law and subject and universality and individuality*

In the next chapter we shall see that Dooyeweerd did not properly distinguish between *law* and *lawfulness* (law-conformity) – he simply used these expressions interchangeably. Yet the (universal) conditions for being this or that type of thing must be distinguished from the (universal) way in which particular entities evince their conformity with these conditions (laws). The term 'structure' is therefore ambiguous. It may refer to the *order for* (*structural law* or *structural principle for*) the existence of a specific type of entities, whereas the *structures of* these latter reveal what is correlated with (and therefore distinct from) the said *order for* entities. A *structure for* has the meaning of a *law for*, while a *structure of* represents the universal way in which individual entities reveal their conformity with the given law for their existence (also known as their *law-conformity*).

By identifying law and law-conformity Dooyeweerd strips factual reality of its *universal side*. For that reason he often explicitly speaks of the *individual factual side*. In his response to Van Woudenberg the argumentation presented by Geertsema in this regard exhibits a number of related problems. He speaks of aspects as being *individualized*. This view is incorrect because the modal universality of aspects can only be *specified*, but never *individualized*. If it was possible to *individualize* what is universal the distinction between universality and individuality collapses. The typical way in which entities function within the modal aspects merely specifies the universal modal meaning of an aspect in a typonomical way. In a footnote Geertsema aims at being "more precise than Dooyeweerd often is himself":

> [I]ndividual things have a lawside and a factualside. In the latter they express both individuality (uniqueness in comparison with other individual things) and universality (what they have in common with other individual things). So the subjectside of individual things encompasses their lawfulness (existing according to the laws that hold for them). On this basis we can gain insight in the laws they conform to" (Geertsema, 2004:67).

To my mind this quotation reflects a mixture of what is correct and incorrect. It is incorrect to assert that "individual things have a lawside and a factualside." Created reality has a law side and a factual side. Individual things function at the factual side and therefore, in their being subjected to the law side, they do not themselves display a law side as well. Geertsema is correct by distinguishing between the unique-individual side of entities and their universal side – as long as we remember that the latter relates to law-conformity as a universal property of *subjects*, for then it is correct to say that the "subjectside of individual things encompasses their lawfulness" – although clarity may be gained if one simply says that the law side (law for) is universal and that at the factual side of reality we may discern both an individual side (uniqueness) and a universal side (law-conformity) of entities. In its *atom-ness* (in its *being an atom* – see page 5, 193, 374) this *individual* atom in a *universal* way shows that it is subjected to the *universal* law holding for its existence.

6.20.6 *Are aspects mental constructs?*

This *nominalistic* idea exerted an enormous influence within modern philosophy ever since Descartes declared, as we have mentioned, that number and all universals are modes of thought (*Principles of Philosophy*, Part I, LVII). The overall effect in respect of the ontic status of modal categories is that aspects or modal functions were no longer appreciated as *ontic* modes of being. In addition concrete entitary reality is stripped of its universal side, thus leaving reality unstructured in its unique (chaotic – Kant) factuality. If no universality is recognized outside human understanding there is also no room left for accepting universal (God-given) *laws for* entities. We argued that the negative fruits of this nominalistic legacy motivate the modern humanistic idea that the human being *constructs* its own world.

Nonetheless there is relative merit in Descartes's claim that 'number' is a mode of thought, because it is only when a given, not yet *counted, multiplicity* within reality is recognized by a human being that we encounter the concept of number – as a response to the implicit question how many? This (human) concept of number is made possible by an *ontic mode of multiplicity* that is not created by humankind and that is also not the product of an autonomous *thought-construction*.

The well-known statement of the early-intuitionist mathematician, Leopold Kronecker, namely that God created the integers and that everything else is accomplished by humans, confuses the ontic mode of multiplicity with the human concept of integers (and other kinds of) numbers. God did not create integers, humankind did. God only created the "multiplicity-aspect" of reality.[1]

Although it may be misleading to talk of numbers as "individual things," as it is done by Frege, he did see something of this distinction as early as 1881. In an article on "*Booles rechnende Logik und die Begriffsschrift*" (unsuccessfully submitted for publication) he wrote: "individual things cannot be assumed to be given in their totality, since some of them, such as number for ex-

1 The title of a recent book edited by Stephen Hawking imitates Kronecker's saying (see Hawking, 2005). Kronecker said: "Die ganzen Zahlen hat der liebe Gott gemacht, alles andere ist Menschenwerk." ["The beloved God made the integers, everything else is human work."]

ample, are first created by thinking" (quoted by Dummett, 1995:3).

The ontic conditions making possible the human concept of numbers explains why Bernays rejects the idea that an axiomatic system in its entirety is an arbitrary construction: "One cannot justifiably object to this axiomatic procedure with the accusation that it is arbitrary since in the case of the foundations of systematic arithmetic we are not concerned with an axiom system configured at will for the need of it, but with a systematic extrapolation of elementary number theory conforming to the nature of the matter (*naturgemäß*)."[1] The "nature of the matter" contains an implicit reference to the ontic status of the "multiplicity-aspect" of reality.

We now proceed with an account of the significance of acknowledging the dimension of (natural and societal) entities for an expanded understanding of the relationship between philosophy and the various academic disciplines.

1 "Gegen diese axiomatische Vorgehen besteht auch nicht etwa der Vorwurf der Willkürlichkeit zu Recht, denn wir haben es bei den Grundlagen der systematische Arithmetik nicht mit einem beliebigen, nach Bedarf zusammengestellten Axiomensystem zu tun, sondern mit einer *naturgemäßen systematischen Exstrapolation der Elementare Zahlenlehre*" (Bernays, 1976:45).

Chapter Seven

Things

The transition from aspects to entities

7.1 Modes of explaining material things

Our human experience of reality is related to concrete things, events and societal relationships. From early childhood we are confronted with an awareness of the *universe*; an awareness that not only generates questions about the boundaries of the universe (mentioned in Chapter 1), but also prompts us to reflect on the 'whatness' of natural and social entities and processes.

Already in the early development of Greek culture, this quest has led to attempts to clarify the mystery of what was designated as 'matter'. These reflections were born from the desire to settle not merely the mystery of what *is*, but indeed to proceed to the discovery of some or other mode of explanation or principle of understanding in order to acquire insight into what exists.

The preceding chapters promoted an understanding of the uniqueness and mutual coherence of the various aspects of reality, as these are explored in a way that highlighted the *foundational role* of philosophy in all scholarly activities. It appeared that the modal aspects are not only modes of being or existence; since they also provide *access* to concrete reality, they serve as *modes of understanding* and *explanation*.

We may use the history of our understanding of physical nature to mediate the transition from the dimension of modal aspects to the dimension of (natural and social) entities and processes. On the one hand, such an analysis will show that it is impossible to reflect on the dimension of entities without employing some or other principle of explanation (mode of understanding), and on the other hand, our historical journey will also demonstrate the inherent limitations of every mode of explanation (in the sense that no single mode of explanation is sufficient for a comprehensive understanding of the multimodal or multi-aspectual nature of many-sided entities).

7.1.1 *Points of departure in Greek culture*

At first sight, it may seem quite unproblematic to acknowledge what is given in the aspects of everyday experience. Yet, as soon as an attempt is made to *explain* what is at hand, the difficulties involved in understanding the complexity entailed in the diversity of experience are reflected in the successive and alternative options chosen during the history of theoretical contemplation. How did Greek culture approach this problem?

Greek culture is indeed appreciated as the seat of western civilization and the fertile soil that gave birth to the so-called exact sciences. In the ripened conception of Plato, mathematics acquired a particularly prominent position. His urge to account for the nature of knowledge is closely connected to this view. It should be remembered that Plato was strongly influenced by a student of Heraclitus, called Cratylus. According to these thinkers, *everything changes*. In Plato's dialogue with the same name (Cratylus), we have found (see pages 105 and 164 above) that Plato realized that *if* everything changes, *then* the conceptual knowledge of reality would be impossible – as soon as something is conceived, it has already changed.

Amidst a world apparently dominated by changefulness, an urge towards the incorruptible, the everlasting and eternal, emerged. This gave birth to a search for the assumed underlying principle, considered to be fixed and firm. The early Greek philosophers chose some or other *fluid* element as a principle of origin, such as *water* (Thales), *fire* (Heraclitus), or *air* (Anaximenes). Soon the point of gravity moved towards the idea of static forms of being (ontic forms), considered to be immutable and eternal.

Of course the subsequent development should take into account the significant role of the school of Pythagoras. The contribution of this school is that it articulated the insight that *rational knowledge* cannot be divorced from *numerical relationships*. Naturally, this school went too far in its one-sided claim that *everything is number*. This thesis rests on the conviction that, with the aid of the relation between integers, i.e. by merely using normal fractions, it is possible to describe whatever exists in numerical terms. What eventually became known as *atomism*, as noted above (see pages 60 ff.), should be seen as a continuation of this tendency towards over-emphasizing the numerical aspect as a mode of explanation. Atomism was begun by the Greek atomists (Leucippos and Democritus); it is operative in the social contract theories of the pre-Enlightenment and Enlightenment eras, and even in the *methodological individualism* of the 20th century (Von Hayek, Watkins, Weber, Simmel, Popper and others).

Cassirer points out that, amongst the "fundamental concepts of pure science, the concept of number stands in the first place, both historically and systematically" (Cassirer, 1953:27). He furthermore states that number enables a consciousness of the meaning and value of the formation of concepts as such. Without number, neither things as such nor their interrelations would have been accessible to rational contemplation. Yet the Pythagorean thesis went too far:

> The claim to grasp the substance of things in number has indeed been gradually withdrawn; but at the same time the insight that the substance of rational knowledge is rooted in number, has been deepened and clarified. Even when the metaphysical kernel of the object is no longer observed in the concept of number, it remains the first and truest expression of rational method in general. In it are directly reflected the differences in principle between the fundamental interpretations of knowledge. Through number the general ideal of knowledge gains a more definite form, in which for the first time it is defined with full clarity (Cassirer, 1953:27).

However, soon the developments within Greek culture became sensitive to *spatial configurations* – such as the shape of calyx leafs in nature, for this shape appeared as an instantiation of a *regular pentagram*. An investigation of the geometrical properties of a regular pentagram led to the discovery that it is not possible to express the ratio between any side and any diagonal of the regular pentagram with the aid of normal fractions, i.e. in terms of the *ratio* of two whole numbers / integers: a/b. At this stage, 'number' simply meant *natural number* and the relation between natural numbers (i.e. fractions). Every relation between natural numbers could be presented in a geometric way, but from the example of the regular pentagram, it is clear that it is not possible to present every relation between spatial magnitudes in terms of natural numbers or their relations.[1]

This limitation at once embodied the discovery of 'incommensurable' quantities – something completely unacceptable for the Pythagoreans, because suddenly within the limiting and form-giving nature of number the *apeiron* (the unbounded-infinite) appeared, i.e. irrational numbers were discovered.

Flowing from the inherent tension between what *is limited* and what *is unlimited* (the *peras* and the *apeiron*) in Greek thought, the discovery of irrational numbers (or in modern mathematical terms: real numbers) inspired the search for an alternative principle of explanation – one that can escape from the unbounded (infinite) present in number.

The alternative mode of explanation was found in *space*. This aspect allowed for the acceptance of *static forms* and yielded the possibility of observing any spatial figure *at once*, without any *before* or *after*. The implication was that the acquisition of concepts is enclosed within the *now*, and in the school of Parmenides, this resulted, as we have seen (see page 254 above), in the equation of *thought* and *being*.

It is known that, on the basis of Babylonian observations, Thales accurately predicted an eclipse of the sun in the year 585 B.C. He also had the remarkable geometrical skill to calculate the height of a pyramid from a sun shadow of 45^0 (keeping in mind that a pyramid differs from something like a tree, where it is possible to establish its height perpendicular to its base). Thales also knew that the diagonals of a rectangle are equidistant and divide each other in equal parts. According to Lorenzen the starting point for geometry as a coherent theoretical system was provided by Thales (Lorenzen, 1960:45-46).

The important feature of this development is that the spatial figures of Greek geometry were *idealized*. This means that a straight line, circle and square are not perceptible in a sensory way – they can only be contemplated intellectually. Plato's account of human knowledge reflects this conviction, because he explicitly states that the conclusions do not rely upon "sensory objects":

> Then by the second section of the intelligible world you may understand me

1 Laugwitz remarks: "Jedes Zahlenverhältnis läßt sich geometrisch darstellen, aber nicht jedes Streckenverhältnis arithmetisch" and adds that this provides the foundation for the primacy of geometry (above arithmetic) – with the implication that in the Books of Euclid number theory became a part of geometry ("Das begründet einen Vorrang der Geometrie vor der Arithmetik, und die Konsequenz sind die Bücher des Euklid: Die Theorie der Zahlen ist ein Teil der Geometrie" – Laugwitz, 1986:10).

> to mean all that unaided reasoning apprehends by the power of dialectic, when it treats its assumptions, not as first principles, but as *hypotheses* in the literal sense, things 'laid down' like a flight of steps up which it may mount all the way to something that is not hypothetical, the first principle of all; and having grasped this, may turn back and, holding on to the consequences which depend upon it, descend at last to a conclusion, never making use of any sensible object, but only of Forms, moving through Forms from one to another, and ending with Forms (*Politeia*, 510D).

Plato's dialogue, Meno, where the leader of the conversation used leading questions in order to allow the conversation partner to produce geometrical proof, solicited Oskar Becker's remark that this gave birth to and appreciation of the *a priori* nature of mathematics (Becker, 1965:X).[1]

The effect of the discovery of irrational numbers was not only that mathematics was geometrized, for it also paved the way for a speculative theory of reality, attempting to explain the entire universe in terms of a *spatial perspective* – as a substitute for the outdated arithmetical orientation of the Pythagoreans. The implication was that Greek thought now understood *matter* in terms of spatial extension. An entity is identified with the place it occupies. Something *is* its place.

It should be noted, however, that Parmenides hardly possessed an independent space concept. He also did not contemplate the idea of *empty space*. When something *is* its place, then the absence of something implies that the subject to which the predicate 'place' applies is not present. Herman Fränkel writes: "With the assertion of a complete filling of space ... the existence of a mere empty space is rather denied than acknowledged."[2]

In its denial of movement, the school of Parmenides, and in particular the arguments of Zeno (see Chapter 5, pages 172 ff.), merely formulated the consequences of over-emphasizing the spatial aspect as a mode of explanation: if something indeed *is* its place, then it can never move, for passing from one place to another entails a change of essence.

The metaphysical over-extension of the *static* nature of space even motivated a remarkable denial of the spatial whole-parts relation. In Ryle's *Dilemmas* the arguments are based on the assumption of the "uncompleted infinite." In addition, he introduces the *whole-parts relation* (with reference to the classical Aristotelian slogan that the "the whole is more than the sum of its parts."[3]

1 On the next page (note 2) Becker proceeds by pointing out that the Aristotelian theory of the infinite and the continuum is still relevant to an adequate foundation of mathematical analysis: "So ist die aristotelische Theorie des Unendlichen und des Kontinuums in ihrer eigenartigen Fragestellung noch von aktueller Bedeutung für das Problem einer wirklich adäquaten Begründung der höheren Analysis" (see also Becker, 1964:69.)

2 "Mit der Behauptung einer volständigen Ausfüllung ... wird der bloße, eventuell leere Raum eher geleugnet als anerkannt" (Fränkel, 1968:181, note 4).

3 "Therefore the state, according to its nature, is prior to the family and the individual, since the whole must precede the part." *καὶ πρότερον δὲ τῇ φύσει πόλις ἢ οἰκία καὶ ἕκαστος ἡμῶν ἐστίν. τὸ γὰρ ὅλον πρότερον ἀναγκαῖον εἶναι τοῦ μέρους* – Aristotle, *Politica*, 1253 a 19-20.

Ryle says that the question, "how many parts have been cut off from an object?" must be distinguished from the question "in how many parts did you divide it?" (Ryle, 1977:61). The first point proceeds from a notion of wholeness containing all its (finite) parts, whereas the second reverts the perspective and explores an ongoing process of division.

This distinction actually imitates B Fragment 3 of Zeno, in which he argues as follows (in the translation of Guthrie): "if there is a plurality, it must contain both a finite and an infinite number of components: finite, because they must be neither more nor less than they are; infinite, because if they are separate at all, then however close together they are, there will always be others between them, and yet others between those, *ad infinitum*" (Guthrie, 1980:90-91). Therefore, assuming a plurality leads to the contradictory conclusion that it contains at once "a finite and an infinite number of components." But given the fact that Parmenides and his school, as an effect of the discovery of irrational numbers, switched from an *arithmetical* mode of explanation to a *spatial* one, we may once more briefly explain the *spatial* whole-parts relation in order to understand what is at stake here. Our argument was that whatever is continuously extended is a coherent whole in the sense that all its parts are connected – therefore we concluded that the terms *coherence* and *connectedness* are synonyms for the terms *wholeness*, *totality* and *continuity*. A whole (or totality) contains all its parts. Paul Bernays affirms that 'wholeness', i.e. the *totality-character* of spatial continuity, will resist a "perfect arithmetization of the continuum" (see Bernays, 1976:74 – see page 61 above). If the plurality of Zeno's first argument refers to a perspective from the parts to the whole, then the number of these parts must be limited, because they constitute at once the world as a whole (the universe). In contrast, if the argument proceeds from the whole to the parts, the *infinite divisibility* operative in this move entails that "there will always be others between them" and so on indefinitely. Fränkel explicitly employs the whole-parts relation to explain the meaning of Zeno's B Fragment 3 (see Fränkel, 1968:430).[1] Perhaps Zeno's B Fragment 3 could be seen as the first "two-directional" discussion of the spatial whole-parts relation.[2]

In order to understand this properly, we have to keep in mind what we have explained earlier, namely that whatever is continuously extended in a spatial sense allows for an infinite divisibility. The spatial whole-parts relation turns the original numerical meaning of succession – the successive infinite – 'inwards', embodied in the successive infinite divisibility of a continuum. In terms of the inter-modal coherence between number and space and in the light of the foundational role of number, it indeed belongs to the meaning of the

1 Guthrie has a positive appreciation of this article by Fränkel. He refers to an English translation: *Zeno of Elea's Attacks on Plurality* (see Guthrie, 1980:88 ff., 512).

2 In response to the phrase "If they are just as many as they are, they will be finite in number" Russell states: "This phrase is not very clear, but it is plain that it assumes the impossibility of definite infinite numbers" (Russell in Salmon, 2001:47). Clearly Russell did not explore a "two-directional" use of the spatial whole-parts relation suggested above.

spatial whole-parts relation that it contains the possibility of endless divisions.

In connection with Zeno's paradoxes, we advanced an alternative approach, holding that, only when the *uniqueness* and *mutual coherence* of number, space and movement are observed, is it possible to avoid the threat of antinomies inherent in Zeno's arguments against plurality and motion.

The spatial metaphysics of Parmenides, inspired Zeno to defend a view of *unitary wholeness* that *excludes* plurality. In other words, Zeno wants to *deny* the 'part' element of the spatial whole-parts relationship, while at the same time holding on to the 'wholeness' that entails it.

Comment: This is a quasi-Wittgensteinian position. Whereas Wittgenstein had to throw away the ladder after climbing it (*Tractatus*, 6.54), Zeno started on top, with wholeness, and then discarded the ladder of infinite divisibility supporting it. The reverse took place in intuitionist mathematics, which started with the original spatial whole-parts relation, but then distorted it by accentuating the part element (with its implied infinite divisibility) at the cost of the element of wholeness (with its givenness all at once). The intuitionistic theory of the real numbers and the continuum followed a similar kind of Wittgensteinean approach – it used the "spatial ladder of wholeness," but immediately discarded it while preserving the infinite divisibility it implied.

Zeno holds that reality is that it is both *one* and *indivisible*. Yet, in order to *argue* for this position, he explores the whole-parts relation in an argument aimed at the *denial* of plurality! The reason for Zeno's consideration of plurality as self-contradictory is that it requires a *number* of (indivisible) units, and because of its implication that reality is *divisible* (see Guthrie, 1980:88). But divisibility threatens the wholeness of a *unit*, since anything divisible has to be a magnitude that is infinitely divisible. The supposed *indivisibility* of a *unit* clashes with its *infinite divisibility*. "Hence, since plurality is a plurality of units, there can be no plurality either" (Guthrie, 1980:89).

Therefore, by denying the *foundational meaning* of multiplicity, Zeno not only distorts the meaning of *number* (plurality), but also misrepresents the meaning of *space*. The infinite divisibility of a spatial whole analogically reflects the original and primitive numerical meaning of the successive infinite. For that reason we argued that through spatial continuity the endlessness of the numerical infinite is "turned inwards." But divorced from its connections with number, the meaning of space collapses (also compare our analysis of the spatial subject-object relation – see pp.222 ff.). The original numerical meaning of the number *one* appears in a non-original mode within space, for within the latter the magnitude of an extended spatial figure (as a whole) provides a different context for unity – a unity that is *infinitely divisible* in its totality. The speculative (metaphysical) notion of a *unitary whole* excluding plurality robs both number and space of their meaning and mutual connections. Moreover, within the domains of space, movement and the physical we can discern numerical analogies – just think of the *length* of a line (specified by a number); consider the *speed* of a moving body as expressed by a number; and the *mass* of a physical body.

Zeno's antinomies (including those of Achilles, the tortoise and the flying arrow) represent the starting-point of a long speculative tradition in which the meaning of space was metaphysically explored within the context of a speculative theory of *being* that finds in God – as the Highest Being (*ipsum esse*) – its conclusion.

By exchanging two *modes of explanation* and attempting to strip them of their intrinsic connections, multiple distortions are caused. The school of Parmenides realized that space provides an original mode of explanation, but in an attempt to 'purify' space of number, it challenged a foundational condition of space, given in the nature and meaning of *multiplicity*. By ignoring the foundational role of a numerical multiplicity, Zeno distorted the meaning of number, and also at once skewed the meaning of space by questioning the divisibility of a spatial continuum. The infinite divisibility of a continuous whole within space is always a reminder of the original successive meaning of number at the basis of space (see pages 163 and 221 above). Just as it is impossible to separate space from other aspects of reality, neither can it be separated from the numerical aspect. Even in the most extreme examples of arithmeticism in modern mathematics, aiming at reducing space to number, certain key features of space were essential. In the case of axiomatic set theory, the use of undefined ('primitive') terms derived from the spatial whole-parts relation, such as *set* or *membership* (see Fraenkel et al., 1973:21 ff.) turned out to be unavoidable. And likewise even the quasi-arithmeticism of intuitionism had to speak of a numerical *domain*.

The original numerical meaning of the number one as an integer analogically appears within the spatial aspect. The *unity* of a spatial subject is found in its *wholeness*. In other words, a spatial unit is constituted as a genuine *whole* or *totality*, a unitary whole, allowing an *infinite divisibility*. The speculative (metaphysical) idea of a unitary whole *precluding* multiplicity robs both number and space of their unique meaning and their mutual coherence.

In respect of the nature of material things, the most important consequence is that the Greek-Medieval legacy only acknowledges *concrete material extension*.

7.1.2 *Transition to the modern era – extension challenged*

Within the Aristotelian legacy, it was believed that celestial bodies obey laws that are different from those for entities on earth. In addition, it was believed that the movement of anything required a *cause*. The problem of motion acquired an increasingly prominent position, although it did not mean that the powerful influence of classical space metaphysics immediately lost its hold. The power of this spatial orientation is indeed still evident in the thought of Descartes (1596-1650) and even Immanuel Kant (1724-1804). In their understanding of nature, both philosophers assign a decisive role to *spatial extension*. For Descartes, *extension* serves as the essential characteristic of material bodies – *res extensa*, for he writes: "That the nature of body consists not in weight, hardness, colour, and the like, but in extension alone" (Descartes,

1965a:200 – Part I, IV). Kant's characterization of material bodies is also oriented toward space. When our understanding leaves aside everything accompanying their representation, such as substance, force, divisibility, etc., and likewise also separates that which belong to sensation, such as impenetrability, hardness, color, etc., then this empirical intuition leaves something else, namely extension and shape.[1]

It should not surprise us therefore that Descartes straightaway applies the feature of (mathematical) continuity to material things and even to atoms that, since Greek antiquity, were supposed to be the final indivisible material particles. He holds that there cannot be atoms or material particles that are inherently indivisible.

> We likewise discover that there cannot exist any atoms or parts of matter that are indivisible of their own nature (Descartes, 1965a:209; Part I, XX).

In this context (XX), he even introduces the idea of God in order to promote the infinite divisibility of matter. He argues that, although God can make a particle so small that no creature can divide it, this does not set any limits to the divine capacity to divide. Therefore, it should be assumed that matter is indeed infinitely divisible:

> Wherefore, absolutely speaking, the smallest extended particle is always divisible, since it is such of its very nature.

That Descartes continues to hold on to *extension* as the essential trait of matter, embodies his connection with the long-standing Greek-Medieval tradition. However, what he has to say regarding the infinite divisibility of matter breathes the spirit of the early modern *functionalistic* orientation.[2] This new functionalistic attitude soon attempted to explain concrete things entirely in *functional* terms, i.e. in terms of *relations*. Yet at the same time Descartes pays attention to motion, which he defines as the "action by which a body passes from one place to another" (Descartes, 1965a:210; Part I, XXIV). This new point of view finds itself at the crossroads of the transition from the medieval to the modern era.

7.1.3 *Motion as the new principle of explanation*

Although Buridan (early 14th century) contributes to uprooting the dominant position of spatial extension and the transition to the modern era, it should be kept in mind that the *impetus* idea itself cannot be equated with the nature of *inertia*. There is a conception that the mechanics of Buridan and classical physics are fundamentally similar, entailing that the *impetus theory* has already practically brought to expression the *law of inertia*.

This convergence is sought first of all in the supposed correlation between

1 "So, wenn ich von der Vorstellung eines Körpers das, was der Verstand davon denkt, als Substanz, Kraft, Teilbarkeit usw., imgleichen, was davon zur Empfindung gehört, als Undurchdringlichkeit, Härte, Farbe usw. absondere, so bleibt mir aus dieser empirischen Anschauung noch etwas übrig, nämlich Ausdehnung und Gestalt" (Kant, 1781/1787-B:35).

2 Functionalism reduces entities to functions, while substantialism reduces functions to entities.

the scholastic view of impetus and the dynamic element of inertial motion. In the second place, it was believed that the assumption of *permanence* in Buridan's view of impetus has already discovered the law according to which completely undisturbed motion will be everlasting (cf. Maier, 1949:142).

It is indeed striking that an *impetus* transferred in a celestial sphere (following a *circular* path) was supposed to be free from any resistance. Therefore it seemed proper to compare it with the underlying idea of inertia. Yet inertia concerns uniform (rectilinear) motion; not *circular* motion. In the case of the impetus theory, there was a difference between what happened in the heavens and what happened on earth. According to the Medieval Scholastic understanding, a force without resistance cannot produce motion. But since *impetus* is artificially and forcefully superimposed upon some or other obstacle (i.e. an interfering force of motion), such an obstacle could only be overcome when the impetus itself was altered in the process of overcoming resistance. However, the decisive difference between these two views lies in that element of inertial movement from which one cannot abstract, namely the inertia of a mass-point.[1] It is possible to abstract from external obstacles and forces, but it is impossible to abstract from that which is crucial for inertia, namely the *mass* of whatever moves. According to classical mechanics, the latter is the true factor in the continuation of movement (inertia). In the case of the impetus theory, inertial mass serves as resistance (obstacle) for the movement of the impetus that caused it. Consequently there is an unbridgeable gap between the impetus theory and the basic idea of inertia, namely the possibility of an everlasting rectilinear motion.

In various contexts, we mentioned Galileo's formulation of the law of inertia on the basis of his *thought experiment* concerning a moving body on a friction-free path extended into infinity (see page 72 above). Opposed to the traditional Aristotelian-Scholastic conception according to which the movement of a body is dependent upon a *causing force*, the law of inertia implies that motion is something *given*, and that therefore, instead of trying to deduce or explain it, it should be accepted as a mode of explanation in its own right. Motion is original and unique; indeed, it embodies a distinct *mode of explanation* that is different from those used by the Pythagoreans (number) and the Eleatic school of Parmenides (space). If motion does not need a causing force, then at most, it is possible to speak of a *change* of motion (*acceleration* or *deceleration*) – which does need *physical force*. A well-known German physicist remarks:

> Since the law of inertia has shown that no force is required for a change of place the most natural thing to do is to accept that force causes a change of speed, or, as Newton says, the magnitude of motion ('Bewegungsgröße') (Von Weizsäcker, 2002:172).[2]

1 In classical mechanics the simplest subject is a mass-point.

2 "Da das Trägheitsgesetz gezeigt hat, daß keine Kraft nötig ist für eine Änderung des Orts, ist es das natürlichste, anzunehmen, die Kraft verursacht eine Änderung der Geschwindigkeit, oder, wie Newton sagt, der Bewegungsgröße."

The idea of a uniform (rectilinear) motion on the one hand expands the inherent limitations attached to number and space as modes of explanation, and on the other, it at once opens the way to consider another problem that has already captured Greek thought. This problem concerns the relation between *persistence* (think about the nature of inertia) and *dynamics* (consider the change of motion requiring a physical force).

In our discussion of the order of the modal aspects of reality we appreciated the contribution of Heraclitus and Plato to our understanding of the relationship between constancy and dynamics (change – see pp. 163 ff. above). Plato's important insight is that change can only be established on the basis of constancy (persistence) – i.e. without an enduring subject, there is nothing to "hold on to;" nothing to which the alleged changes can be attributed. Of course this insight does not force us to join the speculative account Plato gave in his metaphysical theory of static, super-sensory ideal forms – although it is true that his solution does form a lasting attraction for many scholars. We have seen that even Frege said that, without something lasting guaranteeing knowability, everything would collapse into confusion.[1]

The proper elaboration of Plato's insight, namely that change presupposes constancy, is found in Galileo's formulation of the law of inertia and in Einstein's theory of relativity. The core idea of Einstein's theory is after all the constancy of the velocity of light in a vacuum. Although he often merely speaks of "the principle of the constancy of the speed of light,"[2] he naturally intends "the principle of the vacuum-velocity" ("das Prinzip der Vakuumlichtgeschwindigkeit" – see Einstein, 1982:30-31; and 1959:54). It follows that Einstein primarily aimed at a theory of constancy – whatever is in motion moves relative to an element of constancy. It was merely a concession to the historicistic *Zeitgeist* at the beginning of the 20th century that he gave prominence to the term 'relativity' – all movement is relative to the constant *c*.

However, a certain ambiguity can still be found in the thought of Descartes and his followers, for in spite of the fact that they viewed *extension* as the essential property of matter, they simultaneously pursued the kinematic ideal to explain everything that exists and happens exclusively in terms of movement (cf. Maier, 1949:143).[3] It is generally known that Thomas Hobbes took the full step towards the exploration of movement as a principle of explanation in his intended rational reconstruction of reality. According to the newly established natural science ideal, he first demolished reality to a heap of chaos in order subsequently to build up, step by step, a new rationally ordered cosmos, guided by the key concept of "moving body." His acquaintance with Galileo's mechanics enabled him to exceed the limits of space as a mode of explanation. Galileo himself embodies the long history of our understanding of matter up

1 See page 165 above, where the German quotation is given.

2 "das Prinzip der Konstanz der Lichgeschwindigkeit" – cf. Einstein, 1982:32.

3 Maier remarks that "Descartes und seine Schule" indeed pursued a "rein phoronomisches Ideal" and attempted to explain "alles Sein und Geschehen in der Welt lediglich aus Bewegungen."

to this phase of its development, because he explicitly explores the three modes of explanation thus far highlighted in our discussion. He accounts for arithmetical properties (countability), geometrical properties (form, size, position and contact), and kinematic features (motion).[1] Leibniz continues this legacy in his belief that physical events can be explained mechanistically in terms of magnitude, figure, and motion. On October 9, 1687 Leibniz writes in a letter that we "must always explain nature mathematically and mechanically" (Leibniz, 1976:38). In a footnote the Editor of Leibniz's work writes that the latter's approval of Boyle's corpuscular philosophy ought to be understood as "any philosophy which explains physical events mechanistically or in terms of magnitude, figure, and motion" (Leibniz, 1976:349, note 14).

Writing on the foundations of physics, David Hilbert refers to the mechanistic ideal of unity in physics, but immediately adds the remark that we now finally have to free ourselves from this untenable ideal (cf. Hilbert, 1970: 258).[2]

As we have noted, as soon as the kinematic mode of explanation is acknowledged in its own right, the necessity of finding a cause for motion disappears. The classical opposition between *being at rest* and *moving* is therefore untenable because, from a kinematic perspective, 'rest' is a state of movement (cf. Stafleu, 1987:58). Unique and irreducible modes of explanation are not *opposites* – for they are mutually cohering and irreducible.[3]

7.1.4 *Force and energy-operation: another mode of explanation*

Although Descartes and Newton employ the concept of force, it may in general be said that modern physics since Newton is mainly characterized by a *mechanistic* tendency. The mechanistic view consistently attempts to reduce all physical phenomena to a *kinematic perspective*. However, in the course of the 19th century, modern physics already started to explore the nature of energy. The founder of physical chemistry, Wilhelm Ostwald, developed his so-called *Energetik* (enegetics), which even influenced the later views of Heisenberg. Vogel refers to Heisenberg's work "Wandlungen in den Grundlagen der Naturwissenschaft" (Stuttgart, 1949), where the latter explicitly speaks of energy as the basic stuff that constitutes matter in its three-fold stable forms: electrons, protons and neutrons (Vogel, 1961:37). Yet Ostwald's *Energetik* did not exert a lasting influence upon the physics of the 20th century, probably because it was attached to a specific view of continuity that is opposed to an atomistic approach. Niels Bohr in particular mentions the ex-

1 "G. Galilei zählt als primäre Qualitäten der Materie arithmetische (Zählbarkeit), geometrische (Gestalt, Größe, Lage, Berührung) und kinematische Eigenschaften (Beweglichkeit) auf" (Hucklenbroich, 1980:291)

2 It is therefore strange that the contemporary physical scientist from Cambridge, Stephen Hawking, still writes: "The eventual goal of science is to provide a single theory that describes the whole universe" (Hawking, 1988:10).

3 For this reason, we have noted that also number and space also ought not to be seen as *opposites*, as asserted by Lakoff and Núñez (2000:324) owing to their inability to appreciate the unique and mutually cohering nature of these aspects.

cessive scepicism found in Mach's thought regarding the existence of atoms.[1]

The last prominent physicist who consistently adhered to the mechanistic approach was Heinrich Hertz. We have seen that, soon after Hertz's death in 1894, the work in which he attempted to restrict the discipline of physics to the concepts of mass, space and time (see page 264), reflecting the three most basic modes of explanation of reality, namely the modes of number, space and movement, appeared: "The Principles of Mechanics developed in a New Context." This caused him (and Russell) to view the concept of *force* as something intrinsically antinomous.

The Latin designation of *mass* during the medieval period was "quantitas materiae" (see Maier, 1949:144). It consequently appears that number (quantitas) plays a key role in the concept of *mass*. Mass concerns a *physical* quantity, but it is also possible to observe the quantity of energy from the perspective of the *kinematic* modality. In this case, the technical expression *kinetic energy* indicates the action capacity inherent in a moving body (see Maier, 1949:142).

Comment: Von Weizsäcker, Kant and Einstein

> More recently, it is remarkable to see that Von Weizsäcker, in his treatment of the Kantian conception of matter and his view of pure physics, still appreciates the concepts of motion and speed as the basic concepts of all genuine science. (Von Weizsäcker, 2002:196, note). Of course one is immediately reminded of the famous statement by Kant: "However, my claim is that in every particular theory of nature only that much science is found as the amount of mathematics present in it."[2] In contrast, Einstein apparently opts for the *opposite* view when he says, "Insofar as the propositions of mathematics are related to reality they are not certain and in so far as they are certain they are not related to reality."[3]

As soon as the physical aspect of reality surfaced, it opened up the way for 20th century physics to explore it as an independent mode of explanation and to arrive at an even more nuanced understanding of reality. For example, in his *protophysics*, Paul Lorenzen distinguishes four units of measurement reflecting the first four modes of explanation: *mass*, *length*, *duration* and *charge* (Lorenzen, 1976:1 ff.). It is noteworthy to mention that Heisenberg, accepting two universal constants (Einstein's postulate of the velocity of light and Planck's quantum of action), was looking for a third universal constant, namely a universal length. He claims that one has to have at least *three* units – be they *length*, *time* and *mass* or replaced by *length*, *velocity* and *mass* or even *length*, *velocity* and *energy* (Heisenberg, 1958:165). An analysis of the first

1 See Niels Bohr, Atomtheorie und Naturbeschreibung (Berlin, 1931:60 and 12, quoted by Vogel, 1961:35). It should be kept in mind that Mach's views ought to be understood against the background of Ostwald's position.

2 "Ich behaupte aber, daß in jeder besonderen Naturlehre nur soviel eigentliche Wissenschaft angetroffen werden könne, als darin Mathematik anzutreffen ist" (Kant, 1786:IX).

3 "Insofern sich die Sätze der Mathematik auf die Wirklichkeit beziehen, sind sie nicht sicher, und insofern sie sicher sind, beziehen sie sich nicht auf die Wirklichkeit" (Einstein, 1921:124).

four modal aspects would have helped to realize that *four* are needed. Clearly these four units of measurement reflect the meaning of the first four aspects of reality, namely number ('mass'), space ('length'), the kinematical aspect ('duration') and the physical aspect ('charge'). Weinert mentions even that usually physicists "distinguish fundamental constants from *conventional* units" – and he then lists the *kilogramme* (number), the meter (space), the second (the kinematic) and temperature (the physical) (Weinert, 1998:230; and see also Lorenzen, 1989).

A decade after Max Planck discovered his 'Wirkungsquantum', he explicitly addressed the intrinsic untenability of the mechanical understanding of reality.

> The conception of nature that rendered the most significant service to physics up till the present is undoubtedly the mechanical. If we consider that this standpoint proceeds from the assumption that all qualitative differences are ultimately explicable by motions, then we may well define the mechanistic conception as the conviction that all physical processes could be *reduced completely to the motions* (my italics – DS) of unchangeable, similar mass-points or mass-elements (see above page 164).[1]

Einstein is equally explicit in his negative attitude towards "the mechanistic framework of classical physics" (see Einstein, 1985:146).

During the first decades of the 20th century, this growing awareness of the limitations in the formerly dominating mechanistic approach paved the way for a more generally accepted view into the nature of the fourth principle of explanation operative in the history of our understanding of material things.

Eventually the distinction between the kinematic and physical aspects of reality thus became common knowledge. We have argued for the importance of a proper understanding of the relation between *constancy* and *change* (see pages 17 and 164 above). The critical insight is that one cannot refer to (physical) change without acknowledging the presence of persistence at its basis. This insight is confirmed by a thinker who is unfamiliar with the reformational philosophical tradition. Undeniable states of affairs transcend the confines of any particular theoretical paradigm!

According to Janich, the scope of an exact distinction between *phoronomic* (subsequently called *kinematic*) and *dynamic* arguments can be explained by means of an example. Modern physics has to employ a dynamic interpretation of the statement that a body can only alter its speed continuously. Given certain conditions, a body can never accelerate in a discontinuous way, that is to say, it cannot change its speed through an infinitely large acceleration, be-

---

1 "Diejenige Naturanschauung, die bisher der Physik die wichtigsten Dienste geleistet hat, ist unstreitig die mechanische. Bedenken wir, daß dieselbe darauf ausgeht, alle qualitativen Unterschiede in letzter Linie zu erklären durch Bewegungen, so dürfen wir die mechanische Naturanschauung wohl definieren als die Ansicht, daß alle physikalischen Vorgänge sich vollständig auf Bewegungen von unveränderlichen, gleichartigen Massenpunkten oder Massenelementen zurückführen lassen" (Planck, 1973:53).

cause this would require infinite force.[1]

The idea of an attracting force, initially conceived of in connection with magnetism, eventually brought Newton to the insight that magnetism is a force that cannot be explained through motion, although, foundational to the physical aspect, motion is a mode of explanation in its own right. Stafleu points out that the rejection of the Aristotelian distinction between the physics of celestial bodies and the physics of things on earth paved the way, in the footsteps of Galileo and Descartes, to realize that the same physical laws apply to both domains, i.e. that physical laws display modal universality (i.e. they hold universally) (Stafleu, 1987:73). He also remarks that Newton (like Kepler) has indeed already appreciated *force* positively as a principle of explanation that is distinct from motion as an original principle of explanation (see Stafleu, 1987:76). Stafleu summarizes this process, through which the physical aspect emerged as an equally original mode of explanation, as follows:

> In Newtonian mechanics, a force is considered a relation between two bodies, irreducible to other relations like quantity of matter, spatial distance, or relative motion. Though an actual force may partly depend on mass or spatial distance, as is the case with gravitational force, or on relative motion, as is the case with friction, a force is conceptually different from numerical, spatial or kinematic relations (Stafleu, 1987:79).

From the introduction of Niels Bohr's atom theory in 1913, and actually since the discovery of radio-activity in 1896, and of the energy quantum *h*, modern physics had already realized that matter is indeed characterized by physical energy-operation. It is therefore understandable that 20th century physics eventually had to reach a general acknowledgement of the decisive significance of energy-operation for the nature and understanding of the physical world, as it is strikingly captured in Einstein's famous formula:

$$E = mc^2$$

Another realization was that physical processes are *irreversible*. (We have to keep in mind that the time order within the first three modal aspects is reversible.) In itself, this observation justifies the distinction between the kinematic and physical aspects of reality. Both Planck and Einstein knew that, in terms of a purely kinematic perspective, all processes are *reversible*. Einstein refers

---

1 "Die Tragweite einer strengen Unterscheidung phoronomischer (im folgenden kinematisch genannt) und dynamischer Argumente möchte ich an einem Beispiel erlautern, das ... aus der Protophysik stammt. Die Aussage 'ein Körper kann seine Geschwindigkeit nur stetig ändern' kann von der modernen Physik nur dynamisch verstanden werden. Geschwindigkeitänderungen sind Beschleunigung, d.h. als Zweite Ableitung des Weges nach der Zeit definiert. Zeit wird von der Physik als ein Parameter behandelt, an dessen Erzeugung durch eine Parametermaschine ('Uhr') de facto bestimmte Homogenitätserwartungen geknüpft sind ... Bezogen auf den Gang einer angeblich so ausgewählten Parametermaschine kann eine Körper seine Geschwindigkeit deshalb nicht unstetig, d.h. mit unendlich große Beschleunigung änderen, weil dazu eine unendlich große Kraft erforderlich wäre" (Janich, 1975:68-69).

to Blotzmann, who realized that thermodynamic processes are irreversible.[1] In 1824, Carnot already discovered irreversible processes – as we have noted (see page 319). Since 1850 Clausius and Thompson independently developed the second main law of thermodynamics, now known as the law of non-decreasing entropy. This law accounts for the fundamental irreversibility of natural processes within any *closed system*. Clausius introduced the term *entropy* itself only in 1865. In 1852, Thomson explained that, according to this law, all available energy strives towards uniform dissipation (see Apolin, 1964:440 and Steffens, 1979:140 ff.). Planck remarks that "the irreversibility of natural processes" confronted "the mechanical conception of nature" with "insurmountable problems" (Planck, 1973:55).

It is only on the basis of an insight into the foundational position of the kinematic aspect in respect of the physical aspect that an appropriate designation of the first law is made possible. Although we are used to employing the familiar designation of the law of *energy conservation*, we have noted that there is an element of ambiguity attached to the term 'conservation' – as if energy is "held onto." When, on the law side, the retrocipation from the physical aspect to the kinematic aspect is captured by the phrase *energy constancy*, this ambiguity disappears, and we have a concise and precise formulation of this law (see pp.319 ff. above).

7.2 The mystery of matter

The preceding historical sketch makes it clear that, although each of the four modes of explanation opened up a legitimate angle of approach, none can claim to be the *exclusive* and/or *exhaustive* source of our knowledge of material things. Whatever their worth, they merely provide us with a *partial* perspective, one that will always be co-determined by a *totality view* exceeding the scope of any specific mode of explanation (see page 179 above). Such a totality perspective actually exceeds the scope of any special science, since it inevitably rests upon one or another philosophical view of reality. A number of issues that we have discussed within a different context will recur in our present analysis; particularly the problem of individuality (and what is individual) and the necessity of employing modal terms in concept-transcending ways.

These considerations are intimately connected with the modal universality of each modal aspect (in connection with the nature of modal universality, see page 68 above). Acknowledging the modal universality of the different modal aspects is constitutive for an account of typicality, individuality and concept-transcending knowledge (idea-knowledge).

The impressive power of theoretical thinking first of all derives from exploring the *modal universality* of specific modal aspects. The philosophically informed physicist Von Weizsäcker implicitly draws upon this insight when he appreciates quantum theory as the *central theory* of contemporary physics.

1 "Er hat damit das Wesen der im Sinne der Thermodynamik 'nicht umkehrbaren' Vorgänge erkannt. Vom molekular-mechanischen Gesichtspunkte aus gesehen sind dagegen alle Vorgänge umkehrbar" (Einstein, 1959:42).

His explanation highlights the modal universality of the physical aspect, for this modal universality is not restricted by the typical nature of any (type of) entity – it cuts across all typical differences. We have noted that Von Weizsäcker says:

> Quantum theory, formulated sufficiently abstractly, is a universal theory for all classes of entities (Von Weizsäcker, 1993:128).[1]

In addition to this appeal to modal universality, Von Weizsäcker explicitly articulates the fundamental philosophical insight that *everything coheres with everything*: ("Alles hängt mit allem zusammen" – 1993:134).

The modal universality of the four modes of explanation discerned in their successive decisive roles during the history of our understanding of physical nature, entails that the scope of each is *unspecified*. This means that whatever concretely exists, functions within each mode of existence. We have referred to the law of gravity (see page 68 above) as an example of the unspecified modal universality of the physical aspect.

But as soon as the typicality of things is acknowledged, the unspecified universal meaning of the aspects acquire a *typical specification*, for in this case the *effect* of specific *type laws* is manifested within the modal aspects themselves.[2] Type laws are still universal, but no longer in an *unspecified* sense. For example, although the law for being an atom (see page 5) holds for *all atoms* (its *universality*); it does not hold for all kinds of entities (like molecules, planets, animals or states – revealing its specification - for it holds for atoms *only*).

No single entity 'escapes' its (typically specified) function within each modal aspect of reality. Naturally, this applies to the first four modal aspects in particular. As such, this insight already explains why, as we saw from the history of reflection on the nature of material things, every one of these aspects was elevated to become the sole explanatory principle of physical reality during some or other historical phase.

Modal universality and the specificity of type laws are intimately intertwined. An understanding of the former paves the way towards an appreciation of the nature of type laws, while holding on to the insight that all possible kinds of entities in principle function in all modal aspects. However, it is not yet possible to identify what is *uniquely typical* in respect of any particular kind of entity, merely in terms of the universal scope of the modal aspects.

For example, although Newton's formulation of the law of gravity highlights the universality of the physical aspect, his entire *Principia Mathematica* is still permeated by geometrical language and arguments. This shows its de-

1 "Die Quantentheorie, hinreichend abstrakt formuliert, ist eine universale Theorie für alle Gegenstandsklassen."

2 An initial explanation of this important distinction between modal laws and type laws is found in Chapter 3 (see pages 25 ff.). We have pointed out that it can be said in general that modal laws hold for all possible classes of entities, whereas type laws hold for a limited class of entities only.

pendence upon the Greek-Medieval space metaphysics, preventing Newton from employing his own calculus. Although these concepts do take motion and dynamics into account, the general framework continues to be *geometrical* in nature.

In the case of the modal universality of the physical aspect, the implication is that every (modal-functional) physical law holds for whatever actively (i.e. as a subject) functions in the physical aspect. In connection with the way in which Helmholtz formulated the first main law of thermodynamics – the law of *energy constancy*, as we prefer to designate it – Steffens remarks that for "... Helmholtz the law of the conservation of force" includes "all known physical phenomena" (Steffens, 1979:137).

In connection with the question regarding the nature of *matter*, this insight leads us towards an understanding of the coherence between the first four modes of explanation.

From the earliest phase of the development of Greek thought, we find attempts towards an understanding of *matter*. Initially, this urge was closely related to the principle of origin sought by some of the prominent pre-Socratic philosophers. Just recall, for example, the choice of Thales – *water*, Anaximines – *air*, Heraclitus – *fire* or the infinite-unbounded (the *apeiron*) chosen by Anaximander. These elements were thought of as flowing, dynamic principles of origin, because at this early stage in Greek thought, the motive of form, measure and harmony played a subordinate role. Nonetheless, the dialectical struggle between form and the formless was played out in terms of the same basic ontic modes of reality that provided merely a *partial* perspective on material things. Perhaps this is one way to understand the mystery entailed in the question of matter, for if the aspects as modes of explanation enable only *partial answers* from functionally distinct angles of approach, then merely making an appeal to these angles of approach will never solve the problem. It is therefore not surprising that Stegmüller believes that one of the most difficult questions facing science in the 20th century was indeed the concept of matter, which he considers to be mysterious to an utmost degree.[1]

Of course, what Stegmüller has in mind is first of all the old adage that here are no jumps in nature; that it is *continuous* ("*natura non facit saltus*"). Yet we have seen that there is an important difference between physical space and mathematical space – the former is discontinuous, since it is bound to the quantum-structure of energy, while the latter is continuous and therefore allows for infinite divisibility. Stegmüller writes:

> In the preceding sections we have repeatedly established how much precisely those disciplines that are focused on the largest bodily structures as such, astronomy, astrophysics and cosmology, remained dependent upon knowledge of the smallest. In fact presently we often cannot even say if the scientific puzzle or theoretical dilemma that here appears poses a challenge merely for the

1 "Und daß auf der anderen Seite ausgerechnet der Materiebegriff der schwierigste, unbewältigste und rätselhafteste Begriff überhaupt für die Wissenschaft dieses Jahrhunderts blieb" (Stegmüller, 1987:90).

> disciplines concerned with the largest or also at once a challenge for the disciplines concerned with matter. One can defend the mean assertion that the contemporary "matter experts" in a certain sense are forced into a worse acknowledgement than Goethe's Faust. They are not only "not wiser than before," namely in respect of the time when they commenced their research, but they simply have on no accasion become wiser than those first thinkers who more than 2000 years ago attempted to provide a speculative foundation for matter (Stegmüller, 1987:91).

When Stegmüller continues his explanation of the problems attached to an understanding of the nature of matter, the first four aspects of reality suddenly acquire a new actuality. In the first place, he distinguishes two global basic conceptions regarding the nature of matter and points out that these conceptions once again, as previously, occupy a prominent place in current discussions. He calls these two basic conceptions the *atomistic conception* and the *continuity conception*.[1] Laugwitz also points out that, insofar as physics subjects itself to auxiliary means from mathematics, it cannot escape the polarity between continuity and discreteness.[2] The same applies to d'Espagnat, who characterizes classical physics "to be a multitudinist worldview" favouring a conception of nature in which reality basically is constituted by "myriad simple elements – essentially localized 'atoms' or 'particles'." He believes that the more general "quantum field theory is radically at variance with it" with its alternative "notion of a wholeness of some sort." He states: "But theoretical as well as experimental advances gradually made people realize that it [wholeness] constitutes an inherent part of the very quantum formalism and has quite specific experimental consequences" (d'Espagnat, 2006:17).

Suddenly the question concerning the infinite divisibility of matter once again occupies a central position, thus highlighting anew the important distinction between *physical space* and *mathematical space*. It is clear that this distinction between 'atomism' and 'continuity' is based upon number and space as the two most basic modes of explanation of reality. But this is not yet the end of the dependence upon unique modes of explanation, for according to Stegmüller, the two concepts were designed to provide a solution to the following two problems (Stegmüller, 1987:91):

(i) The apparent *indestructibility* of matter, and
(ii) The apparent or real limitless transformability of matter.

When these two problems are assessed in their coherence, it is immediately clear that they depend upon the third and fourth ontic modes of explanation in reality, namely the meaning of kinematic persistence ('immutability') and physical changefulness ('transformability'). The physicist Rollwagen holds

1 "Selbst die beiden großen Grundkonzepte über die Natur der Materie stehen heute nach wie vor zur Diskussion, wenn auch mannigfaltig verschleiert hinter Bergen von Formeln. Diese beiden Grundkonzepte kann man als die atomistische Auffassung und als die Kontinuumsauffassung der Materie bezeichnen" (Stegmüller, 1987:91).

2 "Die Physik, insofern sie sich mathematischer Hilfsmittel bedient oder sich gar der Mathematik unterwirft, kann an der Polarität von Kontinuierlichem und Diskretem nicht vorbei" (Laugitz, 1986:9).

the view that the 'dualism' of wave and particle introduced a new dimension, namely the "possibility of the ... mutual transformation of elementary energy structures" (Rollwagen, 1962:10).

At this point, the key moments in the preceding overview of the history of our understanding of material things recur, owing to the fact that all four modes of explanation still play a decisive conditioning role in our theoretical reflections. The "thing-ness" of material entities once and for all transcends the limited nature of the unique angles of approach (modes of existence and modes of explanation) that served our understanding of matter. Things function at once within all these modes, and yet, in spite of this aspectual many-sidedness of things, their existence is never exhausted by any of the modal aspects. It appears that the *mystery* surrounding material entities derives from this multi-aspectual *but-at-once more-than-merely* aspectual nature of such entities.

It is precisly this more-than-merely-aspectual-nature of material things that sheds a negative light on any *monistic* ideal towards developing a "theory of everything." We saw that with reference to Einstein's search for a unified field theory, a specialist in "super string theory" Greene still believes that physicists will find a framework to combine their insights into a "seamless whole," into a "single theory that, in principle, is capable of describing all phenomena" (see Greene, 2003:viii and page 260 above). He indeed presents "super string theory" as the "Unified Theory of Everything" (Greene, 2003:15; cf. also pp.364-370, 385-386). However, Greene does not realize that, although he has a *purely physical theory* in mind, the *meaning* of the physical aspect of reality inherently points beyond itself to its inter-modal coherence with other aspects, first of all with those aspects that are foundational to the physical aspect (namely the aspects of number, space, and movement). Even the way in which he phrases his goal cannot escape from terms that have their original seat within some of these aspects. Consider for example his reference to a "seamless whole" as well as his use of the quantitative meaning of a "single theory," i.e. *one* theory and the term 'all'. We have repeatedly explained that the core meaning of space (i.e. *continuous extension*) underlies our awareness of *wholeness* and *coherence* (seamless).

The undeniable interconnectedness of all aspects of reality disqualifies each and every claim to (modal) *purity* – at least when this 'purity' is meant to refer to the meaning of an aspect stripped of its coherence with other aspects.

7.2.1 *Kant's synthetic a priori and the distinction between modal laws and type laws*

In his *Critique of Pure Reason*, Immanuel Kant asks: How are synthetical propositions *a priori* possible? (Kant, 1787-B:19). We have seen that, according to Kant, the thought categories of our understanding are not *derived from* nature, but rather *prescribed to* nature in an *a priori* way (Kant, 1783 § 36). Although misdirected by the rationalistic assumptions of his epistemology, Kant, in his search for the synthetic *a priori*, struggles with the nature of *modal universality* (see page 68 above). Yet it is only through the acknowledgement of *type laws* that it is possible to understand in what way the *typical nature* of an entity *specifies*[1] the modal meaning of the aspects in which it functions. This *typical* nature of entities provides a peculiar 'colouring' to their *modal functions*. But most importantly, type laws do not hold for every possible entity – they only apply to a limited class of entities, as Stafleu explains (1980:11, cf. pp.6 ff.).

In order to appreciate Kant's position in this regard, we must remember what has been said about the difference between *modal laws* and type laws. The former hold for whatever there is, while the latter apply to a limited class of entities only. Through modal abstraction, one has access to the (unspecified) universality of modal-functional relationships. But since modal aspects are not concrete entities or events, they cannot be treated *as if* they are entitary in nature, because this would simply amount to a *reification* of modal functions. If one really wants to gain an understanding of the *type law* of any particular *kind of entity* one has to investigate such an entity *empirically*. One cannot derive the *typical nature* of different kinds of physical entities from modal analysis or abstraction – empirical testing *through experimentation* (the relative merit of positivism) is required.

This explains why even Kant was compelled to make a distinction between his (supposedly universally valid *a priori*) *thought categories* on the one hand and so-called *empirical laws of nature* on the other:

> We rather have to distinguish empirical laws of nature, which always presuppose particular perceptions, from the pure or general natural laws, which, without having a foundation in particular perceptions, only contain the conditions of their necessary connection in an experience. In respect of the latter nature and possible experience are entirely the same; and since within these the law-conformity of the necessary connection of appearances in an experience (without which we are totally incapable of knowing any object of the world of the sense), actually is based upon the original laws of the understanding, so it initially does sound strange, but it is nonetheless certain, when I state with respect to the latter: understanding creates its laws (a priori) not out of nature, but prescribes them to nature (Kant, 1783:320; § 36).

This distinction clearly runs parallel with that between *modal* laws and typical laws (*type laws*). Whereas Kant ought to receive credit for wrestling with the dimension of *modal universality*,[2] positivism and neo-positivism ought to be acknowledged for their emphasis on *experimental testing* (not the same as: *verifying*!). Only by studying the *orderliness* or *law-conformity* of entities is it possible to arrive at an understanding of the *type laws* holding for that limited class of entities subject to their peculiar type laws. In the case of physics, it requires empirical research through experimentation.

1 Note that we do not say *individualizes*, because universality does not exist on one end of a continuum, with what is *individual* at the other end.

2 Clouser calls modal universality the "principle of aspectual universality" (Clouser, 2005: 254).

By making an appeal to Dilthey's sketch of seeing the formation of natural scientific theories as constructing reality via the logical mathematical elements of consciousness (and thus asserting the power over nature of this sovereign consciousness as an effect of the autonomy of the human intellect),[1] we are reminded of the limitations of sensory perception (see page 39 above) where we mentioned the views of Weyl and Dingler. Weyl actually followed Hugo Dingler's conception of the principle of *symbolical construction*. He is convinced that the "constructive character of the natural sciences, the situation that their individual propositions do not have a verifiable meaning in intuition (*Anschauung*), but that *truth* builds a *system* which can only as a whole be assessed" (Weyl, 1966:192) has been explained. We also mentioned the view of Planck who believes it is "impossible to find any physical question which can be ascertained directly through measurements without the aid of a theory" (Planck, 1973:341).

Weyl affirms the correctness of Dingler's definition of physics as the discipline in which the principle of symbolical construction is fully carried through and then adds a statement that appeals to the above-mentioned distinction between modal universality and typicality: "But what is connected with the *a priori* construction is *experience* and an *analysis of experience through the experiment*" (Weyl, 1966:192).

Discussing the nature of an *a priori* synthetic element in the "empirical sciences," Stegmüller raises the following possibility – also alluding to this issue (Stegmüller, 1969:316):

> Surely, this cannot imply that the totality of law-statements present in a natural science could be of an *a priori* nature. Much rather, such an apriorism should limit itself to the construction of a limited number of *a priori* valid law relationships, while, furthermore, all more specific laws of nature should be dependent on empirical testing.[2]

Keeping in mind that we must distinguish laws in an *ontical* sense from our *hypothetical* law statements in *scientific formulations*, we also have to note the similarity between Stegmüller's just-mentioned statement and the following explanation by Stafleu (related to the distinction between modal laws and typical laws):

> Whereas typical laws can usually be found by induction and generalization of empirical facts or lower level law statements, modal laws are found by abstraction. Euclidean geometry, Galileo's discovery of the laws of motion ..., and thermodynamic laws are all examples of laws found by abstraction. This state of affairs is reflected in the use of the term "rational mechanics," in distinction from experimental physics (Stafleu, 1980:11).

It must be clear that the intention with the distinction between modal and typi-

1 Weyl refers to the second volume of the 1923 edition of Dilthey's Collected Works (260). Cf. Weyl 1966:192.

2 Fales entertains, in a different context, "the possibility that there are synthetic *a priori* truths; truths about abstract entities may express facts that are not merely the result of linguistic convention" (Fales, 1990:148).

cal laws has indeed captured the reflection of prominent thinkers. To mention one last example: C.F. von Weizsäcker states that, although the basic assumptions of quantum theory could be written on one page (for the mathematically trained reader!), the number of known experiences conforming to this theory runs into the billions – and not a single one is found contradicting quantum theory in a convincing way. He then says, alluding to the universal validity of Kant's thought forms: "I use an idea of Kant and conjecture that quantum theory therefore holds universally in experience, because it formulates the conditions for possible experience."[1]

Stegmüller discerns five distinct properties in respect of synthetic *a priori* propositions:

> (1) all these propositions are true;
> (2) we are in the position of establishing their truth definitely;
> (3) these propositions cannot be proved with the help of the apparatus provided by elementary formal logic. Propositions of this kind were called 'synthetic' by Kant;
> (4) the propositions in question have a factual content, i.e. they tell us something about the real world and thereby about those things and events which are the object of our sense perceptions and observations;
> (5) despite the fact mentioned in (4) the truth of these propositions is not based on experience, i.e. sense perceptions and observations are not necessary to get an insight into their truth. Kant therefore called them '*a priori propositions*' (Stegmüller, 1977:74-75).

If we consider our suggestion, namely that what Kant has in mind with his idea of synthetic *a priori* propositions actually unveils some core features of *modal universality*, then it is immediately clear that synthetic *a priori* propositions are nothing but statements of modal universality. Stegmüller summarizes the mentioned features in the following way. First, he points out that they are "true statements about matters of fact whose truth can be established definitely despite the fact that formal logic is not sufficient to get this insight and observations are not necessary to achieve it." To this, he adds that "most philosophers of Kant's time as well as of the present would react to this characterization by saying that there certainly are no synthetic a priori propositions." The implication of this denial is that most philosophers actually deny the ontic universality of modal aspects – just recall the nominalistic legacy with its rejection of universality outside the human mind, and Descartes's belief that number and all universals are modes of thought.

The fact that Stegmüller is uncomfortable with Kant's view that propositions like these "exist and play a fundamental role in scientific knowledge," and even says that it "is very surprising, if not incomprehensible" in the sense that "it seems to be completely unfounded if not substantiated by additional

1 Von Weizsäcker, 1993:93. "Laws capable of mathematical formulation finally form the hard core of natural science: not the important detail, but the form of universal validity" (1993:113). In another context, he writes that the quantitative results of astronomy are based upon physical laws, and that we postulate, as a working hypothesis, a universal validity for these laws (1993:25).

arguments," demonstrates that he still has difficulty with the nature of (ontic) modal universality.

Our argument regarding the Achilles' heel of positivism (see page 45 above) articulates the insight that modal aspects cannot be observed in a sensory way. We have asked whether they can be *weighed*, *touched*, *measured* or *smelled*, and the negative response demonstrates that they are not *things*, but *aspects* of things (or aspects within which things function). Consequently, we must realize that a theoretical account of what has been 'observed' transcends "sense data." Therefore the restriction of knowledge to *sense data* cannot be maintained.

What Stegmüller says about synthetic statements *a priori*, namely that they are "*true statements about matters of fact whose truth can* be established definitely despite the fact that formal logic is not sufficient to get this insight and observations are not necessary to achieve it" (Stegmüller, 1977:75), highlights the way in which the modal universality of aspects is depicted.

The universality of a modal aspect is a "matter of fact," and establishing its truth transcends formal logic.[1] Moreover, since modal aspects cannot be observed (like things), acquiring insight into their modal universality is not dependent on 'observations', or, to use Stegmüller's words, "observations are not necessary to achieve" this insight.

Recalling what we have said about natural laws and combining it with the importance of modal universality, Popper's statements about natural laws should not surprise us: "I consider it both useful and fruitful to regard natural laws as synthetic and strictly universal statements ('all-statements'). This is to regard them as non-verifiable statements ..." (Popper, 1968:63). Although Popper confuses human law statements with the ontic status of natural laws, he comes close to an understanding of the modal universality of aspectual natural laws.

7.2.2 *Material subjects mistakenly labeled as 'objects'*

As we have argued in connection with the subject-object relation (see pages 345 ff.), material things first of all ought to be understood in terms of their *subject functions*, i.e. in terms of those aspects in which they actively function as subjects. As reflected in the one-sided emphases in the history of our understanding of material things, these functions concern the way in which material things operate as *subjects* within the aspects of number, space, motion and the physical. We have pointed out that the effect of overestimating the human subject was that physical subjects were designated as 'objects'.

Alternatively, by primarily acknowledging the *subject functions* of material entities, we argued that it should be realized that the quantitative, spatial, kinematic and physical properties of material things concern their *subject functions* within these aspects. In this sense, such entities are indeed *physical sub-*

1 Recall our remark that it is the *principle of sufficient reason* that points beyond the limits of logic to the ontic state of affairs that provide the required *grounds* for an argument (see page 184).

jects and not 'objects'. Yet we have argued that every material entity also displays *latent* object functions that may be disclosed by subjects actively functioning within every post-physical aspect. Consequently, we concluded that insofar as material things are objectified in some or other *post-physical* aspect – such as the analytical (where their analytical object function is made manifest in that they are identified and distinguished), the lingual (having been assigned a name, as sign objects), the jural (as legal objects, having become the property of a legal subject), and so on – they are no longer appreciated according to any one of their *subject functions*.

Against the background of the foregoing, we can now proceed by considering one of the classical problems of philosophy: our experience of the *identity* of things amidst a *world of change*.

7.3 The problem of identity

In a striking way, the following Kellogg's advertisement highlights multiple fundamental relational problems: "Eat it again for the first time." This statement is remarkable in *two* respects:

(i) it highlights the apparently dialectical tension of *alteration* and *continuation*, *dynamics* and *persistence* in the human experience; and
(ii) it captures something about *ubiquity* and *contingency*, *universality* and what is *individual*.

Although the phrase "eat it *again*" incorporates an awareness of *succession* and *continuity*, the end of the phrase jumps to something *new* and *discontinuous* from what has been experienced *before*, "for the *first* time." A 'soft' interpretation may suggest that, literally speaking, we are indeed eating the *same* thing every time, namely *Kellogg's*, and that every time we eat the same thing (*identity*, *continuity*) it tastes as good as *new*, as if we are eating it for the *first time* (*non-identity*, *discontinuity*).

It therefore seems as if the advertisement introduces a gap or, even stronger, a *dialectical tension* between the first part and the second part, for what is *granted* in the first part, namely that we are successively eating the *same thing*, is *denied* in the second part, which *affirms* that what we eat is something we have *not* eaten before, i.e. we are eating it *for the first time*.

Yet, when it comes to the problem of *identity*, language does not resolve the issue, partially because lingual units are not mere *copies* of reality. Quine realizes that "The utility of language lies partly in its very failure to copy reality in any one-thing-one-name fashion. The notion of identity is then needed to take up the slack" (Quine, 1958a:208). However, he does qualify this statement by pointing out the real extra-linguistic entities as identical:

> and not the names with one another; the names stand in the statement of identity, but it is the named objects that are identified (Quine, 1958a:208).

This view is defended, because earlier on Quine already insisted that it is incorrect to assert that the laws of logic hold purely by virtue of language, and that they are therefore *analytic* in character.[1]

Quine does realize that the attempt to define analyticity[2] in terms of meaning runs into a circularity, since meaning is defined in terms of analyticity.

7.3.1 *Identity, entity and property*

Let us consider Frege's well-known example of the "evening star" and "morning star." Frege uses this example in order to distinguish between 'Sinn' and 'Bedeutung' (*connotation* and *denotation*). Although there are two *different lingual signs* involved, their reference (denotation) is the *same*. Van Woudenberg proposes to speak of an *identity relation* in cases similar to the "evening star" and the "morning star," and of the relation between entity and property when assertions like "the hat is white" are at stake (Van Woudenberg, 2000:12).

What Quine highlights as identical, namely the real extra-linguistic entities, calls for an account of the *persistence* of such entities. The mode of speech involved in addressing this issue ultimately appeals to our awareness of the meaning of *constancy*, reminding us of the fact that, in his special theory of relativity Einstein gives prominence to the meaning of constancy – and we noted that the crux of Einstein's theory of relativity is in fact to be found in the nature of the (kinematic) order of *constancy* that it presupposes.

Quine considers 'identity' to be "such a simple and fundamental idea that it is hard to explain otherwise than through mere synonyms" (quoted by Grau, 1999:77). It is clear that this view testifies to the fact that Quine is implicitly aware of two things: (i) the core meaning of the kinematic mode that can only be explained by synonyms, and (ii) that our awareness of the identity of entities depends crucially upon what we have called a concept-transcending use of modal kinematic terms. Both Grau and Quine view identity in relation to entity: "No identity without entity. Nonentities are not there to be the same or different" (see Grau, 1999:80).

Bryon and Spielberg correctly emphasize that Einstein's theory concerns 'invariance' – i.e. constancy – but unfortunately, they confused this with the terms *absolute* and *unchanging*:

> Indeed, Einstein originally developed his theory in order to find those things that are invariant (absolute and unchanging) rather than relative. He was concerned with things that are universal and uniform from all points of view (Bryon and Spielberg, 1987:6).

1 "Carnap maintained, and Frege before him, that the laws of logic held by virtue purely of language: by virtue of the meanings of the logical words. In a word, they are analytic. I have protested more than once that no empirical meaning has been given to the notion of meaning, nor, consequently, to this linguistic theory of logic" (Quine, 1973:78). On the next page, however, he does open the door for *analyticity*, but within a perspective that hinges on "social uniformity" – "a sentence is analytic if *everybody* learns that it is true by learning its words" (Quine, 1973:79).

2 We have noted that a sentence is supposed to be analytically true if it is true only on the basis of its meaning (see page 171 above).

The term *unchanging* is simply the denial (negation) of *change* – a *physical* term. We have seen that the physical aspect must not only be distinguished from its foundational kinematic aspect, since there is also an indissoluble coherence between these two aspects. For this reason, within the physical aspect, we discern a structural moment that reminds us of the foundational kinematic aspect – captured by our alternative formulation of the first law of thermodynamics, *energy-constancy* (also see Strauss, 2000).

Implicit in our experience of persistence (identity) is of course also the effect of a structural principle or a law determining the enduring existence of something. Therefore the law of *energy constancy* not only demonstrates the distinct uniqueness of the kinematic and physical aspects, but also the distinction between the law side and factual side. The distinction and mutual coherence of constancy and change render the statement that change is the only constant meaningless and at once accounts for the possibility of the *identity* of entities.

7.3.2 *Modes of explanation making identity understandable*

The idea of an entity transcends any particular mode of explanation through which knowledge of entities is obtainable. One can explore different modes of explanation to demonstrate this perspective.

First, an entity can be seen as *individual*, that is, as *distinct* from other entities. Secondly, the description of an entity may explore – in addition to the just-mentioned numerical mode of explanation – the spatial mode by adding another qualification, for instance when an entity is designated an *individual whole* (totality).

But it is only when we explore the points of entry offered by the kinematic and physical aspects of reality that we are capable of articulating a well-founded intuition of *identity*, for then we can add even further qualifications by saying that an entity is an *enduring* individual whole in spite of any *changes* to which it may be subjected.

We use this insight in order to transcend the static logic of identity in the thought of Von Kibéd (see page 186 above). Von Kibéd does not acknowledge the uniqueness and coherence of constancy and dynamics. This reminds us of the fact that Einstein's theory is one of *constancy* rather than relativity.

In German, the phrase for "uniform motion" is "gleichförmige Bewegung" (see Glahn, 2006:133). "Das Gleiche" refers to what is the *same*, i.e. identical or constant, from which it is clear that the phrase "gleichförmige Bewegung" indeed accentuates *constancy* as the core meaning of the aspect of motion.

7.3.3 *Mechanistic biology and the identity of living entities*

The immense successes of organic chemistry and biochemistry in unraveling the intricate physico-chemical structure and functioning of living entities tempted modern biologists inclined to a *physicalistic* approach to over-accentuate the dynamic flow of the physical-chemical constituents of living entities. Consider for example the fact that all the atoms in the human body, even those in the bones, are exchanged at least once every seven years; all the atoms in the face are renewed every six months; all the red blood cells every four months; and 98% of the protein in the brain in less than a month. Add to this that the white blood cells are replaced every ten days and one-thirteenth of all tissue proteins are renewed every 24 hours (see Jones, 1998:40), then we may justifiably ask: what guarantees the *identity* of a living entity if all its components are in constant flux?

From a purely mechanistic or physicalistic point of view, a living thing is explicable only in physical terms. But then it must display a physico-chemical identity constituted by its atoms, molecules, and macro-molecules. The problem, however, is that all these constituents are constantly changing! Which of these physico-chemical components could then constitute this supposedly purely physico-chemical identity of living things? Will it be the atoms, molecules, and macro-molecules currently present within it, those present years ago, or those that will be present a few years hence!?

When living things are *physicalistically* reduced to their *material constituents*, their *biotical identity* is necessarily lost – since the supposed elements of identity vary continually.[1]

Yet, once the unique biotical function of living things is taken into account, it is even possible to claim that a living thing, considered in terms of the biotic mode of explanation, is in a *stable state* (referred to as *health*), while at the same time the claim can be made – without any contradiction – that, in terms of a physico-chemical mode of explanation (with a view to the flowing equilibrium of its physical-chemical constituents), it exists in an *unstable state*.[2] If the physical-chemical substratum of living things approaches a state of *higher statistical probability*, biotical *instability* increases as a sign of the final process of *dying*.

Important consequences follow from these considerations.

1) Whenever an idea of *identity* is formulated, the decisive clue is always given in the *mode of explanation*. For example, compared to the *physical* identity of material entities, living entities display a *biotical* identity.
2) Furthermore, whenever *identity* is the theme, at least an implicit awareness of the foundational relationship between *constancy* and *dynamics* is entailed.
3) Finally, since an entity is more than the sum of its different modes of explanation, no single (entitary) identity-claim can exhaust the uniqueness of any entity.

Sometimes a distinction is made between living and non-living entities: the first category applies to entities that retain their identity in spite of changes,

1 Plato discusses the example of a wooden ship of which all the constitutive parts are replaced at sea (Phaido, 58; see Van Woudenberg, 2000:28).

2 We should remember that Von Bertalanffy [on the basis of his idea of the dynamic equilibrium (Fliessgleichgewicht)] generalized the second main law of thermodynamics to open systems (see Strauss, 2002a).

whereas the second category refers to material things. Van Woudenberg remarks that the latter groups of entities can only be maintained through *external* intervention. It is clear that he does not realize that von Bertalanffy's *flowing equilibrium* (*Fleissgleichgewicht*) – exemplified in a fire or glacier – does not require external maintenance (such as the case of Plato's ship). Consequently, there ought to be no objection to the identity of physical entities either. But endurance over a period of time may be questioned in terms of what is known as the "principle of the indiscernability of identicals" of Leibniz, implying that, only when we restrict ourselves to a specific moment in time is it possible to affirm the identity of entities. However, as Van Woudenberg remarks, the idea of temporal parts does not preclude the awareness of identity over time, but presupposes it (2000:43).[1]

7.4 Societal identities

Philosophy of mind and social philosophy also struggle extensively with issues of *personal identity* (traditionally, the supposed relationship between body and soul, and more recently the question regarding the identity of mind and brain), and with the complex relationships between the 'individual' and 'society' (*collective identity*). Am I in the first place someone with an *ethnic identity* (such as an Englishman, Zulu, Afrikaner, or Sotho), and only in the *second instance* partaking in a specific political identity (such as South African, American, etc.)?

The basic and important perspective in this context is that no single societal identity – not even when considered merely a "social construction" – can escape either from fundamental ontic modes of explanation co-conditioning human society (such as the lingual,[2] the economic, the social, the aesthetic, the jural and the moral), or the various *possible identities* human beings can construe (and assume) in the process of what Simmel calls *sociation*. Simmel holds that a number of individuals are only transformed from a spatial aggregate or a temporal sequence into a society when (through sociation) these individuals exercise a mutual influence upon each other: "If, therefore, there is to be a science whose subject matter is society and nothing else, it must exclusively investigate these interactions, these kinds and forms of sociation" (see Levine, 1971:24-25).

This means that a person may at once be a father, citizen, club member, and so on, and participate in these distinct social identities without fully being absorbed by any one of them. The mystery of the human being is certainly connected to the fact that no single differentiated social identity can fully and exhaustively encompass the nature of being human.

1 See in this connection also Fine 2003.

2 Owing to what we have discerned as the modal universality of all aspects (see page 68 above) it should be noted that every mode of explanation has a universal scope in the sense that, whatever there is in principle, has a function within every ontic mode of reality.

7.5 Identity and concept-transcending knowledge

We have argued that terms residing within a particular modal aspect may be applied in a twofold sense, either as referring to phenomena functioning *within* the aspect concerned, or employing particular modal terms in such a way that they serve an understanding *transcending* the limited context of the specific mode of explanation in which they have their original seat (see page 176 above). When the awareness of uniform motion is applied to the description of a uniformly moving body in a purely (abstract) kinematic sense, we explore a *conceptual* use of motion. However, the moment we expand our scope, using the term 'constancy' in order to refer to the identity of an entity over time, in spite of the changes it may experience, then the intuition of constancy is applied in a *concept-transcending* way manifest in speaking about the *identity* of such an entity.

We have also equated the phrase *concept-transcending knowledge* with *idea-knowledge*, where it is assumed that a *form of thought* is needed in order to capture that kind of knowledge *transcending* the limits of concept formation. In general therefore, our talking of the *identity* of things rests upon the basis of a *concept-transcending idea* of the *transmodal reality* of entities. It entails that entities belong to a distinct dimension of reality, intimately cohering with another dimension of reality, namely that of modal functions (aspects). We only have access to entities because these aspects not only serve as modes of being and modes of explanation, but also as *experiential points of entry* to entities. But there is no substitute for the acknowledgement that the dimension of entities presents us with an undeniable bottom layer of experience, distinct from but at once intimately cohering with the dimension of modal aspects. Those who deny either of these two dimensions end up, as we saw, in the one-sidedness of a *functionalistic* or *substantialistic* view. The former functionalizes entities, and the latter reifies functions.

7.6 The problem of what is individual

In our discussion of concept-transcending knowledge, references to the *individuality*, *uniqueness*, and *being distinct* of entities (see pages 176 ff. above) turned out to be instances of an idea use of modal arithmetical terms.

From Greek philosophy onward, a key problem was to find what became known as the *principle of individuality*. Aristotle, for example, claimed that *matter* in combination with number is the principle of individuality: "But all things many in number have matter" (Aristotle, 2001:884; *Metaph*. 1074 a 34 – see page 301 above). From the opening sections of Aristotle's work on *categories*, it is clear that he took his starting point from the idea of a *strictly individual* primary substance. Here he defines substance "in the truest and most definite sense of the word" as "that which is neither predicable of a subject nor present in a subject" (Aristotle, 2001:9; Cat. 2 a 11-13). The secondary substances, however, are universal (Aristotle, 2001:9; Cat. 2-4). Species and genera are universal.

Later on, in neo-Platonism, we find that priority is assigned to what is universal, while acknowledging that what is individual cannot be conceived (see Plotinus, *En.* VI, 3,9,36 and *En.* VI, 2,22).

Interestingly, Simplicius already distinguished between the *numerical one* and what is *individual* (Kobusch, 1976:302) – thus implicitly highlighting the difference between a conceptual use of numerical terms (the "numerical one") and a concept-transcending use of numerical terms (the idea of an individual thing). Ammonios Hermeiu influenced the distinction between four 'complexions' found in the thought of Boethius: *substantia universalis*; *substantia particularis*; *accidens universale*; *accidens particulare*. Where *singularity* indicates similarity for Richard von St. Victor, individual substantiality is found in one individual only and therefore cannot be shared by multiple substances, and for this reason it is 'incommunicable' (see Kobusch, 1976:303).

It is striking that the battle field between *universality* and *individuality* is served by our basic intuitions of number and space. The idea of *being distinct* (at least partially) presupposes the *discrete* meaning of number, while *universality* presupposes the spatial awareness of *everywhere*. Of course the numerical point of entry – or mode of explanation – can be complemented in yet another way by the spatial angle of approach, namely when its articulation at once also highlights *numerical analogies* within space. This occurred during the later middle ages, during which the commonly held view (shared by Bonaventura, Thomas Aquinas, Henry of Ghent and Duns Scotus) was that an individual is undivided (the literal meaning of *individuum*), while at the same time being separated from everything else (see Oeing-Hanoff, 1976:306). The term *undivided* reflects the wholeness (*one*-ness) element of the spatial whole-parts relation. In addition to the 'one', it entails the 'many' separated undivided ones (wholes). In other words, the one and the many are situated within the context of the idea of spatial wholeness and distinctness – once again underscoring the unbreakable interplay of the numerical and spatial modes in the articulation of the idea of individuality.[1]

The employment of the spatial whole-parts relation acquired a closer specification in the thought of Boethius, who distinguishes between *homogeneity* and *heterogeneity* – every part of an individual drop of water is still water (physical homogeneity), whereas it is not true that every part of a horse is a horse (biotic heterogeneity) (Oeing-Hanoff, 1976:306).

Leibniz continued the view of Aristotle by inverting the idea that individuality falls under general concepts. Rather, one should say that what is universal is contained or embraced in what is particular and individual (see Borsche, 1976:310). When he determines the individual substance in his *Monadology*, Leibniz assumes an original self-activity prevailing in a state of *continuous change* (appetition) (Borsche, 1976:311).

In close connection with the early Romantic switch from rationalism to

1 Yet more recently, Trincher had to remark that the property of having a geometrically defined surface only applies to macroscopic structures, for even the largest macro-molecule is not delimited by a surface (see Trincher, 1985:336).

irrationalism, Herder affirms that the "deepest foundation of our existence is individual" (Herder, 1877 Vol.II:207). In a letter to Lavater, Goethe mentions the saying that "Individuality is ineffable" (Borsche, 1976:312). According to Fr. Schlegel, individuality is never completed, since it is always involved in continuous becoming ("beständiges werden"). What is essentially incompleted is infinite, and therefore individuality is eternity within the human being and only this can be immortal (Borsche, 1976:315). We shall return to the distinction between *individuality* – which is actually still a universal feature – and what is *individual*.

Goethe plays with the inseparable connection between individuality and universality – in terms of one of our examples: *this chair* (individual side) is *a chair* (universal side). His answer to *what is universal?* (*Was ist das Allgemeine?*) is: the individual instance; and his answer to *what is particular?* (*Was ist das Besondere?*) is: millions of instances (see Von Weizsäcker, 2002:212).[1] Yet this does not mean that Goethe actually maintains a *balance* between universality and individuality, because according to him, in the words of Von Weizsäcker, the *Gestalt* is not rooted in the law, for the law is rooted in the *Gestalt* (Von Weizsäcker, 2002:209).

Clearly these historical lines all converge with the successive modes of explanation that we have discovered as decisive in the history of the concept of matter. Ultimately, no view on individuality can avoid some or all of the first four modes of explanation of reality – in both cases we encounter a mixture of conceptual terms and concept-transcending usages of modal terms.

7.7 The concept of a natural law

Initially the idea of 'law' (nomos) had a broader scope than *law* in a jural sense. Hesiod accepts the idea of *nomos* as given by Zeus, and towards it, the human being in general has to direct the self as an art of life (Hesiod, Erga 276 ff. see Plumpe, 1974:494). Heraclitus advances the idea of divine law (*θεῖος νόμος*) as a principle of the *universe*; the world logos (*λόγος)* (Diels-Kranz, Heraclitus, B. Fragm. 30 ff). Democritus adds the idea that the commands of nature are *necessary* (Plumpe, 1974:494). Cassirer mentions another element in the ancient Greek understanding of *law* – namely that *nomos* constitutes a principle of *ordering* that arranges motion and the diversity within reality (Cassirer, 1911:375).

During the early modern period, a reaction to the traditional Aristotelian-Thomistic view led to a natural scientific orientation that treated *law* predominantly in a *relational coherence* (i.e. a *functional coherence*).

This new focus is an effect of a fundamental switch; one in which the focus is no longer on the *substance* of things (their *essence*), but on the *way* in which

1 In his "Die Vernunft in der Geschichte," Hegel says that the universal can only enter reality through the particular: "Das Allgemeine muss durch das Besondere in die Wirklichkeit treten" (quoted by Van der Hoeven, 1997:149). Pannenberg states: "Laws and law-conformities can only be established in what is contingent" (Pannenberg, 1973:125). Cassirer also acknowledges the inseparability of what is individual and universal where he holds that "judgments concerning individuals can be completely universal" (Cassirer, 1953:226).

we experience them. Galileo is therefore no longer interested in the 'essence' of things, but instead asks *how* they appear to us. What is revolutionary in his view, according to Herold, is that in the absence of thinking about essences (that proceeded from configurations of motion with distinct degrees of perfection), everything is in principle equal in the face of the law – amply demonstrated in the remark that he did not study the pedigree of geometrical figures (see Herold, 1974:502).

Remark: Thus far the following features of natural laws surfaced, that they are *necessary*, that they constitute an *ordering* in the sense of a *relational coherence* regulating *motion* and *diversity* within reality and that they are concerned with the *how* and not the concrete *what* of things.

During the early modern era, these ideas developed within the context of the dominance of the modern natural science ideal. In the thought of Hobbes, science understood as philosophy opened the way of viewing individual things in relation to what is *universal*. His emphasis on the *truth* entailed in universal propositions (see Herold, 1984:503) reveals his nominalistic affinity that projects a human element into the universality of natural law. Particularly, Newton (in his *Principia Mathematica*, 1, 15) explores the view that a law must be understood as a *mathematically conceivable rule* more extensively. While taking distance from the idea of a God-given law, the French Enlightenment, and particularly D'Alembert, derives from the relations between bodies governed by law its necessary *validity* (Herold, 1974:505).

In the thought of Kant, the feature of necessity (*Notwendigkeit*) is accompanied by the *universal-validity* of a law. Insofar as rules are objective, they are designated as laws. These laws are derived *a priori* from the human understanding furnishing the phenomena their lawfulness (*Gesetzmäßigkeit*) (Kant, 1781-A:126). Kant's aim is to render comprehensible the "objective validity of the pure concepts a priori" of the categories of understanding (Kant, 1781-A:128).

Hegel explores a further dimension in his science of logic when he focuses on determining law (*Gesetz*), as what remains the *same* in what *changes* (Hegel, 1957-2:122). Cassirer assumes "ultimate logical invariants" that are not affected by their changing material content. He speaks of "identity and permanence" "at the basis of scientific laws" (Cassirer, 1953:325). He actually comes quite close to an understanding of the conditioning role of the first four modal aspects of reality when it comes to an articulation of the nature of natural laws: "There is no objectivity outside of the frame of number and magnitude, permanence and change, causality and interaction: all these determinations are only the ultimate invariants of experience itself, and thus of all reality, that can be established in it and by it" (Cassirer, 1953:309).

The "ultimate invariants of experience itself" in the final analysis refer to the conditioning role of the most basic modal aspects of reality – they are indeed those 'determinations' co-responsible for the way in which we experience reality.

Without acknowledging the ontic structural configuration of reality and in particular the ontic order of successive modal functions, we are left afloat without an anchoring guideline in our attempts to define the nature of natural laws. In our analysis of the compound (or complex) concept of modal principles (see page 297 above), we point out that such a concept is only possible on the basis of the *constitutive* modal points of entry at the foundation of our theoretical considerations. The first part of our characterization of a principle – namely that it is a *universal* and *constant point of departure* that can only be made valid through human intervention – overlaps with the constitutive elements that ought to be incorporated in an account of natural laws. Let us first of all restrict ourselves to *physical* laws.

Van Fraassen refers to Pierce, who argues that if the 'uniformity' intended by Mill merely meant *regularity* without any real connection between events, then his argument is destroyed (Van Fraassen, 1991:22). The phrase used by Van Fraassen in this context, however, states that a law cannot be "the mere uniformity or regularity itself," for a "law must be conceived as the reason which accounts for the uniformity in nature" (Van Fraassen, 1991:22). The use of the word 'reason' may be interpreted to suggest that laws result from the intellectual endeavors of human beings. Nonetheless he continues by claiming that a "law must be conceived as something real, some element or aspect of reality quite independent of our thinking or theorizing – not merely a principle in our preferred science or humanly imposed taxonomy" (Van Fraassen, 1991:22-23). Within the above-mentioned context, this implies that the word 'reason' means 'cause' – in the sense that a law is the (extra-mental) cause that accounts for the uniformity or regularity of nature.

When Van Fraassen discusses the views of Davidson, he points out that, although Davidson does not attempt to define laws, the latter nonetheless says "that laws are general statements which are confirmed by their instances" (Van Fraassen, 1991:33). In this case, the distinction between *ontic laws* and human statements intended to capture conceptually what such laws entail, collapses. The acceptance of ontic laws needs 'markers', i.e. terms to articulate their ontic nature. The fact that we speak in the *plural* about such laws already reflects the constitutive role of the meaning of number (the one and the many) in our understanding of ontic laws. Furthermore, without the conditioning role of the spatial aspect, it cannot be asserted that laws hold *everywhere*, i.e. that they apply *universally*. Although Van Fraassen acknowledges *universality* as a "mark of lawhood" (Van Fraassen, 1991:26), he also mentions with reference to Armstrong and Lewis that the "criterion of universality" is "no longer paramount" in the "discussion of laws" (Van Fraassen, 1991:28).

Of course one should keep in mind that the distinction between modal laws and type laws (see page 25 above) entails the important difference between *unspecified* and *specified* universality – a modal physical law holds for all kinds of physical entities without specification, whereas a physical type law only holds for a limited class of entities, namely those belonging to that type. Such a type law, for example the law for being an atom, holds universally in the sense that it applies to all atoms, but as we have seen, this universality is specified, since it only holds for atoms, and not for every kind of (physical) entity.

In the case of modal laws and type laws, the reverse of universality is found in the *distinctness* of different laws, specified by using the idea of *delimitation* derived from the primitive meaning of space. Every law has its own *domain* of application; a specific and distinct *sphere* within which it obtains. But only when these two elements are combined with the *constancy* (or *uniformity*) of a law and with its *effect* (*force*), is it possible to account for the constitutive elements of the compound basic concept of a natural law.

Dooyeweerd did not apply his own transcendental-empirical method of analysis to the idea of 'law' in general, for he frequently simply (intuitively) states that a law *determines* and *delimits* whatever is subjected to it.

If the distinct *scope* of laws delimit their unique areas of validity, to allow the concept of a natural law to degenerate into an amorphous collection of predicates, such as found in Stafleu's proposal is not to be recommended. He says that a law is sometimes hidden behind the name axiom, constant, proposition, rule, relation, thesis, symmetry, theorem, design, pattern, connection, prohibition, compassion, phenomenon or prescription (see Stafleu, 2002:39). This list contains elements referring to the law side and the factual side, as well as a mixture of ontic phenomena and the product of human activity. For the sake of convenience for example, Stafleu calls a *mathematical law-conformity* ("wiskundige wetmatigheid") such as the theorem of Pythagoras, a natural law (Stafleu, 2002:39). We shall return to this issue in the next sub-paragraph.

Intrinsic to a natural law is an *order for*, and this mode of speech makes an appeal to the unity in the multiplicity of different laws, for without such *unity*, laws will clash and not be able to constitute an order of laws. The constitutive role of the numerical mode is clear in this concept of order. Furthermore, a law entails its correlate, namely that which is factually subjected to it – this insight entails the inherent universal scope of a law – dependent upon the spatial notion of *everywhere* (at all places). Without the spatial (dimensional) distinction between *above* and *below*, the assumed correlation of law and subject does not make any sense. That the validity of a natural law is not something incidental is captured by saying that it holds *constantly* – demonstrating the constitutive role of the kinematic mode. The notion of *validity* (*being in force*) derives from the core meaning of the physical aspect, and it has to be incorporated in the concept of natural law, because otherwise the ability to say that a law *determines* what is subjected to it would collapse.

The compound or complex basic concept of a natural law may therefore be formulated as follows:

> As a unique, distinct, and universally valid order for what is factually correlated with and subjected to it, a natural law constantly holds (either in an unspecified way as in the case of modal laws or in a specified way as in the case of type laws) within the domain within which it conditions what is subjected to it.

However, what is particularly striking in reflections on the relation between *law* and *factuality*, is the wide-spread confusion of *law* and *law-conformity*. This confusion is found in the last-mentioned statement by Stafleu, as well as in his general characterization of a natural law: "I view every firm natural regularity in the broadest sense of the word as a natural law" (Stafleu, 2002:12).[1] In a similar fashion Clouser conflates *law* and *orderliness* (lawfulness), when he says that " 'law' is our term for the orderliness God has embedded in his creation" (Clouser, 2005:243). In this respect, both Stafleu and Clouser continue the after-effect of nominalism within Dooyeweerd's thought (see pp. 446 ff. below). Clouser says that creational laws "constitute the orderliness He has built into creation and by which it is regulated" (Clouser, 2005:243). It is not 'orderliness' that regulates creation – the orderliness of creation does not regulate for it is just the way in which creation shows that it is regulated by God's creational laws.

In another respect, Stafleu also maintains an older stance in Dooyeweerd's intellectual development, for the former consistently refers to the "subject side" of reality. Dooyeweerd eventually realized that, from a systematic point of view, it is more effective to use a *generic designation* for whatever is correlated with a *law*, for otherwise one will always have to add an additional statement, explaining why the term *subject* has two meaning-nuances: (i) the correlate of what is governed by a law, and (ii) one of the two elements of the subject-object relation found in all (post-arithmetical) aspects – but then one has to add that this subject-object relation is still found on the subject side [taken in sense (i)] of an aspect. We have pointed out earlier that, in his *Encyclopedia of the Science of Law*, Dooyeweerd distinguishes between the law side and *factual side*, and within the latter between the factual subject-side and the factual object-side – making it clear that the subject-object relation is found at the *factual side* of reality (see the references on page 76).

7.7.1 *Law and subject in relation to universality and what is individual*

In our discussion of the all-pervasive influence of nominalism since the Renaissance, the relationship between universality and individuality was placed in the center of our analysis, in close connection with the distinction between conceptual knowledge and concept-transcending (idea-)knowledge. This approach is in line with most of the classical analyses and expositions of the controversy between realism and nominalism, where the difference between universality and what is individual (the general and the particular) is employed as the crucial distinction.

When the nature of law and subject is considered in terms of the relationship between what is universal and what is individual, the most striking differ-

1 "Elke vaste natuurlijke regelmaat in de ruimste zin van het woord beschouw ik als een natuurwet." This confusion is also present in his earlier works (see Stafleu, 1980:6, 8, 13).

ence of opinion is found in connection with the difference between the universality of a law and the way in which the determining role of a law is reflected by what is factually subjected to it.

7.7.1.1 *Abstract and concrete: Stegmüller*

In order to highlight the remarkable systematic convergence of thought between Stegmüller and Dooyeweerd, we begin with Stegmüller's disqualification of the traditional universal-individual distinction in the understanding of realism and nominalism.

Stegmüller claims that the real issue centers in the difference between *abstract* and *concrete* (Stegmüller, 1965:53, 61, 62, 84, 116).[1] This distinction, according to him, is so basic that it is not possible to reduce it to something more fundamental. For this reason, he gives a few examples in order to illustrate the difference between 'concrete' and 'abstract': atoms, molecules, cells, rain-worms, human beings, houses, drops of water, stars, and similar things instantiate the former category, while classes or sets, conceptual extensions, numbers, properties, relations, functions, propositions and similar notions exemplify the latter (Stegmüller, 1965:53). Expressions that assert general properties are not appropriate for differentiating between *Platonism* and *nominalism*, since each can give its own account of them. Platonism takes them for names that refer to "abstract ideal objects," whereas nominalism considers them to be 'syncategorematic' statements (*Synsemantika*) that are, as such, meaningless outside a meaning-giving context. A predicate such as 'human' is an open sentence, expressible as "x is a human being" ('x' is the free variable). Only when this free variable is bound, either by the universal or existential quantifier, is it possible to obtain a meaning for the original open sentence (for instance in the false sentence: "everything is human," or in the true sentence: "there are human beings") (cf. Stegmüller, 1965:49-50).

Clearly, the crucial point is in the kind of bounded variables that are allowed: only "concrete individuals" (the choice of nominalism), or also "abstract objects" (the choice of Platonism) (Stegmüller, 1965:51). This perspective caused Quine to observe that "to be is to be the value of a bound variable," indicating that the bound variables used by someone show what that person considers to be real (cf. Stegmüller, 1965:51). In terms of Stegmüller's analysis, the whole medieval controversy about "universalia ante res" and "universalia in rebus" (cf. Stegmüller, 1965:63, 71, 117) is nothing but the result of an unacceptable entitary-like treatment of ideal objects. This 'Verdinglichung' of ideal objects converted them into objects belonging to the same type as concrete ones (Stegmüller, 1965:62) – a treatment that erroneously identified ideal being with the general (Stegmüller, 1965:59 ff.). Another implication is that conceptualism is not a sub-category of nominalism, but rather of Platonism (Stegmüller, 1965:71). Stegmüller follows the distinction be-

1 A translation of this article is found in Stegmüller, 1977: "The problem of universals then and now" (Stegmüller, 1977:1-65).

tween Platonism and nominalism drawn by N. Goodman and W.V. Quine (Stegmüller, 1965:53). According to their view, conceptualism does allow for abstract ideal (non-concrete) objects (Stegmüller, 1965:69-71).

What is the relation between abstract-concrete and universal-individual?

The above-mentioned examples by Stegmüller, in order to differentiate between abstract and concrete, strictly correspond to the distinction between modality and entity.[1] Within the dimension of modalities, we must distinguish between the meaning of the aspects of number and space, which mutually cohere on the basis of their uniqueness (irreducibility).

We have already argued that speaking of the individual side (or uniqueness) of a concrete thing entails that numerical notions are employed – not to referring to modally abstracted arithmetical states of affairs – but rather to the concrete reality of some entity transcending the numerical function of such an entity. Similarly, when we speak of universality (for example as displayed by a law), we are actually using the original spatial meaning of place (position) within the context of an extended spatial concept – *everywhere* (universally). The conceptual use of the term universality, deriving its meaning from the primitive spatial time-order of simultaneity (the 'everywhere' is thought of "at once," without succession), is nevertheless intimately connected with the deepened number idea of the *at once infinite* (actual infinity), which is easily seen in interchangeable expressions such as: universally = everywhere = at all places. This'all', in a strictly modal sense, expresses the numerical antecipation to the "at once" of the spatial time order (i.e. the idea of the *at once infinite*).

In other words, if we consider the concept 'universality' from the perspective of the deepened (disclosed) arithmetical modality, we have to use the idea of 'all' in the sense of the at once infinite. However, if we consider different concretely existing entities in reality functioning within the numerical mode, and never qualified by it (as the Pythagoreans thought), we are only entitled to employ the primitive meaning of the potential infinite to interpret the 'all', or even strict finiteness (as in the case of finitistic nominalism, which accepts only a finite multiplicity of concretely existing things). In comparison to the irreducibility of the distinction between modality-entity (abstract-concrete), we have to stress the fact that what is *universal* and what is individual (intimately connected with the irreducibility of number and space) are also irreducible to each other![2]

Given this dual irreducibility, it stands to reason that we might investigate

1 Stegmüller later also includes other ideal objects in the former category, such as "being a horse" (Stegmüller, 1965:56.) At least partially, therefore, the opposition abstract-concrete coincides with the distinction between modality and entity – or modality and type law.

2 Sometimes nominalism prefers to use the spatial term whole (totality) to refer to the concrete existence of entities, instead of the expanded numerical term individuality. This change of modal perspective (still using modal terms in an idea context) does not dispense with the basic emphasis on the entitary dimension as the only real one.

the different possible combinations between them. Formally seen, there are four:

(a) abstract universal;

(b) abstract particular (individual);

(c) concrete universal; and

(d) concrete individual.

Stegmüller's distinction between concrete singular terms (such as proper names like 'Sokrates'; Stegmüller, 1965:59), concrete general terms (such as 'blue' and 'father'; Stegmüller, 1965:57), and abstract singular terms (such as 'blueness' (Bläue), "being human" (*Menschsein*), as well as "the class of blue things" and "the class of human beings" (Stegmüller, 1965:57, cf. p.83), only represent possibilities (d), (c), and (b) (in this order). What about possibility (a)? To Stegmüller, an acknowledgment of this possibility would amount to accepting a myth he explicitly rejects; the one of "identifying ideal being with general being" (Stegmüller, 1965:58). Insofar as the notion *abstract* refers to the dimension of modalities, we already encounter problems with this prohibition, because we have seen that the meaning of any aspect can only be analyzed in terms of its coherence with different other aspects. Generally speaking, we may say that every modal aspect universally underlies whatever functions within it. This implies that every modal property of any entity ultimately presupposes the underlying mode concerned in its conditioning universality.[1]

7.7.1.2 *Cross-cutting systematic distinctions*

In the light of our preceding analysis, we must distinguish between:

(1) The law side (resp. norm side) of the different modal aspects as a universal order for (i) entitary subjects and (ii) modal subjects. For example, the natural numbers are modal numerical subjects, displaying their structuredness (orderliness) in their successive numerical values and interrelations – determined by the arithmetical time order on the law side of the numerical mode. The numerical orderliness of concrete entities is seen from their actual (distinct) functioning within the arithmetical modality – i.e. in their being one amongst a multiplicity. Note that (1)(i) concerns modal properties (attributes) of concrete entities.

(2) The typical laws (type laws) for entities (such as atoms, plants, states, humans, books, and so on).

(3) The orderliness of entities, i.e. the universal way in which concrete entities show that they are conditioned by the specified universal entitary law that determines their factual existence – think of being human, being an atom, being blue, being a horse, and so on.

(4) The individual side of every concrete entity (for example, this woman, this atom, this blue ball, this horse, and so on).

The underlying distinctions used in this classification are given in (a) the dis-

1 This is why modal laws apply to all kinds of entities, whereas entitary laws are only applicable to a limited class of entities.

tinction between modality and entity; (b) between law (order) and factuality (that which is subjected to the relevant order), and (c) universality, specified universality (particularity) and what is individual. Since the formulation of (c) is co-determined by the irreducibility of the numerical and spatial modes that co-condition the way in which we speak both of modalities and entities, distinction (c) cross-cuts distinctions (a) and (b).

The systematic distinctions Involved in our discussion
[Points (1) - (4) represent the points on the previous page just below 7.7.1.2]

(1) law-side (norm-side)	—	(b.1) universal type-law (a.1) universal modal law-order
(i) entitary subjects	—	(a.3) modal specificity
(ii) modal subjects	—	(b.3) subject-subject and (a.2) subject-object relations
(2) type-laws	—	(b.2) type-laws
(3) orderliness the universal side of entities	—	(b.4) universal structuredness
(4) individual side of entities	—	(b.5) individual entities

With respect to (a), we might refer to (a.1) as the universal modal order at the law side of every aspect, to (a.2) as the universal modal subjects (which display both a *modal distinctness* and *modal identity*), and to (a.3) as the entitary specified universal modal properties of particular entitary subject-functions in different modal aspects (an instance of particularity or *modal specificity*).[1]

With respect to (b), we must refer to the difference between a universal modal law (b.1) and a universal entitary type law (typical law) (b.2) and, in connection with the factual side, we must distinguish between universal modal subject-subject and subject-object relations (b.3). For example arithmetical addition is applicable in the context of numerical subject-subject relations, whereas the relation between an extended spatial subject (such as a one-dimensional line-stretch) to its boundary (in the one-dimensional case: a point) exemplifies a subject-object relation and, furthermore, between the universal structuredness (orderliness) of entities [(b.4) as the 'image' of (b.2)] and the concrete individual side of entities (b.5).

When we compare the categories obtained from considering the interrelationships between the three sets of underlying distinctions with the four mentioned at first, we obtain the following correlations:

Categories (a.2) and (a.3) correspond to the first and second examples (concerning the natural numbers and *being a multiplicity* for concrete entities) mentioned in brackets after (1) (ii).

7.7.1.3 *Historical connections*

We have seen that according to Aquinas universals have a threefold existence: universalia *ante rem* (the real existence of ideas before the creation in God's

1 For example, when we want to refer to the specified universal modal numerical property of human beings, we mention the class of human beings.

Mind); *in re* (in the things as their universal substantial forms) and *post rem* (their subjective existence in the human mind as universal concepts) (see pages 177, 371 above). The distinction between *ante rem* and *in re* (*ante res* and *in rebus*) is a distorted metaphysical account of the difference between *order for* and *orderliness of*, intimately related to the distinction between law and (universal) factuality. In spite of the fact that Stegmüller does see that, what he calls "the idea of redness" is "not itself red" (Stegmüller, 1965:61), he still asserts that the whole controversy between universalia *ante rem* and *in rebus* is devoid of any logical meaning – it is a fictitious issue pertaining to no logical distinction (1965:117, see page 63). Evidently, he does not realize that it is completely dependent on the meaningful logical distinction between being red and the conditions for being red.

If we can answer this question with a clear yes or no, we would have made a return to the earlier attempt (already rejected by Stegmüller in 1965) that aimed at characterizing nominalism in terms of the opposition between the universal and the individual. Still, as we saw, Stegmüller's own approach is not free from problems. He attempts to explore the possibility of using the abstract/concrete opposition as a defining criterion for distinguishing between Platonism and nominalism (with conceptualism as a sub-category of Platonism). If we accept this criterion as a consistent norm, we have to place conceptualism on the side of Platonism. The only problem is that the meaning of the term Platonism then experiences an implicit shift – it no longer exclusively relates to a separated transcendent realm of ideas (or ideas in the divine Mind), since it now also refers to subjective universal concepts in the human mind.

Occam and Thomas fully agree on the nature of universal concepts in the human mind. They disagree about the implied criterion of truth, because Occam rejects the *universalia ante rem* and *in re*. The point of agreement between Occam and Thomas could not be used to classify Occam as a realist (Platonist) (cf. Stegmüller, 1965:69-71), because the difference in this point of agreement could only be stated when their divergent evaluation of concrete entities is included in the characterization – true realists always see concrete entities as the correlate of some transcendent 'Urbild' or some inherent universal substantial form (the traditions of Plato and Aristotle). Even those variants of nominalism considered to be true nominalists in terms of the abstract/concrete opposition (such as Berkeley and Brentano; cf. Stegmüller, 1965:75 ff.), still have to use linguistic (instead of conceptual) universals, such as words or at least the (universal) rules that regulate the use of words in communication (cf. Stegmüller, 1965:77).

7.7.1.4 *Once again the complex nature of nominalism*

The problem is that all proponents of nominalism want it both ways – they want to accept subjective universals (such as concepts, words, or linguistic rules) and at the same time reject any form of universality outside the human subject – "out there" only individual entities exist. The 'subjective' side of nominalism (including the position of Berkeley and Brentano), is surely *abstract* in terms of the opposition between concrete and abstract – implying that a consistent characterization in terms of this criterion won't be able to find a single true nominalist, for they will all have to be qualified as Platonistic.

To complicate the situation further, we might reverse the same approach and say that anyone accepting nothing but concrete-individual entities is truly a nominalist. However, in this case we have simply introduced a new name for a familiar *ism*, namely irrationalism, which only wants to acknowledge what is unique and individual. This implies that we need nothing more than the distinction between rationalism and irrationalism; a distinction (due to its dependence on the universal-individual opposition) that, according to Stegmüller, is not appropriate to characterize the abstract-concrete opposition (we shall presently return to the reasons why this is the case in Stegmüller's analysis).

We have argued that the complex nature of nominalism is only fully accounted for when we realize that it is rationalistic and irrationalistic at once (see pages 25 and 370 ff. above). In respect of the typical nature of entities, nominalism does not accept any conditioning order for (universal structure for), or any orderliness (universal structuredness) of such entities – every entity is strictly individual. In terms of our definition of irrationalism, this element of nominalism is undeniably irrationalistic. The other side of the coin lies in nominalism's acknowledgement of universals in the human mind (be it as concepts, words or linguistic rules) (compare in this respect the notion of meaning in Wittgenstein's later philosophy: *the meaning is the use* (Wittgenstein, 1968:54; § 139), and Antal's statement: the "rule which regulates the use of signs and their mode of application is called 'meaning' " (Antal, 1963: 26, cf. pp.34, 42, 54, 62, 79). This restriction of knowledge to universals is typical of rationalism in the sense defined by us (note that therefore *both* realism and nominalism are rationalistic). This explains why Berkeley could confess:

> It is, I know, a point much insisted on, that all knowledge and demonstrations are about universal notions, to which I fully agree: but then it does not appear to me that those notions are formed in the manner premised – universality, so far as I can comprehend, not consisting in the absolute positive nature or conception of anything, but in the relation it bears to the particulars signified or represented by it (Berkeley, 1710:55; *Introduction* §15).

Consequently, it is inevitable that we must characterize nominalism with the mentioned distinction: it is rationalistic in terms of the universals (concepts, words or rules) in one's mind, and irrationalistic in terms of its emphasis on the individual side of entities.

7.8 Shortcomings in Stegmüller's analysis of nominalism

Stegmüller denies the existence of *abstract general terms*, because he only accepts *concrete singular terms*, *concrete general terms*, and *abstract singular terms*. These distinctions are dependent on the diagonal opposition of abstract/concrete (A/C) on the one hand, and the opposition of universal/individual (U/I) on the other. Schematically:

Stegmüller accepts the triangles 123, 234, and 341, but rejects 412. His first examples to demonstrate what he means by abstract and concrete, correspond with our own distinction between *modality* and *entity*. If we restrict ourselves to this case, triangle 412 simply states that there are *universal modal aspects*. A closer analysis of Stegmüller's approach reveals that we have to add the distinction between law and factuality in order to supplement our "square of opposition." In a modal context, properties (*Eigenschaften* – cf. Stegmüller, 1965: 53) refer to the modal functions of entities that appear at the factual side (either as subject-functions or as object-functions). He mentions *sets* and *numbers* as examples of "abstrakten Gegenstände" (abstract objects) (cf. Stegmüller, 1965: 53). Although both sets and numbers presuppose the determining numerical time-order for number, they are distinct from this order. Furthermore, modal numerical subjects and their valid arithmetical time order display a distinct *universality*. The distinct universality of modal numerical subjects is nevertheless not as such acceptable to Stegmüller, because he allows for nothing but *abstract singular terms* (for example: the class of blue things; Stegmüller, 1965:57). Thus he only acknowledges the distinctness of the modal numerical property of *being a class*, but not for its universality. In the case of the entitary dimension, this negative approach (rejecting the conditioning role of the spatial term universality in his account of the category *abstract*) is even more striking. *Being a horse* (*Pferdheit*), *being human*, and *being a father* are abstract singular terms, whereas horse, man, and father are considered to be *concrete general terms* (Stegmüller, 1965:56-57). The term 'general' is meant (in the case of concrete general terms) to be applicable to an arbitrary number of entities (the individuality of every entity is decisive for its concreteness) (cf. Stegmüller, 1965:57).

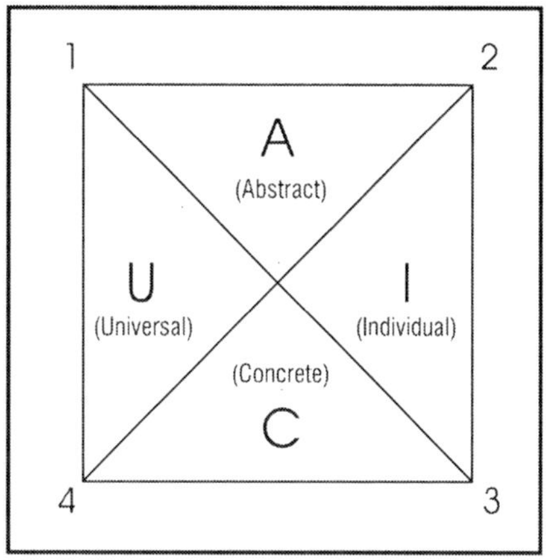

In a later context, Stegmüller uses the following example: "When I imagine my friend Hans" then this representation would be more or less indistinct; "it would be incorrect, however, to call this image of him a general image" (Stegmüller, 1965:77). The crucial question is: is the fact that Hans is a human being included in this representation of my friend Hans or not? If it is excluded, Stegmüller is correct to call it indeterminate (*unbestimmt*). But then it is doubtful whether we still have a distinct representation, because I cannot imagine any of my friends without acknowledging them at the same time *as*

human beings – how else would I distinguish them from pieces of land I adore? Why would it contradict the nature of language to speak of a general representation of some particular friend? We have mentioned that we can refer in a lingual way both to the individual side and the universal side of entities (such as friends), simply by using the respective articles *this* and *a*: *this* human being (friend Hans) is surely *a* human being! This example about "my friend Hans," given in the context of the exposition of Brentano's conception, is in line with Stegmüller's identification of *concrete* with *individual* (cf. Stegmüller, 1965:63). His remark that only that which itself is concrete could appear as a concrete object says nothing against the acknowledgement of the concrete universal side of entities (lingually reflected in the article *a*), but, in the context of his intention, only affirms the tautology: something concretely individual cannot be anything but concretely individual!

This identification of the concrete with concrete individual, while simultaneously disregarding the universal side of concrete entities, is not only typical of nominalism, as we have seen, since it also relativizes the difference between the two pairs of distinctions mentioned in the "square of opposition." If the universal/individual distinction is irreducible (as far as I can see an insight not questioned by Stegmüller), then the identification of what is individual and concreteness must rule out any combination of universality and concreteness. What then about Stegmüller's reference to *concrete general terms*, such as *blue*, *man*, and *father* (Stegmüller, 1965:57)? To him, this combination does not refer to the universal side of concrete entities, but much rather (and true to nominalism in this respect) to an arbitrary number of concrete individual entities ('objects'). In actual fact, therefore, he not only disregards triangle 412, but also triangle 341.

An abstract singular term such as being human (*Menschsein*; Stegmüller, 1965:57) is immediately connected with the class of human beings, which is obviously a numerical feature. In other words, these terms are meant to designate a particular modal property implied in the orderliness of being human. This corresponds with the category (1)(i) (= a.3), where we mentioned the specified universal modal numerical property of human beings, indicated as the *class* of human beings. The only difference is that Stegmüller does not accept the term *universal*, since he uses the qualification singular in an *exclusive* sense. As a substitute for the term singular, we use the term *distinct* in order to accentuate that this modal arithmetical term does not contradict the meaningful additional qualification of *universality* (a term referring to the spatial mode). Being a multiplicity (a *class*) is a *specified universal* numerical property *distinct* from other specified universal modal properties (such as being extended, or being alive – respectively referring to spatial and biotical properties).

The conditioning role of the modal aspects in theorizing implies that an analysis of 'abstract' modal properties is compelled to simultaneously use terms that have their seat in different modal aspects. Stegmüller consistently wants to rule out the conditioning role of the *spatial* aspect by restricting himself to the *numerical* perspective: in order to avoid the 'myth' of "identifying ideal being with general being" (Stegmüller, 1965:58), he prefers to call "abstract objects" *singular* (cf. Stegmüller, 1965:57, 59, 76, 83). Their *singularity* (*distinctness*), however, cannot contradict their universality for two reasons: firstly, because distinctness (singularity) is different from what is concretely individual (even in the latter case, universality is not ruled out), and secondly, because the terms singularity and universality refer to different modal points of entry (only if these terms had the same modal 'origin', a contradiction would result).[1]

Due to this preoccupation with *singularity* at the cost of universality (generality), Stegmüller fails to appreciate the important distinction between a universal *order for* and its correlate found in the universal orderliness of things, as well as the correlation between the universal side and the individual side of concrete entities (the latter shortcoming is co-determined by his nominalistic identification of *concrete* with *individual*). In our pre-scientific experience of reality, we are constantly confronted with its orderliness (structuredness) as reflected in our everyday concepts. Philosophy and the special sciences attempt to account for this orderliness of reality with an analysis of the implied order for this structuredness of reality, and they do this by means of modal analysis (abstraction).

At this point, we also have to recall that analysis and abstraction are synonyms. The logical activity of analyzing or abstracting amounts to the identification (lifting out) of some feature by simultaneously disregarding it (distinguishing it) from non-relevant properties. An implication of the difference between the dimension of modalities and entities is that we must distinguish between modal analysis and entitary directed abstraction (analysis) ("concrete abstraction"). In both cases, we at least have to refer to the orderliness of that which is abstracted, showing that neither the conditions nor results of abstraction (analysis) can do away with universality.

Ultimately these shortcomings are the result of not carefully distinguishing between (a) modality and entity; (b) law (order for) and factuality; and (c) universality, specified universality (particularity) and what is individual. If we consider the intersections of distinctions of distinction (c) with distinctions (a) and (b), we may use Stegmüller's systematic framework ('paradigm') to compare the perspective thus obtained in the following way:

(1) Stegmüller's concrete singular terms (proper names such as 'Sokrates') actually refer to the individual side of concrete entities (category b.5 above).
(2) Stegmüller's concrete general terms (such as blue, man, father) create the impression that he acknowledges our category b.4, i.e. the universal structuredness (orderliness) of entities – an impression that is finally ruled out by his identification (in a typical nominalistic fashion) of con-

1 Stegmüller's usage is reminiscent of the "Logische Untersuchungen" of Husserl, and especially of his 'Ideen' where we read about "eidetischen Singularität" (cf. Husserl, 1913:25).

crete entities with their *individual side*, implying that the expression concrete general is used to refer to an arbitrary number of concrete individual 'objects' (entities).

(3) Stegmüller's abstract singular terms (such as "being a horse" and "the class of human beings") initially also suggests our category b.4 (of course with the exclusion of universality), but eventually they prove comparable to our category a.3, i.e. to the *entitary specified universal modal properties* of particular entitary subject- (or object-)functions in different aspects (an instance of particularity) for example when we want to refer to the specified universal numerical property of human beings by referring to the *class* of human beings.

(4) What is not accounted for in Stegmüller's analysis is the universal modal order on the law side of every aspect (such as the numerical time-order of succession or the spatial time-order of simultaneity) (a.1); universal modal subjects (such as numbers) (a.2); the universality of a.3 and b.4; and finally the distinctions indicated by b.1, b.2 and b.3.

Because what is concretely individual only appears at the factual side of reality, and because the orderliness of entities actually pertains to a form of *specified universality* (typical structuredness) that is still universal and therefore not to be seen as individual (in which case the irreducibility of universality and what is individual would have been eliminated), Stegmüller was not able to allow for abstract general terms. In the case of concrete general terms, he still had a multiplicity of concrete individuals to "split up" the meaning of the predicate *general*, but in his realm of "abstract objects," there is no room for concrete individual entities that can serve as "splitting up agents" for the term *general* in the expression "abstract general".

7.8.1 *The impact of nominalism on Dooyeweerd's thought*

As we have seen, nominalism does acknowledge universality – but only within the subjective human mind (either as universal concepts or as universal words), for outside the human mind, strictly *individual* things exist. In addition, we have seen that modern nominalism strips reality both of its *order determination* and its *orderliness* – degrading factual reality into a non-structured chaos. This is where Kant started, for we have seen that, according to him, the material of experience (sense impressions) is *disordered* and *chaotic*. The lack of determination created by denying the order for and orderliness of reality generated the need for an alternative law-giver that Kant finds in human understanding, for the natural order is (formally) made possible by the categories as *forms of thought*. Thus seen, the concepts of understanding in Kant's view indeed function as the *formal law-giver* of nature. We reiterate one of Kant's frequently mentioned statements: "*Categories are concepts, which prescribe laws a priori to phenomena, and thus to nature as the totality of all phenomena*" (Kant, 1787:163 – see page 382 above).

The remarkable after-effect of nominalism in Dooyeweerd's failure to appreciate the universal side of factual reality – similar to Stegmüller's views. Although Vollenhoven is not a victim of this legacy he does not develop a theory of *type laws* or of *modal specificity* (owing to the influence of nominalism erroneously designated by Dooyeweerd as *modal individuality*).

Of course Dooyeweerd does not adhere in any way to the nominalistic rejection of God's universal law for what is subjected to it. It is also incorrect to say that Dooyeweerd *denies* the reality of law-conformity (*wetmatigheid*), for he consistently speaks of the orderliness of reality. Only insofar as he does not realize that *law-conformity* is a feature of whatever is subjected to God's *law*, can one discern the influence of nominalism in his thought. Nominalism causes him to "misplace/displace" the orderliness of reality (its law-conformity), which results in his identification of *law* and *law-conformity*.

Nonetheless, these qualifications should not close our eyes to the fact that Dooyeweerd's lack of appreciating the universal side of concrete entities functioning at the factual side of reality truly continues a core element of the irrationalistic side of nominalism, given in the view that concretely existing entities ("outside the human mind") are purely individual. This explains why he continues to identify the concreteness of entities with their *individual* side, expressed in his frequently used saying that things function in a *concretely individual way* within universal modal aspects (Stafleu followed him in this regard – see Stafleu, 1980:8 and 2002:12).

7.8.2 *Dooyeweerd and Vollenhoven*

We have noted that neither Vollenhoven nor Dooyeweerd realized that modal terms may be employed in both a conceptual way and a concept-transcending manner (see page 176 above). As soon as the nature of concept-transcending knowledge is acknowledged, the apparent opposition between Vollenhoven and Dooyeweerd regarding the 'place' of God's law is relativized. Vollenhoven's emphasis on *validity*, is represented to a degree in Dooyeweerd's idea of *universal validity*, and in his view that God's law *determines* whatever is subject to it. From the perspective of the nature of concept-transcending knowledge, the choice between the terms *above* (Vollenhoven in his concept-transcending use of the spatial opposition above and below) and *side* (Dooyeweerd in the similar concept-transcending use of another spatial term) is insignificant.

Vollenhoven consistently pays attention to what he calls *diversity* (*verscheidenheid*) and *connection* (*verband*) (see Vollenhoven, 2005:19 ff.). The conceptual use of unity and multiplicity (or: the *one* and the *many*) is focused on the numerical aspect. When this numerical state of affairs is explored in the service of a concept-transcending mode of speech, we normally encounter references to *unity* and *diversity*. Similarly, the spatial meaning of *continuity*, which is equivalent (synonymous) to *connectedness* and *coherence* (see Strauss, 2002), underlies the concept-transcending use of the (original) spatial term *coherence*. In other words, Vollenhoven consistently explores a mode of explanation in which the core meaning of number and space is employed in a concept-transcending manner.

This insight in fact supports an understanding of Vollenhoven and Dooyeweerd that highlights the underlying *convergence* or even *agreement* in their thought. It appears that they differ seriously, because Vollenhoven views God's law as intermediate between God and creation, and characterizes it in terms of 'validity'. Yet he maintains that God's law is 'above' creation. Nonetheless, Vollenhoven and Dooyeweerd share the idea of 'subjection' (Tol introduces the term 'subjectity' – Tol, 1995:102) – an idea that cannot be divorced from the basic spatial intuition of 'above' and 'below'! An understanding of the meaning of *subjected* cannot avoid the ideas of *distinctness*, *delimitation* or the *effect* of what holds (what is *valid*) for whatever is subjected to God's law. The Dutch equivalent for "being subjected to," namely 'onderworpenheid', more explicitly reveals the energetic connotation of *subjection*. [Vollenhoven, for example, mentions the "onderworpen zijn aan dezelfde wet" (Vollenhoven, 2005:34 – "subjected to the same law").] When the assumed *difference* between Dooyeweerd and Vollenhoven is mentioned with reference to the validity of God's law (intermediate between God and creation – Vollenhoven) and the idea of the *law side* of creation (Dooyeweerd), then we are arguing at cross-purposes.

In terms of their concept-transcending modes of speech, both Vollenhoven and Dooyeweerd, in respect of their idea usage of *physical* terms, believe that God's law has an *effect* on what is *subjected* to it – God's law is *valid*, it *holds* in the sense of *determining* creatures (Dooyeweerd: *bepaal*; Vollenhoven: *bepaaldheid* – *determines* and *determination*). From the perspective of an idea usage of spatial terms, there is also a fundamental agreement in terms of the ideas 'above' and 'below' (being *subjected*). The basic *shared* distinction is that between "law and subject"; in addition, Vollenhoven and Dooyeweerd also share similar (almost identical) modes of speech, exploring physical and spatial terms in a concept-transcending way. Given this convergence, adding or omitting the suffix 'side' does not seem so decisive as to be divisive – the *effect* of what is actually shared out*weighs* the diverging emphases in saying that God's law is *above* creation or constitutes the law *side* of creation. (The argument that the law "comes from God" does not disqualify creation (in the sense of what is subject to law – a la Vollenhoven), because whatever is subjected to law also "comes from God." Furthermore, the fact that Dooyeweerd's idea of law entails the "above-below" distinction, one can just as well equate the expression "law side of creation" with the "above-side of creation.")

Clearly, throughout their respective philosophical systematic philosophies, Vollenhoven and Dooyeweerd therefore maintain the basic (shared) distinction between *law* and *subject* (implying, amongst others, an idea usage of spatial terms). Recalling Vollenhoven's "negative theological" remarks concerning his use of the term *boundary* to the exclusion of any connotation of spatiality (instead of recognizing in his use of the term *grens* a concept-transcending employment of a spatial term), we may add that his distinction between

God and creation is entirely dependent upon an idea usage of spatial terms, for he holds that there is nothing "divine that stands *under* the law or anything created that stands *above* the law" (the italics are mine – DS; this once more highlights Vollenhoven's concept-transcending use of spatial terms).

Vollenhoven uses the terms *universalism* and *individualism* to indicate an overestimation of individuality and universality, respectively. Dooyeweerd, on the other hand, uses the correlation of law side and individual factual side to characterize the difference between *rationalism* and *irrationalism*. He denies universality at the factual side of reality and consistently identifies the orderliness of entities, their lawfulness or law-conformity, with the law side of reality. Yet as soon as it is realized that law-conformity is a feature of factual reality, then it is no longer possible to deny the universal side of factual reality. We argued that, in the atom-*ness* of an atom, this (individual) atom, in a (universal) way, is subject to the universal (law) conditions for *being an atom* (where *being an atom* represents the universal side of an atom).

Of course rationalism then cannot any longer be defined as an absolutization of the *law side* of reality, as Dooyeweerd does, because universality is also present at the *factual side* of reality. Furthermore, it is then also insufficient to define irrationalism as an absolutization of the factual side of reality. In other words, whereas *rationalism* leaves no room for true concept-transcending knowledge, irrationalism leaves no room for genuine conceptual knowledge. Vollenhoven explicitly rejects the identification of law and law-conformity: "The law is therefore not regularity, etc.; processes subjected to the law are regular and irregular."[1]

These considerations inspire an alternative definition of rationalism and irrationalism – where the former is defined as an absolutization of *conceptual knowledge* and the latter as an absolutization of *concept-transcending knowledge*[2] – thus placing the relation between individuality and what is universal at the center of this distinction.

7.8.3 *Ambiguities in Dooyeweerd's idea of individuality structures*

What is particularly ambiguous in Dooyeweerd's thought is his mature conception of the law for individual things, captured in his use of the expression *individuality structures*. Already in 1931 the influence of nominalism is apparent in Dooyeweerd's thought, for at this stage of his intellectual development, he frequently straight-forwardly speaks of "individual structure" (beginning on the first page and continued throughout the work). Sometimes the idea of *meaning* is combined with the notion of an *individual structure*, for example when he refers to the "meaning-individual thing structure of organized human communities" (Dooyeweerd, 1931:111 ff.).

1 "Wet is dus niet regelmaat enzovoorts; regelmatig en onregelmatig zijn processen aan de wet onderworpen" (Vollenhoven, 2005:16).

2 That Dooyeweerd's denial of universality at the factual side of reality actually follows from his indebtedness to an element of nominalism, has been argued above.

Insofar as Dooyeweerd does not use the term *structure* in the sense of *construction* (or what has been *constructed*), but in the sense of "law for," his aim is to account for the *law for* concretely existing *individual things*. Yet he does speak of an *individual structure* and sometimes of the 'construction' (Dutch: 'bouw' – see Dooyeweerd, 1936:41) of an entity (in English translated as 'structure' – see Dooyeweerd, 1997-III:60). But as soon as the factual side of reality is at stake, he employs the qualifier *individual* or expresses the conviction that a concrete entity *individualizes* the universal meaning of modal aspects.

One particular meaning intended by the phrase *individuality structure* in Dooyeweerd's thought still echoes the mentioned initial (1931) terminology: *individual(ity) structure*. An individuality structure may either be a *universal law* for an individual entity or it may be an "individual structure." In the latter phrase the word 'structure', for Dooyeweerd, cannot designate something *universal* because it is explicitly qualified as an *individual structure*. For this reason this phrase *individual structure* creates an additional problem because it is truly meant to designate what is *individual*. However, if the term *structure* in the phrase "individual structure" is meant to refer to the law side of reality, then it must be universal.[1] The alternative option is to hold that Dooyeweerd intends to use the phrase individuality structure to describe the above-mentioned (factual) *construction* of an individual entity. However, in this case he would still need to refer to the *structuredness* of such an individual entity – and we have seen that such a *structuredness* remains universal. Although it highlights *being this* or *that* (a universal trait), the ultimate qualification given by Dooyeweerd to the factual side is to say that it is *individual*.

The irony is that *being-an-individual* is a universal trait shared by all individuals, just like *being-a-chair* is a universal trait shared by all chairs or *being-an-atom* one shared by all atoms. This does not entail in any way that we have grasped *what-is-individual* in a concept, pointing at the necessity of arriving at concept transcending knowledge in this regard. Because Dooyeweerd does not acknowledge universality at the factual side of reality (*wetmatigheid*) he constantly confuses what is *individual* with *individuality* – by not realizing that the latter is a universal feature shared by all individuals. One consequence of this is, as mentioned, that Dooyeweerd holds that the (universal!) meaning of modal aspects are *individualized* – contrary to his claim that the functional structure of modal aspects is not affected by "modal individualizations" (Dooyeweerd, 1997-II:423-424). On page 423 (paragraph 3) Dooyeweerd commences by stating that "[M]odal meaning must be individualized" but then – just in the next paragraph – he explains that the "process of individualization, however, does not affect the fundamental functional *structure* of the modal aspect." Also compare his remark on page 422: "Formalism

1 Note that *universality* refers to the *scope* of a law and not to its own (distinct) identity. From another perspective one can distinguished laws, count them and speak of them in familiar numerical terms – such as *one* amongst *many* laws, of *distinct laws*, i.e. of laws that are 'individually' different. But this never means that the scope of a law applies to a single (modal) subject or to a single individual only.

knows nothing about individualizing the modal meaning of law." If the modal (functional) meaning of an aspect displays the trait of universality – as Dooyeweerd holds – then something *does change* when this universal meaning is claimed to be *individualized*. Cf. also Dooyeweerd, 1997-III:76 where he explains that the "structure of individuality" of a linden tree "individualizes" its modal functions and "groups them together in a typical way within the cadre of an individual whole."

Still, the universal modal meaning of all the aspects cannot be *individualized* – they can only be *specified*. The reason given for this is that what is universal and what is individual are not two ends of a *continuum*. These two realities are irreducible. For that reason Dooyeweerd should have investigated instances of *modal specificity* and not of *modal individuality*. But once again remember that Dooyeweerd does not deny law-conformity – he simply identifies it with the law side of reality – also seen in his habit of freely using these expressions interchangeably. What then about the term *individuality* (as contained in the expression individuality structure)? First we have to highlight once more the important difference between the (i) individual and (ii) individuality.

(i) Dooyeweerd consistently understands 'individual' to mean "strictly individual," i.e. without any discernable universal features.
(ii) From a systematic point of view, however, *individuality* refers to a *trait* shared by whatever is *individual* – which means that *individuality* is a *universal feature*. This implies that every individual entity displays this universal trait of "being individual" (= having individuality). Goethe realizes this quite clearly for his quoted answer to "what is unique?" is: "millions of instances " (see above page 432).

Although Dooyeweerd had the intention of dealing with what is *individual*, he did not realize that his use of the term *individuality* entails universality. Discussing *modal individuality* is also intented to account for concrete *individual* functions of individual entities within the modal aspects, but in fact it deals with *concrete universal functions* (because *individuality* is a universal trait of individual entities).

The ambiguity in Dooyeweerd's account is such that we can opt for two alternatives: either interpret the idea of individuality structures (i) as *structures for* what is individual or (ii) as *structures of* what is individual.

Re. (i): In this case the law is universal and the factual side individual.
Re. (ii): In this case it seems as if there are (universal) *structures of* at the factual side – contradicting his idea of the *individual* factual side.

The issue of what is universal and what is individual also permeated Dooyeweerd's view of the relationship between his "individuality structures" and the modal aspects of reality. The reason for this is that he (correctly!) defends the idea of *modal universality*, but incorrectly holds the view, as we saw, that the universal can be *individualized*. Modal laws are not limited to some or other group (or: type) of entities, for every entity in principle functions in all aspects (for that reason they are, strictly speaking, not merely aspects of things). *Type laws*, by contrast, only hold for a limited class of entities, such atoms, states or mammals.

Dooyeweerd holds that transient individual entities ought to be distinguished from the ordering types to which they are subjected:

> The structural types of plants and animals as such are indeed not individual subjects that originate in the temporal process of becoming for much rather they are ordering types belonging to the law side and not the factual side of our empirical world. They can only realize themselves in transient individual living beings, but as ordering types they necessarily bear a constant and foundational character in the time order (Dooyeweerd, 1959:132); and

> Implicitly I have already pointed out earlier that the phyla (in the sense of the highest genotypes [stamtypen] of the realm of plants and animals) cannot come into being and pass away in the process of becoming because they are not themselves "living beings" but much rather *order types* of individual totalities (Dooyeweerd, 1959:141).

Dooyeweerd's own expression "individuality structure" may be interperted to approximate the simple systematic distinction between *universal law* and the universal side of individual things at the factual side of reality. However, as soon as Dooyeweerd attempts to account for something *structural* at the factual side, he reduces (or qualifies) it to be *individual*. For that reason he states what we referred to above, namely that the structural principle (individuality structure) of something (such as the linden tree), "individualizes the modal functions and groups them together in a typical way within the cadre of an individual whole" (Dooyeweerd, 1997-III:76). Elsewhere he writes: "Here we grasp reality in the typical totality-structures of individual things, concrete events, concrete social relationships, etc., in which all modal aspects are typically individualized and integrated in unbroken coherence, grouped together as a whole without any analytical distinction between the modal aspects themselves" (Dooyeweerd, 1996:9). Also compare his qualification of 'subjectivity' by the use of the word 'individual': "The origin of Law and of individual subjectivity , ..." (Dooyeweerd, 1997-I:507).

Henk Hart speaks of universality and individuality as *traits* and proposes "to see" universality "as the way in which conditions relate to subjective existence, while individuality I propose to see as the way in which subjective existence meets conditions" (Hart, 1984:72). On the next page he elaborates this distinction: "universality is simply the relationship between conditions and what is conditioned viewed from the point of view of the conditions. The same relationship going in the other direction is one of individuality. Subjects meet conditions uniquely, even if it is the same conditions they meet." What is lacking in Hart's view is an acknowledgement of universality at the *factual side* of reality – evinced in *being this or that* – which is always correlated with the factual individual side of reality. As explained, pointing at *this chair* (i.e. highlighting its *individual side*) is dependent upon its *being a chair* (its universal side). Although Hart does acknowledge that "being a sparrow" is the way in which "an individual will bear evidence of universality" (Hart,

1984:72), he does not clearly distinguish between the universality of conditions and universality at the factual side.

Zuidervaart appreciates Hart's position in this regard positively and consequently also does not realize that one has to distinguish between universality (and individuality) at the factual side (the distinction between what is *concretely individual* and *concretely universal*. He writes: "Finally, I agree with Hart that nomic conditions (Dooyeweerd's 'law side') hold 'universally for the subjectivities for which they do hold,' and that subjectivities "meet conditions uniquely, even when it is the same condition they meet.' (Hart 1984, 73) This approach promises to clear up confusions in Dooyeweerd's account of structures of individuality, which various authors have pointed out"[1]

Just as Dooyweeerd's notion of an individuality structure is burdened by the idea of an *individual* structure (the underlying issue of what is universal and what is individual), his view of the relation between an *individuality structure* and (universal) modal aspects gets entangled in the untenable view that universal modal aspects or functions can be *individualized*. We have consistently argued that the *typical* way in which an individual entity functions within the modal aspects solely *specifies* that meaning in accordance with the type law concerned. The simplest example of a the modal function of an individual entity (therefore appreciated according to its concrete individual side), is that *this* (one) chair is not *that* (other) chair.

We may repeat the underlying issue once more succinctly: what is *universal* and what is *individual* are not the two ends of a continuum – for if they were, the distinction between them would collapse. Therefore, neither (universal) individuality structures nor (universal) modal aspects could be *individualized*. Yet, appearing at the factual side of reality, concrete entities function in a twofold way within modal aspects:

(i) In a *concrete individual way* (*this* entity and not *that* entity);
(ii) In a *concrete universal way* (this *type* of entity and not that *type* of entity).

We have seen that Vollenhoven does realize that universality and what is individual both appear at the 'subject' side of reality. He also properly distinguishes between law and lawfulness. Yet he did not develop an articulated idea of *type laws*. Dooyeweerd in turn does distinguish between modal laws and type laws but he blurred this distinction by accepting two untenable assumptions: (i) that whatever appears at the factual side of reality must be *individual* and – as a consequence of (i) – (ii) that law and lawfulness (law and law-conformity) are the same.

My explanation for this confusion is that Dooyeweerd did not escape from the influence of *nominalism* in this regard – remember that *nominalism* does not accept any universal traits outside the human mind – in reality threre are solely *individual* entities, without any universal characteristics.

1 What Zuidervaart refers to is found on page 73, but the formulation itself is not found there – "'universally for the subjectivities for which they do hold."

7.9 The distinction between various dimensions of reality

Van Riessen emphasizes that modal aspects serve as *points of entry* (*gateways*, 'toegangspoorten') to reality. In the course of our analyses, we have also employed alternative designations for the aspects of reality, such as *facets*, *ways* or *manners* in which things exist, *modes* (*modalities*) within which (natural and social) entities *function*, and so on. A closer analysis of these designations reveals something inevitable. The notion of a "point of entry" ('gateway') is a *metaphor* employed in order to explain the nature of modal aspects. We have seen that metaphors have meaning on the basis of similarities and differences (intersecting relations between aspects and entities, entities and aspects and between different entities). The reference to *facets* (and synonyms like 'sides' or 'aspects') explores our spatial awareness, which is attached to concretely existing entities functioning within the spatial aspect.

So in fact we speak in terms of *entities* about modal aspects and in terms of *aspects* about entities!

7.9.1 *Complexities involved in characterizing the dimension of modal aspects*

Phrases such as "angle of approach," 'gateway', "point of entry," 'viewpoint' and so on are all *metaphoric* ways in which we allude to the unique dimension of modal aspects. The term 'dimension' is also derived from the meaning of the spatial aspect. One way in which one can explain what the meaning of a 'dimension' is, is that it constitutes the 'horizon' of human experience. The various *dimensions* of human experience[1] therefore form the *experiential horizon* of being human. Once again, the meaning of *horizon* is derived from the *limits* (*boundaries*) of my purview – it serves as an indication of "how far one can see."

From the nature of modal abstraction as the distinctive feature of scholarly (scientific) activities (see Chapter 2, pp. 48 ff.), we have derived the important insight that theoretical thinking inevitably finds its foundation in an all-encompassing and basic understanding of reality, accounting for the unity and diversity of aspects, entities and processes. The main developments within the domain of the philosophy of science during the 20th century, in the footsteps of Kuhn, gave prominence to the metaphor of a *paradigm*. In Greek, the word *παράδειγμα* (*paradeigma*) refers to a 'pattern' (or 'example') and its *copy* – compare the German terms *Urbild* and *Abbild*. The multiple nuances with which the term *paradigm* is used by Kuhn highlight something of the suggestive richness of meaning entailed in this metaphor. Within scholarly disciplines, however, no single useful application of the term *paradigm* can by-pass the idea of a theoretical view of reality and the latter, in turn, always has to give an account of the uniqueness and coherence of the diversity found

1 The dimension of cosmic time, the dimension of modal aspects, the dimension of entities and the central (root) dimension of reality, also known as the religious dimension – on condition that 'religious' is not misunderstood as one facet of life alongside others, but is indeed seen as the core dimension of reality, giving direction to all issues of life.

in our everyday experience. Emphasizing this core element does not exclude other and related meaning nuances, such as the idea of a successful theory (for example the classical particle mechanics of Newton, further analyzed by Stegmüller in the footsteps of the mathematical physicist Sneed – see Stegmüller, 1976).

At this point it is important to realize that, while metaphors are *replaceable*, modal terms and modal analogies are *irreplaceable* – at most one can provide synonyms for modal terms and analogies.[1]

In our analysis of the meaning of *identity*, the distinction between conceptual knowledge and concept-transcending knowledge is applied. When the core kinematic intuition of *continuous flow* is stretched beyond the limits of the motion aspect, we meet the idea of identity amidst change (something continues to be what it is in spite of the changes it undergoes). If we expand this perspective on identity and explore statements in which the core meaning of the first four aspects of reality are stretched beyond the limits of these aspects, it will be possible to articulate the four most basic concept-transcending statements that philosophy can formulate about the universe.

(i) Exploring the meaning of the numerical aspect in a concept-transcending way provides a foundation for the statement that *everything is unique*.

(ii) Stretching the meaning of space beyond its boundaries leads to the statement that *everything coheres with everything else*.

(iii) An idea use of the kinematic aspect underlies the statement that *everything remains identical to itself*.

(iv) Finally, the physical intuition of change may be stretched beyond its boundaries, yielding the claim that *everything changes*.

Only if these statements do not rest upon distinct irreducible modal points of entry, would they be *contradictory*.[2] Therefore, these statements do not *contradict* each other, for they *entail* and *complement* each other. By the same token they illustrate what it means to say that the modal aspects are gateways ('toegangspoorten' – Van Riessen), for by using these four aspects as points of entry, statements about the universe are made possible. Although we have employed two metaphors (*gateway* and *point of entry*) in our reference to modal aspects, a more precise characterization of aspects is possible once we revert to a concept-transcending use of modal terms derived from the first four modal aspects.

That *everything is unique* translates into what has become known (since

1 Recall for example synonyms for the core meaning of space (continuous extension), found in terms such as coherent, connected and the whole-parts relation.

2 In connection with the second statement, we may note that, in spite of his physicalist emphasis on the law of gravitation, the German sociologist Ratzenhofer considers gravitation simply as a modification of the universal basic law of the "mutual dependence of all things" (the "gegenseitige Abhängigkeit aller Dinge") (Ratzenhofer, 1898:84). Ratzenhofer articulated his monistic sentiments in the work Positive Monism (1899) – see also Dooyeweerd (1926:88, note 15).

Groen van Prinsterer and Abraham Kuyper) as the principle of sphere-sovereignty. Within the theory of modal aspects, this principle implies that each aspect is *sovereign within its own sphere*. Likewise, the fact that everything coheres with everything else is expressed in the theory of modal aspects through the retrocipatory and antecipatory analogies within each aspect – known as the *sphere-universality* of every aspect (see page 300 above). The *constant* (enduring) structure of the modal aspects underlies the concrete *functions* that natural and social entities and processes have within each aspect (in some cases subject functions, and in others subject and object functions).

With a slightly different focus, we are here once more fully dependent upon the possibilities provided by the first four modal aspects. However, whereas our formulation of a *law of nature* (or of the pre-positive starting point of a norming principle) embedded these terms within *compound* basic concepts (i.e. these terms are employed in a conceptual way), our characterization of the structure of an aspect had to use *concept-transcending* articulations of terms derived from the same foundational aspects of cosmic reality. A succinct formulation capturing the structure of a modal aspect reads as follows:

> Modal aspects are both *unique* (sphere-sovereign) and *mutually cohering* (their sphere-universality) while *constantly* conditioning (making possible) the *functions* that natural and social entities and processes have within them.

It is therefore *constitutive* for an understanding of modal aspects that a *regulative* (concept-transcending) use be made of the meaning of the aspects of number, space, movement and energy-operation.[1] These elements are said to be constitutive, because they cannot be *replaced*. By contrast, given the possibility of substituting any image or metaphor with a different one, it is understandable that alternative designations for the nature of an aspect emerged. For example, although Dooyeweerd already employed the term 'modality' in 1923, during the first phase of his intellectual development (1918-1928) he also referred to modal aspects as "domain-categories" (*gebiedskategorieën*) (see Verburg, 1989:54 ff., 67).[2] In 1931, Dooyeweerd wrote on the crisis in the humanistic political theory; throughout this work, he avoided the expression *modal aspect*. Rather, he consistently employed the term *function*, although he also used the phrases such "law sphere" (*wetskring*) and "meaning-side" (*zin-zijde*).

More recently, Stafleu introduced the term "relation frame" ('relatiekader') for an aspect (Stafleu, 2002:16 ff.). The second element of this phrase, the

1 This situation is analogous to the conclusion I came to in 1973 in respect of the relation between a concept and an idea of a modal aspect: the idea of an aspect, in a regulative sense, is foundational to the concept of an aspect, while the concept of an aspect is foundational to an idea of it in a constitutive sense (see Strauss, 1973:179).

2 At this stage – under the influence of neo-Kantianism – Dooyeweerd held the view that logic had to use basic categories designated as modal categories of logical thinking. He intended a logical system, embracing "modal categories" such as "identity, diversity, concept, judgement, relation, and necessity" (quoted by Verburg, 1989:57). Surely these "modal categories" are not his eventual modal aspects!

Dutch term 'kader', may mean 'framework', 'cadre', or 'skeleton', and it alludes to the whole-parts relation.[1] The first element, namely the term 'relation', is derived from the spatial meaning of *connectedness*, echoing a basic idea of his 1980 work on the foundations of physics, in which he pointed out that the modal aspects are not merely *modes of being*, since they are also *modes of explanation* and "universal modes of temporal relations" (Stafleu, 1980:15). Clouser suggests that we translate the idea of a modal aspect – law sphere – by referring to a sphere of properties and laws: "aspects are not to be confused with types or classes of things, but are kinds of properties and laws true of all things" (Clouser, 2002:254).[2]

Whatever our choice of images, metaphors or modal terms in order to designate the nature of the dimension of modal aspects, none of these can substitute or eliminate the constitutive elements highlighted above – describing the meaning of a modal aspect by using modal terms derived from the first four cosmic aspects in a concept-transcending way.

Within the physical aspect, with its core meaning of energy-operation, the term *function* has its seat, showing that a succinct designation can employ terms from diverse aspects (*function* – the physical aspect; *side*, *sphere* – the spatial; and so on). At the same time, figurative speech may serve the same purpose, for we noted that an aspect can also be designated as a *gateway*, a *point of entry*, an *angle of approach* and so on. Finally, we see that any combination of modal terms and images is also permissible – such as "domain category," "relation cadre" or "sphere of properties and laws." A collective mode of speech actually amounts to a staggering of modal terms, for then we speak of the *dimension* (spatial) of modal *functions* (physical) of reality.

The nature of the dimension embracing the concrete *what* of reality, i.e. the multiple things, plants, animals, cultural artefacts, societal collectivities and human beings, caused from its inception problems of its own; it required from the outset consideration of the relationship and difference between universality and individuality.

7.9.2 *The dimension of entities*

From the perspective of our everyday experience, the reference to diverse kinds of things (entities) is as natural as it is prior to articulated theoretical reflection. Once theoretical reflection commences, one has to account for a dimension of reality characterized by its concrete *whatness*. Although it intimately coheres with the modal dimension of the *how*, it remains irreducibly different from the latter. We have seen that access to the dimension of entities is mediated by the *point of entry* provided by the various modal aspects of re-

1 "Fit in with the whole scheme of things."

2 The use of the word 'true' does not mean that aspects are statements about things, but rather alludes to the fact that modalities co-condition whatever there is in an ontic sense, and they hold for or apply to everything. Clouser also defines an aspect as "a basic kind of properties and laws" (Clouser, 2005:367).

ality. In addition, one can articulate an understanding of this dimension by using metaphors – such as the italicized one in the previous sentence. Both modal terms and metaphors, nonetheless, fail to fully embrace the *dimensional uniqueness* of (natural and social) entities and processes. For this reason, it is impossible to find a so-called *principium individuationis* (principle of individuation) within any modal aspect, such as the aspect of number. Things are not individual, because they are *many* – they are many because they are *individually distinct*. Naturally *being individual* does not exclude the co-conditioning role of the quantitative aspect of reality.

In a certain sense, one may view the problem of individuality as the crux of the dimension of entities.[1] However, in the light of the nominalistic denial of the law side of reality, as well as its elimination of the factual orderliness of things, employing the term 'individuality' needs to be qualified in order to prevent falling prey to this misunderstanding. Dooyeweerd settled for the designation *individuality structure* as the *law for* entities (from 1935 onwards). The traditional concept of *substance* attempts to account for this dimension of reality by considering an entity in terms of a self-contained existence, independent of any connections or relations beyond itself. The definition in Descartes's "The Principles of Philosophy" (LI) is representative of this legacy: "By substance we can conceive nothing else than a thing which exists in such a way as to stand in relation to nothing beyond itself in order for its existence" (Descartes, 1965:184).

Yet in 1931, the influence of nominalism is still apparent in Dooyeweerd's thought, for we saw that at this stage of his intellectual development, he frequently speaks of "individual structure." Sometimes the idea of *meaning* is combined with the notion of an *individual structure*, for example when he refers to the "meaning-individual thing structure of organized human communities" (Dooyeweerd, 1931:111 ff.).

Since he does not use the term *structure* in the sense of *construction* (or what has been constructed), but in the sense of "law for," one may argue that he aims to account for the *law for* concretely existing *individual* things in speaking of an *individual structure*. Yet, taken *literally*, the phrase *individual structure* creates an additional problem, for then it means that *structure* – in the sense of *universal law for* – *is* individual. In other words, it then asserts that a *universal* law is *individual*. The ripened designation of the dimension of entities, namely that it is the dimension of *individuality structures*, causes further difficulties for Dooyeweerd – once more demonstrating the problematic after-effect of nominalism in his thought. The composite phrase *individuality structure* suggests that the term *structure* represents the "law for," while the term *individuality* represents what is 'subjected' to the law. For Dooyeweerd, the factual side of reality is strictly *individual* – he repeatedly refers to the *individual factual side* of reality. As we have noted, this means that, for him, there is nothing universal (no *universality*) at the factual side of reality.

1 In 1931 Dooyeweerd says that, for theory, individuality is the key problem of temporal reality (see Dooyeweerd, 1931:26).

Van Riessen suggests the phrase "identity structure" as designation of the law for entities (Van Riessen, 1970:158), while another alternative might have been to speak of "entitary structures." Perhaps the best designation is the one employed in the distinction between *modal laws* and *type laws*.[1] Although a derivation from Greek would suggest the word *typonomy* (typos = *type* and nomos = *law* – see page 26 above), its intuitive appeal within the English language is not as clear as that of *type law*.

The use of alternative metaphors may open this domain of nomenclature in a fresh way, but since metaphors lack the precision required by the use of (irreplaceable) elementary and compound basic concepts of disciplines, a further specification will always remain necessary.

As a substitute for his former use of the phrase "typical laws" (or the term 'typicality' – see Stafleu, 1980:6, 11), Stafleu introduces the term 'character' as a general designation of a "cluster of laws of similar things events or their relations" (Stafleu, 2002:9). The only connotation that may help to strengthen the intuitive appeal of the term *character*, is given in its apparent implicit reference to the 'nature' of things, allowing for classifying certain groups of things, distinct from other groups or classes of things.[2]

Even so, Stafleu does not intend to introduce the term *character* as "the essence or nature of things" or processes, for he wants to emphasize that a *cluster of laws* determines the mutual *relations* between things and processes (Stafleu, 2002:9). It seems that Stafleu, in his fear for what he calls 'essentialism', underplays the *thingness* of things by focusing on *relations*.[3] This emphasis comes dangerously close to *functionalism* – the *ism* that functionalizes entities (the opposite of hypostatization or reification – treating a function as if it were an entity, such as when biologists speak of the origin of 'life' instead of the origin of living things).

Stafleu's new emphasis on *relations* – also reflected in the title of his 2002 work – substantiated by 'degrading' a mode of speech in which it will be possible to mention the *nature* of things, may actually continue a longstanding *functionalistic* approach particularly prominent in the entire intellectual development of the natural sciences since the Renaissance. He mentions, as an example of a 'mistaken' understanding of 'zin' (*meaning*), Dooyeweerd's designation of the meaning-kernel of the numerical aspect as *discrete quantity*. For Stafleu, the only acceptable meaning nuance of *meaning* ('zin') is its reference to origin. This referentiality of meaning – i.e. that meaning is relational and ultimately refers to the origin of creation – brings to expression his preference for relations and relational concepts.

1 Clouser prefers to use this phrase – see Clouser, 2005:244, 262-263.

2 Stafleu does point out, however, that scientific classification is something different from typifying characters on the basis of universal "relation cadres" (modal aspects) (Stafleu, 2002:31).

3 The fact that Stafleu sets himself off against 'essentialism' supports his emphasis on relations (see Stafleu, 2006:169).

For example, when it is asserted that the meaning nucleus of the arithmetical aspect is *discrete quantity*, the way in which Stafleu understands *zin* disqualifies it as *essentialistic*, because it does not contain a reference to origin.[1] In fact, Stafleu does not hesitate to call upon the development of modern natural science in its reaction to the essentialistic philosophy of Plato and Aristotle. He says that the question regarding the essence disappeared from modern natural science and that therefore it also should not find shelter in a "relational philosophy."[2] This is nothing but a *deviation* (or even: *derailment*) of Dooyeweerd's original idea. From the early 1920s, the latter took a principled stance in opposition to both the *substantialistic* ('essentialistic') orientation of Greek-Medieval philosophy and the functionalistic ('relationalistic') orientation of modern natural science. An *integral* cosmonic idea, i.e. an encompassing idea of creation in its unity (coherence / relatedness) and diversity (uniqueness / irreducibility) has to affirm both sides of the coin – *uniqueness* and *coherence*. In our discussion of the *meaning* of the modal aspects, we have repeatedly emphasized that this meaning comes to expression in the *coherence* ('relation') between distinct (unique) aspects of reality (see pages 24, 55, 76, 87, 103, 188, 234, 279, and 353).[3]

Reference (relatedness / relation) depends on *uniqueness*, which depends on *coherence*. In the sense of concept-transcending knowledge, the ideas of *uniqueness* and (inter-modal) *coherence* explore modal numerical and spatial terms stretched beyond the boundries of these aspects. It is not a sign of 'essentialism' when the uniqueness of aspects and entities is acknowledged. However, not being willing to speak of the 'nature' of things does not avoid references to 'de-natured' things, explaining why Stafleu nonetheless still has to speak of the relations of (or between) *things*![4]

The (early 20th century) neo-Kantian philosopher, Rickert, continues the mentioned functionalistic tradition with his view that the natural sciences have to proceed in a *generalizing* fashion, in opposition to the *individualizing*

1 "Zo noemt Dooyeweerd voor het eerste aspect de zinkern 'discrete kwantiteit', maar het is onduidelijk in hoeverre dit de zin van dit aspect aangeeft, als de zin een verwijzing naar de oorsprong bedoelt te zijn. Het lijkt er eerder op dat hij 'zin' hier gebruikt alsof discrete kwantiteit de essentie van het getalsaspect aan zou geven" (Stafleu, 2002:282).

2 "De vraag naar de essentie past wel in de essentialistische filosofie van Plato en Aristoteles. Uit de moderne natuurwetenschap is de essentie echter verdwenen en zij hoort ook in een relationele filosofie niet thuis" (Stafleu, 2002:282).

3 Dooyeweerd gives an in-depth treatment of the untenability of this modern functionalism within the science of law. In particular, he discusses the application of "modern functionalistic-mathematical thought" by the Marburg school of neo-Kantianism (Stammler and Kelsen – see Dooyeweerd, 1926:22 ff.).

4 In an e-mail of September 27, 2008 Stafleu remarks that an analysis of the modal aspects precedes that of the "individuality structures" but that this does not make his approach functionalistic. However, his attempt to avoid references to the 'nature' of things as well as the statements quoted from his 2002 work does not escape from the functionalistic mode of thought, even though his *intention* may not be functionalistic.

mode of thought predominant in the (historical) humanities (Rickert, 1913:68-69, 173). Rickert initially develops this perspective by binding the natural sciences to the ideal of transforming all *thing concepts* into *concepts of function* (explicitly designated as *concepts of relations*). This neo-Kantian view of the natural sciences remains completely faithful to the aim of the classical science ideal, namely to reduce all reality to some or other *modal aspect, function* or *relation*. According to Rickert, the (functionalistic) logical ideal of the natural sciences finds its limit in the uniqueness (individuality) of experiential reality itself.[1]

Rickert holds that:

> Whatever the role the category of a thing may fulfill in a theory of the thing world, envisaged as closed, at bottom there is no doubt that the natural sciences have to strive to resolve the rigid and fixed things increasingly, ... and this means nothing else but transforming as far as possible all thing concepts into relation concepts. ... Our theory is valid for the logical ideal of natural scientific concepts, because this ideal solely concerns relation concepts (Rickert, 1913:68-70).[2]

Moreover, highlighting the functionalistic background of an emphasis on relations is further supported by the fact that Stafleu views laws as *timeless*. "Individual things and events are intrinsically temporal, ... The timeless character conditions the existence of individuals concerned in their temporal circumstances" (Stafleu, 2002:14).[3] (Rickert holds the view that values have an *ideal, timeless being*.)

As soon as Stafleu has to articulate more precisely what *characters* are all about, he takes recourse to the precision provided by *modal terms*. He then offers a description that appears as a quasi-compound basic concept: "A character determines an unlimited complete class of temporal subjects" (Stafleu, 2002:14). The term 'determines' is derived from the modal meaning of the physical aspect, the terms 'unlimited' and 'complete' from the spatial mode, and 'class' from a combination of the numerical and spatial aspects. The use of a metaphor, such as figuratively designating a type law as a *character*, in the final analysis requires *modal terms* in order to obtain a precise meaning.

Stafleu says that he defines a *character* as a cluster of *immutable* ('onveranderlijke') natural laws, instead of speaking of their *constancy*, because

1 "That which poses an inaccessible limit to natural scientific concept formation is nothing but unique empirical reality itself, as we intuitively experience it in the immediacy of its individuality" (Rickert, 1913:197).

2 "Welche Rolle also auch die Kategorie des Dinges in einer abgeschlossenen gedachten Theorie der Körperwelt noch spielen mag, so unterliegt es doch jedenfalls keinem Zweifel, dass die Naturwissenschaften danach strebt und streben muss, die starren und festen, und das heist nichts anderes, als die Dingbegriffe so weit wie möglich in Relationsbegriffe umzuwandeln. .... Für das logische Ideal der naturwissenschaftlichen Begriffe in einer 'letzten' Naturwissenschaft ist unsere Theorie gültig, denn es handelt sich bei diesem Ideal nur noch um Relationsbegriffe."

3 "A class is just as timeless as the natural laws determining the class" (Stafleu, 2002:14).

when antecipatory meaning moments are disclosed on the law side of an aspect, then the law side itself indeed *changes*.[1]

Before we focus on some reasons why our theoretical understanding of physical reality is dependent upon philosophical assumptions, we conclude this section with a final remark about the various dimensions of reality.

Foundational to the dimension of modal aspects is the dimension of cosmic time (see our brief discussion of this dimension above, pp.206-211). These two dimensions, in turn, are foundational to the dimension of natural and social entities and processes (subject to type laws). Finally, these three cosmic dimensions find their focal point in what might be called the *directional dimension* of reality. We are more accustomed to refer to the (direction-giving) *central religious dimension* of reality.[2] In the concluding remark of Chapter 2 (page 66; see also page 366 above), we have remarked that human rationality is embedded in a *more-than-rational commitment*, ultimately giving *direction* to all of life.

7.10 The apparent "ontological circle" involved in chacterizing entities

It has already been argued that the dimensions of entities and aspects are mutually interdependent also in respect of what can be said about both. In order to talk about the modal aspects, one inevitably employs metaphors, while at the same time the modal aspects serve as gateways to whatever needs to be said about entities.

On the one hand, it seems that theoretical thought has to realize that the *individuality* of an entity exceeds the limits of concept formation. This insight entails an acknowledgement that the *distinctness* of this unique entitary dimension of reality entails that the individuality and uniqueness of all entities express themselves at once within all the modal aspects of reality in which these entities function. Yet, as soon as it is required to give a theoretical *account* of the individuality of an entity, two perspectives ought to be kept in mind:

(i) the individual side of an entity transcends concept formation, and
(ii) the universal side of an entity (its *being* this or that, i.e. its *orderliness*) points beyond itself to the universal *order for* its existence. In other words, there is no *individuality* without *typicality* – and the latter always reflects the *type law* to which entities are subjected.

The starting point of a theoretical account of entities is therefore always found in the (concept-transcending) *idea* of an individual entity. The second step is to explore the access provided by the modal aspects. As soon as this is done,

1 In the discussion between Van Peursen and Dooyeweerd, the latter responds to the accusation that his conception of law is static by explaining that the meaning dynamic of reality also manifests itself on its law side (see Dooyeweerd, 1960:109).

2 The directional dimension concerns the radical, central and total (RCT) orientation of human life, and should therefore not be confused with any modal aspect (such as the certitudinal mode) or any differentiated, peripheral and partial (DPP) societal connection.

theoretical concept formation comes to the aid of such a theoretical analysis, for now it is possible to discern what has been called the *typical qualifying function* of an entity, such as when it is said that material things are physically qualified, i.e. stamped or characterized by the physical aspect of reality. The phrase *typical qualifying function* refers to the *type law* for an entity. It is important to note that the *idea* of a qualifying function is indeed an *idea* in the technical sense of *concept-transcending knowledge*. The constitutive structure of a modal aspect, identified as the qualifying function within the type law for entities, does allow conceptual knowledge, but as a (theoretically discontinuous) *approximation* of the many-sided individuality of entities, the indication of their qualifying function, on the basis of the *constitutive* knowledge of the structure of the modal aspect under consideration, is *regulatively* deepened by the idea of an individual whole with its characteristic guiding function. Likewise, the *idea* of a foundational function equally manifests this interplay of constitutive and regulative elements.

If this distinction is not made, a *vicious ontological circle* (analogous to the hermeneutical circle) threatens our analysis. The initial statement claims that *individuality* expresses itself within each modal aspect. But then the question arises: how can we explain this "aspect-transcending uniqueness" phenomenon? The answer seems to take recourse to the modal aspects, for suddenly we have to introduce *foundational* and *qualifying* functions.[1] It appears that we are doomed to settle for the vicious circle by simultaneously stating that the individuality of an entity exceeds any modal aspect in which it expresses itself, *and* that an entity *owes* its uniqueness to having a specific foundational and qualifying function!

Formulating the circle in this way does not distinguish between the *regulative* perspective of the (concept-transcending) idea of an entity based upon the *constitutive* employment of the concept of a modal aspect. This is a consequence of the mutuality prevailing between the two dimensions underscored above – the only access (beyond the primary idea of individuality) we have to the dimension of (natural and social) entities is through the gateway of the modal aspects.

In spite of being hampered by the (implicit) nominalistic legacy within his thought, in one of the first accounts of his theory of individuality structures, Dooyeweerd anticipates the argument above. He says that the "individuality of things permeates ... all meaning-functions," but that the "individuality structure can only be understood" from the perspective of "the guiding function" (Dooyeweerd, 1931:110).

7.11 Physical entities exceed the limits of physics

Our analysis of the development of the theoretical understanding of physical reality makes it clear that, to find an *arithmetic*, *spatial* or *kinematic* qualification for *physical entities*, necessarily runs into theoretical antinomies. Al-

1 Dooyeweerd initially believes (1935-1936) that natural things do not have a typical foundational function. In 1950, he relinquishes this position (see Dooyeweerd, 1950:75 note 8).

though material entities without any doubt do function within these three (pre-physical) aspects, they are still always qualified by the *physical* aspect. Approaching (material) entities from any angle, by definition, owing to the multi-aspectual nature of all entities, is incomplete and limited in scope. Complexity theory is sensitive to the incompleteness of knowledge related to complexity. Cilliers writes: "When dealing with complex phenomena, no single method will yield the whole truth. Approaching a complex system playfully allows for different avenues of advance, different viewpoints, and, perhaps, a better understanding of its characteriztics" (Cilliers, 1998:23). If we interpret the "avenues of advance" or "viewpoints" as related to the ontic modal aspects of reality, complexity theory can indeed benefit from the insight that concrete entities and processes transcend the perspective of any aspect. The idea – in the sense of concept-transcending knowledge – of (natural and social) entities and processes implies that they in principle exceed conceptual comprehension. Consider for a moment the many-sided existence of an atom.

In Rutherford's atomic theory, as early as 1911, the hypothesis was posed that atoms consist of a (electrically positive) nucleus and negatively charged particles moving around it (a view inspired by the planetary system). In the following year (1912), Niels Bohr set up a new theory, with two important new ideas: (i) the electrons move only in a limited number of discrete orbits around the nucleus, and (ii) when an electron moves from an orbit with a high energy content to one with a low energy content, electromagnetic radiation occurs. In 1925, Pauli formulated his exclusion principle (Pauli-exclusion).[1] According to the division of charges of electrons, corresponding electron-shells exist, and in each peel, there is room for a 'maximum' number of electrons. This maximum number is yielded by the simple formula: $2n^2$. In the first peel (known as the K-peel), there is room for 2 electrons; in the following L-peel, there is room for 8; in the M-peel for 18; in the N-peel for 32; and so on. Within a peel with a quantum number n, (where there is room for $2n^2$ electrons) sub-orbits are identified so that each sub-orbit with a quantum number l has room for $2(2l+1)$ electrons.

Multiple elementary particles are integrated in the unified functioning of the atom as an individual *whole*. When physicists talk of the nature of these particles, the original meaning of space, combined with numerical analogies within space, is prominent. In other words, the aspects of number and space are first explored in what we say about elementary particles.

The nucleus of the atom (constituted by its protons and neutrons) has a certain *size*, and its diameter, multiplied by 100,000, specifies the distance between the nucleus of an atom and its (circling) electrons. The current physical view is that quarks are the ultimate "building blocks" of these elementary particles. A distinction is drawn between an up quark (with a charge of +2/3) and a down quark (with a charge of -1/3). Apparently there are no free quarks. The

1 It applies to fermions, i.e. elementary particles with a semi-integral spin (1/2, 3/2, 5/2, etc.), for which the statistical laws of Fermi-Dirac are formulated.

proton, for example, consists of two up quarks and one down quark. The size of electrons and quarks is smaller than 10^{-18}; they are so small that they are described as point-like (see Kiontke, 2006:27). Hydrons include those fermions and bosons designated as mesons. Furthermore, hadrons are constituted by quarks. Those known as baryons in turn include nucleons (neutrons and protons) and hyperons. Whereas the hadrons are 'heavy', the leptons are small, including the electron and particles such as the muon, tauton and their corresponding neutrinos. More information on this micro-dimension is found in Penrose (2005:645 ff).

The matter of an atom is concentrated in a volume of less than a 0.00000000000000000001 part of the volume of the atom – which amounts to saying that atoms are more than 99.9999999999999999999% empty (19 zeros before the one – Kiontke, 2006:27). Some facts about the way in which atoms function within the kinematic aspect are equally astonishing. According to wave mechanics, we find quantified wave movements around the atom, and the electron of a hydrogen atom (in its lowest orbit) moves around the nucleus at a speed of about 6.8 million km per hour (Kiontke, 2006:27).

From this, it is evident that the *distinct number* of elementary particles within the internal atomic structure are joined into a typical *spatial* and *kinematic* order of electronic orbits that configure the atom as an individual *physical-chemical micro-totality*. The special spatial configuration manifest within the internal arrangement of the parts of an atom reflects the typical *foundational function* of atoms. Biochemistry discovered many *isomeric* forms; that is, they have identified *chemical configurations* that are constituted by the *same* atoms, viewed from a purely numerical perspective, but that nonetheless, owing to different *spatial* arrangements, differ *chemically*.

The formula C_3H_6O may yield the following (chemically distinct) configurations: $CH_3.CH_2.CHO$ or $CH_3.CO.CH_3$. Another example is $C_4H_4O_4$. That the chemical difference between maleic acid and fumaric acid has its foundation in alternative spatial arrangements is self-evident – just as clear as it is that the number of atoms as such cannot account for this chemical difference. In other words, it is intuitively clear that such molecules have a spatial foundational function, and not a numerical foundational function.

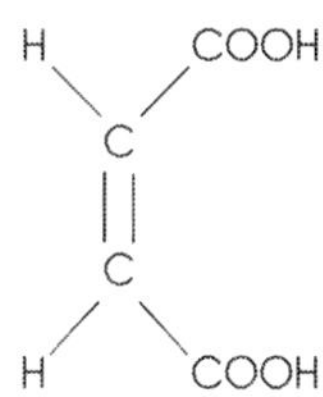

Maleic acid
cis

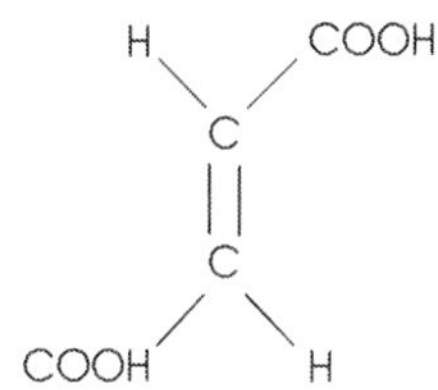

Fumaric acid
trans

The point-like nature of small particles may suggest that, in their case, the *numerical* function is decisive. One may be tempted to argue that such particles have their foundational function in the numerical aspect, but the problem is that in the case of electrons and quarks, their actual size is unknown (Kiontke, 2006: 27). Nonetheless, as Stafleu remarks, the electron is characterized by exactly determinable values for its charge, rest mass, the magnetic moment and the lepton number (Stafleu, 1989:91). A detailed explanation of primary and secondary foundational relations is found in Stafleu, where, for example, energy, force and current are respectively related to quantitative, spatial and kinematic concepts (Stafleu, 2002:26-28, 128-171). His analysis explores the possibility to locate within the first three modal aspects the foundational function of different kinds of physical entities. In doing this he demonstrates how fruitful philosophical distinctions are for a special science such as physics. Quantum electro-dynamics has to take into account the interaction of the electron with its own quantified surrounding field – charge and field are inseparably connected (see Rollwagen, 1962:10).

The problem of interweaving different kinds of entities first of all challenges the limitations of the whole-parts relation – a relation that appears in its original modal meaning within the spatial aspect (see page 87 above).[1] Suppose we ask whether or not Sodium and Chlorine are genuine *parts* of table salt. Surely every division of table salt must continue to display the NaCl structure of table salt. But what happens when the process of division reaches a single salt molecule? Once such a molecule is divided, one is left with a Sodium atom and a Chlorine atom – it is evident that real parts of salt will still possess the same chemical structure of salt, namely NaCl. The critical question is whether Sodium and Chlorine each has a salt structure, i.e. are Sodium and Chlorine true parts of salt? The answer is self-evident, because neither on its own has a NaCl structure!

In this case, the internal sphere of operation of the atoms remains intact although, through a chemical bond, they were taken into the table salt molecule. Dooyeweerd developed an alternative theoretical approach in order to account for the retention of the internal nature of entities that are interlaced – as we mentioned in Chapter 4 in our account of the interlacement between the various intertwined structures within the human body (cf. Dooyeweerd, 1997-III:627 ff., 694 ff. and pages 139 ff. above). When the internal sphere of operation of interwoven entities is retained, the term *enkapsis* is employed (see page 356 above). When one kind of entity is foundational for another kind of entity, the situation is captured by speaking of a *one-sided enkaptic foundational relationship* – which is found in NaCl.

Within the realm of physically qualified entities, we encounter different geno-types. Atoms are, for instance, geno-types within the radical type (realm) of material things. Within different bonds, the same atom displays *variability types*. When an atom engages in chemical bonding, a characteristic enkaptic totality emerges: (i) besides the internal sphere of operation of an entity there is (ii) an external enkaptic sphere of operation in which the enkaptically-bound entity serves the encompassing enkaptic totality.

1 Also recall our discussion of the difference between mathematical space (continuous and infinitely divisible) and physical space (discontinuous and therefore not infinitely divisible). Both kinds are extended (see page 221) – the similarity between them; but within the moment of similarity, the difference between them manifests itself, thus demonstrating the nature of a (modal) analogy.

The factual configuration of a water molecule thus exists on the foundation of the geno-type of the chemical bond between the oxygen and hydrogen atoms. Without these atoms, a water molecule cannot exist. They therefore serve water in the sense of a unilateral foundational relation. Does this imply that the atoms become part of the chemical bond that exists within the molecule? Not at all, because the bond applies only to the binding electrons and not to the whole atom. Besides, the atom nucleus is not just a specific characteristic of the atom, but precisely that nuclear part of an atom that determines its physical-chemical geno-type (compare the atomic number = the number of protons of the nucleus), as well as the atom's place within the periodic table.

The fact that the atom nucleus remains structurally unchanged in chemical bonding secures the internal operation sphere of the atom. Because the electrons cannot be disengaged from the nucleus, these atoms function as a whole in the water molecule. Note that it cannot be said that the atoms function in the chemical bond, for the bonding does not encompass the atomic nuclei. Nonetheless, the atoms (with their nuclei, electron shells and bonding electrons) are present *as a whole* in the water molecule, which embraces them *enkaptically*. The enkaptic interweaving of the atoms within the molecule does not make them intrinsic parts of the molecule, since this would abrogate the internal action sphere of the atoms.

The external enkaptic function of the oxygen and hydrogen atoms in the water molecule indicates the functioning of the atoms within the molecule as a totality via the intermediate role of the chemical bond. This presents us with a three-fold distinction:

(i) Firstly, we must identify the internal action sphere of the atom.

(ii) Secondly, we find the chemical bond that leaves the atom nucleus unchanged, because it only affects the outer electron shells, so that the atom nuclei can in no way be part of the chemical bonding.

(iii) Thirdly, we find the enkaptic structural whole of the water molecule that enkaptically embraces the atomic nuclei – binding them and ascribing to each of them its typical place within the enkaptic whole.

This view avoids the one-sidedness of an atomistic understanding of a molecule (over-emphasizing the continued existence of atoms within the chemical bond at the cost of the totality character of the resulting whole), and of a holistic view (that over-emphasizes the totality-character of the molecule in such a way that the foundational atoms are seen as *integral parts* of the whole). Van Melsen highlights these two extremes: "In modern theories atomic and molecular structures are characterized as associations of many interacting entities that *lose* their own identity. The resulting aggregate originates from the converging contributions of all is components. Yet, it forms a new entity, which in its turn controls the behaviour of its components" (Van Melsen, 1975:349 – quoted on page 62 above).

It should be noted that, in this context also, there is an ambiguity in Dooyeweerd's terminology, because he talks of the interlacement of *individuality structures* (i.e. of *laws*) – instead of clearly stating that the issue concerns the interwovenness of *entities* subject to their type laws (individuality structures). The intention is not to say that *laws* are enkaptically intertwined, but simply to account for the factual interlacements found between different kinds (types) of entities in their subjection to type laws.[1]

Particularly the well-known duality of a wave-particle description of physical entities once more highlights the distinctness and coherence between the dimensions of modal aspects and entities.

7.11.1 *Complementarity – limits to experimentation*

There are also remarkable *limits* to physics in the sense of experimental exactitude and determination. By introducing his principle of uncertainty Heisenberg showed that it is impossible simultaneously to measure the *impulse* and *position* of an electron. The Copenhagen interpretation of quantum physics employs the notion of *complementarity* in order to account for the impossibility of establishing both at once – thus allowing for two irreducible (and complementary) modes of description, in terms of 'place' and 'impulse' respectively. In following some ideas of Mario Bunge the physicist Henry Margenau defends a "moderate reductionism" (see page 261 above) – which he understands to be "the strategy consisting of reducing whatever can be reduced without however ignoring emergence or persisting in reducing the irreducible.[2]

7.11.2 *Wave and particle: the typical totality structure of an entity*

After Einstein reverted to a particle theory regarding the nature of light,[3] it turned out, on the basis of *interference phenomena*,[4] that it is always possible to ascribe a wave-character to elementary particles. Conversely, the *Compton-effect* – regarding the interaction of a photon and an electron – supplied evidence to support the idea of *distinct* particles. De Broglie broadened the perspective by showing that with each and every moving particle (atoms, molecules and even macro-structures) one can associate a wave (cf. Eisberg, 1961: 81, 151).

Although it turned out to be impossible to establish experimentally at *the same time* both the particle and the wave nature Bohr claims that these two perspectives are *complementary* (cf. Bohr, 1968:41 ff.).

1 Stafleu also speaks of the mutual interlacement of laws ('characters' in his terminology, as clusters of laws) (see Stafleu, 2002:150).

2 Cf. Margenau, 1982:187, 196-197.

3 Light quanta are called photons and similar to the neutrino they possess a zero mass.

4 Interference phenomena were established after Michelson – round 1880 – designed an interferometer capable of cutting light and afterwards recombining it. Thus one ends up with the same light beam – with slightly less energy. The remarkable result was that the sum did not produce light but darkness! However, when one of the two halves was blocked with a piece of black paper the other halve still appeared. Seemingly the only way to explain what happened here is to assume that the interference of the split light-waves cancel out each other when re-united.

In the light of the generalization provided by De Broglie one may ask: if it is possible to describe or explain entities qualified by energy in terms of two mutually exclusive experimental perspectives, namely as *particles* and as *waves*, is it then still meaningful to speak about their *unitary structure*? This question pinpoints exactly on that point where the special scientific description reaches its limits and needs to fall back upon a perspective transcending the confines of special scientific inquiry. What is here required is some or other philosophical account transcending the mere combination of one or more (modally delimited) special scientific points of view. The *idea* of the unity and identity of an entity could never be provided to us by theoretically explicating various modal functions, simply because this underlying unity is presupposed in all theoretical explanations. In a strict and technical sense this idea of an entity in its totality – preceding the analysis of its modal aspects – refers to an individual whole embedded in the inter-modal and inter-structural coherence of reality, to an entity emerged in the depth-layer of an all-embracing *temporality* transcending genuine concept-formation and only to be approximated in a *concept-transcending idea*.

A deepening of this basic (transcendental) idea occurs when – through theoretical reflection and investigation – the dimension of micro-structures is unveiled (the micro-world with atoms and sub-atomic particles). It is important in this context, however, to realize that concepts such as *particle*, *field*, and *wave* are not *type concepts* but *modal functional concepts* (sometimes referred to as elementary basic concepts of physics). Consequently, the terms particle and wave analogically reflect retrocipatory structural moments within the structure of the kinematic aspect, namely *movement multiplicity* (numerical analogy) and *movement extension* (spatial analogy). These facets are deepened in physically qualified entities and could be approximated in physical theory from the perspective of mathematical antecipations to the physical aspect – compare Shrödinger's wave function formulated in terms of differential equations.

Since number, space and movement remain irreducible aspects regardless of the nature and type of entities functioning within them (their modal universality), it is also from this perspective understandable why the functionally distinct concepts *particle* and *wave* cannot be reduced to each other – a state of affairs supported by experimental data. Irreducible modal perspectives may also serve as modes of scientific explanation.

Born, Pyrmont and Biem reject the struggle with a *dualism* in this context. They hold that it increasingly becomes clear that "nature could neither be described by particles alone, nor solely through waves," because a proper understanding cannot toggle between a "particle image [*Teilchenbild*]" and a "wave image [Wellenbild]." This leads to a **unitary view** of physical systems.[1] What we have called *modes of explanation* these authors designate as 'Darstellungen' (representations) – and they specifically mention three distinct (but present at once) modes of explanation: an 'Ortsdarstellung' (spatial position),

1 "Mit der Quantentheorie erfaßt man so alle Systeme einheitlich, ..."

a 'Wellendarstellung' (impulses or velocities – kinematic explanation) and an 'Energiedarstellung' (energy – the physical mode of explanation) (Born, Pyrmont, Biem, 1967/68:416-417).

7.12 Living things

Within modern biology, diverging appreciations of living entities are found. Analogous to our analysis of different modes of explanation operative in the history of our understanding of matter, alternative principles of explanation are also found in the biological thinking of the 19th and 20th centuries.

Charles Darwin undoubtedly irrevocably changed the face of modern biology. Before he published his now well-known and famous *On the Origin of Species by Means of Natural Selection or the Preservation of favoured races in the struggle for life* in 1859, biological thought was largely dominated by the Platonic idea that living things are mere *copies* of unchanging, static, eternal (super-sensory) ideal forms, as well as the *vitalistic* Aristotelian tradition, with its emphasis on *purposefulness* (finality/teleology). The biological systematic classification of Ray (1627-1705) and Linnaeus (1707-1778) continued the Platonic legacy. It should be noted that the legacy of idealistic morphology still had representatives in the 20th century. These include Dacqué (cf. 1935, 1940, 1948), Troll (1951 and 1973), Wolf (1951) W. Leinfeller (1966) and Heitler (1976). Troll's work of 1973 is a standard (more than 1000 page) botany textbook. According to Troll, the foundation of comparative morphology is to be found in ideas (in the Platonic sense), which serve to order the "inner articulations of our intuition" by means of which types as "Urbildliche Einheiten" (primordial image-like units) become the subject matter of biology (cf. Ungerer, 1966:232).

During the past 150 years, diverse trends of thought emerged within biology – constantly introducing alternative *modes of explanation*. We once more list these in connection with the above-mentioned main representatives of diverse biological orientations (see Chapter 1, page 6 – here amended by the required references to sources) before we continue our analysis of the nature of living entities.

> The *mechanistic* orientation (Eisenstein – 1975), the *physicalistic* approach (neo-Darwinism), *neo-vitalism* (Driesch – 1929, Sinnott – 1963, 1972; Rainer-Schubert Soldern – 1959, 1962, 1962a; Haas – 1959 1968; Heitler – 1976); *holism* (Adolf Meyer-Abich – 1964, 1965); *emergence evolutionism* (Lloyd-Morgan, Woltereck, Bavinck – 1954; Polanyi – 1967, 1968, 1969); the *organismic biology* of Von Bertalanffy (1973); and *pan-psychism* (Teilhard de Chardin, Bernard Rensch – 1959, 1968, 1969, 1971); recent complexity theory (Behe's notion of "irreducibly complex systems" – see Behe, 2003) and the idea of "intelligent design" that surfaced more recently – not on the basis of lacking sufficient factual knowledge, but supported by scholars with highly specialized natural scientific competencies (see Dekker, Meester, and Van Woudenberg, 2006).

Likewise, the legacy of biological thinking *preceding* the *Origin of Species* is also present in a specific (long-standing) theoretical biological orientation. It is embodied in the tradition of the above-mentioned *idealistic morphology*, combined with Aristotle's *vitalistic* thought up to the *neo-vitalist* biology of Hans Driesch and his neo-vitalist followers. This orientation was accompanied by the idea of (supposedly) an immaterial *vital force* (*entelechie*).[1] Since theory formation always explores certain *modes of explanation*,[2] the effect of elevating one mode of explanation normally results in the theoretical perspective of a *monistic* orientation.

Darwin opposes the tradition of idealistic morphology and its vitalistic belief that living entities are guided by an inherent purposefulness. He rather opted for the idea that living entities are *intrinsically changeful* and subject to *chance* processes. But we shall see that his eventual acceptance of the principle of uniformitarianism (derived from his acquaintance with Lyell's work in the field of geology)[3] did continue a feature formally similar to an element of idealistic morphology.

The best known heritage from this variety of viewpoints is probably what we find in their shared reference to *living matter*. Of course denying the existence of "living matter" does not eliminate an acknowledgement of the *unique* way in which living entities also function within the physical aspect of reality. Karl Trincher identified *four* macroscopic characteristics from which the *physical uniqueness* of a living cell is evident, which include the following (Trincher, 1985:336):

1) spatial macroscopy, which defines the cell as a spatially delimited surface;
2) temporal macroscopy, which determines the finite time in which the energy cycle of the cell occurs;
3) the isothermal nature of the cell, which is responsible for the constancy of temperature throughout the cell; and
4) the persistent positive difference between the higher internal temperature of the cell and the lower external temperature of the environment adjacent to the cell surface.

In addition he argues, also on the grounds of this physical autonomy of living things, that the living cell could not in principle be artificially manufactured. He believes that the opposite position, namely that living things are merely a

1 Driesch argued for his vitalistic position in terms of the remarkable regenerative phenomena found particularly in animal life (he did experimental work on sea urchins – Echinus microtuberculatis) and by opposing the machine model of the mechanistic approach (since Descartes): it is inexplicable how machines could continue to be divided (through cell division) and still remain intact wholes (see Weber, 1999:266 ff., 270 ff.).

2 Such as the numerical (the Pythagoreans with their claim that everything is number), the kinematic (the main tendency of classical physics from Galileo and Newton up to Heinrich Hertz), or the vitalistic trend in biology (Driesch and his followers – exploring the biotic aspect of reality as decisive mode of explanation mentioned in the previous footnote).

3 Principles of Geology (1830-1833) and Elements of Geology (1838). Henslow advised Darwin to take Leyll's first work with him to the Cape Verde Islands – but not to believe it.

developmental product of non-living matter, is responsible for the moral irresponsibility of the contemporary natural sciences.

7.12.1 *The many-sided nature of living things*

Although the habit of speaking of "living matter" is placed within alternative contexts by the different tendencies in modern biology, it still reflects unsolved problems for each of these viewpoints.

For the mechanistic (physicalistic) approach, everything is material in principle, and physically determined, which implies that any terms that appeal to the actual biotic aspect of living things are problematic. Conversely, vitalism searches for immaterial life plans in order to account for the actual nature of 'life'. These life plans are sometimes designated as *formative factors* or *central instances*. It also makes it difficult to speak of "living matter" from the perspective of this approach – a problem that a vitalistic biologist like Haas admits with his emphasis on the fact that physical substances maintain their "being and working" "subsequent to their assimilation" in living things. Understandably, therefore, Haas is also critical of the habit of speaking of "living matter" – according to him, the biochemists and cell physiologists do not know of any "living matter" with "secret vital characteristics" (Haas, 1968:24). He prefers to speak of the material substratum of organisms (Haas, 1968:20-40).

This approach rejects what Haas sees as Aristotle's "monistic vitalism" – and at the same time he draws conclusions about his own approach: "Organisms therefore consist essentially of two realities which are distinguished from each other, a material and a non-material component; it consequently possesses, viewed ontologically, a dualistic constitution" (Haas, 1968:39).

We mentioned the striking way in which Hans Jonas once gave a typification of the *monistic* forms of vitalism and mechanism. A monistic approach does not, like the dualist, reduce reality to two basic principles, because it wants to find *one* all-inclusive and all-clarifying principle. This leads to the two extremes of *pan-vitalism* and *pan-mechanism* highlighted on pages 172 ff. above. Already in the earliest Greek philosophy of nature, we find *hulèsoism* (*zoè* = life; *hulè* = matter): one of Thales's indirectly preserved statements would be that *everything lives*. From this point of view, it is unthinkable that 'life' is not the normal, governing rule in the universe. Jonas remarks: "In such a worldview, death is a puzzle which stares humankind in the face, the antithesis of the natural, self-explanatory and understandable, that which is the common life" (Jonas, 1973:20). The paragraph containing this remark occurs under the heading: *Pan-vitalism and the problem of death* (Jonas, 1973:19). Those who think pan-mechanistically, on the other hand, emphasize the idea that the phenomenon of life is actually a borderline case in the encompassing homogeneous physical worldview. Quantitatively negligible in the immeasurability of cosmic matter, qualitatively an exception to the rule of material characteristics, epistemologically the unexplained in the explainable physical nature – that is how life has become a stumbling block for pan-mechanism: "Conceiving life as a problem here means that its strangeness in the mechanical world, which is reality, is recognized; explaining it means – on this level of the universal ontology of death – denying it, relegating it to a variant of the possibility of the lifeless" (Jonas, 1973:23). This paragraph occurs under the heading: *Pan-mechanism and the problem of life* (Jonas, 1973:22ff.).

We have already repeatedly emphasized that a first step towards solving this problematic situation lies in distinguishing different modalities. The fundamental modal nature of the physical and biotic aspects remains only a functional condition for concrete entities, which still function in these (and other) aspects of reality in a *typical* way. What is needed in this regard is the ever-important basic distinction between the dimension of modal aspects and the dimension of entities. Yet the predominant reifying mode of speech in biological literature refers to 'life' as if it is an entity, for example when biologists speak of the "origin of life" or the "history of life." However, such a mode of speech runs into difficulties the moment it is confronted with what we know about living things.

The first important question should be phrased as follows: is any living thing fully alive, *living* in its *totality* and in *all it parts*?

Of course this question presupposes an insight into another issue – namely the problem of specifying what may be considered genuine *parts* of a living entity. In the light of our argument that the whole-parts relation is derived from the meaning of the spatial aspect within post-spatial aspects (see pages 61, 87, 236, above), this relation will take on an analogical character (see page 221). A spatial whole is homogenous – every part is of the same nature. Even in the case of physical entities, their parts may be similar, as observed from the fact that a part of salt is still salt (with a NaCl chemical structure – see our discussion above on page 466). Yet we have seen that a *biotical* whole is heterogenous, for although a part of a horse is a "horse-part," it is nonetheless *not* a horse (see once more Oeing-Hanoff, 1976:306). It must be acknowledged, however, that from a biotic perspective, the different organs of a horse all exhibit the organic feature of being integrated in the living organism of the horse. For this reason, the heart of a horse differs from that of a cow. Although the different organs of a horse differ amongst themselves and from similar organs found in other animals, they nonetheless all share the same biotic property of being (part-) organs of a *horse*. This feature therefore represents, within the heterogenous biotic whole-parts relation, an element of homogeneity.[1]

In terms of the question regarding the true parts of a living entity, it is clear that the living organism of a living entity indeed evinces such an organic

1 Luhmann understands this issue when he writes in connection with systems: "The system then stamps as a whole through its own nature the nature of the parts, which, nevertheless, can differ in their particulars, although they are qualitatively equal in their being-parts" (Luhmann, 1973:173-174).

whole-parts relation. Holistic approaches employ the idea of integrated organic parts in this regard, designated in German as 'Glieder'.

Nonetheless, this does not exhaust the complexity of what is at stake, because every organ of a living entity is constituted by a complex arrangement of atoms, molecules and macro-molecules – and the latter are certainly not alive. As noted earlier, the neo-Darwinian paleontologist, Simpson, explicitly states that, since molecules are not alive, the expression "molecular biology" is self-contradictory (Simpson, 1969:6).[1]

Therefore, it cannot be denied that, within living things, non-living entities are also present, namely, atoms, molecules and macro-molecules. The conclusion is inevitable: a living entity appears to be at once a 'mixture' of the 'living" and the "non-living" (see page 91 above). In other words, it is not alive through-and-through – it seems to be 'alive' and 'dead' at the same time.

The predominant neo-Darwinian tradition, with its physicalist inclination, opts for a view in which the biotic side of living entities is completely reduced to the interaction between atoms, molecules and macro-molecules. Vitalistic, holistic and organismic approaches, by contrast, do acknowledge what is sometimes referred to as the irreducibility of 'life' – without realizing that this is still a reifying mode of speech, talking about 'life' as if it is a thing, instead of merely being one amongst many aspects of living entities.

No one can deny the *unity* ('oneness') of living things – the unity in the multiplicity of their organic functioning. The phrase "organic functioning" entails the existence of a *multiplicity* of organic activities. In our everyday experience, this is intimately connected to an awareness of the biotic process of growth (differentiation and integration), maturation, ageing and dying.[2]

As an aspect of reality, life pertains to the *how* of entities and not to their concrete *what*. In addition we must emphasize the fact that vital phenomena are always connected to living entities, which cannot, as entities, be totally enclosed in the biotic aspect of reality. Particularly in the vitalist tradition – which sees life as independent variations of an immaterial vital force – this becomes a problem. That the biotic aspect of living entities cannot be seen on its own, i.e. separated from the inter-modal coherence in which it is fitted, is still confirmed by the analogies intrinsic to the structure of the biotic aspect. Even the expression *life force* (vital force), which is so often chosen by vitalism (but remarkably enough, has been replaced with other terms like *Gestaltungsfaktor* or *Zentralinstanz* in the second half of the 20th century), can never indicate or typify the alleged separated existence of the biotic aspect – simply because it unmistakably represents a physical analogy within the modal structure of the biotic aspect. In Chapter 3 we saw that *force* is a term finding its

1 It is therefore slightly surprising that a biologist so critical of neo-Darwinism as Michael Behe still falls into a physicalistic mode of speech when he refers to "molecular life" (Behe, 2003:5). Just like Darwin Behe and other Intelligent Design theorists also do not acknowledge genuine biotic laws – see the penetrating analysis and critique of Zylstra (2004).

2 Even dying is a biotic process, for in the cell, the organelles known as *lysosomes* have a function in the decomposition of living entities.

original, i.e. non-analogical, modal home in the physical aspect of energy operation.

With the help of the theory of an enkaptic structural whole, this problem is placed within a new context. The physical-chemical structure of the constitutive physical components of living things is foundational for their enkaptic (i.e. biotically directed) functions. When this perspective is accepted, the task of organic chemistry can be seen to be foundational for biochemistry in a similar way. Biochemistry ought to focus on the disclosed enkaptic functions of the material structures that are investigated by organic chemistry. This foundational relationship confirms the close interweaving of the structure and functions of the physical constituents of living things. Today it is virtually universal that biochemists do not restrict themselves to an analysis and study of the biotically directed functions of macromolecular material structures, for they are mainly concerned with the structural configuration of these constituents themselves.

Within the context of the ordered (centered) structure of the cell, we find – seen from a biotic angle – the different organs (organelles) of the cell that are parts of a living whole. Because the cell embraces non-living material ingredients as well as the organic functioning present within it, we cannot simply say that these *organelles* are parts of the cell. In order to explain the vital biotic functions taking place within the cell, the term *cell-organism* is preferable. In other words, the different organs in the cell are all parts of the cell-organism. But since the different *organelles* in the cell have their foundation in the physical-chemical constituents, the totality structure of the cell embraces both the living cell-organism and the non-living material constituents present within the cell. This is an example of a unilateral enkaptic foundational relationship.

Consequently, the cell organism is a specific *biotically* qualified layer that can only exist on the basis of enkaptically bound physical-chemical constituents. Because these physical-chemical entities are not biotically qualified, but still function in the living cell, we are obliged to make a threefold distinction in order to give an account of the complex interlacement found in the structure of the living cell (the neovitalist, J. Haas, ops for a dual structure – 1968:39).

(i) Firstly, there are the physical-chemically qualified constituents that as such represent enkaptic structural wholes.
(ii) Secondly, we find the cell's living organism that is biotically qualified and that can only function on the basis of the enkaptically bound building material.
(iii) Thirdly, we find the cell body as encompassing whole which enkaptically embraces both above-mentioned parts. We need the distinction between *concept* and *idea* in order to explain the apparent ambiguity in this context. The idea of enkapsis is used as a substitute for the *whole-parts relationship*, but we still refer to *part structures*! Within the framework of the theory of enkaptic interlacements, the spatial whole-parts relation is no longer employed in a conceptual sense, but in the sense of an idea usage, referring beyond the limits of the spatial aspect to the structural integrity of enkaptically interwoven entities. Without a concept-tran-

scending use of modal spatial terms, the *idea* of an enkaptic structural whole cannot be explained in a non-contradictory way (cf. also Zylstra, 1992:126 ff.).

7.12.2 *The classification of living entities*

The traditional theoretical understanding of reality (the experiential world) generally lacks a proper account of the modal aspects as irreducible functional points of entry to our experience and explanation of the world. For that reason we introduced modal aspects not merely as *modes of being*, but also as *modes of explanation*. The distinctness of the dimension of aspects and entities, on the law side of reality reflected in the distinctness of (unspecified) universal modal laws and (specified) type laws, apparently led to an emphasis on the dimension of entities at the cost of keeping in mind what the implications of the modal dimension were for a meaningful classification of entities.

In our discussion of the distinctive feature of scholarly thinking, our conclusion was that *modal abstraction* provides us with this criterion. One implication of realizing that diverse special sciences are demarcated by distinct modal points of entry to reality is that more than one aspect is required in order to identify one specific modality. Another obvious implication is that the answer to the question regarding the nature of any special science always exceeds the confines of the discipline concerned. For example, the definitions of plant science and animal science do not belong to these disciplines themselves. Saying that "plant science is a study of plants" does not say anything specific about plants, for this definition is solely focused on the *discipline* investigating plants. But let us retreat one more step and ask the question: what is a plant? (or: what is an animal?). A first reaction may be to say that only botany as an academic discipline can tell us what a plant really is. But is this the case?

Suppose there has never been a botanist and for the first time someone commences with a scholarly investigation of the nature of plants. How does that person *know* what a plant is if there is no textbook on plant science? Are there any guarantees that our first botanist indeed investigates *plants*? If there is no plant scholar who can tell her what *plant-ness* is, what would prevent our first botanist from investigating physical entities or animals while being under the impression that they are plants?!

Clearly, without a *prior* knowledge of the nature of plants, not even a (first) botanist will be able to study plants – and this *prior* knowledge does not have any other basis than our everyday, pre-scientific acquaintance with the world we experience. In other words, ultimately not even the discipline of plant science can operate by *negating* our non-scientific knowledge about the world.

The basic experiential awareness of the difference between physical things, plants and animals is correlated with the irreducibility of the relevant aspects, namely the physical, biotic and sensory modes of reality. The implication of a non-reductionistic ontology is that any given entity is either *this* or *that* – it is either physical or biotic (alive), it is either alive (biotic) or sensory (a sentient creature), with no "in-between." For example, the speculative story concerning a long process of a-biogenesis (a-biotic evolution) allegedly stretching over millions of years and aimed at accounting for the emergence of the first truly living entity (in reified mode referred to as the "origin of life") actually camouflages the critical point of embarrassment underlying the entire argument. At every moment of this on-going process, one can ask the meaningful and decisive question: is the 'developing' constellation alive or non-living? Obviously, the answer can only be an affirmation or denial. Nonetheless, the truly critical point is condensed into that unique, abrupt *moment* in which the transition is assumed to take place: at the previous moment, the constellation is non-living, and at the next moment, it is alive. The millions of years are irrelevant; what is required is an account of the *abrupt moment.* A possible escape route is to take recourse to a physicalistic view that asserts that living entities are "nothing but" a complex interaction of physical-chemical elements and processes. Yet, as soon as this view is advanced, the 'problem' evaporates, for then the "non-living" constellation was an interacting physical-chemical system all along, i.e. already '*alive*'! It reminds one of the way in which Simpson 'explains' large gaps in the paleontological record by claiming that if we really recovered all the intermediate forms, it would be clear that there are no gaps. Thus, instead of explaining the gaps, he simply denies that they exist.

What used to be known as a 'kingdom' (and later, in terms of inclusive language, as a 'realm') is directed at the scope of a category of entities conforming to a general *type law*. To the extent in which modern biology (and related disciplines, such as genetics, biochemistry and biophysics) explores the micro-dimensions of living entities, the traditional and familiar classification merely distinguishing between things, plants and animals appear to be unsatisfactory. For example, a group of amoeboid animals, the *Acrasiales* (the sole order within the class of cellular slime molds known as *Acrasieae*), were previously classified as plants. Likewise, the euglenoids, usually classified as plants, are considered to be (protozoan) animals by other zoologists. There are also living entities (amongst unicellular *flagellates*) that display features normally *differentiating* plants and animals. Chlorophyll, usually a characteristic of plants, is found in *Euglena* (one-celled living entities). Yet, at the same time they move and absorb food like animals.

It seems as if one way out of this problem is to introduce a more refined classification of realms ('kingdoms'). The following five 'kingdoms' feature in this regard: "Monera (bacteria, blue-green algae); Protista (protozoa, chrysophytes); Fungi (slime molds, true fungi); Plantae (algae and higher plants); and Animalia (multicellular animals)" (see Bock, 1989:102). Currently, regarding the most basic levels of living entities, preference is given to the distinction between *Prokaryotes* and *Eukaryotes*. The former are living entities *without* a nucleus, mostly *unicellular* and lacking *organelles* bound by membranes, whereas *Eukaryotes*, that can be uni- or multicellular, have a well-defined nucleus (owing to an embracing membrane) and a large number of sub-cellular *organelles*.[1]

Plant cells are identified by their direct use of light energy or owing to the fact that they are parts of a living entity using light energy. They have a cell wall (of *cellulose* – absent in the case of *Animals*, *Fungi* and *Protists*), and within their green parts, they have *plastids* (particularly *chloroplasts*). As an alternative to a two-domain classification, Carl Woese proposes a *three-domain* system: *Eukaryota*, *Bacteria*, and *Archaea*.

However, investigating these distinctions soon makes it clear that, in terms of *modal functions*, the basic range of categories (classifications) pertain to living entities actively (subjectively) functioning (in addition to the aspects of number, space, movement, and the physical) within the *biotic* aspect. Only animals, as sentient creatures, also subjectively function within the *sensory* mode.

In the absence of a clear distinction between modal functions and concretely existing (and functioning) entities, modern biology does not have a univocal idea of a realm ('kingdom') at its disposal. For this reason, it proceeds in respect of *living* things by ultimately making classifications merely within the context of biotically qualified entities! This explains why comparisons and distinctions on this level remain bound to vital (i.e. *biotic*) functions of living things without ever referring to the modal meaning of what is presupposed, namely the *biotic function* of reality. Instead, one reads about *structural* comparisons (always with presupposed similarities in the background). For example, in spite of the absence of a well-defined nucleus, *Prokaryotic* DNA basically functions in the same way as *Eukaryoric* DNA. Modal properties only surface with respect to the aspects of number and space – *counting* sub-cellular organelles, noting differences in *size* and so on. But without alternative *qualifying functions*, the assumed *biotic* domains (realms / 'kingdoms') simply do not transcend the scope of living things (biotic things).

Stafleu attempts to justify the current classification by distinguishing between alternative *foundational* functions ("secondary characteristics") – such as the arithmetical (*Prokaryotes*), the spatial (*Eukaryotes*, *colonies* and *tissues*), and the kinetic (differentiated organisms) (Stafleu, 2002:186-188).Yet even these distinctions do not transcend the qualifying role of the biotic aspect, and at most may serve to arrive at sub-categories (genotypes rather than radical types) within the realm of biotically qualified entities. His emphasis on the modal universality of the irreversible biotic time order of *fertilization* (conception), *germination*, *growth*, *reproduction*, *ageing* and *death* (Stafleu, 2002:173), simply underscores the uniform scope of the biotically qualified realm of living entities. Within the micro-dimensions, we merely encounter living entities displaying either plant-like (*biotically qualified*) or animal-like

1 Next to the nucleus with its pores and nucleoles we may mention organelles such as the endoplasmic reticulum (either smooth or rough), the endomembrane system, the Golgi apparatus (constituted by dictysome membranes), lysosomes, mitochondria, ribosomes, centrioles and peroxisomes.

(sensorily qualified) features. The inability of current biological scholarship to clearly distinguish between biotically qualified features and sensory features – underlying the distinction between the realms of plants and animals, appears primarily to testify to our lack of criteria and not to the basic and limited options provided by two irreducible modal functions available for serving as qualifying functions of distinct realms of entities – plants and animals.

7.13 Does change dominate the bio-world?

Since change presupposes constancy it must be clear that any theory that does not recognize this foundational coherence may end up in serious theoretical difficulties, particularly when entities and processes are contemplated that *enter* into cosmic later modalities, for example when it is contemplated that material entities (having their highest subject function within the physical aspect), either *produce* the biotic aspect of reality or *assume* an additional subject function within the biotic aspect (as Klapwijk argues – see Klapwijk, 2008). Can a particular function *change* into another function? Just consider the conviction that *physical* entities were transformed into *biotic* (i.e. *living* entities).

The problem here is a quite serious theoretical issue, for if we accept that the physical function can *change* (can be 'transformed') into the *biotic* aspect, the next problem is if there will still exist a physical aspect of reality *after* the change of the *physical* into the *biotical*? This seems to be impossible if the physical aspect turned into the biotical aspect! A less rigorous version may contemplate the question whether or not it is possible for one aspect to give rise to the existence of another aspect? For in this case the continued existence of the initial aspect may be affirmed. Yet, if this transition does not *eliminate* the initial (or primary) aspect, it is incorrect to claim that it *changed* into a different aspect. While holding on to the idea of 'transformation' the only other option seems to be to defend some or other view of *emergence* in terms of which it is claimed that an on-going process eventually gives rise to various *new aspects of reality*. It is often asserted that once these additional aspects *emerged* (came into existence) they are *irreducible*. Emergent evolutionists (such as defended by Lloyd-Morgan, Whitehead, Alexander, Woltereck, Bavinck and Polanyi) indeed want to have it both ways: *continuity* in descent (in the *process* of origination) and *discontinuity* in existence (in *structure*). *Structure* thus becomes the product of the *genetic process of becoming*.

Emergence evolutionists openly admit that this position is burdened by an inner antinomy. Richard Woltereck does so in his *Ontologie des Lebendigen* (1940:300ff.), while Michael Polanyi writes:

> We have reached the point at which we must confront the unspecifiability of higher levels in terms of particulars belonging to lower levels, with the fact that the higher levels have in fact come into existence spontaneously from elements of these lower levels. How can the emergent have arisen from particulars that cannot constitute it (Polanyi, 1968:393).

Some neo-Darwinists tend to approximate this emergent-evolutionistic position. Simpson states:

> Man has certain basic diagnostic features which set him off most sharply from any other animal and which have involved other developments not only increasing this sharp distinction but also making it an absolute difference in kind and not only a relative difference of degree (Simpson, 1971:270).

Th. Dobzhansky calls the origination of a new level, i.e. discontinuity, "evolutionary transcendence" (Dobzhansky, 1967:44 – the term "transcendence" is derived from the theologian Paul Tillich):[1]

> The flow of evolutionary events is, however, not always smooth and uniform; it also contains crises and turning points which, viewed in retrospect, may appear to be breaks of the continuity. The origin of life was one such crisis, radical enough to deserve the name of transcendence. The origin of man was another (Dobzhansky, 1967:50).

Furthermore, Dobzhansky holds that "the phenomena of the inorganic, organic, and human levels are subject to different laws peculiar to those levels" (Dobzhansky, 1967:43). At this point something intriguing could be noticed. This quote from Dobzhansky sounds very much like the reformational philosophical idea of *sphere-sovereignty*, embedded in the perspective of an unbreakable correlation between law and what is subjected to law. One important implication of accepting this principle, as applied both to the various aspects of reality and the diverse (kinds of) entities found within it, is the idea of *irreducibility*.

However, it appears there are two perspectives possible when it comes to an account of irreducibility. Dobzhansky, arguing from the "bottom" to the "top" immediately relativizes the idea of different laws peculiar to different levels: "It is unnecessary to assume any intrinsic irreducibility of these laws, but unprofitable to describe phenomena of an overlying level in terms of those of the underlying ones" (Dobzhansky, 1967:43). The reverse path is pursued by Van Huyssteen when he looks back on what emerged, for example, at the level of cognitive and cultural evolution. On the one hand he refers to Darwin who stressed the *continuity* between species in respect of instincts or rational abilities (Van Huyssteen, 2006:81) as well as to the continuity of organic evolution from "unicellular organisms to humans" (Van Huyssteen, 2006:87). But on the other hand he discerns something *unique* within human cognition, culture and religious worldviews, something *irreducible*:

> On the one hand, then, organic evolution – particularly the evolution of the human brain – can be seen as the basis of cultural evolution. On the one hand, the latter can never be reduced to the former. Cultural evolution requires explanations beyond the biological theory of evolution in its strictest sense. Therefore the term 'evolution' applies to both the development of the organic world, from unicellular organisms to humans, and the development of culture. Or in Wuketits's words, biology offers the necessary conditions of culture, but it

1 Van Huyssteen refers to Stewart who argues on the basis of the "notion of *emergence*": "Life is flexible, life is free, life seems to transcend the rigidity of its physical origins. And it is this kind of transcendence that is called 'emergence' " (Van Huyssteen, 2006:55).

> does not offer the sufficient conditions. Cultural evolution (including the evolution of ideas, scientific theories, and religious worldviews) cannot be reduced to biological evolution (Van Huyssteen, 2006:86-87).

He continues to explain his view that within (as part of) the "grandiose universal natural history" once *cultural evolution* commenced it "obeyed its own principles"! This almost sounds like what Dobzhansky had to say about *different laws peculiar to different levels*. However, whereas Dobzhansky rejects *irreducibility* Van Huyssteen upholds it: *culture is not reducible to biological entities*.

> If we should ask whether we are justified in speaking of cultural evolution as we do of biological evolution, the answer, as we saw in our discussion of Plotkin's work, should be yes. We are not only justified in doing this, but it is necessary, since there is one common trait here: both organic and cultural evolution can be regarded as complex learning processes, with human cognition as the crucially important mediator between them. Culture can therefore be understood as the most sophisticated learning process requiring particular modes of explanation and a particular type of evolutionary epistemology that goes beyond strict Darwinism. Wuketits, therefore, correctly argues that although there are biological constraints on cultural evolution, culture is not reducible to biological entities. Cultural evolution indeed depends on specific biological processes, and our cultures therefore are part of a grandiose universal natural history, but cultural evolution, once it started, obeyed its own principles and gave human evolution an entirely new direction, even acting back on organic evolution (Van Huyssteen, 2006:98; regarding the irreducibility of human consciousness, see page 78).

While distinguishing between *laws of nature* (Van Huyssteen, 2006:55) and *cultural evolution* with its "own principles," as well as alluding to "particular modes of explanation" (Van Huyssteen, 2006:98), one may ask the question what the *origin* of these (irreducible) laws[1] and principles is? Since laws *condition* (in the sense of *making possible*) what is subjected to them these laws cannot originate in a process presupposing them. Likewise, if there are "unique principles" for culture, these principles, making possible cultural activities, cannot originate in and through cultural processes.

From a different angle Julian Huxley actually supports Van Huyssteen's idea regarding a "grandiose universal natural history" that gave rise to higher levels that cannot be explained in terms of lower levels. He struggles with the inherent tension between *continuity* and *discontinuity* and what he does may be called an "emergentistic retreat." Huxley warns us against the "nothing but" fallacy:

> We begin by minimizing the difference between animals and ourselves by unconsciously projecting our own qualities into them .... Though early scientific thinkers, like Descartes, tried to make the difference absolute, later applica-

1 He also mentions *regularities* (see Van Huyssteen, 2006: 89, 90, 91). We know that this term is equivalent to *lawfulness* and *law-conformity* and that it therefore reflects a feature of whatever is *subjected to law*. Exhibiting the "measure of law" is a feature of whatever is subjected to a law.

> tions of the method of scientific analysis to man have, until quite recently, tended to reduce it again. This is partly because we have often been guilty of the fallacy of mistaking origins for explanations – what we may call the 'nothing but' fallacy: if sexual impulse is at the base of love, then love is to be regarded as nothing but sex; if it can be shown that man originated from an animal, then in all essentials he is nothing but an animal. This I repeat, is a dangerous fallacy (Huxley, 1968:137).

All variants of an emergent-evolutionistic perspective appears to suffer from the problematic tension between *continuity* and *discontinuity* or *reducibility* and *irreducibility*.

Since neo-Darwinism foremost presents itself as a *theory of change* we may return once more the question raised in Chapter 4: are there any constants in the bio-world?

First of all every single living entity persists over time – and in this sense exhibits an element of constancy, making it possible to refer to its (relative) *identity* – and we saw that this identity cannot be *physical* in nature.

However, once the vital (i.e. biotic) subject function of living things is taken into account, it is even possible to claim that a living thing, from a biotic point of view, is in a *stable state* (referred to as *health*), while simultaneously claiming – without any contradiction – that from a physico-chemical perspective (with a view to the flowing equilibrium of its physical-chemical constituents), it exists in an *unstable state*. If the physical-chemical substratum of living things approaches a state of higher statistical probability, biotical instability increases as a sign of the final *process of dying*.

If change cannot be detected accept on the basis of persistence or constancy, then the question arises what are the conditioning constants underlying the (neo-)Darwinian encompassing theory of *change*? Are there indeed any constant conditioning elements making possible this emphasis on *change*? Darwin proposed one such a conditioning element, namely *natural selection*. With "natural selection," Darwin had in mind the constant struggle in which only the fittest will survive. He explains what he has in mind in the beginning of Chapter 4 as follows:

> On the other hand, we may feel sure that any variation in the least degree injurious would be rigidly destroyed. This preservation of favourable individual differences and variations, and the destruction of those which are injurious, I have called Natural selection, or the Survival of the fittest. Variations neither useful nor injurious would not be affected by natural selection, and would be left either a fluctuating element, as perhaps we see in certain polymorphic species, or would ultimately become fixed, owing to the nature of the organism and the nature of the conditions (Darwin, 1859a:131).

It should be kept in mind that Darwin did not know anything about the genetics of Mendel. The latter's insights into the process of genetic inheritance, after they were rediscovered by Hugo De Vries and Carl Correns in 1900, in fact did not receive a positive response amongst Darwinists, mainly because Mendel's laws acted in a predetermined way. The inherited features discovered by

Mendel proceed according to specific rules and occur as variations within set limits – thus excluding the idea of *random variations*.

The latter was introduced on the basis of the growing knowledge of *mutations*. It appeared that mutations occur indeed at *random*. After about twenty years of controversy between the saltationist and biometric schools of thought T.H. Morgan established a laboratory in which he attempted to show that new species in fruit flies (*Drosophila melanogaster*) could be produced through mutations. Morgan in fact began his career in genetics as a saltationist. However, the experimental work at his lab with *Drosophila melanogaster*, which helped establish the link between Mendelian genetics and the chromosomal theory of inheritance, demonstrated that rather than creating new species, mutations increased the genetic variation in the population (see WEB: *Modern Synthesis*, 2009).

In almost all cases mutations affect *genes* – and Darwinian biologists thought that the variation thus caused by it provides the basis upon which *natural selection* operates. This possibility turned the tide and as an effect we witness the origination of the *Modern Synthesis* which incorporated the role of mutations in the subsequent development of neo-Darwinism. Huxley invented the phrase *the modern synthesis* (see Huxley, 1942) and names such as R. A. Fisher, Theodosius Dobzhansky, J.B.S. Haldane, Sewall Wright, E.B. Ford, Ernst Mayr, Bernhard Rensch, Sergei Chetverikov, George Gaylord Simpson, and G. Ledyard Stebbins are associated with it (see WEB: *Modern Synthesis*, 2009).

Since mutations appeared to be defective in 99% of the instances where they occur, maintaining the idea of mutation as an independent condition for evolution would uproot the foundation of the entire Darwinian idea of "organic transformation." We saw that no one less than Dobzhansky, the well-known neo-Darwinian geneticist, had to observe that "[M]utation alone, ..., could only result in degeneration, decay, and extinction" (Dobzhansky, 1967:41 – see page 113). The remarkable fact is that there is no single genetic mechanism that introduces, regulates or controls mutations (see Scheele, 1997:49). Yet, even though a mutation may cause damage to a gene or even may eliminate it, it may produce an advantage in the chances of survival (Scheele, 1997:50). Mutations therefore differentiate into negative ones (99% result in degeneration, decay, and extinction), neutral ones (no effect on survival), and in a few cases changes that may be advantageous in terms of survival. At this point the modern synthesis opted for the *combined effect* of mutation and natural selection. Subsequently mainstream (neo-)Darwiniain evolutionary theory holds that these two phenomena, namely *mutation* and *natural selection*, always act in combination.

When natural selection is invoked, neo-Darwinism holds that the (mutationally) disadvantaged living entity may emerge as being advantaged – in the sense of having a better chance to survive. At the same time the combined operation of mutation and natural selection serves the claim that new kinds of entities emerged – as we have seen, from the level of unicellular entities up to human beings.

By and large neo-Darwinians acknowledge that mutation alone results in *devolution* ("degeneration, decay, and extinction"). There is nothing *creative* attached to this devolutionary process, it is predominantly destructive. Nonetheless neo-Darwinism is convinced that the *magic wand* that can turn devolution into evolution is given in the operation of *natural selection*. In the above-mentioned quotation from Dobzhansky regarding the degeneration, decay, and extinction caused by mutations alone, a part of the sentence was left out – the full quotation reads:

> Mutation alone, *uncontrolled by natural selection* (my italics – DS), could only result in degeneration, decay, and extinction" (Dobzhansky, 1967:41).

His account of variability has difficulties answering the following question: If living entities during the past three or four thousand million years have been governed by this universal evolutionary law that developed (without any purpose!) towards the human being, it cannot be explained why there still are, apart from the highly evolved animals, such primitive entities as bacteria, algae, mosses, amoebae, worms, etc. – why did the evolved animals not also remain stuck on these original levels? Eisenstein writes: "The simultaneous co-existence of the greatest variety of life forms, from amoeba to humans, anyway proves that from the perspective of nature these are all equitable and equally viable (existenzfähig = able to exist), without any necessity of further development" (Eisenstein, 1975:245).[1]

In order to understand the alleged co-operative action of natural selection and mutation we have to provide some background information. In 1896 the Buchners discovered alcoholic ferments which serve a catalytic function in cells. Initially referred to as "zymase," it gradually became apparent that it is a mixture of enzymes and co-enzymes. Protein refers to macro-molecules consisting of 20 different amino acids. When an amino group (NH_2) of one amino acid is linked with a carboxyl-group (COOH) of another amino-acid, a peptide bond (NH-CO-) is formed – coupled with the release of water (H_2O). Multiple amino acids are bonded in this way into a macro-molecule – a polypeptide. Enzymes have a protein structure built up out of amino acids and occasionally occurs in their thousands in a particular cell. This promotes chemical reactions in the cell, although each kind of enzyme catalyzes only a limited number of reactions. Enzymes are very sensitive to abnormally high temperatures – unlike inorganic catalysts, which normally perform better under warmer conditions. The entire metabolism of the cell depends on the functioning of enzymes.

In the nucleus of the cell nucleotides are formed through the bonding of a sugar and a nitrogenous base on the one hand and a phosphorous acid-remnant on the other. In this way polinucleotide chains are formed. In DNA four

1 We shall see below that the fossil record dominantly evinces forms that (abruptly) appear and then continue to exist basically unaltered for millions of years.

nucleotides are found, namely Adenine (A), Guanine (G), Cytosine (C), and Thymine (T). These spontaneously associate through hydrogen bonds in the links A-T and G-C. Out of this mutual attraction emerges two polinucleotide-strings with various possibilities. A series like ATG ACG is complemented by a series TAC TGC. The so-called genetic code concerns the rule in terms of which a polipeptide series is linked to a given polinucleotide series. This linkage is made possible by RNA – a nucleonic acid differing from DNA in that the T is replaced by U(racil). To transfer the matrix of DNA to protein it appears that a combination of three letters is necessary in the DNA for every amino acid to be formed. This means that some amino acids are correlated with more than one triplet of nucleotides – i.e. different triplets are occasionally attached with only one amino acid. The triplets UAA, UAG, and UGA appear to be inoperative, since they are not correlated with any amino acids.[1]

It is estimated in mammals that "uncorrected errors (= mutations) occur at the rate of about 1 in every 50 million (5×10^7) nucleotides added to the chain." But with 6×10^9 base pairs in a human cell" this implies "that each new cell contains some 120 new mutations." However, since up to "97% of our DNA does not encode anything" this should not be a cause for concern" (Mutations, 2009). Scheele points out that in respect of the amino acid sequence of certain critically specialized proteins no change is allowed because it will compromise the functionality of the protein fatally, i.e. after one single change in such a protein it loses its total functionality and natural selection immediately eliminates it. The effect is that the bearer of the mutated gene cannot continue to exist or to survive. Moreover, two thirds of the genes belong to this category, which means that two thirds of the genes do not change or (d)evolve (Scheele, 1997:90-92).

Whereas *somatic mutations* – that occur within somatic cells by damaging them, making them cancerous or killing them – disappear when such a cell dies, *germline mutations* may be passed down, given certain conditions, to a subsequent generation of gametes (Mutations, 2009). Given the multiple-one mapping between nucleotides and amino acids, a change of *one* letter of the genetic code may be neutral. Alternatively it may change the correspondence and therefore negatively affect the message of the code (these mutations are known as *point mutations*). The scenario gets worse when, in a sequence like ATGACG, one letter drops out (*deletion*) or one is added (*insertion*), because then the entire coding is disrupted. Losing the orderliness of the code should be appreciated against the background of the fact that the possibilities of *randomly* combining amino acids – say in the case of a protein constituted by 300 amino acids – amounts to 20^{300}, a significantly larger number than the esti-

1 There are 20 amino acids and if we consider only two possible combinations of 4 DNA nucleotides only 16 amino acids can be explained: 4x4 = 16. The mentioned nucleotides of A, G, C, and T can however be arranged in 64 combinations of triplets: 4x4x4 = 64.

mated total number of atoms in the universe – 10^{80}![1] Of course the presence of DNA (and protein) merely represents some of the required (macromolecular) physico-chemical substructures of living entities, and as macromolecules they are not alive.

Apparently the role of mutation and natural selection supports the main thesis of neo-Darwinism, namely the assumption that the sole reality within "living" nature is that of *change*. Yet in fact the contrary is true, for if no *constancy* is attached to the (combined) *conditioning role* of *mutation* and *natural selection*, no single evolutionary change is coneivable. In Chapter 4 we have already concluded that only when mutation and natural selection are operating as a constant (conditioning) law for the existence and variability of living entities, is it possible to account for the changefulness of the latter. For that reason we argued earlier that it is not true to claim that everything *changes* within the bio-world, for the "law-effect" of mutation and natural selection is supposed to remain *constant*!

As far as the original position of Darwin is concerned, as presented in 1859, we have to return to his (above quoted) view of *natural selection*, while keeping in mind that – as Dobzhansky emphasized – mutation alone, uncontrolled by natural selection, could only result in degeneration, decay and extinction. The fact that almost all mutations are defective and harmful, make them solid candidates for elimination by natural selection. Darwin explicitly stated that we "may feel sure that any variation in the least degree injurious would be rigidly destroyed" by natural selection (see page 482 above). For all practical purposes the combination of Darwin's original view of natural selection with the neo-Darwinian understanding of mutation therefore rules out all Darwin's hope for evolution, what is left is nothing but *devolution*. The fact of the matter is that natural selection is a conservative process in the sense that it cannot produce or create anything – it can merely select from what is "presented." As Mortenson puts it: "natural selection can explain the survival of the fittest, but not the arrival of the fittest" (Mortenson, 2006).

What modern genetics opened up is an understanding of the variation that is possible within the *same* genetic theme. Mutations move within the same framework, except that in the 99% plus instances they cause degeneration – there is no single example known for a leap into a functional adoption of new genes (Scheele, 1997:110-111). For that reason natural selection actually *conserves* the scope of variation within a population, which explains why it is indeed also described as a *conservative factor*.

John Davison states that natural selection, "the cornerstone of the Darwinian myth, never had anything to do with creative evolution. It served in the

1 The chance of randomly selecting the letters in a specific order (at five thousand flips per second) of *one page* of the genetic code is calculated to take 10 billion years. Since there are actually 500,000 such pages what it will take to "assemble" all of them, in their correct arrangement, is nothing less than 5 million billion years! (Keep in mind that present-day physicists and astronomers estimate our own universe to be less than 14 billion years old.)

past as it does now only to prevent change. That is why every chickadee looks like every other chickadee. … There is not a single extant diploid organism on this planet that will ever become anything basically different from what it already is." Davison, apart from himself, calls upon a number of significant natural scientists in support of his claim that natural selection is "a conservative rather than a creative element."[1] John Davison points out that there "is not a shred of evidence that any prokaryote ever evolved into anything but the same species and certainly not into any eukaryote" (Evolgen Archive, 2009). A similar verdict applies to the experiments done by Dobzhansky with Drosophila, which did not produce a new species – and "it is his credit that he admitted defeat."

> What I am saying is that allelic mutations are either deleterious or neutral and have little or nothing to offer in the way of advantage to the organism. More important, the experimental attempts to demonstrate speciation through selection for such changes have failed. I also do not regard prokaryotes as models for eukaryote evolution. Neither Lamarckian nor Darwinian models have received laboratory support and neither can be reconciled with the fossil record (Evolgen Archive, 2009). In addition Davison refers to the "the total failure of the Darwinian fairy tale to explain anything in evolution beyond the production of varieties and subspecies (Evolgen Archive, 2009).

7.14 Constancy and change within paleontology

What is known as " phyletic gradualism" is basically the 19th century idea of Darwin that species evolve incrementally at a more or less steady rate. How did Darwin assess the paleontological record in the light of his gradualist starting point?

He states that

> Geological research, …, yet has done scarcely anything in breaking down the distinction between species, by connecting them together by numerous, fine, intermediate varieties; and this not having been affected, is probably the greatest and most obvious of all the many objections which may be urged against my views (Darwin, 1859a:307).[2]

The above-mentioned pre-occupation with change, embedded in the overall idea of slow alterations of living entities over vast periods of time, prompted paleontologists under the spell of this paradigm not only to search for these "numerous, fine, intermediate varieties," but also to interpret whatever they found in advance in terms of gradual change. It is understandable that the revolutionary effect of Darwin's new ideas at once also generated substantial respect for his views amongst his followers and inspired them to help to remedy

1 He mentions St George Jackson Mivart, Henry Fairfield Osborn, Leo Berg, Reginald C. Punnett, Pierre Grasse and Otto Schindewolf.

2 In 1859b this reads: "What geological research has not revealed, is the former existence of infinitely numerous gradations, as fine as existing varieties, connecting together nearly all existing and extinct species. But this ought not to be expected; yet this has been repeatedly advanced as a most serious objection against my views" (Darwin, 1859b:207).

this shortcoming of missing transitional forms. Fossil findings were therefore constantly under the pressure to "bridge the gaps." Nonetheless, in spite of the sustained hope that transitional forms will be found, paleontologists were all along very well aware of the discontinuities within the higher systematic categories. Just over 100 years after Darwin's 1859 book Simpson explicitly states that "every paleontologist knows, that *most* new species, genera, and families and that nearly all new categories above the level of families appear in the record suddenly and are not led up to by known, gradual, completely continuous transitional sequences" (Simpson, 1961:360).

One may suspect that this acknowledgement might have resulted in some caution and an attitude of tentativeness and uncertainty. Unfortunately the opposite appears to be the case, because Simpson never doubted for one single moment that the sought after intermediate forms actually did exist. The question is: how does one account for these discontinuities? Simpson proceeds: "Almost all paleontologists recognize that the discovery of a complete transition is in any case unlikely. Most of them find it logical, if not scientifically required, to assume that the sudden appearance of a new systematic group is not evidence for special creation or for saltation, but simply means that a full transitional sequence more or less like those that are known did occur and simply has not been found in this instance" (Simpson, 1961:360). That means that Simpson responds to the mentioned question – namely how does one explain the discontinuities – not by explaining them, but by *denying* them!

About a decade later paleontologists started to realize that they have to take these discontinuities serious. The web article on the *Synthetic Theory of Evolution* says that by the beginning of the 1970's "Stephen J. Gould, Niles Eldredge, and other leading paleontologists" challenged the kind of approach exemplified in what we just quoted from Simpson. It states that they "asserted that there is sufficient fossil evidence to show that some species remained essentially the same for millions of years and then underwent short periods of very rapid, major change. Gould suggested that a more accurate model in such species lines would be *punctuated equilibrium*" (Synthetic Theory of Evolution, 2009).

Particularly significant is the way in which this statement is formulated: "some species remained essentially the same for millions of years and then underwent short periods of very rapid, major change." What is certain is what is embodied in known fossil findings – exhibiting *sameness* over millions of years, or as can also be said, *constancy* over millions of years. What is documented is *stasis*, *constancy* – what is *not* documented is *rapid change* – which is invoked to avoid the phenomenon of the *sudden appearance* of whatever new type that then remains constant over millions of years. The mere abrupt appearance does not reveal *anything* about what happened *before* the unanticipated appearance took place. The attempt to say something about it remains completely *speculative*. In principle there is not really a difference between what Simpson says and what the theory of *punctuated equilibrium* holds, for in both cases the *absence* of transitional forms is actually denied, either by claiming that the transition "did occur and simply has not been found in this instance" (Simpson), or by holding that these (quickly changing) forms do exist but simply was not recorded because, for whatever reason, they did not fossilize (Gould).

However, if we consider one of the authentic statements made by Gould in the seventies, the distinction between fact and fiction more clearly comes to the fore: "The extreme rarity of transitional forms in the fossil record persists as the trade secret of paleontology. The evolutionary trees that adorn our textbooks have data only at the tips and nodes of their branches; the rest is inference, however reasonable, not evidence of fossils" (Gould, S.J. 1977:14). What Eldredge said is even more revealing in this context: "We paleontologists have said that the history of life provides support for the interpretation of gradual development through natural selection while all the time we knew that it was not true" (see Van den Beukel, 2005:105).

After Gould and Eldredge first made their new idea public a significant article appeared in the neo-Darwinist Journal *Evolution* under the title "Paleontology and Evolutionary Theory." In it the paleontologist D.B. Kitts points out that the spatial distribution and temporal sequence of organisms with which paleontology works is founded in the ordering principles of *geology*, and can therefore not be incorporated in any biological theory: "Thus the paleontologist can provide knowledge that cannot be provided by biological principles alone. But he cannot provide us with evolution. We can leave the fossil record free of a theory of evolution. An evolutionist, however, cannot leave the fossil record free of the evolutionary hypothesis" (Kitts, 1974:466). According to him the danger continues to exist that biologists are convinced of the acceptability of the evolutionary hypothesis by a theory which is already inherently evolutionistic: "For most biologists the strongest reason for accepting the evolutionary hypothesis is their acceptance of some theory that entails it" (Kitts 1974:466). His final verdict is devastating: "Evolution requires intermediate forms and paleontology does not provide them" (Kitts, 1974:467).[1]

Eldredge underscored the fact that stasis (constancy) is dominant in the "fossil record": "Gould and I claimed that stasis (= immutability, stand-still), and not change, is the dominant theme of the fossil record" (quoted by Van den Beukel, 2006:106). Most species "enter the evolutionary order fully formed and then depart unchanged" (Berlinski, 2003:158). Eldredge adds the remark: "and this destroys the backbone of the most important argument of the modern theory of evolution" (as quoted by Van den Beukel, 2005:106).[2]

Dawkins strictly continues the epistemic ideal of Enlightenment rationality (see Sterelny, 2001:14) – the scientific description of the universe is "true ...

1 To this he adds the remark: "But most of the gaps are still there a century later and some paleontologists were no longer willing to explain them away geologically" (Kitts, 1974:467).

2 Gould quotes Prothero and Shubin, who wrote, in connection with the supposed evolution of the horse: "This is contrary to the widely held myth about horse species as gradualistically varying parts of a continuum, with no real distinctions between species. Throughout the history of horses, the species are well-marked and static over millions of years" (see Gould, 1997:68).

beautiful and complete" (as Sterelny formulates it – 2001:13). Dawkins adheres to the orthodox Darwinian view believing that the power of selection slowly and incrementally build the exquisite and intricate outfit of living organisms. This conviction clearly shows a prejudiced and premature pre-occupation by Darwin and his followers with *change* which prevented modern (neo-)Darwinian biology to come to terms with the fact that change always presupposes something constant. Steven Stanley raises a legitimate concern in this regard: "Since the time of Darwin, paleontologists have found themselves confronted with gradualism, yet the message of the fossil record has been ignored. This strange circumstance constitutes a remarkable chapter in the history of science, and one that gives students of the fossil record cause for concern" (Stanley, 1981:101). The one-sided emphasis on change actually denied constancy its rightful place. What Gould and Eldredge designated as the *dominant theme* of the fossil record, namely stasis (non-change), highlights the other neglected view point. On the one hand change presupposes the constancy of *conditions* (in the sense of a *law* that determines and delimits those entities subject to it), and on the other those subjects may display, notwithstanding variability within certain boundaries, a relative persistence (or constancy/identity). The latter phenomenon is supported by the mentioned *dominant theme*.

Suppose it would turn out that important systematic units of animals appeared, instead of going through an extremely slow process, at the *same time* without common ancestors, what would it imply for the position taken by Darwin? He explicitly holds: "If numerous species, belonging to the same genera or families, have really started into life at once, the fact would be fatal to the theory of evolution through natural selection. For the development by this means of a group of forms, all of which are descended from some one progenitor, must have been an extremely slow process; and the progenitors must have lived long before their modified descendants" (Darwin 1859a:309).

Ironically enough this is exactly what happened during the "Cambrian explosion." Sterelny is therefore justified in his assessment that the standard (neo-)Darwinian story runs "slap-bang into a nasty fact," the fact namely that about 530 million years ago most "major animal groups appeared simultaneously". He continues:

> In the 'Cambrian explosion', we find segmented worms, velvet worms, starfish and their allies, mollusks (snails, squid and their relatives), sponges, bivalves and other shelled animals appearing all at once, with their basic organization, organ systems, and sensory mechanisms already operational. We do not find crude prototypes of, say, starfish or trilobites. Moreover, we do not find common ancestors of these groups (see Sterelny, 2001:89-90).[1]

1 Schroeder remarks that "530 million years ago in the Cambrian era, with no hint in earlier fossils, the basic anatomies of all life extant today appeared simultaneously in the oceans" (Schroeder, 1998:29-30). He continues: "Following this explosion of multicellular life at the start of the Cambrian era, for some puzzling reason, no other new phyla (basic anatomies) ever appeared. Classes of animals developed within each phyla but they always retained the basic body plan of their particular phylum (Schroeder, 1998:29-30).

Since paleontologists working *within* the neo-Darwinian tradition started to concede that *constancy* dominates the fossil record (living entities persisted basically unaltered over millions of years), this phenomenon requires closer scrutiny. Particularly Schindewolf was extremely critical of the wedding of population genetics and paleontology (he is the author of a standard text book on paleontology, *Grundfragen der Paläontologie*, 1980). He developed his own theory, known as *typostrophism*, through which he accounts for the abrupt appearance of types (typo-genesis), their stabilized (and continued) existence (typo-stasis) and their final disappearance from the paleontological horizon (typo-lysis).[1] He holds that the mode of thought and terminology of population genetics cannot be transferred to the fossil material that forms the chief foundation of phylogeny. He emphatically states that genetics as a discipline falls short of the actual process of evolution. The real issue by far exceeds what is accessible through experimental genetic research on recent organisms. "Evolution in its proper sense is a historical process that occurred in a bygone past" (Schindewolf, 1969:61-62).

Diverging theoretical paradigms caused alternative and even mutually contradicting assessments of fossil findings. As an example we may briefly refer to the account given of the *Archaeopteryx* (discovered already in 1861), which has both reptilian and avian characteristics. Although G.G. Simpson and O.H. Schindewolf largely concur with regard to the discovered state of affairs, they approach the factual information from radically divergent points of departure. Schindewolf is of the opinion that the transition from the *class of Reptiles* to *the class Aves* found expression in the appearance of *Archaeopteryx*. This animal was a *bird* with wings which could fly, the first representative of a new class – the *Aves* (birds). M. Grene characterizes Simpson's approach as follows: "Simpson says Archaeopteryx was a species like any other, originating by normal speciation from other reptilian species; only when we look back over the whole vista of evolution do we say, this particular species was the first of what turned out to be a new class" (Grene, 1974:130).

Grene points out that Simpson and Schindewolf accuse each other of essentially the same or similar mistakes, making use of unnecessary and mystifying presuppositions. She believes that each accepts as premise the negation of the other's conclusions – while hardly if at all differing with regard to the facts:

> Simpson, wedding paleontology to the statistical methods of population genetics, sees a gradual change in populations such that the sharp divisions of traditional morphology become false. Schindewolf, basing his theory on the logical priority of morphology, concludes that the gradualist, statistical picture of

---

1 Without mentioning Schindewolf's theory, Schroeder refers to the data underlying the theory of typostrophism as follows: "Intra-phylum evolution presents a further puzzle. Consistently new organisms, whether among plant groups or animals, make their first fossil appearance highly specialized and fully developed, last their time, and disappear. Rarely if ever are there fossil indications that competition by a new and better-adapted class of animals outstripped an outdated ancestor in the race for food, shelter, and survival, thus driving the ancestor to extinction. The fossil record regularly fails to give any hint at the basic anatomical levels that a change in morphology was in the offing" (Schroeder, 1998:29-30).

> neo-Darwinism is false. To put it very schematically; Simpson argues: the neo-Darwinian theory is true; morphology implies that neo-Darwinism is not true; therefore morphology is wrong. Schindewolf argues: morphology must first be accepted as true; morphology implies that the neo-Darwinian theory is wrong; therefore the neo-Darwinian theory is mistaken. Or to put the matter another way, they agree on the major premise: traditional morphology and neo-Darwinism are incompatible (Grene, 1974:132).

Schindewolf 's insistence that morphology is the key to an understanding of the past is based upon his idea of the *structural design* of particular *types* of entities. He employs the German term *Bauplan* which is meant to capture what we have designated as *type laws* in an earlier context. It should be noted that Darwin, in one of his last letters, doubted that one can avoid the assumption of a *plan* (design) in nature (see Eisenstein, 1975a:412). We have seen that Von Weizsäcker acknowledges the modal universality of the physical aspect of reality (quantum theory applies to all "Gegenstandklassen" – Von Weizsäcker, 1993:128), to be distinguished from type laws holding only for a limited class of entities (such as the type law for being an atom). The biotic aspect of reality displays its own *modal universality* since it holds for all *kinds* of living entities (plants, animals and human beings).

The most fundamental basic concepts of biology as a scholarly discipline constitute this modal universality of the biotic aspect. As analogical basic concepts they actually reflect, within the biotic aspect, structural moments referring back to the four aspects preceding the biotic aspect, namely the numerical, spatial, kinematic and the physical aspects. In Chapters 5 and 6 the distinction between the one and the many is identified as being derived from the numerical meaning of unity and multiplicity. We argued that this arithmetical feature analogically appears in all the post-arithmetical aspects. Just recall, for example, the concept of *mass* employed by the discipline of physics. It represents a numerical analogy within the modal structure of the physical aspect, since it relates to a "quantity of matter." Similarly, a biologist may speak about biomass (see pages 355 ff.).

A spatial figure displays a spatial *unity* amidst its multiple parts – for example the three sides (*multiplicity*) of a triangle (*unity*), the triangle as a spatial unity and multiplicity. The moment the expression *organic life* is employed we encounter an example of a *biotic unity and multiplicity*. Every living entity (note the implied modal universality of the biotic aspect present in this way of addressing the issues) has *multiple organs* that are *united* in its biotic functioning. Every living entity is a *biotic whole* with different *parts*. This statement explores the *spatial analogy* within the structure of the biotic aspect, because as such it appears first in the aspect of space. The biotic endurence or persistence of any living entity analogically reflects the core meaning of the kinematic aspect within the biotic aspect. Although the vitalist tradition intended to reify the biotic mode into an *immaterial vital force* it did not realize that the term *force* is derived from the core meaning of the physical aspect, for when energy operates certain forces come into play, causing certain effects.

The *biotic strength* of a living entity sercures its *survival* potential. For that reason the widely known term *survival* displays the inter-modal coherence between the biotic aspect and the foundational physical aspect.

In other words, the modal universality of the biotic aspect comes into sight in each instance of a biotic function that applies to all possible living entities. Every living entity is bound to the biotic time order of birth, growth, maturation, ageing and dying. The moment the distinct type laws of living entities are included in our perspective we have to take notice of significant *typical* differences, for example displayed in the different life expectations of different kinds of living entities – from one day up to more than thousand years. *Biotic growth* is the embodiment of *increasing subdivisions* accompanied by the *specialization* of distinct organs (or organelles in the case of unicellular entities). Note that the term *increase* analogically reflects the numerical meaning of *more and less* while the term *subdivisions* analogically reflects the meaning of the spatial whole-parts relation (with its implied divisibility). The numerical and spatial foundation of biotic entities therefore make possible basic biological concepts such as *biotic growth* that can be further specified by the expressions *biotic differentiation* and *biotic integration*. The thermodynamics of physically *open systems*, that allow for a continuous interaction with a given environment (known as a *steady state*), is analogically reflected in the modally universal biotic trait of *feeding* (consider the metabolism – anabolism and catabolism taking place within the cell). In spite of the building up and breaking down that take place, every living entity continues its existence, bound to its *typical* life-span.[1]

The legacy of idealistic morphology twisted the idea of a type law (*Bauplan*) into something supra-sensory and static (an element of the thinking of Ray and Linnaeus). Schindewolf does not fall into this trap, but still got accused of it by Mayr in 1963 in his book on *Animal Species and Evolution* (page 673). In 1969 Schindewolf pointed out that typological thinking should not be identified with the Platonic tradition (see Schindewolf, 1969:66), but as recent as 2002 we still find the same one-sided and misguided definition of 'typological' in Mayr's work: *What Evolution Is* (see Mayr, 2002:319).[2]

1 Where Scheele employs the term *typological* he opts for both *constancy* and *variation*: "Our conclusion therefore must be that much rather we have to mention that there is a typological *variation* or *differentiation* that cannot exceed its own boundaries, but which does provide within those limits a large number of possibilities" (Scheele, 1997:117). Also Eisenstein holds that the within the concept of *constancy* the concept of variability is enclosed (Eisenstein, 1975:278).

2 In passing we may note that on page 30 of this work from 2002, Mayr still uncritically included the original (1874) illustration of vertebrate embryos by Haeckel – which turned out to be fraudulent. Wells mentions the dishonesty in Haeckel's sketches: "Haeckel entirely omitted the earliest stages of development in which the various classes of vertebrates are morphologically very different. Biology teachers should be aware that Haeckel's drawings do not fit the facts … it ignores groups that did not neatly fit into Haeckel's scheme" (Wells, 2003:179, 181).

Schindewolf quotes Van Valen who claims that Archaeopteryx is "precisely intermediate in structure between birds and reptiles" and then provides arguments why it in fact belongs to the Aves, including the fact that some of the 35 bird species known since the cretaceous period (derived from 13 families and 8 orders) existed up to the present (about 80 million years) (Schindewolf, 1969:73).[1] He is adamant that Archaeopteryx is a genuine bird and does not reside somewhere in "no man's land" (Schindewolf, 1969:79).

On a more general level he holds, against the nominalistic orientation of neo-Darwinism, that the systematic types (from the species and genera up to orders and classes) are real, concrete entities open to *morphological research*, for otherwise it would be untenable nonsense (*unhaltbarer Unsinn*) (Schindewolf, 1969:83). He reminds us that amongst paleontologists it has long been known that the development of phyla did not proceed in a uniform, continuous tempo (Schindewolf, 1969:88). During the last 50 million years nothing essentially new occurred with the mammals (Schindewolf, 1969:89). But the abrupt appearance of new types must be representative of an exceptional acceleration, for otherwise not enough time would be available. Schindewolf gives the example of *Chiroptera* (the Order of bats). Its *Bauplan* is complete and known since the middle Eocene, but during the subsequent 45 million years no significant changes took shape. If the large distance between this radically new type and its predecessors would have been accounted for in terms of the minute, small and slow changes envisaged by Simpson, then the beginning of this development would have been pushed back to long before the origination of the earth (Schindewolf, 1969:90).

He then continues his argument by mentioning numerous examples of forms that remained constant over millions of years. For example: *Rodentia* – since the Paleocene they did not change for more than 50 million years. In all the cases of enduring existence the essential features remain typically constant (Schindewolf, 1969:115).

The *Coelacanth*, that was supposed to have died out 65 million years ago, until it was found in 1938 off the coast of Madagaskar – still was identical to the fossils of 65 million years ago. Scheele mentions sharks that have not changed over millions of years (Scheele, 1997:93). Army ants have not changed during 100 million years.[2] *Pleisiosaurus* exhibited few changes during a period of 135 million years (Dixon, *et al.*, 1988: 76-77). Sea turtle, 110 million years old, did not change during this period.[3] Blue-green algae (*Cyanobacteria*) are known as the oldest "living fossils," dated to be 3.5 billion years old, yet "they are essentially identical to the blue-green algae that are still living today."[4]

1 He refers to the fact that the wing of Archaeopteryx had the same number of "Handschwingen" than the modern flying birds (Schindewolf, 1969:75).

2 This example and the following 3 are also mentioned by Mortenson, 2006.

3 "Fossils Shed Light on Sea Turtle Evolution." Animal.discovery.com Feb. 24, 2005.

4 See the web site of the *Museum of Paleontology* of the University of California, Berkeley.

What is of course presupposed in the relative constancy of the fossil record is the invariance (constancy) of physical laws. The supposition that the laws and law-conformities currently in force also apply to the past is known as "actualism" (see Schindewolf, 1969:106).[1] However, the overall problem presented to paleontology and biological theory is that the *assumed* rapid changes (such as those intended by the idea of *punctuated equilibria*) are not currently observed and therefore would remain pure speculation.

Gould mentions Charles Oxnard who "studied the shoulder, pelvis, and foot of australopithecines, modern primates (great apes and some monkeys), and *Homo* with the rigorous techniques of multivariate analysis" and who concluded "that the australopithecines were 'uniquely different' from either apes or humans, and argues for 'the removal of the different members of this relatively small-brained, curiously unique genus *Australopithecus* into one or more parallel side lines away from a direct link with man' " (Gould, 1992: 60).[2]

This instance is just one of the many others known to us, which Gould in general describes as "the 'sudden' appearance of species in the fossil record and our failure to note subsequent evolutionary change within them" (Gould, 1992:61). It sounds pretty strange to designate the constancy in terms of a negative qualification, namely "our failure to note subsequent evolutionary change within them"! Apparently the paradigmatic commitment to *change*, at the cost of *constancy*, still burdens neo-Darwinian thought.

The embarrassment with *constancy* as dominant pattern of the fossil record caused Gould to take refuge in the allopatric theory according to which at "another place" "new species arise in *very small* populations that become isolated from their parental group at the *periphery* of the ancestral range." This gives rise to "[S]peciation in these small isolates" that occurred in a way that is "*very rapid* by evolutionary standards" (Gould, 1992:61). Without any *factual support*, this speculative claim then serves to side-step the said embarrassment with the "sudden appearance of species in the fossil record and our failure to note subsequent evolutionary change within them" as "the proper prediction of evolutionary theory as we understand it" (Gould, 1992:61).This assumption of Gould is similar to the embarrassment with the origination of the first living entity. Wilhelm Troll categorically states, in his standard text book on Botany, that the question concerning the origination of life on earth, owing to its speculative nature, does not belong to the domain of biology as an *empirical* science (Troll, 1973:8-9).

Clearly, the general pattern of the fossil record, *discontinuity* and the *stasis*

1 Gould points out that this assumption formed the basis of Leyll's 1830 work on the *Principles of Geology* where it is presented as Leyll's uniformitarian principle (Gould, 1992:148).

2 Gould concedes that "we must recognize three coexisting lineages of hominids *(A. africanus*, the robust australopithecines, and *H. habilis)*, none clearly derived from another" and "none of the three display any evolutionary trends during their tenure on earth: none become brainier or more erect as they approach the present day" (Gould, 1992:60).

or *constancy* of whatever appeared, leave us without any data on the basis of which we can account for the *gaps* that can no longer be explained away geologically (Kitts). The answers to the most burning questions therefore lie hidden below the surface of abrupt appearance and continued (unaltered) existence, and if we do not want to become victims of idle speculation, intellectual honesty may help natural scientists to sincerely confess our *docta ignorantia* (learned ignorance) – acknowledging that in the natural scientific sense of the term *we do not know what really happened.* We have to conclude that Thorpe was right: *fixity* (*constancy*) indeed sticks out like a sore thumb in neo-Darwinian evolutionary theory.

7.15 Concluding remarks

While nominalism proceeds from the assumption of a structureless continuum (each organism is wholly unique and cannot be forced into some or other universal ontic form), idealistic morphology accepts "primal types" (e.g. a primal leaf, a primal plant, or a primal animal) which serve as genuine Platonic models with reference to which any empirically observed living thing or fossil has to be judged.

The idea of a type law that holds, as a typical total structure, as the law for the entities subject to it, represents a structural theory that aims at overcoming the one-sidedness in both a realistic (idealistic) and nominalistic approach. The structureless continuity of a nominalistic vision simply does not allow for relatively constant structural types. Just as a modal-physical law cannot be identified with any subject function or concrete subject, the structural types of plants and animals cannot be identified with particular concrete plants or animals. However, entity structures are types that are embedded in the cosmic dimension of time, still finding their correlate in the succession of transient individual living creatures that appear on the paleontological horizon during the course of earth's history. The psychic-sensitive qualifying function of animals is expressed in their total life orientation. As mentioned, Portmann typifies animals aptly when he says that they are *instinctually-secured* and *milieu-bound* (Portmann, 1969:86).

In spite of its pre-occupation with change neo-Darwinian biology is confronted with the fact that both within biology and paleontology the acknowledgement of constancy is of equal importantance.

Chapter Eight

Human society

Unavoidable philosophical presuppositions

8.1 Individual and society

In Chapter 2, our brief discussion of the problem of unity and diversity within various academic disciplines mentioned, amongst others, the opposition between *atomism* (individualism) and *holism* (universalism) within the discipline of sociology. We argued that these two *isms* respectively over-extend the discrete meaning of number and the original whole-parts relation found within the aspect of space.

In our questioning of key elements within the broader intellectual legacy of the West during the past five centuries, we argued in particular that the all-pervasive influence of modern *nominalism* should be taken into account, for this intellectual orientation effectively *eliminated* every *order for* and *orderliness of* (natural and social) entities within the universe (i.e. *outside* the human 'mind'). We pointed out that this *nominalistic* heritage reached its rationalistic peak in the thought of Immanuel Kant, who elevated human understanding to become the (*a priori*) *formal* law-giver of nature. But then, during the 19th and early 20th centuries, the *irrationalistic* leg of nominalism gained the upper hand in the emergence of *historicism* and the *linguistic turn*. Irrationalism plays an important role in the way in which Smart analyses postmodern social theory. It is supposed to challenge the charisma of modern reason, criticize identity thinking and question the conceptualisation of totality (Smart, 2000:447 ff.).

The effect of this development was that the ontic normative conditions for human societal relationships were *transposed to* and *viewed as* the *product* of *human construction*. Theorising about human society transformed the Kantian and neo-Kantian dualism between 'is' and 'ought' (*Sein* and *Sollen*) into a *separation* of (a-normative) *factual social reality* ('structures', 'systems') and the (*subjectively constituted*) domain of *meanings*, *norms*, *values* and *beliefs* – captured by the general basket-category *culture*. Whereas the initial position of the Baden School of neo-Kantianism (Windelband, Rickert, Weber, and others) still held on to supposedly *ideal* and *timeless* values, the latter soon became fully *historicized* and *relativized* through the emerging idea of *changing lingual* and *social constructions*. Within this picture, the radically opposing perspectives of *sociological individualism* and *sociological universalism* continued an intellectual dilemma dating back long before modernity.

Although Callicles[1] and Protagoras[2] pursued the path of *individualism* already in Ancient Greek culture, the societal wisdom of Greek philosophy culminated in the (*universalistic*) ideal of the city-state, the *polis*, which was supposed to be the all-encompassing *totality* of society, leading the citizen to *moral* perfection. In the course of the fifth century AD, the Roman empire experienced the invasion of the Germanic tribes that eventually caused the final collapse of the western part of the Roman Empire.

> It was the Huns who drove the Goths over the borders of the Roman Empire. The latter defeated the army of Valens at Adrianople in 378. Elton remarks: "Once one group of barbarians had entered the Empire, the Romans could not muster the military strength to keep others out. Vandals, Alans and Suevi crossed the Rhine in 406 and barbarians went on to settle all over the western Empire. Visigoths, Alans and Suevi took land in Spain, Vandals in Africa and Burgundians, Visigoths and Franks in Gaul. Elsewhere, Saxons invaded Britain and at the end of the fifth century, Ostrogoths occupied Italy" (see Elton, 2007). Of course the Eastern part lasted until 1453.

The initial Roman idea of a holy empire (*sacrum imperium*) was continued in the Byzantine Empire, and since Charlemagne (800) and his successors, it returned in the shape of the idea of the *Corpus Christianum*, as the *perfect society* (*societas perfecta*). In all of this, medieval society persisted in a relatively undifferentiated state, further enhanced by the rise of the feudal system. In the process of conquering many countries, the Frankish king laid claim to unoccupied land and then started to hand it out to servants and the nobility as a reward for their support during the wars.

This gave birth to the *feudal system*, in which the owners of large pieces of land acquired within their own domain an exclusive authority in military, judicial, and political affairs. Yet governmental authority was still viewed as a *private entity*. As such, this view precluded a *territorial monopolization* of governmental power – which was essential for the rise of the modern (idea of

1 In his "aristocratic nominalism" Callicles derives from nature the right of the strongest. He opposes the legal order of the state, because it only serves to suppress the strongest through making laws. He therefore admires the tyrant, because the latter breaks through positive laws and subjects the weak to his power as law (in a sense anticipating ideas about 'superman' formulated much later by Nietzsche in the nineteenth century). The tyrant alone is entitled to rights – all the citizens are deprived of any rights and subject to the arbitrariness of the tyrant – this time anticipating the position taken by Hobbes in his *Leviathan* (1651).

2 Protagoras elaborated the Greek nominalistic movement by viewing the human person as being in a constantly changing state that cannot be grasped in any fixed form or measure (every individual is his or her own measure). Only the polis, as bearer of the Greek motive of form, measure and harmony, is capable of supplying the human being with a cultural garb through education and obedience to positive laws. This explains why he holds that human beings, coming from a condition in nature where the state is absent, have the properties necessary for the formation of a state – but not on the basis of a "social contract" (see Menzel, 1929 and 1936). Although Protagoras proceeds from a nominalistic individualistic starting point, his conception of the state does not acknowledge any material boundaries for the competence of the state – even morality and religion are viewed as products of the state.

the) state later on. Portions of governmental authority were spread over cities, guilds and market communities. The Frankish empire of Charlemagne viewed itself as the successor of the Roman Empire, but its division in 843 paved the way for the powerful counts and dukes – in combination with the church – to develop into the real bearers of governmental authority during the subsequent medieval period. Within the guilds, religious and professional interests were integrated in a relatively undifferentiated unit. It was only during the twelfth and thirteenth centuries that international traders (merchant guilds) become influential. Nonetheless, they were succeeded by craft guilds and various traders of a particular craft.

During the medieval era, the attempted synthesis of Greek culture and biblical Christianity led Roman Catholicism further on the universalistic path by superimposing upon the state (with its ideal of moral perfection) the church as a supernatural institute of *grace* – thus not only expanding the idea of the *societas perfecta* (perfect society) but also incorporating it in the *Corpus Christianum* view of the church, which is supposed to complement moral perfection (the natural portal) with (supernatural) *eternal bliss*. The persistence of this view is still found in the famous papal encyclical, Quadragesimo anno (15 May 1931), which explicitly states: "Surely the church does not only have the task to bring the human person merely to a transient and deficient happiness, for it must carry a person to eternal bliss" (cf. Schnatz, 1973:403).

At this point, the Renaissance emerged influenced by the rise of modern *nominalism* (John the Scott, William of Occam). The latter movement challenged the conception of eternal (Platonic) forms in God's mind, as well as the hierarchical structure of the ecclesiastically unified medieval culture. Secular Humanism enthroned the supposedly autonomously free human personality, asserting its authority by implementing a natural scientific (mathematical-physical) mode of analysis – proceeding step-by-step from the simplest elements to the more complex levels. The upshot of the initial successes of the natural science ideal was that human life was viewed as subject to inviolable natural laws, foremost that of cause and effect, i.e. the law of causality (see pages 35 ff. above). It also gave birth to the idea that not only the universe, but also human society ought to be constructed from its simplest elements, the *individuals*. Von Bertalanffy sketches the overall picture when he writes, "First came the developments of mathematics, and correspondingly philosophies after the pattern of mathematics – *more geometrico* according to Spinoza, Descartes and their contemporaries. This was followed by the rise of physics; classical physics found its worldview in mechanistic philosophy, the play of material units, the world as chaos, ..." (Von Bertalanffy, 1968:66).

The political philosophy of John Locke (based upon his atomistic contract theory) and the ideas of the classical school in economics (Adam Smith and his followers) were both in the grip of the natural science ideal. The influence of the science ideal is clear from Viner's characterization: "The claim to fame of Smith in the first place therefore appears to have a foundation, because he has applied the conception of a uniform, natural order just as comprehensively to the world of economics; an ordering that functions on the basis of a natural law and, if left to its own functioning, will be beneficial to humankind (Viner, 1956:92)." Rousseau was the exception, because his contract theory started in an *atomistic* way, but then allows the contract to produce a moral-collective whole in a typical universalistic fashion (the "body politic," the "volonté générale"). With law turned into an expression of the general will manifesting itself only within the state, Rousseau – in spite of his apparent intention to secure individual and societal freedoms – succumbed to a totalitarian and absolutistic view in which those who are not conforming to the general will (which is supposed to be their own will – since freedom is defined as obedience to a law that we have prescribed to ourselves) will be "forced to be free" (see page 510 below).

Yet the dominant spirit of the 18th century, the era of the *Enlightenment*, was one of *rationalistic individualism*. Early Romanticism reverted to an irrationalistic individualism – but the anarchistic consequences of such a position soon inspired an irrationalistic universalism, where each transpersonal community is viewed as a law just for itself (the so-called transpersonalist, freedom-idealism of Schelling, Fichte and Hegel). Western civilization here witnessed the modern ideology of community for the first time – further explored in the 20th century by *Nazism* and *Fascism*.[1]

In the meantime the extremely historically significant *industrial revolution* took place. Yet, owing to the guidance of the classical liberal idea of the state (John Locke) and the classical school of economics (Adam Smith) – *laissez-faire, laissez-passer* – the newly emerging industrial societies abstained from protecting the *economic legal interests* of the workers, and thus gave birth to the labour movements. These eventually entered the political scene in the form of political parties (labour parties) directed at the *sectional interests of labourers* through trade unions – and thus rendered a 'service' to the genesis of the prominent totalitarian régimes[2] of the early 20th century (with their universalist ideologies – in the Italy of Mussolini and Germany of Hitler), finally resulting in the *Second Word War* that forced those Western states with a democratic legacy to enter into this war as well.

The mechanistic worldview paved the way for the eventual emergence of a *technicistic* worldview. The preceding era of handcraft (trade) had a focus on trade tools, but the technical activities of the craftsman, the smith and the carpenter, were still encapsulated within a non-technical context (see Schuurman, 1993:191-192). Within the guild system, there was not yet a differentiation between 'capital' and 'labor'. However, the eventual 'division' of 'capital' and 'labor' gave rise to the labourer that emerged as being stripped of all but one of its former 'assets': *labour power* (*skills*). The development of modern industrial technology during the "industrial revolution" therefore played a decisive role in shaping the social complexity of modern Western societies. Backed up by the ideal of progress emerging from the Enlightenment era,

1 See Koyzis 2003 for a critical analysis of contemporary political ideologies.

2 See the remark on absolutism and totalitarianism on page 547 below.

modern technology soon turned into something upon which trust for the future can be built. The ideological expectation that technology can serve as the redeemer from the defects within society accompanied the rise of modern technology. Schuurman characterises technicism as follows:

> Technicism entails the pretension of human autonomy to control the whole of reality: man as master seeks victory over the future; he is to have everything his way; he is to solve all the problems, including new problems caused by technicism; and to guarantee, as possible consequence, material progress. ... Technicism reduces science to its instrumental use. The economy, as is obvious in Western culture today, is also interpreted technicistically, with utilitarian economics as a complement (Schuurman, 1995:140).[1]

From the preceding sketch, it is clear that the opposition of *individualism* and *universalism* not only constitutes more than a theoretically contested issue, cutting across the significant contours of the development of Western civilization and societies themselves,[2] because it is also intimately connected with the rise of modern *technicism*.

Clearly, the relationship between 'individual' and 'society' constitutes a theoretical and a practical concern. Therefore it is all the more important to attempt to arrive at a satisfactory theoretical understanding of being human within human society. At the same time, it will become clear that every attempt to come to terms with this relationship illustrates the decisive role of foundational philosophical assumptions.

8.2 A false opposition: individuals versus supra-individual totalities

Atomistic approaches deny anything beyond concretely existing *individuals*, whereas holism promotes the idea of societal wholes as *supra-individual totalities*, as *more-than-merely-individual* social realities.

The German sociologist, Ferdinand Tönnies, absorbed both these extremes in his *genetic* distinction between *Gemeinschaft* and *Gesellschaft*.[3] The *British* sociologist, Herbert Spencer, articulated an *individualistic organicism*. By viewing an organism in individualistic terms he believes that it is much easier to see society also as an organism: "On thus seeing that an ordinary living organism may be regarded as a nation of units that live individually, and have considerable degrees of independence, we have less difficulty in regarding a

1 He points out that the process of making everything technological is evident almost everywhere: "in the dehumanisation of labour, in the devastation of nature, in the pollution of the environment, in problems related to nuclear energy, in making society information-compatible, in problems related to farming and meat production, and in problems surrounding genetic engineering " (Schuurman, 1995:86). See also the extensive discussion of the nature of modern technology in Schuurman (2009:131-136).

2 Elements of the above historical contours are also discussed by Min-Sun (2002:10-13)

3 A period of *Gesellschaft* (community) is supposed to follow a period of *Gemeinschaft* (society) (Tönnies, 1965:251). In his *Introduction to Sociology*, he introduces a systematic (non-genetic) distinction, namely between *Samtschaften* ('collectives'), social relationships (*Verhältnisse*), and social 'bodies' or organizations (*Körperschaften*) (Tönnies, 1965:XLV ff.).

nation of human beings as an organism" (Spencer, 1968:54-55). The French thinker, Emile Durkheim, gave priority to his *holistic* understanding of the collective conscience (*conscience collective*).[1] In particular, we noted that the neo-Kantian thinker, Max Weber, continued the *nominalistic* tradition without any hesitation, in terms of which it is always possible to understand "communal human actions" as being reducible "to the actions of the individual human beings concerned" (see Weber, 1973:439).[2] In direct opposition to this *atomistic* (nominalistic-individualistic) orientation, Alexander finds it perfectly meaningful to refer to "structures separate from the individuals who compose it" (Alexander, 1987:10-11).

Those involved in reflecting on the nature of human society meet this problem in the opposition between 'action' and 'order'. Yet an analysis of the (apparent) opposition between 'individual' and 'society' soon proves more complex than it may seem at first. Given the *atomistic* emphasis on the *one* and the *many* (individuals), and the *holistic* focus on an encompassing *whole* (*totality*), employing terms that reveal the *interconnections* between the different *facets* or *aspects* of reality is inevitable. If the antithesis between atomism and holism hinges upon the respective emphases on *multiplicity* and *wholeness*, then the first basic question for understanding theoretical views of society boils down to whether there is a 'domain' in reality where the awareness of the *one* and the *many* (*unity* and *multiplicity* / *being distinct*) finds its seat; and similarly: is there an 'area' where the primary (original) meaning of *wholeness* (or *totality*) is located?

The fact that we argued that these terms respectively are derived from the quantitative and spatial aspects (see pages 60, 87, 236-237 ff.) once more underscores the scope and significance of the theory of modal aspects for theoretical endeavours, also within the domain of the humanities. Reflection on the nature of human society cannot escape a basic level of theorising involving terms with a universal scope (thus displaying modal universality). Ultimately they refer to constant *ontic conditions* making possible historically variable structures and events. Johnson *et al.* did discerned an element of this insight when they state:

> while these concepts do represent universal, constant features of human action, the particular values or contents they have vary historically, and are problems of empirical research (Johnson *et al.*, 1984:72).

A fundamental layer of reality is unveiled here – those conditions that *make possible* whatever we can experience in *variable historical contexts*. Of particular importance is the modal universality of the social aspect. Peter Berger indeed explores an element of its basic structure when he accounts for the delineation of the field of study of the discipline of sociology:

1 Durkheim views society as a whole that may be assessed from the perspective of two angles: the *mechanical* or *organic solidarity* approach (Durkheim, 1972:138).

2 The views of these sociologists are discussed at length in Strauss (2006 – see pp.135 ff., 205 ff., 253).

> The sociologist finds his subject matter in all human activities, but not all aspects of those constitute this subject matter. Social interaction is not some specialised sector of what men do with each other. *It is rather a certain aspect of all these doings* (my italics – DS). Another way of putting this is by saying that the sociologist carries a special sort of abstraction (Berger, 1982:39-40).[1]

The *social aspect of reality* evinces its own *modal universality*. As a "certain aspect" of "human activities," it has a *universal scope* that is unlimited by any *type* of social interaction in particular – underscoring the difference between modal laws and type laws (see p.25). That is to say, the modal (functional) meaning of the social facet of reality holds universally for whatever there is (in a concrete entitary sense – including all kinds of *events*). Although material things, plants and animals are not *accountably free* agents (like human beings), they have an *object-function* within the social aspect that could be opened up (disclosed), although always correlated with the activities of *social subjects* (*individual human beings* or *societal collectivities* according to their *social subject-function*).

In addition to acknowledging the modal universality of the social aspect as being constitutive for being human, the multi-aspectual nature of being human is also extremely important for a deepened understanding of the juxtapositioning of 'individual' and 'society': an individual is *never* exhausted by any aspect of reality in which human beings function. Stated differently: every *individual* totally exceeds the confines of every modal aspect co-conditioning the individual's existence. Social contract theories of the early modern period – amongst others Hobbes, Thomasius, Pufendorf, Locke and Rousseau – proceed from the fictional abstraction of 'isolated' individuals postulated in order to give a *hypothetical* (and therefore *non-historical*) account of the existing order within known societies – as if human individuals are only incorporated in social interaction in a derived sense.[2] With good reason, George Herbert Mead reacts to this abstraction by emphasizing that the social context (co-)determines human existence from the very outset (Mead, 1967:144 ff.).

Once the existence of the *social aspect* of reality is acknowledged and it is understood that alongside all other modal aspects, the social mode also has an ontic meaning – in the sense that it conditions and makes possible whatever we can experience as *social relationships* and *social interaction* – the humanities (implicitly or explicitly) have to come to terms with an understanding of the multiple different *ways* in which human beings interact within society. What is required, is a theoretical account of different forms of social interaction, articulated in terms of what we designated as the *complex* or *compound* basic concepts of sociology.

1 We argued that lifting out a certain aspect as point of entry delimiting the angle of approach of a particular discipline – while disregarding other aspects as modes of explanation – indeed constitutes the distinctive feature of scholarly activities, designated as modal abstraction (see pp. 48 ff.).

2 Taylor is fully justified in appreciating the theories of Hobbes and Locke as an offspring of late medieval nominalism (cf. Taylor, 1989:82 197).

Different elementary basic concepts should be employed at once in order to perform the classificatory task at hand.

A consistent individualism, which would explore the perspective of 'individuals-in-interaction', proceeds from a notion of action that *denies* the inherent *social function* of any human action. The only option left on the basis of this assumption is to *add* the *social* dimension afterwards, as something *foreign* and *different*. Nontheless when we acknowledge the embracing *transcendental* nature of every modal aspect of reality, we have to start from a notion of the social function of reality that is *co-constitutive* for all the *actions* of *individual* human beings. Less durable relationships are therefore already entailed in the fact that individuals *function* in the social aspect as well. As a structural element co-conditioned by the *way* in which human beings function within reality, inter-individual interactions are, in a truly *transcendental* sense, just as 'social' as the existence of any (supra-individual) *societal whole*.

With reference to the increasing complexity involved in an analysis of the elementary basic concepts of a discipline, the following elements were highlighted (see pp. 242 ff.). The existence of societal wholes enables an acknowledgement of the *identity* of social collectivities over time, for notwithstanding the coming and going of their individual members, a given society may experience social growth (social differentiation and integration). This mode of speech analogically reflects the meaning of thermo-dynamically open systems accounting for constancy amidst change.

Thus the *unity* of a specific societal collectivity obtains another specification when it is asserted that, as a social *whole*, it maintains its *identity* in spite of the coming and going of its members. Furthermore, in terms of the biotic analogy within the structure of the social aspect, every societal collectivity is a *differentiated* social whole capable of *integrating* its social activities over time. This persistence may require the operation of social *organs* competent to accomplish this through an ability to structure societal relationships by means of exercising their *social ordering will* in service of an integrated social solidarity and awareness (consciousness). Clearly this formulation, in addition, employs the (analogical) meaning of the sensitive mode of reality. The degree to which *social consensus* or *social conflict* prevails in a particular social sphere (logical-analytical analogy) is often dependent upon the way in which office bearers *control* the situation on the basis of a proper *interpretation* of the *signs* of conflict.[1]

8.3 Classifying social interaction[2]

Taking into account the foundational coherence between the kinematic and the physical retrocipations, analogically reflected within the structure of the

1 Notions of social power, authority and control analogically reflect the coherence with the cultural-historical function, whereas social signs, symbols and interpretation analogically refer to the sign mode (lingual mode).

2 Compare Strauss, 2006:248 ff.

social aspect, enables the introduction of the property of a *solidary unitary character* of particular social forms of life. In spite of the *constant flow*, i.e. coming and going of individual members of a societal whole, the persistence and identity of the social life form concerned are not eliminated. Ryan (1970:174) says:

> There are regularities and constancies in the behavior of groups of people which allow us to talk about groups having a stable structure in spite of fluctuating membership, and about the existence of social roles which can be filled by different people at different points in time.

Within the context of the *spatial* analogy, one can differentiate between coordinational social relationships (*next-to-each-other*) and social relations of super- and subordination (*above and below*). When both features are present, i.e. a durable organization and a relation of super- and subordination, we meet what in German and Dutch are known as *Verbände/Verbanden*. In the absence of a suitable translation of this term, its intended meaning is captured by employing the familiar phrase *social collectivity*. Social collectivities are well-known. Consider the state: it has a durable relation of super- and subordination, and maintains its identity over time in spite of the fact that individual citizens come and go. The same applies to the *church*, the *firm*, the *school*, the *university*, the (nuclear) *family*, the *art association*, the *sports association*, the *cultural association* and the *language association*.

However, when a societal life form possesses only one of these characteristics, we employ the term *community*. A cultural entity, such as a *nation* ('volk' or 'people') as well as the *extended family* only evince a *solidary unitary character*. By contrast, although men and women are equal in a civil legal sense and within the domain of public law, a *marriage community* appears to possess a permanent authority structure, but lacks a solidary unitary character.[1] In the case of *coordinational* relationships, both properties are absent, for they neither display a solidary unity character nor a relation of super- and subordination. The expression *coordinational relationships* therefore aims to reflect what is meant by the Dutch term '*maatschap*'. This Dutch term also does not have a suitable English equivalent. It concerns the social interaction of friends, partners, fellows, mates (pals etc.), peers, and the liberty human beings have to *associate* with an accountable freedom of choice. It should be kept in mind that coordinate relationships also encompass interactions between diverse communities and social collectivities.

As stated, this distinction between societal collectivities, communities and coordinational relationships represent an example of exploring compound or complex basic concepts, in this case within the discipline of sociology (delimited by the social aspect as its angle of approach to concrete reality). As modal totality concepts, they incorporate constitutive modal analogies within the structure of the social aspect, but they do not contain an additional criterion

1 Alan Cameron suggests a relation of mutual super and sub-ordination – to each other rather than the traditional interpretation of the male as head (E-mail comment – 14-03-2008).

enabling further differentiation. Having identified a state, church, firm, university and social club as societal collectivities, on the basis of the fact that they all display the two characteristic features of societal collectivities (a solidary unitary character and a durable relation of super- and subordination), as such merely focuses on what they share without also highlighting their differences.[1]

The strict correlation between collective and communal societal relationships on the one hand, and coordinational relationships on the other provides a deepened and more precise characterization of the extremes of an atomistic and holistic view of human society. When the latter is reified, we meet a universalistic (holistic) view of society, and when the former is deified, individualism (atomism) surfaces. These distinctions are important in overcoming the opposition of individual and society.

8.4 'Socializing' the individual and 'de-totalizing' society

We noted that an effect of the modern natural science ideal, with its urge to enthrone the human person in supposed autonomous freedom, is that human society was reduced to its simplest elements in order to construct a new world out of these atoms in a rational way. The 'atoms' of human society turned out to be (abstracted and isolated) *individuals*. We also noted that the most important assumption (simultaneously also the most fundamental mistake) in this construction of the science ideal, is that the 'social' dimension of reality does not inherently belong to the world of "being an individual." It was particularly the motive of *logical creation* that transposed universal ontic conditions into the assumed constructive power of the human mind. Thus the *ontic universality* of the social aspect was also eliminated – and we saw that eventually, after the rise of *historicism* and the *linguistic turn*, the theme of the *social construction* of reality surfaced.

As an alternative, we emphasized that being human is always co-constituted and co-determined by the social aspect of reality and that the idea of an *individual* inevitably also at once *exceeds* the confines of any modal aspect – including the *social aspect.* In a sense, this perspective complements the insight that every human person necessarily *functions* within each modal aspect (as articulated more extensively in Chapter 4), because 'exceeding' or 'transcending' entails that being human is never *exhausted* by functioning within any modal aspect.

Nor is it exhausted by being a member of any societal collectivity. In the conclusion of *After Virtue* (1981), MacIntyre claims that three centuries of moral philosophy and one of sociology have failed to provide us with a satisfactory "coherent rationally defensible statement" of a "liberal individualist

1 Stafleu's proposal to introduce a "political aspect" of authority and discipline (see Stafleu, 2004:130) does not recognize that the relation of super- and subordination is partially a spatial analogy alongside the next-to-each other as a different partial spatial analogy. These spatial analogies exist within every normative aspect, and therefore cannot serve the idea of a "political aspect." (Also see the critical discussion of Basden, 2005:70 ff.).

point of view" (also see MacIntyre, 1988:ix). The alternative Aristotelian legacy put forward by MacIntyre unfortunately opts for the (universalistic) view that, when a person is separated from social ties, such a person also lacks justice (*themis*) – with reference to Homer and Sophocles, as well as Aristotle himself (see Aristotle, 1894:1253a6 and Homer, *Iliad* IX, 63) (MacIntyre, 1988:96).

This alternative to the atomistic or individualistic idea of an isolated individual is that *functioning* within the social aspect (albeit in a norm-conformative or antinormative way) *inherently* belongs to being human in the sense that no single human being *lacks* a social function, not even when such a person is *alone*.

Note also the difference between *a-social* (= *non-social*) and what is considered socially *antinormative*. Within the jural aspect, for example, we find the opposing forces of *legal* and *illegal*, and within the logical-analytical aspect, those of *logical* and *illogical*. Unfortunately, there is no English equivalent for the mentioned 'il-' forms in the context of the social aspect. Since the opposition of *polite-impolite* does reflect the normative meaning of the social aspect, a more general characterization is needed. The word 'unsociable' is too close to "a-social," although the word "un-social" (similar to *illogical* or *illegal*) does exist in the English language. Therefore we should contemplate for a moment the equivalent of the opposition of *economic–un-economic* in the form of *social – un-social* in order to understand the difference between "a-social" and "un-social." Just as little as *illogical* concepts (see page 93) and *illogical* arguments cease to be cases of *thinking*, do antinormative social actions cease to function *within* the social aspect of reality. To be *impolite* continues to be a form of social action, whereas to buy something or to love someone are clearly *non-social* activities which of course, as activities, still have a function within the social aspect.

In addition, we have shown that the modal structure of the social aspect has its foundation, amongst others, in the numerical and spatial aspects of reality. For this reason, the function of human beings in the social aspect is simultaneously connected with both *social unity and multiplicity* and with *social wholeness* or *social totalities*. Instead of artificially separating these two analogical structural moments and changing this separation into the two mutually exclusive *isms* of atomism and holism, they ought to be understood in their *uniqueness* and *mutual coherence*.

For that reason, we have had to explore the nature of the compound basic concepts of sociology as a discipline in order to distinguish between different *ways* of social interaction (compare the above-mentioned distinction between collective, communal and coordinational relationships). Combining all these considerations effectively amounts to an acknowledgement of the inherent and constitutive human function within the *social aspect of reality* (alongside all other aspects in which a person concretely functions) – they imply the intrinsic *socialization* of the *individual*. In addition, we still have to address the basic and irredeemable flaw in the overestimation of the spatial whole-parts relation (or its analogies within post-spatial aspects), for this distortion consistently resulted in elevating some or other societal collectivity (or even 'society' itself) to become the all-encompassing *whole* or *totality* of society.

The positive side of this universalistic legacy, from Aristotle onwards, is that it did accept the "built-in" social nature of "being human" – just recall Aristotle's "political animal." This mode of thought invariably results in a 'totalizing' view of human society – a perspective in terms of which the *state*, the *church*, or the *nation* (in the ethnic sense of a cultural community – compare Nazism) is seen as the all-encompassing *societal totality*, such that every other societal entity is denatured to become merely a part of the privileged whole.[1]

What Münch calls the distinct "inner laws" of integrated spheres of life (see page 533 below), should indeed be appreciated as a key to transcend the deficiencies in the distortion of all universalistic views of human society. This idea indeed questions the assumption of an all-embracing and all-encompassing totality within society. The key question concerns the *limitations* of the idea of a whole and its parts (compare the example of *NaCl* mentioned above – see page 466).

For a proper understanding of differentiated societies and the place of human beings functioning within them, it is necessary to further explore the difference between modal laws, type laws and their relation with the whole-parts concept. We investigate these issues by first considering some problematic distinctions in the thought of John Rawls.[2]

8.5 Rawls's view of justice and the basic structure of society

Insofar as Rawls understood the opposition between *just* and *unjust* in its universal scope, applying to individuals, associations and the "*basic structure of society*," he clearly came close to an acknowledgement of the *modal universality* of the jural aspect. In its modal universality, the jural aspect underlies the typical function of every societal collectivity within this aspect, and it also embraces every individual human being that functions within this aspect.

8.5.1 *The background of Rawls's theory of justice*

On the one hand, John Rawls's theory of justice is indebted to diverse sources in (political) philosophy, and on the other it produced a configuration in which many key elements within this tradition are transformed. From Greek political thinking, he inherited his view of justice as a *moral virtue*. An awareness of norms and principles leads him to a characterization of the human being as a *moral person* – closely related to the Kantian view of moral autonomy and the categorical imperative (the issue of *constructivism*). From early mod-

1 See Strauss, 2002:102-113 for a more detailed analysis of the role of the whole-parts relation in the thought of some prominent sociologists.

2 Talisse quotes the philosopher Thomas Nagel, who says that John Rawls is "the most important political philosopher of the twentieth century" (Talisse, 2001:5).

ern theories of a social contract, he takes the idea that the initial agreement is *hypothetical* and *non-historical* (Rawls, 1996:271). The way in which Rawls deviates from the radical atomistic (individualistic) orientation of modern contract theories (Hobbes, Pufendorf, Locke, Rousseau and Kant), is found in his assumption that the contracting parties are *heads of families*: "For example, we can assume that they are heads of families and therefore have a desire to further the well-being of at least their more immediate descendants" (Rawls, 1999:111).[1]

This deviation points in another direction. Rawls does not want to proceed merely on the basis of *individuals* in his theory of justice, for he claims repeatedly and emphatically that the *basic structure of society* is the *primary subject* of justice. The *basic structure of society* is acknowledged throughout his argumentation[2] – almost always understood as being the *subject* of justice. Yet occasionally it does happen that he inverts this relation of priority, for example when he states that a "theory of justice depends upon a theory of society" (Rawls, 1978:84).

The importance of the basic structure of society is also seen in the fact that the veil of ignorance and the original position are to deprive those participating in the contract of any knowledge of their particular position within society. This provision only makes sense if the entire argument rests on the background assumption that the basic structure of society has room for diverse positions. This contract theory therefore seems to be "well-informed" by background assumptions regarding the *basic structure of society*, since the latter underlies his emphasis that participants should not know anything about their own position in it.

Antecedent to the contract, there are no "principles of justice" in force. Justice is also not simply an extension from an individual to society as a whole, because it can only emerge from a *joint decision* by rational individuals: "Instead of supposing that a conception of right, and so a conception of justice, is simply an extension of the principle of choice for one man to society as a whole, the contract doctrine assumes that the rational individuals who belong to society must choose together, in one joint act, what is to count among them as just and unjust" (Rawls, 2001:132).

1 In the 1978 edition of A Theory of Justice, it is said that "we may think of the parties as heads of families" (Rawls, 1978:128, 146). This immediately reminds us of Aristotle's view of the family as the "germ cell" of society. A critical assessment of the assumption regarding the heads of families is found in Brennan and Noggle (2000:48-50). Regarding Aristotle's view, see his Politica, Book I (1252a ff., Aristotle, 2001:1127 ff.).

2 Compare the following references to the basic structure of society: Rawls, 1999: 3, 6, 7-8, 9, 10, 17, 18, 32, 38, 39, 47, 48, 49, 53, 57-58, 72-73, 81-82, 82, 89, 91, 93; 95, 95, 95-96, 99, 136, 154, 156-157, 222, 229, 293-94, 308, 365, 409, 511; Rawls, 1996:223, 229, 286, 296, 301, 309, 321, 322, 330, 376, 391, 412, 413, 416, 417; Rawls, 2001:130, 134, 156, 164-165, 167, 174-175, 226, 229, 232-233, 234-235, 235, 250, 255, 256, 278, 308-309, 317, 324, 337, 337-338, 339, 356, 362, 365, 367-368, 390, 391-392, 397, 402, 413, 417, 426, 474, 482, 486, 493, 493, 575, 577-578, 583-584, 584-585; 595-596.

In the current context, we leave aside the problematic consequences of this approach, which is actually indebted to the position taken by Rousseau, who saw in the social contract the basis of all rights obtained in the post-contractual condition. At the same time, Rousseau claims that the social contract assigns to the body politic (the *general will*) an absolute power over all its members. Combined with his definition of 'freedom' as "obedience to a law which we prescribe to ourselves" (Rousseau, 1975:247), this "absolute power" results in the known impasse of his thought, for any person who deviates from the general will is actually disobedient to that person's own will, and as a result must be forced to obey it in order to be free.[1] Not having knowledge of one's position does not entail that the structured multiplicity of *possible positions* within the social system are *eliminated* as well, because this "basic structure of society" is the presupposition of the entire argument.

> No one deserves greater natural capacity nor merits a more favorable starting place in society. Of course, this is no reason to ignore, much less eliminate, these distinctions. Instead, the basic structure can be arranged so that these contingencies work for the benefit of the least fortunate. Thus, we are led to the difference principle if we wish to set up the social system so that no one gains or loses as a result of his arbitrary place in the distribution of natural assets or his initial position in society without giving or receiving compensating advantages in return (Rawls, 1999:87).

What is assumed about society in the background becomes manifest in the two principles of justice emerging from the contract (what Rawls "believe[s] would be agreed to in the original position" – Rawls, 1999:52). The first principle of justice embodies the idea of basic (free and equal) liberties: "First: each person is to have an equal right to the most extensive scheme of equal basic liberties compatible with a similar scheme of liberties for others." The second is specified as "social and economic inequalities are to be arranged so that they are both (a) reasonably expected to be to everyone's advantage, and (b) attached to positions and offices open to all" (Rawls, 1999:53). In his *Political Liberalism*, Rawls explains that his "statement of these principles differs from that given in *Theory,*" for in this work, it reads: "a. Each person has an equal claim to a fully adequate scheme of equal basic rights and liberties, which scheme is compatible with the same scheme for all; and in this scheme the equal political liberties, and only those liberties, are to be guaranteed their fair value" (Rawls, 1996:5); and: "b. Social and economic inequalities are to satisfy two conditions: first, they are to be attached to positions and offices open to all under conditions of fair equality of opportunity; and second, they are to be to the greatest benefit of the least advantaged members of society" (Rawls, 1996:6). In the *Preface* to the revised edition of *Theory*, Rawls succinctly formulates his ripened conception by referring to the "the principle of the equal liberties and the principle of fair equality of opportunity" (*Preface* to *Revised Edition*, Rawls, 1999:xiv).

1 "... ce qui ne signifie autre chose sinon qu'on le forcera à être libre" (Rousseau, 1975:246)! ["... This means no less than that such a person would be forced to be free."]

It is clear that the entire contract theory and its accompanying (agreed upon) principles of justice presuppose the underlying idea of the *basic structure of society*, i.e. of a differentiated society with its inherent social and economic stratification. It should be noted, however, that Rawls himself never uses the phrases "differentiated society" and "social stratification." This is remarkable, particularly in the light of the fact that in both *A Theory of Justice* and his *Collected Papers*, he frequently employs the accompanying idea of the "social system."[1] In sociological systems theory (for example that of Parsons), the concepts of a "differentiated society" and "social stratification" are indispensable (see Strauss, 2006:146 ff., 173 ff.). This idea is frequently articulated by employing the phrase: a *well-ordered society*. What is also intimately connected to these ideas is his notion of the *primary goods* of society, for the latter concerns both his principles of justice and their implication for the basic structure of society. He actually differentiates between two domains of application: "... the two principles of justice assess the basic structure of society according to how its institutions protect and assign some of these primary goods, for example, the basic liberties, and regulate the production and distribution of other primary goods, for example, income and wealth" (Rawls, 1996:309). His intention is to account for the way in which a well-ordered society is "effectively regulated by a shared conception of justice":

> The publicity of the rules of an institution insures that those engaged in it know what limitations on conduct to expect of one another and what kinds of actions are permissible. There is a common basis for determining mutual expectations. Moreover, in a well-ordered society, one effectively regulated by a shared conception of justice, there is also a public understanding as to what is just and unjust. Later I assume that the principles of justice are chosen subject to the knowledge that they are to be public (§23). This condition is a natural one in a contractarian theory (Rawls, 1999:49).[2]

The works and articles published by Rawls are generally seen as making a contribution to *political philosophy* – but they do not reflect an in-depth analysis of issues belonging to the domain of *legal philosophy*. For example, the word *fair/fairness* employed by Rawls in the phrase "justice as fairness" is not related to the legal or jural context of its semantic domain (including synonyms such as *just*, *reasonable*, *impartial*, *evenhanded* and *non-discriminatory*). His aim is rather to develop a conception designated as "justice as fairness": "The central ideas and aims of this conception I see as those of a philosophical conception for a constitutional democracy" (*Preface* for the *Revised Edition*, Rawls, 1996:xi). His emphasis on *free and equal moral persons*

1 In *A Theory of Justice* 51 times and in *Collected Papers* 66 times – the more specific focus of *Political Liberalism* reduced the occurrences of this phrase in this work to merely 6 occasions.

2 Rawls mentions the fact that Locke restricts the right to vote to those who own property and points out that Locke does not accept equal political rights amongst citizens for he "assumes that not all members of society following the social compact have equal political rights" (Rawls, 1996:287).

shows an affinity with the Kantian idea of *moral autonomy* – where being a moral person entails "having a conception of their good and [being] capable of a sense of justice" (see Rawls, 1999:17). A similar remark applies to his use of the word *justice* itself, because one does not find a systematic analysis of the relation between law and morality, and in particular of what is known as justice in the context of *legal ethical principles* or *principles of juridical morality* (such as *fault*, *bona fides*, *good faith*, and *equity*).[1]

It seems to be crucial for Rawls that his political philosophy is built upon an understanding of the 'principles' or 'conceptions' of 'justice', and what is primarily correlated with them, namely the *basic structure of society*. What does this mean?

8.5.2 *Rawls's 'justice' and its 'primary subject'*

Freeman explains:

> Rawls undertakes to show how citizens in a well-ordered society of justice as fairness can come to acquire a *sense of justice*, a disposition to act not simply according to, but also for the sake of justice, as defined by the principles of justice and the legal and social norms that satisfy them (Freeman, 2003:24).

The closing section of this quotation construes a correlation between 'justice' and the assumption that "legal and social norms" may 'satisfy' the former. A first observation is that Rawls in general always *directly* relates *justice* (or: *principles/conceptions of justice*) to something *subject* to it – in the current context: the *basic structure of society*. Indirectly, one may of course argue that the arrangement within the basic structure does (or does not) conform to the regulating *principles* or *conceptions of justice*. Yet, putting alongside each other legal norms and social norms is problematic in a different sense, because Rawls employs the phrase "social norms" only once, in a footnote on pages 442-443 of *A Theory of Justice* (1999): "Thus justice as fairness has the characteristic marks of a natural rights theory. Not only does it ground fundamental rights on natural attributes and distinguish their bases from social norms, but it assigns rights to persons by principles of equal justice, these principles having a special force against which other values cannot normally prevail. Although specific rights are not absolute, the system of equal liberties is absolute practically speaking under favorable conditions."

However, Rawls does not hesitate to refer to "social justice" as an alternative expression for the more generally employed term, 'justice'. It appears that the phrase "social justice" is normally related to social arrangements pertaining to the division of advantages and the proper distributive shares: while the term 'justice' without any qualification normally intends to designate what he has in mind with his first principle of justice: "A set of principles is required for choosing among the various social arrangements which determine this division of advantages and for underwriting an agreement on the proper distributive shares. These principles are the principles of social justice: they provide a way of assigning rights and duties in the basic institutions of society and

1 Only in his Theory of Justice does Rawls mention the classical idea of justice: "give everyone his due" (Rawls, 1999:35, 275-276).

they define the appropriate distribution of the benefits and burdens of social cooperation" (Rawls, 1999:4). Compare his second principle.

The first specification of what Rawls actually means when he employs the expression: the *basic structure of society*, is found early in *A Theory of Justice*. It is articulated by explicitly distinguishing between persons who are capable of behaving *justly* or *unjustly* and institutions and social systems that are said to be *just* or *unjust*.[1] However, his topic is *social justice*, which is concerned with the *basic structure of society*, i.e. the major institutions responsible for the distribution of fundamental rights and duties, as well as the "division of advantages from social cooperation" (Rawls, 1999:6). He now proceeds to explain his understanding of the "major institutions" of society:

> By major institutions I understand the political constitution and the principal economic and social arrangements. Thus the legal protection of freedom of thought and liberty of conscience, competitive markets, private property in the means of production, and the monogamous family are examples of major social institutions (Rawls, 1999:6)

From this it is clear that the *social system* or the *basic structure of society* embraces more than merely the major 'political' and 'economic' institutions of society, for "the monogamous family" is explicitly mentioned (reminding us of the *heads of families* participating in the social contract). Yet he is hesitant to ascribe an unqualified universality to the regulating role of justice. His constructivist approach, proceeding from justice as fairness, does not assume universal first principles having authority in all cases. "In justice as fairness the principles of justice for the basic structure of society are not suitable as fully general principles" (Rawls, 2001:532). In fact his remarks evince an awareness of structural differences within a specific society, as well as between different societies: "They do not apply to all subjects, not to churches and universities, or to the basic structures of all societies, or to the law of peoples" (Rawls, 2001:532). Yet in a different context, Rawls speaks of "the nonpublic reasons of churches and universities and of many other associations in civil society" – from which it is clear that churches and universities are part of "civil society" (Rawls, 1996:213). In the next chapter, a different preference surfaces, where Rawls contends that "different principles" are "more suitable" for churches and universities – he nonetheless still places them within the *basic structure*: "because churches and universities are associations within the basic structure, …" (Rawls, 1996:261).

8.5.3 *Rawls's idea of the basic structure of society*

Although Rawls normally does not use a qualifying term when *justice* is at stake, such qualifying terms do emerge as soon as he focuses on structural differences within a particular society, for then the initial reference to social justice is expanded in the use of a phrase such as "political justice": "Typically, a

1 "Many different kinds of things are said to be just and unjust: not only laws, institutions, and social systems, but also particular actions of many kinds, including decisions, judgments, and imputations. We also call the attitudes and dispositions of persons, and persons themselves, just and unjust" (Rawls, 1999:6).

constructivist doctrine proceeds by taking up a series of subjects, starting, say, with principles of political justice for the basic structure of a closed and self-contained democratic society" (Rawls, 2001:532). What is unclear and ambiguous in his mode of speech is the relatively *undifferentiated* way in which he frequently employs the idea of *justice* and the (relatively differentiated) correlated *subjects*.

However, what is also considered to be *subject* to 'justice', appears to be portrayed in questionable terms. The "principles of justice" seem to be capable of expanding their scope by adapting to different kinds of subjects: "Rather, they are constructed by way of a reasonable procedure in which rational parties adopt principles of justice for each kind of subject as it arises" (Rawls, 2001:532). At the same time, he applies the whole-parts relation to society (the "social structure"). The formulation of principles of justice "presupposes that, for the purposes of a theory of justice, the social structure may be viewed as having two more or less distinct parts, the first principle applying to the one, the second principle to the other" (Rawls, 1999:53). Suddenly, the *social system* is said to display different aspects – in the sense that what is usually designated as the domain of political justice represents one aspect while social and economic relations constitute the other aspect.[1] In a different context, another view is advanced, namely when he characterizes the *basic structure of society* as the primary subject of "political justice."[2]

At this point it should be noted that the political philosophy of Rawls indeed exhibits the influence of conflicting views of society, for on the one hand he wants to maintain continuity with *atomistic* (early modern) theories of the social contract, evident in his construction of an a-historical, hypothetical "original position" covered by a "veil of ignorance," and on the other his emphasis on the (presupposed) *basic structure of society* and on society as a *social system* alternatively opts for a *holistic* (or universalistic) view. The effect of this ambiguity is that his understanding of *state* and *society* often exchange roles, for frequently society itself is depicted as being *democratic*. This leveling of structural differences results in the portrayal of citizenship as related to *society* and no longer merely to the state.[3] It is therefore not surprising that the common contemporary practice of referring to *democratic societies* is also

1 "Thus we distinguish between the aspects of the social system that define and secure equal basic liberties and the aspects that specify and establish social and economic inequalities" (Rawls, 1999:53).

2 "... The primary subject of political justice is the basic structure of society understood as the arrangement of society's main institutions into a unified system of social cooperation over time" (Rawls, 2001:596).

3 Rawls speaks of "a democratic society of free and equal citizens" (see Rawls, 1996:30). Later on he remarks: "... we must distinguish between particular agreements made and associations formed within this structure, and the initial agreement and membership in society as a citizen" (Rawls, 1996:275). Note that citizenship depicts "membership in society" and not just state membership. In his work on the law of peoples, Rawls also speaks of "citizens of liberal societies" (Rawls, 2002:58).

amply present in Rawls's thought.[1]

The underlying notion of the social system, understood in terms of the whole-parts relation, dictates an encompassing societal assignment of "rights and duties within the basic institutions of society"[2]. It is also consistent with an embracing understanding of citizenship exceeding the boundaries of the political community as such, for without hesitation, Rawls speaks of citizenship within the basic structure of society: "This fundamental political relation of citizenship has two special features: first, it is a relation of citizens within the basic structure of society, a structure we enter only by birth and exit by death; and second, it is a relation of free and equal citizens who exercise ultimate political power as a collective body" (Rawls, 2001:577). The conception of the *basic structure of society* increasingly turns out to be an encompassing whole embracing its (subordinate) parts, manifest in its "political and social institutions."[3] Seen from this vantage point, it should not be surprising that subordinate roles are assigned to specific institutions by the *basic structure of society*. For example, Rawls sets out to investigate "a particular political conception of justice" by "looking at the role that it assigns to the family in the basic structure of society" (Rawls, 2001:595). What is of particular significance here, is that a "political conception of justice" assigns a role to the family in the basic structure.[4] Rawls also (on the same page) straight-forwardly asserts that the "family is part of the basic structure."

Rawls holds the view that the equal liberties required by the first principle precedes the way in which the *basic structure of society* arranges the inequalities of wealth and authority.[5] He consistently asserts "the priority of the first principle over the second" (Rawls, 1996:291; cf. Rawls 1999:447). In addition, he defends a serial ('lexical') ordering (or prioritization) in the way in which principles are successively realized, even combining the quality of being *absolute* and *holding without exception* to the principles earlier in the order:

> A principle does not come into play until those previous to it are either fully met or do not apply. A serial ordering avoids, then, having to balance princi-

1 In *Political Liberalism*, this expression occurs on pages: 10, 13, 15, 24-25, 30, 33, 36, 38, 40-43, 61, 65, 70, 79, 90, 95, 134, 136, 154, 175, 177, 198, 205-206, 214, 221, 223, 243, 292, 303, 307, 320, 335, 344, 346, 369, 376, 387, 390, 414, 418, 424, 432 (see also Rawls, 1999:249, 280, 320, 326, 335).

2 "A set of principles is required for choosing among the various social arrangements that determine this division of advantages, and for underwriting an agreement on the proper distributive shares. These principles belong to social justice: they provide a way of assigning rights and duties in the basic institutions of society and define the appropriate distribution of the benefits and burdens of social cooperation" (Rawls, 1999:4).

3 "... in parts of the basic structure and its political and social institutions" (Rawls, 2001:585).

4 However, on the next page, he does back-paddle slightly: "In order for public reason to apply to the family, it must be seen, in part at least, as a matter for political justice" (Rawls, 2001:596).

5 "This means, in effect, that the basic structure of society is to arrange the inequalities of wealth and authority in ways consistent with the equal liberties required by the preceding principle" (Rawls, 1999:38-39).

ples at all; those earlier in the ordering have an absolute weight, so to speak, with respect to later ones, and hold without exception (Rawls, 1999:38).

8.5.4 *Rawls's justice: universal or limited in scope?*

We have seen that, proceeding from justice as fairness, Rawls does not assume "universal first principles having authority in all cases" (Rawls, 2001:532). Our last quotation highlighted his conception regarding the successive realization of principles of justice, where former principles have "an absolute weight." Since Rawls employs the term 'just' in a way that is intimately related to "what is just" – for instance when he speaks of *just institutions* (Rawls, 2001:94, 105) or *unjust institutions* (Rawls, 2001:125) – it is worthwhile to compare the scope of 'just' and 'justice'. Whereas for Rawls, justice does not apply universally, we have seen that what is *just* or *unjust* extends across laws, institutions, the social system, actions, decisions, judgments, imputations, as well as the attitudes and dispositions of persons, and even persons themselves (see the quotation on page 513 above).

Clearly, the opposition between *just* and *unjust* encompasses different *kinds* of entities, configurations and properties. In this regard, it closely approximates the equally familiar opposition of *legal* and *illegal*. Principles of justice, by contrast, have the *basic structure of society* as their primary subject, and in this sense their scope is restricted to only one *kind* or *type* of entity – the *basic structure of society*. In terms of Rawls's own emphasis on the strict correlation between "principles of justice' and their primary *subject*, he is on the brink of discovering a number of fundamental *ontological* distinctions operative in the history of Western scholarship explored in our fore-going analyses.

The first is the strict correlation between *law* and *subject* (a principle and what is subjected to it). A law or principle always *determines* and *delimits* what is subjected to it (see our more detailed account – pages 432 ff.). The second insight concerns the nature and scope of the different *aspects* of reality; what we designated as their modal universality. We observed that Rawls holds the view that the *social system* displays different aspects, where political justice represents one aspect, while social and economic relations constitute another. Remember that the neo-Kantian schools of thought equates function concepts with relation concepts (see Rickert, 1913:68-70). In the third instance, his remarks about the *limited scope* of principles of justice, lacking an unqualified universality, approximates an insight into the dimension of many-sided (natural and societal) entities.[1] If the distinction between aspects and many-sided entities is combined with the first one, it is just one additional step to arrive at the distinction between *modal laws* (*modal principles*) and *type laws* (*type principles*) – where both kinds of laws hold for correlating subjects.

1 Earlier, we mentioned the work of the well-known neo-Kantian philosopher, Ernst Cassirer, on the theme Substance and Function (German: *Substanzbegriff und Funktionsbegriff*) – see Cassirer 1910 and 1953.

When we broaden our perspective by directing our theoretical attention towards other modal aspects or functions of reality – such as the social function, the economic facet or the jural mode – we are still not yet involved in the *classification of societal entities* according to the *kinds* or *types* to which they belong. The mere distinction between *economic* and *un-economic* actions, for example, is not specified in any *typical* way. A consequence of the distinction between modal laws and type laws is that the phrase *typical way* intends to refer to various *types* of entities governed by their type laws. The business enterprise, the state and the university represent different *kinds* or *types* of societal collectivities. Each functions within the economic aspect of reality in a way that reflects their *typicality* – yet abstracting the general structure of the economic aspect always disregards their *typical differences*. Likewise, thermodynamics as a general functional physical discipline abstracts from the typicality of physical entities – it is not interested in the gaseous, solid, or fluid state as such.

Both a state and a business enterprise can *waste their money* (and thus act *uneconomically*) and both ought to function in a way that is guided by economic considerations of *frugality*. But it is only possible to say this when the economic aspect is understood in its *modal universality*, i.e. when the *typical nature* of the business and the state is disregarded. For this reason, we pointed out that modal laws hold universally without any specification. The implication of this modal universality is that universities, businesses, states, families and sport clubs all have to observe the general meaning of economic norms, insofar as they function within the general modal structure of this aspect. In other words, this is an instance of the general perspective that the modal universality of every aspect embraces all possible kinds of entities ('objects') functioning within all modalities.

By contrast, we argued that a valid law for a specific kind or type of entity does not hold for every possible kind or type of entity. Such a type law nonetheless still has its own universality, although its universality is *specified* and *typified*. In Chapter 1 (see page 25), it was stated that the type law for being a state is universal in the sense that it holds for *all states*.[1] Yet not everything in the universe is a state, its type law is specified in the sense that it applies to states only.[2] The other side of this coin was also highlighted, for the uniqueness of the state and a business enterprise is seen in the typically different ways in which they respectively function within the economic aspect.[3]

Moreover, the inherent limitations acknowledged by Rawls in respect of

1 A more familiar designation would be to refer to the structural principle for being a state.

2 Recollect the example of the type law for an atom (see pages 5, 193): The type law for being an atom holds for all atoms (its universality), but since not everything in the universe is an atom, the universality of this type law is restricted to (and therefore solely specified for) a limited class of (physical) entities, namely atoms – plainly because not everything in the world is an atom.

3 For this reason, we formulated the general conclusion that modal laws encompass all possible kinds of entities, whereas type laws only hold for a limited class of entities.

his idea of the "principle of justice" and the lack of universality attached to these principles and their primary subject (the *basic structure of society*) also show how closely he approximated the idea of a *type law*.[1]

However, without explicitly entering into an analysis of these distinctions, Rawls did not benefit from the liberating thrust entailed in them. The first benefit would be a realization that merely referring to "principles of justice"[2] is confusing, for although this phrase appears to designate *modal* jural principles, he actually aims at what we designated as *type laws*. This is derived from the fact that the primary subject of justice is the *basic structure of society* – and the latter concerns the major (political, social and economic) societal institutions.

In the second place, Rawls could have benefited from the insight that every societal institution in principle functions in all aspects of reality, for it entails that the major 'political', 'and 'economic' institutions equally function within the jural (and other) aspects of reality. At the same time, this state of affairs contains a fundamental critique of Rawls's entire approach.

The fundamental problem of his political philosophy is enclosed in the *single* qualification attached to the principles governing the *basic structure of society* – they are said to be principles of *justice*. There are two options: (i) understand 'justice' in a modal (functional) sense or (ii) attach an institutional meaning to it. If option (i) applies, the problem is that, even in his own explanation of the parts of the basic structure of society, he had to take recourse to other *qualifying* or *differentiating* terms, such as *political*, *social* and *economic*! If the *law for* the basic structure is claimed to be found in "principles of justice" then their *primary subject* ought to display the same character. Consequently, falling back onto the said alternative qualifying terms (*political*, *social* and *economic*), highlights an inability on his part to arrive at an efficient characterization of what should be designated as type laws for distinct societal collectivities. If option (ii) is assumed, by attaching an institutional meaning to the principles of justice, the latter take on the role of type laws, generating the problem that every type law determines "multi-aspectual" entities, i.e. many-sided entities that function in more than one aspect *simultaneously*. In addition, the different 'parts' of the *basic structure* (or the *social system*) appear to require alternative qualifying functions in spite of the fact that they all invariably also functioning within the *jural aspect* of reality.[3]

This problem is exacerbated to the extent that Rawls accepts the idea of a social system, for within the context of such a view, the whole-parts scheme prevails and the intrinsic uniqueness of societal collectivities cannot be ac-

1 In Political Liberalism he writes: "The first principles of justice as fairness are plainly not suitable for a general theory" (Rawls, 1996:261).

2 The fact that he sometimes talks about conceptions of justice regulating the basic structure of society demonstrates the ambiguous status of the modern (Kantian and post-Kantian) idea of human autonomy. The impasse in this view is analyzed in Strauss (2006a:71-73).

3 As we shall argue in more detail below, different societal collectivities are qualified by different modal functions.

counted for in a proper way. This gives rise to the question regarding an alternative to the limitations of the whole-parts relation, and whether such an alternative can dispense with the misguided distinction between individual and society.

8.6 Society: towards an alternative to the whole-parts relation

Traditional universalistic views of human society also expanded the meaningful use of the whole-parts relation beyond its limits. Both the ancient Greek idea of the *polis* (the state) and the medieval Roman Catholic idea of the *perfect society* (the church) in principle subsumed all other societal relationships. This universalistic trait was continued via Rousseau and post-Kantian freedom idealism (Schelling, Fichte and Hegel) into the 20th century – it is found in Nazism and Fascism, but also particularly evinced in the idea of a *social system* with its *subsystems*. The ever-recurring difficulty with such a universalistic view is that, at least in its "being-a-part," every part derives its structural sameness (in the sense of equally *being-a-part*) from the *whole*.[1]

Although some of the most prominent contemporary social theorists are oriented to a systems approach, it is remarkable to see that almost all of them at the same time reveal an understanding of the uniqueness of different social forms of life (societal collectivities).

8.7 A critical appraisal of some contemporary theoretical approaches to human society

In an interview Giddens indicated that his theory of structuration must be seen as an attempt to come to terms with its implications for the problem of *individual* and *society* (see Giddens, 1998:75). Rather than starting either with the 'individual' or with 'society', Giddens opts for the *dynamic flow* of "recurrent social practices" – "I wanted to place an emphasis on the active flow of social life" (Giddens, 1998:76).

8.7.1 *A dynamic social field theory*

Since a similar sentiment is found in the theory of a *dynamic social* field developed by Sztompka, we start with a brief discussion of his stance, for through this theory, he aims to surpass the limitations of the systems model (Sztompka, 1993:9 ff.) and to develop a *sociology of social change* that supersedes the doubtful validity of "organic-systemic models of society," as well as the very "dichotomy of social statics and social dynamics" (Sztompka, 1993:9). His aim is furthermore to explore Whitehead's "processual image" which claims that "change is inherent in the very nature of things" (Sztompka, 1993:9).

> Ontologically speaking, society as a steady state does not and cannot exist. All social reality is pure dynamics, a flow of changes of various speed, intensity,

1 Just recall what Luhmann said: "The system as a whole impregnates through its essence (= Wesen) the essence of its parts. Although the latter may then differ in their accidental properties, essentially, in their being-parts, they are qualitatively similar" (Luhmann, 1973:174).

> rhythm and tempo. It is not by accident that we often speak of 'social life', perhaps a more fitting metaphor than the old image of a hide-bound, reified super-organism. Because life is nothing but movement, motion and change, there is no more life when these stop, but rather an entirely different condition, which is nothingness or death, as we call it (Sztompka, 1993:9).

Yet Sztompka does not analyzse the *primitive meaning* of **change**. Such an exercise would have cautioned him in his extreme 'dynamistic' approach. Without something *persistent* or *constant*, it is impossible to detect any *changes* (see pages 163, 186, 234 ff. above). He correctly rejects the old dichotomy of "social statics and social dynamics," but does not see that one cannot avoid the mutual coherence between *constancy* and *change*. Nonetheless, that change presupposes something constant is implicitly acknowledged by his introduction of the expression "social field." For example, he distinguishes four levels within(!) the "socio-cultural field" (*ideal*, *normative*, *interactional* and *opportunity*) and then affirms that each "is undergoing perpetual change" – thus implicitly affirming the *constancy* of each level allowing for the changes taking place *within* them (cf. Sztompka, 1993:10-11). Without constancy (and identity), no meaning can be attached to the word *change*. Sztompka nevertheless still thinks that the only reality this new approach deals with is the dynamic of *constant* (my emphasis – DS!) changes!

Asserting that life is "nothing but movement, motion and change" is tantamount to a denial of the reality of phenomena stamped by the biotical aspect, that is, of anything alive. Therefore, "when those stop," to reverse Sztompka's claim, we already have 'nothingness' and 'death'!

Furthermore, if 'life' is really "nothing but" "movement, motion and change" – why not be consistent and say the same about the 'social'? By maintaining the qualifying role of the term *social* in expressions such as the "social field" and "social life," Sztompka implicitly acknowledges the (ontic) *constancy* of the social dimension (aspect) of reality. If the structure of this aspect itself is subject to change (that is, inherently transient), then its qualifying role has to be substituted by whatever non-social phenomenon it changed into! Such a one-sided emphasis on change cannot but end in insurmountable antinomies.

8.7.2 *The dualism between action and system (order): Habermas*

Habermas distinguishes *three* worlds: "1. The *objective world* (as the totality of all entities about which true statements are possible); 2. The *social world* (as the totality of all legitimately regulated interpersonal relations); 3. The *subjective world* (as the totality of the experiences of the speaker to which she has privileged access" (Habermas, 1984:100 – my italics, DS). The "objective world" is understood in terms of (objective) *entities*, while the *social world* and the subjective world are described in terms of *relations* and *experiences*, respectively.

Implicit in this ontology of Habermas fails to acknowledge *ontic*[1] *modes of being* embracing all three worlds distinguished. Therefore he never relates the acknowledged *social relations* to an ontic social mode of being, making possible all the kinds of social interaction one can distinguish. This observation entails that, even what Habermas would recognize as uniquely *human* acts, in addition to their functions within the normative aspects of reality, also function within the natural sides of reality (his first world). In passing we may note that in an amazingly uncritical way he adheres to a metaphysical legacy reaching back to Greek antiquity. In early Greek philosophy, within the school of Parmenides, emphasis was laid upon unity and truth.

Eventually, in the thought of Socrates and Plato, the beautiful and the good were added, leading to the four transcendental determinations of medieval metaphysics: unity, truth, beauty and goodness (supposed to embrace God – as the highest being or ipsum esse – and creatures analogically). Habermas simply continues three of these, namely the cognitive (truth), the aesthetic and the moral (see for example Habermas, 1995-2:374). The anthropological view that accompanied this heritage considered the human being as a so-called rational-moral being. But Habermas's own view is dependent upon the basic *dualism* between *instrumental actions* and *communicative actions*, since this dualism sets apart the domain of *subject-object* relations (being 'instrumental') from that of *communicative actions* (restricted to *subject-subject* relations). Habermas differentiates rationality in two domains: from "one perspective, the telos inherent in rationality appears to be instrumental mastery; and from the other, communicative understanding" (Habermas 1984:11).

Since Habermas does not have a theory of *modal functions* belonging to a(n ontic) dimension of experiential reality *distinct* from (but intimately cohering with) the dimension of (natural and social)[2] *entities*, he does not realize that all subject-subject relations are *founded* in subject-object relations[3] and that it is therefore unsound to attempt to separate *instrumental* and *communicative actions*. Furthermore, although he constantly speaks of "social entities" such as tribal societies ('Stammesgesellschaften'), institutions, families, and even societies organized by the state,[4] his only recourse to a *theory* of social entities is found in a fusion of his theory of communicative action with the systems theory of Parsons – keeping in mind that he distinguishes between *system* and *life-world*.

Waters summarises this move as follows:

Habermas now proceeds to integrate his own arguments about communicative

1 Regarding the meaning of the term ontic see pages 17 ff. above.

2 Clearly, Habermas excludes 'entities' from the social world, for the latter is characterized by 'relations' only.

3 The only exception is the numerical aspect, where only subject-subject relations are found. Addition, multiplication, subtraction and division operate on numerical subjects.

4 "... staatlich organisierten Gesellschaften ..." ["societies organized by the state"] (Habermas, 1995-2:253).

> action with Parsonian systems theory. First, he argues that Parsons's 'systems within systems' elaboration is not much more than a semantic exercise. Action is not an environment for society but its content – culture, personality and social interaction are the substance of the life-world. Moreover, such constructs as the 'relic system' and 'ultimate values' are abstractions from life-world activity. The only 'real' systems are the structural responses to the AGIL[1] imperatives originally proposed by Parsons and Smelser – economy, polity, societal community, and fiduciary. These are now reinterpreted in the terms of the system/lifeworld couplet ... The economy and the polity are steering agencies, focused on system integration and organized along the lines of strategic action. Societal community and fiduciary are the public and private sectors of the lifeworld, focused on social integration and characterized by communicative action. Note that in undertaking this reinterpretation, Habermas moves system integration to A/G from I and renders I/L the shared location for social integration (Waters, 1994:163).

The mere fact that Habermas speaks of "societies organized by the state" reveals the implicitly *totalitarian* consequences of his own affinity to *systems theory*. In this frame of mind, the functions of society in its totality[2] are viewed in terms of a differentiation between political and non-political *subsystems* as diverse action systems.[3] Through the medium of *money* and *money-exchange*, the economic function of society as a whole is handed over to the capitalistic economic system that forms the specialised foundation for a subsystem extending beyond the normative context of the state: "The functions relevant for society in its totality differentiates into different action systems. Through administration, the military and jurisprudence, the state apparatus specializes in the realization of binding decisions and collective aims." (According to Parsons, the 'polity' is also geared towards "collective goal attainment" – DS.) "Other functions are depoliticized and handed over to non-state subsystems. The capitalist economic system signifies the break-through to this level of system differentiation; it owes its genesis to a new mechanism, the money as tax medium. This medium specializes in a function of society in its totality delegated to the economy and it constitutes the foundation of a normative context of a separated subsystem."[4] From this, it clearly follows that Habermas

1 Regarding Adaptation, Goal-attainment, Integration and Latency (AGIL) see below page 525.

2 "Die gesamtgesellschaftlich relevanten Funktionen" ("Those functions relevant for society in its totality"), Habermas, 1995-2:255.

3 "... nicht-staatliche subsysteme" ["non-state subsystems"], Habermas, 1995-2:255.

4 "Die gesammtgesellschaftliche relevanten Funktionen verteilen sich auf verschiedene Handlungssysteme. Mit Verwaltung, Militär und Rechstprechung spezialisiert sich der Staatsapparat darauf, über bindende entscheidungen die kollektiven Ziele zu verwirklichen" ... "Andere Funktionen werden entpolitisiert und an nicht-staatliche Subsysteme abgegeben. Das kapitalistische Wirtschaftssystem markiert den Durchbruch zu dieser ebene der Systemdifferenzierung; es verdankt seine entstehung einem neuen Mechanismus, dem Steuerungsmedium Geld. Dieses Medium ist auf die vom Staat abgegebene gesamtgesellschaftliche Funktion des Wirtschaftens spezialisiert und bildet die Grundlage für ein normativen Kontexten entwachsenes Subsystem" (Habermas, 1995-2:255-256).

does not acknowledge the *modal universality* of the *economic aspect* of reality.

Giddens evinces a more articulated intuition in this regard when he gives a description of the 'economic', closely approximating the modal universality of the economic aspect of reality: "Rather, the sphere of the 'economic' is given by the inherently constitutive role of allocative resources in the structuration of societal totalities" (Giddens, 1986:34). Rephrased in terms of the idea of the many-sided functioning of societal entities, one can say that Giddens has uncovered the fact that every societal collectivity has a function within the economic modal aspect.

The differentiation of economic life and the emergence of the business enterprise merely *specify* the general modal meaning of the economic aspect – without monopolizing it exclusively, for alongside the non-political spheres of societal life (where each one of these spheres in its own way continues to function within the economic aspect), the state also continues to specify the meaning of the economic mode in a way *typically* different from that of the firm and other societal forms of social life. The striking additional perspective that Habermas wants to advance, is that through the medium of money-exchange a whole domain of "norm-free sociality" is institutionalised.[1]

Although Habermas does react critically to the whole-parts scheme *as such* (see Habermas, 1998:65), he appears not to realize that social systems theory indeed also gives shelter to this scheme. The least one can say is that his thought is ambivalent in this regard, because he does not proceed from a strict modal delimitation (what we will designate as the qualifying function below) of the structural task of the state. This explains why he considers it the main burden of *law* within modern societies to bring about *social integration* (Habermas, 1998:60). Every societal form of life has to attain its own specific "social integration." For example, the *juridical integration* of a multiplicity of legal interests into a public legal order (within the territory of the state) differs in a *typical* way from the *social integration* of changing *fashions*.

There is another subtle 20th century intellectual background, practically intersecting all the different sociological schools of thought of this century – and still found (in some or other variant) in the thought of Weber, MacIver, Parsons and Habermas (to mention just a few). In spite of peripheral differences, they all adhere to the transformation, which the Kantian dualism between 'Sollen' ('ought') and 'Sein' ('is') acquired in the Baden School of

1 "Die kapitalistische Wirtschaft läßt sich nicht mehr wie der traditionale Staat als institutionelle Ordnung begreifen – institutionalisiert wird das Tauschmedium, während das über dieses Medium ausdifferenzierte Subsystem im ganzen ein Stück normfreier Sozialität darstellt" ["The Capitalist economy cannot be understood as an institutional order, like the traditional state. What has been institutionalised is the medium of exchange, while the subsystem embracing this medium manifests an overall domain of sociality free from norms"] (Habermas, 1995-2:256). The phrase "normfreier Sozialität" suggests that economic activities are not subject to economic norms – contradicting the ever-present normative contrary economic – un-economic.

neo-Kantian thought, where it resulted in an assessment of society in *factual terms*, while *norms*, *values* and *beliefs* are located within the sphere of 'culture'. We noted that one of the founding philosophers of the Baden school, Heinrich Rickert, started with ideal values having a timeless validity (culture originates through relating 'nature' to 'values', i.e. through 'Wertbeziehung') – but soon the effects of 19th century historicism and the linguistic turn relativized this view (up to contemporary postmodern views, according to which every individual is supposed to choose her own values at will – in their mission formulation contemporary organizations are sometimes prompted to consider the acceptance of "new values"). The latter lacks an ontic character, for 'culture' is the result of (autonomous) *human construction*.

8.8 The step from modal to typical concepts

In order to further differentiate between societal collectivities, displaying both a solidary unitary character and a durable relation of super- and subordination (such as the state, the firm, the club, the nuclear family, and so on), one has to exceed the confines of the social mode of reality by looking at the specific modal aspect, stamping, characterizing or *qualifying* the type law under consideration. The idea of a *qualifying function* rests upon the *uniqueness* of such a modal function, as well as a continued acknowledgement of the modal universality of every modal function as such. Furthermore, every societal entity also has a *typical foundational function* (compare the above-mentioned example of maleic and fumaric acid explained in the previous chapter – page 465).

In order to distinguish between the collectivities mentioned in the previous paragraph (namely the state, the firm, the club and the nuclear family) one may start by specifying their respective *qualifying functions*. The state, for example, is qualified by the *jural* aspect (it is a public *legal* institution destined to harmonize and balance the multiplicity of legal interests within its jurisdiction on the basis of the monopoly over the "power of the sword" within its territory) – internally, the police force and against external threats/attacks, the military force (infantry, navy and air force). The firm, by contrast, is qualified by the *economic aspect*, while the nuclear family finds its qualifying function in the *moral aspect* of *love*. Because each of these societal collectivities still functions simultaneously in *all* (other) aspects of reality, the acknowledgement of their respective qualifying functions does not terminate any other modal function they may have.

The importance of this distinction between a foundational function and a qualifying function becomes apparent as soon as one wants to account for the distinct nature or "inner laws" governing diverse societal collectivities and communities. Although some of the most prominent contemporary social theorists are oriented towards a systems approach, at the same time some of them reveal a concern for the uniqueness of different social forms of life. Unfortunately, they lack a consistent account of what distinguishes each social form from the others.

8.8.1 *The AGIL scheme of Parsons*

The inadequacy of the Parsonian systems theory is therefore obvious, because (as explained in Strauss, 2006:192-195), the AGIL scheme is constituted as an *ontological design* based upon a *four function paradigm* (*adaptation*, *goal-attainment*, *integration* and *latency*), exploring only three elementary basic (i.e. *analogical*) concepts of sociological theorising – namely the *spatial* opposition of *inner* and *outer*, the analogy of thermo-dynamical open systems (*physical analogy* – "pattern maintenance") and the *biotical analogies* are reflected in the terms *adaptation*, *goalattainment*, and *integration*.

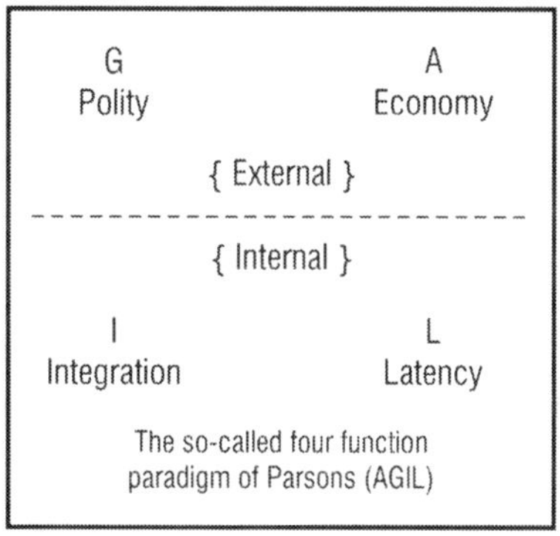

The so-called four function paradigm of Parsons (AGIL)

Unfortunately, these aspects (modal functions) are shared by *all* communal and collective social entities, making it impossible to distinguish them theoretically in this way. Parsons attempts to solve this problem by speaking about *collective* goal attainment – regarding the 'polity' – in order to distinguish it from the 'economy' (supposedly concerned with 'adaptation'). However, both the 'economy' and the 'polity' have to cope with the functional problems of adaptation and goal attainment in their own *typical* ways – entailing that what is *typical* about them as societal collectivities is presupposed and cannot be derived from accommodating these two problems.

Prior to these shortcomings, systems theory has elevated the original *spatial whole-parts relation* in a universalistic way into an exclusive principle of explanation, such that society is captured in this reductionistic, holistic scheme. As long as priority is given to such a whole-parts scheme, the 'individual' will always be sacrificed to whatever societal reality is chosen as the *encompassing whole*. Therefore, anyone buying into systems theory and at the same time trying to uphold a theory of agency, is inevitably confronted with a conflict between atomistic and holistic views of society (cf. Waters, 1994:168 ff.). Waters aptly remarks:

> The intransigent theoretical failures which inhere in the specification of system have generally forced a retreat in the direction of action. In the case of Parsons, the so-called systems theory has an actional foundation. The theoretical prototype of the system, the unit act, precisely specifies a voluntaristic orientation in which purposes are located at the level of the individual. In his middle period (the three-system model) the conditions which constrain such voluntarism disappear from the formula and the governing system, culture, is an analytic abstraction from the concrete orientations of individuals. The pattern variables remain as a classification of human intentions. Only in the later, cybernetic phase does individual intention disappear from the social system en-

> tirely, but only, as Habermas expertly shows, by artificially placing intentions in the cultural environment. Alexander's sympathies with Parsons' middle period indicate a similar orientation in his theoretical approach, assertions to the contrary notwithstanding. This brings us to Habermas himself. His arguments about the mediatization and colonization of the lifeworld would tend to suggest a materialistic orientation in which system dominates action. However, the distinction between lifeworld and system is founded on distinctions at the level of action between strategic and communicative practices. The formation of autonomous steering organizations and the transformation of linguistic human relationships into juridified and monetarized forms is not the consequence of ineluctable system requirements. These are merely vehicles for the strategic intentions of individuals. Habermas' fundamental concern is less to do with the rise of the system than with the progressive eclipse of communicative action by strategic action (Waters, 1994:171-172).

Do we find a solution for the tension between *individualism* and *universalism* in the theory of *structuration* developed by Giddens?

8.8.2 *Giddens: the theory of structuration*

In order to understand the approach of Giddens, two systematic distinctions employed earlier ought to be kept in mind, namely (i) the idea of ontic normativity, that is, the acknowledgement of underlying principles that are not the result or product of human action, but its very *condition*, and the implication of this view for (ii) the distinction between a *principle* and its *application* (giving it a positive form, *positivizing* it). Habermas also explicitly uses this term, for example where he speaks about "the positivization of law" (Habermas, 1996:71 – regarding this idea of positivization as found in the thought of diverse thinkers, see page 94 above).

We saw that the concept of a *principle* is also a *complex basic concept* of those scholarly disciplines involved in a study of reality from the perspective of some or other normative aspect. This consideration entails that, in order to articulate the meaning of a principle, one has to employ terms derived from multiple aspects. An example from everyday life will elucidate the difference between a principle and its application. Consider the social principle of *showing respect*. This principle is *universal* in the sense that there is no single human society in which one does not encounter *some or other* form of showing respect, which is also *constant*. We referred to the example used by Hart concerning showing respect in various cultures – from taking off one's hat, just lifting it slightly or just raising the hand – where the principle of showing respect remained constant through historical changes (see page 294 above).[1]

Moreover, as we saw, principles are not *per se* valid (in force), for they need human intervention *to make them valid*, or *enforce* them (see Derrida,

1 We also mentioned that he is justified in rejecting the extremes of *conservatism* and *chaos*: "Either only lifting one's hat all the way counts as greeting, or anything I choose is greeting. The recognition of 'greeting in principle' makes it possible to avoid both conservatism and chaos" (Hart, 1984:62).

2002:233 ff. and page 291 above). Therefore, we concluded earlier (see page 297) that a principle is a *universal, constant, starting-point for human action that can only be made valid (enforced) by a competent organ with an accountable (free) will, capable of giving a positive shape to such a starting-point in varying historical circumstances, in the light of an appropriate interpretation of the relevant circumstances, and resulting in a norm-conformative or antinormative positivization of the underlying principle.*

Comment: The attentive reader will notice that this circumscription involves the following modal aspects: the numerical (a starting-point); the spatial (universal); the kinematic (constant); the physical (making valid, enforce); the biotical (organ); the sensitive-psychical (will); the logical-analytical (norm-conformative or antinormative); the cultural-historical (shaping, giving form to, positivization); and the sign mode (interpretation). Since these terms are derived from multiple aspects, we pointed out that the concept of a principle is a *compound* or *complex* basic concept within the various disciplines.

Against this background, a distinction ought to be drawn between the principles norming human activities, and the norm-conformative (or: antinormative) ways in which human beings can respond to underlying principles. In the case of societal entities and processes, there is always a difference between "structures for" and "structures of." For example, the scope of the structural principle for being a state encompasses all past, present and future states, wherever they may be found – whereas any concretely existing state, in *being a state*, exhibits the reality that it *is* a state. *Being a state* is the universal way in which *this* state shows that it is subject to the *structure for* being *a* state, i.e. the *structural principle* norming every state. The modern idea of *autonomy*, as well as the idea of the *social construction of the world* reify the *human freedom to positivize*. At the same time, it denies the existence of universal and constant principles underlying every human act of shaping and form-giving (*positivization*).

Anthony Giddens wrestled with these issues in his own way when he introduced his theory of *structuration* in order to emphasize the *actuality* of temporal societal processes through which societal structures are produced and reproduced. According to him, a "double hermeneutic" is implied in all forms of sociological theorizing, because the scholar is simultaneously participant and analyst (see Calhoun *et al.*, 2002:222). The acknowledgement of the "subject-dependency" of societal structures explains why Giddens prefers to speak of 'structuration' instead of merely 'structure'. Although he alludes to "primitive terms," he does not have a clear understanding of what we designated as the indispensable elementary (analogical) basic concepts of sociology as a discipline. In fact, his sociological thought shows a remarkable ambiguity towards the use of modal analogies.

On the one hand, for example, he wants to discard altogether any reference to the phrase *social adaptation*, but nonetheless concedes that there may be legitimate usages of this expression – then he once again fears that such usages may be too *vague* and *diffuse* (Giddens, 1986:233-236, 270-271). In different contexts, he simply uses the terms (social) *differentiation* and *integration* without realising that they reflect *biotic analogies* within the structure of the social aspect (Giddens, 1986:181 ff.). He has a fairly *negative* assessment of the meaning of "social causation" (see Giddens, 1983:80 and 1996:65).

In the absence of well-known 'abuses' of modal analogies comparable to those found in 19th century "organic models," the structure of the logical aspect – and its possible analogies within the context of sociological analysis – apparently does not pose a similar threat to Giddens. Consider his understanding of the "concept of contradiction":

> *Comment*: It is commonly remarked that the concept of contradiction should remain a logical one rather than being applied to social analysis. ... Given that it is used with some care, however, I think the concept to be an indispensable one in social theory (Giddens, 1986:193).

What a *mixed* picture in the thought of one of the leading sociologists of our day regarding the admissibility and status of the elementary basic concepts of sociology: from *rejection* ("social adaptation"), *hesitance* ("social adaptation"; "social causation") and *unawareness* ("social differentiation and integration") to *positive acceptance* ("social contradiction")! Of course there is no *systematic* treatment of analogical concepts in the sociology of Giddens. Where he does refer to the 'primitive' terms of "social science," only three are highlighted: *meaning*, *norm* and *power*:

> *Processes of structuration involve an interplay of meanings, norms and power.* These three concepts are analytically equivalent as the 'primitive' terms of social science,[1] and *are logically implicated, both in the notion of intentional action and that of structure*: every cognitive and moral order is at the same time a system of power, involving a 'horizon of legitimacy' (Giddens, 2002:230).

Although the term 'meaning' resides in the sign mode of reality,[2] which, in a modal-functional sense presupposes *choice* and requires *interpretation*, it is usually (in modern sociological theory) 'loaded' with connotations that are related to *beliefs*, *values* and *norms*.

Comment: In Chapter 4, we saw that animal communication does not refer to the *past* or *future* – it always refers to the vital *here* and *now*, and it is strictly univocal, requiring no interpretation. Animal signs have a single content for every sign. All human utterances, by contrast, can signify a number of different things, depending on the context, intention, or even, in the case of written language, the punctuation. Human language therefore presupposes *freedom of choice*, and the concomitant *multiplicity of meaning*, requiring *interpretation*, which,

1 In a different context, Giddens suggests "that structure, system and structuration, appropriately conceptualised, are all necessary terms in social theory" (Giddens, 1983:62).

2 The phrase "social meaning" in the first place reflects an analogy of the sign mode within the structure of the social aspect and therefore belongs to the domain of elementary basic concepts of sociology.

> in turn, needs further interpretation from the addressee (cf. Nida, 1979:203; De Klerk, 1978:6). It presupposes the responsible free human activity, which requires *accountable choices*. Eibl-Eibesfeldt categorically states that that "which, by contrast, regarding animals, is generally designated as 'language', exclusively moves within ... the domain of interjection, of the expression of moods lacking insight" (Eibl-Eibesfeldt, 2004:214).

It is therefore not surprising that Giddens does indeed relate *meaning* and *norm*. But in doing this, he does not realize that, whereas the *primitive* sense of the term *meaning* marks an *analogical sociological concept*, the complex nature of the term *norm* represents a complex or compound basic concept of sociology. Furthermore, since the core meaning of the cultural-historical mode of reality is given in *power* (*control*, *mastery*), the notion of *social power* (social control, social mastery) also refers to a modal historical analogy within the structure of the social aspect, and therefore belongs to the domain of the *elementary basic concepts* of sociology.

In terms of a non-reductionist ontology, the discipline of sociology is based upon the use of many more elementary basic concepts. Consider numerical analogies (social order – the unity in a multiplicity of social norms, correlated with the unity in the multiplicity of social roles assumed by human subjects within human society),[1] spatial analogies (social super- and subordination, social stratification, social distance), kinematic analogies (social constancy, social stability), physical analogies (social dynamics, social causes and effects), biotic analogies (social life, social differentiation, social integration, social growth, social adaptation), sensitive analogies (social awareness, social sensitivity, social desires/will), logical-analytical analogies (social identification – 'we' and 'they' – social consensus, social conflict, social contradiction, social antinormativity). In addition, sociology (implicitly or explicitly) employs complex or compound basic sociological concepts, constituted by peculiar configurations of elementary basic concepts. We explicated two of them, namely where we analyzed different possibilities of social interaction (coordinational, communal and collective modes), and where the nature of a norm or a principle was explained. In addition, the meaning of *social subjects* and *social objects* can also only be accounted for in terms of *complex* basic concepts. In passing, it should be noted that Giddens is not consistent with his correct emphasis on the *subject-dependent* nature of (continuously functioning) social entities. If this were the case, he would have realized that 'society' (or any social form of life intended by coordinational, communal and collective relationships) is not a *social object*, because social entities in this sense always function *subjectively* within all aspects of reality, including the social aspect. Note that the term 'subjectively' does not convey the connotation of

1 In line with Simmel and Von Wiese, the sociologist Znaniecki views sociology as a special science next to other 'cultural sciences'. Sociology is delimited by the concept of order; "sociology as the science of order among social actions" (Znaniecki, 1963:385 ff.). Fukuyama continues this approach in his primary focus on social order, and his conviction that sociology is "a discipline devoted to the study of social norms" (Fukuyama, 2000:147).

arbitrariness, but simply refers to one pole of the ever-present subject-object relation within human societal interaction. Giddens writes: "According to this conception, the same structural characteristics participate in the subject (the actor) as in the object (society)" (Giddens, 1983:70). Social objects, strictly speaking, are those products brought into being through the cultural-formative activities of human beings (think about the furniture of a living room – social objects; test tubes in the laboratory – scientific objects; tools – technical objects; jails – juridical objects; etc.) – always correlated with "societal-human-subjectivity."

Finally, an account of the inevitable *type concepts* used by sociologists requires the concept of *type laws* as well as that of communal and collective subjectivity (where social entities are understood in terms of their typical totality character, distinctly demarcated by their respective *foundational* and *qualifying* functions – see below, pp. 554 ff.).

O'Brien mentions that Giddens wants to avoid the term 'structure' and rather articulates the idea of "the 'structuring properties' of social interaction" (O'Brien, 1998:10). For Giddens, the concept of *structuration* involves what he calls the "*duality of structure*" which, according to him, "relates to the *fundamentally recursive character of social life, and expresses the mutual dependence of structure and agency*" (Giddens, 1983:69):

> By the duality of structure I mean that the structural properties of social systems are both the medium and the outcome of the practices that constitute those systems. The theory of structuration, thus formulated, rejects any differentiation of synchrony and diachrony or statics and dynamics. The identification of structure with constraint is also rejected: structure is both enabling and constraining, and it is one of the specific tasks of social theory to study the conditions in the organization of social systems that govern the interconnections between the two. ... Structure forms 'personality' and 'society' simultaneously – but in neither case exhaustively: because of the significance of unintended consequences of action, and because of unacknowledged conditions of action (Giddens, 1983:69-70).

The analysis of social relations concerns both "the patterning of social relations in time-space involving the reproduction of situated practices" and "a virtual order of 'modes of structuring' recursively implicated in such reproduction" (Giddens, 1986:17). Moreover, according to the theory of structuration, "the production and reproduction of social action" draw upon "rules and resources" (Giddens, 1986:19). Calhoun remarks:

> Giddens defines *structure* as the 'rules and resources' that act as common interpretive schemes in a particular social system. Giddens argues that structures are related to practices as language is related to speech – in fact, language is an example of what Giddens means by structure. Structures organize practices, but at the same time, structures are enacted and reproduced by practices. Although we experience structures as forces external to us, they have only a 'virtual' existence – they cannot be directly observed except through their effects on practices (Calhoun *et al.*, 2002:223).

These distinctions relate closely to what we said about "structures for" and "structures of" – although the main difference lies in our above-mentioned account of (modal and typical) principles in their *ontic givenness*. As universal, constant starting-points of human action, modal and typical principles *make possible* what is 'structured' through formative human actions of positivization, *giving form* and *shape* to the typical principles for diverse coordinational, communal and collective forms of social life within human society. The relative constancy (*stability*) of these forms (acknowledged by Giddens)[1] reflect (in a universal way) that societal entities are *conditioned* by underlying (modal and typical) *principles*.[2] The *orderliness* exhibited by the "production and reproduction" of societal practices is a feature that we designated as "norm-conformative" *factual* realities, for only that which is subjected to a universal (natural) law or (normative) principle can be appreciated as functioning in a 'law-conformative' or 'antinormative' way.

Understood in terms of these distinctions, *structuration* coincides with the on-going process of *positivizing* those (modal and typical) principles, making possible all coordinational, communal and collective forms of social life. Unfortunately, as argued earlier, since the Renaissance and the rise of modern *nominalism*, the intellectual legacy of the West departed from the idea of *ontic normativity* and implicitly or explicitly adheres to the modern ideal of an *autonomously free personality* – in the spirit of Jean Jacques Rousseau who, as we noted, describes *freedom* as obedience to a *law that we have prescribed to ourselves* (Rousseau, 1975:247). We have shown that as a result of the motive of *logical creation* – initially advanced by Hobbes and brought to its ultimate *rationalistic* consequences by Kant in his view of understanding as the (*a priori*) formal law giver of nature – historicism and the linguistic turn explored the *irrationalistic* side of nominalism by elevating the human subject to be the sole (re-)source of culture and society (see pages 370-379 above). In the thought of Giddens, this constructive capacity of being human is at most related to *nature* transformed into *culture*. Giddens writes:

> Sociology is not concerned with a '*pre-given*' universe of objects, but with one which is constituted or produced by the active doings of subjects. Human beings transform nature socially, and by 'humanizing' it they transform themselves; but they do not, of course, produce the natural world, which is constituted as an object-world independently of their existence. If in transforming that world they create history, and thence live *in* history, they do so because the production and reproduction of society is not 'biologically programmed', as it is among the lower animals (Giddens, 2002:229).

1 Calhoun remarks that Giddens acknowledges stable patterns that are "observable in interactions" and "also suggests that structures are generally quite stable, although they can be changed" (Calhoun, 2002:223).

2 In his distinction between social (face-to-face interaction) and system integration (regarding connections between "those who are physically absent in time and space"), Giddens speaks of 'systemness' (Giddens, 1986:28), similar to our earlier example of 'atomness' as the universal side of an atom. See also Strauss 2006:21 ff., 268-270.

Instead of considering the ontic status of normative principles, a view is developed regarding the transformation of nature into the world of history that is equated with the production and reproduction of *society*! Traditionally, it was affirmed that *culture* arises from a transformation of nature, but for Giddens, the transformation of nature results in 'society'. Furthermore, if the "production and reproduction of society is not 'biologically programmed', as it is among the lower animals," how do we account for the *normative accountability* of human beings in a way that transcends "biological programming"? Surely, the relationship between a *principle* and its *application* concerns an issue that is completely different from the relationship between 'individual' and 'society'. Yet Giddens believes that the main purpose and outcome of his theory of *structuration* is to resolve the problem of 'action' and 'structure'. Giddens explains:

> In the past it was usually seen as a dualism between individual and society, or the actor and the social system. Thinking about this traditional question of the relationship between the individual and society lay at the origin of the idea of structuration (a remark in the interview published by Pierson, 1998:75).

Although there are constructive elements in service of effectively resolving the issue of 'individual' and 'society' in the thought of Giddens, the absence of an articulated account of the elementary, complex and typical basic concepts of sociology as a discipline prevents him from arriving at a satisfactory alternative.

8.8.3 *Habermas and Rawls: acknowledgement of the "inner nature" of distinct societal spheres*

The *complexity* and *mystery* of being human is that it *transcends* the confines of just *being an individual* – just as being human transcends the spheres of operation of distinct *social entities*. This uniquely human transcendence precludes every attempt to characterize or qualify being human merely in terms of one functional quality or of participating within only one societal collectivity. Consequently, acknowledging this fundamental transcendence precludes opting for either an atomistic or a holistic view of society.

However, both in the more recent developments of systems theory within the discipline of sociology and within the domain of political theory, crucial and important starting-points can be found for an alternative understanding of human society.

Once the untenable bias of an atomistic or holistic approach to society is rejected, it becomes possible to appreciate diverse societal institutions and collectivities in terms of their own intrinsic natures; in terms of their own distinct spheres of operation. What is particularly striking in this regard, is that the atomist-holist (individualist-universalist) dilemma held sway for the larger part of 2000 years. Perhaps the first scholar who effectively questioned the whole-parts scheme inherent in universalist theories of society was the German legal scholar Johannes Althusius. He realized that not every societal entity (such as families, churches etc.) is *part* of the state – true parts are solely provinces and municipalities (see Althusius, 1603:16). This insight was accompanied be a clear understanding of the inner structural principles governing distinct societal collectivities – Althusius holds that there are *proper laws* (*leges propriae*), according to which "particular associations are ruled," required by their nature (Althusius as translated in Carney, 1965:16). In Chapter 1, it was briefly mentioned that two 19th century Dutch politicians explored this idea further by designating it as *sphere-sovereignty*, namely Groen van Prinsterer and Abraham Kuyper. Within the context of contemporary political theory, it reminds immediately of Walzer's "spheres of justice" (see Kuyper, 1880 and Walzer, 1983).

The revival of Parsons's structural-functional approach to society in the neofunctionalism of Alexander gave birth to an equally significant emphasis on the "inner laws" of differentiated societal spheres of life. Münch in particular holds that the starting point of the 1980s theoretical debate is found in "Weber's theory of the rationalization of modern society into spheres that are guided to an increasing extent by their inner laws" (Münch, 1990:442). In particular, he mentions the "political system" with "its inner laws" (Münch, 1990:444). Similar points of connection for the acknowledgement of the *sphere-sovereignty* of social entities are present in the political philosophy of John Rawls. In spite of the fact that his thought is torn apart by atomistic and holistic tendencies, he does succeed in transcending this opposition when he shows a sensitivity for *different principles* applicable to *distinct kinds of subjects*: "But it is the distinct purposes and roles of the parts of the social structure, and how they fit together, that explains there being different principles for distinct kinds of subjects" (Rawls, 1996:262). When he continues on the same page by mentioning the distinctive autonomy of elements in society, where principles within their own sphere fit their peculiar nature, his entire mode of formulation approximates the idea of sphere-sovereignty – Rawls says: "Indeed, it seems natural to suppose that the distinctive character and autonomy of the various elements of society requires that, within some sphere, they act from their own principles designed to fit their peculiar nature"!

In his discussion of issues of globalization, Habermas refers to the "classical doctrine of the state," and then mentions private spheres of life: "the state maintains law and order within the borders of its own territory and guarantees security for citizens within their own private spheres of life" (Habermas, 2001:81). On the next page he speaks of "differentiated forms of life." In his extensive work on communicative actions, he also explicitly speaks of the "laws" of "specific social spheres."[1]

In passing, we note that someone like Dunleavy approaches the nature of the state by subdividing the "non-state" domain as follows: the "state and civil society" on the one hand, and the "state and the individual" on the other (Dunleavy, 1987:320). From Aristotle to Kant, the idea of a *civil society* was

1 "den sogenannten Eigengesetzlichkeiten einzelner sozialer Sphären" (Habermas, 1995-2: 437).

identified with the state (see Riedel, 1976:88).[1] It was only after Hegel posited his own (historically non-sensitive) account of "civil society" (in his "Philosophie des Rechts," § 182), that the distinction between *state* and *civil society* became an issue. However, we will not here enter into a discussion of "civil society."

8.9 Contours of a differentiated society

The crucial place of the state within modern society is underscored by the role it played in the ongoing process of societal differentiation.

During the later middle ages the struggle between Emperor and Pope dominated the scene. The superior power of the Emperor is clearly expressed in a letter sent to Pope Leo III by Charlemagne in the year 796. This letter breathes the spirit of Caesaro-Papism. In it, Charlemagne indicates his right and duty to provide material support for the Church, but also to strengthen the Catholic faith: "It is our part with the help of Divine holiness to defend by armed strength the holy Church of Christ everywhere from the outward onslaught of the pagans and the ravages of the infidels, and to strengthen within it the knowledge of the Catholic Faith" (quoted from Ehler and Morrall, 1954:12).

Although by the eighth and ninth century, papal Rome could match the provisions of ancient Rome for its citizens, it was in need of military aid, provided by the Carolingians. Of course, long before the Church attempted to take on political rule, it "was a large and sophisticated enterprise with a huge staff, vast estates, and wide-ranging responsibilities" (Noble, 1984:254). During the eighth and ninth centuries, the Frankish kingdom and during the tenth and eleventh centuries, the Anglo-Saxon kingdom managed to maintain themselves for a "respectable length of time" (according to Strayer, 1970:14). The sweeping changes that occurred before 1000 afterwards became fairly rare, particularly in the chief surviving kingdoms, namely that of England, the West Franks (eventually known as France) and the East Franks (the core of Germany) (see Strayer, 1970:16-17). Strayer also notes that even "the most primitive feudal lordship was a more sophisticated political unit than a primitive Germanic tribe" (Strayer, 1970:18).

During the eleventh century the so-called *Investiture Conflict* emerged against the background of the intertwined relationship between the 'secular' and 'religious' authority that prevailed during the previous centuries. During this period, the Kings had a great influence on Church affairs by appointing abbots, bishops and even popes (Strayer, 1970:20). This conflict lasted for almost fifty years, and eventually the Pope Gregory VII (1073-1085) emerged victorious. Not only did the Church gain almost complete control over European society; it also separated itself from the 'secular' authorities (Strayer, 1970:21-22). What is significant for the modern idea of the state, is that the

1 For Aristotle state (*polis*) and civil society (*koinonia politike*) are synoymous. "The state or political community, which embraces all the rest, aims at good in a greater sense than any other, and at the highest good" (*Politica* 1252a3-5; see Aristotle, 2001:1127: *πόλις καὶ ἡ κοινονία ἡ πολιτική*). Cicero translated it with "societas civilis."

first steps in this process were in the direction of a *differentiation* of the sphere of competence of the Church and the authority of what subsequently became known as the state. Since it is certainly not meaningful to speak of a genuine *state* during the later middle ages, it is also not proper to designate the *Investiture Conflict* as a "struggle of Church and State" (Stayer, 1970:22). Increasingly, the lay ruler was considered to take responsibility for guaranteeing and distributing *justice* through the law and its enforcement (Strayer, 1970:23).

The initial political power claims of the church were based on its relatively *differentiated* position, which enabled it to integrate the relatively *undifferentiated* substructures of medieval society under its umbrella. But through the developments after the *Investiture Conflict*, conditions favouring the rise of a differentiated state increasingly emerged. For example, the rapid growth of educated people during the 12th century was a necessary requirement for an effective state-administration. Unfortunately the schools of education mainly focused on Roman law whereas England, Germany and northern France used customary law not taught in the schools (Strayer, 1970:25). Strayer summarises the situation as follows:

> Thus in the centuries between 1000 and 1300 some of the essential elements of the modern state began to appear. Political entities, each with its own basic core of people and lands, gained legitimacy by enduring through generations. Permanent institutions for financial and judicial business were established. Groups of professional administrators were developed. A central coordinating agency, the chancery, emerged with a staff of highly trained clerks. These professional administrators were not numerous at the time, and they therefore could not be highly specialised. They had to be assisted by short-term or part-time agent-clerks, whose main career was aimed at the Church; as minor barons and knights, or wealthy burgesses. Many of these men worked a few years, or part of each year, as estate-managers, financial agents, local administrators, or judges. In this way, they could gain royal favour and increased income, even if they did not plan to serve the government permanently. Everywhere, there were men who spent most of their lives as professional administrators, and their number increased markedly in the thirteenth century (Strayer, 1970:34-35).

Although Johannes von Salisbury (1115/1120-1180) defends an organic conception of the state (thus affirming Aristotle's view), he designates *justice* as its principle of life (Barion, 1986:48). When Thomas Aquinas entered the scene in the thirteenth century, his account of medieval society was based on an attempted synthesis between Aristotle's philosophy and biblical Christianity. He accepts the dual teleological order of Aristotle with its hierarchy of *substantial forms* arranged in an order of lower and higher. It was designated as the *lex naturalis* (natural law), which is related to the transcendent *lex aeterna* (eternal law), as contained within the *Divine Intellect*. By virtue of its substantial form, the human being depends on the community for the satisfaction of its needs.

The state (both the *polis* and the *Holy Roman Empire*) is at this point still viewed, in line with the conception of Aristotle, as *the* all-encompassing, self-sufficient community (*societas perfecta*). The provision is that Thomas Aquinas applies this only to the *natural terrain*. As the highest community within the domain of nature, the state embraces all other temporal relationships. These lower communities possess a *relative* autonomy, subsumed under what is known as the *principle of subsidiarity*. However, this principle does not eliminate the *universalistic* starting point operative in St Thomas's view of society, since the so-called *relative autonomy* of these lower communities remains connected with the nature of the state as *parts* of a larger *whole*. The parts of a whole share the structural principle of the whole.[1] As a result, the view of St Thomas does not allow for the acknowledgment of societal collectivities that are *structurally different*. In line with the conception of Aristotle, the family for Thomas also remains the *germ-cell* of society. The hierarchical ordering of these communities coheres with each other according to the mutual relationship of a *means* to an *end*, and of *matter* to *form*. Barion explains that, according to Thomas Aquinas, the human being is not fully absorbed within the state, because he also partakes in the supernatural ordering.[2]

When King Philip the Fair decided to tax the French clergy, Pope Boniface VIII considered this an infringement on the freedom of the Church, and responded with his famous Bull, the *Unam sanctam* (November 18, 1302). Gregory already claimed that spiritual power is superior to temporal power. King Henry IV reacted with his theory of two swords, but in his *Unam sanctam*, Boniface VIII states that there is "no salvation or remission of sins outside" the "one Holy, Catholic and Apostolic Church" (Ehler and Morrall, 1954:90). To this, he adds: "But it is necessary that we confess the more clearly that the spiritual power exceeds any earthly power in dignity and nobility, as spiritual things excel temporal ones" (Ehler and Morrall, 1954:91).[3]

In his work, *Monarchia*, Dante (1265-1321) assumed an intermediate position on this issue. In reaction to the claim by Pope Bonaface VIII in the *Unam sanctam*, namely to be the highest spiritual and temporal authority, he holds the view that the worldly authority of the Emperor is not dependent upon the authority of the pope (*Monarchia*, III, 13 ff., quoted by Zippelius, 1980:68). The authority of the Emperor, just like that of the pope, derives directly from God. Yet concerning the overall assessment of the relation between *temporal* happiness vis-à-vis *eternal* bliss, Dante perpetuates the standard Roman Catholic view. There are two aims, requiring two-fold guidance; one by the pope, which has to lead humankind to eternal life, according to Revelation, and the other one where the Emperor has to direct humankind to earthly happiness,

1 Zippelius states this point succinctly: "Parts exist in a proper relation to the whole when one and the same principle rules them" (Zippelius, 1980:67). ["Teile stehen aber dann in einem rechten Verhältnis zum Ganzen, wenn ein und dasselbe Prinzip sie regiert."]

2 "... daß der Mensch nicht mit seinem ganzen Sein und Haben auf die staatliche Gemeinschaft hingeordnet sei, er stehe auch in der Ordnung der Übernatur" (Barion, 1986:50).

3 A part of this conviction is manifest in the incorporation of marriage as a sacrament of the Church. The implication is that "marriage could only be validly contracted in facie ecclesiae, as indeed laid out by the Eastern Church" (Smith, 1964:62).

according to the doctrine of philosophy.[1]

Krüger explicitly refers to Occam in respect of the process of secularisation and regarding his significance for the separation of the natural and supra-natural world.[2] It is remarkable that modern Humanism, such as was the case in the thought of Machiavelli (1469-1527), Bodin (1530-1596), and Hobbes (1588-1679), initially explicitly advocated power-state theories.[3] However, various thinkers gradually attempted to develop a theory of the state, in which it was possible to guarantee various (civil) rights; that is, theories of the "just state" ('regstaatsteorieë').

An important element in the differentiation of modern society can be found in the employment of the idea of sovereignty introduced by Jean Bodin. In 1576, Bodin still employed the word 'republic' for the state while only using the word 'etat' for specific *forms* of the state. The second edition of the Shakespeare-Lexicon shows that the Bard of Avon frequently used the word *state*. In Germany, early 17th century philosophers and politicians already know the phrase "status republicae," but it was not until the latter 18th century that the transformation of territories into states permeated the general consciousness. In Austria, the words *kingdom* and *state* were used adjacent to each other in 1804. [(i)"Königreiche und Staaten"; (ii) "Königreiche und anderen Staaten"]. Particularly phrase (ii) demonstrates that the denotation of both terms intended the same context.

Mager points out that the state concept proceeds from two distinct meaning complexes; one relating to the *Ruler* and the other to the *Body Politic*. The former concerns words such as *status*, *stato*, *estado*, associated with the personal status of the governor (*status regalis*, *regis*, *ducalis*, *ducis*, *principis*), while the latter relates to the constitution of the body politic (here *status* means *species*, *forma politicae*, *reipublicae*). Mager subsequently speaks of these two in terms of the phrases *status regalis* and *species politicae* (Mager, 1968:416). By the end of the 16th century, Machiavelli and Guicciardini used the words *status* and *stato* to designate both the collection of subjects and what is intimately connected with it, namely the *territory* (Mager, 1968:429). During the 16th century the phrase *ius publicum* referred to the *law of the state* (Mager, 1968:486). Particularly in the expression *status regalis*, these elements are fused in a way that produced something completely *new*. The distinction be-

1 The pope who "nach der Offenbarung das Menschengeschlecht zum ewige Leben hinführen soll; und des Kaisers, der nach den Lehren der Philosophie das Menschengeschlecht zu irdische Glück" (*Monarchia*, III, 16, quoted by Zippelius, 1980:68 – translation in text).

2 He speaks of the 'Bedeutung' of Ockham "für die Trennung einer natürlichen und übernatürlichen Welt" (Krüger, 1966:33, note 3).

3 In German, Dutch, and Afrikaans the appropriate designation is 'magstaat' (literally: "power-state"). For lack of a better descriptive translational equivalent, we will sometimes employ the phrase "power-state" as the opposite of 'regstaat'. The latter may be transliterated with the phrase "just state". As an alternative for the idea of a constitutional state under the rule of law, one can simply speak of a *just state* in the terms used by Ronald Dworkin (see Burley, 2004:376). Jim Skillen also employs the phrase *just state* (Skillen, 2001:121 ff.).

tween person and office forms part of this fusion. The moment the distinction is drawn between the person and the official capacity of that person as ruler, the final step to a proper concept of the state is given, for only now is it clear that the authoritative public power of the state ought to be distinguished from the person occupying the office of ruler (Mager, 1968:488). The *regnum republica* is no longer conceived apart from the *sovereignty* of the state (its authority, power, coercion and stature) (Mager, 1968:488). Even the term *state* emerged fairly recently. Traditionally, the term *Politeia* (with its Latin equivalent: *regnum*) served the same purpose. It was only during the 16th and 17th centuries that the term *state* appeared in English, French and German writings (cf. Jellinek, 1966: 132-135).

Although Bodin introduced the concept of sovereignty in order to capture the authority of a state's government, he did not succeed in liberating his thought from the traditional universalistic perspective that proclaims the state to be the encompassing whole of society. In Book III (Chapter 7) of his work on the state, he characterizes the relationship between the family, corporations and colleges to the state as that between the whole (the state) and its parts.[1]

Unfortunately modern political theories focused on the *power* of state-authority to such an extent that their basic problem was captured in the question: does the highest power within the state belong to the people or the monarch? This gave rise to either popular sovereignty or the sovereignty of the monarch. As a consequence, since Machiavelli, a distinction was drawn between kingdom and republic.

8.9.1 *The distinction between kingdom and republic*

Whereas a *kingdom* belonged to a king – as his private property[2] – the state is a *public legal institution* that is destined to serve the *public interest*. This is the authentic meaning of the Latin expression *res publica*. The state in this public legal sense of the term only came into existence, as we pointed out above, through a long and gradual process of cultural development and societal differentiation. This process of *differentiation* in human society brought into existence distinct societal institutions (collectivities) such as the firm, the club, the nuclear family, the school and university, the church and, of course, also the state itself. Through this process of differentiation, a diversity of distinct societal forms of life emerged *alongside* the state. Each of these non-political societal collectivities had its own *form of organization* and its typical internal *sphere of law* (in the jural sense of the term). Together, they co-constitute a diversity of societal interests that ought to be united and integrated within the public legal order of the state.

This public legal character of the state entails that the state is *by definition* a

1 "... sowie zwischen diesen und dem Staat verhält es sich ähnlich wie mit dem Unterschied zwischen dem Ganzen und seinen Teilen" (Bodin, 1981:521). Note that the original text does not use the word *République*.

2 Governmental authority was therefore a private item, something that could be traded (a res in commercio).

republic; a *public legal institution*. Therefore, strictly speaking, it is not correct to employ the term 'republic' as designating some or other *form of organization* of the state. By referring to the *republican* nature of the state, no specific *form of government* ought to be envisaged. In the case of monarchies (distinct from republics) Machiavelli believed that the king has the power (referred to as a *kingdom*), and in the case of the latter, where sovereignty belongs to the people, it is designated as a *republic*. This reminds us of Rousseau, who calls the public person (the people when active) the state and (when passive) the sovereign. As was the case with the former communist "people's republics," a state can be organized as a *totalitarian* and *absolutist* state ("power-state"; 'magstaat'), or as a "just state" ('regstaat'), which is neither totalitarian nor absolutist.[1]

8.9.2 *Can society be 'democratic'?*

In spite of the hypothetical (and therefore a-historical) conjecture of a pre-societal condition found in modern theories of the social contract, human beings never lived 'outside' *society*. The oldest human ways of co-habitation are known as the extended family. Although it is sometimes asserted that segmented (or acephalous) societies existed without 'rulers',[2] it cannot be denied that all known undifferentiated societies display an internal social structure. Kammler distinguishes between *undifferentiated* and *differentiated* societies. The former lacks a significant technology, and within them the realization of political and administrative, economic, juridical, cultic religious, and educational functions, initially fused in the family bond is still *absent*, or they are only present in a *rudimentary* form. Although differentiated societies (designated as "complex societies" by Kammler) do reflect a social stratification and unilateral relations of super- and subordination, relations of super- and subordination are not *absent* in *undifferentiated societies*. Some undifferentiated groups of hunters and collectors already knew the institution of *slavery*. Kammler therefore concludes that even the lowest level of technological and economic development displays elements of *social ordering* (Kammler, 1966:30). Without acknowledging authority and control within the earliest undifferentiated societies, it is not possible to explain how such societies succeeded in protecting themselves against external threats. In spite of the difficulty of identifying the central instance of control, the presence of such a defence organization testifies, according to Kammler, in the light of ethnographical material, that the "political element" everywhere presents itself in undifferentiated societies (Kammler, 1966:31).

The blood ties connecting the members of an extended family continued to play a role in the stronger organization of the *sib* and *clan*, although in many instances the common line of descent became fictitious. However, Lowie

1 South Africa, for example, has a parliamentary democratic republic, distinct from the monarchical republic of the Netherlands.

2 Van Creveld mentions the Australian aborigines, the Eskimo of Alaska, Canada and Greenland and the Kalahari Bushmen (Van Creveld, 1999:2).

points out, in opposition to Morgan's school, that instead of a dull uniformity, primitive society exhibits "mottled diversity; instead of the single sib pattern multiplied in fulsome profusion, we detect a variety of social units" (Lowie, 1921:414). The same applies to the more prominent political organization of the *tribe*. Somewhat larger communities with a slightly more sophisticated political organization, according to Creveld, are found in "some East African Nilotic tribes such as the Anuak, Dinka, Masai, and Nuer" as well as "the inhabitants of the New Guinea highlands and Micronesia; and most – although not all – pre-Columbian Amerindian tribes in North and South America" (Van Creveld, 1999:2). The sociologist, Münch, says that what "we call 'primitive societies' are societies in which action approaches to a higher degree than in any other society the model of action taking place only within the boundaries of a closed community" (Münch, 1990:448).

A shared characterization of these three forms of early human societies is found in the fact that they are all instances of an *undifferentiated society*, in the sense of Kammler's use of the word. Social entities that eventually, through a long process of societal differentiation, emerged as relatively independent units functioning within their own distinct social orbits, such as business enterprises, social clubs, cultural associations, schools and universities, religious denominations, and so on. Within undifferentiated societies, they are still absorbed within an encompassing whole. This does not mean that similar activities did not take place within undifferentiated societies, but merely entails that the undifferentiated totality as such alternatively acts as a political unity, as a business enterprise or as a cultic entity. Yet, unlike what is found within a differentiated society, all these activities are bound by one undifferentiated form of organization. It is only within a differentiated society that distinct societal collectivities operate on the basis of their own form of organization.

A prominent German sociologist of the early 20th century, Othmar Spann, prefers to capture this opposition as one between *individualism* and *universalism* (cf. Spann, 1930:66 ff., 97 ff. and Siegfried, 1974: 51). Another way to designate it is to refer to *atomism* and *holism* (or: *collectivism* – see O'Neill, 1973 and also pages 60-60 above, where these terms are introduced). According to the atomistic (individualistic) view, society is ultimately explicable in terms of the interaction of (autonomous) individuals, whereas holism (universalism) postulates some or other social entity to be the highest and all-encompassing whole of society – such that every other social entity stands in relation as a part to the larger whole.[1]

Given the pervasive influence of Rousseau's idea of popular sovereignty, we have to highlight yet another impasse in his thought, because the issue continues to accompany diverse views on the nature of democracy. The holistic whole-parts scheme in Rousseau's understanding of the sovereign actually replaces the relation of super- and subordination between government and sub-

1 In his critical rejection of the holistic concept of a totality (Ganzheit), Habermas refers to the individuals as incorporated (eingegliedert) in a societal whole (gesellschaftlichen Ganzen) (see Habermas, 1998:65).

ject (the citizens).

Montesquieu already struggled with the ambiguity entailed in the idea of popular sovereignty, for he states that, within a democracy, the people are the monarch in a certain respect, and in another, they are the subject.[1] Rousseau likewise realized that the people can only be either the sovereign or the subject (Rousseau, 1975:305). Cunningham correctly points out that, within the body politic created by the contract, "governed and governors are identical" (Cunningham, 2002:127). Rousseau holds that "[T]he essence of the body politic is given in the reconciliation of obedience and freedom, and the words *subject* and *sovereign* are the identical correlates united in the single word citizen" (Rousseau, 1975:299).

It is therefore not surprising that Habermas as a social philosopher employs a view of democracy in which the emphasis on free and equal members of a legal association is prominent (see Habermas, 1998:141). He holds that, alongside a system of rights, a language must be created in which people can understand themselves as a community of freely associated free and equal legal partners (Habermas, 1998:143). His assumption is that the principle of democracy not only has to determine a legitimate procedure of positing law, since it also has to guide the jural medium itself (Habermas, 1998:142-143). What is absent in Habermas's view, is an account of the legal authority and jural competence entailed in the office of government, for the latter irrevocably implies a relation of super- and subordination between government and subject. Implicitly, Habermas derives this relation from the self-organization of a legal community of free and equal legal partners (*Rechtsgenossen*), that is to say, by implication he attempts to square the circle by deriving the relation of super- and subordination from the equality between free *Rechtsgenossen*.

The idea of *autonomy* pervading the modern world since the Renaissance at most allows for *equality*, but not for any form of super- and subordination. In his portrayal of the freedom prevailing in the state of nature, John Locke employs the significant phrase: "for all being kings as much as he, every man his equal" (see Locke, 1690:179; Chapter IX, § 123). According to the sociologist, Von Wiese, the social sciences frequently trace their problems back to the "last abstraction" given in the "relationship of the one to the many" (Von Wiese, 1959:19). This view underlies his claim that the *next-to-each-other* of people is the most fundamental trait of the social, and that the vertical structuring can never serve as the basis for sociological analysis (Von Wiese, 1959:76-77).

8.9.3 *Problems inherent in the notion of a "democratic society"*

It is only in more recent times in the history of the West that the term democracy has acquired its central importance (cf. Meier, 1970). From Aristotle onwards, it continued to be used as an indication of a *state-form* – and even in

1 Cf. Book II, Chapter 2 of his work on the Spirit of Law – see the translation contained in Bierens de Haan (1947:319-320): "In de democratie is het volk dus in zeker opsicht de monarch, in ander opzicht de onderdaan."

the seventeenth century, Pufendorf still speaks about the three regular forms of the state.[1] The term *republic* appears more frequently. During the period between 1780 and 1800, the term *democracy* acquired its decisively modern meaning (see Maier, 1972, 52 ff.). The French Revolution indeed changed the modern political scene for ever. The idea of popular sovereignty (*a la* Rousseau) now became intimately associated with the idea of *democracy*. During the twentieth century, the most common connotation attached to democracy is that of *popular sovereignty*, i.e. *majority rule*.

Since the noun *democracy* is meant to attach to the idea of the state, its expanded application to *society* implicitly elevates the state to function as the encompassing whole of society. The subtle switch from a *democratic state* to a *democratic society* is therefore more significant than it may initially appear. Ultimately, it reveals a lack of a proper appreciation for the distinctness and uniqueness of non-political institutions and collectivities within a differentiated society.

The practice of speaking of *democratic societies* is truly widespread in contemporary political theory. We merely have to consider how frequently this phrase appears in some of the works of a thinker who, as mentioned above, is considered the foremost political philosopher of the 20th century – John Rawls.[2]

It is quite normal to relate the notion of democracy to the nature of power, because the (democratic) majority is also supposed to be the source of political power. Unfortunately, contemporary discussions of (political) power largely bypass one of the most basic issues in legal philosophy, namely that regarding a theory of the *sources* of law.

The (consensus) approach of the sociologist, Parsons, with his employment of the whole-parts relation (*systems* and *subsystems*), also attempts to avoid relations of super- and subordination in respect of power.[3] According to him, the issue is not power over others, but the "capacity of a social system to get things done in its collective interest" (Parsons, 1969:205).

Within the tradition of a conflict approach, Dahl, by contrast, emphasizes that according to him power entails an element of *conflict*. In a different context, I extensively argued why the conflict-theoretical tradition confuses the

---

1 About the 'civitatis regularis tres formae' (Maier 1972:52).

2 For example, the phrase "democratic society" appears on the following pages of Rawls's works, *Political Liberalism* and the revised edition of his *A Theory of Justice*: See Rawls (1996):10, 13, 15, 24-25, 30, 33, 36, 38, 40-43, 61, 65, 70, 79, 90, 95, 134, 136, 154, 175, 177, 198, 205-206, 214, 221, 223, 243, 292, 303, 307, 320, 335, 344, 346, 369, 376, 387, 390, 414, 418, 424, 432; and Rawls (1999):249, 280, 320, 326, 335.

3 Power should be understood in the sense of a normative calling, allowing for norm-conformative and antinormative usages of power. By contrast, Morriss considers power to be an a-normative, purely descriptive term (Morriss, 1987:199). Already in the thought of Rousseau the relation of super- and subordination is collapsed into the whole-parts relation (the general will embracing every individual as an indivisible part of the whole – "partie indivisible du tout," Rousseau, 1975:244).

cultural calling entailed in power and all forms of the abuse of power.[1] His well-known definition of power focuses on the fact that the power of one person over another is manifest when one person can persuade another to do something that person would not otherwise have done. Dahl's initial article on "The concept of power" appeared in the journal Behavioural Science 2 (1957). It was reprinted in the work edited by Roderick Bell (Edwards and Wagner) in 1969 (see Bell, Edwards and Wagner, 1969:80). This definition covers too much, for even in the case of co-ordinational relationships, i.e. relations on equal footing without any super- and subordination, a polite response to a kind request may cause a person to do something that the person would not otherwise have done. In addition, it completely ignores the elements of *competency* and *legitimacy* attached to the function of a government.

Within the context of state-law, a competent organ is required in order to make valid law (i.e. in order to positivize legal principles). Our assumption is that legal principles are supra-individual and non-arbitrary – a view questioning all forms of the idea of *individual autonomy*, for the latter never succeeds in bridging the gap between the will of an individual and what is supposed to apply to all individuals alike. The validity of law only makes sense when pre-positive jural principles are acknowledged, which are both universal and constant, making possible the changing positive shapes and forms in which they are made valid (see pages 297 ff. above).

However, within a differentiated society, there are multiple instances in which the relations of super- and subordination prevail, i.e. situations in which office bearers have power over others. On a more general level, the basic distinction involved is that between *power over things* and *power over persons*. This distinction concerns *subject-object* and *subject-subject* relations. In the latter case, the meaning of *office* is implied, because power over persons, such as that of a government over citizens, requires a *competent organ* occupying a (legitimate) *office*.

The *constitution* of modern states serves as the originating source of distinct (and original) state legal spheres of competence. In spite of the fact that it is a *formal* source of law, the constitution contains stipulations pertaining to (materially) different legal spheres – such as the domain of *civil private law* (common law – compare human rights), *constitutional law* (compare the procedure specified for putting a government in office on the basis of an election procedure), and so on. The most important trait of the role of the modern state within a differentiated society lies in its *limited* legal competence, leaving open legal spheres distinct from the domain of public law[2] and in the fact that it is the only institution that integrates the multiplicity of legal interests within its territory in one *public legal order*. Derrida is correct in observing in the 'plurality' of languages, cultures and so on, a condition for the state (to which he could have added that that the plurality entailed in this diversity still ought

1 See Strauss, 2006a:207 ff., 216 ff.

2 The domain of public law embraces constitutional law, administrative law, international public law (the "law of nations"), penal law, and the law of criminal procedure.

to be united in a public legal order). Derrida says: "Thus, a state must be attentive as much as possible to plurality, to the plurality of peoples, of languages, cultures, ethnic groups, persons and so on. That is the condition for a state" (Derrida, 1997:15). The public legal order of the state is a typical *ius commune* that intersects all the other societal ties of its citizens. The internal legal sphere of every non-political societal collectivity is always bound to a particular private concern, and therefore constitutes at most a *ius specificum* (belonging to the domain of non-civil private law).

It must be clear that the idea of a *just state* is much more encompassing than what is intended by the expression *democratic state*. In fact, the constitutions of modern states always stipulate in which way, and through which process, a government is put into the office of government. Whatever type of electoral system is used, the *role* of universal suffrage is to put a government in office on the basis of what is known as fair and free elections. It is to this restricted, subservient, but indeed very important element of constitutional law, that *democracy* refers. Yet, putting a government in office[1] does not settle any issue involved in the actual governing task of a government. The entire judicial system of a just state proceeds in accordance with the typical jural principles guiding constitutional law, administrative law, criminal law and criminal procedure – and in none of these legal domains is it the case that the majority of citizens exercise their 'democratic' vote. If a person is found guilty of murder beyond any reasonable doubt (after due process in a criminal court), it would be absurd to subject the validity of this judicial decision to the majority-vote of all the citizens.

The implication is clear: amongst all the public legal freedoms – such as the right to attend and organize political gatherings, the right to organize freely within the domain of public interaction, to express political opinions, the right to establish public media, the right to criticize, and the right to protest – the capstone of our public freedoms lies in the active and passive right to vote and be elected. The adjective *democratic* solely captures this competence of the public-legal co-determination and co-responsibility of the citizenry.

Unfortunately, the adjective *democratic* acquired new life in its use as a noun, for in this case, the state itself was identified with the substantive *democracy*, and subsequently it was combined with a wide array of adjectives. Just contemplate composite phrases such as *direct democracy*, *representative democracy*, *social democracy*, *consociational democracy*, *liberal democracy*, *deliberative democracy*, and so on.

We argue that the adjective *democratic* has a very limited scope, for it merely refers to a restricted element of constitutional law. Therefore, instead of speaking of a *democratic state*, it is preferable to use an adjective with a

1 Since a constitution merely stipulates that, when a political party (or parties – in the case of a coalition government) gained the majority of votes at the polls, it (they) can put the government in office, it never means that the party or parties govern the state – no party is ever a state organ; it remains an organization distinct from the state. Strictly speaking, it is therefore mistaken to say that some or other party governs (referred to as the ruling party).

much wider scope, as applied in the phrase *just state*. This phrase not only captures an acknowledgement of the domain of *public legal freedoms*, but also those of civil private law (*civil freedoms*) and the non-civil private law (*societal freedoms*). At the same time, the idea of a *just state* also avoids the problematic idea of popular sovereignty (majority rule – Cunningham strikingly discusses the issue of the "tyranny of the majority" (Cunningham, 2002:119 ff.) and "majority tyranny" (Cunningham, 2002:134 ff.) – with its implied confusion of the relationship between government and subjects – the government is relegated to become the *servant* of the true sovereign, the 'people'. Thus the authority structure of the state, and its inherent relation of super- and subordination, is turned upside down.

Once this relation of 'above' and 'below' has been restored, it is important to keep in mind that governance within a just state always rests on a jur(idic)ally delimited *legal power* (*competence*), attached to an *office*. On the one hand, it accounts for the formation of positive law through competent state organs and on the other, it precludes interference within the internal domain of freedom (jural spheres) of those social collectivities that are different from the state.

What is the upshot of our argument?

> The adjective *democratic* turned out to have a much more modest domain of application, merely restricted to the election process as stipulated in the constitution of a just state. The constitutionally granted competence of citizens to decide by *vote* who will occupy the office of government, *presupposes* this office with its inherent public legal competencies destined to integrate the multiplicity of jural interests within the territory of the state into one public *legal order* in such a way that, whenever an infringement occurs, the government is entitled to harmonize and balance these interests in a truly *retributive* sense. The real concern of the government in office within a *just* state is therefore to maintain a public legal order in which political freedoms, civil freedoms and societal freedoms are guaranteed.

Yet none of the non-political social collectivities and institutions within a differentiated society is *democratic* in the sense in which this adjective is applicable to (a subdivision of the constitutional law of) the state. Consider the following arguments against the idea of a *democratic society*.

In our discussion of the notion of *society*, it appeared that prominent contemporary thinkers revealed an intuitive awareness of the uniqueness and inner laws of distinct societal spheres within a differentiated society. The implication of this intuitive insight is that the *structural type* of each kind of societal collectivity evinces something peculiar; something underlying its uniqueness. Consequently, the forms of co-responsibility and co-determination in a *business firm*, a *local congregation*, a *university*, a *sports club* and even within the *nuclear family*, are all different from the *state*. Within the context of the state, the adjectival meaning of the term *democratic* contains the idea of a *majority decision* – as stipulated by the applicable constitution. But not even within the normal functioning of a government – in fulfilling its public legal task of balancing and harmonizing a multiplicity of legal interests – is it possible to take recourse to the majority principle.

The reason why the majority principle has a very precise and restricted constitutional place within the state is straight-forward: the majority is never the source of what is *just* (or *true*). Every attempt to attribute these qualities to the majority terminates in a self-defeating *regressus in infinitum*: did the majority decide that what the majority decides is just (or true)? And: did the majority decide *that* the majority decides that *what* the majority decides is just (or true)? ...[1] However, from the fact that it is impossible to determine justice and truth merely on the basis of the *majority principle*, it does not follow that there is not a legitimate place for decisions taken by the majority within a "democratic state." Yet this principle remains purely *formal* in respect of the content (material nature) of legal principles and principles of justice.[2]

Just as it is impossible to settle matters of justice through a *majority vote*, so it is also impossible to settle matters of truth through such a vote. For example, there was a stage in modern history in which the majority of people thought that the surface of the earth is *flat* – but this majority stance did not change the spherical nature of the earth at all! Scholarly disciplines do not function on the basis of "majority rule" – it is only within the institutionalization of science and scholarship in universities that a form of organization emerges in which there are elements of academic co-determination and co-responsibility. However, this does not imply that *students* obtain the competence to appoint their teachers! The presupposition of the teaching situation within academic institutions is that there are learned people and there are learners – an academic relation of super- and subordination. Tests and examinations are conducted by competent scholars appointed in the office of *teachers* – and no "democratic vote" can change things, transforming the learners into the sovereign academic authority with the teachers as their less learned subjected servants.

Within church denominations, the 'democratic' principle normally found within a just state is also absent. Church denominations split on the basis of incompatible *confessions of faith*. Whereas a just state enables the existence of alternative political parties, each with its own *political* confession of faith, a similar situation within the church would immediately lead to charges of heresy, simply because the church by its nature as a community of faith can only allow for *one* confession of faith. Moreover, in the case of the nuclear family (father, mother and children) 'democracy' is plainly absurd for it would entail that the children were there before the (to-be-elected) parents!

Of course, voluntary associations, such as sport clubs, do show important similarities with the structure of co-determination and co-responsibility within a just state, but given the fact that they form part of *non-civil private law* (restricted only to those who are members of this *ius privatum*), this precludes

1 Amongst the "rhetorical ploys and fallacies" discussed by Bowell and Kemp the "fallacy of majority belief" is also mentioned (Bowell and Kemp, 2005:131 ff.).

2 Lambrecht incorporates this reference to the formal character of the majority in his general description of the meaning of the term democracy (Lambrecht, 1999:215).

an identification with the democratic process within the state (state-law and citizenship cut across all non-political ties – explaining its true character as a universal *ius commune*).

By and large, the non-political spheres of a differentiated society therefore cannot be meaningfully characterized as *democratic*. The most important reason why contemporary political theorists still speak of democratic societies, is that they implicitly elevate the state to the level of an encompassing whole, embracing all dimensions of a differentiated society as its *integral parts*. The common practice to refer to *democratic societies* therefore actually contains an implicit *totalitarian* understanding of the state.

Comment: Sometimes the terms *absolutist* and *totalitarian* are seen as synonyms. However, the first term is older than the second, since it is applied to the *absolute monarchies* of the period between 1648-1789 – the age of *absolutism*. The rise of the modern state during the 18th and 19th centuries indirectly benefits from the consolidation through the enhancement of the authority of the monarch. The most important feature of the rise of the modern state, as we noted, lies in the breakthrough of the idea that the state by its very nature is a *res publica*, a public concern. Whereas a monarchy belongs to its King, the state, as a *public* legal institution, belongs to the citizens (the public).

The reformational political philosopher, H.J. Strauss (1912-1995), draws a distinction between *absolutism* and *totalitarian* with reference to the difference between the public domain and the private. The term *absolutistic* characterizes a state in which there is no political co-determination or co-responsibility (i.e. no political freedoms), while the term totalitarian concerns the absence of civil and societal freedoms (i.e. the freedoms of civil and non-civil private law). A true *power state* (German: *Machtstaat*) is therefore both absolutistic and totalitarian.

The **concept** of an all-encompassing societal reality, whether it may be the church, the state, the 'people', the 'Bewegung' (Hitler and his party) or whatever, is practically as old as holistic or universalistic theories of human society. That an alternative name or phrase eventually emerged in order to designate this legacy, such as "total state" or 'totalitarianism', merely added a *new name* to refer to a familiar theoretical stance. Designating our awareness of wholeness with the term *totality* already surfaces in the political philosophy of Hegel and Adam Müller at the beginning of the 19th century; Marx spoke of a "total revolution." In addition, it should be pointed out that before 1931, Germans and Italians started to employ the term *totalitarian/totalitarianism* (Totalitarismus; Totalitario). Apparently as correspondent A. Paquet used the term *Totalitarismus* in the *Frankfurter Zeitung* in 1919. (Paquet wrote a Book: *Im kommunistischen Rußland*, that appeared in 1919.) Likewise, G. Amendola mentioned the "systema totalitario" of Mussilini in 1923. A theoretical explanation of the totalitarian characteristics of Fascism is found in a work by L. Sturzo – *Italien under Faschismus*, published in 1926 (see Kapferer, 1998:1296).

The impasse in this mode of speech can only be transcended if the idea of structurally different societal collectivities is taken seriously – each with its own "inner laws," its typical *sphere-sovereignty*. Although we mentioned that an acknowledgement of this idea surfaced in the thought of prominent contemporary scholars (such as Rawls, Habermas, Münch and Walzer), it should also be noted that this promising intuition calls for a more detailed and well-founded analysis of the distinct nature of all the societal realities of a differentiated society, i.e. of their sphere-sovereignty.

8.10 The typical nature of the state exceeds the scope of any discipline exploring only one modal perspective

Acknowledging the multi-aspectual nature of the state ultimately indicates the type law for being a state. This type law concerns more than searching for some or other distinctive property or properties of the state.

Since Bodin (1530-1596) entered the scene we have been confronted with his idea of *sovereignty* as a unique feature of the state. He developed this view in answer to the question of how we should designate the highest authority in society. For this purpose, Bodin introduced the term *sovereignty*. He differs from Machiavelli in that he regards the government to be bound to both natural and divine law (cf. Mayer-Tasch, 1981:35; Bodin, 1981:211). For this reason, he supported the classical principle of natural law: *pacta sunt servanda* (contracts ought to be kept). Moreover, according to Bodin, the *sovereign authority* and *absolute power* of the government is clearly seen from the fact that it is entitled to make laws to which the subjects are bound without their consent (Bodin, 1981:222). His understanding of sovereign power as "summa ... legibusque soluta potestas" reminds us of the view of Occam (1290-1350) regarding the supposed absolute, despotic arbitrariness of God (*postestas Dei absoluta*). Mayer-Tasch characterizes this position of Bodin as a choice for the "classical formula of juridical-political absolutism" (1981:35).[1]

According to Bodin, sovereignty is not only *absolute*, but also *indivisible*. As a result, he is convinced that, within its territorial boundaries, the state displays an absolute and original competence to the formation of (statutory) law. Of course this view is connected to the relatively undifferentiated medieval society, dominated by the Roman Catholic Church as supra-natural institute of grace. The striking fact about the feudal communities, with their multiple relations of super- and subordination, is that all these forms of organization are characterized by an absence of a *monopoly* over governmental sword power of the government. This makes it clear why Bodin interpreted every original claim to the formation of law as a threat to the idea of the state of a *res publica* – in the sense that it aimed at acquiring original sword power.

Although the guild system obstructed the realization of a genuine state-organization, breaking down the artificial hold of power of the Roman Catholic

1 In reaction to this, Calvin adhered to the conviction that although God is elevated above the law for creatures, he is faithful to them and therefore not arbitrary (see the analysis of Bohatec, 1940:13 ff.).

Church after the Renaissance it led to a process of societal differentiation that was decisive for the emergence of the modern state, because this process generated the distinct legal interests that were eventually bound together within the one public legal order of the state.

The irony is that the intention of Bodin's theory of sovereignty, namely to establish an absolute monarchical power by means of a *monopolization* of the sword power of the government (with its exclusive competence to form positive law), is that such an integration of governmental authority necessarily forms part of the *differentiation* of society. He did not realize that this process of differentiation contradicts the idea of such an exclusive competence to the formation of law, because differentiation gives rise to societal collectivities distinct from the state with their own internal spheres of law.

How then are we to understand the idea of sovereignty when it is ascribed to the authority of a state? It should indeed capture the *legal competence*, embedded in the office of government – a competence enabling the *formation* of positive law. This legal power represents an anology of the cultural-historical aspect within the structure of the jural aspect (jural competence – see page 95 above), and it embodies the official power of a state-government over its citizens (power over persons – an instance of subject-subject relations). In this way, typical principles for being a state are given form, i.e. they are positivized.[1] Moreover, it must be remembered that legal power – representing the cultural-historical analogy on the norm side of the jural aspect – is not the same as the *original function* of the state within the cultural-historical aspect. The latter is found in the *power of the sword*.

As a result of the unbreakable coherence between the qualifying jural aspect and the foundational cultural-historical aspect of the state, both functions ought to be recognized as intrinsic to the multi-aspectual nature of the state. When Habermas distinguishes between *positive law* and *political power*, or when he says that law serves the organization and guidance of state power, it is clear that he does not sufficiently recognize the intrinsic qualifying role of the jural aspect in the state.[2] Although he acknowledges a conceptual kinship between *Rechtsetzung* (law-making) and *Machtsbildung* (power formation), he does not realize the difference between the jural power entailed in law-making (a retrocipation to the cultural-historical aspect within the jural) and original power formation in a cultural-historical sense.[3]

1 In our analysis of principles, we argued that within certain societal collectivities, power over persons concerns subject-subject relations, a competence to positivize principles and also a subjective moment on the norm side of reality (see page 249).

2 Habermas distinguishes between the "Rechts- und Machtskode," between "Recht und politischer Macht," and in this context he holds: "Das Recht erschöpft sich keineswegs in verhaltenssteuernden Normen, sondern dient der Organisation und der Steuerung staatlicher Macht" (Habermas, 1998:180; "Law by no means exhausts itself in behavioural norms, but serves the organization and regulation of state power" – cf. Habermas, 1996:144).

3 "Die konzeptuelle Verschwisterung von Rechtsetzung und Machtbildung ..." (Habermas, 1998:185; cf. Habermas, 1996:149).

Considering the legal principles involved in erecting and maintaining a state focuses on an all-important aspect of reality, in which the state as a societal collectivity functions, namely the jural aspect. Accounting for the legal competencies entailed in the office of government, however, adds a complication to the picture, because it shows that functioning within a particular aspect implies that such an aspect cannot be understod in isolation from all other aspects, as clearly seen in the nature of *legal competence*, which reveals an interconnection between the jural and the cultural-historical aspects of reality. Before we explore this complication further, we first briefly look at the many-sidedness of the state.

8.10.1 *The multi-aspectual nature of the state*

Natural and social entities, as well as all events (or processes) within reality in principle function in *all* the (ontically given) distinguishable modal aspects. The state, for example, is a social entity comprising a multiplicity of individuals (designated as *citizens*). As such, the state has a function within the *quantitative aspect* of reality. Stating that the state functions within the numerical aspect of reality, affirms *one* of its *many* modes of being (existence). Yet the existence of the state is not exhausted by its arithmetical functioning. Regarding the spatial function of the state, one contemplates the *territory* of a state. On the basis of this *locality*, a state not only embraces the existence of its citizens in a specific way, since it is also dependent upon their connection to the state in spite of their relative movements (a term stemming from the kinematic aspect). Through its juridical organization, the "sword power" of the state is capable of using the required *force* whenever necessary – in service of restoring law and order when certain legal interests are encroached upon (think about the *actions* of the police or the defense *force*). The term *force* stems from the physical aspect of energy-operation and in this context, it elucidates the function of the state within this aspect.

The state as a public legal institution binds together the *lives* of its citizens in a specific way – in the sense that a certain portion of one's life time actually belongs to the state (insofar as work for the part of one's income destined for tax-paying is concerned) and also in the necessity that the state can only maintain its territorial integrity against possible threats from outside if citizens are integrated within the defense force[1] – even running the risk of being *killed* in military action. Clearly, *life* and *death* assume their own roles within the state as an institution – and it undeniably testifies to the fact that the state does function within the *biotic aspect* of reality as well. Jim Skillen correctly points out: "Likewise, a political community exhibits biotic functions by the fact that its citizens function biotically, and many of its laws deal with public health and natural environmental regulations" (Skillen, 2008:12). The *nation* of a state (transcending diverse ethnic communities without eliminating their right of continued existence) operates on the basis of a national *consciousness* and an

1 Not to forget the role of the police force in protecting of the legal interests of citizens.

emotional sense of belonging. Although it does not apply to all citizens, a worthwhile state should succeed in making its citizens *feel* at home (the notion of a *Heimat*). These phenomena clearly cannot be divorced from the *sensitive-psychic function* of the state. Furthermore, once we realise that citizens ought to *feel* at home within the state, they can also positively *identify* with it (consider ID documents in this regard) – the political content of what sociologists would call the 'we' and the 'they' – those belonging to this state and those not belonging to it. Since the core meaning of the logical-analytical aspect is captured in the reciprocity of *identification* and *distinguishing* – whoever identifies something is also involved in distinguishing it from something else (see the sketches on pages 14 and 151 above), the *national identity* of the citizens of the state testifies to the fact that this identity cannot be understood apart from the function of the state within the logical-analytical aspect. When we take into account the argumentative possibilities entailed by functioning within the logical-analytical aspect of reality, we discover that the nature of the *public opinion* within any particular state in a broader sense manifests the function of the state in the said aspect.

The historical aspect of reality concerns *formations of power*, since it brings to expression the basic trait of *culture*, namely the uniquely human calling to have stewardship of the earth and to disclose the potential of creation in a process of cultural development. Such a process goes hand-in-hand with an on-going development of human society in which – through increasing differentiation and integration of specific societal zones (spheres) – distinct societal collectivities, such as the state, eventually emerge. It is only on the basis of its "sword power" that the state can function as a public legal institution, because maintaining a public legal order requires a monopoly of the "sword power" over the territory of the state. Of course, the function of the state in the historical aspect is also clearly evidenced in the actual *history* of every distinct state. That the state has a function within the sign mode of reality is obvious from its national *symbols* (anthem, flag, etc.) and from its official *language(s)*. Similarly, the function of the state within the social aspect of reality is evident in the way it binds together its citizens within a public legal institution. It thus determines a specific kind of social interaction. Participating in a general election, acquiring an ID, observing traffic rules, respecting the rights of fellow citizens – and many more forms of social interaction exemplify the function of the state within the social aspect of inter-human interaction.

Raising taxes not only affects the financial position of the citizen, but also enables the state to fulfill its legal obligations in governing and administering a country – bringing to light a facet of the *economic function* of the state. Although a state is not an artwork, it is a typical task of a government to *harmonize* clashing legal interests. Establishing balance and harmony amongst the multiplicity of legal interests within a differentiated society is always guided by the *jural function* of the state. In addition to this internal coherence between the jural and aesthetic aspects of the state the latter also has an external (i.e. original) function within the aesthetic aspect, displayed in the characteristic format of published (promulgated) state laws, in the aesthetic qualities of governmental buildings (houses of parliament, jails), and so on. The idea of *public justice* is impossible without the function of the state within the jural aspect of reality. The state also requires an *ethical integrity* amongst its citizens, for without this *loyalty*, the body politic will fall apart (of course the government must also conform to standards of public decency and integrity). It is therefore appropriate that the extreme of disobedience to this loyalty is punishable if a citizen is found guilty of *high treason*. The nation of a state must share in its vision, its convictions regarding establishing a *just* public legal order, and give each citizen its due. It is only on this basis that the highly responsible task of governing a country could be en-*trust*-ed to those in office. Terms like 'trust', 'certainty' and 'faith' are synonymous. The *certitudinal* or *fiduciary* aspect of reality – the faith aspect – is therefore *not* foreign to the existence of the state. Just like all the other aspects this one intrinsically co-conditions the existence of every state and also explains why no single state can exist without functioning within the faith aspect of reality as well.

8.10.2 *The type law for being a state*

Perhaps the most striking fact concerning the history of political theories is that they always return to the nature and interconnection of 'might' and 'right'. Sometimes the state is portrayed as an institution endowed with absolute *power*, while and at other times it is seen as protecting what is right. Litt notes that all reflection on the nature of the state oscillates between the two poles of state acitivity; *Macht* and *Recht* (*might* and *right* – see Litt, 1948:23).

Power within the state is easily identified with the *military strength* at is disposal. The earlier development of modern political theories as we noted, confined themselves to choosing between the two extremes of popular sovereignty and the sovereignty of the monarch. By the turn of the 20th century an alternative theory of sovereignty emerged, namely that of *state-sovereignty*. Prominent political theorists in this tradition are Gerber, Laband, Jellinek and Otto Von Gierke (1841-1921). By identifying the state with its power, the demands of 'right' are excluded – the state turns into a pure power institution. For this reason, Gierke considers 'might' and 'right' as two independent and specifically distinct sides of communal life (see Von Gierke, 1915:105).

The jural is thus turned into something completely *external* to the state. This raises a question regarding the *jural competence* involved in the formation of law. Is it possible for an institution that is characterized by the non-juridical feature of cultural-historical power to play a role within the domain of law formation? Parsons echoes this tradition in his view that the political organization ought to obtain effective control over the internal organization of 'force', and it should be integrated with the "legal system." "Because of the problems involved in the use and control of force, the political organization must always be integrated with the legal system" (Parsons, 1961:47). The doctrine of the sovereignty of law (legal sovereignty) is also misguided, because the judiciary constitutes merely one of the three central government functions of a state (the other two being the legislative and the executive). The office of government is not occupied by any one of these three state-functions. Of course performing these functions are constantly bound to and normed by the type law of a just state.

The integral coherence between the dimensions of modal aspects and multi-aspectual entities (including societal collectivities) first of all calls for the acknowledgement of both the cultural-historical and the jural aspects as intrinsically belonging to the structural principle of the state. 'Might" and 'right' cannot be separated.[1] In fact, they serve as the two characteristic functions of the structural principle of the state, its type law. Already in Chapter 7, we used the phrases *foundational function* and *qualifying function* to capture the structural uniqueness (typicality) of natural entities. In the present context, a similar task confronts us.

The classification of social forms of life with the aid of compound basic concepts, such as specified in the distinction between *societal collectivities*, *communities* and *coordinational relationships*, does not contain any criteria enabling the discernment of typical differences such as between a church denomination or a state. A university and business firm both display the distinctive features of societal collectivities, namely the presence of a durable relation of super- and subordination and a solidary unitary character. However, these two features do not provide us with criteria to distinguish between societal collectivities as such. The only theoretical access we have to the *typical differences* exhibited by these forms of life is the point of entry provided by the meaning of distinct sphere-sovereign modal aspects.

The decisive element in applying the idea of sphere-sovereign modal aspects to distinguish between different kinds (types) of societal collectivities, is to realize that the *foundational function* and the *qualifying function* of such social forms of life are mutually dependent and co-determinative. This becomes evident as soon as one realizes that most societal relationships have their foundational function within the cultural-historical aspect, since most are based upon some or other *type* of *power formation*. Without taking the qualifying function of these social forms of life into account, it will not be possible to differentiate the types of power formation exemplified by them. Of course this view assumes the (ontic) modal universality of all aspects, including the jural. It opposes the nominalistic legacy that does not accept ontic universality. Jellinek, for example, explicitly rejects the idea of the jural as independent (in an ontic sense) from the human being. According to him, law is an ingredient of human representations in the human mind, and coming to a closer determination of what law is, amounts to establishing which part of the

1 Habermas correctly emphasises the intrinsic coherence (interne Zusammenhang) between law (Recht) and the democratic-constitutional origination, acquisition and application of political power. ["... der interne Zusammenhang zwischen dem Recht und der demokratisch-rechtstaatlichen Organization der Entstehung, des Erwerbs und der Verwnedung politischen Macht ..." (Habermas, 1998:70) Later, he mentions the mutually constitutive coherence between "Recht und politischer Macht" ("law and political power" – Habermas, 1998:208).]

contents of human consciousness should be designated as law.[1] Compare the equally nominalistic stance of Descartes regarding universals as modes of thought, as mentioned in Chapter 1 (page 19).

In the case of the state, guided by the jural aspect as its qualifying function, the specific type of power formation lies in the integration of the "sword power" within the territory of a state. The idea of the state does not allow for original sources of "sword power" within its territory distinct from what has been monopolized by the state. The government of a state has to have sole jurisdiction (monopoly) over the power of the sword. Of course this entails that the existence of a state presupposes that all undifferentiated forms of life ought to be left behind, such as those in the medieval guild system and the feudal system.

Power formation is not something purely factual, and even less is it a-normative. However, since power originally stems from the historical aspect of reality, it must be clear that the foundational function of the state lies in the historical aspect. Human beings are called to explore different kinds of power formation, both in the context of subject-subject relations and subject-object relations. Depending upon the typical qualifying function of other societal collectivities, alternative types of power formation are found within a differentiated society, such as capital (economic power) in the firm or the power of God's Word in religious life. Power formation within the state should always be guided by its leading or *qualifying function*. For that reason, the term *power* ought not to be taken in the negative sense of untamed, brutal force. Much rather, according to the cultural mandate to humankind, it must be seen as a cultural calling; an assignment given to humankind, which places a peculiar task and responsibility on the shoulders of office-holders within the power formations of the state. Naturally, owing to the effects of sin, this task can be accomplished to a better or worse extent.

The jural aspect as qualifying function of the state stamps the state as a *res publica* and explains why no other societal collectivity has the calling to function as a *universal integrator* of diverse *legal* interests in a truly *public* legal sense. None of the non-political societal collectivities, such as the firm, university, free association, or the church has a *juridical qualifying function*. This follows from the fact that, in their formation of law, they are restricted to *specific* private legal spheres. By contrast, the internal law of the state, in a typical universal-juridical way, transcends the boundaries of all classifications on the basis of societal institutions and communities distinct from the state. This diversity of persons is united as *citizens* of one and the same *state*.

1 "Entweder man sucht die Natur des Rechtes als einer vom Menschen unabhängigen, in dem objektiven Wesen des Seienden gegründeten Macht zu erforschen, oder man faßt es als subjektive, d.h. innermenschliche Erscheinung auf. ... Das Recht ist demnach ein Teil der menschlichen Vorstellungen, es existiert in unseren Köpfen, und die nähere Bestimmung des Rechtes hat dahin zu gehen, welcher Teil unseres Bewußtseinsinhaltes als Recht zu bezeichnen ist" (Jellinek, 1966:332).

8.10.3 *The unique position of the state within society*

The juridical unity binding together government and citizen within the state *abstracts* from ethnic differences, racial differences, religious differences, language differences, differences in economic position and social rank, family differences and differences in dispositions or intelligence. However, the public legal integration achieved by the government ought not to be appreciated in a *universalistic* sense, as if it amounts to leveling or negating all the mentioned differences. On the contrary, it belongs to the nature of the state that it creates a *juridical unity*, leaving the diversity *intact*. The outcome is that all people in the territory of the state, by means of this juridical unitary organization, receive a function within the state as collectivity.

> The state therefore *disregards* non-political ties in order to care for the *legal interests* entailed by them!

We saw that all collective and communal relationships are strictly correlated with coordinational relationships. The implication of this insight is that the full meaning of being human can never be *exhausted* by being a citizen within the state.

The internal public law of the state (such as constitutional law, criminal law, criminal procedural law, administrative law and whatever public jural interests there may be) is guided by the *typical principle* of the *public interest*. It concerns the *res publica* or public law in a broad sense, also embracing international public law.

The internal public law of the state contains the internal order arrangements according to which the *governance* and *practice of law* of the different legal organs (such as the legislative, judicial, and executive competencies) take shape. It also relates to the *political way* in which government and subjects participate in the legal intercourse occurring within the state in the public sphere where the public opinion is articulated.

At this point, we may recall the distinction between constitutive and regulative (legal) principles (see pages 313, 315, 317 ff. above). Constitutive legal principles could be compared to bricks in a wall – without them the house will collapse.[1] Standing on the roof supported by the walls may enable us to observe a purview transcending the limits of the house. Similarly, those modal analogies within the legal aspect referring back to the aspects *preceding* the jural mode are *constitutive structural building blocks* within the jural aspect. They are found in *every possible legal order*. Yet the meaning of the jural aspect may also be *disclosed* under the guidance of the two aspects succeeding the jural aspect, namely the ethical and the certitudinal aspects. In other words, whereas *constitutive* structural elements within the meaning of jural relationships are presupposed in every possible legal order, the *regulative* structural elements only come into view when the meaning of legal relation-

1 Plato's understanding of justice in fact only pertains to constitutive elements within the functional structure of the jural aspect – it accounts for the foundational moments of harmonizing through avoiding what is excessive (the jural indeed presupposes the economic and aesthetic aspects).

ships is *deepened* and *disclosed* (opened up) by the moral and fiduciary aspects. Fault, as we saw (see pages 229 and 289 above), refers to a disclosed principle of criminal law. It requires that the punishment should fit the crime (taking into account both forms of fault; *intent* and *negligence*).

This principle of punishment proportionate to fault is a *deepened* legal-ethical principle of *justice* that is fundamentally different from the strict responsibility for outcomes evident in undisclosed legal systems (e.g. the talio-principle in the Old Testament, known as the "eye for an eye" or "tooth for a tooth" principle). In the *talio*-principle, the ethical aspect of moral love had not yet deepened the meaning of the jural aspect of reality, since the attitude of the actor was neglected, and only the consequences of the act taken into account (see pages 219 ff. above).

In an ethically deepened or disclosed legal system, life imprisonment is only one *application* (positive expression) of the underlying principle of *punishment according to fault*. Another application of the same principle, for example, could be a shorter term imprisonment, depending upon the degree of possible mitigating circumstances.

Other *deepened* principles of justice in disclosed legal systems are, for example, the principle of *equity*, good faith (*bona fides*), and the *dignity of the human person*. Disclosing the meaning of the jural aspect therefore opens up regulative *legal-ethical principles* of *justice*. Of course this *deepened meaning* of the jural aspect cannot be grasped *conceptually* – it can only be approximated in terms of knowledge exceeding the boundaries of conceptual thought. When Jacques Derrida speaks about justice, he implicitly alludes to the deeper meaning intended here. He says: "Justice, if it has to do with the other, with the infinite distance of the other, is always unequal to the other, is always incalculable. You cannot calculate justice. Levinas says somewhere that the definition of justice – which is very minimal but which I love, which I think is really rigorous – is that justice is the relation to the other. That is all. Once you relate to the other as the other, then something incalculable comes on the scene, something which cannot be reduced to the law or to the history of legal structures" (1997: 17-18).

8.10.4 *The salus publica as regulative typical principle*

The fact that the state is a *public* institution that concerns itself with the interests of the public (*res publica*) reveals the public interest as *typical principle* in the *regulative* sense of the word. The regulative modal legal principles receive a *typical specification* in the idea of the *public legal interest*.

The specific content of the principle of the *salus publica* (public good) derives its meaning from the *sphere-sovereignty* of the state. Often, those principles referring to the internal structural nature of a societal bond are also captured by calling them *material* principles. Acknowledging *material principles* gives recognition to the sphere-sovereign, entitary-structural nature of the societal life form concerned, while at the same time differentiating itself from *formal* principles that cut across the typical principles of different life forms (or: spheres of law).

The *idea of public justice* therefore does not only exhibit the *typically deepened meaning* of the qualifying jural aspect of the state, since its entire focus presupposes the *material sphere of competence* of the state as a *public legal institution*. Only this focus provides a meaningful portrayal and delimitation of the idea of the *salus publica*, because it entails that the state in principle cannot assume an *absolute sovereignty* over those societal spheres that *differ in principle* from the structural nature of the state, although it was often factually attempted by absolutist and totalitarian states in their abuse of power.

Dooyeweerd rightly points out how diverse and unstructured the appeal to the idea of the *salus publica* could be if divorced from an insight into the sphere-sovereignty of the qualifying jural function of the state:

> The idea of the "*salus publica*" displays a genuine Protean character in political theory. It was subordinated to the ancient universalistic-organic theory of the State, to the doctrine of the "reasons of State," to Wolff's natural law theory of the police State, to Hobbes' and Rousseau's natural law construction of the Leviathan State, to the classical liberal doctrine of the constitutional State (Locke and Kant), and also to the modern totalitarian political theories.
>
> For the sake of the public interest, Plato and Fichte defended the withdrawal of children from their parents to entrust their education to the body politic. With a further appeal to the public interest, Plato wanted to abolish marriage and private property as far as the ruling classes of his ideal State were concerned. Aristotle wanted education to be made uniform in "the public interest"; on the same grounds, Rousseau wished to destroy all the particular associations intervening between the State and the individual. Wolff desired the body politic to be part of all human affairs and, at least for the Protestant churches, he wanted the government to determine the confession. The idea of the "*salus publica*" was the hidden dynamite under the humanistic natural law theories of Hugo Grotius and S. Pufendorff.[1] In Chr. Wolff's doctrine of natural law, this idea resulted in a frankly admitted antinomy with his theory of innate natural rights.[2] The slogan of the public interest was the instrument for the destruction of the most firmly established liberties because it lacked any juridical delimitation.
>
> The terrible threat of Leviathan is audible in this word as used in a juridically unlimited sense. The universalistic political theories could conceive of the relation between the State and the non-political societal structures only in the schema of the whole and its parts. This is why they could not delimit the idea of "the public interest." (Dooyeweerd, 1997-III:442-443.)

Summarising what is at stake in this context, one can say: *Those structural principles guiding the functioning of the state therefore ought to*

1 Cf. Dooyeweerd's In den Strijd om een Christelijke Staatkunde, I, XV (Anti-Revolutionary Politics, Year 1, three-monthly Journal) pp.142 ff.

2 Chr. Wolff, Jus Naturae VIII, 1, § 117; here he speaks of a real "collisio legum" between his principles of natural law and the basic principle of his theory of the State: "Salus publica supreme lex esto." He cuts the Gordian knot with his construction of an emergency law of the State: "Necessitas non subditur leg."

be seen as the universal, relatively constant starting points that can only be made valid by competent organs within the material sphere of competence of the state as public legal institution.

Constitutional law in a restricted sense encompasses the principles related to the *organizational form* of the state and with the arrangement of the mutual relationship between the internal spheres of legal competence, as well as the obligations of the central and lower legislative administrative bodies (such as courts, parliament, provincial councils, municipalities and state departments).

Constitutional law also embraces the various public legal freedoms of citizens, such as the right to political gatherings, the right to organize freely, to express political opinions, establish public media, to criticize, to protest and, of course, the capstone of our public freedoms: the active and passive right to vote and to be elected.

The *right to vote* represents the principle of *political co-determination* within the constitutional state. The principle of *differentiated governmental functions* requires that, *even* if the same organ performs more than one of these functions – namely the legislative, the judicial and the executive –these functions should be distinct.[1]

In the execution of its typical state task, the government itself is also subject to the law in a *formal sense*. Within the British tradition, this is known as the "rule of law." In his famous work, *Law of Constitution*, the well-known constitutional scholar, A.V. Dicey, states that the "rule of law" entails that no one is elevated above the law. According to this rule, no person is punishable unless an existing law is violated, and the case was brought to court and handled in accordance with the applicable procedural rules. The third feature posits that the general principles of a constitution – for example the right to personal freedom and the right to organize public meetings – are the result of judicial decisions that are determinative for the rights of private persons brought to court in specific cases (Dicey, 1927:183 ff., 191 ff.)

However, in spite of this formal connection to the law, we still find in Dicey's thought a view of *parliamentary sovereignty* that practically assigns an *unlimited competence* (in the formation of law) to parliament – clearly evinced in the subsection discussing the "Unlimited legislative authority of Parliament" (Dicey, 1927:39 ff.). According to Dicey, the *sovereign power* of the parliament can be seen clearly in its ability to legislate what is unjust: "The point to be now noted is that such enactments being as it were the legislation of illegality are the highest exertion and crowning proof of sovereign power" (Dicey, 1927:48).

The mere fact that the word *illegality* is employed, demonstrates that a competent state organ such as the parliament, can also act in *antinormative* ways!

In spite of his consideration of *possible restraints*, his view does not in princi-

1 Montesquieu's theory of the *trias politica* actually suggested that there are *three* governments within *one* state – instead of one government performing three distinct governmental functions!

ple succeed in arriving at an *inner delimitation* of governmental authority (cf. Dicey, 1927:74, 77). In this regard, he does not transcend the shortcomings in the prevailing humanistic dogma of sovereignty.

8.10.5 *Political aims presuppose the internal structural principle of the state*

In the light of peculiar circumstances, any government may pursue various, sometimes widely diverging *goals*. These purposes, aims or goals do not necessarily have to fall *outside* the typical sphere of competence of the state – although it is often the case that they in fact do precisely that. Any societal collectivity, including the state, may pursue certain goals on the basis of being *this* or *that*. Without an intrinsic, sphere-sovereign structural nature (being *this* or *that*), this would be impossible. As a consequence, certain practical aims can never *qualify* the state. Rather, being uniquely qualified by the jural aspect, we may distinguish between *typical* and *a-typical* aims of the state. For example, it is the task of the state proper to maintain civil and criminal law courts. One cannot imagine a state for which this is not the case – at least not if the state aims at *being a state*. To establish a kind of (economically) qualified enterprise (a firm), although it may be an a-political goal of the state, does not however contribute to the definition of the state. In fact, even if a state abstains from establishing any kind of firm or business enterprise, but manages to function as a public legal institution, it can still be a *good* state.

8.10.6 *Non-civil private law*

We have seen that the state can only emerge if various (sphere-sovereign) societal collectivities and communities that are distinct from the state simultaneously originate. These non-political societal life forms distinguish themselves from the state in that none of them is qualified to be a *public* legal institution. The church, the firm, the university, the sports club, and many other non-political communities, in other words, do not have a qualifying *jural* function.

> Does this imply that these forms of social life do not have a jural function at all?

The answer must be negative; because we argued that the distinctive trait of the dimension of entities is that they have functions within *each* aspect of reality, without distinction. It therefore follows that each will also function within the *jural aspect*, a fact that accounts for their respective *internal* legal arrangements. The church has its own internal jural function – known as ecclesiastical law; the same applies to marriage, the school, the business enterprise, and so on. The important point is to realize that the juridical functioning of any non-political form of social life is always characterized by the peculiar (non-jural) qualifying function of the life form concerned. The internal law of marriage as community is qualified by *love* as its guiding function; the internal law of the firm is stamped by the *economic* aspect as its qualifying function, and so on.

Each of these non-political societal forms of life therefore contains its own internal and original jural competence to form (non-civil) private law. In a negative way, these forms of life demonstrate the *limitations* of the state's competence to form law as well.

Dicey's view of "absolute parliamentary sovereignty" or the conviction of Rousseau and Kant, namely that the *general will* of the people within the state, is the *only* source of law, in principle consistently eradicates the differences between the *materially* distinct spheres of law within a differentiated society. All these internal spheres of law, qualified by some or other non-jural aspect, belong to the domain of *non-civil private law*. This designation naturally implicitly refers to the domain of *civil law*.

8.10.7 *Civil law*

The public law-sphere of the state characteristically displays a relation of super- and subordination. By contrast *civil law* has a *coordinational* structure – it concerns the actions of individuals or societal wholes on equal footing next to or opposed to each other. Within this legal sphere, coordinational legal relationships are arranged while *disregarding* the non-jurally qualified contexts within which they may occur.

The Greek view of the city state as all-inclusive bond did not leave any room for the freedom and equality of the individual. Stoic natural law particularly posited the natural freedom and equality of all people. The original Roman *ius civile* was a *tribal law* which, as such, was *undifferentiated* and *exclusive*. Participation in this legal sphere was connected to being a *member* of the Roman community. Those not belonging to it had no rights (they were *exlex*, *hostis*). The Roman power motive, which indeed permeated all Roman culture, separated the power of the *gentes* (with their ancestral cults) from the power sphere of the Roman tribe (the *civitas*). During the rise of the Roman republic, the power of the *gentes* was broken and transposed to the smaller Roman family community (*familia*). This *familia*, which still displayed an *undifferentiated* character, was guided by a head (*pater familias*) who possessed absolute power and even the power of life and death over members of the *familia*.

Without cancelling this exclusive and totalitarian legal sphere of the *pater familias* during the gradual expansion of the Roman Empire, there emerged a need to make legal provision for *all* people within the Roman Empire, including those who were not Roman citizens. This development gave birth to the Roman *ius gentium* which, as a kind of private international law, appreciated every person (slaves excluded) as a *legal subject*. The classical Roman jurists connected this view to the Stoic natural law (*ius naturale*), according to which all people are naturally *free* and *equal*. In this way, as complement of the public law of the state, an inter-individual legal domain emerged, in which every free person can function as a *legal subject*, irrespective of what collective or communal links such a person may have. The development of the Roman *ius gentium* actually provides the historical starting point of today's *civil private law*.

Yet the Roman power motive, with its aim for *world dominance*, constantly threatened this sphere of freedom of the individual. It also explains why the Byzantine Empire (commencing by the end of the 3rd century A.D.), oriented to the Greek-Eastern conception of a *holy empire* (*sacrum imperium*), destroyed the domain of personal freedom. It was the later Byzantine Emperor, Justinian I (reigning from 527-565 A.D.), who largely cleared up the prevailing uncertainties through the comprehensive *codification of law*, known as the *Corpus Juris Civilis* (528-534 A.D.). This codification provided a link that was explored by the eventual developments in later Roman law and its reception in Roman-Dutch law.

Since civil private law observes *coordinated legal relationships* between individuals while abstracting from the non-jural qualification of the societal context in which these legal relationships appear,[1] civil law ought to be seen as the reverse side of both public law and non-civil private law.

The link with the state is of fundamental importance – not only because civil private law has a jural qualification as well, but also because the state is required to handle civil cases through an impartial *civil jurisprudence*. Within civil law, the *juridical value* of the human personality is protected regardless of the person's socially differentiated ties (related to differences in social rank, language, race, religion, or economic position).

However, in order to abstract from the multiplicity of these differentiated ties, it is supposed that such a differentiated multiplicity of non-political societal forms of life are *ab initio* in existence. In these forms of life a person functions as member of a larger whole and therefore cannot be considered as a person with its own individual private sphere. In the UK tradition, this sphere of individual freedom and the principles of justice protecting it are found within the *common law*.

Dooyeweerd's significant contribution to the distinction between *public law*, *non-civil private law* and *civil (private) law* may be explained by adding a rather extensive quote from him in this context (Dooyeweerd, 1997a:98).

> The inner delimitation of the legal power of the state is determined by the internal structural principle of the societal institution. The *ius publicum*, constitutive of the internal law of the state as public legal institution, does not permit service to group interests external to the jural qualifying function of the state.
>
> Therefore, the nature of the state is irreconcilable with the allocation of *privileges* to specific persons or groups. Similarly, no individual or group may withdraw from the public legal power of the government within the *sphere of life of the state*.
>
> For this reason, the state had to commence its entry into the world scene by

1 In his *Philosophie des Rechts* Hegel characterizes civil society (*bürgerliche Gesellschaft*) in similar terms, for in the latter a human being counts "weil er Mensch ist, nicht weil er Jude, Katholik, Protestant, Deutscher, Italiener usf ist" (§ 209; see § 190). ("because such a person is a human being, not because this person is a Catholic, Protestant, German, Italian etc.")

starting to do away with the undifferentiated spheres of authority of private lords and societal collectivities which withdrew their subjects from the legal power of the state.

In order to achieve this aim, the *public legal principle* of *freedom* and *equality* has to be pursued. It also forms the basis upon which *civil legal* private freedom and equality are to be attained. As long as it is possible for private lords and private societal collectivities to exercise an exclusive and undifferentiated power over their subjects, there is no room for a truly *ius publicum* or for a truly *civil ius privatum*.

It is only the *state*, on the basis of its public legal power, that can open up to the individual person a *civil legal* sphere of freedom, providing that person with a guarantee against the over-exertion of power by specific private communities, or by the public legal power itself, as long as the public office bearers remain aware of the inner limits of their competence.

The state, in view of the inner nature of the *ius publicum*, does not have the competence to bind the exercise of civil private rights to a specific socio-economic destination, simply because the *ius publicum* intrinsically lacks any specific *economic qualification*.

8.10.8 *Criminal law and civil law*

Because penal law belongs to the sphere of public law, the super- and subordinational structure of the state also typically characterizes it. Criminal law arranges the legal protection of *public legal interests*, such as the security of the state and the public legal interest of every citizen regarding that citizen's *property* and *bodily integrity*. A fundamental principle of criminal law is that no sentence could be enforced without a pre-existing promulgated law. In addition, the innocence of the accused is accepted until the guilt of that person is established and a legitimate conviction determined (a person cannot be prosecuted twice for the same crime). The burden of proof, residing with the state, requires a high degree of probability – the guilt of a person must be proven *beyond all reasonable doubt*. (Some of these principles belong to *criminal process law*.)

Civil private law, as a truly coordinational law that provides for every person within the territory of the state, regulates the mutual action of free and equal individuals standing next to, and sometimes opposed to, each other. The principle of *civil freedom* reveals itself in various forms, from the choice of marriage partner, the free exercise of property rights, and contractual freedom, up to the freedom to initiate a civil legal action. The principle of *civil equality* manifests itself in the fundamental equality of everyone before the civil law, which means that all *royal rights*, according to which someone on behalf of owning certain properties may be assigned certain public legal privileges, are irreconcilable with the nature of civil law as a differentiated legal system. As in the case of criminal law, the civil legal process also displays a *typical* nature with a *material structure* that cannot be defined in a purely *formal* way. In the civil legal process, the required standard of proof to establish the liability of the defendant is only on a preponderance of probability.

At this point we may summarize our analysis of the position of the state within a differentiated society by briefly highlighting again the distinctions introduced.

A differentiated society does not only present itself to us in diverse, unique (sphere sovereign) life forms (church, state, business, family, school, etc.), but also in three indissolubly coherent legal spheres.

(i) *The public legal sphere*
This sphere encompasses the relations within the state between government and subject, as well as the legal order among nations (international law), with its accompanying coordinational nature. As such, it encompasses *international public law*, *constitutional law*, *penal law*, *penal procedural law*, and *administrative law*. The political rights of citizens are circumscribed by this legal sphere: the right to political assembly, organization and opinion, as well as the rights to criticism and protest, with the right to vote and be elected as their capstone.

(ii) *Civil private law*
Civil private law abstracts from all non-state relationships in which a citizen may take part. This legal sphere protects citizens in their position as *free individuals* within differentiated legal interaction, and as such, it is a guarantee of *individual personal vindication* in legal life. It is distinct from constitutional law, in which there is a relationship of super- and subordination between government and subject. *Civil private law* maintains a coordinational legal relationship among individuals and institutions. Both public law and civil private law are *jurally* qualified.

(iii) *Non-civil private law*
This is the sphere that encompasses the internal law of the various non-political societal forms of life. This means that, in every instance, the law is *differently* qualified. Internal business law is qualified by the economic function of the firm; internal church law (ecclesiastical law) is characterized by the certitudinal function of the church as a faith-collectivity, etc. This sphere of law delimits the legal competence of the state *externally* – that is, apart from the internal delimitation of government action as guided by the juridical qualification of the state as such.

Although Kalsbeek is acquainted with the state as a public legal institution, his explanation in this regard may be misunderstood in respect of the relationship between the state and the non-political societal collectivities within a differentiated society. He writes: "The state is public in the sense that every person living within its territory and every community and association having its domicile therein is subject to the state's legal jurisdiction and has a right to its legal protection" (Kalsbeek, 1975:226). Unless "subject to the jurisdiction" is interpreted as the jural way in which the state integrates non-state entities within its public jural integrating function, it may suggest an *elimination* of the sphere-sovereignty of these non-state entities. According to their inner structural principles they are strictly speaking *not* subject to the "state's legal jurisdiction" because that would suspenjd their own inner sphere of law. The latter phrase, for that matter, lacks any precision: the moment it is properly specified, the inner limitation of the competence of the state surfaces. For example, if this phrase intends the above-mentioned domain of public law, encompassing constitutional law, the international law of nations, administrative law, criminal law and criminal procedure, then it is clear that none of these spheres embraces anyone of the internal spheres of law of any non-political societal collectivity.

If these fundamental differences between the legal spheres are ignored, assessing the diverse legal interests within the territory of the state in a constitutionally correct manner would be impossible. This approach emphasizes above all else the *jural* task of the state as a public legal institution, called upon to maintain a public legal order. Only in this way can proper care be provided for the *legal protection* to which all citizens are entitled, knowing that the government is called upon to secure their particular (state and non-political) legal interests, and to harmonize these interests in one *public legal order*.

However, without the foundational aid of a theory of modal aspects, it will not be possible to realize that the modal universality of the jural aspect transcends any societal collectivity or community as such. For example, the jural aspect cannot be identified with the "legal system" of the state. Before we proceed by discussing the shortcomings of Luhmann's understanding of the legal system, Chaplin's analysis of "public justice" as political norm requires our brief attention.

8.11 Chaplin: "public justice" as critical political norm

Although I share Chaplin's preference for a positive designation of the core meaning of the jural aspect – *tribution* instead of *retribution* (Chaplin, 2007:130, note) – it should be kept in mind that Dooyeweerd does use the term *retribution* in a positive sense, for he warns against the misunderstanding inherent in equating retribution with criminal law, or even worse, with a response to a legal wrong. The legal measure of proportionality entailed in retribution is applicable to every jural fact with its accompanying legal consequences (effects): "Retribution is not only exercised *in malam* but also *in bonam partem*" (Dooyeweerd, 1997-II:130).

8.11.1 *State and society: differentiated spheres of law*

The jural aspect is of central importance for Dooyeweerd's political philosophy, because the sphere-sovereignty of the different modal aspects provides the decisive perspective needed to avoid a totalitarian view of the state within a differentiated society. Yet he does not distinguish between *state* and *society* as such, as Chaplin alleges (Chaplin, 2007:131). He solely employs the idea of a *differentiated* society and then delineates within it the different spheres of law – public law, civil and non-civil private law. Chaplin is therefore mistaken in attributing to Dooyeweerd the view that the "civil-law sphere" of the state is a "part of public law" (Chaplin, 2007:134). To Dooyeweerd, civil law is civil *private law*, never *public law*. There is only one place in *A New Critique* where Dooyeweerd employs the combined phrase *civil and non-civil private law* (Dooyeweerd, 1997-III:692), although the implied distinction does occur in other contexts as well. Slightly differently formulated, the same distinction surfaces on one of the pages from which Chaplin quotes Dooyeweerd, namely page 446 of NC-III.[1] Under the heading "The civil law-sphere of the State" Dooyeweerd explains: "The internal public law-sphere of the State has its typical correlate in the sphere of civil law as a private *common law*" (Dooyeweerd, 1997-III:446).

8.11.2 *Internal function and external relations*

Chaplin highlights a certain ambiguity and inconsistency in Dooyeweerd's thought regarding the distinction between "internal functions" and "external relations" (Chaplin, 2007:130-133). According to him, one finds on NC-III page 483 an instance where the "state's regulation of private economic structures" is "cited as an example of the internal economic functioning of the state" (Chaplin, 2007:131). By contrast, Chaplin says that, in "raising its own revenue, the state requires its citizens to fulfill the proper duties of membership in the political community" (Chaplin, 2007:131). The heading of the section to which Chaplin refers concerns the integrating function of the state in respect of its *internal political economy* (Dooyeweerd, 1935:420 ff.;[2] see Dooyeweerd, 1997-III:482 ff). Yet it should be kept in mind that initially, *political economy* emerged as a study of the *economies of states*, with the tax obligation included in its agenda (see Myrdal, 1932:7, 86-87).[3]

In the context of his analysis, focused on the way in which the state expresses itself within the diverse modal functions of reality, Dooyeweerd was here supposed to analyze the internal function of the state within the economic aspect. But instead of doing this, he embarked upon a treatment of the integrating function of the state regarding *political economy*. Surely performing "a political integrating function" (NC-III:482) forms a part of the *political task* of the government, and therefore ought not be confused with the original function of the state within the economic aspect. Dooyeweerd is fully aware of this, because he immediately adds that this task of the government is totally different from the integrating function of "economically qualified societal relationships" (NC-III:482). The term 'internal' in Dooyeweerd's exposition therefore refers to the government's *political* integrating task, and not to its "internal economic function." The fact that the political task of integrating economic legal interests presupposes independent economically qualified, organized communities (*verbanden*) *external* to the state (sphere-sovereign), does not mean that this task itself is *external* to the state.

The criticism therefore should have been that, instead of discussing the intrinsic function of the state within the economic law sphere (such as observed

1 On page 132 of his article, Chaplin quotes Dooyeweerd with reference to NC-III:416, while in fact this quotation is actually derived from page 446.

2 The Dutch text reads: "De integreeringsfunctie van den staat in de interne politische economie ..."

3 The 19th century witnessed the rise and development of political economy, although a professorship was already established at the University of Vienna in 1763.

in tax revenue), Dooyeweerd actually discusses an element of the government's task of integrating (economic) legal interests. The content of what Dooyeweerd explains is correct; the location however should have been different.[1]

In this context, Chaplin gives a positive assessment of Dooyeweerd's notion of "political enkapsis" as an alternative to "both individualism and universalism" by emphasizing that the state does not have "any original competence in non-political structures." This 'simultaneously' affirms that the "state has the competence to regulate *externally* any non-political structures insofar as their activities have public-legal consequences" (Chaplin, 2007:132). However, the full context of the partial quotation that Chaplin gives here shows that his own qualification is incorrect: "insofar as their activities have public-legal consequences." Dooyeweerd says that this "harmonization process should consist in weighing all the interests against each other in a retributive sense, based on a recognition of the sphere-sovereignty of the various societal relationships" (Dooyeweerd, 1997-III:446). The very fact that civil private law never originated independently of the body politic, shows that the harmonization of legal interests cannot be restricted to "public-legal consequences," as Chaplin alleges. Encroaching upon the sphere-sovereignty of one non-political social entity by another also does not transform the private legal interests involved into something with a public legal nature. The integrative task of a government embraces both upholding those state institutions concerned with *public legal interests* (criminal courts), and those involved with *personal and societal freedoms* (civil courts).

The two most basic public legal interests of importance for the individual citizen concern personal bodily integrity and the public-legal side of property rights. For this reason, the state takes the initiative when assault or murder occurs, or when someone's property rights are violated (compare the classical phrase from Locke: life, liberty and property – Locke, 1966:119). Entering into a contract and terminating it belong to the domain of civil law. Note that the relevance of the three distinguished spheres of law within a differentiated society are not restricted to any specific or privileged *place* within the territory of the state. When someone is killed during a church service, criminal law is immediately activated, and when the minister defames a member of the congregation from the pulpit, the sphere of civil private law can be activated.

8.11.3 *A different idea of internal and external coherence*

Of course, there is a different way to distinguish between what may be designated as the *internal* and *external* coherence between different aspects; in particular between the qualifying aspect of an entity or societal structure and the original function of that (natural, cultural or societal) entity within all other aspects of reality. Although Dooyeweerd did not make this distinction, Hom-

1 This remark does not invalidate the distinction highlighted by Chaplin between "a balance of payment deficit and a budget deficit" (Chaplin, 2007:131).

mes, his successor, did. A work of art, for example, is qualified by its aesthetic aspect and this aspect shows its *internal* coherence with the other cosmic aspects through its retrocipatory and antecipatory analogies. Earlier, we discussed fundamental modal norms, reflected on the law side (norm side) and based upon modal retrocipations and antecipations (see page 343 above, regarding modal aesthetic principles). Consider the difference between the function of an art work within the numerical aspect (it is *one*) and the retrocipation within this aspect to number, given in *aesthetic unity and multiplicity*. Similarly, there is a difference between the original sensory function of an art work (one *sees* a sculpture, *reads* a book or *listens* to music) and the *aesthetic sensitivity* displayed by a work of art for the nuances within reality. The history of such a work concerns its function within the historical aspect of reality, while the *aesthetic form and style* reflect the historical analogy within the aesthetic aspect. We furthermore argued (see page 230 above) that a work of art ought to do *aesthetic justice* to what it portrays, it ought to evince *aesthetic care* and *integrity* and it ought to be *aesthetically convincing* – showing that there is also an antecipatory coherence between the aesthetic aspect and the jural, moral, and certitudinal aspects of reality (to be distinguished from the original function of an artwork within these aspects – as a legal object, as something worthy of love (moral object-function) and as something trustworthy (reliable) in its own right (as a certitudinal object).

The retrocipatory coherence between the qualifying aesthetic and the other aspects of an artwork constitutes the intrinsic (inner) coherence between these aspects, whereas the original (authentic) function of such an artwork within every modal aspect gives expression to the extrinsic (outer) coherence between the qualifying aspect and the other functions of the work of art.

Similarly, one can discern an inner and outer coherence between the qualifying jural aspect of the state and the other original modal functions of the state. The fact that a state also functions within the economic aspect, implies that every state has to observe economic normativity in its task of managing all its resources in a *frugal way*. Therefore, the economic actions of the state are subject to economic principles. By contrast, the economic retrocipation within the structure of the jural aspect on the law side constitutes the jural principle of *avoiding what is excessive*. (In particular, Dooyeweerd refers to the concept of *jural economy* and its implied elementary basic concepts of *legal interest* and *legal proportionality* – see Dooyeweerd, 1967:27.) Public opinion manifests the original function of the state within the logical-analytical aspect, but it differs from the jural accountability of natural persons or legal entities, because the jural accountability analogically reflects the logical principle of sufficient reason within the jural aspect. (This logical principle, in turn, reflects the causal physical analogy within the structure of the logical-analytical aspect on its norm side.) The legal power vested in the office of government is nothing but the *competence* to form positive law (to positivize jural principles). This office, with its competence (legal power) is different from the original function of the state within the cultural-historical aspect, for the latter concerns the original *power of the sword.*

Finally, in order to highlight one further instance of the difference between the inner and outer coherence, we mention the deepened principles of jural morality (designated as legal-ethical principles – see page 289 above). These disclosed principles come into view through the antecipatory coherence between the qualifying jural aspect of the state and the moral (ethical) aspect of love. The love for a country by its citizens shows the original function of the state within the moral aspect, and therefore differs from legal-ethical principles such as the fault principle, equity and *bona fides*.

8.11.4 *Justice and the distinction between constitutive and regulative structural elements*

The distinction between constitutive and regulative (legal) principles already received our attention (see pages 229, 315, 317 ff. above). Before the term *justice* can acquire a well-delineated meaning, it is necessary to come to terms with the constitutive structural elements within the jural aspect, as well as the regulative (deepened or disclosed) jural principles. Dooyeweerd phrases this distinction in terms of the difference between retrocipatory and antecipatory analogies within the structure of the jural aspect. From the perspective of analysing the basic concepts of the discipline of law, Dooyeweerd raises the question: "Is it possible to grasp in concepts all the analogies that we discern in the jural mode of experience?" His answer reads:

> No, this is only possible in relation to the retrocipatory analogies. The antecipatory ones only reveal themselves when law is opened up and starts to anticipate later aspects on the basis of a historical disclosure of culture. The jural elements of fault (or guilt) bonos mores, bona fides, equity, etc. are not found in a closed legal order. And so they fall outside the ambit of the concept of law, for that concept can only encompass those modal moments that are found in all legal orders, including therefore a primitive legal order (Dooyeweerd, 1967:2).

Whereas concepts such as a *legal order* (elementary basic concept expressing the numerical analogy within the jural aspect) and *jural causality* (elementary basic concept reflecting the physical analogy within the jural mode) apply to distinct (constitutive) structural elements within the jural aspect, the *concept of law* embraces all these retrocipatory analogies at the same time.

In order to appreciate Dooyeweerd's view of the *idea* of public justice, a proper understanding of his *concept of law* is required. Chaplin approaches the idea of public justice by relating it to sphere-sovereignty, for "public justice involves *harmonizing the various interests which arise from the legal sphere-sovereignty of various social structures*" (Chaplin, 2007:134). He then focuses on the term 'harmonizing' which, according to him, Dooyeweerd employed for the sake of "systematic consistency," because the jural aspect has its direct foundation in the aesthetic aspect. He then interjects: "But I suggest that the *economic* foundation of the legal aspect illuminates more clearly what Dooyeweerd actually has in mind here" (Chaplin, 2007: 134). He continues: "We might more felicitously speak of a 'frugal' or non-excessive balancing of legal interests. The state's responsibility to render justice to each legal interest could then be described, more evocatively, as preventing the excessive satisfaction of each of these interests at the expense of others. When justice is done, there will be such an element of 'frugal (re-)tribution'; or, more elegantly perhaps, 'balanced rendition' " (Chaplin, 2007: 134). On the previous page, Chaplin provides a quote by Dooyeweerd, in which it appears that Dooyeweerd sees a close link between "public social justice" and the "harmonizing of interests." Dooyeweerd writes: "The internal political activity of the State should always be guided by the idea of public social justice. It requires the harmonizing of all the interests obtaining within a national territory, insofar as they are enkaptically interwoven with the requirements of the body politic as a whole. This harmonizing process should consist in weighing all the interests against each other in a retributive sense, based upon a recognition of the sphere-sovereignty of the various societal relationships" (Dooyeweerd, 1997-III:446).

Chaplin is correct in pointing out that the phrase "public social justice" does not properly render the meaning of the original Dutch text. The Dutch text reads: "Steeds behoort de interne politische activiteit van den staat onder *typische leiding* te blijven van de idee *der publiek*[*e*][1] *verbandsgerechtigheid*, welke een evenwichtige harmoniseering in den zin der vergeldende afweging eischt van alle belangen, welke zich binnen het landsgebied geldend maken, inzooverre zij enkaptisch verlochten zijn met de eischen van het staatsgeheel, en welke ook de *juridische* souvereiniteit in eigen kring eerbiedigt" (Dooyeweerd, 1936-III:401-402).

Chaplin suggests that the correct translation should be: "public *communal* justice" (Chaplin, 2007:133). However, translating the original Dutch text is more complicated, because Dooyeweerd distinguishes between *gemeenschap* (normally rendered as *community*) and *verband*. According to him, only historically founded *gemeenschappen* are to be designated as *verbanden*. Whereas the Dutch term *gemeenschap* could be translated with the term *community*, the term *verband* (= historically founded *gemeenschap*) is normally translated (in NC) as *organized community*. Because Dooyeweerd's terminology confuses the modal totality concept regarding different ways of social interaction with the issue of a foundational function of societal entities, we opted for a different translational equivalent for verband: soci(et)al collectivity (see page 242 above). The phrase "publieke verbandsgerechtigheid" therefore rather requires a circumscription in English, such as: "The internal political activity of the State should always be typically[2] guided by the idea of the

1 In quoting the Dutch phrase "publiek verbandsgerechtigheid" the author accidentally omitted the "e."

2 Note that the English translation omits the crucial term typische! We shall return to the distinction between modal and typical below.

public justice of the state as an organized community."[1]

When Dooyeweerd says that public justice *requires* the harmonization of interests, he highlights a constitutive (retrocipatory) element within the structure of the jural aspect. But his understanding of the constitutive structure of this aspect embraces *all* retrocipatory analogies, not only *one* of them, such as the aesthetic or the economic. Chaplin apparently did not realize that the concept of law is a compound or complex concept, embracing every one of its foundational analogical moments. Dooyeweerd specifies what this entails both with respect to the law side and the factual side of the jural aspect.

> 1. The modal meaning of the juridical aspect on its law-side is: the unity (the order) in the multiplicity of retributive norms positivized from super-arbitrary principles and having a particular, signified meaning, area and term of validity. In the correlation of the interpersonal and the communal functions of the competency spheres, these norms are to be imputed to the will of formative organs, and they regulate the balance in a multiplicity of interpersonal and group interests according to grounds and effects, in the coherence of permissive and prohibitive (or injunctive) functions by means of a harmonizing process preventing from any excess, in the meaning nucleus of retribution.
>
> 2. The modal meaning of the juridical aspect on its subject-side is the multiplicity of the factual retributive subject-object relations imputable to the subjective will of subjects qualified to act, or per *repraesentationem* to those not so qualified. These subject-object relations are bound to a place and time, in the correlation of the communal and the interpersonal rights and duties of their subjects. In their positive meaning – in accordance with (or in conflict with) the juridical norms – these subject-object relations are causal with respect to the harmonious balance of human interests in the meaning of retribution (Dooyeweerd, 1997-II:406).

From this comprehensive account, it is clear that the concept of law does not allow for any *trade-off* by means of which one retrocipatory analogy (such as the aesthetic) could be replaced by another (such as the economic), because they are all equally constitutively and mutually involved.

The normative meaning of the qualifying jural aspect is sometimes depicted as the jural norm of integration, and following some of Dooyeweerd's analyses, it is specified as a government's task to establish balance and harmony amongst the multiplicity of legal interests within its territory, and to restore this balance in a retributive sense whenever it is disturbed. Yet a proper articulation of the function guiding the government ought to include every distinct analogical moment within the jural aspect. Let us therefore attempt to give an account of this guiding normative jural meaning that incorporates all retrocipatory analogies on the law side of the jural aspect.

A government, as a legally competent state-organ, is called upon to maintain a balance and harmony in the multiplicity of collective, communal and

1 Literally, publieke verbandsgerechtigheid should be rendered as: "public, organized communal justice."

coordinational legal interests within its legal domain by effectuating a durable positivization of underlying typical legal principles that can only be made valid (enforced) if the government has an accountable free, legal will, enabling it to correctly interpret juridically relevant events in order to, without any abuse of power, act by harmonizing the legal interests involved. The distinct modal analogies of the jural aspect embodied in this formulation are: multiplicity (number), legal domain (space), made valid (enforce/legal validity – physical), legal organ (biotic), legal will (sensitive-psyhical), legal accountability (logical-analytical), positivization (cultural-historical), legal interpretation (sign), legal intercourse (collective, communal and coordinational – social), avoiding abuse of power (economic), jural harmonization (aesthetic). It is therefore not meaningful to consider any constitutive structural element within the jural aspect in isolation from all the others.

This analysis runs parallel with the all-encompassing significance of the (elementary and compound) basic concepts of the science of law. Dooyeweerd remarks that none of these belongs to any specific division of the science of law (such as "civil law, commercial law, law of civil procedure, law of criminal procedure, and administrative law").[1]

Only once the *full scope* of the foundational structure of the jural aspect is taken into account, is it possible to consider its deepening and disclosure under the guidance of the post-jural aspects. The most obvious legal-ethical principles are those of fault, *bona fides* and *equity* (see page 229 above).

Insofar as the term *justice* is employed in Dooyeweerd's thought, two things should be kept in mind. First, it presupposes the entire constitutive structure of the jural aspect, capable of being grasped in the concept of law (we used the image of *bricks in a wall* – see page 555 above). Secondly, it is meant to take into account the deepened or disclosed structure of the jural aspect.

As an effect of the process of societal differentiation, the deepened modal personality principle (*persoonlijkheidsbeginsel*), as well as the sharp distinction between public legal power and private property right materialized (see Hommes, 1972:481). This distinction intimately coheres with the rise of the modern state, supported by a process through which governmental power changed from a private property right (*patrimonium*) into a *public office* in service of the idea of the *public interest*, having as its counter-pole the differentiation of private property rights. The juridical personality principle is guided by the regulative jural principle of the *value of the human being* (*dignitas humana*) (Hommes, 1972:487) and it acquired a significant role in re-

1 Dooyeweerd lists the following concepts analysed in his Encyclopedia of the Science of Law: "legal norm, legal subject and legal object, legal fact, subjective right and legal duty, area of validity and the locus of a legal fact, lawfulness and unlawfulness, jural attribution and accountability, jural will, jural causality (legal ground, legal consequence as to the law side; the subjective or objective causality of, respectively, a legal transaction or objective legal fact, as to the subject-side), jural positivizing and the originating jural form (formal source of law), legal organ and jural competence (legal power), jural interpretation and legal significance, jural fault or guilt, good morals, good faith" (Dooyeweerd, 2002:199-200).

spect of the articulation of the public spheres of jural freedom of the human personality within the state[1] as well as within the non-political spheres of life.[2]

With reference to NC-III:445, Chaplin correctly remarks that "the principle of the public interest binds the state to the norm of public justice" (Chaplin, 2007:136).

The regulative principle of the *salus publica* brings to expression the deepened meaning of the jural aspect and the typical nature of administrative law. The crucial question is how the idea of the *salus publica* is to be limited, because we saw it has been used for diverse extremes – from the separation of parents and children (Plato and Fichte) in the name of public interest to Locke's state nihilism (*laissez-faire, laissez-passer* – see Locke, 1690, § 158:197). There is simply no criterion in the idea of the public good or pubic interest that can curb the extremes of state nihilism and state absolutism and totalitarianism.

The only way to break through this deadlock is to revert to an explication of the meaning and role of the jural aspect as the qualifying function of the state. This crucial insight concerns the fact that the body politic finds its stature within the context of the public-*legal* nature of the state as a public-*legal* institution (a *res publica*). Apart from acknowledging the guiding role of the jural aspect, the idea of the *salus publica*, as well as that of the state as a *res publica*, do not obtain a structural delimitation. As a societal collectivity ("organized community" in Dooyeweerd's sense) the *typical* public-legal nature of the state differs from every other legal sphere within society in that the non-political social forms of life are not *qualified* by the jural aspect. Therefore, since their jural function is not the leading or guiding function, their internal law remains bound only to their specific spheres of competence. It remains a *ius specificum*, a specific law; limited to a *section* of the population of the state. It is only state citizenship that cuts across all other societal ties, because it is united into a truly universal collective communal law (*ius commune*). This *universality* is not *unlimited*, for it does not authorize the state to engage in the usurpation of any specific private legal sphere – as if the state would be called to pursue a private religious integrative task (by establishing a universal church denomination), or to involve itself in elevating a particular cultural legacy to become normative for the body politic as a whole. Particular private concerns from the non-jural qualifying function of non-political social collectivities and communities, contradict the universal public-*legal* nature of the state, for the jural qualification of the task of the state cannot transform itself simultaneously into a *universalistic* totality embracing every non-political

1 Think of the freedom of expressing an opinion in the form of the press, confidentiality of letters [briefgeheim], prohibition of arbitrary arrest (habeas corpus), religious freedom, an so on.

2 Freedom of association and meeting, of religious gathering, the freedom to educate children in accordance with their own convictions, and so on. See Hommes, 1972:504 and, in respect of educational freedom – see Chaplin, 2007:135.

sphere as an integral part.

The term *justice* captures all the *deepened* and *disclosed* moments in the opened-up (antecipatory) structure of the qualifying jural aspect *requiring* the constitutive building blocks as their irreplaceable foundation. For this reason, the idea of *public justice* does not receive its content from any retrocipatory moment as suggested by Chaplin – who nowhere in his article attempts to explain the nature of justice in terms of the disclosure of the jural aspect. The ethically deepened respect for the *dignity* of the human being indeed guides the different spheres of law – public law (including constitutional law, administrative law, criminal law and procedure), and civil private law. The regulatively deepened jural principles of *equity* and *bona fides* cut across the different legal spheres that are intertwined with the complex and multi-aspectual functioning of the state. It is only on behalf of these disclosed jural principles that the word *justice* acquires its full meaning. We stated earlier that the deepened principles of jural morality are all principles of *justice*.[1]

8.11.5 *sphere-sovereignty: typical and a-typical tasks*

We noted that, through societal differentiation and integration, distinct sphere-sovereign social collectivities emerged in the course of a long historical developmental process. The type law for each distinct social collectivity delimits its inner sphere of operation, guided by its typical qualifying modal function. Once the inner structural principle of social entities is recognized, it is possible to distinguish between *typical* and *a-typical* tasks of specific social forms of life.

Chaplin points out that this distinction caused severe difficulties for reflection on the task of government. How does one conceive of a government regarding what is supposed to fall *outside* its sovereign sphere of operation? With reference to De Ruiter, he notes that it seems as if Dooyeweerd cannot specify any structural norms for the a-typical tasks of the state (Chaplin, 2007: 141). In respect of a "nationalized industry," he explores a good intuition, saying, "Here too the state must respect economic sphere-sovereignty if the industry is to function properly. But it is not essential to the nature of the state that it operate any nationalized industries" (Chaplin, 2007:141).

Of course, the picture is complicated if the government has to weigh the interests of diverse non-political entities.

> It is certainly true that, if the government does any of these things, it must do so in manner which treats the juridical interests of the relevant persons or structures equitably. It should not, arguably, dole out huge subsidies to opera companies while starving community theatres of funds, or bail out loss-making car companies while driving efficient farmers out of business (Chaplin, 2007: 142).

In a more general sense, Chaplin reiterates the criterion of respecting the

1 The deepened modal jural principles obtain a typical specification through the type of law of the state.

sphere-sovereignty of non-political social entities. "And it is also clear that the government must act in a manner which respects the sphere-sovereignty of the parties involved" (Chaplin, 2007:142). In South Africa, for example, the landscape made it very difficult for a private company to build a railway network. Therefore, the state took over this responsibility. In a similar way, *Yskor* was erected as a steel industry. The Dutch Reformed Church also established an industry, namely farming (Kakamas), in order to serve as a resource supplement in difficult times. Chaplin's discussion could benefit by taking into account an article written by H.J. Strauss in *Philosophia Reformata* (1965). In this article, Strauss engages in a consideration of these issues and suggests an alternative that transcends the criteria that may be deduced from the sphere-sovereignty of the state. He affirms the positive stance advanced by Chaplin, regarding the sphere-sovereignty of non-political societal entities, including a brief account of their typically different tasks guided by their diverse qualifying modal aspects (see Strauss, 1965:198-199).

In order to assess a-typical governmental tasks, a broader perspective is required; one in which the *genesis* of a differentiated society is kept in mind. Such a process of differentiation basically follows the historical norms for civilizational development, amongst which the principles of historical continuity, historical differentiation and historical integration are prominent (see Strauss, 1965:200-203). Tasks of an a-typical nature within any community or societal collectivity therefore should be assessed in terms of the normativity of the principles guiding meaningful cultural unfolding and development in a general civilizational sense.

It should not be difficult to determine whether or not a specific task is *typical* or *a-typical*. The question is simply: if one imagines a situation in which the task under discussion is not performed, will that mean that the societal life form involved can no longer perform those tasks that typically belong to its calling and sphere of competence? For example, is it possible to say that a state not running a railway network or a steel industry is still a proper state? The answer is obviously *yes*, for states do not need to be involved in such tasks in order to be full-blown states. Likewise, a church denomination that is not running a farming industry will continue to function as a church.

Therefore it seems that two considerations ought to guide and direct an institution in taking on *a-typical tasks*.

1. Treat the a-typical domain (sphere) in accordance with its own inner structural principle (inner *sphere-sovereignty* or *type law*). A business firm ought to be managed *as* a business firm – not an integral part of the state; a farming industry managed by a church denomination does not become ecclesiastical – it is not transformed into a religious service or a part of the congregation.
2. Always work towards a situation where the relation of dependence could be terminated – support the a-typical sphere to regain its independence and thus to realize its own internal sphere-sovereignty in the ongoing process of differentiation and integration of society. This second consid-

eration retreats to a more general perspective, focused on the *dynamic development of society*, subject (amongst others) to the historical principles of historical continuity, historical differentiation and historical integration. This implies that the process of taking on a-typical tasks and allowing the spheres involved to eventually once again reach a situation where they can take responsibility for and give shape to their own distinct callings. In other words, neither the emergence of a-typical tasks nor their termination proceeds in an a-normative way. But if the norms for meaningful historical development are not applied, the reactionary effect may be a return to an earlier undifferentiated condition mediated by an ongoing process of de-differentiation.

8.12 The shortcoming in Luhmann's system-theoretical conception of the legal system

Modern *systems theory* is justified in its rejection of atomistic (individualistic) interpretations of human society. In his systems theory, Niklas Luhmann employs the importance of the term *differentiation* with reference to the whole of society as a "differentiated unity." The latter has to be "distinguished from that of its parts" (Luhmann, 1990:410). He draws a distinction between two 'levels' – that of the *whole* and that of the *parts*. Since these two levels mutually "mirror each other without being reduced to each other" this 'constitutes' a 'paradox' (Luhmann, 1990:410).

Within the same systems-theoretical tradition as Luhmann, we may note the promising view on distinct social systems advanced by Münch. He points out that a meaningful account of social order cannot be divorced from differentiation. He then mentions Luhmann's view: "differentiation is itself the only possible way of ordering action in modern societies" (Münch, 1990:443). Weber's theory of rationalization of modern society prompted the idea of spheres that are increasingly guided by their *inner laws* (see page 533 above). Münch points out that this "theory of rationalisation has been combined – by Schluchter and Habermas – with the theory of functional differentiation as it was formulated by Luhmann" (Münch, 1990: 442). In particular, Münch refers to Luhmann's notion of *autopoieisis*, according to which society "is compartementalized into a growing number of autopoietical – that is, self-regulating – systems which treat each other as environments to which they have to adapt actively" (Münch, 1990:444). Maturana acknowledges Francisco Varela as the co-author of the idea of autopoeisis. In general terms, Maturana provides the following circumscription: "Living systems are autonomous entities, even though they depend on a medium for their concrete existence and material interchange; all the phenomena related to them depend on the way their autonomy is realized. A perusal of present-day biochemical knowledge reveals that this autonomy is the result of their organization as systems in continuous self-production" (Maturana, 1978:36).

Although Münch on the one hand maintains a potentially holistic view by employing the whole-parts (system-subsystem) scheme, his above-mentioned idea of the *inner laws* of *differentiated spheres of life* opens up an alternative understanding entailed in accepting the principle of sphere-sovereignty. We highlight the implications of this principle with specific reference to Luhmann's conception of the *legal system*.

In his sociological theory of law, Luhmann captures the rapidly increasing complexity of society during the modern age by referring to different *meaning spheres* (Luhmann, 1985:147). He has a clear sense of the fact that the "pre-modern high cultures presume the *incomplete functional differentiation of the societal system*" (Luhmann, 1985:133). In those "pre-modern high cultures," the "legal principle takes on a generalized moral form in the idea of what is equitable and just" (Luhmann, 1985:145).

When Luhmann speaks of the tautological character of the normativity of the legal system (*Rechtssystem*), he highlights something of the indefinability of the jural aspect of reality.[1] He relates the notion of *retribution* to the jural aspect (Luhmann, 1985:122) and frequently speaks of a *legal order* (cf. Luhmann, 1985:133-134). Note that the concept of a *legal order* is on the same level as that of the *legal validity* of positive law (see Luhmann, 1985:132, 146, 160). A legal order concerns the *unity in the multiplicity* of jural stipulations and legal validity concerns their enforcement. As we saw Kant already realized that (state) law inherently entails (legal) force, which caused Derrida to confirm that "there is no law (*loi*) without enforceability" (Derrida 2002:233 – see page 291 above). Luhmann evinces an ambiguous attitude towards the *force of law*, because on the one hand, he associates it negatively with *violence*, and on the other he speaks of enforcement in a non-pejorative way (Luhman, 1985:168-169; 122, 129, 130-131, 133, 162, 168-170). In a different context he mentions ordinary events – such as a plane crash, a couple that divorces, someone accidentally pushing a vase from the shelf of a shop – and asks the question of how such events cohere from a *jural point of view* (*Rechtsgesichtspunkt* – Luhmann, 1986:34).

Although this may suggest that Luhmann indeed recognises the *jural* as *an aspect* of reality, his systems orientation explores an opposite path, for he distinguishes between less and more highly developed archaic societies (see Luhmann, 1985:135), and refers to "pre-modern high cultures [that] presume the *incomplete functional differentiation of the societal system*" (Luhmann, 1985:133). According to him, the *legal system* emerged as a differentiated functional subsystem of society.

His understanding of the *legal system* is closely related to his views on *law* as the *structure of society* (see Luhmann, 1985:103 ff.). His analysis here is guided by the idea of "the institutionalization of the scheme of societal system differentiation" and he emphasizes that "law must be seen as a structure that defines the boundaries and selection types of the societal system" (Luhmann,

1 "Das Rechtssystem selbst ist gehalten, die Normqualität seiner Entscheidungen tautologisch zu definieren. Es kann die Worte wechseln, zum Beispiel Normen durch 'Sollen' definieren, aber das dient nur der verbalen Verschleierung der Tautologie. Rechtsnormativer Sinn ist das, was ein Element als zugehörig zum Rechtssystem auszeichnet" (Luhmann, 1986:20).

1985:105). He holds the view that the insights pertaining to self-referential systems ought to be applicable to the *legal system* as one of the differentiated functional systems discernable in the evolution of society.[1]

However, by viewing the legal system as a product of differentiation, one runs into difficulties when it comes to an account of the complexities of modern differentiated societies with their distinct spheres of law. At the heart of these complexities, a proper understanding of the jural guarantees for societal freedoms is required. For this purpose, it is necessary to amend the views of Luhmann on the legal system by underscoring the distinction between *aspects* and (natural and social) *entities* that guided our analyses throughout this work.

It is unfortunate that Luhmann's view of the *Ausdifferenzierung* (differentiating out) of the legal system still lacks a proper understanding of the *multiple legal spheres* within a differentiated society, as explained above. Distinguishing the various aspects of reality makes it easier to characterize differentiated societal collectivities and spheres. Within the 'economy', the business enterprise as a social collectivity is indeed qualified by its economic aspect – and within the domain of politics, the state as societal collectivity is qualified by its jural function, whereas science is normally practiced within academic institutions such as universities, qualified by theoretical thought, that is to say that the logical-analytical aspect of reality characterizes the university.

Since, as explained in Chapter 3, the various aspects of reality are not themselves concrete entities, they constitute universal *modes of being*, within which every distinct (differentiated) societal entity has a function. Furthermore, in the light of the modal universality of each aspect (transcending whatever functions within it), it is insufficient, as Luhmann does, to speak of the *legal system* as an *undifferentiated whole*. Every societal collectivity, including the university, social club, art association, state, nuclear family, confessional denominations, and so on, has a function within the jural aspect of reality. This insight rules out applying the jural aspect merely to *one* social entity functioning within it, such as the state, identified with the legal system of society. The alternative view developed in our preceding analysis explores the sphere-sovereignty of different modal aspects of reality, as well as the idea of type laws in order to avoid problems like those involved in Luhmann's thought.

8.12.1 *The fundamental distinction between a power state ('magstaat') and a just state ('regstaat')*

Against this background, we can now conclude by distinguishing the *two basic state types* – (i) a "just state" and (ii) a "power-state":

"Just state" (a constitutional state under the rule of law):

1 "Wenn diese Ausgangsannahmen zutreffen, dann müßten die Einsichten über typische Strukturen und operative Probleme selbstreferentieller Systeme auf unseren Fall des Rechtssystems anwendbar sein; dann müßte die Evolution der Gesellschaft zu Ergebnissen geführt haben, die in ausdifferenzierten Funktionssystemen, unter anderen im Rechtssystem, die typischen Merkmale selbstreferentieller Systeme aufweisen" (Luhmann, 1986:12-13).

A "just state" involves civil and non-civil private freedoms, as well as public legal freedoms.

"Power state" (an absolutist and totalitarian state)

In a "power-state," there are no such freedoms.[1]

If these fundamental differences between the various legal spheres are ignored, it would be impossible to value the diverse legal interests within the territory of the state in a constitutionally correct manner.

This approach emphasizes above all else the *jural* task of the state as a public legal institution called upon to maintain a public legal order. Only in this way can proper care be taken of the *legal protection* to which all citizens are entitled, knowing that the government has the calling to protect their particular (state and non-state) legal interests, and to harmonize all these interests in one *public legal order*.

8.13 Justice and legal validity (the force of law): Derrida, Dooyeweerd and Habermas

Habermas addresses the reality that positive law may not be just. He phrases it in terms of the tension between validity and facticity (positivity) (Habermas, 1998:171). "This initially appeared in the dimension of legal validity (as the tension between the positivity and the legitimacy of law) and inside the system of rights (as the tension between private and public autonomy)" (Habermas, 1969:136). Although Habermas speaks of "the horizontal association of consociates," he also has to acknowledge the vertical relation (of super- and subordination) within the state. "With the conceptual move from the horizontal association of consociates who reciprocally accord rights to one another to the vertical organization of citizens within the state" (Habermas, 1996:135). However, this acknowledgement of verticality creates the problem of a *duplicated sovereignty*. He is compelled to take account of the super- and subordination of government and subject, but this does not liberate him from the idea of *popular sovereignty*.[2] According to him, "[A] popular sovereignty that is internally laced with individual liberties is interlaced a second time with governmental power" (Habermas, 1996:135). "Political power" or "state power" has to be "*authorized* by legitimate law" (Habermas, 1996:136). Therefore, although Habermas does speak of *law and the state*, he does not introduce the idea of *state law*. Law appears to be *external* to the state, for the latter manifests itself only as (state) *power*. He also employs the phrase *governmental power*. The "collective binding decisions of an author-

1 A *totalitarian state* extends its power beyond its own confines in order to exercise control over all societal collectivities that are distinct from the state, – such as economic, religious and educational institutions (see Hayes, 1969, 94-95), i.e. it denies its citizens their *personal* (*civil*) and *societal* freedoms. A state is *absolutist* if it does not leave any room for the co-determination and co-responsibility of its citizens (i.e. it denies their *political freedoms*) (see page 547 above).

2 Popular sovereignty in the sense that "all governmental authority derives from the people" (Habermas, 1996:135).

ity" must utilize 'law' to "fulfill its own functions."[1]

When Habermas refers to *law*, the *legislature* or *legislative bodies*, he does not have a modal function or aspect in mind. He says, "the legislature is constituted as a branch *within* the state" (Habermas, 1996:135). That he does not have the *jural aspect* in mind is confirmed by his reference to the *legal system* (see Habermas, 1996:39 and Luhmann's notion of the legal system, page 577 above). Furthermore, a "branch *within* the state" is a *part* of the state – and no modal aspect can ever be a part of the state.

Habermas came very close to an understanding of the foundational and qualifying functions of the state. He does mention, in connection with the views of philosophers of natural law, that the "administrative power" has "a monopoly on the instruments of legitimate force" (Habermas, 1996:137). What is nonetheless absent, is the idea of the cultural-historical foundational *function* directed by the jural as qualifying *function* of the state. 'Power' and 'law' are treated as two *entities* that ought to be *connected*: "It is not the legal form as such that legitimates the exercise of governmental power but only the bond with *legitimately enacted* law" (Habermas, 1996:135).

It is nonetheless noticeable that the distinction between law and morality causes Habermas to distinguish between different kinds of "normative principles and rules," and that the latter may refer to the "same problems in different ways" (Habermas, 1996:106). It is a small step from here to acknowledge that human actions function in "different ways" within distinct modal aspects. The subsequent issue will then be to observe the *interconnections* between different aspects, expressed in composite phrases such as *legal validity*, through which the relationship between two different aspects is expressed.

Although Derrida does not employ an articulated theory of modal aspects, he does approximate a theory of intermodal connections, for we saw that he places *credit* against the background of acknowledging the universality of 'faith'. We noted his emphasis that "faith is absolutely universal" (see Derrida, 1997:22 and page 101 above). This paved the way for his implicit acknowledgement of the coherence between the economic aspect and the certitudinal aspect, given in the reality of *credit* (as *economic trust*).

Likewise, when Habermas speaks of a *legal order* (Habermas, 1996:106) or when Derrida mentions a *juridical order* (*Rechtsordnung*) in his discussion of Benjamin's views, it should be clear that the qualifying part of the phrase *juridical* (*legal*) *order* is derived from the root *ius*, i.e. from the *jural aspect*. But what about the term *order*? Is it possible to relate it to an aspect that is *different* from the jural?

If the multiplicity of (positivized) legal norms within a specific legal sphere is not integrated into a unity, it will be impossible to speak of a *legal order*. For that reason we argued that the phrase *legal order* represents a numerical

1 "The idea behind government by law requires that the collectively binding decisions of an authority that must make use of the law to fulfill its own functions are not only cast in the form of law, but are for their part legitimated by statutes enacted in accordance with that procedure" (Habermas, 1996:135).

analogy within the structure of the jural aspect. The possibility of speaking of juridical norms (or a legal order) therefore evinces the correlation between the norm side and the factual side of the jural aspect. What Habermas calls "law making" (Habermas, 1996:258) concerns the formation of law, i.e. an analogy of the cultural-historical aspect within the jural aspect.

In order to appreciate what the expression "the force of law" (*legal validity*) entails, it can be compared to the phrase *legal order*. A legal order concerns a unity in the multiplicity of juridical norms, i.e. it brings to expression the numerical analogy on the law side of the jural aspect. Dooyeweerd explains it as follows:

> Every differentiated legal order is constituted as a unity of public law, civil private law, the totality of internal law of all non-state societal collectivities (for example ecclesiastical law), as well as coordinational law (the internal contractual rules regulating agreements between parties). All these legal rules that belong to intrinsically different societal spheres are interwoven into a jural unity, that is to say, they ought to be brought into harmony with each other in accordance with retributive criteria, so that no cluster of legal norms threatens another (hence the internal law of the contracting parties should never violate the basic and mandatory legal principles and norms) (Dooyeweerd, 1997-II:11).[1]

Correlated with and subjected to the juridical norms of a specific legal order, every legal subject is also a subjective *jural unity in the multiplicity* of legal relationships. According to Dooyeweerd, the legal relationships are "juridically harmonized in a retributive sense," because a "legal object is also an objective jural unity in a multiplicity of legal relationships."[2] We can now focus on the nature of the composite expression *legal validity*.

8.13.1 *The force of law: legal validity*[3]

The issue is to realize that the terms 'force' and 'validity' are both derived from the physical aspect of energy-operation. When energy operates the forces generated cause changes (effects), i.e. they embody what is known as

1 Compare with this the way in which Habermas speaks of "legal domains/spheres of law" ('Rechtsgebieten') – clearly pointing at a spatial analogy within the modal structure of the jural aspect (see Habermas, 1998:59). [Incidentally, the English translation, "different subject areas of law," is not a proper rendering of 'Rechtsgebieten' (see Habermas, 1996:40).]

2 "The same piece of land may be the property of A, the object of B's usufruct and of C's mortgage" (Dooyeweerd, 1967-II:12-13).

3 Within the circles of legal philosophy and philosophy in general during the first part of the 20th century, the idea of validity played a significant role. Whereas the initial position of the Baden School of neo-Kantianism (Windelband, Rickert, Weber, and others) still held on to the validity of supposedly ideal and timeless values, the latter soon became fully historicized and relativized through the emerging idea of changing lingual and social constructions. Alan Cameron mentioned that, during the latter half of the same century, the same concept of validity features in Anglo-American jurisprudence, especially from the appearance of H.L.A. Hart's The Concept of Law (1961), where he addresses the conditions of legal validity. See Simmonds, 1986:85, 92.

physical causality.[1]

Derrida explores examples of analogies of *force* in his discussion of the "differential character of force," aimed at avoiding "the risks of substantialism or irrationalism" (Derrida, 2002:234). He does this in connection with his notion of *différance* because he does not offer a systematic theory regarding the intermodal connections (analogies) between various aspects of reality. He does however mention the relation between *force* and *signification* (in various forms, such as " 'performative' force, illocutionary or perlocutionary force, of persuasive force and of rhetoric" (Derrida, 2002:235). He concedes (on the same page) "that [he has] ... always been uncomfortable with the word *force* even" though he "often judged it indispensable." Perhaps his hesitance stems from the fact that he did not contemplate the fact that modal analogies form a constitutive element in the (normative) meaning of the jural aspect – as indeed given in the physical analogies of jural *force*, jural *being-in-force*, jural *en-forcement*, having jural *validity*, as well as *jural causation* (causality).

Habermas attempts to define "Geltung" (validity), but he does this in a circular way and in a manner that misses the connection with the physical aspect (see Habermas, 1998:138). His main reference on this page is to *action norms* as behavioural expectations, embedded in a context of the "predicate 'gültig' " (*valid*), which (tautologically) expresses a non-specific meaning of normative *Gültigkeit* (validity).[2]

Derrida does comprehend the constitutive meaning of the physical analogy within the jural aspect: "Applicability, 'enforceability', is not an exterior or secondary possibility that may or may not be added as a supplement to law. It is the force essentially implied in the very concept of *justice as law*, of justice as it becomes law, of the law as law [*de la loi en tant que droit*]" (Derrida, 2002:233).

Configurations such as *legal order*, *legal sphere*, *legal constancy* and *legal validity* are expressions of constitutive structural moments within the modal structure of the jural aspect. For this reason, they are treated as *elementary basic concepts* by Dooyeweerd, and also designated as *analogical basic concepts* of the discipline of law. To reiterate:

> The other elementary basic concepts analysed in his five-volume *Encyclopedia of the Science of Law*, are the following: *Legal life* and *legal organ* [biotic analogy within the jural]; the juridical *will-function* [sensitive analogy within the jural]; *legal accountability*, *legal conformity* and *legal contradiction* [logical-analytical analogy within the jural]; *legal power* and the *formation* (positivization) *of law* [cultural-historical analogy within the jural]; *juridical meaning* and *juridical interpretation* [lingual analogy within the jural]; *legal*

1 Of course one does not have to sell the term causality out to either its deterministic or indeterministic interpretations (compare the difference between Einstein, Planck, von Laue and Lenard on the one hand, and Schrödinger, Heisenberg and Jordan on the other). A formulation avoiding both extremes (see page 226 above) states that, although nothing happens without a cause (partially conceding the view of determinism), whatever the outcome may be does not need to be fixed in advance (partially conceding what indeterminism appreciates).

2 Of course one can relate the term "action" to the core meaning of the physical aspect.

> *intercourse* in the correlation of *jural coordinational* and *communal relationships* [social analogy within the jural]; *juridical economy* and avoiding *jural excesses* – such as an *abuse of power* [economic analogy within the jural]; the juridical harmonization of interests [aesthetic analogy within the jural]. See also the extensive analysis of these basic concepts by Hommes (Hommes, 1972:106-480 and compare the explanation on page 570 above).

It should be noted, however, that the neo-Kantian interpretation of validity exerted an influence on the views of Dooyeweerd. It is first of all found in his use of the expression *universally valid* in connection with the nature of (structural) principles (see Dooyeweerd, 1997-I:160). Pre-positive legal principles are not yet positivized for, as we noted (see page 94 above), they are dependent upon human intervention in order to be *made valid*, to *have effect*. It is only when principles are given a positive shape that they exhibit *validity*. Consequently, insofar as principles are given as universal and constant starting-points for human action, they are not yet valid, and insofar as they are positivized (made valid), they are no longer universal in an unspecified sense. It is secondly found in the fact that Dooyeweerd did not realize that *legal validity* is actually synonymous with *legal force* (*rechtskracht*) (see Dooyeweerd, 1967-II:14). In passing, we may note that, although Kelsen upholds the neo-Kantian view of validity, his understanding of the nature of a norm reflects an element of *modal universality*, for example when he affirms that the essential content of a norm holds everywhere (universally) and always: "But because those essential moments of the contents of a norm are a priori unlimited in respect of space and time, the norm holds (*gilt*), insofar as it does not posit limits for its contents in respect of space and time, everywhere and always (*überall und immer*)" (Kelsen, 1966:137).

From Kant, we inherited a distinction between law and morality in terms of the following opposition: law is said to be *universal*, *external* and *compulsory*, whereas morality is supposed to be *individual*, *internal* and *voluntary*. Merely recollecting the distinction between collective and communal societal relationships on the one hand and coordinated societal relationships on the other, it will be clear what happened here. By describing *law* as universal, external and compulsory, the jural is identified with typical collective jural relationships – such as those found within the state, where the 'external' authority of the government applies 'universally' and in a coercive (compulsory way) to all subjects. When morality is defined in terms of ethical coordinational relationships, it is characterized as individual, internal and voluntary. But if we choose the nuclear family (a moral societal collectivity), then the authority of the parents is valid for all children ('universal'), external and compulsory, whereas any jural coordinational relationship displays the opposite features. It is therefore clear that the general modal meaning of law and morality ought to be understood in abstraction from all collective, communal and coordinational relationships. Habermas has this confusion in mind when he warns us that one must not succumb to what he calls an ingrained prejudice, namely that "morality pertains only to social relationships for which one is personally responsible, whereas law and political justice extend to institutionally mediated spheres of interaction" (Habermas, 1996:109).

From the perspective of the universality of the jural aspect the modal meaning of "legal validity," in its abstraction from any social collectivity (such as the state), is *unspecified*. But, within the state, where it is connected to the reality of "law-enforcement," this modal universality acquires a typical meaning, and not a merely modal jural meaning. Entering into a sale and purchase agreement entails that a positive shape is given to legal (and other)[1] modal principles. Yet in this case, no "sword power" is assigned to either contracting party, simply because the coordinational relationship on equal footing lacks a relation of super- and subordination, although the ability to enforce contracts presupposes the sword-power of the state.[2]

8.13.2 *Law and justice*

We argued that a true state requires that the government unite the "power of the sword" in such a way that it acquires *monopoly* over it (internally, the police force and as safeguard against possible hostilities from outside the defense force, traditionally differentiating in the air force, navy and military force). This monopoly ought to manifest itself over a spatially delimited cultural area – the *territory* of the state. But we have to repeat that the term *power* ought not to be taken in the negative, a-normative sense of untamed, *brutal force*. When Derrida refers to Benjamin, who speaks of the monopolization of Gewalt (violence), it is clear that the normative task incorporated in state power is identified with an abuse of power, i.e. with violence that is anti-normative in character.[3] Legal power is therefore not authorized violence (see Derrida, 2002:262, where he discusses the difficulties in translating *Gewalt*). Although it could be employed in norm-conformative and norm-violating ways, it cannot by definition be identified with either. Much rather, according to the cultural mandate to humankind, it must be seen as a cultural calling, as an assignment given to humankind, which burdens the power formations of the state with a particular task. Naturally this task could be accomplished in a better or worse way, in norm-conformative or in norm-violating (anti-normative) ways. Moreover, the power of the state, as manifested in its function within the cultural-historical aspect, must be distinguished both from *legal power* (the cultural-historical analogy within the jural aspect mostly captured by calling it *legal competence*) and from *force*, because, as we saw, the latter is derived from the core physical meaning of energy-operation.[4]

Derrida does acknowledge the historical foundation of law – the fact that it constantly changes: "There is a history of legal systems, of rights, of laws, of positive laws, and this history is the history of the transformation of laws." These changes (deconstruction) form the "condition of historicity … and

1 Such as showing respect (social norm), expressing clearly what one wants (lingual principle), distinguishing properly (logical-analytical norm), and so on.

2 Alan Cameron points out that there are also *unenforceable* contracts.

3 See Derrida, 2002:267 where he quotes the following words from Benjamin: "Interesse des Rechts an der Monoplisierung der Gewalt."

4 In his discussion of Pascal, Derrida uses the terms force and power interchangeably (Derrida, 2002:241).

progress" (Derrida, 1997:16). Yet for Derrida, "justice is not the law. ... Without a call for justice we would not have any interest in deconstructing the law" (Derrida, 1997:16; see also Derrida, 2002:254). Derrida is pleased with Levinas, who says that "justice is the relation to the other" (Derrida, 1997:17).[1] Derrida often explains the difference between law and justice in terms of the difference between what is *calculable* and *incalculable* (see Derrida, 1997:17 and 2002:235, 244, 249-250, 257).

On the one hand, Derrida emphasizes that there must be laws: "Law is the element of calculation, and it is just that there be law" (Derrida, 2002:244). This affirmation that it is 'just' that there are laws becomes all the more significant when Derrida at the same time stresses that "[L]aws are not just in as much as they are laws." This view causes Derrida to depreciate the nature of law even further, for he argues that the reason why we obey laws is not because they are just, but because they have 'authority': "The justice of law, justice as law is not justice. Laws are not just in as much as they are laws. One does not obey them because they are just but because they have authority" (Derrida, 2002: 240).

If it is "just that there be law" this 'justness' links what is just with the universality (and calculability) of law. Yet in general, Derrida opposes the universality of law on the one hand and the direction of justice towards what is unique and singular on the other. He introduces the idea of an address (in the sense of *direction*), which "says something about law [*droit*] and about what one must not miss when one wants justice, when one wants to be just," but "the address always turns out to be singular" (Derrida, 2002: 245). Having initially stated that *laws are not just in as much as they are laws*, Derrida here proceeds to speak of "justice, as law": "and justice, as law,[2] seems always to suppose the generality of a rule, a norm or a universal imperative" (Derrida, 2002:245).

The opposition between *law* and *justice* that Derrida has in mind is clearly captured in the following significant question:

> How to reconcile the act of justice that must always concern singularity, individuals, groups, irreplaceable existences, the other or myself *as* other, in a unique situation, with rule, norm, value, or the imperative of justice that necessarily have a general form, even if this generality prescribes a singular application in each case? (Derrida, 2002:245).

Note the subtle distinction between the "act of justice" – that 'must' concern singularity and a unique situation – and the "imperative of justice" – that "nec-

1 Of course, this is a fairly undifferentiated designation, for the relation to the other may come to expression in different ways or modes of being, not merely the jural within a context of justice. The relation to the other also comes to expression in economic actions, social contexts, lingual interaction, and so on – and all of these modal functions differ from the jural aspect that is also discernable in those (many-sided) interrelationships.

2 In the same context Derrida mentions the possibility of " apply[ing] a just rule." If a rule is taken to be universal (and calculable), and if justice does not follow from applying rules, how then is it possible for Derrida to refer to "a just rule"?

essarily have a general form." This distinction may give the (mistaken) impression that Derrida considers justice to obtain when a universal imperative (namely the imperative of justice) is *applied*. There seems to be an inherent dialectic between universality and singularity (uniqueness / individuality) in Derrida's exposition . A few pages further on, he says: "One must know that this justice always addresses itself to singularity, to the singularity of the other, despite or even because it pretends to universality" (Derrida, 2002: 248).

What rips apart the homogeneous fabric of law making, of a previously founding law, a preexisting foundation, is a *decision* (Derrida, 2002:241), where "the *decision* between just and unjust is never insured by a rule" (Derrida, 2002:244). When a law is obeyed, even in the sense of autonomy, i.e. the "freedom to follow or to give" to oneself "the law," such an application of a rule (the effect of a calculation) may 'perhaps' be seen as 'legal' – in the sense "that it conforms to law" – "but one would be wrong to say that the *decision* was just. Simply because there was, in this case, no decision" (Derrida, 2002:251). A decision opens the way to justice, but as soon as one attempts to interpret a decision as conforming to a law, there is *no decision*!

It is only in respect of a being that is free and responsible in a given act that one can say "its decision is just or unjust" (Derrida, 2002:251). As long as a "programmable application or the continuous unfolding of a calculable process" is at stake, it "might perhaps be legal" but "it would not be just" (Derrida, 2002:252) – "only a decision is just" (Derrida, 2002:253). Discourses may be (obliquely) seen as discourses of justice in reflecting "the undecidable, the incommensurable or the incalculable, on singularity, difference and heterogeneity" (Derrida, 2002:235).[1]

However, according to Derrida, "the undecidable is not merely the oscillation between two significations or two contradictory and very determinate rules, each equally imperative … between two decisions." What is undecidable delivers itself to "the impossible decision while taking account of law and rules" (Derrida, 2002:252).[2] And "at no time can one say *presently* that a decision is just, purely just (that is to say, free and responsible)" (Derrida, 2002:252). Justice emerges from a decision through the ordeal of the undecidable, for otherwise it would not be a *free* decision: "A decision that

1 "It goes without saying that discourses on double affirmation, the gift beyond exchange and distribution, the undecidable, the incommensurable or the incalculable, on singularity, difference and heterogeneity are also, through and through, at least oblique discourses on justice."

2 The full text reads as follows: "Yet, the undecidable is not merely the oscillation between two significations or two contradictory and very determinate rules, each equally imperative (for example, respect for equity and universal right, but also for the always heterogeneous and unique singularity of the unsubsumable example). The undecidable is not merely the oscillation or the tension between two decisions. Undecidable – this is the experience of that which, though foreign and heterogeneous to the order of the calculable and the rule, must [dolt] nonetheless – it is of duty [devoir] that one must speak – deliver itself over to the impossible decision while taking account of law and rules" (Derrida, 2002:252).

would not go through the test and ordeal of the undecidable would not be a free decision; it would only be the programmable application or the continuous unfolding of a calculable process. It might perhaps be legal; it would not be just" (Derrida, 2002:252).

The position Derrida takes on law and justice therefore boils down to a particular view of an *act*, a *decision* and *freedom* (from law) in his thought. In his view that justice is incalculable, one finds an anti-mathematical trait, similar to what forms a part of Goethe's understanding of the nature of a law, supposedly "free from mathematical ingredients" (see page 272 above). Of course the difference between Goethe and Derrida is that the latter does not combine his understanding of 'incalculable' justice with any notion of law, owing to the distinction he makes between the universality of law and the singularity of justice.

The guiding and refining role of an act, a decision and freedom in Derrida's thought about justice (what is just and unjust) presupposes his acknowledgement of the rightful place of law ("it is just that there be law" – Derrida, 2002:244). The universality of law appears to play a *constitutive role*, whereas the guiding and refining role of justice serves a *regulative purpose* by being focused on unique and singular historical circumstances. This orientation shows remarkable similarities with Dooyeweerd's view of the relationship between law and justice. According to Dooyeweerd, those aspects that are foundational to the jural aspect are indeed constitutive for the meaning of the jural and for every imaginable legal system or juridical order. The government of every constitutional state under the rule of law has to bind together (i.e. *integrate*) the multiplicity of legal interests within its territory, thus erecting and maintaining one public legal order. Such a state is called upon, through appropriate juridical organs, to establish balance and harmony amongst these legal interests. Wherever an infringement of rights takes place, such a state ought to restore the resulting imbalance in a retributive manner. Those modal analogies within the legal aspect referring back to the aspects *preceding* the jural mode, are therefore what we designated as *constitutive* structural building blocks within the jural aspect. They are found in every possible legal order.

The task of the juridical integration of legal interests entails a differentiation between the legal spheres of a differentiated society distinguished above, namely the domain of *public law* (correlated with public freedoms – like the freedom to express political views, to organize political parties, and to participate in the capstone of political freedom: the right to vote), the domain of *personal-individual freedom* (common law or civil law) and the domain of *societal freedoms* (non-civil private law). We argued that the nature and existence of these legal spheres are *constitutive* for the existence of a *just state* (*Rechtsstaat*). The specification *constitutive* implies that the legal order of a state cannot function properly except on the basis of the presence of all three these legal domains. A more refined analysis may proceed by specifying the constitutive legal principles in the subdivisions of the above-mentioned legal spheres. Consider for example the fundamental *procedural principle* of civil law,[1] namely that the other side must also be heard (the *audi et alteram partem* rule). Or contemplate the basic (constitutive) nature of criminal law, evinced in the principle that a person can only be found guilty if the sentence is based upon evidence enabling a conclusion *beyond any reasonable doubt*. In addition, it must be based upon *due process*. Within the domain of administrative law, the requirement of *due care* obtains, and so on.

However, as we saw (see pages 228 ff. above), the meaning of the jural aspect may also be disclosed under the guidance of the two aspects succeeding the jural aspect, namely the ethical (core meaning: love) and the certitudinal aspect (core meaning: trust, certainty). Traditional (undifferentiated) societies lack the disclosure of the forward-pointing (antecipatory) coherence between the jural aspect and the moral and certitudinal aspects. In other words, since the jural aspect is foundational to the moral (ethical) aspect, it is quite possible that the structure of this aspect may still only appear in its *restrictive*, i.e. "not yet disclosed," meaning (see Hommes, 1972:481-546). A system of penal law, in which the meaning of the jural aspect is undisclosed, will display all the constitutive structural moments within the jural aspect, such as that it is constituted as a *legal order* with its own *domain of jurisdiction*, its own form of *jural causality* (*Erfolgshaftung*), and so on. These constitutive elements are accounted for in the concept of law (see page 570 above). The disclosure and deepening of the jural aspect, however, open up the antecipations to the moral and certitudinal aspects. The disclosed meaning of the jural aspect exceeds the *concept* of law, it can only be approximated in an idea (in the sense of that which exceeds the confines of a concept, i.e. in terms of concept-transcending knowledge). Dooyeweerd writes: "the moment of fault cannot be grasped in a basic concept of legal science. Within the jural aspect jural fault anticipates the moral mode of experience. It only reveals itself when we encounter a structure of legal life that has been opened-up. It is therefore theoretically encapsulated in the theoretical idea of the jural aspect" (Dooyeweerd, 1967-II: 75-76).

At this point, one can compare the following explanation with what has been said about the deepened idea of the jural aspect in Dooyeweerd's legal philosophy: "This excess of justice over law and calculation, this overflowing of the unpresentable over the determinable, ..." (Derrida, 2002:257). What one can designate as idea-knowledge (concept-transcending knowledge) in Dooyeweerd's thought is portrayed by Derrida as an "excess of justice over law and calculation," as an overflow "of the unpresentable over the determinable." Recalling Aristotle's view of equity (see page 229 above) it must be clear that the field of application of equity concerns unique historical situations, transcending the *generality* of statutory law. This direction of equity towards what is 'singular' (to employ the term used by Derrida) is an instance of

1 Alan Cameron points out that this principle of course also serves as a procedural principle of administrative law which is classified as (internal) public law, e.g. when administrative and quasi-public tribunals review the conduct of the state (personal communication).

a deepened legal-ethical principle. We designated all the deepened principles of jural morality collectively as *principles of justice* – such as the mentioned fault principle, the principle of equity, that of *good faith* (*bona fides*), *legal certainty*, the *dignity of the human personality* and so on.

The similarity between Derrida and Dooyeweerd is that they both distinguish between the concept of law and what exceeds this concept. The difference between them is that, whereas Derrida considers the 'exceeding' to leave behind every form of calculability and universality, Dooyeweerd holds on to the idea of deepened legal *principles* – where justice serves as the embracing term capturing all the disclosed principles of juridical morality (the mentioned legal-ethical principles). Although this may appear to be a substantial difference, closer investigation reveals a further remarkable similarity. In order to understand this similarity, we have to recall that modern nominalism denies the universal features of factual reality – concretely existing entities (individuals) and events are strictly individual because, according to nominalism, there is no universality outside the human mind.

We saw that Dooyeweerd accepts ontic universality (for example the universal scope of the different aspects of reality). Derrida, as we have seen, also does not hesitate to affirm universality. He claims that faith is universal, deals with *messianicity* and he speaks of the "historicity of law" (Derrida, 2002:266). The historicist intends to strip historical reality of its universality by claiming that every historical event is unique and non-repeatable. However, the opposite of the aim is achieved – *uniqueness* and *non-repeatability* are two universal traits of *historicity* applying to *all* historical events. In other words, factual reality in its assumed uniqueness, singularity or individuality cannot escape from displaying *universal* features. *Messianicity* and *historicity* share in this quality. In *being historical*, every unique historical event evinces the fact that it is subjected to universal conditions for being historical in a universal way. Historicity exhibits the measure of the law for being historical, and in its orderliness (or even disorderliness), it reflects the correlation between conditions (law) and what is conditioned (what is subjected to law).

In a typical nominalistic fashion, Dooyeweerd denies the universal side of factual reality (its law-conformity). This does not mean that he denies law-conformity as such. He simply identifies law and law-conformity, thus handing factual reality over to what is purely *individual*. Derrida accepts universal principles and the universality of law-conformity (by talking about *messianicity* and *historicity*), but when it comes to the meaning of justice, he wants to be *free*, liberated from universality and focused on what is unique and 'singular' – as if factual reality in this case is also suddenly stripped from any universality. The nominalistic element in Dooyeweerd's thought, given in his denial of the universal side of factual reality, is mirrored in Derrida's thought in respect of his peculiar view of justice "as the experience of absolute alterity" of what "is unpresentable" (Derrida, 2002:257) and as concerned with "singularity, individuals, ... irreplaceable existences, ... in a unique situation" (Derrida, 2002:245). If one employs our definition of rationalism as an overestimation of (universal) conceptual knowledge and irrationalism as an over-estimation of concept-transcending (idea-)knowledge, then it is clear that both Derrida and Dooyeweerd are partially in the grip of the irrationalistic side of nominalism.[1]

The starting point for overcoming this nominalistic element may be present in a two-fold way. In reaction to Dooyeweerd, one may say that he has to acknowledge universality at the factual side of reality – evinced in the atomness of an atom and the houseness of a house. How should this be done in respect of what Dooyeweerd and Hommes call the legal-ethical principles of justice?

By maintaining the idea of *positivization* it should be kept in mind that not even a deepened legal principle can ever be "individualized" – for then the basic distinction between universality and individuality will collapse. Through disclosed legal-ethical principles, the deepened meaning of law acquires ever-increasing *specified* forms of justice, without ever converting justice into something purely and solely unique and individual. This last remark can liberate Derrida from his misguided attempt to bridge the gap between the universal and the unique. By conceding that whatever is given in its factuality always displays both an individual side and a universal side – just compare the statement that *this* state (individual side) is *a* state (universal side) – one can avoid squaring the circle, because universality and what is invidual are *irreducible*. Von Ranke once said: "Consider aristocracy according to all its features, and never one could fathom the existence of Sparta" (cf. Landmann, 1973:81).[2] Sparta is similar to the instances of justice in Derrida's thought (connected to *acts*, *decisions* and *freedom*), but although it cannot be fathomed (*ahnen*), Sparta does not fall outside the (universal) conditions for *being-an-aristocracy*, it is merely one of many possible specified instantiations of these conditions. Likewise, what Derrida calls justice does not need to fall outside universal conditions or evincing in *being-just* that there is also *universality* attached to just situations (acts or free decisions) – without denying the truly factual uniqueness (individual side) of such situations.

8.14 Equity and transformation – a case study

In South Africa, the new constitution makes provision for redressing the injustices of the past (*Apartheid*) in terms of equity (in addition to the constitution, also see the employment equity Act – No.55 of 1998). The interpretation of 'transformation' requires that societal institutions in South Africa ought to 'reflect' the (numerical) composition of its population. The problems involved in these views remind us of the traditional distinctions between republic, monarchy, aristocracy, and democracy, where the role of numerical relationships also appears to be decisive (cf. Arendt, 1953, 133 ff.). However, a closer scrutiny of these views will show that the role of numerical relations and calculations can easily have (albeit unintended) *totalitarian* conse-

1 Since nominalism does acknowledge universal concepts or words within the human mind, we pointed out that it is rationalistic at the same time.

2 "Denke dir die Aristokratie nach allen ihren Prädikaten, niemals könntest du Sparta ahnen."

quences, because quantitative considerations *alone* do not contain any safeguards against totalitarian and absolutistic views and practices. These consequences may need further reflection on what has been designated as the 'total state' (see page 547 above).[1]

Chapter 2, 8(1) of the Constitution, stipulates that "[T]he Bill of Rights applies to all law, and binds the legislature, the executive, the judiciary and all organs of state." Of particular importance are Sections 9(2) and 9(3) of Chapter 2. Section 9(2) states: "Equality includes the full and equal enjoyment of all rights and freedoms. To promote the achievement of equality, legislative and other measures designed to protect or advance persons, or categories of persons, disadvantaged by unfair discrimination may be taken"; and 9(3) holds that the "state may not unfairly discriminate directly or indirectly against anyone on one or more grounds, including race, gender, sex, pregnancy, marital status, ethnic or social origin, colour, sexual orientation, age, disability, religion, conscience, belief, culture, language and birth."

The affirmative action entailed in section 9(2) seems to contain an implicit tension in relation with the meaning of section 9(3). Disadvantages caused by "unfair discrimination" ought to be rectified through legislation and other measures, thus leaving room for what is called "fair discrimination."

Section 9(1) states that: '[E]veryone is equal before the law and has the right to equal protection and benefit of the law'. Taken together, these three sections contain the central problem of the current constitutional dispensation of South Africa.

On the one hand, equality and non-discrimination represent the application of a *constitutive legal principle* prevalent within the constitutions of all modern constitutional states. In a state where these constitutive jural principles do not find application, all sorts of injustices are possible. The legal order of a just state therefore depends crucially upon the validity of this principle in a concretised form, i.e. in the shape of positive law.

Speaking of "unfair discrimination" at once sanctions "fair discrimination." Since *fairness* is merely a synonym for *equity*, and since equity, as we argued, is one of the *regulative* jural principles[2] guiding the actions of competent jural organs in a refined, deepened and specified way in the assessment of unique (and formally and materially) unforeseeable situations, events or circumstances, these sections balance the constitutive demands of equality and non-discrimination with the morally deepened jural guideline of a regulative equity perspective.

If the difference between constitutive and regulative principles is not prop-

1 Leibholz points out that in the total state, not only all human beings in all walks of life are subjected to the grasping power of the state, since the educational institutions are also absorbed in the firm hold of a totalitarian state or "ruling party." The merciless control of this total state embraces human life from birth to death (Leibholz 1954, 1125-126).

2 Regarding the distinction between constitutive and regulative , see pages 228, 311, 315, 390, ff.

erly observed, one may ultimately interpret affirmative action in a *constitutive* sense – on the same level as the (constitutive) appeal of modern *Bills of Rights*. But as soon as these are both viewed as constitutive, our understanding becomes a victim of an inherently contradictory view, for in a *constitutive sense*, it is contradictory to authorize discrimination and condemn it at the same time.

The implicit confusion of a constitutive and a regulative perspective is embodied in Section 9(5): "Discrimination on one or more of the grounds listed in subsection (3) is unfair unless it is established that the discrimination is fair." The statement that "[D]iscrimination on one or more of the grounds listed in subsection (3) is unfair" echoes the constitutive thrust of the general scope of a legal order, which means that the term 'unfair' should have been replaced by a constitutive legal term such as *unlawful*. But by not having done this, the second statement, entailing that it can be established that 'discrimination' is 'fair', contradicts the first by literally asserting that discrimination can be both fair and unfair.

Acknowledging that this section in fact does employ the term 'unfair' (instead of *unlawful*), we do have a way of avoiding this dilemma. The problem may be resolved by applying the distinction between *constitutive* and *regulative*, for in doing this, one can argue that the first part of Section 9(5) ought to be interpreted in a constitutive sense, while the second part should be seen in regulative terms, in which case it is no longer contradictory, owing to the difference between a constitutive and a regulative context.

It is striking to consider the position of the *Constitutional Court of South Africa* in this regard. In a particular case[1] the absence of the distinction between constitutive and regulative perspectives in many instances haunts the deliberations of the judges concerned. This is clear from the fact that judges tend to interpret Section 9(2) in separation from Section 9(3) – as if both have a constitutive jural meaning. In one instance, Mokgoro supports the conclusion drawn by Moseneke, but where the latter argued in terms of Section 9(2), the former holds that the facts of this case ought to be decided in terms of Section 9(3) (see CCT, 2004:67). With reference to the *Pretoria City Council Case v. Walker*[2] Sachs focuses on the situation "where the measure advances the disadvantaged but in so doing disadvantages the advantaged," and then points out that "members of the advantaged group are not excluded from equality protection" (CCT, 2004:150). In addition, he mentions that the same judgment holds the following: "Courts should, however, always be astute to distinguish between genuine attempts to promote and protect equality on the one hand and actions calculated to protect pockets of privilege at a price which amounts to the perpetuation of inequality and disadvantage to others on the other" (CCT, 2004:150).

As a rule, "constitutional democracies" take a *Bill of Rights* to serve as a

1 Case CCT 63/03, decided on 29 July 2004.

2 1998 (2) SA 363 (CC); 1998 (3) BCLR 257 (CC).

constitutive building block for their legal order. In the case of South Africa, where the injustices of the past provide unique circumstances in which, on the grounds of equity, affirmative action serves a disclosed jural purpose, we have to realize that the constitutive legal order is regulatively deepened. Moreover, because it embodies a regulative principle, employment equity has to be appreciated in terms of a time limit, for if it is interpreted as a *permanent* task, it not only turns into a constitutive stipulation, but at the same time it changes our constitution into something flawed, with an inner contradiction, sanctioning non-discrimination and discrimination at once and forever.

It is unfortunate that the regulative role of equity is currently eroded by interpreting it in quasi-constitutive terms, namely in the service of *transforming* the entire South African society into something that ought to reflect the *composition* of its population. If transformation is not interpreted in terms of equity, this alternative will cause it to decay into a new *Leviathan*, the *dictatorship of numbers*. Such a totalitarian view is not democratic in any sense of the word, and is irreconcilable with the idea of the *just state* as well as a *differentiated society*, which is a pre-supposition of the idea of the constitutional state, for in practice it eliminates the principal structural differences between the state as a public legal institution and the various non-political societal spheres.

Our preceding considerations regarding the nature and place of the state in a differentiated society broadens the perspective of all those social sciences that have a limited field of investigation in the sense that they proceed from the perspective of some or other (normative) aspect of reality (such as the disciplines of history, economics, sociology and ethics).

Let us now turn to the domain of a non-political societal collectivity belonging to the sphere of non-civil private law by analyzing, as an example, the structural principle of the *university* as an *academic institution*.

8.15 The position of the university within a differentiated society

Modern highly differentiated industrial societies are all co-dependent upon the role of universities as academic institutions. Even the initial genesis of modern Western universities during the late medieval period and the early modern era, faculties such as those of divinity, law and medicine intended to provide society with pastors, lawyers and doctors. During the past five hundred years, the scholarly scope of the academic enterprise broadened its reach to such an extent that there is almost no single domain within society that does not have intimate links with important competencies and skills acquired through some or other course at a university.

Soon after the rise of universities by the end of the 12th century, it became clear that modern society will have to take into account three future powers: the church, the king, and the academic podium (*sacerdotium*, *imperium*, *studium*) (cf. Rashdall, 1936:2, 573; Romein, 1947:2 and Stellingwerf, 1971:136). We noted that at this stage, the all-encompassing grip of the church started to fade (see pp. 499 ff. above), particularly through developments during the early 14th century. Late scholastic nominalism challenged the authority of the pope and the church, and the Renaissance opened up new vistas, exploring the possibilities of intellectual pursuits as an infinite task – thus continuing views advanced by thinkers such as Cusanus, Giordano Bruno and eventually Galileo, Descartes and Newton.

The differentiation of church and university opened up a domain of freedom for intellectual pursuits. Yet liberation from the totalitarian Greek-Medieval grip of the state and the church on society does not necessarily mean that ideological distortions disappeared, although the liberation of cultural life from control by the church and the state opened up the freedom of scientific research and theorizing to such an extent that these liberties were acknowledged in the constitutions of modern states. However, as Topitsch argues, the 'ideology' of (neo-)liberalism had to struggle against the spiritual climate in which traditional ideologies maintained themselves – dating back to Medieval and Greek influences (Topitsch, 1969:24).

During and after the early modern era, and particularly after the rise of positivism as a philosophical orientation dictating the appreciation of 'science' (restricted to physics as model of science), modern universities claimed to have moved ahead by taking a second liberating step. Whereas the state had to free itself from dominance by the church, science and the university had to rid themselves of the interventions of both the church and state.

Unfortunately, the beginning of the 20th century was dominated by the positivist belief in the neutrality and objectivity of 'science'. As such, this view confused religious beliefs within the context of ecclesiastical communities with scholarly (scientific) convictions that are inherent to the scientific enterprise. However, an insight into the nature of the latter had to wait for extensive developments within the new philosophy of science, worked out during the second half of the 20th century (discussed in Chapter 2 above).[1] Positivism was severely challenged by Popper and also by the Frankfurt school of neo-Marxism. The slogan of 'objective' and 'neutral' science had to face the protesting students of the late nineteen sixties. They claimed that Western societies actually surrendered to ideological distortions with an appeal to "objective (neutral) science." Suddenly the appeal to "academic freedom" no longer appeared convincing at all. The solemn academic, exercising his or her personal (individual) academic freedom, increasingly had to face claims to accountability. Already during the seventies, specific scientists and disciplines began to give an account of their 'relevance' for 'society' at large. The ivory tower of the isolated academic was broken down, and institutions for higher learning had to 'position' themselves within society in terms of what then, i.e. during the eighties of the previous century, emerged as the necessity of formulating a *vision* and *mission*.

Instead of being *liberated* from society, the contemporary university now

1 The most important insights of this new philosophy of science were anticipated by Dooyeweerd during the three decades preceding 1950.

had to face the issue of accountability. Suddenly a 'third' function of the university emerged, adjacent to *teaching* and *research*, namely *community service*. The outcome of this process was that *collective accountability* (vision and mission formulation) and *additional community service* globally became an integral part of contemporary university life. What was not realized is that the need to formulate a mission and vision actually emerged from the theory of *instrumental organizations* (cf. Luhman, 1973:55 ff.). This theory, which is intimately linked to modern *systems theory*, views the possible *aims* of a social collectivity and its *inner nature* as interchangeable. As a consequence, it is no longer capable of distinguishing between *typical* and *a-typical* aims. Only when the typical structure, the unique intrinsic nature of an institution, is accounted for, will it be possible to come to a meaningful distinction of what *typically* does and does not belong to the task of that institution.

If the university is an *academic* institution to begin with, this very basic character should play a guiding role in all its activities. Stellingwerf remarks that the oldest idea of the university is that of an *universitas magistrorum et scholarium*, i.e. a *community of teachers and students* (Stellingwerf, 1971: 35). This led to the idea that the various disciplines are not disconnected, but should be appreciated within the perspective on the unbreakable *coherence* that prevails between them (the idea of an "universitas scientiarum"). Only after the Enlightenment, particularly in Germany, the importance of combining teaching with research was brought into practice. In England, the training and education of a responsible 'gentleman' impregnated the university with a distinctive twist (cf. Stellingwerf, 1971:36-37).

It is important to clarify the *academic* character of the university *before* any attention is given to its possible 'aims' or 'goals'. Of course any social collectivity can also pursue *goals* or *aims* that do not typically belong to its basic nature. We mentioned the example of South Africa where the state was involved in many a-typical endeavours, such as erecting a railways, establishing a steel industry (Iscor) as well as a Post Office. Since it is possible to envisage a state not involved in economic businesses such as running a railways, managing a steel industry, or providing a country with a postal service, shows that these tasks exceed the range of typical governmental obligations. However, a state that is not involved in the *harmonization* of the *legal interests* within its territory – on the basis of the "power of the sword" – is no longer a state in the proper sense of the term.

Of course this does not mean that it may not be justified to be involved in such tasks – as argued above (see page 574). The guiding perspective should always be that, whatever is undertaken, should not lead to a greater measure of dependance upon the state, but rather enhance an increasing sense of responsibility and independence within non-political domains of society. For otherwise the a-typical tasks of the state may lead to a *de-differentiation* of society and an eventual return to an undifferentiated totalitarian societal condition.

Similarly, a university may also undertake a-typical tasks, brought together under the flag of "community service." Once again, these tasks should be recognized for what they are: *a-typical*. This qualification simply means that there may be (and **are**) universities that restrict themselves to their typical (and therefore: *primary*) 'service' to society, namely to provide society with high quality teaching and research. When universities – for the sake of relieving poverty, uplifting communities and opening opportunities for formerly disadvantaged people – involve themselves in a-typical tasks ("community service"), the aim should always be to help various sectors of society to grow and mature to a point where they can properly function within their social orbits. This should be the *Leitmotif* of all "community service" of universities – observing the principle of general cultural civilizational development.

The same applies to the "entrepreneurial" activities of universities. They are all geared towards the ideal of making the university financially independent – such that the typical task of the university as an academic institution may flourish. If this a-typical side of modern universities is mistakenly understood to be a typifying characteristic, it would be impossible to identify the difference between the university and a firm.

If it is true that an institution such as a university cannot be divorced from its social and economic environment, the question remains why it cannot be identified with it either.

8.15.1 *University or multiversity*

The history of Western universities started with the Greek *episteme* and Latin *scientia* and culminating in the English *science* and German *Wissenschaft*. We have already noted that the science ideal and positivism accounts for the predominantly natural scientific contents assigned to the term 'science' within English-speaking countries. However, it is preferable to continue the German tradition, according to which 'Wissenschaft' encompasses all academic disciplines, including the humanities (such as economics, linguistics, political science, the discipline of law, theology and aesthetics).

Many competent scholars point out that the initial *unity* of knowledge and science (Wissenschaft), constitutive for a *uni*-versity, eventually disintegrated to such an extent that there were no longer any visible signs of an integrated perspective. The (philosophical) claim of positivism, namely that the disciplines operate without any philosophical presuppositions whatsoever, contributed substantially to the emerging reality of a *multi*-versity. The well-known defense of the "value-free" nature of the academic enterprise advanced by Max Weber did not challenge the positivistic emphasis on 'factuality'. In respect of the positivistic pre-occupation with 'facts', Topitsch objects that this 'Tatsachenfetischismus' serves to justify existing social relationships (Topitsch, 1969:45).

It should be kept in mind that initially various faculties maintained what is known as a *studium generale*, in which a unifying perspective amidst the diversity of scholarly disciplines prevailed. The nature of this *studium generale* did not oppose a differentiation of disciplines (such as that between the *triv-*

ium and *quadrivium*). In fact, the initial crucial impulse was not even directed towards *training* for specific *occupations*. Rather, the central concern was scholarly reflection *itself* (cf. Grundmann, 1960: 39 ff. 63 ff.).

It must therefore be seen as a serious loss that our modern and contemporary universities succumbed to the practice of a *multiversity* in spite of the unity that is still present in their organizational structures, and which is brought to expression in their vision and mission formulations. What is notably *absent* in most vision and mission statements, is a constructive and integrating ideal of the unity of all disciplines, guided by a particular life and worldview.

In practical terms, this more recent disintegrating legacy left universities open to the threat of specialized 'idiotism'. A victim of this ailment, in popular parlance, is a person who "knows more and more about less and less, until that person knows everything about nothing." Already in the thirties of the previous century, Ortega Y. Gasset highlighted this negative tendency.

Ironically enough, this disintegrating effect of the philosophy of positivism found a counter-force in the newly emerging philosophy of science that once again emphasizes the necessity and inevitability of a directing *paradigm*. Hand-in-hand with this development, neo-Marxism advanced a radicalization of the biblical account of *sin* by proclaiming that even every *new order* is corrupt by definition – in the words of Ernst Bloch: "what is cannot be true."

The underlying historicistic orientation and hermeneutic inclination of the contemporary 'Zeitgeist' enhance the idea that the ongoing changes the university experienced in its historical development preclude the idea of a *constant underlying structural principle*. Rossouw points out that, in spite of all the differences entailed in the idea of the university, one can nonetheless identify elements of continuity as well: "A diversity of paradigmatic views regarding the nature and function of the university developed in different eras and cultural environments. This diversity of paradigmatic views, or sets of presuppositions and convictions, do represent competing but nonetheless not totally irreconcilable manifestations of the idea of the university" (Rossouw, 1993: 31). To this, he adds that these manifestations constitute "in reality a continuous conceptual tradition in which already established meanings repeatedly at once are conserved and transformed, in which they at the same time transcended and maintained a relative validity" (Rossouw, 1993: 31). Although one can support this recognition of both constancy and *change*, the absence of acknowledging not merely a *conception* or *idea* of the university, but indeed a *normative structural principle* at the foundation of variable historical manifestations, does highlight an alternative orientation – one that is clearly skeptical about *ontic normativity* (see page 290 above). The recognition of the *historical context* of scholarly thinking combined with the necessity of always *interpreting* data does not eliminate the underlying relevance of (universal) logical principles, but *presuppose* them.

Moreover, although scientific thinking in all its manifestations operates under the guidance of (deepened) logical thinking, a university is much more than merely its function within the logical-analytical aspect of reality. Particularly, the modernist faith in the rationality of the human intellect created an overestimated expectation, namely that the intellectual endeavours of universities ought to give *guidance* to all of life. Hazel Barnes's 1970 book on *The University as the New Church* therefore did not come as a great surprise.

8.15.2 *External and internal intermodal connections*

When Kuyper made a plea for the internal autonomy of the university (its sphere-sovereignty) he aimed at a university that should be free from interference by state and church. The pursuit of scholarly activities nonetheless continues to depend upon a supra-theoretical commitment that transcends all the differentiated societal ties a person may have.

The characteristic theoretical-analytical function of scholarly activities within the university displays both an inner and an external coherence with other facets of reality.

Think for a moment of the nature of the credo of an academic institution – the formulation of its vision and mission. One can compare it with the confession of a community of faith, for in both cases we encounter the way in which a university and a church function within the certitudinal aspect of reality. But as little as one can identify a political party with a church merely on the basis of its political convictions (credo), is it possible to identify a university and a church merely because they both have a function within the faith aspect of reality. This coherence between the logical-analytical function and the certitudinal (faith) function of a university is *external* and should therefore be distinguished from the inner connection between these two aspects, manifested in the *logical trust* (logical certainty) guiding scientific thinking in various contexts.[1]

In a similar way, one can distinguish between an internal and external coherence between the logical and the historical aspect of a university. The former is observed in the *logical mastery* or *logical control* over a given field of study (knowledge-material – internal antecipating coherence), while the latter is found in the concrete function of the university in the historical aspect (its history). The logical principle of thought economy mentioned earlier (Occam's razor), displays the inner connection between the logical and the economic aspects of reality. It ought to be distinguished from the original function of the university within the economic aspect (the external coherence – consider the university budget).

8.15.3 *University, state and law*

During the late Medieval period, the emerging universities manifested their "sphere-sovereignty" in an extremely independent way, simply because in these institutions, lecturers and learners did not have any assets. It was strong

1 See page 314 ff. above for a brief analysis of the relationship between 'reason' and 'faith'.

because it was not bound to any *place* or *possessions* (Stellingwerf, 1971:143). This phase soon passed. Buildings, capital and libraries proved to be cultural objects, essential for a *structured* functioning of the university. Already before 1500, the Pope in Italy turned 64 of the 70 existing universities into internationally acknowledged institutions. This led to a restriction of universities to the *national* domain, but the most important element was that, whereas three faculties served specific occupations, liberal arts were to be organized in a basic faculty responsible for the general development of the student. This general development within the faculty of the humanities had a 'propaedeutic' function with respect to the other three faculties (cf. Stellingwerf, 1971: 145).

Even in those cases where the government does make a contribution to the functioning and continued existence of universities, this does not mean that the state can prescribe to a university what its *scientific orientation* ought to be. At most, the state can apply a general *formal* criterion, stipulating the minimal formal requirements for an acceptable university. The *directional choice* of a university – manifested in its statement of vision and mission – in principle lies beyond the grasp of governmental interference. If the state transgresses in this regard, it has irrevocably set foot on the path of a totalitarian practice, disregarding the original sphere-sovereignty not only of the university, but in principle of all non-political societal collectivities.

The fact that the government of a constitutional state under the rule of law (a *just* state) has to integrate and harmonize a multiplicity of legal interests within its territory, does not entail that these legal interests emerge or derive from the state or are created by the state. Much more, for the state to *be* a state a differentiated society is presupposed, within which there are original spheres of competence not reducible to state law. For this reason, a university, once erected, has the right to determine its own character (its directional choice) – a right not granted by the state, but merely *acknowledged* (and to be *respected*). Therefore, in spite of the fact that both the state and the university function within the legal aspect of reality, they still maintain their respective *juridical sphere-sovereignty*.

The last part of this chapter will be dedicated to an analysis of some implications of distinction between different kinds of (sphere-sovereign) societal relationships and activities, focused on the growing importance of leisure against the background of structural changes within modern industrial society.

8.16 The importance of recreation and leisure in a differentiated society

Leisure in the widest sense of the term is qualified by the *social aspect* of reality. This encompasses derived social activities such as *tourism*, *play*, *sport* and *recreation*. It often occurs that leisure is negatively defined as the *time left after work* (see Parker, 1976:12). While *work* becomes the decisive determinant, non-work implicitly derives its nature from work. Parker discusses work-leisure relationships in general descriptive terms such as 'identity', 'contrast' and 'separateness' (Parker, 1995:29 ff.). He concludes by confessing, with reference to the relationship between "work and non-work," that "we do not yet know much about the pathways underlying such relationships, their direction of causality, or relative strength" (Parker, 1995:36). Clearly he does not bring into play the *modal irreducibility* of the social, the economic and other aspects of reality as a guideline – for then he would have realized that *different aspects* cannot be juxtaposed in a cause-effect relation (compare our earlier remark concerning Kant's distinction between *succession* and *causality* on page 83).

Work normally occurs within an *economic* context – careers and jobs generate the income required to survive economically within society. Yet this does not mean that the economic motive involved in *work* entails that non-work could be captured in (negative) economic terms as well. Rather, one has to consider other possible qualifying functions as well, in order to understand the nature and context of *free time*.[1] In addition, one should realize that leisure activities within a differentiated society presuppose the distinction between collective and communal relationships on the one hand, and coordinational relationships of the other. Leys's argument regarding the proposed bigger role of the state within third-world countries is therefore understandable, because "most of the population is still caught up in pre-capitalist production relations, not fully proletarianized or urbanized, relating more directly with other (small commodity) producers, related more directly to nature, and living a correspondingly different symbolic life" (Leys, 1982:306).

In differentiated societies, there are various social forms of life that bind members together for the greater part of their lives in a way that is independent of their will. The state, for example, does not originate in a hypothetical "social contract" – which explains why it can organize the political life of its citizens independent of their will (consider for example tax obligations).

All forms of life that embrace the lives of their members partially or fully for the greater part of their life span could be called *institutional*. Marriage exhibits an institutional nature, because it is meant to constitute the marriage relationship for the duration of the partners' lives. A person is born within a family and a circle of relatives and grows up without having a choice regarding this. Not all collective forms of life possess an *institutional* character. Think for example of a *firm*, a *university college* or a *sport club* – these are all examples of collectivities based upon completely *voluntary* membership.

It is nonetheless impossible that a person merely lives within collective or communal relationships without also participating in coordinational relationships. Two families, for example, stand in a *coordinational* relationship

1 Tribe emphasizes "free time" as an element of *leisure*: "A common element in many definitions of leisure is that of free time". "Leisure = discretionary time." "Recreation = pursuits undertaken in leisure time." "Tourism = visiting at least one night for leisure and holiday, business and professional or other tourism purposes" (Tribe, 1995:2).

("inter-collective"), two married couples in an inter-communal *coordinational* relationship, and so on. Furthermore, every individual in a differentiated society is taken up in countless inter-individual coordinational relationships, where she informally relates with fellow human beings in co-ordinate relations. Conversely, no life is exclusively involved in coordinational relationships, because the counterpart of the latter is found in institutional and non-institutional collectivities and communities.

It is amazing to consider the strength of the naturalistic understanding of the human being – as a mere 'organism' – continued within sociological reflections on leisure. Geba, for example, advocates a new leisure model, focused on the difference between "being lived" (nature) and "living life" (the symbolic realm) (Geba, 1982:93). He calls it the lifestyle "Attitude Model" (Geba, 1982:91). "Attitude, then, encompasses the whole organism and its interaction with the environment" (Geba, 1982:98). "The lifestyle attitude of the human organism is made up of the sum [100] total of its motions, sensations, emotions, and thoughts. All are movements, and together they give the life of an individual its personal style" (Geba, 1982: 99-100). The view that "all are movements" is reminiscent of an early phase of the modern humanistic science ideal, where Thomas Hobbes attempted to analyze all reality in terms of the basic concept of the "moving body."

The Greeks already conceived of *work* as the precondition for *leisure*. This explains the long-standing legacy of defining *leisure* as *freedom from work*. Jarvie captures an element of this heritage when he writes:

> Focusing on the perceived fusion or polarity between work and leisure Parker argues that the former occurs when people refuse to divide up their lives between work and leisure. When polarization does occur the corresponding functions of leisure are identified as 'spillover' or compensatory. Work may be said to spill over into leisure to the extent that leisure is the continuation of work [22] experiences and attitudes. In some instances leisure activities 'compensate' for the dissatisfactions felt in work (Jarvie, 1994:21-22).

Since the Latin root *licere* originally means *to be permitted* (freedom) and to be *regulated* (constraint), it also becomes clear why leisure cannot be divorced from *normativity* (norms permit certain actions, while regulating/prohibiting others). This link with *normativity* also implies why the *time* inherent in leisure can never be reduced to mere (a-normative) *physical* time – although we have to acknowledge that all forms of *social time* are founded in *physical time*.

A closer look at what could be called *unoccupied* (spare) *time* or *free time* soon reveals that the opposite kind of time, namely *occupied time*, actually relates to different *kinds* of functioning for individuals – collective, communal or coordinational. Within each collectivity or community, individuals only come into focus as *parts* (members) of a larger *societal whole* and never as *individuals in their own right*. Consequently, the societal domain where so-called "free time" can surface *par excellence* is within the area of *coordinational relationships*.

Since one crucial characteristic of coordinational relationships is that they lack a relation of super- and subordination, they do not have office bearers exercising *authority* or *power*. The inherent (coordinational) social freedom characteristic of leisure therefore excludes the nature of *force* (the exercise of power and control – or relations of super- and subordination), which of course does not mean that leisure falls outside the matrix of normativity. Just as little as one can *force* someone to be *free* (see Rousseau, 1975:246 and page 510 above), is it possible to *enforce* leisure.

Since *work* normally takes place within some or other collective or communal relationship, and since leisure by nature belongs to the domain of coordinational relationships, it must be clear from the start that there can never be an *either/or* choice between the two. It is just as meaningless to opt for a life merely and solely constituted by work as it is to opt for a life of leisure without any communal and collective demands whatsoever.

It is therefore not meaningful to assign a *privileged position* to *work* over *leisure*, or vice versa. This would be tantamount to a distortion of the unbreakable correlation between collective and communal relationships on the one hand, and coordinational relationships on the other.

Murphy mentions that, in terms of a "structural-functional paradigm ... leisure is primarily residual instead of an integral part of life" (Murphy, 1987:11). Alternatively, he advocates a holistic approach: "Emerging holistic models based on enlightened research into the human condition suggest that the needs of the person determine what will be intrinsically motivating and therefore what constitute the leisure experience" (Murphy, 1987:16). We have to be careful not to allow "human needs" to level the structural differences between different kinds of societal relationships, for as soon as *coordinational relationships* are acknowledged as *basic* (*constitutive*) for social life, it is no longer possible to define leisurely activities as a *derivative* of work obligations or as a *luxury* that competes with collective and communal control.

Embedded in the very nature of coordinational relationships, leisurely activities constitute an *inherent* and *integral part* of the *social well-being* of all human beings. This position is one of *principle*, which appreciates leisure in *its own right*.[1] That is to say, it simply states the *originality*, *uniqueness* and *irreplaceability* of the *coordinational nature* of *leisure* without committing itself to any *specific positive social form* it might assume within any particular society. In fact, the simple ratio of time for leisure and work constantly varies throughout the history of humankind.

This feature decisively places all forms of leisure first of all within the scope of the *social sciences* (the humanities in a broader sense). Not without good reason did Lanfant, already in 1974, say that sociology is obsessed in its search for a definition of *leisure* (Lanfant, 1974:180).

1 Cooper advances a "conception of leisure as activity desired for its own sake" (intrinsic desire)" (Cooper, 1989:66).

Whereas leisure is seen as embedded in *socially qualified* coordinational relationships, *commerce* on the other hand should be seen as embodied in *economically qualified* coordinational relationships. *Leisure* and *commerce* are therefore structurally differentiated, owing to their respective *modal foci*. Of course when one proceeds from an *economistic* perspective, as Karl Marx did, the only option left is to transform leisure into an *extension* of *capital* (wealth). He defines leisure in terms of *wealth*: "Wealth is disposable time and nothing more" ("*Reichtum ist verfügbare Zeit und nichts weiter*" Marx, 1983:311).

The opposite (capitalist) extreme is involved when Johnson rejects the term *leisure* altogether (as an ideological notion) and instead prefers to "talk about sites of consumption and reproduction of labour power" (quoted by Tomlinson, 1981:65-66).

8.17 Structural changes within modern industrial society

Before the industrial revolution and the institution of machine technology and automation since the later 18th century, early modern societies were, to a large extent, governed by the dictates of the necessities of life. The irony of this development in Western society is that it was accompanied by a derailed understanding of the task of government derived from the classical liberal idea of the state (Locke) and the classical school in economics (Adam Smith and others). It was a direct result of the rise of classical physics with its deterministic worldview, as imitated by economic and political theories.

The classical liberal idea of the state, as advanced by Locke in his *Two Treatises of Civil Government* (1690), did not consider the social contract to represent a radical break with the (hypothetical) initial "state of nature." Locke proceeds from the absolute and inalienable human rights of the individual to *life, liberty* and *property* – which could not be given up through any social contract. These human rights require protection through an organized power in the transition to the civil state. But only two *basic rights* should be given up: (1) the right of individuals to do what they consider to be necessary for their own well-being, and (2) the right to punish an offender (Locke, 1966:181 – §128). Consequently, Locke conceives of the civil state as a mere continuation the state of nature protected by an organized maintenance of the civil freedom rights of *life, liberty* and *property*. In order to *maximise* civil freedoms, there has to be a minimum of governmental interference – the classical liberal doctrine of state withdrawal: *laissez faire, laissez passer*.

Reducing the task of government to a minimum in order to secure a maximum amount of civil freedom, caused political theorists to speak about Locke's *state nihilism*. The classical liberal idea of the state provided subsequent governments with ample grounds for backing off from their *normative* governmental task to protect the legal interests of their citizens within all the different societal contexts in which these interests manifest themselves. Jeremy Bentham (1748-1832) and James Mill (1773-1836) took the individualistic view of society to its peak with their motto to ensure *the greatest happi-*

ness for the greatest number.

The classical school of economic theory founded by Adam Smith (1723-1790), explored the dubious 'fruits' of Locke's idea of the state in the economic sphere of society and combined it with the 'law' of supply and demand. The effect of this 'law', as manifest in the functioning of the market, gave rise to the idea of an "invisible Hand" operative in the domain of economic affairs. Mandeville once referred to this fiction as the "fable of the bees" – as if greater prosperity and happiness would ensue when each individual pursue his or her own *self-interest* and *greed* optimally (this is also the way in which Milton Friedman, a prominent American economist of the 20th century, prefers to explain how the classical school elevated these two *vices* to overall guidelines for living: self-interest and greed).

Clearly, the theoretical 'guidance' provided by the liberal idea of the state (Locke) and the classical school in economics (Smith), **resulted in the Tory government refraining from adopting legal measures to protect** the economic legal interests of labourers. Initially, during the 17th century, the British government managed to protect the labourer by means of several labour laws. As late as 1756, a law was promulgated to make provision for courts of law to determine the wages for piece jobs. The apparently strange phenomenon is that, during the second half of the 17th century, a 'systematic' repealing of "labour-protecting laws" started to take place in Britain. By 1776, when protests against the "Spinning Jenny" were presented to the *House of Commons*, the latter refused to receive them. The extreme consequence of this process is seen in the legal banning of all labour organizations in 1799!

The latter were thus left vulnerable in the hands of the growing class of (exploiting) capitalistic entrepreneurs.

8.17.1 *Consequences for 'labour': the rise of trade unions*

Comfortably forgetting **that the hands-off liberal government** neglected its normative task of integrating the multiplicity of legal interests on its territory into one public legal order of justice, economic theory, motivated by the humanistic science ideal, believed that all economic relations are governed by exact natural laws instead of being ruled by economic norms – notably manifest in the "law of supply and demand." This "law of nature," ultimately, should be held accountable for the miserable situation of wage labourers. Thomas Malthus (1766-1834) and David Ricardo (1772-1823) pursued this path explicitly.

It was within this climate that the modern wage labourer emerged during the industrial revolution. **The humanistic science ideal deprived the labourers of human dignity by reducing them to a production factor** existing alongside buildings, machines and raw material – degraded to the level of an *economic thing*.

Malthus also designated this inexorable law of supply and demand as the "iron wage law," inhibiting any higher wages, since the gross national product (GNP) only has a fixed amount available for wages. The only solution for this problem was in reducing the number of labourers, in order to allow a bigger share for each. By means of birth control, there would be less labourers – eventual suicide for the increasing number of labourers already living below the bread line.[1]

In the mean time, these labourers continued to be *human beings*, capable of exploring the normative possibilities provided to them in the erection of trade unions. The unequivocal aim was to fight for those rights consciously neglected by the government. Owing to the guidance of the humanistic science ideal, societies in Western Europe were thus transformed by an increasing *class struggle*. According to Karl Marx, this situation is characteristic of the uncontrolled state of nature as sketched by Hobbes: a war of all against all. At the same time, these circumstances provided 'fruitful' soil for the communistic prophecies of Marx himself.

It is within the context of this derailment that Marx found ample material to substantiate his concern for the inhuman conditions prevailing within the industrial arena where women and children were maltreated by the capitalist class by means of such abuses as working days of up to sixteen hours. His concern is that, whereas machines were supposed to liberate human beings from routine labour and open up free time, in the hands of the capitalist entrepreneurs, this technology resulted in massive enslavement, simply enhancing the weakening conditions of the proletariat.

The subsequent history reveals the story of the rise of trade unions established to restore the rights of labourers ignored by governments under the spell of Locke and Smith.

Particularly, the radical revolutionary focus of neo-Marxism during the sixties of the 20th century returned to the young Marx in order to transcend the fixed situation in Russia while propagating the *perpetual revolution*. Only through the latter will it be possible, according to Marcuse and other neo-Marxist authors, to secure the road to freedom, eroticism, play and creativity. The student revolts of the late 1960s manifested these convictions, which led to a radical rejection of what was considered to be the "establishment." Neo-Marxism realized that, within the welfare state, the labourer eventually acquired a settled position with constructive benefits from social insurance. Therefore, it directed its focus towards the new generation of students who were 'untied', without any standing duties or responsibilities. The irony is that all the leaders of these revolts were fully settled within the 'establishment' a decade later. My recollection of an article in the *Time* of the late nineteen seventies tells the story of one of these former student leaders, who said that "God is dead, Marx is dead, and I am not feeling too well myself"!

During the first half of the 20th century, significant changes occurred within the world of labour and industry. For example, on May 14, 1921 Belgium promulgated a law shortening the work day to 8 hours and the work week to 48

1 More recently Corijn still refers to the dominating contemporary ideology where economic laws and market laws are 'mystified' as "natural laws" (Corijn, 1998:198).

hours. Already in 1889, the first of May became an international day dedicated to the struggle for the 8-hour working day.

In the course of this struggle, basic human rights slowly but surely made headway and managed not only to secure a shortened work day and limited working hours, but also safety and security within the workplace, and eventually also the thirteenth pay cheque. The effect was that more time and money became available for leisurely activities!

Without the power usurpation during the rise of classical capitalism, the struggle for regaining "leisure time" surely would not have been so heroic. However, the success of this battle during the 20th century highlights the quality and value of the newly acquired *leisure time*.

8.18 The scope of leisure and the quality of life

Although leisure has been portrayed in functional terms (as a means to an end) or as some or other kind of activity, we have seen that an accountable structural location of its place cannot bypass the societal distinction between collective and communal relationships, where the former are always correlated with coordinational relationships – and leisure finds its societal 'place' within the latter kind of relationships.

It is therefore *not* meant to designate a specific *social class* within society – such as suggested by T. Veblen, who referred to the practices of the new "leisure class" of the late 19th and early 20th centuries in a derogatory way . A *class of people* (layer of society) in principle functions within all three modes of social interaction: collective, communal and coordinational, and therefore it can never be identified with only *one* of them.

Depending upon the specific preferences, desires and aspirations of an individual, that person may engage in leisurely activities in numerous ways. This spectrum naturally coheres with the full range of normatively differentiated options open to free human functioning within all the normative aspects of reality. Let us take an example from the world of *tourism*. Tourism as such is a prime example of giving shape to the freedom entailed in coordinational leisurely activities. A tourist is not engaged in a "money-generating" endeavour (economically qualified), although it has become increasingly *expensive* to be a tourist. Naturally, tourism is founded in economic relations, for without the necessary economic means, any touring plan would fail before its inception. But the economic facet in general never qualifies tourist activities.

The tourist is always guided by a rich diversity of *socially differentiated* special *interests*. These may vary over the entire spectrum of aspects: some tourists are mainly interested in art galleries, owing to their aesthetic hobbies or preoccupations; others choose to opt for natural scenery and environments (eco-tourism), flora and fauna, historical or archaeological sites, and so on. Many times, these interests are enhanced by special associations advancing such causes by, amongst others things, organizing extensive tours for members of these associations. By contrast, countries with a rich variety of tourist attractions actively market them, and thus attract prospective tourists.

The same diversity is present in the normal experience of leisure, because the *normative plasticity* of human endeavours cutting across all aspects of reality enables any individual to pursue particular and distinct leisurely goals and to freely and (re-) creatively give shape to the leisure side of coordinational relationships. (Recreation always takes place during leisure time.) It is important to remember that, not only individuals participate in coordinational relationships, but also particular communities (such as married couples) and collectivities (such as the nuclear family or firms, church communities or social clubs). Think about the leisure side of a shopping mall, where an individual, a couple, or a family can shop or simply enjoy an evening out by exploring restaurants, cinemas, theatres, pubs, discos, and so on.

The *quality of life* is not merely constituted by leisure, since the *meaning of life* is inherent in all human endeavours and relationships (including collective and communal relationships). *Happiness* is also not something a person can pursue as a *primary* goal. Rather, when something worthwhile is carried out and done well, then happiness may ensue as a *bonus*. Similarly, when leisure is pursued properly and in norm-conformative ways, it may be transformed into a joyful experience of leisure.

With the immense supply of amusement and recreation, many people are tempted into the leisure role of *passive spectators*, which may lead, in certain respects – think of unhealthy living patterns – to become a threat to the worthwhile quality of life. This simply emphasizes once more that – like all human activities – leisure may become a false ideological seat, a "pastoral haven." By contrast, one should acknowledge that, just like all other typically human possibilities, leisure represents a *normed* dimension of *social* life. Utilizing leisurely options on the one hand therefore leaves open alternative positive (norm-conformative) options, and on the other, anti-normative deviations as well.

8.18.1 *Exercise and sport*

An anthropological perspective on the human being ought to account for the physico-chemical substructure, the biotical substructure, the sensitive substructure and the qualifying, although in itself unqualified normative structure of being human (see Chapter 4). This perspective entails that the original seat of the terms *health* and *illness* is located within the *biotic aspect* of reality. This means that only entities functioning actively within the biotic aspect can be identified as *healthy* or *ill*. Given the peculiar biotical structure of the human body, biotical health is dependent upon the frequent and regular use of the body. When this is done in a conscious and purposeful way, we encounter *exercise*, which ought to be accompanied by the appropriate *nutrition* to be optimally effective. The combination of *nutrition* and *exercise* plays a basic role in the study of sport.

When this situation is approached from the angle of the qualifying normative structure of the human bodily existence, the most general perspective is oriented to the nature of *play*. Though *playfulness* is also observed in animal life (exhibiting the fun of being alive), the latter lacks the *accountable freedom* of human play. This explains why the inherent dynamics and plasticity of human play could be channelled in various (competitive and non-competitive) directions, from ordinary *games* up to and including highly specialized kinds of *sport*.

Understood within a typical human societal context, *play* could be both *aimless* and *purposeful*, *relaxing* or *highly demanding*, and could even be *disinterested* or *serious* (see Riezler, 1981:439-451). Since culture is the *first nature* of being human, *play* is actually a manifestation of this first nature. Consequently, it shares in the *creative imagination* characteristic of all typical human cultural activities. Structured forms of play therefore demonstrate the rich variations of imaginitivity operative as an effect of the creative imagination of human beings. Consider for example the best-known *games* and *sports*.

The concept of a *game* is more encompassing than *sport*. However, distinguishing between game and sport should avoid the (earlier mentioned) physicalistic or psychological definition of *human action* as a *willed muscle movement*. The inherent *normativity* of human behaviour often entails the normative *obligation* to act or not act in a physical way. Our legal practice is acquainted with acknowledging a *commission* and *omission* as *legal acts*.[1]

One cannot argue that chess is just a *game* because no great physical activity takes place when one is involved in playing a game of chess. Biochemical research actually showed that, although the human brain occupies only 2 percent of the mass of the human body, it carries 25 percent of its metabolism (see Plamenac, 1970:444). Playing chess therefore undeniably at least entails physical (brain) activity – not to mention the arm used to move pieces around on the chess board!

A broad definition of sport may even start with *playful movement*. But within sport, *playing* is limited by the *rules* and *aim* of the game. The energetic and vital urge of participants on the one hand constantly challenge the natural abilities of the athlete/player within the constraints of the rules of a particular sport. The element of competition and the will to conquer the opponent or natural conditions induce the importance of *exercise* and *training* upon the presupposition of equal chances of all participants. This opens the possibility of emphasizing competition, a proper organization of matches and well-adapted rules by placing even less weight on movement and bodily development. Thus borderline cases, such as chess and bridge, can be included within the category of sport.

The urge to achieve better results and reach more extreme heights historically led to national and finally international sport. On the one hand, it enhanced more specialized studies of the nature of exercise and the techniques

1 A person who fails in the duty to switch the signal for incoming trains from safe to unsafe may not have moved a muscle, but omission was nevertheless the legal cause of the resulting accident (see pages 24 and 281 above)!

required to improve results through effective ways of exercise, while on the other this led to a situation where the *socially guided* structure of sport partially or totally assumed the nature of a *profession* (the professionalization of sport). All in all, this finally led to the involvement of sport organizations and even those instances where states are investing in the development of sport (for example, contribute to the training of athletes before the Olympic games).

The other end of the continuum, where playfulness becomes dominant, opens up *recreative sport*, which may even be *non-competitive*. No match – only recreational forms that show similarities with genuine sporting activities.

The highly differentiated nature of modern society embodies a vast diversity of sport practices. The many-sidedness displayed in these practices constantly embodies the intimate connections between social and professionally economic orientations within the leisure dimensions of human society. In this sense, *leisure*, although not an all-inclusive enterprise, is co-constitutive for the quality of a life worth living.

8.19 Concluding remar

The difference between the dimension of modal aspects and concretely existing (natural and social) entities at once explains the intimate connection between these two dimensions of reality. Whatever exists functions with all the aspects of reality in principle. In some instances, the possible object functions are dependent on being opened up or disclosed by the objectifying activities of subjects functioning as subjects in cosmic later modalities.

The distinction between modal laws and type laws (see page 25) reveals a certain foundational order between the two dimensions of reality mentioned above, because the unspecified modal meaning of every aspect, displayed in its modal universality (see pages 68, 421), acquires a typicality through the way in which entities and processes, governed by type laws, specify the general modal meaning of all the aspects in which they function.

Since the many-sided existence of entities exceeds the boundaries of any and all aspects, every special scientific analysis of typical entitary functions inevitably causes these disciplines to proceed from an underlying idea regarding the typical totality character of such entities and processes – and this necessity reveals the foundational role of philosophy with regard to the various academic disciplines. No single discipline can escape from an implicit or explicit philosophical view of reality.

Chapter Nine

Philosophy is more than merely the "Discipline of the Disciplines"

The encompassing scope of philosophy

9.1 Philosophy as totality science

The key elements of the expression *totality science* are familiar to us. In Chapter 2, it was argued that the distinctive feature of scholarly thinking lies in *modal abstraction*, where abstraction is based upon the analytical acts of *lifting out* (identifying) and *disregarding* (distinguishing) (see also Sketch 5 on page 151). The word *totality* is the equivalent of *wholeness*, and is derived from the core meaning of the spatial aspect, where we find the whole-parts relation in the first place (see pages 87, 225, 236, 239, 307, 353, 391, 405, 407). It is fashionable nowadays to employ the term *holistic*, where the intention is simply to refer to all the facets or aspects of whatever is discussed. When the term *totality* is used in the compound characterization of "totality science," it is clear that this term extends beyond the limits of the spatial aspect, and that it is therefore applied in a concept-transcending way. It certainly does not aim to subsume all reality in the whole-parts scheme. In fact, we argued against an unqualified application of the whole-parts relation to everything (see pages 466 ff.).

Abstraction presupposes the (quantitative) meaning of the one and the many (multiplicity),[1] while the term *totality* presupposes the spatial whole-parts relation. Since the modal aspects are not just *modes of being*, but also *modes of explanation*, it stands to reason that the first two modal aspects, namely number and space, will play a key role in scholarly parlance, as well as within our non-scientific experience. From childhood, we are exposed to these two most basic and fundamental modes of experience and explanation. For example, when one opens the *Yearbook* of a state, information is provided regarding the population (*how many* citizens) and the *size* of the country – exploring our awareness of multiplicity and geographical size in a specific way. A combination of these two modes of explanation serves the formulation of the basic idea of the *universe* – in Dutch and Afrikaans, the 'all' of the 'whole', designated as the 'heelal'; literally, the *whole* of *everything*.

As innocent and straight-forward as this designation appears, we only need to recall the basic assumptions of modern set theory to realize how problem-riddled the combination of multiplicity and wholeness can be. For example, the basis of Cantor's set theory is fully dependent upon the combination of

1 When this awareness of unity and multiplicity is stretched beyond the boundaries of the numerical aspect, the well-known expression unity and diversity, is applicable – as an instance of the *concept-transcending* use of arithmetical terms.

these two elements, for he defines a *set* as *bringing together* (*Zuammenfassung*) definite, properly distinct (*wohlunterschiedenen*) elements of our intuition or thought into a whole (*zu einem Ganzen*) (see Cantor, 1895:481).

Bertrand Russell and Ernst Zermelo independently discovered the intrinsically problematic nature of the notion of a set and its elements (see Husserl, 1979:xxii, 399 ff.). Consider a set *C*, which has as its elements those sets *A* that do not contain themselves as elements.[1] Now contemplate two options; the one supposing that *C* is an element of *C*, and the other that *C* is not an element of *C*. It should be kept in mind that the condition for a set to be an element of *C* is that it cannot contain itself as an element. The upshot is perplexing.

(i) If *C* is an element of *C*, it must conform to this condition, i.e. that it does not contain itself as an element:

If C is an element of C, then C is not an element of C

(ii) If *C* is not an element of *C*, then it meets the condition for being an element of *C*

If C is not an element of C, then C is an element of C

(iii) Therefore,

C is an element of *C* if and only if it is not an element of C!

that is, $C \in C \Leftrightarrow C \notin C$!

The apparently innocent combination of *multiplicity* and *wholeness* therefore caused havoc within the discipline of mathematics, giving rise to conflicting schools of thought within this special science. While the logicism of Russell held that mathematics *is* actually logic, the intuitionist school and that of axiomatic formalism accepted a pre-logical or extra-logical subject matter. Intuitionism reacted both to logicism and formalism and in particular to what Russell and Zermelo discovered – and in doing that it generated a *whole new mathematics*:

> The intuitionists have created a whole new mathematics, including a theory of the continuum and a set theory. This mathematics employs concepts and makes distinctions not found in the classical mathematics (Kleene, 1952:52 – also compare a remark by Evert Beth, quoted in Chapter 1, page 5).

In 1900, the French mathematician, Poincaré, made the proud claim that mathematics has reached absolute rigour. In a standard work on the foundations of set theory, however, we read: "ironically enough, at the very same time that Poincaré made his proud claim, it has already turned out that the theory of the infinite systems of integers – nothing else but part of set theory – was very far from having obtained absolute security of foundations. More than the mere appearance of antinomies in the basis of set theory, and thereby of analysis, it

1 The set of 120 people is not a person, but a set. Therefore it does not contain itself as an element. By contrast, the set of all imaginable thoughts can be imagined – and therefore does contain itself as an element.

is the fact that the various attempts to overcome these antinomies, ..., revealed a far-going and surprising divergence of opinions and conceptions on the most fundamental mathematical notions, such as set and number themselves, which induces us to speak of the third foundational crisis that mathematics is still undergoing" (Fraenkel *et al.*, 1973:14).

In passing we may remark that, on the basis of Cantor's set theory, modern axiomatic formalism believes itself to have *arithmetized* mathematics completely, but did not realize that this claim begs the question.[1]

From a more general perspective, one can observe that axiomatic formalism tends to be *atomistic* in its mathematical approach, while intuitionism proceeds from an explicit acknowledgement of the (spatial) whole-parts relation[2] – normally functioning as the starting point for a *holistic* stance. Remarkably enough, the two disciplines, delimited by the aspects of space and number (arithmetic and geometry or, in more general terms, algebra and topology), served a distorted understanding of the unity and diversity within the universe. The fundamental issue concerns the coherence of what is unique and irreducible, i.e. the question of how we account for the "coherence of irreducibles" (see pages 8, 13, 157). It is also known as the quest for a *basic denominator*, which traditionally, in its derailed format, assumed the form of an *ismic* orientation, in which some or other aspect of reality is deified and elevated to serve as the sole and encompassing mode of explanation. In Chapter 2, the problem of unity and diversity was illustrated by looking at the opposition between atomism (individualism) and holism (universalism) within various academic disciplines – covering mathematics, physics, biology, psychology, logic, linguistics and sociology (see pages 60 ff.).

In addition to the first two modal aspects, all the other modal aspects (provisionally distinguished in this work) served as a pseudo point-of-rest, an ulti-

1 The circularity present in this claim is discussed in Strauss, 2005 (Chapter 2). If the core meaning of space (continuous extension) entails the idea of wholeness and totality, the idea of an infinite totality (required in employing the at once infinite) presupposes the irreducibility of space. Yet, only when the at once infinite is used (namely in Cantor's diagonal proof of the non-denumerability of the real numbers) does one 'succeed' in 'reducing' space to number. This results in the realization that space can be reduced to number if and only if it cannot be reduced to number! We noted that Paul Bernays clearly understood that it is the totality-character of spatial continuity that will resist a perfect arithmetization of mathematics (see Bernays, 1976:74) Just recall his remark: "It is in any case doubtful whether a complete arithmetization of the idea of the continuum could be justified. The idea of the continuum is anyway originally a geometrical idea" (Bernays, 1976:188).

2 "In agreement with intuition, Brouwer sees the essence of the continuum not in the relation of the element to the set, but in that of the part to the whole" (Weyl, 1966:74). Following Aristotle, Weyl also holds that "[I]t rather belongs to the essence of the continuum that each of its parts is infinitely divisible" (Weyl, 1921:77). In opposition to the notion of a continuum of points, Weyl indicates that the concept of *environment* must still be used to salvage continuity: "To render a continuous coherence of points analysis has (since it decomposed the continuum into a set of isolated points) found refuge in the concept of environment up to today " (Weyl, 1921:77).

mate explanatory device during the history of philosophy and the disciplines.

Yet, before we proceed, another opposition requires attention, namely that between *universality* and *what is individual*. The awareness of *being distinct* presupposes the discrete meaning of the numerical aspect, co-conditioning the nature of individuality (and our reflection on what is individual). Likewise, our understanding of universality presupposes an understanding of the meaning of space and spatial location – in the sense of *everywhere*. The ability to distinguish between the universal side of things (the *atom-ness* of an atom) and their individual side (*this* atom), is therefore also co-conditioned by the first two cosmic aspects. When universality is emphasized at the cost of individuality, we meet the stance of *rationalism*, and when individuality is reified at the cost of universality we meet *irrationalism*. This new understanding of rationalism and irrationalism enabled a deepened view on the ambiguity within nominalism (see pages 370–379), because in respect of the supposedly purely individual outside world, nominalism is irrationalistic, and regarding the universal concepts within the human mind, it is rationalistic.

The fundamental role of the quantitative and spatial aspects in theoretical designs throughout the history not only resulted in *distorting* views – such as those of atomism (individualism), holism (universalism), rationalism and irrationalism, for a positive assessment shows that no single philosophical view of reality (underlying the disciplines) has ever succeeded in escaping from the uniqueness and interconnectedness of *multiplicity* and *coherence*. For this reason, we have to acknowledge the co-conditioning role of these two modes in respect of our characterization of universality and individuality (see pages 430 ff.). Moreover, the nature of concept-transcending knowledge serves as the justification of the two most basic philosophical statements: (i) everything is unique, and (ii) everything coheres with everything else (see page 455). Applied to the modal aspects themselves, these two statements transform into the principles of modal sphere-sovereignty and modal sphere-universality.

The preceding remarks, based upon the more extensive argumentation at various earlier locations in this work, demonstrates that general philosophical concerns – regarding unity and diversity and the quest for a basic denominator – enable a discernment of connections between number and space, and all the post-arithmetical and post-spatial aspects, as well as between the disciplines having diverse modal aspects as their delimiting angle of approach to reality. It is only a *totality science* that will be able to account for these connections. One way to show what this means is to make a brief comparison of the approach in this work with the fairly widespread traditional way of presenting philosophy through a number of philosophical sub-disciplines.[1]

1 Within this context, we may therefore leave aside an analysis of subsequent philosophical issues and their ramifications within the disciplines (such as constancy and dynamics, conceptual knowledge and concept-transcending knowledge), because an in-depth treatment of these is found throughout this work.

9.2 Philosophical sub-disciplines versus an encyclopedic approach

Traditionally, the discipline of philosophy includes sub-disciplines such as metaphysics, logic, ethics, epistemology, philosophy of religion (proof for the existence of God), philosophy of mind, and philosophy of science.

The basic thrust of the approach explored in this work is guided by the ideal of a *non-reductionist ontology*. Since the legacy of metaphysics by and large represents reductionist interpretations of reality (foremost found in the well-known isms, such as arithmeticism, mechanicism, physicalism, organicism, psychologism, logicism, historicism, and so on), one may give preference to the idea of a *non-reductionist ontology*. The latter has to account for all the dimensions of reality – from an comprehensive account of the dimension of cosmic time, the dimension of modal aspects and the dimension of (natural and social) entities, up to the central, direction-giving dimension where ultimate commitments (manifested in communal ground motives) have their seat.

Once this broad ontological scope has been exposed, there is no longer any need for the traditional subdivisions of philosophy. Ethics, which is supposed to serve as the basket for all forms of normativity, is now replaced by an awareness that all the post-sensitive modal aspects are *normed*. Logical principles, cultural-historical principles, lingual, social and economic norms, as well as the demand to be norm-conformative actions within the aesthetic, jural, moral and certitudinal aspects obtain their rightful place within their proper spheres, without being reduced to matters of ethics or morality *per se*. Liberated from the impossible task of accounting for all possible forms of normativity, ethics can now focus on its specific task as a distinct academic discipline with its own modally delimited field of investigation.

Furthermore, including just a few special science philosophies within the scope of philosophy, such as philosophy of religion, philosophy of mind, and philosophy of 'science',[1] actually reduces the intrinsic foundational role of philosophy in respect of *all* the special sciences. Owing to the analogical connections between all aspects and the structural interlacements between diverse entities, every discipline, taking any aspect as its angle of approach, through its elementary, compound and typical basic concepts, shows its dependence upon a philosophical view of reality. The alternative may therefore be designated as an *encyclopedic* approach, because it does not shy away from analyzing the given diversity of intermodal connections and inter-structural interlacements, guided by the principles of sphere-sovereignty, sphere-universality and enkaptic intertwinements.

Even epistemology, by exploring the meaning of the logical subject-object relation, has to perform a task that falls within the context of a non-reductionist ontology, for it has to account for the not-yet-deepened as well as for the disclosed structure of the logical aspect. Moreover, scholarly knowledge proved to be crucially dependent upon those (logically objectified) cognitive configurations combining a multiplicity of features or hallmarks into the unity

1 Largely reduced to the philosophy of physics.

of concepts. The epistemological discernment of the restrictive and expansive boundaries of rationality (see pages 182 ff.) is only meaningful if the distinction between conceptual knowledge and concept-transcending knowledge is utilized.

Both logic and ethics, strictly speaking, are therefore actually *special sciences*, i.e. distinct academic disciplines, that ought not to be taught in philosophy courses. Since a philosophical anthropology investigates the typical totality-structure of the human personality, constituted by the physical, biotic and sensitive layers and qualified by the guiding normative structure (which, in itself, is not qualified by any modal aspect – see Chapter 4), anthropology belongs to the dimension of entities. It encompasses the issue of the "philosophy of mind" (including the distinction between 'mind' and 'brain').

The philosophy of time embraces everything within temporal reality. Physics can therefore not claim monopoly on the question: what is time? For this reason, any account of time has to be primarily *philosphical* in nature. The boundary questions regarding the assumed coming into being of our cosmos exceeds physics, because whatever happens within the universe is determined by some or other conditioning order. Whatever we can *date,* presupposes these underlying conditions, which themselves therefore cannot be dated.

In Chapter 2, a number of issues in the developments within the philosophy of science are highlighted. It appeared that a theoretical view of reality is not something accidental, for it belongs to the constitution of every academic discipline. Moreover, the long cherished capabilities of human reason ultimately became an idol that is clearly visible in the (apostate) *trust* in the rationality of reason (as Popper pointed out). Also recall the insight advanced by Stegmüller, namely that a self-guarantee of human reason, within whatever domain, is impossible, for one always first has to believe in something to be able to justify something else (see page 10).

A key element in the development of 20th century philosophy of science is its recognition of the institutional – and therefore social – side of the academic enterprise. It opened the way to the acknowledgment of thought communities. Viewed from the perspective of the dimension of ultimate commitments, one may also speak of *conviction communities* within diverse scholarly disciplines, analogous to confessional communities that are organized in various (church) denominations.

However, in what has been said, an important aspect of scholarship was not mentioned, namely the *history* of philosophy and the disciplines. When scholars assess the contemporary state of their discipline, they tend to do so in an *a-historical* way. On the one hand, the implicit criterion is that everyone working within that special science ought to be "up to date," but on the other, the current state of the discipline is presented as if what is said about its current state has *always* been the case!

Suppose we say that mathematics is set theory. This claim has to be qualified, for otherwise it becomes absurd. Set theory was developed by Cantor between 1874 and 1899. In the light of this historical reality, one may ask with Reuben Hersh: "What does this assumption, that all mathematics is fundamentally set theory, do to Euclid, Archimedes, Newton, Leibniz, and Euler? No one dares to say they were thinking in terms of sets, hundreds of years before the set-theoretic reduction was invented" (Hersh, 1997:27). It seems that Hersh is also correct in arguing that the only way out of this impasse is to reinforce it: "... their understanding of what they did must be ignored! We know better than they how to explicate their work! ... That claim obscures history, and obscures the present, which is rooted in history" (Hersh, 1997:27).

This example clearly demonstrates the need to distinguish between those conditions making possible a discipline like mathematics and the historically changing responses to those conditions exhibited throughout the history of mathematics. This distinction, which applies to every scholarly discipline, once more accentuates the importance of an acknowledgement of underlying ontic conditions that enable human responses. The practice of any special science is made possible by all four dimensions of created reality, and no response should be appreciated as the last point of reference, because such a mistake will uproot both the historical past, the historical present and even challenge the historical future of a discipline (precluding any significant *new* developments).

Although the main focus of this work is on ontic conditions and their implications for the diverging paradigms (theoretical designs) within the various disciplines, we frequently alluded to the underlying, direction-giving ultimate commitments in every scholarly orientation. It is certainly one of the merits of Dooyeweerd's philosophy that it provided us with detailed analyses of what he designated as the four basic motives – ground motives – of Western culture (see in particular Dooyeweerd, 2003).

A brief overview of the ground motives in the history of Western culture and philosophy follows.

9.3 Ultimate commitments in the history of Western culture[1]

9.3.1 *Greek culture: the urge towards the incorruptible and immutable*

In the course of our analyses, we mentioned aspects of the four ground motives within the history of the West, namely the Greek ground motive of form and matter, the biblical motive of creation, fall and redemption, the Medieval Scholastic motive of nature and grace, and the modern humanistic motive of nature and freedom.

During the 7th and 6th centuries B.C. – the period when Greek philosophy entered the scene – Greek society experienced the turmoil of a transitional phase in all its facets. The reign of the noble patrician clans during the age of

1 Just as philosophy is more than merely the *discipline of the disciplines* the ground motives are more than merely motivating theoretical thought – they drive cultural developments in all their facets, including their societal complexities.

chivalry was now confronted with a process of cultural development and differentiation. The traditional sources of economic income of the nobility - agriculture and stock-breeding – were soon overshadowed by the money aristocracy. A cultural crisis, geared towards the attempt to reconcile the older religions of nature with the Olympic cultural religion, emerged through the theogonic constructions of Homer and Hesiod. This tension eventually materialized in the central direction-giving power of the ultimate longing for incorruptibility amidst a world of change.[1]

This ultimate concern for the aeonian, incorruptible, immutable and imperishable amidst the ever-changing flow of material things directed early Greek philosophy towards a search for the underlying *original principle* of everything – thus revealing the central commitment of Greek philosophy to the basic motive of *form* and *matter* (the *constant* and the *changing*). Anaximander already explicitly introduced the idea of an *Archè* (*'Αρχή*) – which he called the *apeiron*, the *infinite* and *limitless*. Fr. Solmsen explains that Anaximander is the only thinker "for whom the *apeiron* itself was the enduring and all-encompassing entity" (Solmsen, 1962:114). Sweeney also points out that the *Archè* of Anaximander is "by nature the Infinite, the Boundless, the Limitless" (Sweeney, 1972:65). According to this, the *to apeiron* "is indeterminate, inexhaustible, everlasting, untraversable, and without any extrinsic limit" (Sweeney, 1972:62). Before Anaximander, the *Archè* was often identified with the *fluid* (flowing) nature of some element that was considered to be original (for instance, *water* in the case of Thales, *air* with Anaximines and later on, *fire* with Heraclitus). According to Anaximander, however, the *apeiron* does not *age* (Diels-Kranz, B Fr.2), and is without *death* and *corruptibility* (Diels-Kranz, B Fr.3).[2]

Clearly, this concern was an expression of the fundamental aim to account for the underlying *unity* of the *Archè* that reveals itself in multifarious changing forms that are doomed to return to the formless origin (cf. Anaximander's B Fr.1). The *order* (*limited form*) represents the repressed *form motive*, which is, in its dialectically depreciated meaning, the source of *punishable injustice* (*adikias*). However, the dialectics between *form* and *matter*, *constancy* and *change*, are implied by the reciprocal determination of these two opposing moments. Although disqualifying order (and therefore *constancy*) in B Fr.1, the unstructured and indeterminate *Archè* is nevertheless seen as *everlasting*, without *age*, *death*, or *corruptibility*, and therefore *constant*. It is striking that the fundamental dialectic of Greek thought is captured by employing modal kinematic and modal physical terms (namely *constancy* and *change*) in a con-

1 Although Bos questions the way in which Dooyeweerd accounts for the genesis of the motive of form and matter, and also prefers to speak of the titanic meaning-perspective, he believes that the extensive analysis of the development of this motive in Dooyeweerd (1949) (see Dooyeweerd 2003), contains a valid perspective on the inherent dialectic of Greek thought (see Bos, 1994:220).

2 Subsequent references to the Fragments of pre-Socratic philosophers will omit the Diels-Kranz reference.

cept-transcending way. Other modal points of entry, such as the spatial aspect (compare the terms *limited* and *unlimted*), are used in a similar fashion.

The dialectical tension, where the *form* and the *formless* (the *limited* and the *limitless*), not only *oppose*, but also *need* each other, is clearly stated by a follower of Pythagoras, namely Philalaos. According to him, the *world order* is composed of the *limitless* and the *limited* (B Fr.1). Against this dualistic background, the Pythagoreans used the conviction that everything *is* number as starting point. Number functioned as the dialectical unity between the opposing motives of form and matter: as a progressive series, number shows the formlessness of the matter-motive, but simultaneously establishes *order*, *form* and *harmony* in the cosmos (the *form motive*). According to its form-giving function, it has been assumed that the essence of everything could be expressed in terms of the relation between two integers (i.e. in terms of *rational numbers*). In his dealing with a *regular pentagram*, Hippasus of Metapont, however, discovered (± 450 B.C.) that the ratio of certain line segments is *incommensurable* (cf. K. von Fritz, 1965:271 ff., especially pp.295-297; also see Chapter 2, page 44 and page 403 ff. above),[1] marking the discovery of *irrational numbers*. Within the form-giving function of number, the Pythagoreans were thus confronted with an unbounded and infinite series, indicating something *formless*. To escape from the fate of irrational numbers, they translated all their arithmetical problems into *spatial terms* (any spatial figure has a *definite* and *limited form*). This possibility of handling irrational numbers (with their implied unlimited and infinite series of numbers) in a *geometrical* way, caused a fundamental *geometrization* of Greek mathematics. Not only is this an outcome directed by the basic motive of *matter* and *form*, since it also means a shift in the (implicitly present) modal (aspectual) focus: rejecting the numerical point of entry, the *spatial aspect of reality* provided Greek thought with a rich variety of terms to express its view of reality at this stage (even the modern distinction between *analytical* and *synthetic* judgments is dependent upon this development).

Parmenides explored this spatial orientation in his static *philosophy of being*. The only road to truth is via the theorem that whatever is, *is*, since non-being is neither knowable nor expressible (B Fr.2). The terms Parmenides uses to characterize the nature of being, are derived from the modal perspective of the *spatial aspect*. As hallmarks of being, the following are mentioned: since it is unborn, it is imperishable (cf. Anaximander's B Fr.2 and Fr.3), ... it was not and will never be because in hanging together, it is as an indivisible whole in the present – unified, coherent (B Fr.8, 3-6 – see pages 87 and 268 above).

The matter pole of the Greek ground motive dominated the undifferentiated patrician clans who were the *bearers of power* within the Greek city-states (the *polis*). The dark and unforeseeable ever-flowing stream of life was believed to be subject to *blind fate* (*chance*), to the *Anankè*. The popular assem-

1 Hippasus has apperently betrayed a mathematical secret by publishing this (see Riedweg, 2005:107 and Fowler, 1999:289-302).

bly of free citizens was actually a power within the various forms of the Greek *polis*.

The further development of the Greek *polis* terminated the dominance of clans, tribes and brotherhoods. The initial four Ionian tribes were replaced by ten new *territorial* tribes. Upon this basis, the Athenian democracy reached its peak under the reign of Pericles (446-404 B.C.), although it was not capable of maintaining itself after the end of the Persian wars – it collapsed soon after the reign of Pericles.

The flourishing of the *polis* during the golden 5th century B.C. finds its spiritual roots in the shift of primacy from the *matter* motive to the *form* motive, which ran parallel to the transition from the older undifferentiated clans and tribes to the relatively more differentiated legal order of the *polis* (city-state).

While the *pre-polis* period was still strongly influenced by conceptions of *dikè*[1] and *themis* (the internal legal order of the clans) as the guardian of the natural order of things, the 5th century witnesses significantly new developments. *Dikè*, for example, loses its original meaning and acquires a new content, designating the *positive law* formed by the *polis* and the punishment exercised on the basis of these positive laws.

At this stage, a new theory of knowledge surfaces, which acknowledges nothing *universal* outside the human being – only strictly individual entities populate the world. Yet the human mind, according to this theory, does have the ability to refer to such a multiplicity of strictly individual things, but it is only done by employing *concepts* or *words* (*nomina*) referring beyond the human mind to these individual entities. This issue gave birth to the above-mentioned classical philosophical struggle, namely that between *realism* and *nominalism*. The former accepts universality both 'outside' *and* 'inside' the human being, whereas the latter accepts it only *within* the human mind.

We pointed out that the early 5th century thinker, Callicles, derives from nature the right of the strongest (see page 498 above). We mentioned there that he opposes the *legal order* of the state, because it only serves to suppress the strongest through laws and consequently admires the *tyrant*, for the latter breaks through positive laws and subjects the weak to its power as law. The tyrant alone is entitled to have rights – all other citizens are deprived of rights. Since they are subject to the arbitrariness of the tyrant Vollenhoven here speaks of an "aristocratic nominalism" (Vollenhoven, 1933:83).

The *Sophists*, and in particular Protagoras, elaborated this *nominalistic* orientation in connection with an understanding of human persons and their place within society. The view of the former (that is, the human person) is in

1 Heraclitus, although thinking under the primacy of the matter motive, attempted a dialectical synthesis between matter and form, because the word *logos* (reason) reveals itself both as *Anankè* and *Dikè* (justice). Anaximander views the taking on of a limited form as an encroachment against the formless *apeiron* (the formless-infinite), which is therefore doomed – by virtue of the law of justice (*dikè*) – to suffer from punishment and penance and return to its formless origin, according to the order of time (B Fr. 1).

the grip of the *matter* motive: human subjectivity is constantly changing and cannot be grasped in a fixed form or measure (every individual is his or her own measure). Only the *polis*, as bearer of the Greek *form* motive, is capable of supplying the human being with a cultural attire through education and obedience to positive laws – thus demonstrating the *primacy* of the *form motive* in the thought of Protagoras. This explains his view that human beings, coming from a condition in nature where the state is absent, have the properties necessary for the formation of a state – but not on the basis of a "social contract" (see Menzel, 1929 and 1936). Although Protagoras proceeds from a nominalistic *individualistic*[1] starting-point, his conception of the state does not acknowledge any *material*[2] boundaries for the competence of the state – even morality and religion are viewed as products of the state.

In the thought of Socrates, the Greek form motive fully assumes *primacy*. Although he identifies the *state* with *positive law* – a stance currently designated as *legal positivism* – his conception definitely differs from a modern legal positivist such as Kelsen in the sense that he considers the 'justness' of laws to be dependent upon the knowledge the legislators have of a given *world-order*. His life ended tragically. He was accused of misleading the youth and as a result had to drink the poison goblet. When he was given the opportunity to escape, he rejected it to show that he was the *best* citizen of the Athenian democracy. At the same time, he also wanted to show how *corrupt* the Athenian democracy had become – they did not even have a place left for their **best** citizen!

In Plato's greatest dialogue, *Politeia* (*The Republic*) – representing the culmination of the first phase of his theory of ideas – he defends (in preparation of his ideal state with its three classes) a tripartite understanding of the soul (cf. *Politeia*, 436 ff.). These three parts of the soul[3] continued via the Middle Ages to exert an influence on the traditional understanding of the 'faculties' of the soul (even in 20th century Reformed theology): *thought*, *will* and *feeling* – compare also Hitler's estates in Nazi Germany and the *id*, *ego* and *superego* in the depth psychology of Sigmund Freud.

This three-fold conception of the soul provides the basis for his *theory of the state* and lays the foundation for his understanding of the first three cardinal virtues he distinguishes. According to this, *wisdom* (*sophia*) is the virtue of the *rational part* of the soul, *courage* (*andreia*) is the virtue of the *spirited*

1 We have described individualism (atomism) as the view that wants to explain society and societal institutions purely in terms of the interaction between individuals. Universalism (holism), by contrast, postulates some or other all-encompassing societal whole or totality.

2 The term 'material' here refers to the inner nature or the inner structural principle of the state. It is usually distinguished from a 'formal' view. The latter does not intend to account for structural differences or inherently limited spheres of competence. A thinker like Rousseau, for example, advocated the idea of a just state in a formal sense, but materially fell back into a totalitarian and absolutistic theory of the state. Hobbes defended a totalitarian and absolutistic theory in both a formal and material sense.

3 Namely the *logistikon*, *thumoeides* and *epithumétikon*, i.e. *thought*, *fervour* and *desire*.

part, while *temperance* as virtue represents – under the rule of the rational part – the union of the *thumoeides* and the *epithumétikon*. *Justice*, as general virtue, embraces the former three, and thus also has a bearing on the ideal *state as a whole* (cf. *Politeia*, 433A-C). Justice prohibits the transgression of the legal domain of the different parts of the soul, i.e. it demands avoiding any *legal excess* – which also applies to the three *estates* within the state (cf. *Politeia*, 443 ff.).

Both Plato and Aristotle gave primacy to the form motive and opted for the ideal of moral goodness (form perfection) as the ultimate aim of society. Aristotle distinguishes between *natural law* and *positive law* for the first time. Justice in a broad (moral) sense (*dikaion politikon*) embraces all virtues (such as *courage*, *moderation*, *friendliness*, and so on) and manifests itself within the state. Justice concerns *legal norms* and their *obedience* in a *strict sense*. One of the hallmarks of this strict form of justice is *equality*. Equality may be determined according to an arithmetical or a geometrical yardstick. The latter is applied in the case of *distributive justice* – regarding the distribution of "honor or money or such other possessions of the community as can be divided among its members," while the "other kind is shown in private transactions or business deals, where it serves the purpose of correcting any unfairness that may arise."[1] The latter kind of justice, designated as *corrective justice*, operates according to an "arithmetical proportion" requiring from the judge to find (or restore) the mean between what is *too much* and what is *too little* (cf. *NE*, Book V, Chapter 4). Well anticipating what the classical Roman jurists took justice to mean, Aristotle's understanding of distributive justice could be summarized in the guideline: *treat equals equally and non-equals unequally*.

9.3.2 *Transition to Stoic philosophy and the medieval synthesis*

Eventually the Stoa – with Zeno from Cyprus (336-264 B.C.) as their founder – radicalize the idea of *natural law*. The prevailing contrast between Hellenic people and savages is now challenged. The mature Greek conception allows only the free Greek to participate in political decision-making and jurisprudence, and thus to come to a fulfilled life within the *polis*. The Stoa accepts a macropolis, universally embracing all gods and human beings (as *equals*) alike. Certainly, this new view is linked to the empire of Alexander the Great, which terminates the independent existence of the Greek *polis* (city-state). The Stoa attempts to unify the opposing poles of matter and form. Whatever there is, is supposed to participate in the *Divine Reason* (*Nous*) through the *logoi spermatikoi* (germ-like rational particles). The Stoic epistemology (theory of knowledge) anticipates the British empiricist philosophers of the 17th century by assuming that the human spirit is a clean slate (a *tabula rasa*) at birth. Yet, in their view that knowledge acquired through the senses cannot guarantee truth because the latter is dependent upon *inborn concepts*, the Stoics also anticipate the views of Descartes (1596-1650) and Kant (1724-1804).

1 *The Nicomachean Ethics* (*NE*), Book V, Chapter 2.

The Stoa accepts *nomos* as a *universal natural moral law*.

In the thought of Cicero, the *fluid* and *changeful* nature of law and legislation cannot be accepted. We find that he rather adheres to the view already advanced by Plato and Aristotle, namely that all *positive law* must be understood in terms of fixed legal principles, flowing from an ethical world order. The preferential designation of this order is found in the notion of a *natural law*, a *lex naturalis*.[1]

Cicero conceives the individual in a two-fold way: as Roman citizen (*civis Romanus*), an individual participates in the Roman *populus* (the *public*) without having any *claim* to it, or a *subjective right* in respect of it. As a *private person*, an individual disposes over an *inviolable personal legal sphere*. The legal sphere of the *familia*, with its head of the house, the *pater familias*, provides the starting point for the way in which Cicero accounts for subjective private law. The *family* is still an undifferentiated unit that functions as a miniature state.[2] Through the expansion of the Roman Empire, a need for legal provision for non-Romans within the Empire became apparent. This need generated the development of the *ius gentium*, which should not be seen as the starting point of the law of nations, but rather as the point of departure of what we now know, as noted, as *civil private law* – a legal sphere destined to protect the *personal freedom* of the individual within the differentiated legal intercourse within modern society (see pages 282, 545, 561, 562).

The theory of natural law advanced by Cicero strongly influenced classical Roman jurists such as Gaius, Ulpianus, Modistinus and others, as can be seen in their definitions of *justice* and *natural law*.

Neo-Platonism, and particularly the thought of Plotinus, elevated the One above all multiplicity, while reserving the second 'hypostasis' for the *one-many*, the unity in the multiplicity of (Platonic) ideas (known as the *hen-polla*). This second level, emanating from the 'One', is occupied by the *Nous* (*Reason*), the third by the *Soul* (*Psychè*), and the fourth by *matter*.[3]

Although Augustine sincerely expresses the biblical creation motive on the one hand by proclaiming God's sovereignty over everything created *from nothing* (*ex nihilo*), he also combines these two and transposes the Platonic

1 Compare in terms of this, his work "De republica" (Book 3, Chapter 22) as well as his "De Officiis."

2 Dooyeweerd describes the nature of the familia as follows: "Each familia is a family community, an economic unit, a miniature state, and a community of worship. Above all, it is the embodiment of the religious authority of the household gods, who represent the communion between the living and dead members of the familia. The head of the familia is usually the oldest male, the pater familias, who wields the power of life and death over all – his wife, his children, his slaves, and his so-called clients. He also presides as the priest" (Dooyeweerd, 2003:24).

3 The original form-matter dualism acquires an ambiguous form in the thought of Plotinus. On the one hand he claims that through the process of emanation the One turns into its opposite: matter. At the same time he holds that the diminishing radiation coming from the One finds in matter its last form (*εἶδος τι ἔχατον*, En. V,8,7,22-23).

ideas into the Divine Mind. Prior to creation, these ideas have been within the Divine Mind (*divina intelligentia – De diversis questionibus*, 83,46).

In his famous work, *Civitas Dei* (*The City of God*), Augustine observes the biblical distinction between the kingdom of God and the kingdom of darkness, but owing to the neo-Platonic influence upon his thought, he gives it an unbiblical (dualistic) twist. The earthly world is interpreted as the temporal and changeful, which as such already displays an inherent *defect* in relation to God. The earthly state is understood in the sense of the classical Greek *totalitarian* state. In this dispensation, both are related and mixed. Yet, the earthly state is merely a *copy* of the City of God – their relationship is conceived according to the Platonic scheme of the ideal form and its copy. This copy is inherently *bad* – explaining why it is also designated as *Babylon* with a monarch called *Diabolus*. It should also be kept in mind that the *City of God* does not coincide with the temporal church institution, for as a sacramental institute of grace, the *Corpus Christi* (Body of Christ) is elevated above all societal institutions and is intended to encompass the entire life of the Christian. The fact that Augustine confuses *creation* and *fall* not only results in the misrepresentation of the antithesis between sin and redemption by focusing the latter upon two totalitarian spheres of life (the *City of Babylon* versus the *City of God*), but also exerts a significant influence upon the subsequent struggle between church and state.

By viewing the church as a *perfect society* superior to the state, Augustine lays the foundation for the Scholastic ground motive of *nature* and *grace*. This motive was brought to a unique synthesis in the thought of Thomas Aquinas (1225-1274).

When Thomas Aquinas enters the scene in the 13th century, his account of medieval society is based upon an attempted synthesis between Aristotle's philosophy and biblical Christianity. He accepts the dual teleological order of Aristotle with its hierarchy of substantial forms arranged in an order of lower and higher. It was designated as the *lex naturalis* (natural law), which is related to the transcendent *lex aeterna* (eternal law) as contained within the *Divine Intellect*. By virtue of its substantial form, the human being depends upon the community for the satisfaction of its needs.

The state (both the *polis* and the *Holy Roman Empire*) is viewed, in line with Aristotle's conception, as *an* all-encompassing, self-sufficient community (*societas perfecta*). The provision is that Thomas Aquinas applies this only to the *natural sphere*. As the highest community within the domain of nature, the state embraces all other temporal relationships. These lower communities possess a *relative* autonomy, subsumed under what is known as the *principle of subsidiarity*. However, this principle does not eliminate the universalistic starting point in St. Thomas's view of society, since the so-called *relative autonomy* of these lower communities remains connected to the nature of the state as *parts* of a larger *whole*. What is part of a whole shares in the same structural principle as the whole. As a result, the view of St. Thomas does not allow for the acknowledgment of societal collectivities that are *structurally different*. In line with Aristotle's conception, the family for Thomas also remains the *germ cell* of society. The hierarchical ordering of these communities coheres with each other according to the mutual relationship of a *means* to an *end*, and of *matter* to *form*.

As the encompassing community within the natural domain, the state actually only forms the *natural foundation* for the church as overarching *superstructure*, as the *supernatural* institute of *grace*. The state carries human beings to their highest *natural* aim in life, namely *goodness*, whereas the church elevates them to their supertemporal perfection, *eternal bliss* (see page 499 above).

Similar to Aristotle, the view of society in the thought of Thomas Aquinas correlates with his view of the human being. According to him, the essential rational nature of the human person is not radically affected by the fall into sin, that is, the fall does not touch the *root* (radix) of human existence. The fall into sin only caused the *loss of faith*. Redemption therefore means that faith is returned to the person as a *supernatural gift of grace* through the church. Within the natural domain, human reason is *autonomous* – it can even provide natural proof for the existence of God (cf. *Summa Theologica*, 1, 2, 3). Yet the supernatural *revealed truths* have to supplement and *perfect* these insights. The well-known position of Thomas Aquinas in respect of the relation between nature and grace is that grace does not negate nature, but perfects it (*Gratia naturam non tollit, sed perficit*). Pope John Paul II still echoes this legacy in 1998: "Just as grace builds on nature and brings it to fulfillment,[1] so faith builds upon and perfects reason" (John Paul II, 1998).[2]

Thomas's view of the function of the church as a supernatural institution of grace exerted a strong influence upon the official position of the Roman Catholic Church later on. Just recall the papal encyclical *Quadragesimo anno* (15 May 1931), it is explicitly stated:

> Surely the church does not only have the task to bring the human person merely to a transient and deficient happiness, for it must carry a person to eternal bliss (cf. Schnatz, 1973:403 – see page 499 above).

In his understanding of *law*, Thomas Aquinas perpetuates the Greek legacy. *Justice* serves as one of the four moral virtues (alongside *wisdom*, *temperance* and *courage*). Justice 'tributes' to someone that which legally belongs to the person. Aquinas also continues the Aristotelian distinction between *commutative* and *distributive* justice – with equality respectively viewed in terms of an *arithmetical* and a *geometrical* standard.

He adds what he calls *legal justice* (*iustitia legalis*). This form of justice assigns certain legal duties (such as military service). Natural law forms the ba-

1 A reference to *Summa contra Gentiles*, I,7 is given.

2 Pope John Paul II also refers to the classification of faith and reason in terms of *grace* and *nature* by Thomas Aquinas: "Although he made much of the supernatural character of faith, the Angelic Doctor did not overlook the importance of its reasonableness" (John Paul, 1998).

sis of all positive law, and whenever a positive legal stipulation contradicts natural law, it loses its *legal validity*. It is possible to deduce the *objective* natural law (valid for humanity as a whole) from the teleological ethical *basic principle*: "Do what is good and avoid what is bad." Subjective natural law embraces the legal competencies of a person by virtue of objective natural law (such as the right to life, integrity, acquisition of property, etc.).

9.3.3 *Nominalism paving the way for modern Humanism*

In our discussion of the nature of nominalism, we briefly paid attention to the key elements within a realistic understanding of reality, as well as to the rise of the modern humanistic ground motive of nature and freedom (see pages 370 ff.). This nominalistic movement provides the starting point for modern philosophy and modern political thought. It emancipates the modern person from the authority of faith as dictated by the church (and the Pope). In the transition to Renaissance and modernity, the dominance of society by the Pope and the Roman Catholic Church started to fade. This historical process paved the way for various *individualistic* theories of human society and the state, that is, theories denying the reality of supra-individual communities, in their attempt to explain society fully in terms of the actions of individual human beings. Initially, as in the case of Machiavelli (1469-1527), Bodin (1530-1596) and Hobbes (1588-1679), the result of this transformation was explicit *totalitarian* theories.[1] However, various thinkers gradually attempted to develop a theory of the state in which it is possible to guarantee various (civil) rights, that is, theories of the "just state" ('regstaatsteorieë').

The impact of the ground motive of nature and freedom occupied our attention at various occasions earlier in this work. In particular, we mentioned the rise (and initial dominance) of the natural science ideal (nature pole) and that via Rousseau, it was Kant who reverted to the primacy of the personality ideal (freedom motive). The nature motive is embodied in the ideal of an all-encompassing natural science capable of explaining whatever there is in natural scientific categories (such as cause and effect: causality), while the freedom motive manifested itself in the ideal of the autonomous human personality. Although the personality ideal gave birth to the science ideal, the latter turned into a Frankenstein, challenging its maker. If reality in its entirety is understood as subjected to the exact natural laws of cause and effect, then the human being is also reduced to become merely an atom amongst atoms, a cause amongst causes, and an effect amongst effects – totally causally determined, without any freedom.

Post-Kantian freedom idealism carried this primacy shift to its ultimate consequences. However, Auguste Comte once again reverted to the primacy of the science ideal, followed by Marx's dialectical materialism and Darwin's evolutionistic physicalism (see Strauss, 2007). During the second half of the

1 We explained earlier (see page 42) the term used in German, Dutch and Afrikaans, namely that 'magstaat' literally means 'power-state'. The idea of a 'regstaat' was translated as a "just state."

19th century, a number of intersecting lines emerged, perhaps best exemplified in the development of Edmund Husserl's thought.

The transition to the 20th century, and particularly the first three decades of this century, witnessed one of the most amazing and productive periods in the history of philosophy. When Sigmund Freud wrote his work on the interpretation of dreams (1900), it is striking that he no longer maintained a logical focus (analysis), but explicitly opted for *interpretation* (Traum*deutung*). Dilthey already opened the way to the further development of hermeneutics within the context of the linguistic turn that emerged during the transition of the 19th to 20th century. When Wittgenstein published his *Tractatus-Logico Philosophicus*, the shift to language is quite clear, for *language* is portrayed as constituting the limits of the individual's world: "The boundaries of my language means the limits of my world" (5.6). Wittgenstein's thought gave birth to logical positivism and later on to analytical philosophy and ordinary language philosophy – although Wittgenstein would not have considered himself to be a member of any of these trends of thought per se. Karl Jaspers further explored what is now known as existentialism, a trend that Heidegger combined with the phenomenology of Husserl – the first edition of Heidegger's *Sein und Zeit* appeared in 1927 in Husserl's *Jahrbuch für Philosophie und phänomenologische Forschung*. In 1923, Martin Buber published his personalist work *Ich und Du*. The first three decades also witnessed the dominance of the two neo-Kantian schools of thought, known as the Baden school (Windelband, Rickert, Weber and others) and the Marburg school (Cohen, Natorp, Cassirer, Kelsen and others). During the early twenties, the philosophical endeavors of Dooyeweerd and Vollenhoven also emerged, eventually known as the tradition of reformational philosophy.

A brief analysis of the intellectual journey of Edmund Husserl (1859-1938) will provide a slightly more detailed example of the motivating power of the humanistic ground motive within the thought of this significant modern philosopher, who truly wrestled with what he experienced as the crisis of *Europe* and the disciplines. Close to the end of his life Gödel spent much time in studying the works of Husserl (see Yourgrau, 2005:170, 182). The fact that we have seen that Gödel formulated ideas concerning primitive terms and coherence that closely approximate the principles of sphere-sovereignty and sphere-universality may be linked to the fact that Dooyeweerd acknowledged that during his early development he was strongly influenced by Husserl (see Dooyeweerd, 1997-I:v).

9.3.4 *Husserl wrestling with the dialectic of humanistic thought*

Husserl was born on April 8, in the same year when Darwin's "The origin of Species" appeared (1859) – Prossnitz, Moravia. He died on April 27, 1938 (Freiburg). He studied astronomy, mathematics and physics at Leipzig and Berlin. In 1887, he started his academic career as "private lecturer" in Halle. He was a professor in Göttingen from 1901 to 1916, and he taught in Freiburg (Breisgau) from 1916 to 1928.

The connection between phenomenology and intuitionism – one of the trends in modern mathematics – is obvious from the fact that Husserl received his first philosophical impulses from Leopold Kronecker (1823-1891), who is known as one of the leading figures in the intuitionistic mathematics of the late nineteenth century. As one of the leading mathematicians of his time, Kronecker particularly reacted against the idea of infinity as a completed totality, because he was influenced by C.F. Gauss (1777-1855), the 'prince' of mathematics. The latter wrote to Schumacher in 1831, that he "protests against the use of an infinite magnitude as something completed, which is never permissible in mathematics." Early intuitionism understood infinity in the literal sense of succession without end (endlessness). On the basis of this assumption, Kronecker believed that it is possible to reduce mathematics to finite natural numbers.

9.3.4.1 *The early development of Husserl*

The central significance of these problems becomes understandable if we keep in mind that Husserl commenced his academic studies in 1876, and that he studied mathematics in Berlin between 1878 and 1881, with Karl Weierstrass (1815-1897), amongst others. The latter, however, alongside his younger contemporaries, Richard Dedekind (1831-1916) and Georg Cantor (1845-1918), provided a new foundation for mathematics, in which ample use is made of the idea of completed infinitude (we proposed to distinguish between the *sucessive infinite* and the *at once infinite* – traditionally known as the *actual infinite*, as opposed to the *potential infinite* – see page 237 above).[1] It is therefore clear that Husserl's initial philosophical interests challenged him with the problematic relation between the successive infinite and the at once infinite.

Husserl continued his initial mathematical studies in Berlin, and he finished his dissertation on the theme: *Beiträge zur Variationsrechnung* in 1883. During the same year, he worked as assistant for Weierstrass until the latter took ill, upon which Husserl moved to Vienna (1884-1886), where he studied philosophy mainly with Brentano (1838-1917).

With a view to his "Habilitationschrift" he moved to Halle in 1886, where Carl Stumpf (1848-1936) was located. Here he also met Georg Cantor, and they became very good friends. Cantor, who was working out his transfinite arithmetic on the basis of his 'Mengenlehre' (set theory) at this stage, wholeheartedly advocated the employment of the actual infinite (at once infinite) in mathematics. It is remarkable that Cantor's strong opponent was no one less than Kronecker. Cantor himself worked within the platonistic tradition of scholasticism and his Platonism exerted a significant influence on Husserl's thought .

1 For example, Weierstrass defined an irrational number – such as $\sqrt{2}$ – as the actual infinite set of rational numbers smaller than $\sqrt{2}$.

9.3.4.2 *Husserl's Philosophy of Arithmetic*

In 1887, Husserl completed his habilitation with a psychological analysis of the concept of number. In 1891, he followed this with his first large work: *Philosophie der Arithmetik* (published in the *Husserliana* as Volume XII, 1970).

During this period, Husserl particularly struggled with the problem of the infinite. On the one hand, he related the infinite to the psychical ('psychological') nature of a collective synthesis in a psychologistic sense (Husserl, 1970:64 ff.), which coheres with the principle of succession (Husserl, 1970:220). On the other hand, he focused on the question regarding the existence of the actual infinite in view of the finitude of human nature (Husserl, 1970:191). At this stage of his development – and in line with Kronecker – he envisaged the ideal of developing a *finite* arithmetic, and even contemplated a second volume of his *Philosophy of Arithmetik*.

In the *Foreword* to this volume, L. Eley concludes from the fact that this second volume never appeared in print, that the ideal of a finite arithmetic failed. From the unpublished manuscripts of this second volume, it is clear that Husserl did not succeed in providing a foundation for arithmetic without the assumption of the actual infinite.

The effect of this failure was that, during the nineties, Husserl developed a closer affinity for Platonism in its linkage to the acceptance of the actual infinite. According to this Platonism, the entities and relations with which the mathematician operates exist in a Platonic realm of ideal forms.

9.3.4.3 *Platonism in Husserl's thought*

The influence of this mathematical Platonism clearly manifests itself in Husserl's next large (two-volume) work: *Logical Investigations* (*Logische Untersuchingen – LU*) (1900-1901; Husserliana, Volume XVIII, 1975). In this work, he accepts a world-in-itself, with "ideal objects," independent of human consciousness. Whereas he often uses the word 'Idea' in *LU,* he later prefers the terms Form (*Eidos*) and Essence (*Wesen*). In *LU*, the universally valid *essences* are independent of the flowing psychical acts that acquire a direct grip on them through an inner evidence (*LU*, I:190). For Husserl, *evidence* here constitutes the experience of truth (*LU*, I:230). Truth in itself is the correlate of being in itself (*LU*, I:229).

The effect of Platonism in his *LU* causes an ambiguous methodology: on the one hand, a *phenomenological descriptive analysis of essences*, and on the other a *natural scientific explanation*. At this stage of his development, Husserl still views material things as independent from the human consciousness. Only interpreted objects have a mode of being dependent upon the intentional consciousness. In general, one can therefore say that the all-pervasive presupposition of *LU* lies in the acceptance of a world-in-itself, with "ideal objects" that are independent from human consciousness. In his authoritative work on the development of Husserl, we find that De Boer characterizes this position as *realistic* (De Boer, 1966:315). Picker relates this directly to Husserl's mathematical studies (Picker, 1961:289).

9.3.4.4 *The genesis of Husserl's transcendental idealism*

With the exception of his *Five Lectures on Phenomenology*, Husserl did not publish anything for the next decade. For the first time, he now engaged himself in a penetrating study of Immanuel Kant (1724-1804). The outcome of the crisis through which Husserl struggled during this period appeared in his first article in the newly established philosophical journal, *Logos* (1910), under the title: *Philosophie als strenge Wissenschaft* (*Philosophy as an exact Science*). Although Husserl did not support the classical modern ideal of a *mathesis universalis* (i.e. a *mathematical science ideal*), he did remark later (at his 70th birthday celebration) that he wanted to do for philosophy what Weierstrass achieved for mathematics.

He now realized that, although he succeeded in conquering logical psychologism, he failed to liberate himself from *epistemological psychologism*. As an alternative, he presented his so-called *transcendental idealism*.

9.3.4.5 *The intuitionistic core of Husserl's transcendental idealism*

In his 1913 work – *Ideen zu einer reinen Phänomenologie und phänomenologische Philosophie* – the intuition of essences is also designated as *eidetic reduction*. In the natural attitude, material things are simply at hand (Husserl, 1913-I:62-63). The basic thesis of this attitude is: "I and my Umwelt" ("Ich und meine Umwelt"). This attitude views reality as absolutely at hand – with consciousness as a layer of reality itself. In order to alter this attitude, Husserl introduces the *philosophical* (transcendental, phenomenological) *reduction* (epochè: *εποχή*) in addition to his eidetic reduction (Husserl, 1954:153).

Without losing or changing anything, our natural conviction is bracketed (suspended) through the transcendental reduction – which is a matter of "complete freedom" (Husserl, 1913-I:65). The philosophical *epochè* is also constituted by the maxim that we withhold ourselves from all judgments regarding the contents of all given philosophies (Husserl, 1913-I:40-41).

Through this transcendental reduction, I no longer accept the world as presented by the natural attitude and the positive sciences, namely as a universal existential basis for knowledge and as a world existing prior to all knowledge of it: "In the change of the epochè nothing is lost" (Husserl, 1954:179). What this in fact means is that the transcendental reduction does not eliminate the existing world, but simply sets aside a specific *natural interpretation* of it.

In opposition to *LU* – which still accepts the world in a platonistic sense as presupposition – the transcendental reduction reveals that consciousness itself is the sole true ground of the world: the reality of the entire world exists only as the correlate of the *intentional consciousness*. In other words, matter does not serve as the foundation for consciousness, because the reverse is the case: "Reality, both the reality of things taken separately and the reality of the universe essentially lack independence. It is not something absolute which in a secondary sense is connected to something else, for in an absolute sense it is precisely nothing, it does not have an *absolute essence*. It displays the nature of something that in principle *merely* (emphasis from Husserl!) exists intentionally, merely conscious, that is to say, it can only be represented and realized in possible appearances" (Husserl, 1913-I:118).

In all this, Husserl closely approximates the transcendental constituting motive of Kant, holding that understanding prescribes its laws (*a priori*) to nature (Kant, 1783:§36). Kant elevated human understanding to become the formal (*a priori*) law-giver of nature (see page 73 above).

Husserl accounts for the *transcendental motive* as follows: "It is the motive of investigating the final source of all knowledge acquisition, the reflection of the knowing person upon itself and its knowing life ... Worked out radically, it is the motive of a philosophy based purely in this source. Therefore it is a universal philosophy with an ultimate foundation" (Husserl, 1954:10; see *Phänomenologische Psychologie*, 1962:298, where he characterizes complete phenomenology as "universal philosophy").

Although the term *transcendental* highlights an element of similarity with Kant, the difference emerges from their respective views of what *idealism* means. As a "systematic unveiling of the constituting intentionality," philosophy as transcendental idealism does not leave open, in the Kantian sense of a limiting concept (*Grenzbegriff*), a world of "things-in-themselves" (Husserl, 1950:118-119).

The final justification of knowledge reveals that Husserl ultimately takes refuge in an extreme epistemic *intuitionism*. The guiding norm for phenomenology is: "Accept nothing but that which we can master with insight as it is essentially presented within pure consciousness" (Husserl, 1913-I:142). This norm presupposes the authentic intuitionistic basic principle of Husserl's thought. Indeed, he speaks about the "principle of all principles: that every originally given intuition is a legitimate source of knowledge, that everything which is presented to intuition (so to speak in its lived-through reality) simply ought to be accepted as it is given, but that also merely within the boundaries within which it is given no conceivable theory can cause us to err" (Husserl, 1913-I:52). The transcendental phenomenology of Husserl is therefore explicitly *intuitionistic*: what is directly given in our intuition is the final source of all knowledge and within those limits, no theory can dethrone it!

9.3.4.6 *Husserl and the mathematical intuitionism of Herman Weyl*

In a lecture by Weyl at the University of Lausanne (1954), he explains that his belief in positivism was first shaken when he fell in love with "a young singer whose life was grounded in religion and who belonged to a circle that was led philosophically by a well-known Hegelian." The effect of this shock continued until he married a pupil of Husserl's: "So it came to be Husserl who led me out of positivism once more to a freer outlook upon the world" (Weyl, 1969:287).

Husserl's shift to transcendental idealism inspired Weyl to deduce the following intuitionistic ground rule: "The immediate 'seeing' not just the sensory, experiencing sight but seeing in general, as given in ordinary consciousness of whatever kind, is the ultimate source of justification for all reasonable assertions. What offers itself to us in our intuition, is simply to be accepted in the form in which it gives itself, but only within the limits within which it gives itself" (cf. Weyl, 1969:288 and Husserl, 1950:52).

9.3.4.7 *The basic motive at the root of Husserl's phenomenological intuitionism*

Setting aside the natural attitude (Husserl, 1913-I:65) with one stroke (Husserl, 1913-I:67) is a matter of full freedom ("meine volle Freiheit' – Husserl, 1913-I:67; a matter of our complete freedom – Husserl, 1913-I:65). It is indeed through complete freedom that Husserl ensures the validity of his *intuitionistic* (*transcendental-idealistic*) *phenomenological science ideal*. He does not wish to return to the pre-Kantian rationalistic science ideal. In *Krisis*, he disqualifies this "rationalistic science ideal" (Husserl, 1954:119).

The crisis that Husserl discerns in respect of Europe and the disciplines is merely rooted in what he calls a *misguided rationalism* (an "verirrenden Rationalismus") (Husserl, 1954:337). In opposition to such a misguided rationalism, Husserl posits the unlimited possibilities of the *intuitionistic, phenomenological reason*. This trust is fundamentally threatened by the increasing influence of naturalism and objectivism, as well as the irrationalism of his own student, Heidegger (*Sein und Zeit*/Being and Time). Husserl experiences this with a sense of hopelessness – as the crisis of Europe and the academic disciplines. He writes:

> In order to comprehend what is wrong in the present crisis the concept Europe once again has to be viewed by means of the historical directedness towards the infinite aims of reason; it must be demonstrated how the European world was borne from reason-ideas, that is, out of the spirit of philosophy. The crisis will then clearly emerge as the apparent failure of rationalism. The basis of this failure of a rational culture, however, … is not inherent to rationalism, since it is only found in its externalization, in its decay into *naturalism* and *objectivism*. The crisis of European existence provides only two options: the decline of Europe in the alienation from its own rational existential meaning, the decay into an animosity towards the spiritual and a lapse into barbarism, *or* the rebirth of European existence through the spirit of philosophy, particularly through a heroism of reason that will consistently triumph over naturalism (Krisis, 1954:347-348).

His deepest trust in the intuitionistically conceived of (transcendental-idealistic, phenomenological) philosophical reason explains why he compares the "total phenomenological attitude" and its accompanying *epochè* with a *religious conversion*, because it indeed harbours the largest existential change that confronts humankind as a task (Husserl, 1954:140).

Through complete freedom, the European is called to establish the intuitionistic, phenomenological science ideal as the only road to the rebirth of Europe through the spirit of philosophy. The slogan of his article: *Philosophy as a Rigorous Science,* continues to overarch his philosophical endeavors.

Yet he was unseccessful in containing the growing crisis he experienced. For that matter, his phenomenology – in the hands of Heidegger, Sartre and Merlau-Ponty – was turned into its opposite: an irrationalistic and existentialistic *freedom motive*, which derives its motivating power not from an intuitionistic *science ideal*, but from the ideal of an *autonomously free personality*. This development ruined his dream of philosophy as an irrefutable, apodictically certain science:

> Philosophy as science, as a serious, exact, yes apodictic exact science – *der Traum ist ausgeträumt*" (Husserl, 1954:508 – "the dream is dreamed").

9.4 The presence of ultimate commitments within the special sciences

Particularly within the domain of the philosophy of science, various prominent thinkers have acknowledged the presence of supra-rational orientations (see Chapter 2). We focus briefly on a number of disciplines in this final subsection.

9.4.1 *Mathematics*

The Greek ground motive of form and matter directed the way in which Greek mathematics developed. The Pythagoreans believed that the essence of everything could be expressed by integers and their relations (in fractions). (See pages 44, 53, 162, 403.) Yet, the relation between the side and diagonal of a regular pentagram led to the discovery of 'incommensurable' quantities – a discovery that undermined the limiting and form-giving nature of number, because within this function, the unbounded infinite (the *apeiron*) appeared. This discovery of irrational numbers was to remain a secret. Hippasus, who is supposed to have published it, was therefore considered "to have betrayed a mathematical secret" (Riedweg, 2005:107). The effect was that Greek philosophy began to explore an alternative mode of explanation, which led to the *geometrization* of Greek mathematics.

With the rise of early modern philosophy, as seen, the humanistic freedom motive gave birth to the (rationalistic) natural science ideal, which deified mathematical and physical thought. Leibniz merged his mathematical discoveries (his so-called infinitesimal calculus) with his monadology in such a way that the distinct monads represented both discreteness and the personality ideal, while the assumed *lex continui* represented the primacy of the mathematical science ideal in his thought, also expressed in his remark that God is the great Geometer. From Galileo on the common view has been that the book of nature is written in the language of mathematics. In anticipation of the arithmetization of mathematics during the second half of the 19th century, Jacobi stated that "God ever arithmetizes" in the first part of this century (see Bell, 1965-I:16). Bell himself demonstrates the effect of the motive of natural scientific control in a remark directed at the Pythagorean claim that everything is number: "If 'Number rules the universe' as Pythagoras asserted, then number is merely our delegate to the throne, for we rule Number" (Bell, 1965-I:16).

The rationalistic science ideal inspired the leading mathematician of the 20th century, David Hilbert, to attempt to prove that mathematics is consistent (non-contradictory). In his famous oration on logic and knowledge of nature during 1930, upon receiving honorary citizenship of Königsberg (the village where Immanuel Kant lived), he concluded with optimism: "We must know, we shall know."[1] Ironically, Kurt Gödel showed only a year later in 1931, that any formal axiomatic system is incomplete and dependent upon an intuitive insight transcending the formalism of the system (see page 315 above). Therefore the self-insufficiency of human thought also manifested itself within the domain of mathematics.[2]

9.4.2 *Physics*

In our discussion of the development of the concept of matter (in Chapter 7), the successive conceptions not only illustrated diverging views of reality, but also that they have their roots in deeper, supra-theoretical ground motive orientations. From the Renaissance, modern Humanism started to dominate the scene, and its science ideal found its first stronghold in an over-estimation of the disciplines of mathematics and physics. The deterministic thrust of the classical mechanistic tendency in the development of modern physics lasted well into the 20th century, for even Einstein did not wish to accept a dice-playing God (Planck, Von Laue and Lenard also supported a deterministic stance).[3] Yet the element of indeterminacy in the emergence of quantum theory seemingly opened space for human freedom. The assumption is that quantum theory eliminated the physical law of causality. However, Max Planck, wanting to uphold the law of causality, nonetheless advanced arguments supporting the freedom of the human will (see Planck, 1973:302).

9.4.3 *Biology and bio-philosophical anthropology*

Although Darwinism and neo-Darwinism are in the grip of the classical science ideal, with its continuity postulate (speculatively bridging whatever gaps there may be in the line from atoms to human beings), the neo-vitalistic biology of Driesch developed a negative understanding of the (Aristotelian) idea of a *vital force* (*entelechie*). As such, he considers it to be a "system of negations" (Driesch, 1920:513; 459 ff.), that is, it cannot be positively determined: 'entelechie' is non-mechanical; it is not energy, not force, not a constant (Driesch, 1920:460), and non-spatial (Driesch, 1920:513). The difference be-

1 "Wir müssen wissen, wir werden wissen" (Hilbert, 1970:387).

2 Recall the words of Hermann Weyl: "It must have been hard on Hilbert, the axiomatist, to acknowledge that the insight of consistency is rather to be attained by intuitive reasoning which is based on evidence and not on axioms" (Weyl, 1970:269).

3 Heisenberg remarks that "mechanistic thinking" gave rise to a "gradually spreading materialistic world view" that blossomed in the Enlightenment (Heisenberg, 1956:90-91).

tween the atomistic 'Einselkausalität' and the holistic 'Ganzheitskausalität' is also framed in terms of the opposition 'Zufall' and 'Ganzheit' (chance and totality).[1] In the thought of Driesch, determination is opposed to genuine freedom. He declares that the question of freedom is to be considered as a metaphysical question of faith, which cannot be answered by the science of philosophy (cf. Driesch, 1931:93-122). Although Kant and Driesch differ in their views on the nature of philosophy, they agree that freedom is not a question of scientific proof, but rather of (practical) faith.

In his theory of the freedom of the will, Arnold Gehlen continues Driesch's negative description of the 'entelechie'. However, with an explicit appeal to the freedom idealism of Schelling, he immediately transforms it in order to provide a point of entry for *freedom*. At the same time he realizes that Driesch actually brought biotical phenomena under the reign of the deterministic classical ideal of science. Therefore, once again, he wants to restrict causality to *mechanical causality*: "Since causality is only thinkable as mechanical causality, the entelechie is negatively free, i.e. spontaneous and primary in a sense which cannot be subjected to a closer determination" (Gehlen, 1965:60).

The tension between nature and freedom brought Max Scheler to his well-known characterization in terms of what he calls the 'Weltoffenheit' of human beings (Scheler, 1962:38, 40). Against this background, Plessner develops his own perspective on the human being as an eccentric creature, while biologists and anthropologists such as Portmann, Overhage and Gehlen give the notion of 'Weltoffenheit' a prominent place in their writings. Even theology takes advantage of this notion. Wolfhart Pannenberg, for example, interprets it in terms of what he calls "der grenzenlosen Angewiesenheit des Menschen" (*the unlimited dependency of the human being*), while relating it to the "fundamental biological structure of being human" (Pannenberg, 1968:11; cf. also Scherer's treatment of the 'Weltoffenheit' of the human being, 1980:79 ff.). Ultimately, the term 'Weltoffenheit' is used to embody the reaction against the claims of the science ideal, namely that the human being is determined in all respects. In the final analysis, the intention of these authors is to show that the human being is free from being determined by natural causality. Of course the opposite tendency, the continuity postulate of neo-Darwinism, persisted throughout the 20th century. In the mid-sixties of this century Jevons articulates this continuity postulate clearly: "The continuity of the hierarchy offers hope that by systematic, stage by stage comparisons, the gap between molecules and large animals can be bridged ..." (Jevons, 1964:97).

In his PhD thesis, dealing with philosophical aspects in the biology of Portmann, R. Kugler states that Portmann essentially understands the human being in terms of freedom (Kugler, 1967:75). At the same time, Portmann is well aware of the fact that, as a "philosophical idea," freedom withdraws from scientific grasp. Kugler places this approach within the "large tradition" of a

1 Dooyeweerd points out that this distinction reflects the ground motive dialectic of nature and freedom (Dooyeweerd, 1997-III:748).

"philosophical determination of the human being," dating back to Immanuel Kant: "The innermost essence of the human being is freedom, it is the possibility of the human being to transform itself into that what it is" (Kugler, 1967:81). Compare this announcement with the following statement by Plessner: "As eccentrically organized creature the human being must make itself into that what it already is" (1965:309).

Gehlen points out that this mode of expression manifests the logical scheme of a normal teleology. This tradition is influenced by Fichte: "I want to be free ... means: I want to make myself into that what I shall be before I am it, in order to be able to perform it" (cf. Gehlen, 1965:103-104). And we have seen that Fichte himself is dependent on Kant, who introduces teleology as a bridge to human freedom. The philosophical tradition in which "mechanical causality" and 'teleology' (nature and freedom) are always dialectically related, inspired Eduard von Hartmann to remind natural scientists in the following way: "If our natural scientists were philosophically better trained, they would have been aware of the fact that the whole German speculation, from Leibniz to Kant and up to the present, equally decisively rejects a teleology separated from mechanical causality, as it does with a mechanical causality divorced from teleology" (quoted by Haas, 1959a:456).

In the thought of the well-known but independent neo-Darwinist thinker, Stephen Gould, the dialectical tension between nature and freedom is found in his reaction against the biological determinism of the sociobiologist E.O. Wilson (see Wilson, 1975). On the one hand Gould upholds the basic thesis that *humans are animals*. However for him this statement does not "imply that our specific patterns of behaviour and social arrangements are in any way directly determined by our genes" (Gould, 1992:251). For that reason he answers the question regarding the "evidence for genetic control of specific human social behavior" totally in the negative: "At the moment, the answer is none whatever" (Gould, 1992:252). He explicitly states that he rather opts for *freedom*: "Better to stick resolutely to a philosophical position on human liberty: what free adults do with each other in their own private lives is their business alone" (Gould, 1992:267).

Gould mentions a statement of Wolfgang Wickler: "It follows from evolutionary theory that the genes run the individual in their own interest." Gould's reaction is radical: "I confess I cannot regard such a statement as much more than metaphorical nonsense" (Gould, 1992:269). However, the question is: how does Gould reconcile his view that humans are animals with the freedom and liberty of humankind? It appears that the basic motive of nature and freedom settled within the domain of the 'biological'. Gould believes that "the issue is not universal biology vs. human uniqueness, but biological potentiality vs. biological determinism" (Gould, 1992:252). *Potentiality* here represents the humanistic freedom motive and *determinism* the classical humanistic science ideal. In reaction to the meaningless speculations of sociobiologists Gould therefore posits human flexibility with a vast range of potential behaviour.[1] In the final analysis Gould attempts to maintain a relative balance between the dialectically opposed poles of the ground motive of nature and freedom.[2]

9.4.4 *Psychology*

During the 19th century atomistic association psychology represents the domination of the humanistic science ideal in its attempt to explain reality in terms of its simplest elements. In connection with the human will Schopenhauer kept alive the Kantian demarcation of the science ideal and the personality ideal – phenomena and Things-in-Themselves, i.e. nature and freedom. What is indestructible is not the individual being, but the everything-penetrating, everything-shaping and the everything-disrupting *will*. Freedom is impossible within the world of phenomena, for it can only exist in the sphere of the "Thing-in-Itself" (cf. Drewermann, 2006:48). Drewermann points out that in the psychoanalysis of Freud the I does not belong to the intelligible world, in its merely psychical (phenomenal) existence it is *unfree*.

Since 1879 Wundt established empirical psychology in a natural scientific sense. His atomistic approach, aiming at an imitation of the analytical methods of physics and chemistry, continues the primacy of the original science ideal, although some of the successors advanced a holistic approach, also found in the two schools of *Gestalt*-psychology (the Berlin school and the Leipzig schools.[3] Ziehen continues the Wundt tradtion in his work on physiological psychology (11th edition in 1920). However, within the deterministic articulation of the science ideal he switched to a psychical basic denominator in his so-called pan-psychismus (see Zeihen, 1920:536, note 3). Within the field of biology this trend, representing the naturalistic science ideal of modern Humanism, was further explored in the pan-psychistic identism of Rensch (see Rensch, 1968, 1971, 1973).

After Jaspers published his *Algemeine Psychopatologie* in 1912 the freedom motive of Humanism made its own contribution to the development of 20th century psychology. In reaction to behaviourism and depth psychology existentialism acquired an influence within, what is also sometimes designated as "humanistic psychology." This is not meant in the sense of the humanistic world and life view directed by its dialectical ground motive of nature and freedom, but simply to arrive at a more comprehensive understanding of the human person in its totality.

1 "We are both similar and different from other animals. In different cultural contexts, emphasis upon one side or the other of this fundamental truth plays a useful social role. In Darwin's day, an assertion of our similarity broke through centuries of harmful superstition. Now we may need to emphasize our difference as flexible animals with a vast range of potential behavior. Our biological nature does not stand in the way of social reform" (Gould, 1992:259).

2 We saw that Simpson in his own way emphasized the 'absolute' difference in kind between humans and animals (see page 114 above – Simpson, 1971:270).

3 Gestalt psychology gave rise to its own therapy, known as *Gestalt Therapy*. See Perls, Hefferline and Goodman (1986).

Consider, for example, Frankl's *logotherapy* (the search for meaning), inspired by existentialism – while keeping in mind that the latter reflects the primacy of the irrationalist turn within the humanistic personality ideal during the 20th century – a tendency continued within postmodernism.

The well-known French philosopher, Merleau-Ponty, developed his philosophical anthropology on the basis of psychological and psycho-pathological considerations. On the one hand, together with Sartre, he accepts the thesis: "I am my body." On the other hand, however, he also holds the opinion that one's historical existence must repress the bodily organism down to the pre-personal level of an anonymous complex.

Inspired by the nature-pole of the basic motive (ground motive) of humanism, Merleau-Ponty writes: "I cannot understand the function of the living body except by enacting it myself, and except in so far as I am a body which rises towards the world" (Merleau-Ponty, 1970:75). From the opposite motivation he states: "... so it can be said that my organism, as a pre-personal cleaving to the general form of the world, as an anonymous and general existence, plays, beneath my personal life, the part of an inborn complex" (Merleau-Ponty, 1970:84). On the one hand I am my body, and on the other hand my body is seen as a pre-reflexive, pre-personal, anonymous complex by virtue of its being-in-the-world (Merleau-Ponty, 1970:79, 80, 82, 83, 86). *Nature* and *freedom* reciprocally endanger and presuppose each other: "... for most of the time personal existence represses the organism without being able either to go beyond it or to renounce itself; without, in other words, being able either to reduce the organism to its existential self, or itself to the organism" (Merleau-Ponty, 1970:84). The dialectical movement, to and fro, between these poles is best illustrated in his following words: "Man taken as a concrete being is not a psyche joined to an organism, but the movement to and fro of existence which at one time allows itself to take *corporeal form* and at others moves towards *personal acts* (my italics – DS)" (Merleau-Ponty, 1970:88).[1]

9.4.5 *Sociology*[2]

Both Comte and Spencer adhere to variants of the humanistic science ideal. Even a prominent 20th sociologist such as MacIver embraced this science ideal by the end of his academic career. However, the supposed authenticity of the human personality generated a striking reaction to the mechanistic, materialistic and behavioralistic transgressions of the humanistic science ideal in the thought of Sorokin: "Hence the general tendency of the sensate mentality to regard the world – even man, his culture, and consciousness itself – materialistically, mechanistically, behavioristically. Man becomes, in sensate scientific definitions, a 'complex of electrons and protons', an animal organism, a

1 Van Huyssteen refers to the words of Merleau-Ponty mentioned above (see Merleau-Ponty, 1970:75 and Van Huyssteen, 2006:276), but he does not discern the dialectical tension in the thought of Merleau-Ponty.

2 Compare in this connection Strauss, 2006 – particularly Chapters 2 and 3.

reflex mechanism, a variety of stimulus-response relationships, or a psycho-analytical 'bag' filled with physiological libido. 'Consciousness' is declared to be an inaccurate and subjective term for physiological reflexes and overt actions of a certain kind" (Sorokin, 1946:93-94).

The neo-Kantian Baden school – in particular Wilhelm Windelband, Heinrich Rickert and Max Weber – also advanced the cause of the personality ideal (freedom motive) by means of transforming the Kantian domain of *Sollen* (ought) into their idea of *values*. Throughout the 20th century, the dialectic of nature and freedom exerted its influence upon the development of the discipline of sociology. Catton, for example, returned to a consistent *naturalistic* approach (see Catton, 1966), similar to the positivistic legacy of empiricism in sociology (see Neurath, 1973). In his initial action theory and subsequent system theory, Parsons appears to acknowledge the normativity of human life, although his AGIL scheme (Adaptation, Goal-attainment, Integration and Latency) continues to reflect elements of a naturalistic mode of explanation (see pages 525 ff. above).

9.4.6 *Political Theory*

In the first sub-paragraph of Chapter 8, we provided a brief characterization of the role of the form motive in ancient Greece, focused on the city state (polis), which, according to both Plato and Aristotle, had to lead its citizens to moral perfection. During the middle ages, the synthesis motive of nature and grace amended the idea of the perfect society (*societas perfecta*) by subordinating the state, leading to the highest *temporal* fulfillment in human life (moral perfection), to the church as supra-natural institute of grace aimed at eternal bliss.

The humanistic science ideal in its individualistic attire inspired the social contract theories that aimed at a rational (and not historical) reconstruction of society in terms of its simplest elements, individuals. As we have seen, these theories invariably led to a theory of the power state, although some of them hoped to accomplish the opposite, namely to establish a plea for a *just state*. In the case of Rousseau, the picture was further complicated by the fact that he opted for the simultaneous employment of both atomistic and holistic elements in his thought, although this complication simply underscores his ultimate impasse, demonstrated in his internally antinomic conviction that the dissenting minority ought to be *forced to be free*. During the 19th century, the firm hold of the humanistic science ideal captured the spirits of Marx and Spencer – the former with his dialectical materialism and the latter with his organicistic individualism.

Even the political theory of the 20th century continued to struggle with the impasse entailed in the opposing views of atomism and holism. Yet, we noted that key figures within the tradition of system theory, as well as political philosophers like Rawls and Habermas have stumbled upon the inner laws of societal institutions, in this regard bringing them close to the position of reformational philosophy, which eventually, after Althusius for the first time opened the way to escape from the speculative metaphysical way in which the whole-parts scheme was traditionally applied to the whole of society and inspired by the biblical creation motive, articulated an understanding of what has been called the *sphere-sovereignty* of each distinct societal collectivity (Groen van Prinsterer, Abraham Kuyper and Herman Dooyeweerd).

9.4.7 *Theology*

The proliferation of distinct theological stances that came upon the scene in the 20th century cannot be understood except on the basis of acknowledging their underlying philosophical orientations. While traditional (Aristotelian-Thomistic) scholasticism mainly borrowed elements from Greek philosophy to provide content for the nature pole, subordinated to the grace motive, modern scholasticism – neo-Scholastisim – tend to borrow thought schemes from modern Humanism in order to shape the nature motive.

Just consider the remarkable dependence of Barth's understanding of God upon existential (freedom) philosophy.[1] In existential philosophy, the 'essence' of human existence lies in freedom. Likewise, in the theology of Barth, the 'essence' of God lies in the freedom of God. Similar to the way in which the center of human existence coincides with the *existential act* for existential philosophy (the being of the human person is the becoming of the self-constituting act), Barth considers the 'essence' of God to be identical to His 'Act'. The 'being' of God is no resting being, but the self-constituting act of God, *actus purus*. Just as the 'essence' of human existence in existential philosophy is the *act of choice*, the action of the self-electing freedom, so in Barth's theology the 'essence' of God is a an act-of-choice, the action of the self-electing God. Just as the human person is a human-in-community, according to the dualistic wing of existential philosophy (Jaspers), so God is God-in-community in Barth's theology, the Triune God, God in relation, and God is identical to His relation. Just as the human being, according the existential philosophy, always changes, so Jesus of Nazareth must always again become the Son of God, according to Barth. As the *spoken word* in existential philosophy does not reveal its source in the existential human being, but rather conceals it because it belongs to the contingent world (that has a mode of being different from the existential), so Barth holds that God's spoken Word does not give away its source in God's Word, but rather conceals it – for as the already spoken Word, it belongs to a different reality, that of 'Welthaftigkeit', which is foreign and opposed to God. As the human being is *history* according to existential philosophy, so God *is* history in the thought of Barth . The original history [*Ur-Geschichte*] of God in principle must be distinguished from the 'Biblical' history, which, as tradition, is depreciated and designated as having no intrinsic authority, for according to Barth, God *is* history.[2]

Without articulating any detail, we briefly refer to what has been mentioned

1 Zuidema highlighted these similarities – see Zuidema, 1953:96-98.

2 However, the analogy with existential philosophy does find a limit in Barth's scriptural faith (see Zuidema, 1953:97-98).

in Chapter 1 – in this context to be seen as a few more striking examples of the neo-Scholastic legacy. The historicism of Dilthey and Troeltzsch largely influenced the theology of Pannenberg; positivism and neo-positivism (fruits of the humanistic science ideal) influenced the atheistic theology of Altizer and Cox; existentialist and hermeneutical philosophical trends (in spired by the humanistic freedom motive) influenced Fuchs, Ebeling and Steiger, while liberation theology is influenced by the return to the freedom motive in neo-Marxist thought.

During the last two decades two significant and penetrating works focusing on the foundations of theology appeared – see Spykman (1992) and Troost (2004). Spykman writes:

> The day is now past that Reformed dogmatics can afford to be unaware of the importance of a Christian philosophy for its theology. The renewal of Reformed dogmatics now calls for a more explicit delineation of its philosophical foundations and context. To neglect the Christian philosophy associated with the names of Vollenhoven and Dooyeweerd is to impoverish Reformed theology (Spykman, 1992:100).

9.5 Concluding remarks

Philosophy indeed represents an intellectual enterprise with its own intrinsic field of investigation – the cohering diversity within the universe, encompassing all its dimensions, aspects, entities and processes. Although it is almost self-evident that this scope will be of significance for the special sciences, this work attempts to show that the foundational role of philosophy in respect of the various academic disciplines is intrinsic to them. On the one hand, problems and distinctions pertaining to states of affairs that cannot be denied by scholars from diverse backgrounds were discussed, and on the other it should be kept in mind that rationality itself is rooted in ultimate commitments – ground motives – transcending the theoretical sphere of science and scholarship as such, while giving direction to the theoretical thought and the distinctions articulated within every distinct theoretical view of reality.

No single special science can avoid a particular systematic perspective on the distinction between uniqueness and coherence regarding the relation between universality and what is individual, in respect of constancy and change, about conceptual knowledge and concept-transcending knowledge, and so on. In fact, every scholarly paradigm (theoretical view of reality) implicitly or explicitly employs an idea of the diversity of aspects and (natural and social) entities, embedded in an idea of the correlation of law and subject. Within the post-sensitive aspects, the idea of law is specified in terms of the complex or compound nature of a principle (see pages 291 ff.).

The issue is not *whether* a scholarly discipline proceeds from some or other philosophical view. Rather, the question is whether this unavoidable underlying philosophical perspective succeeds in giving a satisfactory theoretical account of reality by articulating a non-reductionist ontology – subject to the ontic principle of the excluded antinomy (see pages 183 ff.). What is peculiar about the idea of a non-reductionist ontology is that it is informed by the biblical appreciation of the unity and goodness of creation, in particular articulated in the New Testament.[1] World history witnessed multiple worldviews torn apart by elevating something in creation while at the same time depreciating something else – the legacy of *holy cows* and *scapegoats*, *golden calves* and *black sheep*.

Besides the task of analyzing and making appropriate distinctions about the diversity of creation, popular philosophy every now and then sees philosophical thinking as a way to holiness, to a virtuous life (Plato), as a lifestyle that ought to lead to the good (Plotinus), as a path to arrive at rational self-perfection (Descartes), to change reality through philosophical thought activities (into a heaven on earth, the worker's paradise – Marx), etc. The role that the many philosophical tendencies fulfill is to localize the source of evil somewhere in reality, and to lead humankind to a domain of safety, integrity and even salvation.

In other words, we are confronted with a way of liberation, a call to move away from one area of creation to "the kingdom of freedom, of virtue, of self-perfection or of goodness and autonomy." This means that the directional antithesis between *good* and *evil* is understood in *structural* terms, for example, is identified with *specific opposed territories*. For Greek philosophers, *evil* is found in the *material* world; for the existential philosopher of the 20th century, it is found in societal structures that threaten the *individual freedom* of a person; for the neo-Marxist and the social conflict theorist (cf. Hegel, Simmel and Dahrendorf), it is found in the *authority structure* of societal life forms; for other thinkers in the supposed inevitability of natural causality, and for others in the appearance of freedom that an individual is supposed to possess.

According to Wolters this apostate style of practicing science – in philosophy and in special sciences – indicates the way to the good, to the meaning of life, in short, the path to salvation – as the escape from one area of creation to another of creation: for example by moving to rationality, to forming, to the collective whole (of the nation, the state or the church), to freedom, etc.

Yet a biblical perspective does not localize evil in a specific area of creation, but in the *apostate direction* of the human heart, while salvation is equally a *directional* matter. If we look at philosophy, and the different existing special sciences, from the depth perspective of worldview, the most remarkable fact is that we are constantly confronted by what may be called a *surrogate salvation appeal*.

Each of these ways to salvation rests on an over-evaluation of a well-created part of reality, which at once leads to a depreciation of something else within creation – already a fundamental characteristic of the ancient heresy of *gnosticism*. At the same time, this attitude idolizes (deifies) something within creation – a point of departure of all idolatrous service, which brings honor,

1 Compare 1 Timothy 4:4: "For everything God created is good."

meant for the Creator, to a creature. Wolters succinctly characterizes this structure-direction distinction: "It is in this feature of traditional philosophy, which I have called the 'metaphysical soteriology' (and which has been blunted but not completely eradicated, in most Christian philosophies) that its religious nature comes most clearly to the fore. In my view, it ought to be a mark of philosophy which seeks to be as radical as the Bible that it renounces this whole enterprise, and simply accepts, as a point of departure, that every creature of God is good, and that sin and salvation are matters of opposing religious direction, not of good and evil sectors of the created order. All aspects of created life and reality are in principle equally good, and all are in principle equally subject to perversion and renewal" (Wolters, 1981:10-11).

* * *

Although many problems, trends of thought and diverging views were covered, there are still equally many themes and sub-disciplines left aside in our analyses. Yet, if what has been considered indeed supports the claim that philosophy is the discipline of the disciplines, this work has achieved its aim.

Literature

Adorno, T.W. & Horkheimer, M. 1973. *Aspects of Sociology*, The Frankfurt Institute for Social Research (Translated by John Viertel; Original title: *Soziologische Exkurse*). London: Heinemann.

Adorno, Th. W. 1970. *Der Positivismusstreit in der deutschen Soziologie*. Berlin: Luchterhand.

Agassi, J. & Cohen, R.S. (Editors) (1982): *Scientific Philosophy Today*, Essays in Honour of Mario Bunge, Boston Studies in the Philosophy of Science, Volume 67. Dordrecht: Reidel.

Agazzi, E. 1991. The Problem of Reductionism in Science. In: Episteme, A Series in the Foundational, Methodological, Philosophical, Psychological, Sociological, and Political Aspects of the Sciences, Pure and Applied, Volume 18, Editor: Mario Bunge, *Foundations and philosophy of Science Unit*, McGill University. Boston: Kluwer Academic Publishers. [Contributions from: E. Agazzi / Introduction; E. Agazzi, Reductionism as Negation of the Scientific Spirit; M. Bunge, The Power and Limits of Reductio; P. Hoyningen-Huene, Theory of Antireductionist Arguments: The Bohr Case Study; M. Stoeckler, A Short History of Emergence and Reductionism; E. Engeler, The Technical Problem of "Full Abstractness" as a Model for an Issue in Reductionism; J. Vuillemin, A Neutral Reduction: Analytical Method and Positivism; P. Weingartner, Reductionism and Reduction in Logic and in Mathematics; R. Morchio, Reductionism in Biology; H. Primas, Reductionism: Palaver without Precedent; G.G. Granger, Must a Science of Artificial Intelligence be Necessarily Reductionist?; P. Suppes, Can Psychological Software be Reduced to Physiological Hardware?; E. Klevakina, On the Problem of Reducing Value-Components in Epistemology.]

Agazzi, E. 2001. Philosophy of Nature and Natural Science. In: *Philosophia Naturalis*, 33(1&2):1-23.

Alexander, J.C. & Seidman, S. (Editors) 1990c. Culture and Society, Contemporary Debates. New York: Cambridge University Press.

Alexander, J.C. 1985. *Neofunctionalism*, London: Sage Publications

Alexander, J.C. 1987. *Sociological Theory since World War II, Twenty Lectures*. New York: Columbia University Press.

Alexander, J.C. 1988. *Action and its Environments*. New York: Columbia University Press.

Alexander, J.C. 1990. *Analytic Debates: Understanding the Relative Autonomy of Culture*. In: Alexander & Seidman 1990c (pp.1-27).

Alexander, J.C. 1990a. *Differentiation Theory and Social Change*. Co-editor Paul Colomy, New York: Columbia University Press.

Alexander, J.C. 1990b. Differentiation Theory: Problems and prospects. In: Alexander, 1990a (pp.1-15).

Alexandroff, P.S. 1956. *Einführung in die Mengenlehre und die Theorie der reellen Funktionen*. Berlin: Deutscher Verlag der Wissenschaften.

Althusius, J. (1603): *Politica Methodice Digesta* [Collection of Systematic Political Views] (1603, 3rd Edition 1614). The Politics of Johannes Althusius; an abridged translation of the 3rd ed. of *Politica Methodice Digesta*, Atque exemplis sacris et Profanis illustrata / Althusius, Johannes; translated, with an introduction by Frederick S. Carney, 1965. London: Eyre and Spottiswoode. See also the complete translation: Althusius, J. 1979. *Politica Methodice Digesta of Johannes Althusius* (Althaus), with an introduction by Carl Joachim Friedrich. New York: Arno Press.

Altner, G. & Hofer, H. 1972. *Die Sonderstellung des Menschen*, Stuttgart: Fischer.

Altner, G. (Editor) 1973. *Kreatur Mensch, Moderne Wissenschaft auf der Suche nach dem Humanen*, München.

Altner, G. 1976. *The Nature of Human Behaviour*. London: Allen & Unwin.

Angelelli, I. 1984. Frege and Abstraction. In: *Philosophia Naturalis*, Vol.21, Part II (pp.453-471).

Antal, L. 1963. *Questions of Meaning*. The Hague: Mouton.
Apolin, A. 1964. Die Geschichte des Ersten und Zweiten Hauptzatzes der Wärmetheorie und ihre Bedeutung für die Biologie. In: *Philosophia Naturalis*, Volume 4.
Appleby, J., Covington, E., Hoyt, D., Latham M. Sneider, A. 1996. *Knowledge and Post-modernism in Historical Perspective*. New York: Routledge.
Arendt, H. 1953. Ideologie und Terror: Eine neue Staatsform. In: Seidel & Jenkner 1968 133-167.
Aristotle 1894. *Politica*, Text edition by F. Susemihl and R.D. Hicks: *The Politics of Aristotle, A revised text*, New York: McMillan & Co.
Aristotle. 2001. *The Basic Works of Aristotle*. Edited by Richard McKeon with an Introduction by C.D.C. Reeve. (Originally published by Random House in 1941). New York: The Modern Library.
Austin, J.L. 1962. *How to Do Things with Words*. New York: Oxford University Press.
Avey, A.E. 1929. The Law of Contradiction: Its Logical Status. In: *The Journal of Philosophy*, (26):519-526.
Ayer, A.J. 1959. *Logical Positivism* (Editor). New York: Free Press.
Ayer, A.J. 1959a. Verification and Experience. In: Ayer, 1959 (pp.228-243).
Ayer, A.J. 1967. *Language, Truth, and Logic*. London: V. Gollancz (First published January 1936; Second edition – revised and reset – 1946).
Barion, J. von 1986. *Grundlinien philosophischer Staatstheorie*. Bonn: Bouvier Verlag Herbert Grundmann.
Barth, K. 1957. *Church Dogmatics*, Vol.II, Part I, Editor: Bromiley and others, Edinburgh: Clark.
Barwise, J. 1986. The Situation in Logic – I. In: Marcus *et al.* 1986:183-203.
Basden, A. 2005. Brief comments on Stafleu's proposal for a new political aspect. In: *Philosophia Reformata*. 70(1):70-75.
Bavinck, H. 1918. *Gereformeerde Dogmatiek*. (4 Vols. unaltered reprint, 1967) Kampen: Kok.
Beardsley, M.C. 1958. *Aesthetics*. New York: Harcourt, Brace & World.
Beck, H. 1999. *Metaphysiche Implikationen im Konstruktiven Realismus*. In: Wallner 1999.
Becker, O. (Editor), 1965a. *Zur Geschichte der griechischen Mathematik*, Wege der Forschung, Band 43, Darmstadt (pp. 271-307).
Becker, O. 1964. *Grundlagen der Mathematik in geschichtlicher Entwicklung*. Freiburg: Alber.
Becker, O. 1965. Preface. In: *Zur Geschichte der griechischen Mathematik, Wege der Forschung*. Volume 43, Darmstadt: Wissenschaftliche Buchgesellschaft.
Beckner, M. 1971. Organismic Biology. In: *Man and Nature, Philosophical Issues in Biology*, New York: Dell.
Behe, M.J. 2003. Design in the Details: The Origin of Biomolecular Machines. In: Campbell and Meyer, 2003 (pp.287-302).
Behe, M.J. 2003a. *Darwin's Black Box. The Biochemical Challenge to Evolution*. New York: The Free Press.
Bell, R., Edwards, D.V. and Wagner, R.H. 1969. *Political Power: a reader in theory and research*. New York: Free Press.
Benacerraf, P. and Putnam, H. 1964 (Eds.). *Philosophy of Mathematics, Selected Readings*. Oxford: Basil Blackwell.
Bennet, J. 1974. *Kant's Dialectic*. Cambridge: University Press.
Berberian, 1994. *A First Course in Real Analysis*. New York: Springer.
Berger, P.L. 1967. *The Social Construction of Reality, A Treatise in the Sociology of Knowledge*, New York: Anchor Books.
Berkeley, G. 1710. *The Principles of Human Knowledge*, Edited and introduced by G.J. Warnock. Oxford: The Fontana Library (1969).
Berlinski, D. The Deniable Darwin. In: Campbell and Meyer, 2003 (pp.157-177).
Bernays, P. 1974. Concerning Rationality. In: *The Philosophy of Karl Popper*, The Library of Living Philosophers, Volume XIV, Book I. Edited by P.A. Schilpp. La Salle. Illimois: Open Court.

Bernays, P. 1976. *Abhandlungen zur Philosophie der Mathematik*. Darmstadt: Wissenschaftliche Buchgesellschaft.

Bernstein, Richard J. 1983. *Beyond Objectivism and Relativism*. Science, hermeneutics and praxis. Philadelphia: University of Pennsylvania Press.

Beth, E.W. 1965. *Mathematical Thought*. New York: D. Reidel Publishing Company.

Bierens De Haan, J.D. 1947. *Politeia, Groote Mannen over Staat en Maatschappij*. Amsterdam: Elsevier.

Birtwistle, G.M 1996. Filosofie van de Kunst en de Aesthetica. In: *Kennis en Werkelijkheid*. Edited by René van Woudenberg and others. Amsterdam: Buijten & Schipperhijn (pp. 342-370).

Black, M. 1949. *Language and Philosophy*. Ithaca: Cornell University Press.

Black, M. 1979. *More About Metaphor*. In: Ortony 1979.

Bock, W.J. 1989. Animal kingdom. In: McGraw-Hill, 1989. *Concise Encyclopedia of Science & Technology*. New York: McGraw-Hill Publishing Company (pp. 102-103).

Böckenförde, E-W., 1976 (Editor). *Staat und Gesellschaft*. Wege der Forschung Volume CDLXXI. Darmstadt: Wissenschaftliche Buchgesellschaft.

Bodin, J. 1981. *Sechs Bücher über den Staat*, Buch I-III. Übersetzt und mit Anmerkungen versehen von Bernd Wimmer, Eingeleitet und herausgegeben von P.C. Meyer-Tasch. München: Verlag C.H. Beck.

Bohatec, J. 1940. Autorität und Freiheit in der Gedankenwelt Calvins. In: *Philosophia Reformata*, 5(1):1-28.

Böhme, G. 1966. Unendlichkeit und Kontinuität. In: *Philosophia Naturalis*, Volume 11, (pp.304-317).

Bolzano, B. 1920. *Paradoxien des Unendlichen* (1851). Leipzig: Reclam.

Born, M., Pyrmont, B. and Biem, W. 1967-1968. Dualismus in der Quantentheorie. In: *Philosophia Naturalis*, Volume 10 (pp.411-418).

Borsche, T. 1976. Individuum, Individualität. In: *Historisches Wörterbuch der Philosophie*, Eds. J. Ritter, & K. Gründer & G. Gabriel, Volume 4 (pp.310-323). Basel-Stuttgart: Schwabe & Co.

Bos, A.P. 1994. H. Dooyeweerd en de Wijsbegeerte van de Oudheid. In: Geertsema *et al.*, 1974 (pp.197-227).

Bousquet, G.-H. 1965. Vilfredo Pareto. In: Recktenwald, 1965:390-406.

Bowell, T. and Kemp, G. 2005. *Critical Thinking, A Concise Guide*. London: Routledge & Kegan Paul.

Boyer, C.B. 1959. *The History of the Calculus and its Conceptual Development*. New York: Dover.

Bracken, J.A. 2001. *The One in the Many, A Contemporary Reconstruction of the God-World Relationship*. Grand Rapids: William B. Eerdmans.

Brand, P.A. 2000. The Architecture of Semantic Domains. A grounding hypothesis in Cognitive Semiotics. In: *Revista Portuguesa de Humanidades*. Volume 4, Issue 1/2 (pp.11-51).

Brennan, S. & Noggle, R. 2000. *Rawls's Neglected Childhood*. In: Davion and Wolf, 2000 (pp.46-72).

Breuer, T. 1997. Universell und unvollständig: Theorien über alles? In: *Philosophia Naturalis*. 34:1-20.

Bril, K.A. 1973. A Selected and Annotated Bibliography of D.H. Th. Vollenhoven. In: *Philosophia Reformata*, Year 38 (pp.212-228).

Brouwer, L.E.J. 1907. *Over de Grondslagen der Wiskunde*. Amsterdam: Maas & Van Suchtelen.

Brouwer, L.E.J. 1911. Beweis der Invarianz der Dimensionenzahl. *Mathematische Annalen*, LXX (pp.161-165).

Brouwer, L.E.J. 1919. *De Onbetrouwbaarheid der Logische Principes*. In: Brouwer, 1919a.

Brouwer, L.E.J. 1919a. *Wiskunde, Waarheid, Werkelijkheid*. Groningen: Noordhoff.

Brouwer, L.E.J. 1924. Bemerkungen zum natürlichen Dimensionsbegriff. In: Brouwer, 1976, (2)554-557.

Brouwer, L.E.J. 1964. Consciousness, Philosophy, and Mathematics. In: Benacerraf et.al., 1964, pp.78-84.

Brouwer, L.E.J. 1976. *Collected Works*. Edited by Hans Freudenthal. Amsterdam: North-Holland.

Brunner, E. 1972. *Dogmatik I, Die Christliche Lehre von Gott* (first edition 1946), Zürich: Theologischer Verlag.

Buri, F., Lochman, J.M. & Ott, H. 1974. *Dogmatik im Dialog, Theologie, Offenbarung, Gotteserkenntnis*, Gütersloh: Gütersloher Verl.-Haus Mohn.

Burley, J. 2004. *Dworkin and his Critics, with Replies by Dworkin*. Oxford: Blackwell Publishing.

Butts, R.E. and Brown, J.R. (Eds.) 1989. *Constructivism and Science*. Dordrecht: Kluwer.

Buytendijk, F.J.J. 1970. *Mensch und Tier*. Hamburg: Rowohlt.

Cahn, S.M., Kitcher, P., Sher, G., and Fogelin, R.J. (General Editor). 1984. *Reason at Work*. New York: Harcourt Brace Jovanovich.

Calhoun, C., Gerteis, J., Moody, J., Pfaff, S., and Virk, I. (Eds). 2002. *Contemporary Sociological Theory*. Oxford: Blackwell Publishing.

Cameron, A.M. 2000. Implications of Dooyeweerd's Encyclopedia of Legal Science. In: Dooyeweerd, 2000:191-238.

Campbell, J.A. and Meyer, S.C. 2003. *Darwinism, Design, and Public Education*. East Lansing: Michigan State University Press.

Cannon, W.B. 1929. Organization for Physiological Homeostasis. In: *Physiological Review*, 9:397.

Cantelon, H., Gruneau, R. 1982. *Sport, Culture and the Modern State*. Toronto: University of Toronto Press.

Cantor, G. 1895. Beiträge zur Begründung der transfiniten Mengenlehre. In: *Mathematische Annalen*, Volume 46 (pp.481-512) and 1897 Volume 49 (pp.207-246).

Cantor, G. 1962. *Gesammelte Abhandlungen Mathematischen und Philosophischen Inhalts*. Hildesheim: Oldenburg Verlag (1932).

Carnap, R. 1922. *Der Raum. Ein beitrag zur Wissenschaftslehre*. Berlin: Reuther & Reichard.

Carney, F.S. 1965. The Politics of Johannes Althusius; an abridged tr. of the 3rd Edition of *Politica methodice digesta* [Collection of Systematic Political Views], with an introduction by Frederick S. Carney. London: Eyre and Spottiswoode.

Cassirer, E. 1910. *Substanzbegriff und Funktionsbegriff*. (Berlin), Darmstadt: Wissenschaftliche Buchgesellschaft, 1969.

Cassirer, E. 1911. *Das Erkenntnisproblem in der Philosophie und Wissenschaft der neueren Zeit*. Volume II. Berlin: Verlag Bruno.

Cassirer, E. 1912. Das Problem des Unendlichen und Renouviers 'Gesetz der Zahl', In: *Philosophische Abhandlungen, Hermann Cohen zum 70sten* Geburtstag (4 Juli 1912) Darmstadt.

Cassirer, E. 1923. *Das Erkenntnisproblem*. Vol.III. Berlin: Bruno.

Cassirer, E. 1928. Zur Theorie des Begriffs. In: *Kant-Studien*, Vol. 33:129-136.

Cassirer, E. 1929. *Philosophie der symbolischen Formen*. Vol.III. Berlin: Bruno.

Cassirer, E. 1944. *An Essay on Man*. New Haven, Conn.: Yale University Press.

Cassirer, E. 1953. *Substance and Function*. First edition of the English translation of *Substanzbegriff und Funktionsbegriff*: 1923; (First German edition 1910). New York: Dover.

Cassirer, E. 1957. *Das Erkenntnisproblem in der Philosophie und Wissenschaft der neueren Zeit*. Stuttgart: Kohlhammer Verlag.

Cassirer, E. 1971. *Das Erkenntnisproblem in der Philosophie und Wissenschaft der neueren Zeit*. Volume Two. 3rd edition, Darmstadt: Wissenschaftliche Buchgesellschaft.

Cathrein, V. 1909 (2nd Edition). *Recht, Naturrecht und positives Recht. Eine kritische Untersuchung der Grundbegriffe der Rechtsordnung*. Freiburg im Breisgau: Herder (first edition 1901).

Catton, W.R. 1966. *From Anamistic to Naturalistic Sociology*. New York: McGraw-Hill.

Chandler, A.D. 1979. *The Visible Hand*. Cambridge MA: Harvard University Press.

Chaplin, J. 2007. "Public Justice" as a critical political norm. In: *Philosophia Reformata*. 72(2):130-150.

Chase, G.B. 1996. *How has Christian Theology furthered Mathematics?* In: *Facets of Faith and Science*, Volume 2, The Role of Beliefs in Mathematics and the Natural Sciences: An Augustinian Perspective, edited by Jitse M. Van der Meer, New York: University of America Press (pp.193-216).

Chihara, C. 2004. *A Structural Account of Mathematics*. Oxford: Clarendon Press.

Cilliers P. 1998. *Complexity and Postmodernism: Understanding complex systems*. London: Routledge.

Clark, M. 2002. *Paradoxes from A to Z*. London: Routledge.

Cloeren, H.J. 1984. Ockham's razor. In: *Historisches Wörterbuch der Philosophie*, Eds. J. Ritter, K. Gründer & G. Gabriel. Volume 6 (pp.1094-1096). Basel-Stuttgart: Schwabe & Co.

Clouser, R.A. 2005. *The Myth of Religious Neutrality: An Essay on the Hidden Role of Religious Belief in Theories*. Notre Dame: University of Notre Dame Press (new revised edition, first edition 1991).

Clouser, R. 2005a. A Critique of Historicism. In: *Relativity and Relativism, Acta Academica Supplementum* 2 (guest editor Strauss, D.F.M.), pp.1-19.

Coase, R.H. 1937. The Nature of the Firm, *Economica*, 4(1937):386-405.

Coetzee, P.H. 1955. *Die Plek van die Staatsadministrasie in Friedrich Darmstädter se Regs- en Staatsbeskouing*. Amsterdam: Teerhuis & Klinkenberg.

Coley, N.G. and Hall, M.D. (Eds.) 1980. *Darwin to Einstein: Primary Sources on Science and Belief*, Harlow, Essex: Longman in association with the Open University Press.

Collins, F. 2007. *The Language of God*. New York: Simon & Schuster.

Cooper, W.E. 1989. Some Philosophical Aspects of Leisure Theory. In: Jackson and Burton 1989 (pp.49-68).

Copi, I.M. 1994. *Introduction to logic*, 9th Edition. New York: Macmillan.

Corijn, E. 1998. *Verkenningen in de Ontwikkeling van de Studie van de Vrijetijd*. Brussel: VUB Press.

Cornford, F.M. 1966. *The Republic of Plato*. Translated with Introduction and Notes. Oxford: Clarendon Press.

Coseriu, E. 1978: *Einführung in die strukturelle Betrachtung des Wortschatzes* (originally published in 1966/1973). In: Geckeler: 1978.

Critcher, C. 1995. *Sociology of leisure: a reader*. London: E & F. N. Spon.

Croce, B. 1953. *Aesthetic as Science of Expression and General Linguistic*. 3rd edition of revised edition (translated from the definitive Italian edition from 1920). New York: Noonday.

Croon, J.H. 1974. Pericles. In: *Grote Winkler Prins*, Volume 15 (pp.248-249).

Cunningham, F. 2002. *Theories of Democracy*. London: Routledge.

Dacqué, E. 1935. *Organische Morphologie and Phylogenie*, Berlin: Walter de Gruyter.

Dacqué, E. 1940. *Die Urgestalt*, Leipzig: Im Insel Verlag.

Dacqué, E. 1948. *Vermächtnis der Urzeit*, München: Leibniz Verlag.

Dahl. R. 1969. *The Concept of Power*. In: Bell, Edwards and Wagner.

Dante, A. 1943. Over de Monrachie, translation contained in Bierens de Haan, 1943 (pp.147-154).

Dantzig, T. 1954. *Number: the Language of Science*. (First edition 1947). New York: The Free Press.

Darwin, C. 1968. *On the Origin of Species by Means of Natural Selection or the Preservation of favoured races in the struggle for life* (1859), Edited with an Introduction by J.W. Burrow. Hardmondsworth: Penguin Books 1968. The version available on the WEB (downloaded on October 29, 2005):

http://www.infidels.org/library/historical/charles_darwin/origin_of_species/Intro.html slightly differs in some respects from the Penguin edition.

Darwin, C. 2005. *On the Origin of Species by Means of Natural Selection or the Preservation of favoured races in the struggle for life* (1859b). WEB version: http://www.infidels.org/library/historical/charles_darwin/origin_of_species/Intro.html (accessed on October 29, 2005).

Davidson, D. 1986. *Inquiries into Truth and Interpretation*. Oxford: Oxford University Press.

Davion, V. and Wolf, C. 2000. *The idea of Political Liberalism: Essays on Rawls*. New York: Roman & Littlefield.

Dawkins, R. 1986. *The Blind Watchmaker*. Burntmill: Longman Scientific & Technical.

De Brugh, M. 1995. Oren om Rechtop te lopen. In: *Wetenschap, Cultuur en Samenleving*, Year 24th, Nr.1/2, January-February (pp.19-22).

De Graaff, A.H. 1980. Psychology: Sensitive openness and appropriate reactions. In: *Journal for Christian Scholarship*. 1980 – 16(3&4)135-152.

De Klerk, W.J. 1978. *Inleiding tot die Semantiek*, Durban: Butterworth.

De Saussure, F. 1966. *Course in general Linguistics*. Edited by Charles Bally and Albert Sechehaye, London: McGraw-Hill.

De Vleeschauwer H.J. 1952. *Handleiding by die Studie van die Logika en die Kennisleer*, Pretoria: Uitgewery J.J. Moerau & Kie.

Dedekind, R. 1887. *Was sind und was sollen die Zahlen*, 10th ed., 1969. Braunschweig: Friedrich Vieweg & Sohn.

Dekker, C., Meester, R., and Van Woudenberg, R. 2005. *Schitterend Ongeluk of Sporen van Ontwerp*? Kampen: Ten Have.

Dennet, D.C. 1978. *Brainstorms, Philosophical Essays on Mind and Psychology*, Sussex.

Dennet, D.C. 1987. *The Intentional Stance*, Cambridge: Mass., MIT Press.

Depew, D. 2003. Intelligent Design and Irreducible Complexity: A Rejoinder. In: Campbell and Meyer, 2003 (pp.441-454).

Depew, D.J. and Weber, B.H. (Editors) 1986. *Evolution at a Crossroads: the New Biology and the new Philosophy of Science*, First Printing 1985. Cambridge, MA.: MIT.

Derrida, J. 1982. Signature Event Context. In: *Margins of Philosophy*, Translated, with Additional Notes, by Alan Bass, Chicago: University of Chicago (pp.307-329).

Derrida, J. 1993. *Aporias*. Translated by Thomas Dutoit. Stanford: Stanford University Press.

Derrida, J. 1997. *Deconstruction in a Nutchell, A Conversatin with Derrida*. Edited with a commentary by John D. Caputo (The Villanova Roundtable). New York: Fordham University Press.

Derrida, J. 2002. *Force of Law*. In: Derrida, J. *Acts of Religion*, London: Routledge (pp.230-298).

Descartes, R. 1965. *A Discourse on Method, Meditations and Principles*, translated by John Veitch, Introduced by A.D. Lindsay. London: Everyman's Library.

Descartes, R. 1965a. *The Principles of Philosophy*. In: Descartes 1965.

Descartes, R. 1976: *Descartes's Conversation with Burman*. Translated by J. Cottingham, Oxford: University Press.

d'Espagnat, B. 2006. *On Physics and Philosophy*. Princeton: Princeton University Press.

Dicey, A.V. 1927. *Law of Constitution*, 8th edition, London: MacMillan Company.

Diels, H. and Kranz, W. 1959-60. *Die Fragmente der Vorsokratiker*. Vols. I-III. Berlin: Weidmannsche Verlagsbuchhandlung.

Diemer, A. 1970. Zur Grundlegung eines allgemeinen Wissenschaftsbegriffes. In: *Zeitschrift für allgemeine Wissenschaftstheorie*. Volume 1/2:209-227.

Dilthey, W. (1927): *Der Aufbau der geschichtliche Welt in den Geisteswissenschaften*. Reprint of the Berlin Edition (Gesammelte Werke, Vol.V, 1927). Göttingen: VandenHoeck & Ruprecht.

Dingler, H. 1967. *Die Grundlagen der Naturphilosophie*. Darmstadt: Wissenschaftliche Buchgesellschaft.

Diwald, H. 1963. *Wilhelm Dilthey, Erkenntnistheorie und Philosophie der Geschichte.* Göttingen: Musterschmidt-Verlag.

Dixon, D. *et al.*, 1988. *The Macmillan Illustrated Encyclopedia of Dinosaurs and Prehistoric Animals.* New York: Macmillan, 1988.

Dobb, M.H. 1969. Kapitalismus. In: *Sowjetsystem und demokratische Gesellschaft*, Vol.III, (Edited by C.D. Kernig). Freiburg: Herder.

Dobzhansky, T. and Ayala, F.J. 1974. *Studies in the Philosophy of Biology. Reduction and Related Problems.* Berkeley: University of California Press.

Dobzhansky, Th. 1967. *The Biology of Ultimate Concern.* New York: New American Library.

Dooyeweerd, H. 1926. *De Beteekenis der Wetsidee voor Rechtswetenschap en Rechtsphilosophie* (*The Significance of the Cosmonomic Idea for the Science of Law and Legal Philosophy*), Inaugural Lecture Free University. Amsterdam October 15.

Dooyeweerd, H. 1928. Het juridisch causaliteitsprobleem in 't licht der Wetsidee. In. *Antirevolutionaire Staatkunde.* 1928(1):21-121.

Dooyeweerd, H. 1930. De Structuur der Rechtsbeginselen en de Methode der Rechtswetenschap in het Licht der Wetsidee (The Structure of legal principles and the method of the science of law in the light of the Cosmonomic Idea). In: *Wetenschappelijke Bijdragen, Aangeboden door Hoogleraren der Vrije Universiteit ter Gelegenheid van haar Vijgtigjarig Bestaan.* Amsterdam: De Standaard (pp.225-266).

Dooyeweerd, H. 1931. *Crisis in de Humanistsche Staatsleer.* Amsterdam: N.V. Boekhandel W. Ten Have.

Dooyeweerd, H. 1932. Norm en Feit, In: *Themis*, 93(E), pp.155-214.

Dooyeweerd, H. 1935-1936. *De Wijsbegeerte der Wetsidee.* 3 Vols. Amsterdam: Paris.

Dooyewerd, H. 1936. The Problem of Time and its Antinomies on the Immanence Standpoint I, *Philosophia Reformata*, Year 1, 2nd Quarter, pp.65-83.

Dooyewerd, H. 1939. The Problem of Time and its Antinomies on the Immanence Standpoint II, *Philosophia Reformata*, Year 4, 1st Quarter, pp.1-28.

Dooyewerd, H. 1940. The Problem of Time in the Philosophy of the Cosmonomic Idea I, *Philosophia Reformata*, Year 5, 3rd Quarter, pp.160-192.

Dooyewerd, H. 1940a. The Problem of Time in the Philosophy of the Cosmonomic Idea I, *Philosophia Reformata*, Year 5, 4th Quarter, pp.193-234.

Dooyeweerd, H. 1941. De transcendentale critiek van het wijsgeerig denken en de grondslagen van de wijsgeerige denkgemeenschap van het avondland. In: *Philosophia Reformata*. 6:1-20.

Dooyeweerd, H. 1948. *Transcendental Problems of Philosophic Thought.* Grand Rapids: WM. B. Eerdmans Publishing Compnay.

Dooyeweerd, H. 1949. *Reformatie en Scholastiek in de Wijsbegeerte*, Volume I, Het Grieksche Voorspel. Franeker: T. Wever.

Dooyeweerd, H. 1958. *Encyclopaedie der Rechtswetenschap.* Part I, Amsterdam: Bureau Studenteraad. [The date 1958 indicates the year in which Judge G.F. de Vos Hugo received his copy. It is identical to the copy that I have purchased in 1970 at the Free University of Amsterdam (available since 1967). It is also identical to an early sixties Volume sent to my father by Dooyeweerd (my father studied with Dooyeweerd in the thirties).]

Dooyeweerd, H. 1959. Schepping en Evolutie. In: *Philosophica Reformata*, 24:113-159.

Dooyeweerd, H. 1960. Van Peursen's Critische Vragen bij "A New Critique of Theoretical Thought." *Philosophia Reformata*, 25(3&4):97-150.

Dooyeweerd, H. 1962. *Verkenningen in de Wijsbegeerte, de sociologie en de Rechtsgeschiedenis.* Amsterdam: Buijten & Schipperheijn.

Dooyeweerd, H. 1967 (to be published): *The Encyclopedia of the Science of Law*, Collected Works of Herman Dooyeweerd, A Series Vol. 9, General Editor D.F.M. Strauss; Special Editor Alan Cameron [The Dutch texts comprise (a) The Introduction, (b) The Historical Volume, (c) The Systematic Volume, (d) The Distinction between Public Law and Private Law and (e)

The Theory of the Sources of Positive Law. References to the Dutch text will be to 1967-I (historical part) and 1967-II (systematic part).] Lewiston: The Edwin Mellen Press.

Dooyeweerd, H. 1996. *Christian Philosophy and the Meaning of History*, Collected Works, B Series, Volume 1, General Editor D.F.M. Strauss. Lewiston: Edwin Mellen.

Dooyeweerd, H. 1997. *A New Critique of Theoretical Thought*, Collected Works of Herman Dooyeweerd, A Series Vols. I-IV, General Editor D.F.M. Strauss. Lewiston: Edwin Mellen.

Dooyeweerd, H. 1997a. *Essays in Legal, Social and Political Philosophy*. Collected Works of Herman Dooyeweerd, B Series Volume 2, General Editor D.F.M. Strauss, Lewiston: Edwin Mellen.

Dooyeweerd, H. 1999. *In the Twilight of Western Thought*. Collected Works of Herman Dooyeweerd, B Series, Volume 4, General Editor D.F.M. Strauss, Special Editor J.K.A. Smith. Lewiston: Edwin Mellen.

Dooyeweerd, H. 2000. *Contemporary Reflections on the Philosophy of Herman Dooyeweerd*, Collected Works of Dooyeweerd, Series C – Dooyeweerd's Living Legacy, Edited by D.F.M. Strauss and Michelle Botting. Lewiston: Edwin Mellen

Dooyeweerd, H. 2002. *Encyclopedia of the Science of Law*, Volume I, *Introduction*, A Series, Volume 8, General Editor D.F.M. Strauss, Special Editor A.C. Cameron, Lewiston: Edwin Mellen.

Dooyeweerd, H. 2003. *Roots of Western Culture, Pagan, Secular and Christian Options*, B Series, Volume 3, General Editor D.F.M. Strauss, Lewiston: Edwin Mellen.

Dooyeweerd, H. 2004. *Political Philosophy*. D Series, Volume 1, General Editor D.F.M. Strauss, Lewiston: Edwin Mellen.

Dooyeweerd, H. 2004a. *Reformation and Scholasticism in Philosophy*, Vol.I, Collected Works of Herman Dooyeweerd, A Series, Volume 5, General Editor D.F.M. Strauss, Lewiston: Edwin Mellen.

Dooyeweerd, H. 2008. *The Struggle for a Christian Politics*, B Series, Volume 5, General Editor D.F.M. Strauss, Lewiston: Edwin Mellen.

Dorfman, R., Samuelson, P.A., Solow, R.M. 1958. *Linear Programming and Economic Analysis*. New York, McGraw-Hill

Drake, S. 1999. *Essays on Galileo and the History of Philosophy of Science*, Volume I. Toronto: University of Toronto Press.

Drewermann, E. 20-06. *Atem des Lebens, Die moderne Neurologie und the Frage nach Gott*. Volume I, *Das Gehirn*. Düsseldorf: Patmos.

Driesch, H. 1920. *Philosophie des Organischen*. Leipzig: E. Reinicke.

Driesch, H. 1931. *Wirklichkeitslehre: ein metaphysischer Versuch*. Leipzig: E. Reinicke.

Drieschner, M. 2004. Die Quantenmechanik – eine Revolution der Naturphilosphie? In: *Philosophia Naturalis*. 41(2):187-225).

Dummett, M.A.E. 1978. *Elements of Intuitionism*. Oxford: Clarendon Press.

Dummett, M.A.E. 1995. *Frege, Philosophy of Mathematics*. Second Printing. Cambridge: Harvard University Press.

Dunham, W. 1994. *The Mathematical Universe*. New York: John Wily & Sons Inc.

Dunleavy, P. and O'Leary, B. 1987. *Theories of the State. The Politics of Liberal Democracy*. London: MacMillan Education LTD.

Durkheim, E. 1972. *Selected Writings* (edited by A. Giddens), Cambridge: Cambridge University Press.

Dworkin, R. 2004. See Burley 2004.

Ebeling, G. 1974. *Der theologische Begriff von Gott*. In: F. Buri, J M Lochman & H Ott (eds.), Gütersloh: Gütersloher Verl.-Haus Mohn.

Ebeling, G. 1982. *Dogmatik des Christlichen Glaubens*, I. Tübingen: Mohr.

Ehler, S.Z. and Morrall, J.B. 1954. *Church and State Through the Centuries. A Collection of historic documents with commentaries*. London: Burns & Oates.

Eibl-Eibesfeldt, I. 1972. Stammesgeschichtliche Anpassungen im Verhalten des Menschen. In: Gadamer and Vogler, 1972 (pp.3-59).

Eibl-Eibesfeldt, I. 2004. *Grundriß der vergleichenden Verhaltensforschung, Ethologie*. 8th revised edition. Vierkirchen-Pasenach: Buch Vertrieb Blank GmbH.

Einstein, A. 1921. Geometrie und Erfahrung, *Sitzungsberichte der Preußischen Akademie der Wissenschaften*, First Half Volume, Berlin, pp.123-130.

Einstein, A. 1959. Autobiographical Notes. In: *Albert Einstein, Philosopher-Scientist*. Edited by P.A. Schilpp. New York: Harper Torchbooks.

Einstein, A. 1980. Herbert Spencer Lecture, Oxford 10. Juni 1933. In N.G. Coley & M.D. Hall (Hrsg.), *Darwin to Einstein: Primary Sources on Science and Belief*. Essex: Harlow.

Einstein, A. 1982. *Grundzüge der Relativitätstheorie*. Reprint of the 1969 Braunschweig edition (original edition 1922). Wiesbaden: Friedrich Fieweg & Sohn.

Einstein, A. 1985. *Relativity, the Special and General Theory*. Bristol: Arrowsmith (reprint of the first 1920 translation).

Einstein, A. 2006. *The Collected Papers of Albert Einstein*. Volume 10 (Translator: Ann Hentschel). Princeton: University Press.

Eisberg, R.M. 1962. *Fundamentals of Modern Physics* (Second edition). New York: Wiley.

Eisenstein, I. 1975. Ist die Evolutionstheorie wissenschaftlich begründet? In: *Philosophia Naturalis*, Archiv für Naturphilosophie und die philosophischen Grenzgebiete der exakten Wissenschaften und Wissenschaftsgeschichte, Vol.15, No.3, Part 1.

Eisenstein, I. 1975a. Ist die Evolutionstheorie wissenschaftlich begründet? In: *Philosophia Naturalis*, Archiv für Naturphilosophie und die philosophischen Grenzgebiete der exakten Wissenschaften und Wissenschaftsgeschichte, Vol.15, No.3, Part 2.

Eldredge, N. 1995. *Reinventing Darwin: The Great Debate at the High Table of Evolutionary Theory*. New York: Wiley.

Elton, H. 2007. Late Antiquity in the Mediterranean: The Collapse of the Roman Empire-Military Aspects, In: ORB Online Encyclopedia.

Engels, F. 1964. *Dialectics of Nature*. Translated from the German by Clemens Dutt. Moscow: Progress Publishers.

Evolgen Archive, 2009. *Random Mutation and Natural Selection*. WEB Site: http://evolgen.blogspot.com/2005/06/random-mutation-and-natural-selection.html (accessed 22-03-2009).

Ewing, A.C. 1962. *Ethics*. London: The English University Press LTD.

Felgner, U. (Editor) 1979. *Mengenlehre*. Darmstadt: Wissenschaftiche Buchgesellschaft.

Fern, R.L. 2002. *Nature, God and Humanity*, Cambridge: University Press.

Ferns, I. 2007. Can morality be regarded as a universal phenomenon? Comments from a psychological perspective. In: *Journal of Humanities* (Tydskrif vir Geesteswetenskappe), 47(2):155-164.

Fichter, J.H. 1968 (Edited by Erich Bodzenta): *Grundbegriffe der Soziologie*. Berlin: Springer-Verlag.

Fine, K. 2003. The Non-Identity of a Material Thing and Its Matter. *Mind*, Volume 112, (pp.196-234).

Fodor, J.D. 1977. *Semantics: Theories of Meaning in Generative Grammar*. Hassocks: Harvester Press.

Fourie, F.C.v.N. 1993. In the Beginning there Were Markets? (pp.41-65). In: *Transactions, Costs, Markets and Hierarchies*. Edited by Christos Pitelis. Oxford: Basil Blackwell.

Fowler, D. 1999. *The Mathematics of Plato's Academy*. Second Edition, Oxford: Clarendon Press.

Fraenkel, A. 1928. *Einleitung in die Mengenlehre*, 3rd expanded edition Berlin: Springer.

Fraenkel, A. 1967. *Lebenskreise – Aus der Erinnerungen eines jüdischen Mathematikers*. Stuttgart.

Fraenkel, A. 1968. *Abstract Set Theory*. Amsterdam: North Holland.

Fraenkel, A., Bar-Hillel, Y., Levy, A. & Van Dalen, D. 1973. *Foundations of Set Theory*, 2nd revised edition. Amsterdam: North Holland.

Fränkel, H. 1968. Zeno von Elea im Kampf gegen die Idee der Vielheit. In: *Um die Begriffswelt der Vorsokratiker*, *Wege der Forschung*, Band IX, Editor Hans-Gerog Gadamer, Darmstadt: Wissenschaftliche Buchgesellschaft (pp.425 ff.).

Fränkel, H. 1968a. *Wege und Formen frügrieschischen Denkens*. Herausgeber F. Tietze. (3rd revised edition). München: Beck.

Freeman, K. 1949. *Companion to the Pre-Socratic Philosophers*. Oxford: Basil Blackwell.

Freeman, S. 2003. *The Cambridge Companion to Rawls*, Cambridge: University Press.

Frege, G. 1884. *Grundlagen der Arithmetik*. Breslau: Verlag M & H. Marcus (Unaltered reprint, 1934).

Frege, G. 1893. *Grundgesetze der Arithmetik*, Vol. I. Jena (Unaltered reprint, Hildesheim: G. Olms, 1962).

Frege, G. 1895. Kritische Beleuchtung einiger Punkte in E. Schröders Vorlesungen über die Algebra der Logik, *Archiv für systematische Philosophie* (hrsg. P. Natorp). Vol. I., Berlin.

Frege, G. 1903. *Grundgesetze der Arithmetik*, Vol. II Jena (Unaltered reprint, Hildesheim: G. Olms, 1962).

Frege, G. 1979. *Posthumous Writings*. Oxford: Basil Blackwell.

Frege, G. 1983. *Nachgelassene Schriften* (ed. Hermes, H., Kambartel, F., Kaulbach, F.), 2nd expanded edition. Hamburg: Felix Meiner Verlag.

Friedrich, H. (Editor) 1973. *Mensch und Tier, Ausdruckformen des Lebendigen*. München: Deutscher Taschenbuch Verlag.

Fukuyama, F. 2000. *The Great Disruption, Human nature and the reconstitution of social order*. London: Profile Books Ltd.

Gadamer, H-G, & Vogler, P. 1972. *Neue Anthropologie*, Vol. II. Stuttgart: Georg Thieme.

Gadamer, H-G. 1967: *Kleine Schriften I, Philosophie / Hermeneutik*. Tübingen: Mohr.

Gadamer, H-G. 1989. *Truth and Method*, Second Revised Edition (first translated edition 1975). New York: The Continuum Publishing Company.

Gadamer, H.-G. 1991. Die Hermeneutik und der Diltheyschule. In: *Philosophische Rundschau*, 38(3), (pp.161-177).

Gadamer, H-G. 1998. *Truth and Method*. Second, Revised Edition. New York: Continuum.

Galilei, G. 1973. *Unterredungen und mathematische Demonstration über zwei neue Wissenszweige, die Mechanik und die Fallgesetze betreffend* (1638). Darmstadt: Wissenschaftliche Buchgesellschaft.

Galileo Gallilei 1638. *Dialogues and Mathematical Demonstrations concerning Two New Sciences*, the German translation. Darmstadt: Wissenschaftliche Buchgesellschaft, 1973.

Galileo, G. 1957. *Discoveries and opinions of Galileo*. Translated with an Introduction by S. Drake. Garden City, N.Y.: Doubleday.

Gardner, M. 1968. *Mathematical Puzzles and Diversions*. Harmondsworth: Penguin Books.

Geba, B.H. 1985. *Being at Leisure, Playing at Life: a Guide to Health and Joyful Living*. La Mesa, Calif.: Leisure Science Systems International.

Geckeler, H. (Editor.) 1978. *Strukturelle Bedeutungslehre*. Darmstadt: Wissenschaftliche Buchgesellschaft.

Geckeler, H. 1971. *Strukturelle Semantik und Wordfeldtheorie*. München: Fink.

Geertsema, H.G., Zwart, de Bruijn, J., van der Hoeven, J., Soeteman, A. (Editors) 1994. *Herman Dooyeweerd 1894-1977. Breedte en Actualiteit van Zijn Filosofie*. Kampen: Kok.

Gehlen, A. 1965. *Theorie der Willensfreiheit und frühe Philosophische Schriften*. Berlin: Luchterhand.

Gehlen, A. 1971. *Der Mensch; seine Natur und seine Stellung in der Welt*. Frankfurt: Athenaum.

Gentzen, G. 1965. *Die Widerspruchsfreiheit der reinen Zahlentheorie*. Darmstadt: Wissenschaftliche Buchgesellschaft (*Unaltered Reproduced Copy* of the original article which appeared in *Mathematische Annalen*, 112, 1935 – pp.493-565).

Giddens, A. 1983. *Central Problems in Social Theory*. London: The MacMillan Press.

Giddens, A. 1986. *The Constitution of Society*. Los Angeles: University of California Press (first published in 1984).

Giddens. A. 1998. *Interview Three – Structuration Theory* (with C. Piersen). In: Pierson, 1998 (pp.75-93).

Giddens, A. 2002. Some New Rules of Sociological Method, Section Reprinted in: Calhoun, C., Gerteis, J., Moody, J., Pfaff, S., and Virk, I. (Eds). 2002 (pp.226-231).

Gierke, Otto von 1915. *Die Grundbegriffe des Staatsrechts und die neuesten Staatsrechtstheorien*. Tübingen: Mohr.

Glahn, I. (Ed.) 2006. *Grosses Handbuch Formel-Sammlung, Mathematik, Physik, Anorganische Chemie*. München: Compact Verlag.

Glas, G. 2002. Churchland, Kandel and Dooyeweerd on the reducibility of mind states. *Philosophia Reformata* (pp.148-172).

Gödel, K. 1964. What is Cantor's Continuum Problem? In: Benacerraf & Putnam, 1964. (This article appeared in 1947 in the *American Mathematical Monthly*, 54, pp.515-525).

Goerttler, K. 1972. *Morphologische Sonderstellung des Menschen im Reich der Lebensformen auf der Erde. In:* Gadamer & Vogler, 1972, Volume 2 (pp.215-257).

Goodfield, J. 1974. Changing Strategies: A Comparison of Reductionist Attitudes in Biological and Medical Research in the Nineteenth and Twentieth Centuries. In: Dobzhansky and Ayala, 1974 (pp.65-86).

Gosztonyi, A. 1976. *Der Raum; Geschichte seiner Probleme in Philosophie und Wissenschaften*. Freiburg: Alber (Vols. 1 & 2).

Gould, S.J. 1977. Evolution's Erratic Pace, in: *Natural History*, Vol.86:5 (May 1977).

Gould, S.J. 1992. *Reflections in Natural History. Ever Since Darwin*. New York: W W Norton & Company.

Gould, S.J. 1994. *Further Reflections in Natural History, Hen's Teeth and Horse's Toes,* New York: W W Norton & Company.

Gould, S.J. 1996. *Life's Grandeur*. London: Vintage (Random House).

Graefe, A. and Parker, S. (Eds.) 1987. *Recreation and Leisure: an Introductory Handbook*. State College, Pa.: Venture.

Grau, A. 1999. "No Entity without Identity" – Schellings Identitätsbegriff im Lichte analytischen Denkens. In: *Kant-Studien*, 90, Heft 1 (pp.75-90).

Greene, B. 2003. *The Elegant Universe*. New York: W.W. Norton & Company Inc.

Grene, M. 1974. *The Understanding of Nature, Essays in the Philosophy of Biology*, (Boston Studies in die Philosophy of Science, Vol.XXIII), Boston.

Grene, M. 1986. *Perception, Interpretation, and the Sciences: Toward a New Philosophy of Science*. In: Depew, D.J. & Weber, B.H.: Evolution at Crossroads, 2nd Print, London.

Grene, M. 1986. Perception, Interpretation, and the Sciences: Toward a New Philosophy of Science. In: Depew & Weber 1986.

Grene, M. 1986a. Philosophy of Biology 1983: Problems and Prospects. In: Marcus *et al.*, 1983:433-452.

Griffioen, S. 1986. De Betekenis van Dooyeweerd's Ontwikkelingsidee, In: *Philosophia Reformata*, 51 (1&2) (pp.83-109).

Grondin, J. 1994. *Introduction to Philosophical Hermeneutics*, foreword by H-G Gadamer, New Haven: Yale University Press.

Grondin, J. 2003. *Hans-Georg Gadamer, A Biography*. Yale: Yale University Press.

Grünbaum, A. 1952. A Consistent Conception of the Extended Linear Continuum as an Aggregate of Unextended Elements. In: *Philosophy of Science*, Vol.19, nr.2, April (pp.288-306).

Grünbaum, A. 1967. *Modern Science and Zeno's Paradoxes*. Middletown: Wesleyan University Press.

Grünbaum, A. 1974. *Philosophical Problems of Space and Time*. Dordrecht (Holland): D. Reidel (second, enlarged edition).

Grundmann, H. 1960. *Vom Ursprung der Universität im Mittelalter*. Darmstadt.

Grünfeld, J. 1983. Euclidean Nostalgia. In: *International Logic Review*, June 1983, 27 (pp.41-50).

Guthrie, W.K.C. 1980. *A History of Greek Philosophy*. Volume II. *The Presocratic Tradition from Parmenides to Democritus*. Cambridge: Cambridge University Press.

Haas, J. 1959. *Das stammesgeschichtliche Werden der Organismen und des Menschen*, Vol.I, Basel: Herder.

Haas, J. 1959a. Naturphilosophische Betrachtungen zur Finalität und Abstammungslehre. In: Haas 1959.

Haas, J. 1968. *Sein und Leben, Ontologie des organischen Lebens*, Karlsruhe: Badenia Verlag.

Habermas, J. 1970. Gegen ein positivistisch halbierten Rationalismus. In: Adorno 1970 (pp.235-266).

Habermas, J. 1970a. *Erkenntnis und Interesse*. Frankfurt am Main: Suhrkamp Verlag.

Habermas, J. 1971. *Theorie und Praxis, Sozialphilosophische Studien*. 4th revised and expanded edition, Berlin: Suhrkamp Taschenbuch Verlag.

Habermas, J. 1981. *Theorie des kommunikativen Handelns*. Volume 1. Frankfurt am Main: Suhrkamp Verlag (Fourth revised edition, 1987 – 'Taschenbuch' edition 1995).

Habermas, J. 1984. *The Theory of Communicative Action: Reason and the Rationalization of Society. A critique of Functionalist Reason*. Volume 2. Boston: Beacon (translation of Habermas, 1981).

Habermas, J. 1987. *Theorie des kommunikativen Handelns*, Volumes I and II, Frankfurt am Main: Suhrkamp Tachenbuch edition 1995 (first edition 1981).

Habermas, J. 1990. Modernity versus Postmodernity, New German Critique 22, 1981:3-14. Reprintied in: Alexander, J.C. & Seidman, S. 1990c (pp.342-354).

Habermas, J. 1994. *The Past as Future*. Interviewed by Michael Heller. Translated by Peter Hohendahl, London: University of Nebraska Press.

Habermas, J. 1995-1 & 1995-2. *Theorie des kommunikativen Handelns*. [Theory of Communicative Action] Volume 1 & Volume 2. Frankfurt am Main: Suhrkamp Verlag (Fourth revised Edition, 1987 – 'Taschenbuch' Edition 1995 – originally published in 1981).

Habermas, J. 1996. *Between Facts and Norms: Contributions to a Discourse Theory of Law and Democracy*, translated by William Rehg; 2nd print, Cambridge: Massachusetts: MIT Press.

Habermas, J. 1998. *Faktizität und Geltung. Beiträge zur Diskurstheorie des Rechts und des demokratischen Rechtsstaats*. Frankfurt am Main: Suhrkamp edition (first edition 1992).

Habermas, J. 2001. *The Postcolonial Constellation, Political Essays*. Cambridge, Massachusetts: Polity Press.

Hacohen, M.H. 2002. *Karl Popper, The Formative Years 1902-1945*. Cambridge: University Press.

Haeffner, G. 1982. *Philosophische Antropologie*, Stuttgart [Appeared in 1989a as: *The human situation: a philosophical anthropology*; translated by Eric Watkins, Notre Dame, (Ind.): University of Notre Dame Press].

Hamilton, W. 1852. *Discussion on Philosophy and Literature*. London.

Hart, H. 1984. *Understanding our World*, Lanham: University Press of America.

Hart, H. 2000. Notes on Dooyeweerd, Reason and Order. In: Dooyeweerd, 2000 (125-146).

Hart, H.L.A. & Honoré, T. 1985. *Causation in the Law*. Oxford: Clarendon Press (third edition).

Hartmann, N. 1926. *Ethik*. Berlin und Leipzig: Walter de Gruyter.

Hartmann, N. 1957. *Kleinere Schriften*, Volume II, Berlin: Walter de Gruyter.

Hasse, H. & Scholtz, H. 1928. Die Grundlagenkrisis der griechischen Mathematik. In: *Kant-Studien*, Vol.33 (pp.4-35).

Hawking, S.W. 1988. *A Brief History of Time*. London: Transworld Publishers.

Hearne, V. 1986. *Adam's Task: Calling Animals by Name*. London.

Hegel, G.W.F. 1807. *Phänomenologie des Geistes*. Würzburg. Hamburg: Felix Meiner Verlag (Der philosophische Bibliothek, Band 114, Sixth Edition 1952).

Hegel, G.W.F. 1830. *Wissenschaft der Logik*, Volume II, (Herausgeber H. Glockner. Gesammelte Werke 1969).

Hegel, G.W.F. 1833. *Vorlesungen über die Geschichte der Philosopie*. Sämtliche Werke, Volume 17, Editor Hermann Glockner. Fourth edition of the 'Jubiläumausgabe', Stuttgart: Friedrich Frommann Verlag.

Hegel, G.W.F. 1957. *Sämtliche Werke*, Volume 2. Stuttgart: Friedrich Frommann Verlag.

Heidegger, M. 1969. *Identity and Difference*, New York: Harper & Row.

Heidegger, M. 1970. *Phänomenologie und Theologie*. Frankfurt: Klostermann.

Heidegger, M. 1992. *The Concept of Time*. Oxford: Blackwell.

Heisenberg, W. 1956. *Das Naturbild der heutigen Physik*. Hamburg: Rowohlt.

Heisenberg, W. 1958. *Physics and Philosophy. The Revolution in Modern Science*. New York: Harper Torchbooks.

Heisenberg, W. 1973. *Wandlungen in den Grundlagen der Naturwissenschaften*. Stuttgart: S. Hirzel (10th Edition).

Heitler, W. 1976. Über Komplimentarität von lebloser und lebender Materie, In: *Abhandlungen der Mathematisch-Naturwissenschaftliche Klasse*, year 1976, Number 1, Mainz/Wiesbaden.

Hemleben, J. 1974. *Ernst Haeckel, der Idealist des Materialismus*. Hamburg: Anthroposophischen Buchhandlung.

Hempel, C. 1959. The Empiricist Criterion of Meaning. In: Ayer, 1959 (pp.108-129).

Henke, W. & Rothe, H. 1980. *Der Ursprung des Menschen*. Stuttgart: Fischer.

Herder, J.G. (Editor Bernhard Suphan). 1877. *Herders sämmtliche Werke*. Berlin: Weidmann, (1877-1913).

Herder, J.G. 1978. *Johann Gottfried Herder, Abhandlung über den Ursprung der Sprache, Text, Materialen, Kommentar*. Herausgeber Wolfgang Proß, München: Carl Hanser Verlag.

Herold, N. 1974. Der Gesetzesbegriff in Philosophie und Wissenschaftstheorie der Neuzeit. In: *Historisches Wörterbuch der Philosophie*, Eds. J. Ritter, K. Gründer & G. Gabriel, Volume 3 (pp.501-514). Basel-Stuttgart: Schwabe & Co.

Hersh, R. 1997. *What is Mathemtics Really?* Oxford: Oxford University Press.

Heyns, J.A. (without date): *Algemene Inleiding tot die Etiek* (*General Introduction to Ethics*). Stellenbosch: University – Theological Faculty.

Heyns, J.A. 1970. *Die Nuwe Mens Onderweg*. Cape Town: Tafelberg.

Heyting, A. 1949. *Spanningen in de Wiksunde*. Groningen & Batavia: P. Noordhoff.

Heyting, A. 1971. *Intuitionism. An Introduction*. Amsterdam: North Holland Publishing Company.

Hilbert, D. 1924. Die Grundlagen der Physik. In: Hilbert 1970 (pp.258-289) [It appeared in *Mathematische Annalen*, Volume 92 (pp.1-32).]

Hilbert, D. 1925. Über das Unendliche, *Mathematische Annalen*, Vol.95, 1925: 161-190.

Hilbert, D. 1970. *Gesammelte Abhandlungen*, Vol.3, Second Edition, Berlin: Verlag Springer.

Hinrichs, C. 1957. *Einleitung*. In: Meinecke 1957.

Hobbes, Th. 1651. *Leviathan*, Harmondsworth: Pelican edition (1968).

Hofer, H. & Altner, G. 1972. *Die Sonderstellung des Menschen*. Stuttgart: G. Fischer.

Holz, Fr. 1975. Die Bedeutung der Methode Galileis für die Entwicklung der Transzendentalphilosophie Kants. In: *Philosophia Naturalis*, Vol. 15 (3):344-358.

Hommes, H.J. 1961. *Een Nieuwe Herleving van het Natuurrecht*, Zwolle: W.E.J. Tjeenk Willink.

Hommes, H.J. 1972. *De Elementaire Grondbegerippen der Rechtswetenschap*, Deventer: Kluwer.

Hommes, H.J. 1981. *Hoofdlijnen van de Geschiedenis der Rechtsfilosofie*, Deventer: Kluwer.

Homo habilis finds 1997. WEB site: [http://www.talkorigins.org/faqs/homs/a_habilis.html] (04/28/1997).

Horneffer, R. 1933. *Die Entstehung des Staates. Beiträge zum öffentlichen Recht der Gegenwart*. Tübingen: Mohr.

Hörz, H. 1967. Contribution on Physics in: *Naturforschung und Weltbild*, Berlin.
Hucklenbroich, P. 1980. Der physikalische Begriff der Materie. In: Ritter, *Historisches Wörterbuch der Philosophie*. Volume 5, Stuttgart: Schwabe & Co Verlag (pp.921-924).
Hume, D. 1962. *A Treatise of Human Nature*. London: Collins, The Fontana Library (First published in 1739).
Hurewicz, W. and Wallman, H. 1959: *Dimension Theory*, 5th edition. Princeton: University Press.
Husserl, E. 1891. *Philosophie der Arithmetik mit erganzenden Texten* 1890-1901. (Expansion of his *Habilitationsschrift*). Den Haag: Nijhoff (1970).
Husserl, E. 1900-1901. *Logische Untersuchungen* (2 Vols.), 2nd edition 1913. Husserliana, Vols. 18 & 19. The Hague: Nijhoff (1975-1984).
Husserl, E. 1913. *Ideen zu einer reinen Phänomenologie und phänomenologische Philosophie*, *Husserliana*, Band III, 1950. The Hague: Martinus Nijhoff.
Husserl, E. 1913a. *Ideen zu einer reinen Phänomenologie und phänomenologische Philosophie*, *Husserliana*, Band IV, 1952
Husserl, E. 1935. *Die Krisis des europaïschen Menschentums und die Philosophie*, contained in: Husserl, 1954. The Hague: Martinus Nijhoff.
Husserl, E. 1950. *Cartesianische Mediationen und Pariser Vorträge*, Edited and introduced by S. Strasser. Den Haag: Martinus Nijhoff, 1950.
Husserl, E. 1954. *Die Krisis der europäischen Wissenschaften und die Transzendentale Phänomenologie*, Husserliana Band VI, The Hague: Martinus Nijhoff.
Husserl, E. 1962. *Phänomenologische Psychologie*, Husserliana Volume IX. The Hague: Martinus Nijhoff.
Husserl, E. 1979. *Aufsätze und Rezensionen* (1890-1910), Husserliana, Edmund Husserl, Gesammelte Werke (Collected Works), Volume XXII, From the *Husserl Archives* in Leuven, guided by Samuel Ijsseling with the aid of Rudolf Boehm, Editor with text additions, Bernhard Rang, The Hague: Martinus Nijhoff.
Huxley, J. 1942. *Evolution: The Modern Synthesis*. New York: Harper & bros.
Huxley, J. 1968. *Evolution in Action*. Hammondsworth: Penguin Books.
Jackson, E.L. and Burton, T.L. (Eds.) 1989. *Understanding Leisure and Recreation: Mapping the Past, Charting the Future*. State College, Pa.: Venture.
Jacobson, T.L. and Servaes, J. (Editors), 1999. *Theoretical Approaches to Participatory Communication*. Cresskill, N.J.: Hampton Press.
Janich, P. 1975. Tragheitsgesetz und Inertialsysteem. In: *Frege und die moderne Grundlagenforschung*, red. Chr. Thiel, Meisenheim am Glan: Hain.
Jarvie, G. and Maguire, J. 1994. *Sport and Leisure in Social Thought*. London: Routledge.
Jaspers, K. 1948. *Philosophie*. Berlin: Springer Verlag.
Jellinek, G. 1966. *Allgemeine Staatslehre* (Dritte Auflage). Berlin: Verlag Dr. Max Gehlen.
Jenkinson, A.G. 1986. Die Diminutief in Afrikaans, Deel 1. In: *SA Journal of Linguistics*, Year 4, Nr.2, April (pp.39-62).
Jevons, F.R. 1964. *The Biochemical Approach to Life*. London: George Allen & Unwin Ltd.
John Paul II 1998. Encyclical Letter: *Fides Et Ratio*. To the Bishops of the Catholic Church on the relationship between Faith and Reason. WEB site:
http://www.vatican.va/holy_father/john_paul_ii/encyclicals/docu-
cents/hf_jp-ii_enc_15101998_fides-et-ratio_en.html (10/01/2009).
Jonas, H. 1973. *Organismus und Freiheit, Ansätze zu einer philosophischen Biologie*, München: Vandenboeck and Ruprecht.
Jonas, H. 1974. *Philosophical Essays: From Ancient Creed to Technological Man*. Prentice-Hall, INC.: Englewood Cliffs, New Jersey.
Jones, A. (Ed.) 1998. *Science in Faith, A Christian Perspective on Teaching Science*, Essex: Romford.
Kalsbeek, L. 1975. *Contours of a Christian Philosophy*. Toronto: Wedge Publishing Foundation.
Kammler, H. 1966. *Der Ursprung des Staates*, Eine Kritik der Ueberlagerungslehre, Köln.

Kant, I. 1781. *Kritik der reinen Vernunft*, 1st Edition (references to CPR A). Hamburg: Felix Meiner edition (1956).

Kant, I. 1783. *Prolegomena einer jeden künftigen Metaphysik die als Wissenschaft wird auftreten können*. Hamburg: Felix Meiner edition (1969).

Kant, I. 1785. *Grundlegung der Metaphysik der Sitten*. Darmstadt: Wissenschaftliche Buchgesellschaft (1968).

Kant, I. 1786. *Metaphysische Anfangsgründe der Naturwissenschaft*. Herausgeber Wilhelm Weischdel. Darmstadt: Wissenschaftliche Buchgesellschaft.

Kant, I. 1787. *Kritik der reinen Vernunft*, 2nd Edition (references to CPR B). Hamburg: Felix Meiner edition (1956).

Kant, I. 1788. *Critique of Practical Reason*. Darmstadt: Wissenschaftliche Buchgesellschaft (1968).

Kant, I. 1790. *Kritik der Urteilskraft* (1790, 1793, 1799). Darmstadt: Wissenschaftliche Buchgesellschaft (1968).

Kant, I. 1797-A-1798-B. *Die Metaphysik der Sitten. Metaphysische Anfangsgründe der Rechtslehre*. Immanuel Kant, Werke in Zehn Bände, Edited by Wilhelm Weischedel, Volume 7. Darmstadt: Wissenschaftliche Buchgesellschaft.

Kant, I. 1798-A-1800-B. *Anthropologie in pragmatischer Hinsicht*, Erster Teil, *Anthropologische Didaktik, Von der Art, das Innre sowohl als das Äussere des Menschen zu erkennen*. Immanuel Kant, Werke in Zehn Bände, Edited by Wilhelm Weischedel, Volume 10. Darmstadt: Wissenschaftliche Buchgesellschaft.

Kant, I. 1961. *Critique of Pure Reason*, Translated by F.M. Müller. New York: Dolphin Books.

Kapferer, N. 1998. Totalitarismus. In: *Historisches Wörterbuch der Philosophie*, Eds. J. Ritter, K. Gründer & G. Gabriel, Volume 10 (pp.1296-1300). Basel-Stuttgart: Schwabe & Co.

Katscher, F. 1970. Heinrich Hertz. In: *Die Grossen der Weltgeschichte*, Volume IX, *Röntgen bis Churchill*. München: R. Oldenburg.

Kattsoff, L.O. 1973. On the Nature of Mathematical Entities. In: *International Logic Review*, Number 7, 1973 (pp.29-45).

Kaufmann, F. 1957. Karl Jaspers und die Philosophie der Kommunikation. In: Schilpp 1957 (pp.193-284).

Kaufmann, F. 1968. *Das Unendliche in der Mathematik und seine Ausschaltung*. Darmstadt: Wissenschaftliche Buchgesellscahft (first edition 1930, Vienna: Franz Deuticke).

Kaufmann, F. 1981. Karl Jaspers and a Philosophy of Communication. In: Schilpp 1981 (pp.211-295).

Kawalec, P. 2006. Atomism, http://72.14.207.104/search?q=cache:oj3RfXVz2ysJ:www.kul.lublin.pl/etk/angielski/hasla/a/atomism.hmtl+Energeticism&en&ct=clnk&cd=9 (visited on 04-04-2006).

Kelsen, H. 1925. *Hauptprobleme der Staatsrechtslehre*. Berlin: Scientia Aalen 1960.

Kelsen, H. 1960. *Reine Rechtslehre*, Mit einem Anhang: *Das Problem der Gerechtigkeit*. Vienna: Verlag Franz Deuticke.

Kelsen, H. 1991. *General Theory of Norms*. Translated by Michael Hartney. Oxford: Clarendon Press.

Keynes, J.M. 1939. *The General Theory of Employment Interest and Money*. London: MacMillan and Co.

Kiontke, Siegfried 2006. *Physik biologischer Systeme, Die erstaunliche Vernachlässigung der Biophysik in der Medizin*. München: Mintzel.

Kitts, D.B. 1974. Paleontology and Evolutionary Theory. In: *Evolution*, 28.

Klapwijk, J. 2008. *Purpose in the Living World*. Cambridge: University Press.

Kleene, S.C. 1952. *Introduction to Metamathematics*. Amsterdam: North-Holland Publishing Company.

Klein, F. 1925. *Elementar Mathematik vom höheren Standpunkte aus, Geometrie*. Third Edition, Volume II. Belrin: Verlag Julius Springer.

Klein, F. 1932. *Elementary Mathematics from an Advanced Standpoint*. London: Macmillan.

Klein, F. 1939. *Elementary Mathematics from an Advanced Standpoint. Geometry*. London: Dover Publications.

Klein, G. 1990. The Atheist in the Holy City. Cambridge MA: MIT Press.

Kline, M. 1980. *Mathematics, The Loss of Certainty*, New York: Oxford University Press.

Kluxen, W. 1971. Analogie. In: *Historisches Wörterbuch der Philosophie*, eds. J. Ritter, K. Gründer & G. Gabriel, Volume 1, pp. 214-227. Darmstadt: Wissenschaftliche Buchgesellschaft.

Kobusch, Th. 1976. Individuum, Individualität. In: *Historisches Wörterbuch der Philosophie*, Eds. J. Ritter, K. Gründer & G. Gabriel, Volume 4 (pp.299-304). Basel-Stuttgart: Schwabe & Co.

Koehler, O. 1973. *Vom unbenannten Denken*. In: Friedrich 1973.

Koestler, A. and Smythies, J.R. (Eds.) 1972. *Beyond Reductionism*. New York: Macmillan.

König, G. 1972. Denkökonomie. In: *Historisches Wörterbuch der Philosophie*, Eds. J. Ritter, K. Gründer & G. Gabriel, Volume 2 (pp.108-109). Basel-Stuttgart: Schwabe & Co.

Körner, S. 1968. *The Philosophy of Mathematics*. London: Hutchinson & Co.

Korzybski, A. 1948. *Science and Sanity. An Introduction to non-Aristotelian Systems and General Semantics*. Connecticut: The International non-Aristotelian Library Publishing Company.

Koyzis, D.T. 2003. Political Visions & Illusions. A Survey & Christian Critique of Contemporary Ideologies. Downers Grove, Illinois: Intervarsity Press.

Kreitzer, M.R. 2007. Towards a Biblical Philosophy of Science. In: *Christianity and Society*, The Biannual Journal of the Kuyper Foundation, Vol. XVII, No.2, Winter 2007 (pp.6-19).

Kremer, K. 1966: *Die neuplatonische Seinsphilosophie und ihre Wirkung auf Thomas von Aquin*, Leiden: Brill.

Kropotkin, P.A. 1903. *Mutual Aid: A factor in Evolution*. Original Publisher: McClure Phillips & Co. New Edition 1972, *Foreword* by Ashley Montagu and *The struggle for existence*, by Thomas H. Huxley. A new introduction for the Garland ed. by Esther Kingston-Mann. New York: Garland.

Kropotkin, P.A. 1995. *Evolution and environment*. Edited with an *Introduction* by George Woodcock (Works, Vol.11). Montreal: Black Rose Books.

Krüger, H. 1966. *Allgemeine Staatslehre*. Zweite, durchgesehene Auflage. Berlin: W. Kohlhammer Verlag.

Kugler, R. 1967. *Philosophische Aspekte der Biologie Adolf Portmanns*. Zürich: Editio Academica.

Kuhlen, L. 1981. Zur Abgrenzung von Recht und Moral. In: *Analise & Kritik, Zeitschrift für Sozialwissenschaften*, hrsg. M. Baurman, A. Leist & D Mans, December, (pp.223-236).

Kuhn, T. 1970. *The Structure of Scientific Revolutions*, 2nd revised edition, Chicago: The University Press of Chicago.

Kuhn, T. 1974. *Second Thoughts on Paradigms*. In: Suppe, 1974.

Kuhn, T.S. 1977. *The essential tension: selected studies in scientific tradition and change*. Chicago: University of Chicago Press.

Kuhn, T.S. 1984. Objectivity, Value Judgments, and Theory Choice. In Cahn *et al.*, 1984 (pp.371-385).

Kushner, B.A. 2006. The Constructive Mathematics of A.A. Markov. In: *The American Mathematical Monthly*, Volume 113, Number 6 (pp.559-566).

Kuyper, A. 1880. *Souvereiniteit in eigen kring*. Amsterdam: Rodopi. (Presentation at the Inauguration of the Free University of Amsterdam, October 20, 1880.)

Laitman, J.T. 1985. Evolution of the upper respiratory tract: The fossil evidence. In: Tobias, 1985.

Lakoff, G. & Johnson, M. 1999. *Philosophy in the Flesh. The Embodied Mind and Its Challenge to Western Thought*. New York: Basic Books.

Lakoff, G. & Núñez, R.E. 2000. *Where Mathematics Comes From, How the Embodied Mind Brings Mathematics into Being*, New York: Basic Books.

Lambrecht, L. 1999. Demokratie. In: Sandkühler, H.G. 1999 217-225.

Landmann, M. 1962 (in Cooperation with G. Diem). *De Homine*. Freiburg: Alber.

Landmann, M. 1969. *Philosophische Anthropologie*, Berlin: Walter de Gruyter.

Landmann, M. 1973. *Philosophie: ihr Auftrag und ihre Gebiete*, Darmstadt: Carl Habel Verlagsbuchhandlung.

Lanfant, M.-F. 1974. *Sociologie van de Vrije Tijd*. Utrecht: Het Spectrum.

Lategan, L.O.K. and Smith, J.H. 2006. *Time and Context Relevant Philosophy. Festschrift* dedicated to D.F.M. Strauss. Bloemfontein: Association for Christian Higher Education.

Laugwitz D. 1986. *Zahlen und Kontinuum. Eine Einführung in die Infinitesimalmathematik*. Mannheim: B.I.-Wissenschaftsverlag.

Laugwitz D. 1997. Mathematische Modelle zum Kontinuum und zur Kontinuität. *Philosophia Naturalis*. 34:265-313.

Leakey, R.E. & Lewin, R. 1978. *People of the Lake. Mankind and its Beginnings*. Garden City, N.Y.: Anchor Press/Doubleday.

Leakey, R.E. 1973. Skull 1470, Discorvery in Kenya of the earliest suggestion of the genus Homo – nearly three million years old. In: *National Geographic*, Vo.143, No.6, June 1973.

Leftow, B. 2005. Eternity and Immutability. In: Mann, 2005 (pp.48-77).

Leibholz, G. 1954. Das Phänomen des totalen Staates. In: Seidel & Jenkner 1968 123-132. It originally appeared in: *Mensch und Staat in Recht und Geschichte*. Festschrift für Herbert Kraus. Kitzingen: Holzner, pp.156-162.

Leibniz, G.W.H. 1965. *Correspondence with Clarke*, Third Paper, published in the translation of M. Morris: *Leibniz, Philosophical Writings*, London: Everyman's Library.

Leibniz, G.W.L. 1976. *Philosophical Papers*. Edited by Leroy E. Loemker. Synthese Historical Library, Volume 2. Dordrecht-Holland: D. Reidel (first print 1969).

Leiminger, K. 1967. *Die Problematik der reinen Rechtslehre*. New York: Springer.

Leinfeller, W. (1966): Ueber die Karpelle verschiedener Magnoliales I, *Oesterreichische Botanische Zeitschrift*, 113, 1966.

Lemmon, E. J. 1968. *Introduction to axiomatic set theory*. London: Routledge & Kegan Paul.

Lennox, J.C. 2007. *God's Undertaker, Has Science Buried God*? Oxford: Lion.

Leys. C. 1982. Sport, the State and Dependency Theory, Response to Kidd. In: Cantelon and Gruneau 1982 (pp.305-315).

Litt, Th. 1948. *Staatsgewalt und Sittlichkeit*. München: Erasmus Verlag.

Locke, J. 1966. *Two Treatises of Civil Government*, London: Everyman's Library (1690).

Lonergan, B. 2000. *Insight: A study of Human Understanding*. Collected Works of Bernard Lonergan. Volume 3. Toronto: University of Toronto Press.

Lorenz, K. 1980. Mereologie. In: *Historisches Wörterbuch der Philosophie*, Eds. J. Ritter & K. Gründer. Volume 5. Basel-Stuttgart: Schwabe & Co (pp.1145-1148).

Lorenzen, P. 1960. *Die Entstehung der exakten Wissenschaften*. Berlyn: Springer-Verlag.

Lorenzen, P. 1968. Das Aktual-Unendliche in der Mathematik. In: *Methodisches Denken*, Frankfurt am Main: Suhrkamp Taschenbuch Wissenschaft (73), pp.94-119 – also published in Meschkowski, 1972:157-165.

Lorenzen, P. 1972. Das Aktual-Unendliche in der Mathematik, In: Meschkowski, 1972:157-165.

Lorenzen, P. 1976. Zur Definition der vier fundamentalen Meßgrößen. In: *Philosophia Naturalis*, Volume 16 (pp.1-9).

Lorenzen, P. 1989.Geometry as the Measure-Theoretic A Priori of Physics, in: Butts and Brown (Eds.), 1989:127-144.

Lotter, M.-S. 2000. Das individuele Gesetz. Zu Simmil's Kritik an der Lebensfremdheit der kantischen Moralphilosophie. In: *Kant-Studien*. Volume 2 (pp.178-203).

Loux, M.J. 2002. *Metaphysics: A Contemporary Introduction* (second edition). London: Routledge & Kegan Paul.

Lowe, E.J. 1998. *The Possibility of Metaphysics. Substance, Identity, and Time*. Oxford: Clarendon Press.
Lowie, R.H. 1921. *Primitive Society*. London: George Routledge & Sons.
Luckmann, T. & Berger, P.L. 1969. *The social construction of reality; a treatise in the sociology of knowledge*, London: Allen Lane.
Luhmann, N. 1973. *Zweckbegriff und Systemrationalität*. Frankfurt am Main: Suhrkamp-edition.
Luhmann, N. 1985. *A Sociological Theory of Law*. London: Routledge & Kegan Paul.
Luhmann, N. 1986. *Die soziologische Beobachtung des Rechts*. Frankfurt am Main: Metzner.
Luhmann, N. 1990. The Paradox of System Differentiation and the Evolution of Society. In: Alexander 1990a (409-440).
Luyten, N.A. (Editor) 1974. *Fortschritt im heutigen Denken*? Publication of the "Görres-Gesellschaft für Interdisziplinäre Forschung"; Series: *Grenzfragen*, Vol. 4. Freiburg: Alber.
Mac Lane, S. 1986. *Mathematics: Form and Function*. New York: Springer-Verlag,
MacIntyre, A. 1981. *After virtue: a study in moral theory*. Notre Dame, Ind.: University of Notre Dame Press.
MacIntyre, A. 1988. *Whose justice? Which rationality*? Notre Dame, Ind.: University of Notre Dame Press.
MacIver, R.M. 1964. *Social Causation*. First Edition, 1942. New York: Harper & Row.
MacKenzie, P. 2005. Elephant Environmental Impact, WEB-Site (accessed 30-07-2005): http://elephant.elehost.com/About_Elephants/Impact/impact.html.
Mäckler, A. (Hrsg). 2000. *1460 Antworten auf die Frage: was ist Kunst*? Köln: DuMont.
Mager, W. 1968. *Zur Entstehung des modernen Staatsbegriffs*. Akademie der Wissenschaften und der Literatur. Abhandlungen der Geistes- und sozialwissenschaftlichen Klasse. Wiesbaden: Akademie der Wissenschaften und der Literatur in Mainz in komission bei Franz Steiner Verlag GMBH.
Maier, A. 1949. *Die Vorläufer Galileis im 14. Jahrhundert*, Roma: Edizioni di Storia e letteratura.
Maier, A. 1964. Diskussion über das Aktuell Unendlichen in der ersten Hälfte des 14. Jahrhunderts. In: *Ausgehendes Mittelalter*, Vol. I, Rome: Roma: Edizioni di storia e letteratura.
Maier, H. 1972. Demokratie. In: *Historisches Wörterbuch der Philosophie*, Eds. J. Ritter, K. Gründer & G. Gabriel, Volume 2 51-55. Basel-Sturrgart: Schwabe & Co.
Maine, H.S. 1986. *Ancient Law: its Connection with the Early History of Society, and its Relation to Modern Ideas*. Tucson: University of Arizona Press.
Majerus, M.E.N. 1998. *Melanism: Evolution in Action*. Oxford: Oxford University Press.
Malherbe, D.F. 1947. Kuns – Selfstandig en Afhanklik [Art, Independent and Dependent]. In: *Philosophia Reformata*, 2nd Quarter (pp.67-85).
Malthus, T.R. 1798. *An essay on population*. The 1914 edition appeared again in 1960 with an introduction by Michael P. Fogarty. London: Dent.
Mann, W.E. (Editor) 2005. *The Blackwell Guide to the Philosophy of Religion*. Oxford: Blackwell Publishing.
Mannheim, K. 1982. *Structures of Thinking*. Edited by David Kettler, Volker Meja and Nico Stehr and translated by Jeremy J. Shapiro and Shierry Weber Nicholson, London: Routledge & Kegan Paul.
Marcus, R.B. and Dorn, J.W. and Weingartner, P. 1986. *Logic, Methodology and Philosophy of Science* VII, Proceedings of the seventh International Congres of Logic, Methodology and Philosophy of Science, Salzburg, 1983. Amsterdam: North Holland.
Margenau, H. 1982. *Physics and the Doctrine of Reductionism*. In: Agassi, J. & Cohen, R.S. (Editors) 1982.
Marion, J-L. 1982. *Dieu sans l'être: hors-texts*. Series: Théologiques. Paris: Communio/Fayard.
Marion, J-L. 1995. *Saint Thomas d'Aquin et l'onto-théologie*. Revue Thomiste, 95:31-66.
Marion, J-L. 1995a. *God Without Being*. Chicago: University of Chicago Press.
Marx, K. 1983. *Grundrisse der Kritik der Politischen Ökonomie*. Berlin: Dietz Verlag.

Maus, H. 1956. Geschichte der Soziologie. In: Ziegenfuss, 1956.

Mayr, E. 2002. *What Evolution Is*. London: Phoenix.

McGrath, A. & McGrath, J.C. 2007. *The Dawkins Delusion. Atheist fundamentalism and the denial of the divine*. London: Society for Promoting Christian Knowledge.

McGrath, A.E. 1999. *The Foundations of Dialogue in Science & Religion*, Oxford: Blackwell.

McIntire, C.T. (ed.) 1985. *The Legacy of Herman Dooyeweerd*. Lanham: University Press of America.

McIntire, C.T. 1985. Dooyeweerd's Philosophy of History in: McIntire, C.T. 1985 (pp.81-117).

McMullin, E. 1983. Values in Science, *Proceedings of the Philosophy of Science Association* (PS), Volume 2, 1983.

Mead, G.H. 1967. *Mind, Self, and Society, From a Standpoint of a social behaviorist*, Chicago: University of Chicago Press.

Meggle, G. 1999. Kommunikation/kommunikatives Handeln. In: *Enzyklopädie Philosophie*, herausgegeben von Hans JörgSankkühler. Hamburg: Felix Meiner Verlag.

Meier, C. 1970. *Entstehung des Begriffs Demokratie: vier Prolegomena zu einer historischen Theorie*. Frankfurt am Main: Suhrkamp Verlag.

Meier-Oeser, S. 1998. Symbol. In: *Historisches Wörterbuch der Philosophie*, Eds. J. Ritter, K. Gründer & G. Gabriel. Volume 10, (pp.710-723). Darmstadt: Wissenschaftliche Buchgesellschaft.

Meinecke, F. 1957. *Die Idee der Staatsräson in der neueren Geschichte*, Works Volume I, Editor Walther Hofer. München: R. Oldenburg Verlag.

Meinecke, F. 1965. *Die Entstehung des Historismus*, Works Volume III, Editor Carl Hinrichs. München: R. Oldenburg Verlag.

Mekkes, J.P.A. 1940. *Proeve eener Critische Beschouwin van de Ontwikkeling der Humanistische Rechtsstaatstheorieën*. Rotterdam: Libertas.

Menzel, A. 1929. *Beiträge zur Geschichte der Staatslehre*. Vienna & Leipzig.

Menzel, A. 1936. Griechische Staatssoziologie. In: *Zeitschrift für öffentliches Recht*, XVI.

Merleau-Ponty, M. 1970. *Phenomenology of Perception*, translated by Colin Smith, 5th Edition. London: Routledge & Kegan Paul.

Meschkowski, H. (Editor) 1972. *Grundlagen der modernen Mathematik*. Darmstadt: Wissenschaftliche Buchgesellschaft.

Meschkowski, H. 1967. *Problemen des Unendlichen*. Braunschweig: Vieweg

Meyer-Abich, A. 1964. The Historico-Philosophical Background of the Modern Evolution-Biology: nine Lectures delivered during October and November of 1960 at the Department of Zoology of the University of Texas in Austin, Texas USA, *Acta Biotheoretica. Supplementum* 10, Leiden: Brill.

Meyer-Tasch, 1981. *Einführung in Jean Bodins Leben und Werk*. In: Bodin, 1981 (pp.11-79).

Meynell, H.A. 2004. The philosophy of Dooyeweerd, a transcendental Thomist appraisal. In: Faith and Philosophy, *Journal for the society of Christian Philosophers*. 20(3) July 2003 (pp.265-87).

Mill, J.S. 1886. *System of Logic, Radiocinative and Inductive*. London: Longmans.

Mills, G.C., Lancaster, M. and Bradley, W.L. 2003. Origin of Life and Evolution in Biology Textbooks: A Critique. In: Campbell and Meyer, 2003 (pp.208-219).

Min-Sun, K. 2002. *Non-Western Perspectives on Human Communication: Implications for Theory and Practice*. Thousand Oaks, California: Sage Publications.

Modern Synthesis, 2009. http://www.answers.com/topic/modern-evolutionary-synthesis (accessed 19-03-2009).

Morriss, P.1987. *Power: A Philosophical Analysis*. Manchester: Manchester University Press.

Mortenson, T. 2006. *Origin of the Species, Was Darwin Right*? DVD Hebron: KY.

Müller, F.M. 1961. Translater of Kant's *Critique of Pure Reason*, Second Edition, Revised. New York: Dolphin Books, Doubleday & Company.

Müller, G.H. 2001: Umwelt, article in: *Historisches Wörterbuch der Philosophie*, eds. J. Ritter, K. Gründer & G. Gabriel, Volume 11, pp.99-107. Darmstadt: Wissenschaftliche Buchgesellschaft.

Munson, R. (Editor) 1971. *Man and Nature, Philosophical Issues in Biology*. New York: Dell.

Murphy, J. Concepts of Leisure. In: Graefe, A. and Parker, S. (Eds.) 1987 (pp.11-23).

Mutations, 2009. WEB Site:http://users.rcn.com/jkimball.ma.ultranet/BiologyPages/M/Mutations.html
(accessed 24-03-2009).

Narr, K.J. 1973. Kulturleistungen des frühen Menschen. In: Altner 1973.

Narr, K.J. 1974. Tendenzen in der Urgeschichtsforschung. In: Luyten, N.A. (Editor) 1974.

Narr, K.J. 1976. Cultural Achievements of Early Man. In: Altner, 1976.

Natorp, P. 1921. *Die logische Grundlagen der exakten Wissenschaften*, first edition 1910, Leipzig: B. G. Teubner (1969 reprint of the 1921 edition, Wiesbaden: Sändig).

Needham, J. 1968. *Order and Life*. 2nd edition, Cambridge, M.I.T. Press.

Neeman, U. 1986. Das Primat der Ontologie vor dem der Methodologie. In: *Philosophia Naturalis*, Band 23, Volume 1.

Neurath, O. 1973. *Empiricism and Sociology*. Edited by M. Neurath and R.S. Cohen, Boston: McGraw-Hill

Nida, E.A. 1979. *Componential analysis of meaning*. The Hague: Mouton.

Noble, T.E.X. 1984. *The Republic of St. Peter. The Birth of the Papal State* 680-825. Philadelphia: University of Pennsylvania Press.

Northrop, E.P. 1964. *Riddles in Mathematics*. Harmondsworth: Penguin Books.

Obojska, L. 2007. "Primary Relations" in a new foundational axiomatic framework. In: *Journal of Philosophical Logic* (2002) 36: 641-657, Springer 2007.

Oeing-Hanoff, L. 1976. Individuum, Individualität – Hoch- und Spätskolastik. In: *Historisches Wörterbuch der Philosophie*, Eds. J. Ritter, K. Gründer & G. Gabriel, Volume 4 (pp. 304-310). Basel-Stuttgart: Schwabe & Co.

Olthuis, J. 1989. The Word of God and Creation. In: *Journal for Christian Scholarship*, 1st and 2nd Quarter (pp.25-37).

O'Neill, J. 1995. *The Poverty of Postmodernism*. London: Routledge.

O'Neill, O. 2003. Constructivism in Rawls and Kant. In: Freeman, 2003 (pp.347-367).

Oparin, A.I. 1953. *Origin of Life*. New York: Dover Publications (1938).

Ortony, A. (ed.) 1979. *Metaphor and Thought*. New York: Cambridge University Press.

Ouwendorp, C. 1994. Het probleem van het universle en individuele. In: *Philosophia Reformata* 59(1):26-57.

Ouweneel, W.J. 1984. *Psychologie – Een Christelijk Kijk op het Mentale Leven*. Amsterdam: Buijten & Schipperheijn.

Ouweneel, W.J. 1986. *De leer van de mens. Proeve van een christelijk-wijsgerige antropologie*. Amsterdam: Buijten & Schipperheijn.

Overhage, P. 1972. *Der Affe in dir*, Frankfurt am Main: Josef Knecht.

Overhage, P. 1973. Die Evolution zum Menschen hin. In: *Gott, Mensch, Universum*. Graz: Styria.

Overhage, P. 1977. *Die biologische Zukunft der Menschheit*. Frankfurt am Main: Josef Knecht.

Pannenberg, W. 1973. *Wissenschaftstheorie und Theologie*. Frankfurt am Main: Suhrkamp.

Pareto, V. 1963. *The Mind and Society; a Treatise on General Sociology*. Edited by Arthur Livingston. Translated by Andrew Bongiorno and Arthur Livingston, with the advice and active cooperation of James Harvey Rogers. New York: Dover Publications (Dover-edition 1916).

Pareto, V. 1966. *Course d'Economie Politique* (1896), § 585, Translated by Finer, Vilvredo Pareto: *Sociological Writings*, London.

Parker, S. 1995. Towards a Theory of Work and Leisure. In: Critcher 1995 (pp.28-37).

Parsons, T. 1961. An Outline of The Social System. In: *Theories of Society*, Vol.I, edited by Parsons, Shills, Naegele and Pitts, New York: Free Press.

Parsons, T. 1967. *Sociological Theory and Modern Society*. New York: The Free Press.

Parsons, T. 1967a. Some Reflections on the Place of Force in Social Process. In: Parsons, 1967:264-296.

Parsons, T. 1969. *Politics and Social Structure*. New Haven: Yale University Press.

Parsons, T. 1977. *Social Systems and the Evolution of Action Theory*. New York: Free Press.

Parsons, T., Bales, R.F. and Shills, E.A. 1953. *Working Papers in the Theory of Action*. New York: Free Press.

Penrose, R. 2004. *The Road to Reality. A Complete Guide to the Laws of the Universe*. London: Vintage Books.

Picker, B. 1961. Die Bedeutung der Mathematik für die Philosophie Edmund Husserls, *Philosophia Naturalis*, Vol. 7, 1961.

Pierson, C. 1998. *Converations with Anthony Giddens: Making Sense of Modernity*. Cambridge: Polity Press.

Plamenac, M. 1970. Bio-physical Analysis of Vital Force of Living Matter. In: *Philosophia Naturalis*, Band 12.

Planck, M. 1910. Die Stellung der neueren Physik zur mechanischen Naturanschauung (Vortrag gehalten am 23. September 1910 auf der 82. Versammlung Deutscher Naturforscher und Ärzte in Königsberg i. Pr.). In: Max Planck, 1973:52-68.

Planck, M. 1913. Neue Bahnen der physikalischen Erkenntnis (Rede, gehalten beim Antritt des Rektorats der Friedrich-Wilhelms-Universität). In: Max Planck, 1973:69-80.

Planck, M. 1926. Physkalische Gesetzlichkeit (Vortrag, gehalten am 14. Februar 1926 in den akad. Kursen von Düsseldorf). In: Planck, 1973: 183-205.

Planck, M. 1973. *Vorträge und Erinnerungen*, 9th reprint of the 5th edition. Darmstadt: Wissenschaftliche Buchgesellschaft.

Plantinga, A. 1993. *Warrant and Proper Function*. Oxford: Oxford University Press.

Plato, 1973. Edited by E. Hamilton and C. Huntington: *The Collected Dialogues of Plato; Including the Letters*. Princeton: University Press.

Plessner, H. 1965². *Die Stufen des Organischen und der Mensch, Einleitung in die philosophische Antropologie*. Berlin: Walter de Gruyter.

Plessner, H. 1975. *Autobiographical article: Helmut Plessner. In:* Pongratz 1975.

Plessner, H. 1975a. Zur Anthropologie der Sprache. In: *Philosophia Naturalis*, Vol.15, Section 4.

Plumpe, G. 1974. Die religiöse und theologische Bedeutung des Gesetzesbegriffs. In: *Historisches Wörterbuch der Philosophie*, Eds. J. Ritter, K. Gründer & G. Gabriel, Volume 3 (pp.494-495). Basel-Stuttgart: Schwabe & Co.

Pojman, L.P. (Ed.) 2000. *Life and Death: a Reader in Moral Problems*. Belmont, Calif.: Wadsworth Publishing Co.

Polak, L. 1921. *De zin der vergelding: een strafrechts-filosofies onderzoek*. Volume 1. Amsterdam: Emmering.

Polanyi, M. 1968. *Personal knowledge; Towards a Post-Critical Philosophy*. Chicago: University of Chicago Press (electronic edition 2003).

Pongratz, L.J. 1975. *Philosophie in Selbstdarstellungen*, Hamburg: Felix Meiner.

Popper, K. 1966. *The Open Society and its Enemies*, Vol. I & II, London: Routledge & Kegan Paul.

Popper, K.R. 1974 Scientific Reduction and the Essential Incompleteness of All Science. In: Dobzhansky 1974 (pp. 259-284).

Portmann, A. 1965. *Vom Urpsrung des Menschen*, Basel: Schwabe.

Portmann, A. 1967. *Probleme des Lebens*, Eine Einführung in die Biologie, Basel.

Portmann, A. 1969. *Biologische Fragmente zu einer Lehre vom Menschen*, 3rd expanded edition, Basel.

Portmann, A. 1969a. *Einführung in die vergleichende Morphologie der Wirbeltiere*, 4th expanded edition, Stuttgart.

Portmann, A. 1970. Der Mensch ein Mängelwesen? Chapter in: *Entlässt die Natur den Menschen?* München.
Portmann, A. 1970a. Ein Wegbereiter der neuen Biologie, Vorwort. In: Jakob von Uexküll 1970 (pp.IX-XXI).
Portmann, A. 1973. *Biologie und Geist*, Frankfurt am Main.
Portmann, A. 1973a. Der Weg zum Wort. In: *ERANOS* Vol 39, Leiden 1973.
Portmann, A. 1974. *An den Grenzen des Wissens, von Beitrag der Biologie zu einem neuen Weltbild*. Wien: Econ.
Portmann, A. 1975. Homologie und Analogie, Ein Grundproblem der Lebensdeutung. In: *ERANOS* Vol.42, Leiden.
Portmann, A. 1977. Die biologischen Grundfragen der Typenlehre. In: *ERANOS* Volume 43, Leiden.
Portmann, A. 1990. *A zoologist looks at humankind*, translated by Judith Schaefer. New York: Columbia University Press.
Prantl, C. 1851. *Über die Entwicklung der aristotelischen Logik aus der platonischen Philosophie*, presentation, April 5, 1851 (made available by the *Wissenschaftliche Buchgesellschaft*, Darmstadt).
Prantl, C. 1855. *Geschichte der Logik im Abendlande*, Volume I, Leipzig: S. Hirzel (Unaltered reprint in 1955; Graz. Akademische Druck- U. Verlagsanstalt).
Prinsloo, E.D. 1989. Logic and culture. In: *South African Journal for Philosophy*, Volume 8, (pp.94-99).
Putnam, H. 1982. *Reason, Truth and History*. Cambridge: Cambridge University Press.
Quételet, L.A.J. 1835. *Sur l'homme et le développement de ses facultés ou essai de physique sociale*. Paris: Bachelier. A new edition appeared in 1869: Quételet, L.A.J. 1869. *Physique sociale, ou: Essai sur le développement des facultés de l'homme*. Bruxelles: Muquardt.
Quine, W. V.O. 1958. *Methods of Logic*. London: Routledge & Kegan Paul.
Quine, W.V.O. 1953. *From a Logical Point of View*. Cambridge Massachusetts: Harvard University Press.
Quine, W.V.O. 1970. *Philosophy of Logic*. Englewood Cliffs: Prentice Hall.
Rashdall, H. 1936. *The Universities of Europpe in the Middel Ages*, Volume I.
Ratzenhofer, G. 1898. *Die sociologische Erkenntnis. Positive Philosophie des socialen Lebens*. Leipzig – reprint Amsterdam: Liberea, 1968.
Rawidowicz, S. 1964 (second edition). *Ludwig Feuerbachs Philosophie, Ursprung und Schicksal*, (first edition 1930). Berlin: Walter de Gruyter & Co.
Rawls, J. 1978. *A Theory of Justice*. Oxford: Oxford University Press (first edition 1971).
Rawls, J. 1996. *Political Liberalism*. Revised Edition. Cambridge: Harvard University Press.
Rawls, J. 1999. *A Theory of Justice*. Revised Edition, Cambridge: Harvard University Press.
Rawls, J. 2001. *Collected Papers*. Edited by Samuel Freeman. Cambridge Massachusetts: Harvard University Press.
Rawls, J. 2002. *The Law of Peoples*. Harvard: Harvard University Press.
Recktenwald, H.C. (Editor) 1965. *Lebensbilder großer Nationalökonomen*, Berlin: Kiepenheuer & Witsch.
Regan, T. 1997. *The Case for Animal Rights*. Los Angeles: University of California Press (first edition 1983).
Reichling, A. 1967. *Het Woord*. Zwolle (1967²) (1935 – Nijmegen: Berkhout).
Reichling, A. 1970. *Wat wel in Van Dale staat, Onze Taal*, Nov./Dec. 1970, Vol.39, nr. 11 & 12.
Reid, C. 1970. *Hilbert*, With an appreciation of Hilbert's mathematical work by Hermann Weyl. New York: George Allen & Unwin.
Reiss, H. 1966. *Politisches Denken in der deutschen Romantik*. Bern: Francke Verlag.
Rensch, B. & Schultz, A.H. (Editors) 1968. *Handgebrauch und Verständigung bei Affen und Frühmenschen*. Symposium der Werner-Reimers-Stiftung für anthropogenetische Forschung. Bern: Huber.

Rensch, B. 1968. Discussion Remarks, attached to Von Bertalanffy: Symbolismus und Anthropogenese. In: Rensch, B. & Schultz, A.H. (Eds.) 1968.

Rensch, B.1971. *Biophilosophy*, New York: Columbia University Press.

Rensch, B. 1973. *Gedächtnis, Begriffsbildung und Planhandlungen bei Tieren*, Hamburg: Parey.

Rice, M. 2000. What is Science? In: Dooyeweerd, 2000:239-270.

Rickert, H. 1913. *Die Grenzen der naturwissenschaftlichen Begriffsbildung*, Tübingen: Mohr (1902) 1913^2.

Rickert, H. 1924. *Das Eine, die Einheit und die Eins*. Tübingen: Mohr.

Rickert, H. 1929. *Zur Lehre von der Definition*, 3rd improved edition. Tübingen: Mohr.

Riedel, M. 1976. Der Begriff der "burgerlichen Gesellschaft" und das Problem seines geschichtlichen Ursprungs. In: Böckenförde, 1976 (pp.77-108).

Riedweg, C. 2005. *Pythagoras: His Life, Teaching, and Influence*. Ithaca: Cornell University Press.

Riezler, K. 1981. Play and Seriousness. In: Lüschen, G.R.F and Sage, G.H. 1981 (pp.439-451).

Ritter, J. 1971-2001 (Editor). *Historisches Wörterbuch der Philosophie*. Volumes 1-12, Stuttgart: Schwabe & Co.

Robinson, A. 1966. *Non-Standard Analysis*. Amsterdam: North-Holland (second edition, 1974).

Robinson, A. 1979. *Selected Papers of Abraham Robinson*. Edited by H.J. Keisler. New Haven: Yale University Press.

Roelofse, J.J. 1982. *Signs and Significance*. Johannesburg: McGraw-Hill.

Rollwagen, W. 1962. *Das Elektron der Physiker*. Munich: Max Hüber.

Romein, J. 1947. *Universiteit en Maatschappij in de Loop der Tijden*. Leiden.

Roper, D. 1992. *The Reformational Contribution to Aesthetic Theory*. In: *Issues*, No.7, November 1992 (pp. 3-31).

Rorty, R. 1989. *Contingency, Irony and Solidarity*. New York: Cambridge University Press.

Rossouw, H.W. 1993. *Universiteit, Wetenskap en Kultuur*, red. Anton Van Niekerk. Kaapstad: Tafelberg.

Rossouw, D. 2003. *Intellectual Tools: skills for the human sciences* (tanslated by Carina Fourie). Pretoria: Van Schaik.

Rousseau, J.J. 1966. *The Social Contract and Discourses*, translated by G.D.H. Cole, London: Everyman's Library.

Rousseau, J.J. 1975. *Du Contrat Social et Autres Oeuvres Politiques*. Paris: Editions Garnier Fréres.

Ruse, M. and Travis, J. 2009. *Evolution: The First Four Billion Years*. London: Cambridge Massachusettes.

Russell, B. 1897. *An Essay on the Foundations of Geometry*. Cambridge: University Press.

Russell, B. 1920. *Introduction to Mathematical Philosophy*. London: George Allen & Unwin Ltd. (First edition 1919).

Russell, B. 1956. *The Principles of Mathematics*. London: George Allen & Unwin. (First published in 1903, Second edition 1937, Seventh edition 1956).

Ryan, A. 1970. *The Philosophy of the Social Sciences*, New York: Pantheon Books.

Ryle, G. 1977. *Dilemmas*, with a *Preface* by René Meyer. Pretoria: J.L. Van Schaik. (Achilles en die Skilpad – pp.50-69).

Salmon, W. (Ed.). 2001. *Zeno's Paradoxes*. New York: Bobs-Merrill. [Contributors: Resolution of the paradox, by A. Shimony. – Introduction, by W. C. Salmon. – The problem of infinity considered historically, by B. Russell. – The cinematographic view of becoming, by H. Bergson. – Achilles and the tortoise, by M. Black. – Achilles on a physical racecourse, by J. O. Wisdom. – Tasks and super-tasks, by J. Thomson. – Tasks, super-tasks, and the modern Eleatics, by P. Benacerraf. – Comments on Professor Benacerraf's paper, by J. Thomson. – Zeno and the mathematicians, by G. E. L. Owen. – Modern science and refutation of the paradoxes of Zeno. Zeno's metrical paradox of extension. Modern science and Zeno's paradoxes of motion. By A. Grünbaum. – Appendix: Sets and infinity, by W. C. Salmon. First published in 1970 by New York: Bobs-Merrill.]

Sandkühler, H. G. 1999. *Enzyklopädie Philosophie*. Hamburg: Felix Meiner.

Sargent, T.D., Millar, C.D. and Lambert, D.M. 1998. The 'Classical' Explanation of Industrial Melanism: Assessing the Evidence. In: *Evolutionary Biology*, 30 (299-322).

Scheele, P. 1997. *Degeneratie, Het einde van de Evolutietheorie*. Amsterdam: Buijten & Schipperheijn.

Scheler, M. 1962. *Die Stellung des Menschen im Kosmos* (1928). 6th edition, Bern-München: Francke.

Schelling, F.W.J. 1968. *Schriften von 1806-1813, Ausgewählte Werke*, Vol.4, Darmstadt.

Scherer, G. 1980. *Strukturen des Menschen, Grundfragen philosophischer Antropologie*, Essen: Ludgerus Verlag.

Schilpp, P.A. 1951 (red.). *Albert Einstein, Philosopher-Scientist*, London, Vol. I.

Schindewolf, O.H. 1969. *Ueber den "Typus" in morphologischer und phylogenetischer Biologie*, Wiesbaden.

Schindewolf, 1980. *Grundfragen der Paläontologie*, (first edition 1950 Stuttgart: Scheizerbart). New York: Arno Press.

Schlick, M. 1925. *Allgemeine Erkenntnislehre*. Second Edition. (Series: Naturwissenschaftliche Monographien und Lehrbücher, Volume 1). Berlin: Springer.

Schlick, M. 1959. The Turning Point in Philosophy. In: Ayer, 1959 (pp.53-59).

Schlick, M. 1974. *General theory of knowledge*. Translated by Albert E. Blumberg ; with an introduction by A.E. Blumberg and H. Feigl. New York: Springer-Verlag.

Schnatz, H. 1973. *Päpstliche Verlautbarungen zu Staat und Gesellschaft, Originaldokumente mit deutscher Uebersetzung*, Edited by H. Schnatz. Darmstadt: Wissenschaftliche Buchgesellschaft.

Schneider, E. 1962. *Einführung in die Wirtschaftstheorie*, IV. Teil, Ausgewählte Kapitel der Geschichte der Wirtschaftstheorie, I. Band, Tübingen: Mohr.

Schopenhauer, A. 1974. *On the Fourfold Root of the Principle of Sufficient Reason*, translation by E.F.J. Payne, La Salle, Illinois: Open Court (Original title: *Ueber die vierfache Wurzel des Satzes vom zureichenden Grunde*, 1813).

Schroeder, G.L. 1998. The Science of God. The convergence of scientific and biblical wisdom. New York: Free Press.

Schröder, E. 1877. *Der Operationskreis des Logikkalkuls*. Unaltered edition. Darmstadt: Wissenschaftliche Buchgesellschaft (1966).

Schrödinger, E. 1955. *What is Life? The Physical Aspect of the Living Cell*, Cambridge: University Press.

Schubert-Soldern, R. 1959. *Materie und Leben als Raum und Zeitgestalt*. München: Pustet.

Schubert-Soldern, R. 1962. *Mechanism and Vitalism: Philosophical Aspects of Biology* Edited by Philip G. Fothergill; foreword to the American Ed. by James P. Doll, Translation was made by C.E. Robin. Notre Dame: Indiana University of Notre Dame Press.

Schutz, H. 1974. *Der sinnhafte Aufbau der sozialen Welt*, Springer Verlag 1932, Frankfurt am Main: Suhrkamp.

Schuurman, E. 1993. Techniek, Technologie en het Technisch Wereldbeeld. In: Van der Ploeg *et al* (pp.191-206).

Schuurman, E. 1995. *Perspectives on Technology and Culture*. Sioux Center: Dordt College Press.

Schuurman, E. 2005. *The Technological World Picture and the Ethics of Responsibility, Struggles in the Ethics of Technology*. Dordt: Dordt College Press.

Schuurman, E. 2009. *Technology and the Future*. 2nd edition. Jordan Station: Paideia Press.

Schwartz, J.H. 1985. Toward a synthetic analysis of Hominid Phylogeny. In: Tobias, 1985.

Seerveld, C.G. 1958. *Benedetto Croce's Earlier Aesthetic Theories and Literary Criticism*. Kampen: J.H. Kok N.V.

Seerveld, C.G. 1968. *A Christian Ccritique of Art and Literature*. Toronto: The Association for Reformed Scientific Studies.
Seerveld, C.G. 1970. A Christian Tin-Can Theory of Man. In: *Journal of the American Scientific Affiliation*, Minnesota, August.
Seerveld, C.G. 1979 (Ed. John Kraay and Anthony Tol). Modal Aesthetics: Preliminary Questions with and Opening Hypothesis. In: *Hearing and Doing*, Philosophical Essays Dedicated to H. Evan Runner, Toronto: Wedge Publishing Foundation (pp.263-294).
Seerveld, C.G. 1980. *Rainbows for the Fallen World: Aesthetic Life and Aesthetic Task*, Toronto.
Seerveld, C.G. 1985: *Dooyeweerd's Legacy for Aesthetics: Modal Law Theory*. In: McIntire, C.T. (ed.) 1985 (pp. 41-79).
Seerveld, C.G. 1985. Dooyeweerd's Legacy for Aesthetics. Modal Law Theory. In: McIntire, 1985 (pp.41-79).
Seerveld, C.G. .1987. Imaginativity. In: *Faith and Philosophy*, Vol.4, Nr.1, January.
Seerveld, C.G. .2001. Christian aesthetic bread for the world. In: *Philosophia Reformata*. 66(2):155-177.
Seidel, B. and Jenkner, S. 1968 (Editors). *Wege der Totalitarismus-Forschung*. Darmstadt: Wissenschaftliche Buchgesellschaft.
Selous, F. 1881. *Hunter's Wanderings in Africa*. 1985 Edition. Alberton: Galago Publishing.
Shapiro, S. 1997. *Philosophy of Mathematics, Structure and Ontology*. Oxford: Oxford University Press.
Shapiro, S. 2005 (Editor). *The Oxford Handbook of Philosophy of Mathematics and Logic*. Oxford: Oxford University Press.
Sikkema, A. 2005. A Physicist's Reformed Critique of Nonreductive Physicalism and Emergence, *Pro Rege*, 33:4 (June):20-32.
Silver, B.L. 1998. *The Ascent of Science*. Oxford: Oxford University Press.
Simmonds, N.E. 1986. *Central Issues in Jurisprudence: Justice, Law and Rights*. London: Sweet & Maxwell.
Simpson, G.G. 1961. *The Major Features of Evolution*. New York: Columbia University Press.
Simpson, G.G. 1969. *Biology and Man*. New York: Harcourt.
Simpson, G.G. 1971. *Man's Place in Nature*, Section from "*The Meaning of Evolution*" (revised edition Yale University 1967), Reprinted in Munson, 1971.
Singer, P. 2000. *All Animals are Equal*. In: Pojman, 2000.
Singh, D. 1985. On Cantor's concept of set. In: *International Logical Review*, Nr.32, December.
Skillen, J.W. 1994. *Recharging the American Experiment. Principled Pluralism for Genuine Civic Community*. Grand Rapids, MI: Baker Book House.
Skillen, J.W. 2001. Politics in one world. In: *Philosophia Reformata*. 66(1):117-131.
Skillen, J.W. 2008. The Necessity of a Non-Reductionist Science of Politics, Presented at the conference sponsored by Metanexus Institute, "Subject, Self, and Soul: Transdisciplinary Approaches to Personhood" Madrid, Spain, July 13-17, 2008 (unpublished version, 17 pp.).
Skolem, Th. 1922. *Einige Bemerkungen zur axiomatischen Begründung der Mengenlehre*. In: Felgner, 1979 (pp.57-72).
Skolem, Th. 1929. Über die Grundlagendiskussionen in der Mathematik. Proceedings of the 7th Scandinavian Mathematical Congress, Oslo 1929, pp.3-21, Reproduced in: Felgner, 1979 (pp.73-91).
Smart, B. 2000. Postmodern Social Theory. In: Turner, 2000 (pp.447-480).
Smend, R. 1930. *Der Staat als Integration*. Berlin und Leipzig: W. de Gruyter & co.
Smith, A.L. 1964. *Church and State in the Middle Ages* (1905). New York: Barnes & Noble.
Smith, K.C. 1992. Neo-Rationalism versus Neo-Darwinism: Integrating Development and Evolution. In: *Biology and Philosophy* 7, 1992:431-451.
Sober, E. 1981. *Holism, individualism and the units of selection*. In: PSA 1980 (2):93-121.
Sober, E. 1987. *The Nature of Selection; Evolutionary Theory in Philosophical Focus*. London: The MIT Press.

Sokal, S. & Bricmont, J. 1998. *Fashionable Nonsense: Postmodern Intellectuals' Abuse of Science*, Picador, New York, 1998; cf. the German edition: *Eleganter Unsinn: Wie die Denker der Postmoderne die Wissenschaften missbrauchen*, München: C.H. Beck 1999.

Solmsen, Fr. 1962. Anaximander's Infinite, *Archiv für Geschichte der Philosophie*, Vol.44.

Sorokin, P. 1928. *Contemporary Sociological Theories*. New York and London: Harper & Bros.

Spann, O. 1937. *Naturphilosophie*, Jena: Verlag Gustav Fischer.

Spencer, H. 1968. *Reasons for Dissenting from the Philosophy of Comte and other Essays*. Berkeley: Glendessary Press.

Spencer-Brown, G. 1969. *Laws of form*. London: Allen & Unwin.

Spengler, O. 1923^2. *Der Untergang des Abendlandes; Umrisse einer Morphologie der Weltgeschichte*. Two Volumes. München: Beck.

Spivak, G.C. 1999. *A Critique of Postcolonial Reason. Toward a History of the Vanishing Present*. Cambridge: Harvard University Press.

Spruyt, C.B.: *Die Geschichte der Philosophie in Holland in den letzten zehn Jahren*. In: Archiv für die Geschichte der Philosophie, Hermann Diels, Wilhelm Dilthey, Benno Erdmann und Eduard Zeller, hrsg. Ludwig Stein, Volume II, Druck & Verlag von Georg Reinen, 1889.

Spykman, G.J. 1992. *Reformational theology: a new paradigm for doing dogmatics*. Grand Rapids: Eerdmans.

Stafleu, M.D. 1968. Individualiteit in de fysica. In: *Reflexies*, Opstellen aangeboden aan Prof. Dr. J.P.A. Mekkes. Amsterdam: Buijten & Schipperheijn.

Stafleu, M.D. 1972. Metric and Measurement in Physics. In: *Philosophia Reformata*, 37(1/2): 42-57.

Stafleu, M.D. 1980. *Time and Again, A Systematic Analysis of the Foundations of Physics*. Toronto: Wedge.

Stafleu, M.D. 1987. *Theories at Work: On the Structure and Functioning of Theories in Science, in Particular during the Copernican Revolution*, Lanham: University Press of America.

Stafleu, M.D. 1989. *De Verborgen Structuur*. Amsterdam: Buijten & Schipperheijn.

Stafleu, M.D. 1999. The idea of a natual law. In: *Philosophia Reformata*, 64 (1): 88-104.

Stafleu, M.D. 2002. *Een Wereld vol Relaties*. Amsterdam: Buijten & Schipperheijn.

Stafleu, M.D. 2002a. Evolution, History, and the individual character of a person. In: *Philosophia Reformata*, Volume 67(1):3-18.

Stafleu, 2004. On the character of social communities, the state and the public domain. In: *Philosophia Reformata*. 69(2):125-139.

Stafleu, 2006. Infinity and Continuity. In: Lategan & Smit, 2006 (pp.163-174).

Stanley, S.M. 1981. *The New Evolutionary Timetable: Fossils, Genes, and the Origin of Species*. New York: Basic Books.

Steffens, H.J. 1979. *James Prescott Joule and the concept of energy*. Folkstone, Eng.: Dawson. New York: Science History Publications.

Stegmüller, W. 1965. Das Universalienproblem eins und jetzt. In: *Archiv für Philosophie*, 6 (1956):192-225; 7 (1957):45-81 (made available combined in 1965 by the Wissenschaftliche Buchgesellschaft, Darmstadt in 1965 – translation available in Stegmüller, 1977:1-65).

Stegmüller, W. 1969. *Metaphysik, Skepsis, Wissenschaft*, (first edition 1954). Berlin/New York: Springer.

Stegmüller, W. 1969a. *Main Currents in Contemporary German, British and American Philosophy*. Dordrecht: D. Reidel Publishing Company, Holland.

Stegmüller W. 1977. *Collected Papers on Epistemology, Philosophy of Science and History of Philosophy*. Volumes I and II. Dordrecht-Boston: D. Reidel Publishing Company.

Stegmüller, W. 1987. *Hauptströmungen der Gegenwartsphilosophie*. Volume III, Stuttgart: Alfred Kröner Verlag.

Stellingwerf, J. 1971. *Inleiding tot de Universiteit*. Amsterdam: Buijten en Shcipperheijn.

Sterelny, K. 2001. *Dawkins vs. Gould, Survival of the Fittest*. London: Icon Books.

Sterelny, K. 2009. Philosophy of Evolutionary Thought. In: Ruse and Travis, 2009:313-329.

Stich, S.P. (ed.). 1975. *Innate Ideas*. Berkeley: University of California Press.

Stone, C.D. 1998a. *Should Trees Have Standing? – Toward Legal Rights for Natural Objects*. In: VanDeVeer, 1998 (pp.148-159).

Strauss, D.F.M. 1973. *Begrip en Idee*. Assen: Van Gorcum.

Strauss, D.F.M. 1980. *Inleiding tot die Kosmologie*, Bloemfontein: VCHO.

Strauss, D.F.M. 1981. Woord, Saak en Betekenis. In: *Acta Academica*, UFS, Bloemfontein (pp.5-34).

Strauss, D.F.M. 1982. The Place and Meaning of Kant's Critique of Pure Reason (1781) in the legacy of Western philosophy. In: *South African Journal of Philosophy*, Volume 1, (pp.131-147).

Strauss, D.F.M. 1983a. Individuality and Universality in Reformational Philosophy. In: *Reformational Forum*, 1(1):23-36.

Strauss, D.F.M. 1984. An analysis of the structure of analysis, (The Gegenstand-relation in discussion). In: *Philosophia Reformata*. 49(1): 35-56.

Strauss, D.F.M. 1985. Taal en Historiciteit als Bemiddelaars tussen Geloven en Denken. In: *Philosophia Reformata*. 50(2):130-148.

Strauss, D.F.M. 1991. The Ontological Status of the principle of the excluded middle. In: *Philosophia Mathematica* II, 6(1):73-90.

Strauss, D.F.M. 1991a. Hoe kan ons wetenskaplik oor God praat? In: Journal for Christian Scholarship, 2nd quarter (pp.23-43).

Strauss, D.F.M. 2000. Kant and modern physics. The synthetic a priori and the distinction between modal function and entity. In: *South African Journal of Philosophy*, 2000, pp.26-40.

Strauss, D.F.M. 2000a. The Order of Modal Aspects. In: Strauss & Botting, 2000 (pp.1-29).

Strauss, D.F.M. 2001. *Paradigms in Mathematics, Physics*, and *Biology – their Philosophical Roots*. Bloemfontein: Tekskor (Revised Edition, 2004).

Strauss, D.F.M. 2002. The scope and limitations of Von Bertalanffy's systems theory. In: *South African Journal of Philosophy*, Volume 21, (pp.163-179).

Strauss, D.F.M. 2002a. Philosophical Reflections on continuity. In: *Acta Academica*, 34(3) (pp.1-32).

Strauss, D.F.M. 2003. Frege's Attack on 'Abstraction' and his Defense of the 'Applicability' of Arithmetic (as Part of Logic). In: *South African Journal of Philosophy*, Volume 22, (pp.63-80).

Strauss, D.F.M. 2003a. Popper and the Achilles heel of positivism. In: *Koers*, Vol.68, Nr. 2 & 3(pp.255-278).

Strauss, D.F.M. 2005. *Paradigmen in Mathematik, Physik und Biologie und ihre philosophische Wurzeln*. Frankfurt am Main: Peter Lang.

Strauss, D.F.M. 2006. *Reintegrating Social Theory – Reflecting upon human society and the discipline of sociology*. Frankfurt am Main: Peter Lang.

Strauss, D.F.M. 2006a. The mixed legacy underlying Rawls's Theory of Justice, In: *Journal for Juridical Science*, 31(1):61-79.

Strauss, D.F.M. 2007. Did Darwin develop a theory of evolution in the biological sense of the word? In: *South African Journal of Philosophy*, Vol.26(2):190-203.

Strauss, D.F.M. and Botting, M. (Eds.) 2000. *Contemporary Reflections on the Philosophy of Herman Dooyeweerd*. Lewiston: The Edwin Mellen Press.

Strauss, H.J. 1965. Nie-Staatlike Owerheidstaak in Beskawingsamehang. *Philosophia Reformata*. 30(2-4):198-204.

Strayer, J.R. 1970. *On the Medieval Origins of the Modern State*. Hew Jersey: Princeton University Press.

Suppe, F. 1974: *The Structure of Scientific Theories*, Edited with a *Critical Introduction* by Frederick Suppe. Urbana: University of Illinois Press.

Sweeney, Leo S.J. 1972. *Infinity in the Presocratics: A Bibliographical and Philosophical Study*, The Hague.

Synthetic Theory of Evolution, 2009. *An Introduction to Modern Evolutionary Concepts and Theories. Micro and Macro Evolution.* WEB site: http://anthro.palomar.edu/synthetic/synth_9.htm (accessed 24-03-2009).

Sztompka, P. 1993. *The Sociology of Social Change*, Cambridge Massachusetts: Blackwell.

Tait, W. 2005. *The Provenance of Pure Reason, Essays in the Philosophy of Mathematics and Its History.* Oxford: University Press.

Talisse, R.B. 2001. *On Rawls.* Belmont:Wadsworth.

Taylor, C. 1989. *Sources of the Self, The making of the modern identity.* Cambridge, Massachusetts: Harvard University Press.

Teensma, E. 1969. *The Paradoxes.* Assen: Van Gorcum.

Titze, H. 1984. Zum Problem der Unendlichkeit. In: *Philosophia Naturalis*, Vol.21 (pp.139-156).

Tobias, P.V. 1985 (Editor). *Hominid Evolution,* New York: Liss.

Tol, T. 1995. Time and Change in Vollenhoven. In: *Philosophia Reformata*, 60(2): 99-120.

Tomlinson, A. 1981. *Leisure and Social Control.* Brighton: Chelsea School of Human Movement.

Tönnies, F. 1957. *Community and Society*, Harper & Row, translated and edited by C.P. Loomis, New York 1957 (a translation of Tönnies 1887/1972).

Tönnies, F. 1965. *Einführung in die Soziologie* (1931), third reprint, Stuttgart: Enke.

Tönnies, F. 1972. *Gemeinschaft und Gesellschaft*, Darmstadt: Wissenschaftliche Buchgesellschaft (first edition, 1887).

Topitsch, E. 1969. *Die Freiheit der Wissenschaft und der politische Auftrag der Universität.* Berlin: Luchterhand.

Traeger, L. 1904. *Der Kausalbegriff im Straf- und Zivilrecht.* Marburg: N.G. Elwart'sche Verlagsbuchhandlung.

Tribe, J. 1995. *The economics of leisure and tourism.* Oxford: Butterworth-Heinemann.

Trier, J. 1973. *Aufsätze und Vorträge zur Wortfeldtheorie.* Edited by von Anthony van der Lee and Oskar Reichmann, The Hague: Mouton.

Trincher, K. 1985. Die Dualität der Materie. In: *Philosophia Naturalis*, Vol.22, part 3 (pp. 329-342).

Troeltsch, E. 1922. Die Krisis des Historismus. *Die neue Rundschau* 33.

Troll, W. 1951. Biomorphologie und Biosystematik als typologische Wissenschaften, *Studium Generale* 4 (376-389).

Troll, W. 1973. *Allgemeine Botanik*, revised and extended edition, Stuttgart: Ferdinand Enke Verlag.

Troost, A. 2004. *Vakfilosofie van de Geloofswetenschap. Prolegomena van de Theologie.* Budel: Damon.

Turner, B.S. 2000. *The Blackwell companion to social theory.* Malden, Mass.: Blackwell Publishers.

Ungerer, E. 1966. *Die Wissenschaft vom Leben*, Band III, Der Wandel der Problemelage der Biologie in den letzten Jahrzehnten. Freiburg: Alber.

Unschuld, P. 2003. *Was ist Medizin? Westliche und östliche Wege der Heilkunst*, München: Beck.

Vaihinger, H. 1949. *The Philosophy of "As If."* London: Routledge & Kegan Paul (translated by C.K. Ogden).

Van Creveld, M. 1999. *The Rise and Decline of the State.* New York: Cambridge University Press.

Van den Beukel, A. 2005. Darwinisme: wetenschap en/of ideologie? In: Dekker, *et al.*, 2005 (pp.101-116).

Van der Hoeven, J. 1963. *Kritische Ondervraging van de Fenomenologische Rede*, Amsterdam: Buijten & Schipperheijn.

Van der Hoeven, J. 1997. Hartstocht en Rede (II). In: *Philosophia Reformata.* 62(2): 145-165.

Van der Ploeg, A., Dulhaart, M.H.J., Vlug, A.E. 1993. *Tastend Zien, Opstellen in dankbare herinnering opgedragen aan dr. F. De Graaff.* Sliedrecht: Merweboek.

Van Heerden, C. 1965. *Inleiding tot die Semantiek*, Johannesburg: Butterworth.

Van Huyssteen, J.W.V. 1997. *Essays in Postfoundationalist Theology*, Grand Rapids: William B. Eerdmans Publishing Company.

Van Huyssteen, J.W.V. 1998. *Duet or Duel? Theology and Science in a Portmodern World*. Trinity Press International. Pennsylvania: Harrisburg.

Van Huyssteen, J.W.V. 1999. *The Shaping of Rationality. Towards Interdisciplinarity of Theology and Science*. Grand Rapids: William B. Erdmanns.

Van Huyssteen, W.J.V. 2006. *Alone in the World? Human uniqueness in science and theology*. Grand Rapids: William B. Eerdmans.

Van Melsen, A.G.M. 1975. Atomism. In: *Encyclopedia Britannica*, 15th edition, London, Volume 2, pp.346-351.

Van Niekerk, M. 1993. *Sin- en Identiteitsproblematiek in die Kuns van George Grosz*, unpublished Ph.D., Potchefstroom University.

Van Riessen, H. 1970. *Wijsbegeerte*. Kampen: J.H. Kok N.V.

Van Stigt, W.P. 1990. *Brouwer's Intuitionism*. Amsterdam: North Holland.

Van Woudenberg, R. 2000. *Het Mysterie van de Identiteit, Een Analytisch-Wijsgerige Studie*. Nijmegen: SUN.

Van Woudenberg, R. 2003. 'Aspects' and 'Functions' of Individual Things. In: *Philosophia Reformata*, 68(1):1-13.

VanDeVeer and Pierce, C. 1998. *The Envirenmental Ethics & Policy Book*, 2nd Edition. New York: Wadsworth Publishing Company.

Veling, K. 2000. *Ruimte voor de rede, Filosofie als Systematische Reflectie*. Kampen: Agora.

Verburg, M.E. 1989. *Herman Dooyeweerd, Leven en werk van een Nederlands christen-wijsgeer*. Baarn: Ten Have.

Verelst, K. 2006. *De Ontologie van den Paradox*. Brussel: University Publication.

Viner, J. 1965. Adam Smith. In: Recktenwald, 1965 (pp.91-98).

Visagie, P.J. 1982. Some Basic concepts concerning the idea of Origin in Reformational Philosophy and Theology. In: *Journal for Christian Scholarship*, 1st and 2nd quarter (pp.1-13).

Visagie, P.J. 1988. Methods and levels of archeological discourse analysis – with special reference to J. Derrida. In: *Interim*, 9:45-71 (Publication of the Interdisciplinary Research Unit for Studies in the Philosophy of the Academic Disciplines).

Visagie, P.J. 2005. Applying some philosophical tools to the theme of relativity and relativism. In: *Acta Academica, Supplementum*. 2005(2):134-158.

Vogel, H. 1961. *Zum Philosophischen Wirken Max Plancks. Seine Kritik am Positivismus*. Berlin: Akademie-Verlag.

Vollenhoven, D.H. 1933. *Het Calvinisme en de Reformatie van de Wijsbegeerte*. Amsterdam: H.J. Paris.

Vollenhoven, D.H. 1948. *Hoofdlijnen der Logica*. Kampen: Kok.

Vollenhoven, D.H. 1967. *Isagoogè Philosophiae*. Amsterdaum: Edition of the Philosophical Institute at the Free University.

Vollenhoven, D.H. 1968. *Problemen van de Tijd in onze Kring*. Presentation to the Circle of Amsterdam (Association for Reformational Philosophy – Chairperson Dooyeweerd). Text based upon a tape recodring by J. Kraay, checked and corrected in a few instances by Vollenhoven, with a six point summary written by Vollenhoven himself (added on page 8).

Vollenhoven, D.H. 2005. *Isagôgè Philosophiae*. Introduction to Philosophy, Translated from the Dutch by John H. Kok, Edited by John H. Kok and Anthony Tol. Dordt: Dordt College Press.

Von Baader, F. 1851. *Gesammelte Schriften zur philosophischen Erkenntniswissenschaft oder Metaphysik, Sämtliche Werke*. Edited by F. Hoffmann & Julius Hamberger, Volume 2 (replicated by Scientia Verlag, Aalen, 1963).

Von Bertalanffy, L. 1966. Mind and Body Re-Examined. In: *Journal for humanistic Psychology*. 6:113-138.

Von Bertalanffy, L. 1968. *Organismic Psychology and Systems Theory*. Massachusetts: Clarke University Press.

Von Bertalanffy, L. 1968a. Symbolismus und Anthropogenese. In: Rensch. & Schultz, 1968.

Von Bertalanffy, L. 1973. *General System Theory*. Hammondsworth: Penguin University Books.

Von Fritz, K. 1945. The Discovery of Incommensurability by Hippasus of Metapontum. *Annals of Mathematics*, 46:242-264 (German text in Becker, 1965:271-307).

Von Kibéd, A.V. 1979. *Einführung in die Erkenntnislehre, Die Grundrichtungen und die Grenzen der Erkenntnis der Wahrheit*. München: Ernst Reinhardt.

Von Königswald, G.H.R. 1968. Problem der ältesten menschlichen Kulturen. In: *Handgebrauch und Verständigung bei Affen und Frühmenschen* (Ed. B Rensch, Stuttgart 1968).

Von Savigny, F.C. 1948. *Grundlagen der historischen Rechtschule*. Frankfurt am Main: Vittorio Klostermann.

Von Uexküll, J. & Kriszat, G. 1970. *Streifzüge durch die Umwelten von Tieren und Menschen, Bedeutungslehre*. Frankfurt am Main: S. Fischer Verlag.

Von Uexküll, Thure, 1970. Die Umweltforschung als subjekt- und objektumgreifende Naturforschung. In: Uexküll, Jakob von, 1970.

Von Uexküll, J. 1973. *Theoretische Biologie* (1928). Frankfurt am Main.

Von Weizsäcker, C.F. 1972. *Voraussetzungen des naturwissenschaftlichen Denkens*, Herderbücherei, Band 415. München: Carl Hanser Verlag.

Von Weiszäcker, C.F. 1993. *Der Mensch in seiner Geschichte*. München: DTV.

Von Weizsäcker, C.F. 2002. *Große Physiker, Von Aristoteles bis Werner Heisenberg*. München: Deutscher Taschenbuch Verlag.

Wachter, H.E. 1975. *Das Unendlichkeitsparadox in Raum und Zeit und seine kosmologischen Konsequenzen*. Kaiserslautern.

Waldenfels, B. 1971. Aporie, Aporetik. In: Ritter Volume 1 (pp.447-448).

Wallner, F. 1992. *Acht Vorlesungen über den Konstruktiven Realismus*. Vienna: WUV Universitätsverlag.

Wallner, F.G. 1999. *Konstruktion und Verfremdung*, Herausgeber Wallner, F.G. & Agnese, B., Vienna: Universitäts-Verlagsbuchhandlung.

Wallner, F.G. 2003. *Konstruktion und Erziehung, Zum Verhältnis von konstruktivistischem Denken und pädagogischen Intentionen*, Herausgeber Wallner, F.G. and Greiner, K., Hamburg: Verlag Dr. Kovac.

Walzer, M. 1983. *Spheres of Justice: a Defense of Pluralism and Equality*. New York: Basic Books.

Wang, H. 1988: *Reflections on Gödel*. Cambridge Massachusetts: MIT Press.

Warren, M.A. 2000. *Difficulties with the Strong Animal Rights Position*. In: Pojman 2000.

Waters, M. 1994. *Modern Sociological Theory*. London: Sage Publications.

Weber Marcel 1999. Hans Drieschs Argumente für den Vitalismus. *Philosophia Naturalis*, 36:263-293.

Weber, Max 1918. *Parlement und Regierung im neu geordneten Deutschland*. München: Dunkler & Humblot.

Weber, Max 1973. *Gesammelte Aufsätze zur Wissenschaftslehre*, 4th edition. Tübingen: Mohr.

Weideman, A. 2006. A Systematically Significant Episode in Applied Linguistics. In: Time and Context Relevant Philosophy, *Festschrift* dedicated to D.F.M. Strauss (Guest Editors L.O.K. Lategan & J.H.Smit), Bloemfontein: Association for Christian Higher Education (pp.231-244).

Weideman, A. 2007. The Idea of Lingual Economy. In: *Koers*, 72(4):1-27.

Weideman, A. 2009. *Beyond expression, A systematic study of the foundations of linguistics*. Jordan Station: Paideia Press.

Weinert, F. 1998. Fundamental Physical Constants, Null Experiments and the Duhem-Quine Thesis. In: *Philosohpia Naturalis*, 35:225-251.
Wells, J. 2003. Haeckel's Embryos and Evolution: Setting the Record Straight. In: Campbell and Meyer, 2003 (pp.179-186).
Wells, J. 2003a. Second Thoughts about Peppered Moths. In: Campbell and Meyer, 2003 (187-192).
West, E.S. and Todd, W.R. 1966 (third edition). *Textbook of Biochemistry*. New York: MacMillan.
Weyl, H. 1921. Ueber die neue Grundlagenkrise der Mathematik, *Mathematische Zeitschrift*, Volume 10 (pp.39-79).
Weyl, H. 1931. *Die Stufen des Unendlichen*: Vortrag, gehalten am 27. Oktober 1930 bei der Eröffnung der Gästetagung der Mathematischen Gesellschaft an der Universität Jena im Abbeanum.
Weyl, H. 1932. *Das Kontinuum*, 2nd edition, Berlin. (1960 Edition, New York: Chelsea publishing co.)
Weyl, H. 1946. Mathematics and Logic. In: *American Mathematical Monthly*, Vol. 53.
Weyl, H. 1966. *Philosophie der Mathematik und Naturwissenschaft*, 3rd revised and expanded edition. Vienna: R. Oldenburg.
Weyl, H. 1969. Insight and Reflection. In: *The Spirit and Uses of the Mathematical Sciences*, ed. T.L. Saaty and F.J. Weyl, London: McGraw-Hill.
Weyl, H. 1970. *David Hilbert and His Mathematical Work*. In: Reid, 1970 (pp. 243-285).
White, A.C. 1949. *The Call of the Bushveld*. Bloemfontein: Whiteco House (first edition 1948).
White, M.J. 1988. On Continuity: Aristotle versus Topology? In: *History and Philosophy of Logic*, 9:1-12.
Whitrow, G. J. 1961. *The Natural Philosophy of Time*. London: Thomas Nelson & Sons Ltd.
Williamson, O.E. 1975. *Markets and hierarchies: Analysis and Antitrust implications*. New York: The Free Press.
Williamson, O.E. 1985. *The Economic Institutions of Capitalism*. New York: The Free Press.
Wilson, E.O. 1975. *Sociobiology*. Cambridge, Massachusets: Harvard University Press.
Windelband, H. 1924. Geschichte und Naturwissenschaft. In: *Präludien*, Vol.2, Tübingen: Mohr.
Windelband, W. 1935. *Lehrbuch der Geschichte der Philosophie*, Mit einem Schlußkapitel: *Die Philosophie im 20. Jahrhundert* und einer *Übersicht über den Stand der philosophiegeschichtlichen Forschung*. Editor Heinz Heimsöth. Tübingen: Mohr.
Wittgenstein, L. 1966. *Tractatus Logico-Philosophicus*. (1921: Third edition. London: Routledge & Kegan Paul.
Wittgenstein, L. 1968. *Philosophical Investigations*. (1953), Third Edition 1968. Oxford: Basil Blackwell.
Wolff, K. 1971. Zur Problematik der absoluten Überabzählbarkeit, in: *Philosophia Naturalis*. Band 13, 1971.
Woltereck, R. 1940. *Ontologie des Lebendigen*. Stuttgart: Ferdinand Enke Verlag.
Wolters, A. 1981. Facing the Perplexing History of Philosophy. In: *Journal for Christian Scholarship*, 17(4):1-31.
Wolterstorff, N. 1987. *Art in Action*. Grand Rapids: Eerdmans.
Yockey, H. 2005. *Information Theory, Evolution, and the Origin of Life*. Cambridge: Cambridge University Press.
Yourgrau, P. 2005. *A World Without Time. The forgotten Legacy of Gödel and Einstein*. London: Penguin Books.
Zeitlin, I.M. 1973. *Rethinking Sociology, A Critique of Contemporary Theory*. New Yersey.
Zeitlin, I.M. 1984. *The Social Condition of Humanity*, Second edition. Oxford: Oxford University Press.
Ziegenfuss, W. 1954. *Gesellschaftsphilosophie*, Stuttgart: Ferdinand Enke Verlag.
Ziehen, Th. 1920. *Leitfaden der Physiologischen Psychologie*. Jena: Verlag von Gustav Fischer.

Zippelius, R. 1980. *Geschichte der Staatsideen*. München: Verlag Beck.

Znaniecki, F. 1963. *Cultural Sciences, Their Origin and Development* (1952), Urbana.

Zuidema, S. U. 1953. Theologie en Wilsbegeerte in de "Kirchliche Dogmatik" van Karl Barth. In: *Philosophia Reformata*, year 18, 2nd and 3rd quarter (pp 77-138).

Zuidervaart, L. 1973. *Systematic Philosophy Paper*, ICS, March 19, 1973.

Zuidervaart, L. 1977. *Explorations into a philosophical aesthetics*. Stenciled Paper. Tontono: Institute for Christian Studies.

Zuidervaart, L. 1995. Fantastic things: critical notes towards a social ontology of the arts, *Philosophy Reformata* (pp. 37-54).

Zuidervaart, L. 2004. The Great turning point: Religion and rationality in Dooyeweerd's transcendental critique. In: *Faith and Philosophy, Journal for the society of Christian Philosophers*. 20(3) July 2004 (pp.65-89).

Zuidervaart, L. 2008. After Dooyeweerd: Truth in Reformational Philosophy. WEB Publication available at: http://records.icscanada.edu/ir/articles/20081007-1.shtml (accessed on June 1, 2009).

Zylstra, U. 1992. Living Things as Hierachically Organized Structures. In: *Synthese*, 91:111-133.

Zylstra, U. 2004. Intelligent-Design Theory: An Argument for Biotic Laws. In: Zygon, Vol.39, no.1, March (pp.175-191).

Index of Subjects

M

Index of Names

Y

Z

Lightning Source UK Ltd.
Milton Keynes UK
25 January 2010

149065UK00001B/241/P

9 780888 152077